The Vertebrate Body
Sixth Edition

Alfred Sherwood Romer

Late Alexander Agassiz Professor of Zoology, Emeritus, Harvard University

Thomas S. Parsons

Department of Zoology, University of Toronto

SAUNDERS COLLEGE PUBLISHING
Philadelphia New York Chicago
San Francisco Montreal Toronto
London Sydney Tokyo Mexico City
Rio de Janeiro Madrid

rs to:
n Avenue
, NY 10017

Address editorial correspondence to:
West Washington Square
Philadelphia, PA 19105

Text Typeface: ITC Garamond Book
Compositor: The Clarinda Company
Acquisitions Editor: Ed Murphy
Project Editors: Joanne Fraser and Robin Bonner
Copyeditor: Ann Blum
Art Director: Carol C. Bleistine
Cover Design: Lawrence R. Didona
Text Artwork: ANCO/Boston
Production Manager: Tim Frelick
Assistant Production Manager: JoAnn Melody

Cover credit: Illustrated by Lawrence R. Didona

Library of Congress Cataloging in Publication Data

Romer, Alfred Sherwood, 1894–1973.
   The vertebrate body.

   Bibliography: p.
   Includes index.

   1. Vertebrates—Anatomy. I. Parsons, Thomas Sturges,
1930– . II. Title.
QL805.R65 1985   596'.01   85-8196
ISBN 0-03-058443-4

THE VERTEBRATE BODY   ISBN 0–03–058446–9

3456   032   987654321

CBS COLLEGE PUBLISHING
Saunders College Publishing
Holt, Rinehart and Winston
The Dryden Press

Preface

In a prefatory Apologia to the first edition of this text, the late Professor Romer listed six features that he considered desirable in such a volume. Now, some thirty-five years later, since they are as important as ever, I repeat his list and comments on each.

1. *Adequate illustrations.* Even though a student may cover considerable ground in the laboratory, he cannot see all types and structures of interest, and may fail to see the forest for the few trees visible to him.
2. *A truly comparative treatment.* Overemphasis of human structure is not desirable, even for the premedical student. Such a course should be, for him, essentially a "cultural" background, to given him better understanding of the peculiarities and seemingly irrational construction of the human body.
3. *Proper paleontologic background.* The known facts of vertebrate history should be utilized to give, not only adequate treatment of the skeletal and other systems to whose evolution paleontology contributes, but also a modern phylogenetic point of view.
4. *A developmental viewpoint.* Embryologic history is crucial in the establishment of homology, a leitmotiv of comparative anatomy. Further, the time element should be kept in mind and in the consideration of any vertebrate body, for an "adult" is merely one stage in a long developmental series.
5. *Inclusion of histologic data.* The wielder of the scalpel is liable to forget the basic materials of which gross structures are composed. Such an organ as a stomach is merely a flabby, rather revolting and uninteresting object unless we consider the varied internal epithelia and minute glands which furnish much of its excuse for existence.
6. *Consideration of function.* The almost complete separation of form and function prevalent in instruction today [1949; the same is probably not now the case] is unnatural and unfortunate. It is doubtful if there is such a thing as a nonfunctioning structure, although mention of function is often taboo in morphologic works. Nor do functions take place *in vacuo* or without purpose in benefiting the structures which compose the organism, despite the contrary feeling that some physiologic treatises imply. Even if attention be held to comparative anatomy in a narrow sense, the study of homology immediately raises the question of the changes of function associated with the changes, often remarkable, undergone by homologous structures.

In preparing this edition I was faced with the problem, as are virtually all authors or revisors of textbooks, of what topics to include or enlarge and what to drop or shorten. I received many recommendations, often mutually contradictory ones, concerning this; not a few of them appear to reflect the opinion that the amount of detail is about right or slightly too much on everything except the special interest of the person doing the recommending—that special interest obviously requires far more detail! I was, therefore, forced to make my own decisions and must take full blame for any glaring omissions or obviously too-detailed sections.

It is always easier to add things than to drop them. Thus I suspect that this edition is slightly, but only very slightly, longer than its predecessor. Prices and resistance to using a text tend to increase with its length, so I have tried to keep any increase minimal. However, I prefer a text that can also serve as a reference. I do not expect my students to learn everything in this book; indeed, there are large sections that I do not cover or assign at all in my course. My aim is to produce a book that is detailed and comprehensive enough to be a useful reference book for a student interested in vertebrate structures and one that is still not so encyclopedic as to be unreadable or useless as a textbook for a second- or third-year university course.

In preparing this edition, I believed that there were four possible ways to treat functional anatomy: (1) leave the book much as it was, by integrating functional considerations rather than having separate chapters; (2) sacrifice some of the comprehensiveness of the coverage to make room for a reasonable consideration of function; (3) add a chapter or more on functional aspects of locomotion and feeding, thus greatly increasing the size of the book; or (4) add a very short, and thus inevitably superficial, summary of selected functional aspects. I chose the first as the most satisfactory course to pursue. I can only hope that those who wish to include more information on such functional aspects will see other redeeming features in this book and use appropriate supplementary material for the topics I have omitted.

I have benefited greatly in the preparation of this edition from numerous comments, suggestions, and criticisms given by many colleagues and friends, notably J. A. Burns, D. G. Butler, C. S. Churcher, A. F. Ellsworth, C. Gans, J. Goodman, R. Hirakow, D. Klingener, J. Lai-Fook, E. C. Minkoff, J. M. Moulton, M. C. Parsons, J. J. Thomason, R. Walker, Y. L. Werner, J. E. M. Westermann, and K. G. Wingstrand, plus anonymous reviewers for Saunders. Numerous students in my course also provided suggestions; M. J. Bazos was especially helpful in this regard. Special thanks are to Professor Wallace E. McLeod, who checked an early draft of the Appendix on Scientific Terminology, correcting some of my more glaring deficiencies in the classical languages. Like the others mentioned, he cannot be blamed for any residual problems or errors. As usual, the staff at Saunders was pleasant and helpful; for this edition my dealings there were mainly with Michael J. Brown. Please note that all those mentioned here helped with *this* edition. Many others, cited in earlier prefaces, made suggestions which have improved all subsequent editions, this one included. Their omission here does not mean that their contributions have, in some mysterious way, lapsed.

Finally, the present edition has benefited immeasurably from the ever cheerful and resourceful help that I have received for the past three summers from Margaret Anne Adair. She has made drawings, tracked down references, helped edit and typed corrections and new sections, and done all the other sorts of odd jobs that are so vital in the preparation of such a volume. Without her presence this new edition would probably not yet be finished, and both it and my temper would certainly be the worse.

Professor Romer closed the Apologia to the first edition by venturing "to render homage to the two men . . . who many years ago instilled in me an interest, which has never flagged, in the story of the vertebrates; Dr. William King Gregory, under whom I studied as a graduate student, and Dr. J. H. McGregor, whom I served as a teaching assistant in vertebrate zoology." Dr. Alfred Sherwood Romer, under whom I so studied and for whom I so served, instilled a similar interest in me, and I wish to render similar homage.

Thomas S. Parsons

Contents

Chapter 1

Introduction

This work is designed to give, in brief form, a history of the vertebrate body. Basic will be a comparative study of vertebrate structures: the domain of comparative anatomy. This is in itself an interesting and not unprofitable discipline. Of broader import, however, is the fact that the structural modifications witnessed are concerned with functional changes undergone by the vertebrates—changes correlated with the varied environments and modes of life found in the course of their long and eventful history. The evolutionary story of the vertebrates is better known than that of any other animal group, and vertebrate history affords excellent illustrations of many general biologic principles. Knowledge of vertebrate structure is of practical value to workers in many fields of zoology. To the future medical student, such a study gives a broader understanding of the nature of the one specific animal type on which his later studies will be concentrated.

For the most part (Chaps. 6–17), the present volume is devoted to a consideration, *seriatim,* of the various organs and organ systems. In this chapter, a "bird's eye" view of vertebrate structure is given, together with certain introductory matters. In other early chapters, general or preliminary topics are discussed, including the evolutionary history of the vertebrates and their kin (Chaps. 2 and 3), cells and tissues as the basic structural elements (Chap. 4), and embryonic development (Chap. 5).

The Vertebrate Body Plan (Fig. 1)

Bilateral Symmetry. A primary feature of the vertebrate structural pattern is the fact that the members of this group are bilaterally symmetric, with one side of the body essentially a mirror image of the other. Vertebrates share this type of organization with most invertebrate groups, notably the annelid worms and the great arthropod phylum, which includes crustaceans, arachnids, insects, and so forth. In strong contrast is the radial symmetry of coelenterates and echinoderms,* in which the body parts radiate out from a central axis like the spokes of a wheel. The degree of activity of animals appears to be correlated with the type of symmetry that is present. The radiate echinoderms and coelenterates are in general sluggish types, slow moving or fixed to the bottom or, if free floating, mainly drifters with the current rather than active swimmers. Vertebrates, arthropods, and marine annelids are, on the other hand,

---

*The echinoderms are not truly radial; they start off life as bilateral animals, but later assume a more or less radial form and the habits that go with it (though some may be active predators).

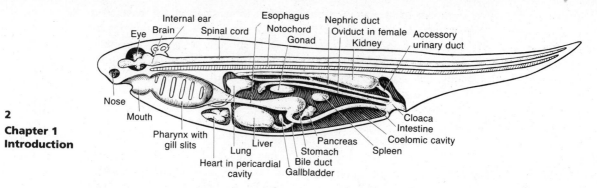

Eye Brain Internal ear Spinal cord Esophagus Notochord Gonad Nephric duct Oviduct in female Kidney Accessory urinary duct Nose Mouth Pharynx with gill slits Lung Liver Stomach Bile duct Gallbladder Heart in pericardial cavity Pancreas Spleen Cloaca Intestine Coelomic cavity

Figure 1. Diagrammatic longitudinal section through an "idealized" vertebrate, to show the relative position of the major organs.

generally active animals. Activity would seem to have been one of the keys to the success of the vertebrates and is in a sense as diagnostic as any anatomic feature.

Regional Differentiation. In any bilaterally symmetric animal, we find some type of longitudinal division into successive body regions—in the annelid worms, for example, a rather monotonous repetition of essentially similar segments, or in insects, a pattern in which such segments are consolidated into head, thorax, and abdomen. Vertebrates, too, have well-defined body regions, although these regions are not directly comparable to those of invertebrate groups.

There is in vertebrates a highly specialized **head**, or cephalic region; in this region are assembled the principal sense organs, the major nerve centers that form the brain, and the mouth and associated structures. In vertebrates, as in all bilaterally symmetric animals (even a worm), there is a strong tendency toward **cephalization**—a concentration of structures and functions at the anterior end of the body.

In all higher, land-dwelling vertebrate groups, a **neck** is present behind the head; this is little more than a connecting piece, allowing movement of the head on the trunk. The presence of a neck is not, however, a primitive vertebrate feature. In lower, water-breathing vertebrates, this section of the body is the stout **branchial region**, containing the breathing apparatus. The appearance of a distinct neck occurs only with the shift to breathing with lungs and the reduction of the gills.

The main body of the animal, the **trunk**, is the next region; this terminates in the neighborhood of the anus or cloaca. Within the stout trunk are the body cavities containing major body organs, the viscera. In mammals, the trunk is divisible into the **thorax** and **abdomen**, the former containing the heart and lungs within a rib basket, the latter enclosing most of the digestive tract; there is, however, no clear subdivision here in lower vertebrates.

In most bilateral invertebrates, the digestive tube continues almost the entire length of the body. Among the vertebrates, however, we find, in contrast, that the digestive tract and other viscera stop well short of the end of the body; beyond the trunk there typically extends a well-developed **tail** or **caudal region**, with skeleton and muscles, but without viscera. The presence of a postanal tail is a basic feature of our group, laid down, it would seem, in an early stage of chordate evolution. The tail is, of course, the main propulsive organ in primitive aquatic vertebrates. In terrestrial animals, it tends to diminish in importance but is often long, stout at the base, and well developed in many amphibians and reptiles. In mammals, it is generally persistent but is merely a slender appendage. In birds, it is shortened and functionally replaced by the tail feathers, arising from its stump; in some forms—frogs, apes, and man—it is, exceptionally, lost completely as an external structure.

Gills.   The presence, in the embryo if not in the adult, of internal gills developed as a paired series of clefts or pouches leading outward from an anterior part of the gut— the pharynx—is one of the most distinctive features (perhaps *the* most distinctive feature) of the vertebrates and their close kin. In higher vertebrates, the gills are functionally replaced by lungs, but branchial pouches are nevertheless prominent in the embryo. In lower water-dwelling vertebrates, gills are the primary breathing organs. Among small invertebrates, many with soft membranous surfaces can get enough oxygen through such membranes to supply their wants. But in forms with a hard or shelly surface, and especially in large forms in which the surface area is small compared with the bulk of the body, gills of some sort are a necessity. Typical invertebrate gills, as seen in crustaceans or molluscs, are feathery projections from the body surface. The vertebrate gill, however, is an internal development, connected with the digestive tube. Water enters the "throat," or pharynx (usually through the mouth) and passes outward through slits or pouches; on the surface of these passages are the actual gills, membranes through which an exchange of oxygen from the water for carbon dioxide in the blood takes place. Quite in contrast is the function of the gills in certain lowly relatives of the vertebrates. There, as we shall see, the system of gills is of primary importance in food collection—a fact tending to explain the unusual vertebrate condition of an association of the breathing organs with the digestive tube.

Notochord.   In the embryo of every vertebrate there is, extending from head to tail along the length of the back, a long, flexible, rodlike structure—the notochord. In most vertebrates, the notochord is much reduced or absent in the adult, where it is replaced by the vertebral column or backbone. But it is still prominent in some lower vertebrates and is the main support of the trunk in certain simply built vertebrate relatives (such as amphioxus) in which no vertebral column ever forms. So significant is this primitive supporting structure that the vertebrates and their kin are termed the phylum Chordata, a name referring to the presence of a notochord.

Nervous System.   Longitudinal nerve cords are developed in various bilaterally symmetric invertebrate groups. These, however, are frequently paired and may be lateral or ventral in position. Only in the chordates do we find developed a single cord, dorsally situated and running along the back above the notochord or the vertebrae. Invertebrate nerve cords are generally solid masses of nerve fibers (and supporting cells) running between equally solid clusters of nerve cells, termed **ganglia**. The chordate nerve cord is, in contrast, a hollow, nonganglionated structure, with a central, fluid-filled cavity. In various invertebrates, the process of cephalization is reflected in a concentration of nerve centers in a brainlike structure. Independently, we believe, the vertebrates have evolved a hollow **brain** with characteristic subdivisions at the anterior end of the hollow nerve cord—the **spinal cord**. Not exactly matched in any invertebrate group is a series of characteristic **sense organs** developed in the head of vertebrates—paired lateral eyes and, primitively, one or two dorsal, nearly median eyes; nasal structures, usually paired; and paired ears with equilibrium as their primary function.

Digestive System. All metazoans (with degenerate exceptions) have some sort of digestive cavity with a means of entrance to and exit from it. In many of the more primitive metazoans, there is but a single opening, serving as both mouth and anus. In vertebrates, as in other more progressive metazoans, there are separate anterior and posterior openings, serving, respectively, for the entrance of food materials and the exit of wastes. The mouth is situated near the anterior end of the body, commonly somewhat to the underside. In arthropods and annelids, the digestive tube reaches to the posterior end of the body. In vertebrates, however, this is not the case; the anus lies at the end of the trunk, leaving, as we have mentioned, a caudal region in which the digestive tube is absent.

In most vertebrates, the digestive tube is divided into a series of characteristic regions serving varied functions—**mouth, pharynx, esophagus, stomach,** and **intestine** (the last variously subdivided). In lower vertebrates, the esophagus may be almost nonexistent, and in some groups, even the stomach may be absent. In mammals and certain other vertebrates, the digestive tract terminates externally at the **anus.** In most groups, however, there is a terminal segment of the gut, the **cloaca,** into which urinary and genital ducts also lead.

A **liver,** which performs to some extent a secretory function but is mainly a seat of food storage and conversion, is present in vertebrates as a large ventral outgrowth of the digestive tube. Somewhat similar but variable structures are present in many invertebrates. In most vertebrate groups, a **pancreas** is present beside the gut as, primarily, an enzyme-secreting gland.

Kidneys. Among invertebrates, some type of kidney-like organ for the disposal of nitrogenous waste and for the maintenance of a proper composition of the internal fluids of the body is often present, typically as rows of small tubular structures termed **nephridia.** Of chordates below the vertebrate level, amphioxus has nephridia of a special type. In true vertebrates, however, the **kidney tubules** serving such a function are of a markedly different type and are characteristically gathered into compact paired kidneys, dorsal in position. **Kidney ducts,** of variable nature, lead to the cloaca or to the exterior, and a **urinary bladder** may develop along their course.

Reproductive Organs. Male and female sexes are almost invariably distinct in the vertebrates, as they are in many invertebrate groups. The tissues producing the germ cells—the **gonads**—develop into either **testis** or **ovary.** In all except the lowest vertebrates, a system of ducts leads the eggs or sperm to or toward the surface (frequently by way of the cloaca); in the female, special regions of the duct may be present for shell deposition or for development of the young.

Circulatory System. In vertebrates, as in many invertebrates, there is a well-developed system containing a body fluid, the blood, with tubular vessels and a pump, the **heart,** to bring about its circulation. The heart in vertebrates is a unit structure, ventrally and rather anteriorly situated. In certain invertebrates, the circulation is of an "open" type: the blood is pumped from the heart to the tissues in closed vessels but is then released and makes its return to the heart by oozing through the tissues without being enclosed in vessels. In vertebrates, as in some of the more highly organized invertebrates, the system is closed; not only is the blood carried by the **arteries** to the various organs, but the return to the heart, after passing through the tissues in small tubes, the **capillaries**, is also made in closed vessels, the **veins**. In most vertebrates, **lymph vessels** form an additional means of returning fluid from the cells to the heart. Many invertebrates contain in their blood streams pigmented metallic compounds in solution, which aid in the transportation of oxygen. Among vertebrates, almost exclusively, the iron compound **hemoglobin** is the oxygen carrier; furthermore, this chemical is not free in the blood but is contained in **red blood cells** (white cells with other functions are also present).

In annelids, the circulation of the blood is in general forward along the dorsal side of the body and backward ventrally in its return to the tissues. The reverse is true of the vertebrates. The blood from the heart passes forward and upward (primitively via the gills) and back dorsally to reach the organs of trunk and tail, and a major return forward—from the digestive tract—is ventral to the gut (although dorsal veins are also important).

Coelom. In certain invertebrates, the internal organs are embedded in the body tissues. In others, however, there develop body cavities—**coelomic cavities**—filled with a watery fluid, in which most of the major organs are found. This latter condition is present among vertebrates. A major body cavity—the **peritoneal cavity**—occupies much of the trunk and contains most of the digestive tract; various other organs (reproductive, urinary) project into it. Anteriorly there is a discrete **pericardial cavity** enclosing the heart, and, in mammals, the lungs are contained in separate **pleural cavities**.

Muscles. Musculature in the vertebrates is of two types, **striated** and **smooth** (or nonstriated)—the two differing sharply in minute structure and in distribution in the body. The former, roughly, includes all the voluntary musculature of the head, trunk, limbs, and tail, and the muscles of the branchial region; the smooth musculature, more diffuse, is mainly found in the lining of the digestive tract and blood vessels. The musculature of the heart is in various respects intermediate in microscopic structure. The striated musculature of the trunk develops, unlike most other organ systems, as a series of segmental units.

Skeleton. Hard skeletal materials are present in all vertebrates, and, in all except certain derived groups, consist in part, at least, of bone. Superficial skeletal parts, the **dermal skeleton**, correspond functionally to the "armor" of certain invertebrates and are typically bony; internal skeletal structures, the **endoskeleton**, are formed as **cartilage** in the embryo but are generally replaced by bone in the adult. Cartilage-like materials are found in some invertebrates. **Bone**, however, is a unique vertebrate tissue. It differs in texture and minute structure from the typical chitinous or calcareous skeletal materials of invertebrates. The salts deposited in bone are mainly calcium phosphate—in most (but not all) invertebrate skeletal structures, carbonate is the common calcium compound.

Appendages.   Two pairs of limbs, **pectoral** and **pelvic**, are found in most vertebrates in the form of fins or legs and become increasingly prominent in higher members of the group. They are, however, little developed or absent in the lowest vertebrates, living and extinct, and hence are not absolutely characteristic. They may also be lost in specialized forms. Their structure (in contrast to arthropod limbs) includes internal skeletal elements, with muscles for their movement arrayed above and below.

Segmentation.   The great invertebrate phyla Arthropoda and Annelida are notable for the presence of **metamerism**: a serial repetition of parts in a long series of body segments. In annelids, this segmentation is readily apparent; in arthropods, the metameric structure may be more or less obscured in the adult but is clearly seen in embryos or larvae.

Vertebrates, too, are segmented, but the segmentation is limited and has obviously developed independently from that of those invertebrate groups in which all structures, from skin inward to the gut, exhibit segmentation. Among the vertebrates, neither skin nor gut is segmented; the metameric arrangement is primarily that of the trunk muscles. In relation, however, to the attachments of these muscles and their nerve supply, much of the skeleton and nervous system has taken on a segmental character.

The Body in Section (Fig. 2).   We have noted some of the more important body features, with particular regard in many cases to their anteroposterior positions. We may now briefly consider the general organization of the body as seen in cross section.

Structurally, the most simple region of the body is the tail, strongly developed in

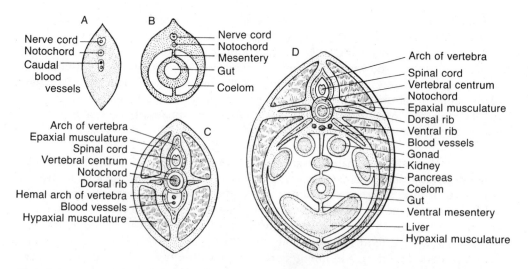

Figure 2.   Cross sections through the body of a vertebrate. *A, B,* Much simplified sections through tail and trunk, to show the essential structure of the trunk as a double tube; in the tail, the "inner tube" of the gut is absent. *C, D,* More detailed diagrammatic sections of the tail and trunk to show the typical position of main structures.

most vertebrate groups. The section of a tail (Fig. 2A, C) is typically a tall oval, the surface skin-covered. Somewhat above the center is seen the notochord, or the central region of the vertebrae that typically replaces it in the adult and, above this, a cross section of the nerve tube; the two structures are invariably closely associated topographically. The body cavity and associated viscera are absent here; representing them (in a sense) are caudal blood vessels lying below the notochord. Almost all the remainder of the tail is occupied by musculature, usually powerful. This musculature is arrayed in right and left halves, with a medium septum dividing them above and below.

A typical section through the trunk is more complicated, even when, as in Figure 2B, this is represented in its most generalized condition. One may consider the trunk as essentially a double tubular system, roughly comparable in structure to the casing and inner tube of an old-fashioned automobile tire. The outer tube in itself contains all the major elements seen in the section of the tail—notochord and nerve cord, and musculature descending on either side beneath an outer covering of skin. Internally, it is as if we had taken the little area below the notochord in the tail, where only the blood vessels were present, and expanded this to enormous proportions as the coelomic cavity of the trunk. With the development of this cavity, the outer "tube" of the trunk now has an inner as well as an outer surface. The surface lining the body cavity is the **peritoneum**, and that part of this lining that forms the inner surface of the outer tube is the **parietal peritoneum**. The part of the outer tube between coelomic cavity and the surface of the body is the body wall.

The "inner tube" is primarily the tube of the digestive tract. The outer lining, facing the coelomic cavity, is peritoneum—**visceral peritoneum**. The inner lining is the epithelium lining the digestive tract. Between the two, analogous to the musculature in the body wall, are smooth muscle and connective tissues. In the embryo, the gut is connected with the "outer tube," both dorsally and ventrally by **mesenteries**—thin sheets of tissue bounded on either side by peritoneum. The dorsal mesentery—that above the gut—usually persists, but the ventral portion frequently disappears for most of its length.

Although we shall treat the arrangement of the organs in the coelom in more detail in a later chapter, here we may go somewhat further in considering the position of the viscera. In Figure 2 D, we have indicated the fact that the digestive tube is not a simple tubular structure but has various outgrowths—most characteristically, the liver ventrally and the pancreas dorsally. These are (in theory and in the embryo, at any rate) median structures and are developed within the ventral and dorsal mesenteries. Further, we may have other organs projecting into the body cavity but arising from tissues external to it. In many groups, the kidneys project into the abdominal cavity at either upper lateral margin, and the reproductive organs—ovaries or testes—typically project into the cavity more medially along its upper border. It must be noted that the relative size of the body cavity is never so great as represented in this and other diagrams; the viscera actually fill virtually all the available space.

Directions and Planes

Although the vertebrate body is essentially a bilaterally symmetric structure, there are many exceptions to this general statement. Organs that primitively lay in the midline may be displaced: the heart may be off center; the abdominal part of the gut—stomach and intestine—is usually twisted, and the intestine may be convoluted in a complicated asymmetric manner. Again, in paired structures, those of the two sides may differ markedly; for example, in birds, just one of the two ovaries (the left) is functional in the adult. Still greater asymmetry is seen in the flounders, where the whole shape of the body is affected by the substitution of the two sides for the normal top and bottom of the animal.

Either in theory or in practice, the body of an animal may be sectioned in various ways at various angles. If the body is considered as sliced crossways, as one would cut a sausage, the plane of section is termed **transverse**. If the line of cleavage is vertical and lengthwise, from snout to tail, the plane is a **sagittal** one. Sometimes this latter term is restricted to a cut actually down the midline—the midsagittal plane—and similar sections to one side or the other are termed parasagittal; but frequently such cuts are considered parts of a series of sagittal sections in a broad sense. The third major plane of cleavage, in the remaining direction, is that of slices cut the length of the body, but horizontally, so as to separate the body into dorsal and ventral parts. Such a plane is termed a **frontal** one—that is, one parallel to the "forehead" of the animal.

Direction within the body is of importance in the description of structural relationships and the naming of the various organs. Terms in this category, fixing a position or pointing out a direction, may be considered.

The head and tail ends of the body are, in most vertebrates, the direction toward which and from which movement of the animal normally takes place. **Anterior** and **posterior** are the common terms of position in this regard; **cranial** and **caudal** are less used but are essentially synonymous. Upper and lower surfaces—back and belly aspects—are reasonably named **dorsal** and **ventral**. Position in the transverse plane is of course given with reference to the midline; **medial** refers to a position toward the midline; **lateral**, a more removed position.

A fourth pair of terms of less exact meaning but of considerable use are **proximal** and **distal**. The former refers in general to the part of a structure closer to the center of the body or some important point of reference; the latter, to a part farther removed. These terms are clearly available for the limbs and tail. Within the head and trunk, their use is less clear, but we may, for example, speak of proximal and distal parts of a nerve with obvious reference to the spinal cord or brain as a center, proximal and distal regions of arteries with reference to the heart as the assumed center, and so on.

For these adjectives of position, there are, of course, corresponding adverbs ending in **ly,** and others (rather awkward) to denote motion in a given direction, ending in **ad**, as **posteriorly, caudad**.

The major directional terms, anterior, posterior, dorsal, and ventral, apply with perfect clarity to almost all known vertebrates. But, in man, we have an exception, an aberrant form that stands erect and hence might have different directional terms applied to him.

It is unfortunate that in the terminology most generally used in medical anatomy this is the case (Fig. 3). The head and "tail" ends of the body are, in the erect human position, above and below rather than fore and aft and are termed **superior** and **inferior**, rather than anterior and posterior. Cranial and caudal could, of course, have been used instead, but medical people seem to like these alternatives no better than many comparative anatomists. To add to the problem, superior and inferior mean

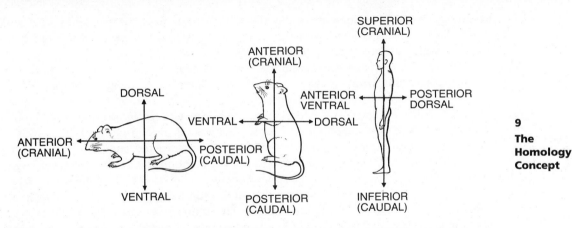

Figure 3. Diagram to show the contrast in positional terms between normal vertebrates and man.

higher and lower in Latin; thus, they are occasionally used as synonyms for dorsal and ventral in comparative anatomy. More confusing, however, is the fact that anterior and posterior have generally been used in man quite needlessly to replace dorsal and ventral, so that the back side of the human has been generally termed **posterior** and the belly surface **anterior**. Thus, this pair of terms may have contradictory meanings in special human anatomy and in more normal usage, causing considerable confusion. For example, each spinal nerve has two roots (cf. Fig. 386, p. 543). In a dissecting room, the two roots in a human cadaver have been generally termed posterior and anterior. But if a neurologist working with (say) rats tries to use the same nomenclature, he is in an obviously absurd position; one root is no more "anterior" or "posterior" than the other. In both rat and man, however, designation of the nerve roots as dorsal and ventral is reasonable and logical. Long-established customs, no matter how illogical, are hard to break down; but in a recently adapted revision of human anatomic nomenclature, the use of dorsal and ventral has been agreed to in such cases.

The Homology Concept

Even in the early days of zoologic research, it was recognized that within each major animal group there was a common basic pattern in the anatomic plan of the body. The same organs could be identified in many or all members of a group, although frequently much modified in size, form, or even function in correlation with changing habits or modes of existence. With the acceptance by biologists of the principle of evolution in the 60's and 70's of the last century, real significance was given to the concept of **homology**: the thesis that specific organs of living members of an animal group have descended, albeit with modification, slight or marked, from basically identical organs present in their common ancestor. For many decades, the tracing of homologies was a leading motif in zoologic research.

Many of the results of such studies were novel and exciting. It was found, for example, that the three little auditory ossicles of our own middle ear (p. 531) were once part of the jaw apparatus of our piscine ancestors and appear even earlier to have been part of the supports of the gills of ancestral vertebrates. The muscles with which we smile or frown are derived from those that once helped our fish forebears to pump water through their gills.

Homologous organs are those that are identical—the same—in the series of forms studied. But what do we mean by "the same"? One tends, unthinkingly, to believe that the same actual mass of material, the very same limb or lung or bone, has been handed down, generation by generation, like an heirloom. This is quite absurd, but such a concept has obviously influenced, unconsciously, the minds of many workers. In reality, of course, every organ is re-created anew in every generation, and any identity between homologues is based upon the identity or similarity of the developmental processes that produce them.

Genetics gives us a firm base for the interpretation of these processes. They are controlled by hereditary units, the **genes**. These tiny structures are present to the number of thousands, at least, in the chromosomes of every animal cell. The development of the individual is directed by the genes transmitted to the fertilized egg by the parents. Each gene may affect the development of a number of structures or parts of a body; conversely, every organ is influenced in its development by a considerable number of genes. If the genes remain unchanged from generation to generation, the organ produced will remain unchanged (apart from environmental effects upon an individual that may be obvious but are not inherited by the next or later generations) and the homology is absolute.

Changes, however, do occur in genes, as **mutations**; these mutations produce changes in the structures to which the genes give rise. If the mutations produce effects of small magnitude and occur in only a few of the genes concerned, the organ will be little modified and its homology with the parent type will remain obvious. If, however, the mutations are numerous and marked in their effects, the organ may be radically modified and its pedigree will be much less clear. In a sense, a study of the homology of organs is merely a study of phenomena produced by genes. If the genetic constitution of all animal types were well known, the determination of homology between structures might well rest upon the degree of identity of the genes concerned in their production. But this is not a matter of practical importance, because there are few animals whose genetic constitution is at all adequately known, and it is improbable that our range of knowledge will ever be broadened to the necessary degree.

What then are the best criteria for the establishment of homology? Function is no certain guide, because organs that are clearly homologous in two animals may be put to quite different uses. Observation shows that the shape, size, or color of a structure gives little positive evidence of identity. Similarity in general anatomic position and relations to adjacent organs is a more useful clue to identification. Best of all is similarity in developmental history. Embryologic processes in vertebrates tend to be conservative, and organs that are quite different in the adult condition often reveal their homology through similarity in early embryonic stages.

Homology is generally applied to structural identity. Some investigators have proposed that the concept be broadened to include functional identity. This suggestion has not, however, met with general acceptance. The term **analogy** is in some regards a parallel, on the functional side, to homology; analogous organs are those that have similar functions. The term is, however, somewhat restricted, for, as generally used, it implies that the organs concerned are not homologous. A lung and a fish gill, for example, are analogous, for both are used for respiration, but the two are quite different, i.e., not homologous, structures.

Adaptation and Evolution

The varied modifications that vertebrate structures have undergone and the varied functions that they have assumed have, of course, come about as the result of evolution. One cannot make a comparative study of the vertebrates without formulating some general concept of the nature of evolutionary processes. Most structural and functional changes in the vertebrate body are quite clearly adaptive modifications to a variety of environments and modes of life. How have these adaptations been brought about? Proper discussion would require a volume in itself; here, we can merely indicate the general nature of the problems concerned and current majority opinion as to their interpretation; the bibliography contains references to more detailed works on this topic.

We sometimes speak, thoughtlessly, of adaptive changes, as if the animal "willed" them or as if its needs or desires in themselves brought new structures or structural changes into being. It would be advantageous, one might say, for a fish to be able to walk on land, and so some fishes made themselves legs; it would be "nice" if the cow's early ancestors developed teeth better able to cope with grain and grass, and so the teeth promptly became larger.

Obviously, such ideas are absurd. They are, however, not far removed from certain theories of evolution that have had, and still have, some acceptance. These theories assume that evolution is an unnatural phenomenon—that changes have been brought about by some "inner urge" within the organism or are the result of the "design" of some supernatural force. Because such theories are nonscientific, they cannot be scientifically disproved; but we are at liberty to look for more reasonable explanations of evolution, based upon known facts. If someone tells us that the operation of an automobile engine is controlled by a little invisible daemon dwelling therein, we cannot prove him wrong. But nothing is gained by adding this hypothetic daemon, and we would prefer an attempt to explain the engine's working in terms of known mechanical principles, the nature of electric currents, and the explosive structure of hydrocarbon molecules.

A more plausible attempt at interpretation of structural evolutionary changes was advocated almost two centuries ago by Lamarck—a belief in the inheritance of acquired characters through the effect of use and disuse. If the giraffe's ancestors stretched their necks to obtain foliage on high branches, the effects of stretching, this theory assumes, would be transmitted to their offspring, generation after generation, and an elongate neck gradually developed in the hereditary pattern. If the snake's lizard ancestors ceased to use their legs in locomotion, the cumulative result of disuse would be the eventual loss of the limbs. This attractive theory seems simple, reasonable, and natural. But its present standing is poor indeed. We may summarize by saying that no one, despite repeated efforts, has been able to furnish any valid proof of the inheritance of an acquired characteristic as proposed by Lamarck. Structures useful to an animal may and often do increase in size or complexity in the course of time, and useless or little used structures may diminish. But there is no evidence that the use or disuse of parts by an individual has any effect upon the build of its offspring.

The science of genetics has clearly demonstrated that evolutionary changes are due to mutations. These may produce effects of some magnitude, but most cause only minor modifications: a mutation in a fruit fly may, for example, have no greater visible result that the splitting of a single bristle. We now have considerable knowledge of the chemical structure of the materials involved and of chemical and physical influences (such as radiation) that play important roles in bringing them about. An understanding of these topics, which are extremely interesting in their own right, is not essential to an understanding of evolution, and discussion of them, therefore, is omitted here. As for evolutionary theories, however, two things stand out clearly: (1) There is no evidence of "design" or "direction" in mutations. They appear to be quite random, rather than tending in any one direction. Some may well be advantageous; most, however, are obviously harmful, and many are lethal. (2) There is no evidence that mutations have any relation whatever to use or disuse of body organs; characters acquired by the individual have no specific influence on the nature of mutations of the genes in its sex cells—mutations whose effect is transmitted to the offspring.

The process of mutation thus seems to be merely one of blind, random change. But vertebrate evolution certainly appears to have resulted in changes both useful and adaptive. How can such results have come about by means of the mutation process?

Our modern ideas on evolution stem from the publication in 1859 of Charles Darwin's On the Origin of Species. Besides giving a mass of evidence that evolution had actually occurred, Darwin presented, for the first time, a convincing theory of *how* it occurred: this is his theory of **natural selection**. Basically Darwin noted that all species of animals and plants vary and that these variations tend to be inherited. These are obvious and generally accepted ideas: like begets like. Also in nature, species tend to reproduce such that their numbers would increase geometrically if there were no checks (a female cod lays from two to nine million eggs a year; this gives a theoretical potential that boggles the mind). Because numbers of most forms do *not* increase greatly, there must be a very high "infant mortality." Darwin postulated that minor differences would affect an animal's ability to survive. Those differences that are evolutionarily "better," that is, those that promote survival, will, if inherited, become more common in succeeding generations. Eventually these differences would become the norm, and "new" characters would have evolved.

Although they had, in fact, been discovered by Mendel by 1865, the basic principles of genetics were not generally known in Darwin's day, and he had no real knowledge of the mechanism of inheritance (in fact, some of his ideas were incorrect). This is, of course, basic information now that is included in all elementary textbooks. Major points include the fact that there is no real "blending"; instead, inheritance is by more or less discrete units or genes. For each "character" (in a genetic sense, a character of an animal may depend on many genetic characters), any animal, with various exceptions that need not be treated here, receives one gene from each parent. When that animal produces eggs or sperm, each will contain one of the two original genes. Genes may occasionally change their structure or properties; such a change is a mutation and the mutated form of the gene will be inherited. Mutations of genes or new combinations of existing genes will be responsible for the new characters that may or may not improve the animal's chances of survival.

Obviously, things are not simple, and many complications must be considered. First, it is rarely possible to say that a character is always "good" or "bad." Its value, or lack thereof, to the animal will depend on the animal's environment. A thick fur coat is an obvious advantage to an arctic mammal; in a tropic mammal, it could be quite disadvantageous. Some characters are usually beneficial (such as a better means of temperature control) and others are not (those that interfere with normal development, for example), but these are probably rare. Thus, as conditions change, characters that were once selected against will become useful and will be selected for.

Second, it is a marked oversimplification to say that a character such as blue eyes is inherited if the proper gene is present. The potential to develop blue eyes may be inherited, but whether or not they actually form will depend on many other factors. A shortage of a particular food could reduce the animal's ability to form the proper pigment, or, as a more extreme example, if something prevented eyes from forming at all, it would obviously not be possible to have blue ones.

Third, that which is "selected" is not a specific character. Animals survive or not as a whole, and *all* the characters in an animal must be such that it can survive and reproduce if the "good" genes it possesses are to be passed on. Individual characteristics must be not only beneficial but also able to work together to form a properly integrated whole. Thus, when statements are made to the effect that such a character is advantageous and thus selected for, remember that several steps are being omitted in the argument. Such short cuts save time and effort and are frequently used in this text and others, but they should be recognized as short cuts.

Fourth, in general, genes are carried in duplicate in the cells of every animal (one gene coming from each parent); if the members of a gene pair differ in their potentialities, one tends to dominate over the other in the structures or functions that it controls. It is obvious that selection can have no influence over the "weaker" of such a pair of genes—technically termed **recessive**—unless by chance both members of the pair of genes concerned are of the same recessive nature. A little consideration makes it clear that, as a result of this situation, it is practically impossible to eliminate completely a recessive mutant, even if highly deleterious, from an animal stock in which it is once established. It is reasonable to believe that with numerous variables of this sort present in a stock, circumstances might arise (particularly in changed environments) in which certain "suppressed" variants or combinations of them might eventually prove highly advantageous if they should come to light in an individual and result in evolutionary change in the population as a whole. Every species, it would seem, has within its "gene pool" an amount of potential variation that might, to a considerable degree, enable it to adapt to a new situation without the introduction of new mutations.

The mechanism of evolution just outlined can be used to explain many of the observed trends in the phylogeny of vertebrates and other organisms. Given a supply of random mutations, natural selection will act powerfully to eliminate unfit types and preserve better-fitted forms in which one or a group of useful mutations have occurred in the germ plasm. It is expected that, with time, species will diverge in structures and other characters. Those in different, geographically separate populations will experience different environments so that different qualities will be selected. Moreover, the mutations that occur and are thus available for selection will probably not be the same. After a sufficient (and variable) period of time, the populations will have diverged so that they are different species, families, and so on. **Parallelism** and **convergence**, cases in which two groups of animals change independently but in a similar way or change to become similar from dissimilar ancestors, may reflect selection of similar traits to solve similar problems. Wings are essential for flight in both bats and birds; they show certain similarities but developed from normal front legs quite independently. Such resemblances are, at least in theory, never perfect. Bats remain mammals and birds remain birds; characters not directly involved in the mechanism of the wings are characteristic of their group and quite dissimilar.

This lack of perfect resemblance is also important in the "law" that evolution is not reversible. Animals can never return to the ancestral condition because the environment can never be exactly the same; just the presence of the descendant form makes it different. Also, the chance that the right mutations will appear in the correct order to retrace perfectly the previous history in reverse is extremely unlikely. Naturally, various structures, having previously evolved, can be lost. Although this is, in a sense, a reversal, it is a very minor one. Some parallelisms can be striking resemblances of reversals; the fusiform shape and the general appearance of sharks, ichthyosaurs (extinct reptiles), and porpoises provide the classic example. However, porpoises have definitely *not* reverted to being fish. They are perfectly good mammals with lungs, not gills, with expanded cerebral hemispheres, not primitive brains, with metanephric, not opisthonephric, kidneys, and so forth. However fishlike a mammal may be or become, it is virtually impossible that it will actually revert to being a true fish.

The differences that are selected may be very minor, and the amount of selection that is needed is surprisingly small. Various complex mathematical formulae have been devised to show the changes, given different degrees of selection or starting frequencies, but a few minutes of playing with pencil and paper (or a pocket calculator) will show that this is, in fact, true. Evolution does not depend on all-or-none chances of success, so arguments assuming that it does ("this difference is too small to be important") are simply invalid.

One final note on this theory and its consequences: Selection is nothing more than a measure of differential success in reproduction; it has no teleologic or "purposeful" component. Therefore, selection can only adapt an animal to the existing conditions in the environment. Despite this, the term **pre-adaptation** is often used. It describes a situation wherein a character or group of characters was selected because they were of immediate adaptive or selective advantage; later, as conditions changed, these same characters turned out to be even more advantageous and allowed the animal to survive under the new conditions or to adapt to them in a different way. For example, lungs are essential for almost all terrestrial vertebrates and must have appeared before the animals became terrestrial. However, they must have been (and were) adaptive in the original aquatic environment to have arisen in the first place. Their subsequent use by the animal on land was, in a sense, a happy accident—a case of evolutionary serendipity.

Surface-Volume Relations

In any group of animals, large and small forms will differ notably in the relative size of various organs or parts. The reason for many of these proportionate differences lies in a geometric principle so obvious that it is often overlooked, namely, that *as the size of an animal (or any other object) changes, surfaces increase (or decrease) proportionately to the square of linear dimensions, while volumes change proportionately to the cube of linear dimensions.*

This principle is of wide application, because surface-volume relationships are to be found in a variety of structural and functional features of vertebrates. We cite obvious examples: (1) The strength of a leg (like any supporting column) is proportional to its cross section, which varies as the square of linear dimensions, whereas the weight that it supports is proportionate to the cube of linear dimensions. In consequence an elephant cannot have gazelle-like legs. (2) The amount of food that an active animal needs is roughly proportionate to its volume,* the amount of food that its intestine can absorb depends on the area of the intestinal lining. In consequence, large animals have a disproportionately elongated intestine or one with a complicated structure, resulting in a greater internal surface area for digestion.

Anatomic Nomenclature

The student of vertebrate morphology is confronted with a bewildering array of unfamiliar names of anatomic structures. This is unfortunate but inescapable. Vertebrate structures are numerous; for many there are no everyday terms. Even where such names are available, they are often vague and not exactly defined in common usage. Further, it is desirable to have some international system of terms understood in the same sense by scientists of every country.

When anatomy was first studied, all learned works were, as a matter of course, written entirely in Latin. In consequence, Latin names, where already in existence, were applied to anatomic structures, and if no term existed, one was manufactured from Greek or Latin roots and cast in Latin form. Some notes regarding the formation of anatomic terms are given in Appendix 2. Today Latin has ceased to be an international language as far as the general text of scientific books is concerned. Latin anatomic terms, however, are still used. We cannot do without them, although we often use them in a somewhat anglicized form—speaking, for example, of the "deltoid muscle" of the shoulder rather than the "musculus deltoideus," or of the "parietal bone" rather than the "os parietalis."

Latin is an inflected language, and thus its nouns and adjectives have a variety of endings to express not merely singular and plural numbers but also a variety of cases and a rather arbitrary system of genders. Until recent decades, some knowledge of Latin grammar was part of the equipment of every college student, and the manipulation of Latin terminology presented no difficulty. Today, this is not the case, rather unfortunately, for a biologist should know at least enough Latin grammar to avoid such gaucheries as speaking of "humeruses" instead of humeri and "femurs" instead of femora. Fortunately, the number of noun and adjectival endings ordinarily used in anatomic terms is limited, and these can be readily learned (cf. Appendix 2).

It is accepted procedure in anatomic nomenclature that where a structure is present in mammals—particularly in man—the name used there is applied to the same structure in other forms.

Thus, for example, man and many mammals have a clavicle, or collar bone, and the equivalent element in the shoulder structure should be called by the same name in reptiles, amphibians, or fishes, even though its appearance is radically different. Sometimes, however, an incorrect identification may be made and a name wrongly applied. Teleost fishes have a bone similar in position to the clavicle, and that name was customarily given to it; we now know, however, that the teleosts lack the true

---

*Emphasis on *active*; basal metabolism in a resting condition is another matter.

clavicle; the bone present there is a different one (the cleithrum, p. 201). If homologies are in doubt, it is better to use a different name for the structure in question. For example, there is a muscle in the thigh of reptiles that may be homologous with the sartorius muscle of mammals; however, because there is some doubt of the homology, it is customary to give the reptilian muscle a different name—the ambiens muscle (p. 297). If a structure encountered in a lower group has no mammalian equivalent, a new name must, of course, be used.

Although, in general, anatomic terminology has been a rather stable and uniform system, there arose, quite naturally, a number of differences in terminology between different schools of work and in different countries. Motivated by the laudable desire to achieve uniformity, the German Anatomical Association, in convention at Basel near the end of the last century, brought forward a comprehensive scheme of terms, which members hoped would receive universal adoption in human anatomy. This terminology, usually referred to as the "BNA," was adopted by medical schools and has been widely used in medical work. Many of the terms in this code were, unfortunately, ineptly chosen, and we noted earlier the conflicts in terms of body position. At an International Congress of Anatomists meeting in Paris in 1955, a modified code was drawn up, improving the situation to a considerable degree, and the revisions embodied in the new Nomina Anatomica Parisiensia, NAP, are now gradually supplanting the older terms in medical school practice. There is also a "standard" nomenclature in use for veterinary anatomy. All these systems are based on mammals and become difficult to use for lower vertebrates. More recently, avian anatomists have thus proposed a "standard" nomenclature for birds.

Taxonomy and Classification

Obviously, it is essential to have names for different animals if we are to discuss them. It is equally clear that the more information that the names provide, the more useful they will be, and that there must be general agreement on the names so everyone can understand them. No system is perfect and the one used to name animals is no exception, but it is a workable and generally accepted system dating back over 200 years to the tenth edition of Systema Naturae by the Swedish botanist Linnaeus.

The basic unit is the **species**. A species can theoretically be defined as a group of interbreeding or potentially interbreeding individuals. However, this definition is rarely used in practice. Sometimes it cannot be applied: Many plants and lower animals and even a few vertebrates do not use sexual reproduction, so there is no breeding. Fossils, too, are excluded from any testing by this means. Even with normal, sexually reproducing animals, it is rarely feasible or even possible to test whether animals can interbreed. Finally, the distinction is not clear-cut, and, especially in captivity, forms that everyone agrees are separate species (lions and tigers, for example) may interbreed. A somewhat cynical definition, but one with some degree of truth, is that a species is a group that is considered to be a species by a competent taxonomist. Actually, the problems are not as great as you may think from this. In most cases, species are relatively distinct, separate, and recognizable with practice. Lions, tigers, leopards, and jaguars are all big cats but are obviously different; they are good species.

Species are grouped in various larger or higher units. All species belong to genera (singular, genus), all genera to families, all families to orders, all orders to classes, and all classes to phyla. Thus, the domestic dog *(Canis familiaris)* is in the genus *Canis*, the family Canidae (all dogs), the order Carnivora (most carnivorous mammals), the class Mammalia, and the phylum Chordata. Sometimes more subdivisions are needed; in such cases the prefixes super-, sub-, and infra- are used. In some cases, still other ranks are inserted, but these need not be considered here. There are no clear-cut rules distinguishing these ranks or definitions for them. All are subjective, though in many cases they correspond to familiar groupings that have common English names. In the example just used, *Canis* includes what one thinks of as dogs in a loose sense: domestic dogs, wolves, and coyotes, and the like. Foxes are dogs in a still broader sense and are in the same family, but not in the genus *Canis.* Similarly, squirrels are a family, and rodents are an order.

Because all groups above the specific level are subjective and because new evidence is constantly being discovered, there is no "correct" classification or universal agreement. Ideally, the classification should reflect the phylogeny of the animals. Such a classification is frequently called a "natural" one. In theory, all the members of any taxon (species, genus, or whatever level) share a common ancestor not shared with the members of any other comparable taxon. Again, this is theory; aside from the common ancestor being unknowable without a perfect fossil record, which does not exist, there is no agreement about whether the common ancestor is theoretically an individual, a species, or some higher taxon. Nevertheless, most people assume that the classification they use is, as far as possible, a "natural" one. Recently, other types of classification, based simply on resemblances, have been proposed; most of these resemble the traditional classifications, but the ideas behind them are not always accepted.

Appendix 1 gives a classification of chordates. Like any other, this is arbitrary in many ways, and most zoologists would disagree with one or more of its features. However, it does serve as a framework and will be used throughout this book. Unfortunately, confusion can easily arise. Not only are there many synonyms (the order including the turtles can be called Testudines, Testudinata, or Chelonia), but also in many cases, the same name is used for different concepts (some workers, for example, exclude the dinosaurs from the Reptilia). The classification here is moderately conservative and standard, but in a few places we have used relatively uncommon terms or arrangements that we believe better reflect the relationships of the animals concerned. There is currently much debate concerning the methods to be used in erecting such a classification; the differences in the results are usually less than those in the methods, but they do exist. For those familiar with these problems, we have not used a cladistic analysis but have followed more traditional lines.

Finally, a few practical matters are worth noting. The formal names of taxa are in Latin (or are used as if they were Latin—many are based on Greek roots, personal names, or various unlikely sources). Names of families, genera, and species are controlled by a complex set of rules administered by an international commission. Basically, these rules concern priority; the first name used, with certain exceptions, must be retained in order to promote stability. The names of higher taxa are not so controlled and may be changed as concepts of the relationships of the animals change.

Generic and specific names are always (or should be) printed in italics (*Canis familiaris* for example); names of higher taxa are not italicized. In writing or typing, italics are indicated by underlining (i.e., <u>Canis familiaris</u>). Generic names are always capitalized; specific names are never capitalized, even if they are based on a proper noun (*Sciurus carolinensis,* the gray squirrel of eastern North America). A generic name may be used by itself, but a specific name must always include the generic name. The generic name *Sciurus* is used *only* for squirrels, but the specific name *carolinensis* may be used in many different genera (besides *Sciurus*, the mammalian genera *Blarina*, *Castor*, and *Clethrionomys* include species or subspecies of this name). Names above the generic level are always capitalized and are plural; thus, "the Canidae are . . . ." The names of families always end in -idae; those of subfamilies end in -inae, and those of superfamilies, in -oidea. There are no standard endings for other ranks, though in some groups many ordinal names end in -iformes. Common names are usually not capitalized, although proper names in them are, and some people, particularly ornithologists, favor using capitals for common names of species.

Chapter 2

The Vertebrate Pedigree

Although this text primarily concerns vertebrates, there exist various animal types lacking a backbone but closely allied to the vertebrates. Study of these more lowly forms contributes to our understanding of vertebrate structure and history. Further, although (as will be seen) we have little certain knowledge of the early ancestry of the vertebrates, the subject of their pedigree deserves consideration. In this chapter, we will first describe, as lower chordates and hemichordates, certain small marine animals definitely allied in some manner to the vertebrates, then discuss possible relationships of the vertebrates to various invertebrate phyla, and finally, attempt to plot out a reasonable vertebrate pedigree.

The vertebrates do not in themselves constitute a major division of the animal kingdom. They are considered merely one subdivision, although by far the largest subdivision, of the phylum **Chordata**, the other members of which are to be considered briefly here. The "**lower chordates**" lack the backbone and many other advanced structures of their vertebrate relatives. They do, however, exhibit the basic features characteristic of the vertebrates and not found elsewhere in the animal kingdom. These features indicate that they are truly related to vertebrates and hence properly included in a common group with them. The term Chordata itself implies that a notochord (or chorda) is present. Again, a dorsal, hollow nerve cord is a common feature. Most characteristic of all is the fact that pharyngeal gills are universally present in chordates.

Amphioxus

Most workers subdivide the phylum Chordata into three subphyla—in roughly ascending order, Urochordata, Cephalochordata, and Vertebrata. Here we change the order and begin our discussion of the lower chordates with the **Cephalochordata**, in which the similarities to the vertebrates are most obvious. The subphylum includes only a few closely related forms, all commonly termed **amphioxus**, a name replaced in formal taxonomic usage by *Branchiostoma* (Figs. 4, 5). They are translucent animals, fishlike in appearance and proportions, found in shallow marine waters in various regions of the world, and sometimes locally abundant. As the shape suggests, they can swim readily, but because of their poorly developed fins, rather ineffectively; for the most part they spend their time with the body buried in the sands of the bottom, with merely the anterior end projecting.

Despite the piscine appearance, it is obvious that we are dealing with forms far

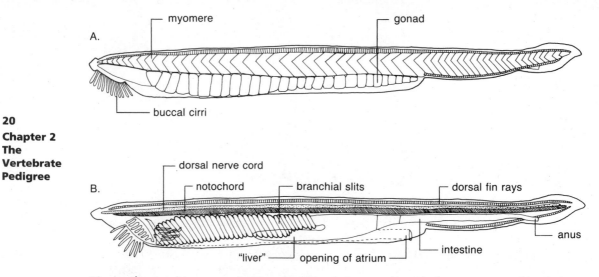

Figure 4. Amphioxus, a primitive chordate. *A,* As seen through the transparent skin. *B,* A parasagittal section to show the internal structures.

more primitive than any fish. There are no paired fins or limbs of any sort. Cartilage-like materials stiffen the gills, dorsal fin, and mouth parts, but no part of the normal vertebrate skeleton of vertebrae, ribs, or skull is to be found. The main skeletal structure is the notochord, which persists through life and (in contrast to the vertebrates) extends clear to the tip of the snout—a feature to which the group owes its name. The notochord of amphioxus is structurally quite unlike that of vertebrates in that some of its cells have become muscular, enabling the animal to vary the tension within and hence the rigidity of the notochord. The details of how it functions are not known, but certainly it prevents telescoping during vigorous swimming and serves as a convenient central "peg" on which to hang the body organs. There are nerves serially arranged along a typical, dorsal, hollow nerve cord; but although the cord is somewhat larger anteriorly, there is no true brain, and there are only dubious traces of sense organs that might correspond to the nose or eye. Much as in fishes, the major musculature consists of a segmental series of blocks of muscle arranged in V's down either side of the body; alternating waves of contraction of these muscles bring about the swimming movements.

For the most part, the digestive tract is very simple. There is a mouth (**buccal cavity**) surrounded by a circle of stiffened projecting cirri. The pharynx is greatly elongated, extending about half the total length of the body. Back of the pharynx, the gut is a straightforward tube with little sign of division into successive chambers, although chemical treatment of food appears to predominate in the anterior part of its length, absorption at the back. A large pouchlike outgrowth is generally compared to a liver, although the homology is dubious. Within this outgrowth, also called a **cecum**, there may occur phagocytosis and intracellular digestion, features found in many invertebrates but not in vertebrates. As in vertebrates, the tube ends at the anus, far short of the end of the body, which thus terminates in a true tail.

The **pharynx** is highly specialized for food collecting. Amphioxus lives on particles gathered from the sea water; these are taken in through the mouth by ciliary action and strained out from the water as it passes out of the body through the

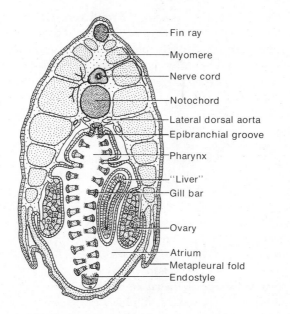

Fin ray
Myomere
Nerve cord
Notochord
Lateral dorsal aorta
Epibranchial groove
Pharynx
"Liver"
Gill bar
Ovary
Atrium
Metapleural fold
Endostyle

Figure 5. Cross section of amphioxus through the pharynx. The peribranchial space surrounding the pharynx, liver, and so on, is, despite its seeming internal position, actually external to the body and is somewhat analogous to the gill chamber of bony fishes. It is formed by the downgrowth around the pharynx of great metapleural folds meeting one another ventrally, and it connects with the exterior through a posteriorly placed opening. Since the gills lie in a diagonal position, such a vertical section as that shown cuts through a number of successive bars. (After Al-Hussaini and Demian.)

branchial slits. These slits are very numerous, far more so than in any vertebrate, for there may be as many as 50 or more pairs of them, and each gill is essentially a double structure, developing in much the horseshoe-shaped manner seen in acorn worms (cf. Fig. 9). A pair of great folds grows downward to meet ventrally and enclose the whole branchial system, serving to protect these delicate organs when amphioxus is buried in the sand. This forms a pocket, the **atrium**, or peribranchial space, opening to the surface only by a pore at the back of the pharynx. But whereas the branchial slits, by their major development, emphasize the animal's relationships to the vertebrates, they differ from those of typical vertebrates in both function and mode of operation. It appears that much of the "breathing" of amphioxus is done through the skin—which, in contrast to that of vertebrates, is quite thin—and that here, as in other lower chordates, the gills are primarily feeding devices. Further, water currents through the gills in vertebrates are effected by muscular pumping; in amphioxus, ciliary action alone is responsible, and cilia are highly developed in the pharynx. A prominent feature is the development of a longitudinal midventral (hypobranchial) groove termed the **endostyle**, running the length of the pharynx; in this, a sticky mucus is abundantly secreted. Ciliary currents carry streamers of this material up the pharyngeal walls, past the branchial slits; trapping food particles on the way, the mucus collects in a dorsal (epibranchial) groove. From this point, cilia carry back the mucus and the enclosed food particles in a continuous slimy band to the intestine; the animal feeds itself by a conveyor belt system.

The major blood vessels of amphioxus are laid out clearly on the vertebrate pattern (Fig. 331), with the blood coursing forward ventrally and back dorsally after passing upward through the gills. There are, however, no blood cells, red or white, or blood pigments and, further, there is no single heart; movement of the blood is accomplished by waves of contraction along some of the principal vessels, together with the contraction of numerous tiny heartlike bulbs situated along the course of the arteries below the gills.

The gonads differ from those of vertebrates—indeed, from those of all other chordates—in being numerous and segmentally arranged. Still more divergent from the vertebrate plan is the nature of the excretory organs. The vertebrate kidney is of a unique type, composed of distinctive water-filtering units that will be described in detail later. In amphioxus, on the other hand, the structures are of a very different sort; they are, as in many invertebrates, segmentally arranged protonephridia, which, in amphioxus, resemble to a degree those characteristic of many flat worms.

Where does amphioxus stand in relation to the vertebrates? A few theorists, for whose ideas as to vertebrate evolution the existence of amphioxus is inconvenient, deny that it is closely related. But the features in which amphioxus resembles the backboned animal are so numerous and so basic that this position is untenable. Strongly in contrast is the suggestion that amphioxus is a secondarily simplified vertebrate (such a condition of secondary simplification is often referred to as "degenerate," but that term has unfortunate moral overtones and is usually best avoided). As we shall see, the young of lampreys, quite unlike the adults, live a life as sedentary filter-feeders and are comparable to amphioxus in many ways. Is amphioxus a lamprey that has, so to speak, never grown up and hence retained, as an adult, the simplicity of structure of the larva? It is probable, as we shall see presently, that a factor to be kept in mind in evolutionary studies is the phenomenon known as **neoteny**, in which larval characteristics are retained to a greater or lesser degree while the reproductive organs become functional. The terminology here is complex, and people, even dictionaries, differ on definitions. One recent and carefully worked out system uses **paedomorphosis** for any retention, in the adult, of characters present in the young of ancestral stages. In theory, this can be caused by **neoteny**, which involves the slowing of somatic development, or by **progenesis** (often called **paedogenesis**, a term with several other meanings, even used for some types of parthenogenesis), which involves the speeding of sexual development; in practice, these are usually impossible to differentiate. For simplicity, we will lump them all together and use neoteny, only because it is shorter and easier to spell than the alternatives. But amphioxus shows many features that are different from those expected in the young of an ancestral lamprey—too many to make this suggestion of relationship by simplification plausible. As a working hypothesis, we shall, like most students of the subject, interpret amphioxus as a specialized and modified survivor of a type of animal ancestral to the vertebrates. As will be seen later, it is important to realize that amphioxus not only is more primitive structurally than any vertebrate but also has a mode of life essentially different from and more primitive than that of vertebrates. In general, vertebrates actively and agressively seek large food objects, eat by muscular movements of jaws or analogous structures, and typically use their gills—operated by means of well-developed muscles—exclusively for breathing. It is often tacitly assumed that the vertebrate ancestor was likewise a vigorous, active swimming form. But amphioxus, although able to swim, is essentially sedentary and, as we have noted, is in contrast a filter-feeder, using cilia rather than muscles in food gathering and utilizing the gills for feeding rather than breathing.

Tunicates (Figs. 6–8)

In seeking further primitive relatives of the vertebrates we may be well advised to look for other filter-feeders that, even if simpler in structure than amphioxus, show at least some of the basic chordate characters of branchial slits, notochord, and dorsal, hollow nerve cord. Such a group is that of the **Urochordata**, the **tunicates** or **sea squirts** and their relatives. These rather common small marine organisms are essentially inactive; the adult does not seek its food but is a highly developed filter-feeder, accepting such particles as it can attract by ciliary action. Many tunicates are found floating freely in the water, singly or in groups, often as tiny barrel-shaped structures; others are attached to the bottom, either as branching colonies or as individuals (Fig. 8 *C–E*). Simplest are the solitary tunicates (Fig. 6 *B*). As an adult, such an animal is an almost formless lump attached to a rock or other underwater object and covered with a leathery-looking "tunic." The only structural features seen externally are an opening at the top, into which water passes, and a lateral opening, through which the water current flows outward. The creature shows no external resemblances to the vertebrates, and internally much of the structure is equally unfamiliar to a student of

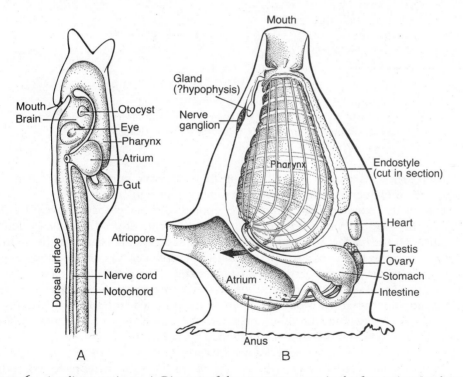

Figure 6. A solitary tunicate. *A,* Diagram of the structures seen in the free-swimming larva (head end above, and only a short section of tail figured). The otocyst is a simple ear structure. *B,* The sessile adult, formed by elaboration of the structures at the anterior end of the larval body. The original dorsal side lies at the left. The large pharynx is attached to the body wall above and below (left and right in the figure); the atrium (corresponding to the peribranchial space of amphioxus) bounds it on either side. Water passing through the latticework gills of the pharynx enters the atrium and, as indicated by the arrow, streams out through the atriopore. (After Delage and Hérouard.)

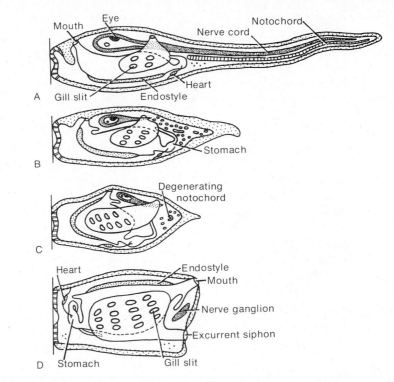

Figure 7. Diagrams of metamorphosis of a solitary tunicate. In *A* the free-swimming larva is attaching to the substrate by anterior suckers. In *B* and *C,* the tail is being resorbed with the loss of the notochord and reduction of the nervous system. The internal organs gradually rotate to their adult positions as shown in *D.* (After Storer and Usinger.)

that group. There is no notochord. Nor is there a nerve cord; instead, there is a simple (and solid) nerve ganglion with a few nerves splaying out from it.

Much of the interior of the animal is occupied by a barrel-shaped structure that serves as the food-gathering device. The water current, created by ciliary action, passes through slits in the sides of the barrel into a surrounding chamber, the **atrium**, which leads to the lateral excurrent opening (or atriopore). On closer examination, it becomes obvious that the barrel is an exaggerated set of internal gills, constituting the pharyngeal region of the animal; there is even an endostyle comparable to that noted in amphioxus, which produces sheets of mucus used to trap food particles, as in amphioxus. Below the enormous pharynx, the digestive tube narrows to form esophagus, stomach, and intestine—all of modest size.

We have here, in the pharyngeal apparatus, a high degree of development of one of the primary characters to be sought in a relative of the vertebrates. For other chordate characters, however, we must turn to the developmental history. In many tunicates, propagation takes place mainly by a process of budding. But in some tunicates, there is a distinct larval form (Figs. *6 A,* 7) that looks like an amphibian tadpole. The "head" of the larva corresponds to the entire body of the adult. The tail is a swimming organ, useful in transporting the young tunicate about in its search for a home. Once the animal attaches and "settles down" to its sedentary adult existence,

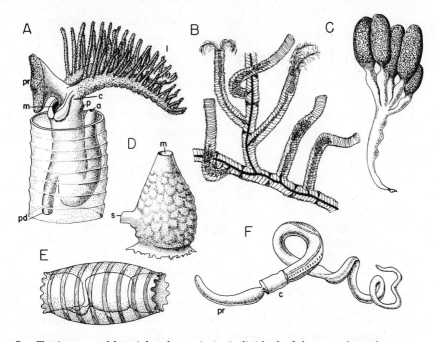

Figure 8. Tunicates and hemichordates. *A,* An individual of the pterobranch genus *Rhabdopleura* projecting from its enclosing tube. *B,* A part of a colony of the same. *C,* A colonial sessile tunicate; each polygonal area is a separate individual of the colony. *D,* External view of a solitary tunicate (cf. Fig. 6 *B*). *E,* A free-floating tunicate, or salp. *F,* An acorn worm *(Balanoglossus). a,* Anus; *c,* collar region; *l,* lophophore; *m,* mouth; *p,* pore or opening from coelom; *pd,* stalk (peduncle) by which individual is attached to remainder of colony; *pr,* proboscis or anterior projection of body; *s,* siphon that carries off water and body products. (Mainly after Delage and Hérouard.)

the tail dwindles and is absorbed into the body. In this tail, however, are to be found major proofs of the vertebrate relationship of the tunicates. In the larval tail (as the group name, Urochordata, implies), there is a well-developed notochord and, above this, a typical hollow, dorsal, nerve cord. These structures are, however, less advanced than those of amphioxus, because here there is no segmentation of the swimming muscles or of the nerves supplying them. Anteriorly, there are, in the larva, a rudimentary brain and sense organs. At metamorphosis, the notochord and the larval nerve cord (unnecessary in the sessile adult) disappear.

The tunicates, thus, are definitely chordates and definitely related to the vertebrates. How do they fit into the evolutionary story? Those who believe that the vertebrate ancestors were from the earliest times actively swimming animals would regard the tunicates as a specialized side branch of the vertebrate ancestral line and would consider that the common ancestor of vertebrates and tunicates was a free-swimming adult, somewhat like the larval tunicate. From this the vertebrates may have "ascended" by an improved continuation of an active mode of life, whereas the tunicates tended to become "degenerate" and lost most progressive structural features except for the food-straining gill barrel—became, in fact, fit subjects for evolutionary sermons on the results of slothful living. There is, however, another interpretation that is more probable, namely, that the chordate ancestor of the vertebrates was, rather, a sessile food-strainer somewhat like an adult tunicate; that the tail first appeared as an adaptation in the larva, aiding it to find a suitable place in which the animal could "settle down"; and that the development of higher forms came about by the retention of the tail and the free-swimming habit in adult life with the elimination of a sessile adult stage. It is reasonable to believe that this is an example of neoteny—reproduction by larval animals, with the result that in their descendants, the old adult stage was obliterated, and there began a new progressive evolutionary series leading to the vertebrates.

Hemichordates

A further series of forms definitely related to the vertebrates is that of the **Hemichordata**. Here, the key chordate characters are little developed, so that we may consider the hemichordates a phylum separate from the Chordata, although related to them.

The best known hemichordates are the **acorn worms**, such as *Balanoglossus* (Figs. 8, 9), termed as a group the class **Enteropneusta** and found not uncommonly in tidal flats. The long, slender body suggests that the acorn worms are active animals. This is not the case; they are essentially sedentary burrowers in mud and are filter-feeders comparable in their general mode of life to the tunicates. The general body shape is wormlike, but there the resemblance ends, for their structure is not at all comparable to that of ordinary annelid worms. Even externally the acorn worms are distinctive. The body terminates anteriorly in a tough yet flexible and muscular "snout" or proboscis, of variable length which serves as a burrowing organ. Behind the proboscis, a distinct thickened section of the body forms the "collar" region; the name acorn worm originated from the fact that in some forms, the proboscis and collar have somewhat the appearance of an acorn in its cup.

In most regards, acorn worms show no special resemblance to the vertebrates or other chordates. For a short distance in the collar region, there is a dorsal nerve cord that is more or less hollow. But over the rest of the body, the nerve cells and fibers are rather diffusely distributed in the skin, although there is some development of solid dorsal and ventral strands of neural tissue. There is no proper notochord, although a stout pouch of tissue at the base of the proboscis has been compared (rather dubiously) to an imperfectly developed structure of that sort.

But, as in tunicates, we find that the vertebrate type of pharyngeal gills are pres-

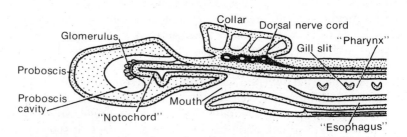

Figure 9. Diagrammatic longitudinal section through the head of *Saccoglossus,* an acorn worm. (After Storer and Usinger.)

ent in a characteristic and highly developed manner. The gills are not so pronounced as they are in tunicates, but there is, behind the collar, an extended pharynx (partitioned off from the food passage to the stomach), from which on either side open out branchial slits quite comparable, even in details of structure and development, to those of amphioxus. Vertebrates are not descended from acorn worms as such, but these forms may reasonably be interpreted as a group not distantly removed from our early chordate ancestors—essentially sedentary food-strainers with, however, some degree of potential motility in the adult.

The acorn worms show, in the branchial slits at least, definite proof of vertebrate relationships. But in a second type of hemichordates, the **Pterobranchia** (Fig. 8 *A, B*), hardly a trace of vertebrate structure is to be found, and were it not clear that they are affiliated with the enteropneusts, one would hardly suspect that they belonged to this general stock. The pterobranchs are tiny, rare marine animals of which only a few genera are known. They form little plantlike colonies, whose individuals project like small flowers at the ends of a branching series of tubes. The short body is doubled back on itself, so that the anus opens anteriorly back of the head. Proofs of relationship to the acorn worms lie in the fact that there is a snoutlike anterior projection beyond the mouth, corresponding to the enteropneust proboscis, and back of this a short collar region. But almost all resemblances to acorn worms—to say nothing of the more highly developed chordates—are lacking. There is little development of a nervous system, no trace of a hollow nerve cord, and not the slightest suggestion of a notochord. Even the feeding mechanisms are of a very different type. True, these plantlike animals feed, as do more typical lower chordates, on food particles drawn in by ciliary action. But there is almost none of the branchial mechanism, which is so important in the filter-feeding of tunicates and amphioxus. One of the two better-known pterobranchs has a single pair of small pharyngeal gills, the other none at all. Instead, there project from the collar region large tentacle-like structures, termed **lophophores**; these are supplied with bands of cilia that collect food particles and bring them to the mouth. Even acorn worms, with good gills and no lophophores, gather almost all their food by cilia and sheets of mucus on the outside of the proboscis and adjacent areas.

So unchordate-like are these small creatures that one is tempted to believe that they are "degenerate," perhaps relatively modern in development. But it is believed that they are a very ancient group indeed. Paleontologists have long been familiar with a variety of small tubelike structures termed **graptolites**, which were abundant in the seas about the time of the appearance of the oldest vertebrates. These tubes are similar to those that shield the modern pterobranchs.

With the pterobranchs, we conclude the series of lowly chordates or near-chordates.* What do they teach us about vertebrate origins? As we have said, one (at first thought) might expect the vertebrate ancestral line to lead through active little forms, paralleling, at least, such invertebrate groups as the progressive arthropods such as crustaceans and insects. But consideration of the lower chordates tends to dim such expectations. For the most part, they are sluggish, sedentary, passive filter-feeders. However, before attempting to reach a conclusion, let us survey the various invertebrate phyla in search of possible chordate relatives.

---

*Possibly allied to chordates are some slender, elongate, deep sea forms termed the **Pogonophora**, but they are poorly known and obviously "degenerate" (the digestive tube is absent), and we may leave them, without loss, in oblivion.

Invertebrate Phylogeny

In recent decades students of invertebrate zoology have come to agree on many points (but not all!) concerning the phylogeny of animals without backbones (Fig. 10). All forms above the level of protozoans and sponges are termed the **Metazoa**. The majority opinion is that the ancestral metazoan was a sessile, attached form, probably appearing like some of the living coelenterates (Cnidaria) such as the sea anemones or little hydras. The living coelenterates, however, are specialized in the possession of stinging cells, which enable them to capture prey of some size. A coelenterate ancestor, lacking such weapons and depending for a living on food particles floating within reach in water currents, seems satisfactory as a truly primitive metazoan.

Despite their specialization in developing stinging cells, the coelenterates appear clearly to represent the ancestral condition in one very important point, the absence of the middle body layer. As presumably was the case in a truly primitive metazoan, the coelenterate body has a simple, two-layered structure, with little between the

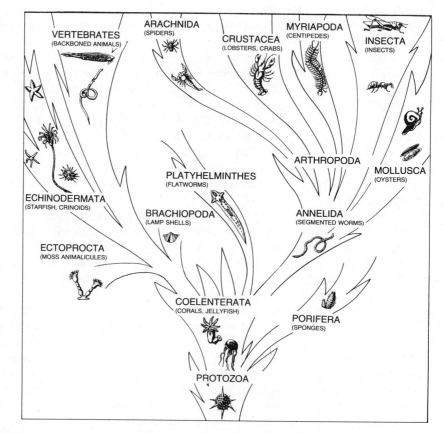

Figure 10. A simplified family tree of the animal kingdom, to show the probable relationships of the vertebrates. (After Romer, Man and the Vertebrates, University of Chicago Press.)

"skin"—the ectoderm—and the lining, termed the endoderm, of the inner gut cavity. Above the coelenterate level, most animals have a third intermediate body layer, the mesoderm, from which muscular, circulatory, and other systems are formed; the mesoderm becomes, in bulk at any rate, the most important of the three tissues. Further, many invertebrates develop a coelom within the mesodermal tissues.

Two contrasting methods of embryonic formation of this third body layer are seen. In one type especially characteristic of echinoderms—starfishes, sea urchins, sea lilies, and the like—the mesoderm arises in the form of pouches growing outward from the walls of the gut; these pouches remain in the adult as closed body cavities. In a second type, the mesoderm arises as solid masses of cells budded off from an area near the posterior end of the body, and the body cavities arise by cleavage within the masses of mesodermal cells. To this second type belong the annelid worms and the molluscs. The great group of joint-legged animals, the arthropods, appear to belong to this second group as well, although their developmental pattern is much modified; and certain other forms, such as the flat worms, appear to be offshoots from the base of this major stock. We thus have the concept that, above the coelenterate level, the invertebrates form two great branches, in Y-fashion, with the echinoderms at the end of one branch and the great host of familiar advanced invertebrates clustered on the other. A few of the less familiar, mainly sessile marine forms, such as the lamp shells (brachiopods) and moss animalcules (ectoprocts), do not fit well on either main branch but are perhaps somewhat closer to the echinoderms.

The two stocks contrast not only in the mode of formation of the middle body layer but also in the patterns of cleavage of the fertilized egg, the method of gastrulation, and the larval development. Both echinoderms on the one hand and aquatic annelids and molluscs on the other grow from the egg into tiny larvae of simple structure, with bands and tufts of cilia arranged in characteristic patterns on the surface of the body. The echinoderm larva has an arrangement of these and other features that differs markedly from those of the larvae of annelid worms and molluscs.

From which point on this family tree of the invertebrates does the chordate (and vertebrate) branch arise? Theories on this subject have been numerous but have provided few positive results.

One solution to the problem might be to suggest a direct origin of chordates from the most primitive metazoans. Here there are no great difficulties to overcome, for such animals have few specialized features that must be lost before starting on the path toward the vertebrates. But in reality, advocacy of such a descent would seem to be begging the question. A number of basic advances are common to almost all invertebrate phyla: presence of a middle body layer, of a true body cavity (coelom), of both mouth and anus in the digestive tract, and so forth. It seems highly improbable that the vertebrates acquired these progressive features entirely independently of other groups. Search seems warranted for possible relatives, if not direct ancestors, at a higher level.

Annelids as Ancestors

The annelid worms offer a possible point of departure; the theory of vertebrate origin from annelids was warmly advocated during the later decades of the nineteenth century. The common earthworm is none too prepossessing as an ancestor, but there are numerous marine annelids of a more progressive and attractive nature. Annelids show

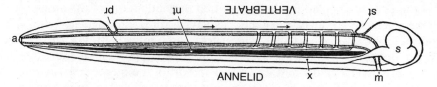

Figure 11. Diagram to illustrate the supposed transformation of an annelid worm into a vertebrate. In normal position this represents the annelid with a "brain" *(s)* at the front end and a nerve cord *(x)* running along the underside of the body. The mouth *(m)* is on the underside of the animal, the anus *(a)* at the end of the tail; the blood stream (indicated by arrows) flows forward on the upper side of the body, back on the underside. Turn the book upside down

and now we have the vertebrate, with nerve cord and blood streams reversed. But it is necessary to build a new mouth *(pr)* and anus *(st)* and close the old ones; the worm really had no notochord *(nt)*; and the supposed change is not as simple as it seems. (From Wilder, History of the Human Body, by permission of Henry Holt & Co., publishers.)

bilateral symmetry, as do vertebrates, and, in correlation with this, some are, like typical vertebrates, active animals in contrast with the sessile types common in many invertebrate phyla. Then too, they are segmented forms, as the vertebrates are to at least some extent. As in vertebrates, the central nervous system is composed of a brainlike mass at the anterior end of the body and a longitudinal nerve cord.

So far so good. But beyond this point, the comparison breaks down. Even the segmentation is a weak argument; because the annelid is segmented in every respect, from skin to gut, whereas the segmentation of a vertebrate is primarily confined to part of the middle body layer. The annelid has, it is true, a longitudinal nerve cord. But it is solid, not hollow, and ventral, not dorsal, in position. This last point is especially troublesome to advocates of this theory. They have "resolved" the difficulties by assuming that a vertebrate is a worm upside down (Fig. 11). This is difficult to accept (even a worm usually knows which way is up) and involves further perplexities. The worm's mouth is on the underside of the head, as is that of a vertebrate. A reversal of surfaces implies that the old mouth of the worm has closed and been replaced, historically, by a new one. Traces of this theoretic old mouth in vertebrate embryos—it should pass upward and forward through the brain to the top of the head—cannot be found.

Even if this difficulty in orientation were solved, there are other problems. In an annelid, there is no trace of a notochord or internal gills. This may be discounted, perhaps, by the fact that (as we have seen) these structures are little developed in some of the simplest chordates. But a crucial difference is that the type of mesoderm formation differs, because in chordates the pouch-type of mesoderm formation is the basic pattern. There is thus almost no positive reason to believe in a descent of vertebrates from annelids, and there are so many difficulties that there is little reason to take stock in an annelid theory.

Arachnids as Ancestors

The arthropods, including crustaceans, myriapods, arachnids, and insects, are, it is thought, descended from annelids or from worm ancestors closely related to them.

The arthropods include the most progressive and successful of all invertebrate animals, and it is natural that they have received considerable attention as possible vertebrate ancestors. Among the arthropods, it is the arachnids that have been selected as the most likely candidates for vertebrate kinship. Spiders are the most common of arachnids, but the scorpions are more generalized types. Still more primitive are aquatic arachnids, including the horseshoe crab (*Limulus*) and relatives of high geologic antiquity; one ancient extinct group, the eurypterids, has been thought by some to be close to the ancestry of the vertebrates.

In arachnids, as in annelids, there is a ventral nerve cord. This poses the same problem as in the worms: the upper and lower surfaces must, it seems, be reversed, a new mouth must be exchanged for an old one, and so on. Again, as in annelids, we find conspicuous differences from vertebrates in segmentation; still again, there is no trace in arachnids of notochord or internal gills. An added difficulty with arachnids as ancestors of the vertebrates is the presence of numerous, complex, jointed legs of the arthropod type. It is impossible that these were transformed into fish fins, and the arthropod legs must be done away with before the development of vertebrate appendages can begin. The old eurypterids were covered with chitinous armor, which in some cases had a superficial resemblance to the bony armor of certain archaic fishes, such as those shown in Figure 18. But even this resemblance is meaningless; because the resemblance is a top-to-top one, whereas, because of the reversal of surfaces, it should be the bottom of the arachnid that resembles the upper side of the vertebrate. In summary, to make a backboned animal from an arachnid, the supposed ancestor must have lost almost every characteristic feature that it once possessed and reduced itself practically to an amorphous jelly before resurrecting itself as a vertebrate.

An amusing variant of the arachnid theory removes a whole series of difficulties caused by the necessity of turning the arachnid over to make a vertebrate. Under this theory, there has been no reversal of surfaces. The arthropod digestive tube has expansions and subdivisions that resemble the cavities of the brain and spinal cord of vertebrates. This particular theory assumes that the original ventral nerve cord of arachnids migrated upward to surround the original digestive tube, which actually became the cavities of the nervous system; meanwhile, a brand new digestive tract developed by a closing over of the furrows present ventrally between the jointed legs. Technically, this theory solves the difficulties encountered in turning the animal over, but it appears to offer new problems as difficult as those solved. There is no adequate explanation for this change from one digestive system to another; the intermediate stages are, to say the least, difficult to imagine, and the embryology provides no support for such an idea.

Echinoderm Affinities

Unlikely as it seems at first sight, the best clues to chordate relationships are to be found in a study of the echinoderms—the starfish, sea urchins, and the like. Several lines of work suggest that, despite the obvious and strong contrasts, the two phyla are nevertheless related. In most vertebrates, mesoderm formation is a complex process, but in amphioxus, we find mesoderm forming as pouches from the gut, just as it does in echinoderms. Further, whereas amphioxus has a long series of body segments, lower chordate types have only three; this is also the case with echinoderms. Also, some hemichordates have a ciliated larva (Fig. 12) of the same type as that of

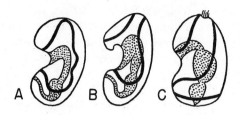

Figure 12. Diagrammatic side views of the larvae of *(A)* a sea cucumber, *(B)* a starfish, and *(C)* an acorn worm, all much enlarged. The black lines represent ciliated bands. The digestive tract (stippled) appears through the translucent body. Views are from the left side; the larvae are bilaterally symmetric. (After Delage and Hérouard.)

echinoderms—so similar, in fact, that until the life history was known, the hemichordate larvae were thought to be those of starfish.

Even biochemistry helps to establish the case. The proteins of blood serum vary greatly from form to form, and it appears, in general, that the more closely related two animals are, the more similar are these proteins. Tests of the sera of acorn worms and lower chordates show their definite relationships to those of echinoderms but not to other invertebrates. Again, muscle chemistry tends to link the two groups. The muscles of animals contain phosphorus compounds that speed up the release of energy for muscle activity. In vertebrates, the material combined with the phosphate is creatine; in most nonvertebrate groups, another compound, arginine, is present instead. But some echinoderms have creatine as well as arginine, and arginine is present in some tunicates and is present (as well as creatine) in some hemichordates—facts that tend further to link the two groups.

These arguments have been thought convincing by most workers and represent the majority view. However, there is not universal agreement, and some workers have recently presented cogent arguments against echinoderm affinities, advocating, instead, such things as nemertine worms as possible vertebrate ancestors. At the other extreme, some consider that some forms usually believed to be odd extinct echinoderms were actually early but already highly complex chordates. In the future, we may have to change our minds, but we still prefer the theory just outlined.

Chordate Phylogeny

What do these resemblances mean? Surely, the vertebrates and their chordate relatives are not derived from echinoderms, with their specialized organs and skeletal plates and their pronounced (though secondary) radial symmetry. Such a form as a starfish or a sea urchin is obviously far from any line leading to a vertebrate. But one important point should be kept in mind. Most echinoderms are free-living and capable of locomotion to some degree, but the fossil record indicates that the ancestral echinoderms were sessile forms. One group of living echinoderms, the crinoids or sea lilies, still constists mainly of such sessile forms. Attached by a stalk to the sea bottom, they spread out above their compact bodies a series of feathery arms. Along these arms are bands of cilia, which filter out from the water the food particles upon which the sea lily subsists and carry them down to the mouth.

Here lies, one may believe, the clue to the whole story. This mode of life is precisely that of the little pterobranchs, which, as we have seen, are unquestionably related to vertebrates despite their simple structure and the absence in them of almost every diagnostic character of vertebrates and even of chordates.

Despite the contrasts between primitive echinoderms and pterobranchs, the two can readily be derived from an ancient common ancestor and, except for the little proboscis, which tends to tie them to the acorn worms, the pterobranchs are certainly close to the pattern expected in this ancestor.

Clearly, the evidence suggests that this ancestor was a sessile bottom dweller, subsisting on food particles gathered and brought to the mouth by outstretched lophophore arms. On this basis, a reasonable theory of chordate evolution can be erected (Fig. 13). Little animals of this character, collecting food by lophophores, are common today, notably brachiopods or lamp shells (so-called because the body and lophophore are protected by a paired shell) and the tiny ectoprocts—the "moss animalcules." These forms are suspected of at least distant relationships to echinoderms and chordates, and the chordate ancestor was rather surely one of a series of such lophophore-bearers. From it, with elaboration of varied specialized organs, may have arisen the echinoderms, and from it, with little change except a thickening of a collar region and the development of a small proboscis, may have come the pterobranchs.

An early development of true chordate characters was a shift in the method of obtaining food particles, the substitution of gill-filtering for lophophores. Even in one of the pterobranchs, a single pair of gill slits has developed, apparently aiding in the flow of food materials into the digestive tract. With the increase and elaboration of the branchial filtering system, lophophores were abandoned. The acorn worms, still essentially sedentary, appear to represent a side branch at this stage of evolution.

Further elaboration of the pharyngeal straining device led eventually, in one higher branch, to the typical tunicates, in which the whole animal seems to be little more than an elaborate food-filter. But as this stage was approached, there appeared, it would seem, a new adaptation that was to alter radically the whole picture of higher chordate evolution.

The embryo or larva of a sessile organism must find a proper place on the sea bottom on which to settle down for adult life. How to reach and select it? Some acorn worms, we have noted, have a ciliated larva, but such a larva's powers of locomotion are limited. Much better is the tadpole-like larva that may have developed before the tunicate level was reached. Here the tail is muscular; a notochord stiffens the structure; a nerve cord and nerves supervise locomotion; and sense organs guide the movements of this new swimming structure toward a proper place for fixation and adult existence.

Once this new larval structure was evolved, a radical change of direction in chordate evolution opened out. A new active type of life became possible. Conservative forms specialized as tunicates. But for others, neoteny appears to have entered the picture. The animal ceased to "settle down." The larval locomotor structures were retained throughout life, even though filter-feeding continued to be the means of substenance; the pharyngeal filtering apparatus could be transported from place to place as favorable opportunities presented themselves. Amphioxus, although a bit off the direct ascending line, represents a more advanced stage in which filter-feeding persists but the ancestral fixed adult condition has been abandoned. But most chordates did not stop at this level. A great burst of evolutionary activity resulted from

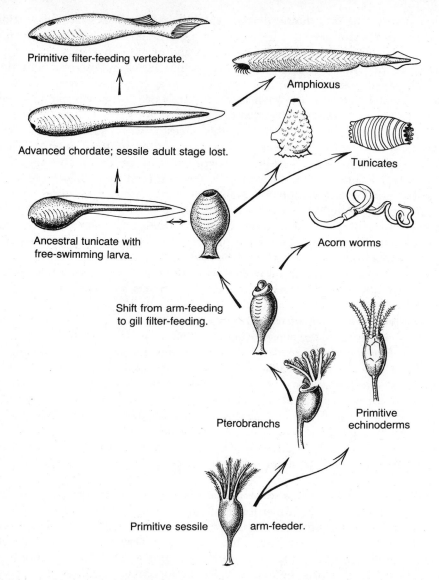

Primitive filter-feeding vertebrate.

Amphioxus

Advanced chordate; sessile adult stage lost.

Tunicates

Ancestral tunicate with
free-swimming larva.

Acorn worms

Shift from arm-feeding
to gill filter-feeding.

Pterobranchs

Primitive
echinoderms

Primitive sessile   arm-feeder.

Figure 13.   A diagrammatic family tree suggesting the possible mode of evolution of vertebrates. The echinoderms may have arisen from forms not too dissimilar to the little pterobranchs; the acorn worms, from pterobranch descendants which had evolved a gill system but were little more advanced in other regards. Tunicates represent a stage in which, in the adult, the gill apparatus has become highly evolved for feeding, but the important point is the development in some tunicates of a free-swimming larva with advanced features of notochord and nerve cord and free-swimming habits. In further progress to amphioxus and the vertebrates, the old sessile adult stage has been abandoned, and it is the larval type that has initiated the advance. (From Romer, The Vertebrate Story, University of Chicago Press.)

the development of these new locomotor potentialities and led to the major story of vertebrate evolution.

"Visceral" and "Somatic"

In later chapters, we will frequently encounter the terms **visceral** and **somatic**— visceral and somatic skeletal structures, visceral and somatic muscles, and visceral and somatic nerves. The visceral structures have mainly to do with the gut (particularly the pharynx) and its appendages; the somatic structures are those of the "outer" tube of the body (Fig. 2 *B*). One might assume that the two terms had a mere topographic meaning and nothing more, but it is highly probable that there is a long phylogenetic history behind visceral-somatic distinctions (Fig. 14).

Such a lower chordate as a solitary tunicate consists of little else than a pharynx and an appended gut, plus such necessary additions as the gonads and a very simple nervous system. Except that it is, of course, sheathed externally by skin or tunic, the whole animal represents essentially the visceral component of the vertebrate. The somatic component is the new, added series of locomotor devices: swimming muscles, notochord and, for their direction, a more highly evolved nervous system and sense organs. At first these somatic structures were for the most part appended posterior to the visceral animal and were mainly larval. As vertebrate evolution progressed, the two became more broadly overlapping and coordinated with one another. But even today, as seen in development and in adult structure, the original distinctions tend to persist. In many ways, one can regard a vertebrate as two distinct animals, visceral and somatic. The two are welded into a single structure, but some traces of the distinctions between them still persist. The weld is an imperfect one.

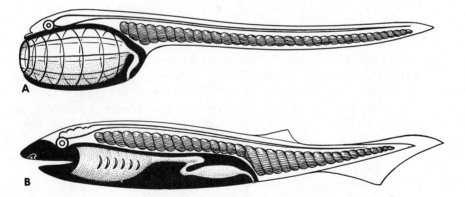

Figure 14. Diagrams to show the contrast between "visceral" and "somatic" components of the chordate body. *A,* A theoretical type of chordate essentially similar to the tunicate larva, but with the somatic component retained in the adult; below, a true vertebrate. The area of the visceral component is outlined in black. In *A,* the somatic animal lies posterior to the visceral animal (representing the ancestral chordate), except that sense organs and anterior part of the nerve cord extend forward dorsally. In *B,* the visceral and somatic components overlap to a considerable degree, and integration of the two is advancing.

Chapter 3

Who's Who Among the Vertebrates

The study of organs and organ systems and their varied forms and functions—the main concern of the present work—provides us with only a one-sided account of the vertebrates. What one should know is not merely the discrete parts, but the total animal, its life and its place in nature. Our present study provides us with no more of a rounded picture of the vertebrates than would the dissection of a cadaver and a course in physiology provide us with a complete knowledge of mankind. It is to be hoped that the student will read some works on the "natural history" of vertebrates and thus gain an idea of the living animals whose bodies are verbally dissected in this book. In this chapter, we shall survey the membership of the vertebrate groups in order to place the forms discussed within a phylogenetic framework.

The Geologic Record

The fossil record and the extinct animals included in it require attention in this regard. In comparative anatomy, one often compares the organs of *existing* members of different groups as though one had descended from the other—as though mammals had descended from the existing reptiles, and these from existing amphibians and fishes. Obviously, however, this is not the case. A turtle is a reptile, but it is not a mammalian ancestor. It has had just as much time to diverge from the common primitive reptilian stock as has the mammal. A frog is an amphibian, but it is definitely not the sort of amphibian from which more progressive terrestrial vertebrates were derived. Only through paleontology, the study of fossils, can we hope to discover the nature of the actual ancestors from which the varied living vertebrates arose. Naturally, the fossils are not always of those actual ancestors, but they are, in many cases, very close to them and are certainly far more similar than any living forms.

In discussing fossils, some notion of the geologic time scale is necessary (cf. Table 1). The earth's history of several billion years is divided by geologists into a few major time units termed **eras**; these are subdivided into a number of **periods**. For the earlier eras, there is little adequate knowledge of life of any sort; the fossil record is almost entirely confined to the last three eras, spanning somewhat over half a billion years of earth history. The first of these three eras, the **Paleozoic Era**, or Age of Ancient Life, covered about 340 million years and is divided into half a dozen periods. The fossil record remaining from the seas of the oldest period (Cambrian)

Table 1. Geologic Periods Subsequent to the Time When Fossils First Became Abundant

(The Carboniferous is frequently subdivided into two periods, Mississippian [earlier] and Pennsylvanian [later]. The time estimates are based on the rate of disintegration of radioactive materials found in a number of deposits.)

Era (and Duration)	Period	Estimated Time Since Beginning of Each Period (in Millions of Years)	Epoch	Life
Cenozoic (age of mammals; about 65 million years)	Quaternary	2+	Holocene (Recent)	Modern species and subspecies; dominance of man.
			Pleistocene	Modern species of mammals or their forerunners; decimation of large mammals; widespread glaciation.
	Tertiary	65	Pliocene	Appearance of many modern genera of mammals.
			Miocene	Rise of modern subfamilies of mammals; spread of grassy plains; evolution of grazing mammals.
			Oligocene	Rise of modern families of mammals.
			Eocene	Rise of modern orders and suborders of mammals.
			Paleocene	Dominance of archaic mammals.
Mesozoic (age of reptiles; about 165 million years)	Cretaceous	135		Dominance of angiosperm plants; extinction of large reptiles and ammonites by end of period.
	Jurassic	190		Reptiles dominant on land, sea, and in air; first birds; archaic mammals.
	Triassic	230		First dinosaurs, turtles, ichthyosaurs, plesiosaurs, mammals; cycads and conifers dominant.

(*Table continues next page*)

Table 1. Geologic Periods Subsequent to the Time When Fossils First Became Abundant (*continued*)

Era (and Duration)	Period	Estimated Time Since Beginning of Each Period (in Millions of Years)	Epoch	Life
Paleozoic (about 360 million years)	Permian	280		Radiation of reptiles, which displace amphibians as dominant group; widespread glaciation in southern hemisphere.
	Carbon-iferous	350		Fern and seed fern coal forests; sharks and crinoids abundant; radiation of amphibians; first reptiles.
	Devonian	410		Age of fishes; first trees, forests, and amphibians.
	Silurian	430		Invasion of the land by plants and arthropods; archaic fishes.
	Ordovician	500		Much as in the Cambrian period, with very few traces of fish armor; brachiopods and cephalopods dominant.
	Cambrian	590		Appearance of all major phyla and many classes; first vertebrates (ostracoderms); dominance of trilobites and brachiopods; diversified algae.

contains abundant representatives of almost every major animal group except the vertebrates, which are represented only by a few scraps of armor. A modest number of archaic jawless fishes have been found in the Ordovician and Silurian periods that followed. In the Devonian period, fishes were abundant and varied in freshwater and marine deposits—so abundant that this period is sometimes termed the Age of Fishes. The continental sediments of the Devonian period indicate to the geologist that some regions were subject to marked seasonal droughts, as are certain tropical regions today. Times of abundant rainfall alternated with seasons when streams ran dry and pools were stagnant. These conditions may have had a major influence on the history of fishes and in the development of terrestrial life; but this is open to considerable debate.

At the very end of the Devonian appeared the first four-footed vertebrates, the amphibians. Primitive members of that group are common in the swamp deposits that characterize the Carboniferous period, the age during which the major coal seams of Europe and North America were formed. Well before the end of that period, the first reptiles had evolved, and early reptile orders were common land animals in the Permian period, with which the Paleozoic Era closed.

The **Mesozoic Era**, the "Middle Age" of the story of life, is frequently termed the Age of Reptiles, because members of that class dominated the land during that era, and many types of reptiles now extinct flourished in the seas and in the air. The highest of vertebrate groups had their beginnings in the Mesozoic; the oldest mammals appeared near the end of the Triassic period; and the oldest known birds appeared toward the end of the Jurassic, but both groups remained inconspicuous until the end of the era.

The **Cenozoic Era** is the Age of Modern Life or the Age of Mammals. At the end of the Mesozoic the reptilian hordes became greatly reduced, leaving that class of vertebrates in its modern impoverished phase. Modern types of birds had appeared by the beginning of the Cenozoic, and, most conspicuously, the mammals rapidly evolved into the varied progressive groups that dominate the land today. Periods may be divided into subdivisions terms **epochs**; in our geologic chart, we have listed the Cenozoic epochs to show, primarily, the stage-by-stage rise of the mammals during Cenozoic times.

Vertebrate Classification

The backboned animals constitute the major subphylum Vertebrata of the phylum Chordata (see Appendix 1 for a tabular presentation of vertebrate classification). There is no general agreement on the classification of vertebrates; indeed, it is currently a hotly debated topic. Some of the disagreement concerns the methods to be used, a conflict that may or may not be reflected in the system finally obtained. Much of the problem, however, is simply disagreement over the course of vertebrate phylogeny and the relationships of various groups, especially extinct ones for which our knowledge is, necessarily, far from complete. In this chapter, we present one coherent scheme. Although we sometimes note areas of particularly contested ideas, many points presented here as "facts" are actually only opinions, even minority ones. Probably no one who is an expert in the field would agree with *all* our decisions—by the time this book is published, *we* will probably have changed our minds on some things. In general, we have been conservative in our classification, have tended to be "lumpers" rather than "splitters," and have used a traditional rather than a cladistic approach.

The first stage in classifying the vertebrates is to divide them into a series of classes. The distinguishing features of certain of these classes are obvious to anyone who has the slightest familiarity with nature. The class **Mammalia** includes the mammals, the familiar warm-blooded, hair-clothed animals among which man himself is to be included; the birds, class **Aves**, are readily distinguished by the presence of feathers and wings and by their possession, equally with mammals, of a high, controlled body temperature. The class **Reptilia**, lacking the progressive features of the birds and mammals, represents a lower level, mainly of land dwellers, with lizards, snakes, turtles, and crocodiles as living representatives. A fourth group is that of the class **Amphibia**, including frogs, toads, and salamanders—four-legged (usually and primitively at least) animals, but reminiscent of fishes in many respects.

One commonly lumps the remaining lower vertebrates as "fish," and these forms (or most of them) are sometimes included in a single vertebrate class—the attitude being that, after all, they seem to be built on a common plan, as water dwellers with gills and locomotion performed by fins rather than limbs. This, however, is a rather personal, human viewpoint. An intellectual and indignant codfish could point out that this is no more sensible than putting all land animals in a single class, since, from his point of view, frogs and men, as four-limbed lung-breathers, are much alike. Actually, when we look at the situation objectively, a codfish and a lamprey, at two extremes of the fishy world, are far more different structurally than an amphibian and a mammal. The fishes are perhaps best arranged in three classes of lower vertebrates: class **Agnatha** for jawless vertebrates, such as the living lampreys and fossil relatives; class

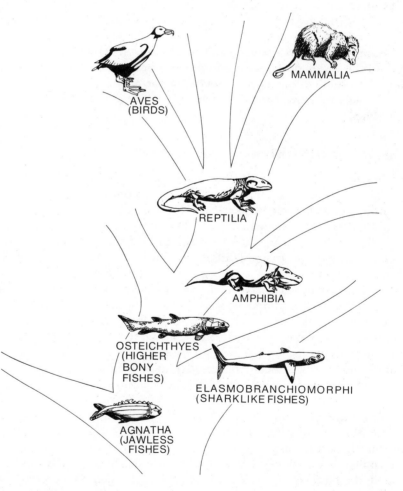

Figure 15. A simplified family tree of the classes of vertebrates. (Mainly after Romer.)

Elasmobranchiomorphi for certain extinct armored fishes and the more modern cartilaginous fishes, sharks and their relatives; and class **Osteichthyes**, the higher bony fishes that today constitute most of the piscine world. Others would recognize up to about a dozen classes of fish.

If we wish to group these seven classes, we may for convenience consider the four higher land groups as constituting a superclass **Tetrapoda**, or four-footed animals, the fishes as making up a superclass **Pisces**:

Superclass Pisces
   Class Agnatha
   Class Elasmobranchiomorphi
   Class Osteichthyes

Superclass Tetrapoda
   Class Amphibia
   Class Reptilia
   Class Aves
   Class Mammalia

This is but one of several alternative methods of grouping the vertebrate classes. Some, placing emphasis on the development of jaws, would contrast with the Agnatha all the remaining vertebrates as the **Gnathostomata**, "jaw-mouthed" forms. Still another grouping is to consider the three highest classes as forming a group termed the **Amniota**, the remaining four constituting the **Anamniota**. This is based upon the fact that the lower types generally have a rather simple mode of reproduction, with eggs laid in the water and young developing there, whereas reptiles evolved a shelled egg, laid on land, within which a complex sort of development (described in a later chapter) takes place. Some reptiles and almost all mammals bear their young alive but have retained the same general pattern of embryonic development; the name Amniota is derived from the amnion, one of the membranes surrounding the growing embryo in mammals and birds as well as in reptiles.

In its simplest form, the phylogeny of the vertebrate classes (Fig. 15) may be diagrammed thus:

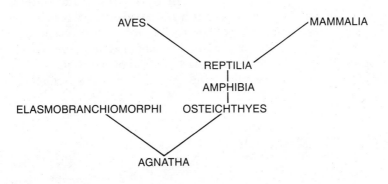

Jawless Vertebrates

Living **lampreys** and **hagfishes**, together termed **cyclostomes** (Figs. 16, 17), are representatives of a lowly group, the class **Agnatha**—jawless vertebrates. Best known is the marine lamprey (*Petromyzon*). This fish is eel-like in appearance but much more primitive in its structure than true eels (which are highly developed bony fishes). The lamprey is soft-bodied and scaleless and, though having a feeble skeleton of cartilage, lacks bone entirely. There are no traces of paired fins, and jaws are totally

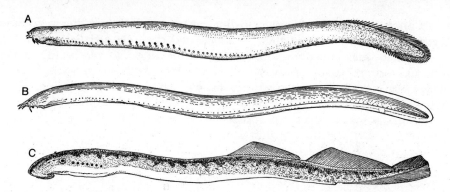

Figure 16. Three types of cyclostomes. *A,* The slime hag, *Bdellostoma; B,* the hagfish, *Myxine; C,* the lamprey, *Petromyzon. A* and *B* are members of the Myxinoidea; *C,* of the Petromyzontia. (From Dean.)

lacking. The adult lamprey is predaceous nevertheless; the rounded oral cup forms an adhesive disc by which it attaches to the higher types of fishes upon which it preys as a bloodsucker, and a rasping tonguelike structure within the mouth is a fairly effective substitute for the absent jaws. There is only a single nostril, opening high on top of the head, and having a hypophysial pouch (cf. p. 501) combined with it. The branchial passages are not slits, as in typical fishes, but spherical pouches, connected by narrower tubes with the pharynx and body surface. In various less obvious structural characters, noted in later chapters, the lampreys likewise show a series of features in which they differ from typical fishes—features that appear to be in part primitive, in part aberrant, and in part highly specialized.

The excessively slimy hagfishes are purely marine in habit and differ in a number of ways. In fact, the differences are so great that we consider the lampreys and hagfishes to be separate orders (**Petromyzontia** and **Myxinoidea,** respectively) and use cyclostomes as a common name only. The rasping tongue is present, but the mouth is surrounded by short tentacles instead of a sucker. The hags are scavengers rather than active predators, burrowing into the flesh of dead or moribund fishes. The nostril is at the tip of the snout rather than atop the head, and the branchial pouches in

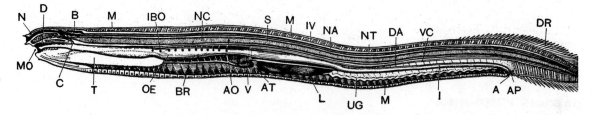

Figure 17. Longitudinal section of a slime hag, *Bdellostoma. A,* Anus; *AO,* ventral aorta; *AP,* abdominal or genital pore; *AT,* atrium of heart; *B,* brain; *BR,* gill pouch; *C,* duct from nasal pit to throat; *D,* horny, toothlike structures; *DA,* dorsal aorta; *DR,* dorsal fin rays; *I,* intestine; *IBO,* internal gill openings; *IV,* septum between muscle plates; *L,* liver; *M,* muscle segments; *MO,* mouth; *N,* nostril sac; *NA,* sheath of spinal cord; *NC,* notochord; *NT,* neural tube (spinal cord); *OE,* pharynx; *S,* sheath of notochord; *T,* extrusible "tongue"; *UG,* urogenital organs; *V,* ventricle of heart; *VC,* posterior cardinal vein. (From Dean.)

some hagfishes do not open directly to the surface but join to a common external opening on either side. The skeletons of hagfish and lampreys are also very different.

Hagfish eggs are laid in the sea, and the young develop directly there. The marine lamprey, in contrast, has a distinct freshwater larval stage. Every spring, lampreys ascend the streams to spawn, and the developing young spend several years of their lives as larvae (ammocoetes), which lie almost buried in the mud of brooks and streams. These larvae are not at all predaceous; there is no rasping tongue or oral sucker. Instead, they are filter-feeders that strain food particles much as does amphioxus. A stream of water is brought into the mouth by muscular action (not ciliary, as in amphioxus), passes through a pharynx that even has a structure comparable to the amphioxus endostyle, and thence flows out the branchial slits. At the end of the larval period, there is a sudden marked change in structure—a **metamorphosis**—and the young lamprey, with adult features fully developed, descends to the sea. It is possible, however, for lampreys to remain in fresh waters for their entire lives—the sea lamprey has successfully invaded the American Great Lakes—and certain small species of lampreys never take up a predaceous life but reproduce and die in their native streams.

It is generally agreed that the absence of jaws, and, probably, of fins, is a primitive feature of cyclostomes. Other characters, however, are more dubiously primitive. There is considerable reason to regard the absence of a bony skeleton as a secondary feature; the predaceous or scavenging habits can hardly have been present in ancestral vertebrates (mutual cannibalism is not, to say the least, advantageous), and the rasping tongue is a lamprey specialty. Cyclostomes represent a primitive level of vertebrate development; they are not, however, in themselves primitive or ancestral vertebrates.

When we look into the fossil record, we find that the oldest and most primitive of fossil vertebrates, found in Cambrian, Ordovician, and Silurian deposits and surviving into the Devonian, were small fishlike creatures known as **ostracoderms** (Fig. 18). These are of several groups (orders), and thus ostracoderm does not appear as

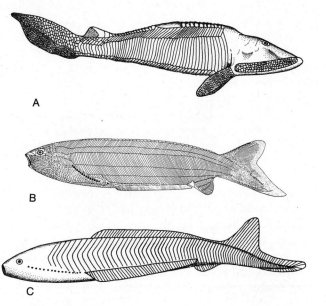

Figure 18. Fossil ostracoderms. *A, Hemicyclaspis,* a member of the Osteostraci; *B, Pharyngolepis; C, Jamoytius,* two members of the Anaspida. In all of these forms, the gills had separate openings to the outside and there was a single median nostril on the top of the head. (*A* and *C* after Moy-Thomas and Miles; *B* after Ritchie.)

a formal term in Appendix 1. Superficially there is little resemblance to the cyclostomes, but study during this century, especially by the Swedish paleontologist E. A. Stensiö, shows that ostracoderms were ancient jawless representatives of the class Agnatha. In many ostracoderms, there was, as in cyclostomes, only a single nostril that was high atop the head.

Many ostracoderms (like cyclostomes) lacked paired appendages, although in some, there were aids to navigation in the form of peculiar flaps projecting from the body behind the head rather comparable to pectoral fins or folds extending out on either side of the trunk as "stabilizers."

A major contrast with the modern cyclostomes lies in the skeleton. All ostracoderms were covered by good bony armor or at least scales, and in some, the head also contained an internal body skeleton. It was formerly assumed that the primitive vertebrates were (like the living cyclostomes and sharks) boneless forms, with a skeleton of cartilage only. This may have been true of the still older ancestral chordates and immature ostracoderms, but the prevalence of bone in the oldest known fossil vertebrates and evidence of reduction instead of increase in ossification in the later history of many fish groups suggest that ancestral vertebrates were armored as adults and that absence of bone in the lower living vertebrates is a secondary rather than a primitive characteristic.

As to the reasons for this early development of bone, we are not certain. One suggestion lies in the fact that we often find in association with these ancient vertebrates remains of eurypterids—ancient water scorpions—and less familiar crustaceans termed ceratiocarids. Both of these arthropod types were voracious and, on the average, considerably larger than the little ostracoderms amidst which they lived. It may be that in their earliest phases the vertebrates were the prey; bony armor may have been a defense against these invertebrate predators. Later, as vertebrates became larger, speedier, and themselves predaceous, the eurypterids vanished from the fossil record and the ceratiocarids shrank to insignificance. Other suggestions are that the armor helped to prevent undue water loss or represented a store of calcium salts; none of these ideas is really convincing.

Ostracoderms are not one small group of similar animals; they are a very mixed group of ancient jawless forms, usually (as here) divided into four distinct orders. Best known are the **Osteostraci**, forms such as *Cephalaspis* (Fig. 18 *A*). These had a greatly expanded "head" region (Fig. 19), most of which was occupied by large branchial chambers. Because they lacked jaws or other biting or rasping structures, it seems obvious that these very old vertebrates, like other chordate ancestors and like the larval lampreys of today, made their living by filter-feeding with their pharyngeal gills. Many of them, although capable of locomotion with a fishy tail, were much flattened and must have been relatively sluggish animals. These forms are especially noteworthy, because the entire head is enclosed in a solid bony shield. Careful dissection of the shields has revealed the detailed structure of the long since rotted gills, brain, cranial nerves, and blood vessels (Fig. 20). In the Osteostraci, as in the modern lampreys, there was a single nostril placed dorsally between the eyes.

A second order, the **Anaspida** (Fig. 18 *B*), shares with the Osteostraci the characteristic of a single median nostril on top of the head. However, these were not flattened bottom dwellers but appear more like active swimmers (although fins were very poorly developed, so we may doubt that they were very agile). The tails tip downward, the reverse of the shark condition, despite some early pictures that show them restored upside down. Instead of heavy armor, these forms possessed a thinner

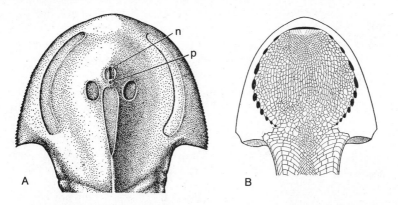

Figure 19. *A,* Dorsal and *B,* ventral views of the head region of a fossil ostracoderm of the *Cephalaspis* type (cf. Fig. 18 *A*). Dorsally are seen openings for the paired eyes, median eye *(p),* and a median slit *(n)* for nostril and hypophyseal sac. Ventrally, the throat was covered by a mosaic of small plates, covering an expanded set of gill pouches. Round openings on either side are for the gill orifices; the mouth is a small anterior slit. (After Stensiö.)

covering of scalelike plates, and bone was restricted to the outside of the head so that internal structures are almost unknown. The gills opened through a series of small, lateral, circular pores. Although some investigators suggest that the anaspids were free-swimming and lived on plankton near the surface of the water, it seems more probable that they, like the osteostracans, were basically bottom dwellers, fil-

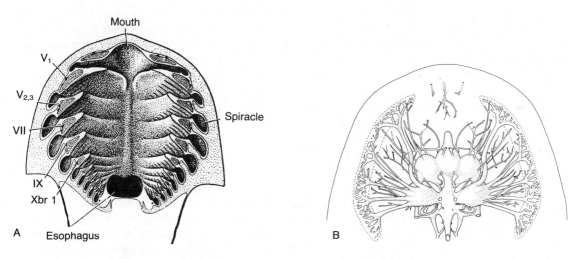

Figure 20. Detailed anatomy of the head of ostracoderms of the order Osteostraci. *A,* A restoration of the pharyngeal region, as seen in ventral view after removal of the bones on the underside of the head. The mouth is very small; the pharynx, an expanded food-straining device. V_1, V_2, V_3, *VII, IX, Xbrl, Sections of branchial nerves associated with successive gills (cf. Fig. 404). By this interpretation, two gill pouches appear anterior to that representing the spiracle, but other workers believe the first pouch here to be the spiracle. B,* Cast of the cranial cavity, the orbits, and canals for the ear, nerves, and blood vessels, as seen in ventral view. The Swedish paleontologist Stensiö has demonstrated the anatomy of these forms in exquisite detail. (After Stensiö.)

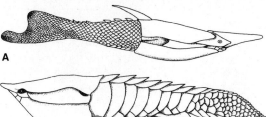

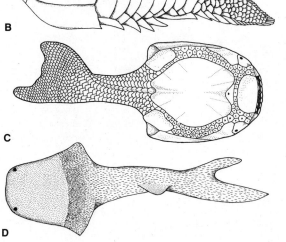

Figure 21. Fossil ostracoderms. *A, B,* and *C* are Heterostraci and *D* is a member of the Coelolepida. The heterostracans were relatively streamlined forms; the true shape of coelolepids is generally not known. *A, Pteraspis; B, Anglaspis; C, Drepanaspis; D, Logania.* (After Moy-Thomas and Miles.)

tering out food particles from the mud. Interestingly, at least one anaspid (Fig. 18 *C*) shows a reduced skeleton and appears, for a variety of reasons, to be close to the ancestry of modern lampreys. The relationships of the hagfish are not known; some investigators think that they are moderately closely related to lampreys and thus, presumably, descended from anaspids, but others claim that they are derived from very different forms.

A third and rather large order is the **Heterostraci** (Fig. 21 *A–C*). These animals are characterized by having an armor of large plates over the head and smaller scales over the rest of the body. They differ from the previously described ostracoderms in lacking a dorsal median nostril. Most investigators believe that they possessed paired nostrils (like us, goldfish, or any other jawed vertebrates), but a few claim that they had a single anterior nostril (like hagfish). In members of this group, the gills possessed a common opening on each side, at the back edge of the "head." Internal ossification was extremely limited, so we know little of their anatomy. Typical heterostracans were rather streamlined-looking, almost like cartoon space ships, but paired fins were lacking, although lateral and dorsal spines may have provided some stabilizing effects. The eyes were lateral, not dorsal as in bottom-dwelling osteostracans, and the bony plates around the mouth may have allowed a small amount of nibbling. Thus, these forms may have been scavengers or even have preyed on soft-bodied animals such as worms; they would not have to have been filter-feeders. Some members of this order were flatter and often showed reduced ossification (Fig. 21 *C*); however, the eyes are lateral and the mouth rather dorsal, so their mode of life is hard to imagine. Still other flattened heterostracans lost the eyes and had tubular mouths.

Finally, there is a small order, the **Coelolepida**, members of which had only small scales in the skin (Fig. 21 *D*). They are very poorly known and need not be discussed here.

Those, then, are the jawless vertebrates—two groups of highly specialized, often "degenerate," forms still extant and four groups of extinct and thus rather poorly known animals. Unfortunately, their relationships with other groups are a mystery. There are no known intermediates between the lower chordates and any of these forms. The earliest fossil agnathans are merely small pieces of bone, not whole animals. Their histology shows them to be heterostracans but tells us almost nothing about them. The living lampreys appear to be highly modified descendants of the anaspids; the origin of the hagfish is disputed.

Presumably, the jawed vertebrates arose from some group of ostracoderms. But again, intermediates are completely lacking, and we know nothing of the actual stages. We cannot even say whether jaws were evolved more than once. When we first meet jawed fishes in the fossil record, they are of two distinct sorts, so that a common ancestor is hard to draw or describe. The number of common features, however, strongly suggests that such a common ancestry was involved.

Elasmobranchiomorphs

Placoderms. The ostracoderms were at the peak of their development during the Silurian. At the end of that period, there appeared somewhat more advanced fish types that were exceedingly prominent in the following Devonian period, but became extinct well before the close of the Paleozoic. These were mostly grotesque forms, quite unlike any fishes living today.

These forms were for a long time (and often still are) considered to form a separate class, the Placodermi, thought to be ancestral to all higher vertebrates. However, more recent work has shown that one group originally placed in this "class" probably belongs with the bony fish to be discussed later; most of them seem more closely related to sharks and their allies so that we may lump the two into one class, the **Elasmobranchiomorphi**. The **Placodermi** are, then, reduced to the status of a subclass of the elasmobranchiomorphs.

All had jaws. This represents a major advance over the ostracoderms, one that opened up new avenues of life to fishes and enabled them to become more active and wider-ranging animals. The term **gnathostome**, "jaw-mouthed," is often applied to placoderms and all higher vertebrates, in contrast with the Agnatha. Originally, the jaws of placoderms were thought to be primitive or aberrant in structure, not supported by the next posterior visceral arch (the hyoid arch; cf. Fig. 165). This is now doubted by most investigators. Unfortunately, the hyoid and other visceral arches were, presumably, cartilaginous—at least we do not find them in fossils—so we cannot be certain. Paired fins, too, were developing, in connection with the new freedom that fishes were acquiring, but these structures were variable and often oddly designed (from a modern point of view) as though nature were still "experimenting" with them. Some of these early placoderms were freshwater dwellers, but during the Devonian a majority of them inhabited the seas.

Best known of the placoderms were the members of the order **Arthrodira**, the jointed-necked fishes (Fig. 22 *A, B*). In these, the head and branchial region was covered by a great bony shield, and a ring of armor sheathed much of the body. The two sets of armor were connected in most by a pair of movable joints (hence the

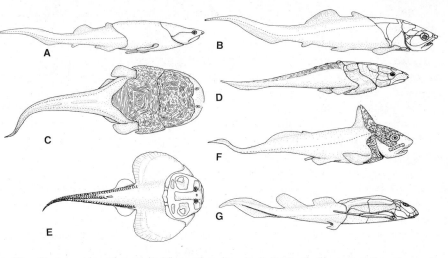

Figure 22. Various placoderms. *A,* The arthrodire *Arctolepis; B,* the arthrodire *Coccosteus;* *C,* the phyllolepid *Phyllolepis; D,* the petalichthyid *Lunaspis; E,* the rhenanid *Gemuendina; F,* the ptyctodontid *Rhamphodopsis; G,* the antiarch *Bothriolepis. C* and *E* show dorsal views; all others are lateral. (After Stensiö.)

name). Peculiar bony plates served the function of jaws and teeth. The posterior part of the body was quite naked in typical arthrodires, but there was a covering of bony scales in the most primitive types. In some forms, true paired fins have been found, but in the most primitive arthrodires, little but a pair of large, hollow, fixed spines projecting outward from the shoulder region—some sort of holdfast or balancing structures—have been found.

Arthrodires were, for the most part, active predators, living on the diverse fish of the Devonian lakes and seas. Indeed, one of them was the largest form of its time, *Dunkleosteus* from the black shales of northern Ohio, a fish similar in shape to the one shown in Figure 22 *B* but reaching a length of 9 m. However, some arthrodires became flattened bottom-dwellers with reduced armor and weak jaws.

All the other placoderms are variants on the arthrodire pattern. The armor that split into separate parts covering the head, lower jaw, and anterior part of the trunk, the presence of some sort of spine in front of the pectoral fins, and tendencies toward bottom dwelling, flattening of the body, and reduction of armor are common, though not universal, characteristics of the different orders. All the trends we mentioned appear in members of the orders **Phyllolepida** (Fig. 22 *C*), **Petalichthyida** (Fig. 22 *D*), and **Rhenanida** (Fig. 22 *E*). They are a mixed (and poorly understood) lot, but appear to be early "attempts" at producing fish comparable to the modern skates and rays. Another order, the **Ptyctodontida**, includes a group that apparently became specialized as crushers of molluscs (Fig. 22 *F*)—comparable and in many ways very similar to the holocephalians discussed below. It is of considerable interest that at least one ptyctodont possessed specialized "claspers" on the pelvic fins of males. These are used in mating and are otherwise known only in the cartilaginous fish or Chondrichthyes. Their presence here is one of the characters linking placoderms and chondrichthyeans.

A last order of placoderms is the **Antiarchi** (Fig. 22 *G*)—grotesque little animals that had armor like that of the arthrodires, but small heads, tiny nibbling bony jaws, extensive armor over the front of the trunk, and, for forelimbs, a pair of jointed "flippers" projecting from the body like bony wings. These were pectoral fins but had dermal armor completely encasing them, so that they appeared more like the appendages of a lobster than any sort of fish fins. The heavy armor, flat ventral surface, and dorsally placed eyes indicate that these forms were bottom dwellers and probably relatively inactive.

Most placoderms were obviously far from the main lines of vertebrate evolution, and few if any of the known types can be regarded as actual ancestors of later vertebrates. As a group, however, they appear to represent one of nature's first essays in the development of jawed vertebrates. Most of these "experimental models" were not, in the long run, successful; but it is highly probable that, with loss of bony armor, some placoderms gave rise to the sharks and chimaeras, the second subclass of the Elasmobranchiomorphi.

Chondrichthyeans. The modern sharks are the typical representatives of a major surviving group of jaw-bearing marine fishes—the subclass (often considered a class) **Chondrichthyes**. The name **cartilaginous fishes** refers to the fact that bone is virtually unknown in any member of the group (there *may* be some bone at the bases of the small scales or denticles). It seems probable that the absence of bone in sharks is due to a process of reduction; the toothlike denticles present in the shark skin and the spines sometimes present on the fins appear to be the last remnants of the armor that once sheathed their placoderm forebears. The main group of cartilaginous fish forms the infraclass **Elasmobranchii**—the sharks, skates, and rays plus some extinct relatives (Fig. 23). Present are well-formed jaws, although, in the absence of bone, there is no formed skull and the upper jaws are independent of the

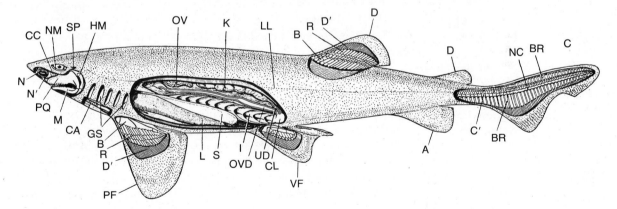

Figure 23. Diagrammatic dissection of a female shark. *A,* Anal fin; *B,* basal elements of fin; *BR,* basal and radial elements, upper and lower lobes of caudal fin; *C,* caudal fin; *C',* centrum; *CA,* conus arteriosus; *CC,* cartilaginous cranium; *CL,* cloaca; *D,* dorsal fins; *D',* dermal rays of fin; *GS,* gill slits; *HM,* hyomandibular; *I,* intestine with spiral valve; *K,* kidney; *L,* liver; *LL,* lateral line; *M,* mandible; *N, N',* anterior and posterior openings to nasal pouch; *NC,* notochord; *NM,* "nictitating membrane" of eye; *OV,* ovary; *OVD,* oviduct; *PF,* pectoral fin; *PQ,* cartilage of upper jaw; *R,* radials of fin; *S,* stomach; *SP,* spiracle; *UD,* urinary duct; *VF,* pelvic (ventral) fin. (From Dean.)

braincase. The gills border slitlike passages, typically five in number, that open separately to the surface, and there is generally a small accessory anterior opening (the spiracle). The numerous teeth are constantly and rapidly replaced. Absent are the peculiarities of the cyclostomes and the curious structural "experiments" seen in the arthrodires. The nostrils are double and placed beneath the tip of the snout.

A feature of sharks and their relatives, which is probably not primitive, is the fact that they produce large eggs containing considerable yolk. These eggs are, in many members of the group, encased in a horny shell before they are laid. To effect this, they must be fertilized before leaving the mother's body, and the male sharks have developed "claspers" projecting from the pelvic fins to aid in introduction of the sperm. Internal fertilization allows the possibility of development of the young within the body of the mother. In various sharks and rays, the fertilized eggs are retained in the mother's reproductive tract and develop there, so that the young are born alive (a limited number of reptiles and almost all mammals have similarly developed this procedure).

Elasmobranchs first appear in the latter part of the Devonian, and *Cladoselache* (Fig. 24 *A*) of that period may be close to the ancestry of later sharklike fishes. It is representative of the order **Cladoselachii**, a group characterized by broadly attached paired fins and the absence of claspers on the pelvic fins. Both of these traits have been considered primitive, but in both cases this is debatable. An aberrant offshoot of the cladoselachians were the **Pleuracanthodii** (Fig. 24 *B*), a small group of Paleozoic (and earliest Mesozoic) forms that lived in fresh water. The unusual symmetrical tail, the fleshy paired fins, and the double anal fin are all unlike those of any normal shark.

There was a variety of shark forms in the seas of the late Paleozoic, and toward the end of the Mesozoic, we find shark types similar to those in modern oceans (Figs. 23, 24 *C*). All these may be placed in the large order **Selachii**, the normal sharks. They are almost purely predaceous in habit, and with few exceptions are purely marine. Except for loss of bone, the sharks appear to be, in general, "proper" fishes of a fairly primitive type. They possess well-developed paired fins and a powerful tail fin, with the tip of the body curving into its upper lobe.

In the Mesozoic, too, there appears the first of the skates and rays, members of the order **Batoidea** (Fig. 24 *D*). These are forms derived from sharks. They have taken to a mollusc-eating diet and a bottom-dwelling mode of life with which their flattened body shape is correlated. In typical rays, the tail and pelvic fins are much reduced but the pectoral fins are greatly expanded and, stretching forward above the gill openings, may meet in front of the head. Locomotion is accomplished by undulatory movements of these broad appendages. Because in a resting position the mouth may be buried in the mud or sand of the sea bottom, the spiracle (small or even absent in sharks) is a large opening behind the eyes through which water enters the pharynx.

A distinct group of cartilaginous fish, as opposed to the elasmobranchs, is the infraclass **Bradyodonti**—forms characterized by slower replacement of teeth (hence their name) and a flap of skin covering the separate branchial slits. Such a flap, or operculum, is also present in placoderms and bony fish. Most of these are extinct, but the chimaeras or ratfish (Figs. 25, 26) include living, though relatively rare, oceanic forms. These are, like the skates, mainly mollusc-eaters; the body is not greatly depressed, and their peculiarities include, among other features, the development of

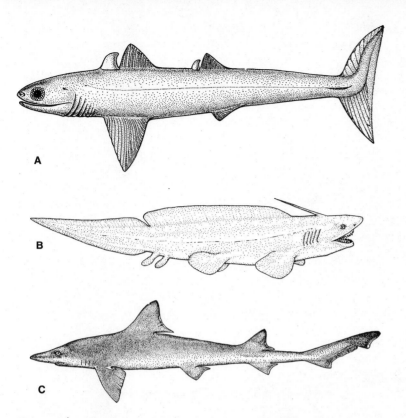
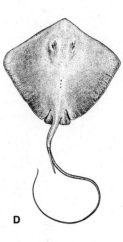

Figure 24. Elasmobranch fishes, members of the Elasmobranchiomorphi, which have cartilaginous skeletons and in which the upper jaw is not fused to the braincase. *A, Cladoselache,* a primitive member of the Cladoselachii; *B, Xenacanthus,* one of the Pleuracanthodii; *C, Mustelus,* a typical shark of the Selachii; *D, Dasybatis,* one of the greatly flattened forms of the order Batoidea, with greatly expanded pectoral fins and a long whiplash tail. (*A* after Dean and Harris; *B* after Špinar and Burian; *C* and *D* after Garman.)

large tooth plates and of upper jaws that (in contrast to those of elasmobranchs) are solidly fused to the braincase. Although the living forms show rather little variety, the extinct ones, mainly Paleozoic and early Mesozoic, are diverse and, as would be expected with extinct forms having cartilaginous skeletons, poorly understood—so poorly that the classification here must be considered provisional.

Approximately ten orders are currently recognized, but we need not consider them all; they may conveniently be divided into two series or superorders. The **Holocephali** consists of those, like the living forms (order **Chimaeriformes**, Fig. 25 *C*), that have the upper jaw fused to the braincase and the dentition represented by a series of large tooth-plates; the extinct orders include various quite bizzare forms (Fig. 25 *B*). The second superorder, the **Paraselachii** (Fig. 25 *A*), are known, and poorly at that, only as fossils. They differ from holocephalians in having numerous teeth (as in sharks); the upper jaw may be fused to the braincase (as in the Holocephali) or free (as in the Elasmobranchii).

Although some investigators have tried to derive the Bradyodonti directly from placoderms, especially ptyctodonts, this idea is not now widely accepted. We assume **51**

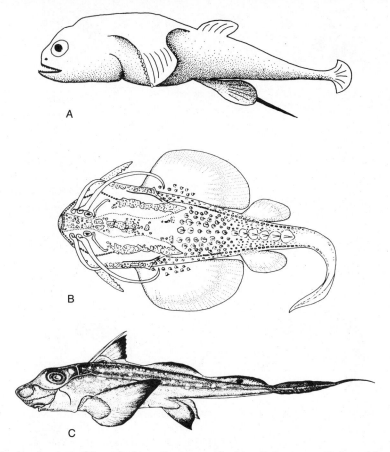

Figure 25. Bradyodont fishes. *A,* An extinct paraselachian *Iniopteryx; B,* the odd extinct holocephalian *Menaspis; C,* the living chimaeriform *Chimaera.* (*A* based on drawings by Zangerl and Case; *B* after Bendix-Almgreen; *C* after Dean.)

that they, like other Chondrichthyes, had a placoderm ancestor but not via a line independent of that leading to elasmobranchs. That is, we consider the Chondrichthyes to be a natural group; if the alternate theory were to be proved correct, the Elasmobranchii and Bradyodonti would have to be raised to the rank of subclasses.

Bony Fishes

The class **Osteichthyes** includes the vast majority of fishes. As the name implies, they are forms in which a bony skeleton has been retained and improved upon. A characteristic pattern is to be found, with variations, in most members of the group, in the bones of the skull, jaws, coverings of the gills, and in a set of bony scales covering the body. It was once believed that these fishes were descendants of sharklike forms and that bone in them was a new acquisition. It now, however, appears more probable that the bony skeleton here is simply a retention with modifications of that which was present in the ancestral vertebrates.

In this section, we shall outline the general evolutionary history of the great groups of fishes that, beginning in the Devonian, are generally agreed to be proper

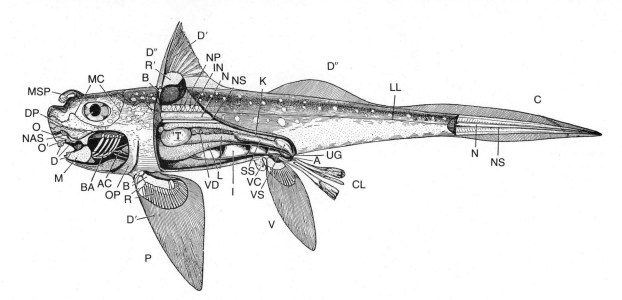

Figure 26. Diagrammatic dissection of a male chimaeroid. *A,* Anus; *AC,* conus arteriosus; *B,* basal element of fin; *BA,* branchial arches; *C,* caudal fin; *CL,* clasper; *D,* dental plates; *D′,* dermal rays of fin; *D″,* dorsal fins; *DP,* small dermal plates of lateral line canal; *I,* intestine with spiral valve; *IN,* interneural cartilages of vertebral column; *K,* kidney; *L,* liver; *LL,* lateral line groove; *M,* lower jaw cartilage; *MC,* lateral line canals (mucous canals) of head; *MSP,* frontal spine peculiar to male; *N,* notochord; *NAS,* nasal pouch; *NP,* neural arch; *NS,* sheath of notochord; *O, O′,* grooves leading to and from nasal cavity, covered by fold of skin; *OP,* operculum; *P,* pectoral fin; *R,* radial cartilages of fin; *R′,* fused radials of anterior dorsal fin; *SS,* sperm sac; *T,* testis; *UG,* urogenital opening; *V,* pelvic (or ventral) fin; *VC,* anterior accessory clasper; *VD,* epididymis; *VS,* sperm vesicle. (From Dean.)

members of the Osteichthyes. But first we must mention the interesting but puzzling little group of very ancient types, the **Acanthodii** (Fig. 27). These were jawed fishes that appeared even earlier than the placoderms, because, whereas the acanthodians, which survived to the Permian, are most abundantly known in the early Devonian freshwater deposits, there are fragmentary remains of them well back in the Silurian. Most acanthodians were small, roughly of minnow size. As in ostracoderms and placoderms, the head and body were covered with bony plates and scales, and there was a fair amount of internal ossification. The tail was strongly tilted upward in a sharklike pattern. Paired fins were developed, but in a most unusual manner. Each fin was supported by a very stout spine; and in addition to the normal two pairs of paired fins, accessory pairs, up to as many as five, may also be present. Similar spines lay in front of the dorsal and anal fins. The acanthodians are often called "spiny sharks," but apart from the tail fin, there is little in their structure that is particularly sharklike (although one authority has recently claimed that they are, in fact, sharks). Neither do they resemble the true placoderms, although they were once included in that group. Recent detailed studies have revealed features, especially in the skull and branchial skeleton, suggesting their relationships to the Osteichthyes. Because of their peculiar paired fins, however, the known acanthodians do not appear to be directly ancestral to the proper osteichthyans. Thus we place them, with some reservations, as a subclass of the Osteichthyes.

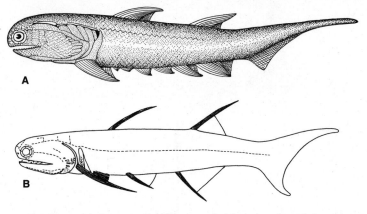

Figure 27. Acanthodians, primitive jawed fishes probably related to the modern Osteichthyes. *A, Climatius,* a form with heavy scales and very large fin spines; *B, Ischnacanthus,* one with greatly reduced armor and thin spines. (*A,* data from Watson; *B* after Moy-Thomas and Miles.)

As yet, the problem of the nature of the truly ancestral jawed vertebrates from which the Placodermi, Chondrichthyes, Acanthodii, and other Osteichthyes were derived is unsolved. Probably these ancestors arose in the Silurian or even earlier, quite possibly in fresh waters; but the preserved record of the rocks of these older times appears to be almost exclusively marine. The biggest gap in the fossil record of vertebrates is our absolute lack of knowledge of the presumed, though possibly nonexistent, common ancestor of the various gnathostome groups.

Leaving aside the puzzling "spiny sharks," the first of the Osteichthyes proper are found in rock of early Devonian age; the class is, thus, somewhat older than the sharks. By the middle of the Devonian, bony fishes were already the dominant forms in fresh waters, where they remain varied and abundant in later Paleozoic periods. Bony fishes are also present in saltwater deposits in the Paleozoic, and, by the end of the Triassic, marine waters appear to have become the headquarters of the class. Lungs appear to have been present in all primitive bony fishes, although today such structures usually have been lost or converted into a hydrostatic organ, the swim bladder. Lungs were long considered an aid to survival under conditions of seasonal drought; such conditions are believed by many geologists to have been present in the Devonian fresh waters in which the ancestral Osteichthyes lived. Various other possibilities have been suggested and many investigators now believe that lungs arose in estuarine or other near-shore marine situations. Later, with climatic changes and particularly with the movement of most surviving bony fish types into the sea, the lung lost its importance.

The phylogeny of the bony fishes is currently much debated; we present here a possible scheme, but not one that all investigators would accept (Fig. 28). At the very beginning of their known history, the Osteichthyes (apart from the problematic acanthodians) were already subdivided into two major groups, termed the subclasses **Sarcopterygii** and **Actinopterygii**.

Sarcopterygii. In considering the descent of terrestrial vertebrates, the Sarcopterygii are the most important of the two, because they contain the order **Crossopterygii**, from which land vertebrates appear to have descended, and the order **Dipnoi**,

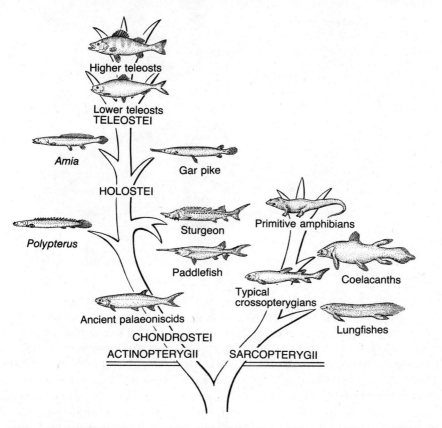

Figure 28. A simplified family tree of the bony fishes (apart from acanthodians), to show their relations to one another and to the amphibians.

the lungfishes, which are surviving cousins of our piscine ancestors. Some of the sarcopterygians have, in contrast to the other subclass, internal nostrils, as do all terrestrial vertebrates (because of the presence of such structures, these fishes have sometimes been called the **Choanichthyes**). In stronger contrast with the actinopterygians is the fact that there are fleshy-lobed paired fins (a feature to which the group name refers) and, as a technical character, scales that in early forms were of a structure quite distinct from those of the actinopterygians (the cosmoid scale, cf. p. 171).

Crossopterygians. In the Devonian the most common of bony fishes were crossopterygians (Fig. 29 *A*)—aggressive, predaceous fishes that show important structural features of a sort to be expected in the ancestors of the amphibians. In the Carboniferous, however, they became relatively rare, and typical crossopterygians, termed rhipidistians (suborder **Rhipidistia**), were extinct before the close of the Paleozoic.

Meantime, however, a peculiar group, the suborder **Coelacanthini** (Fig. 29 *B*), appeared in the late Paleozoic and Mesozoic seas. These forms had stub snouts, feeble jaws, and teeth. The last fossil coelacanths are found in Cretaceous rocks, and it was long taught that our crossopterygian relatives had been extinct since the days of the dinosaurs. In 1939, however, to the surprise of science, a strange fish caught off the coast of South Africa proved to be a coelacanth! This single specimen, scientifically

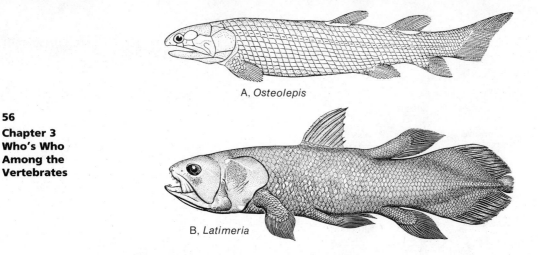

A, *Osteolepis*

B, *Latimeria*

Figure 29. Crossopterygians. *A,* Typical Devonian form; *B,* the only living coelacanth. (*A* after Traquair; *B* after Millot.)

termed *Latimeria*, was, unfortunately, incompletely preserved. Since then, many other specimens have been obtained from deep waters off the Comoro Islands in the Indian Ocean, and their structure is being studied intensively by French scientists.

Knowledge of the structure of this fish is highly important, because we have here the closest living fish relative of the tetrapods. But although the terrestrial vertebrates are derived from crossopterygians, and *Latimeria* is a crossopterygian, one must not expect this survivor to resemble closely the tetrapod ancestors. Even the oldest known fossil coelacanths already differed in various skeletal features from typical members of the order, and the shift from ancestral streams and ponds or coastal waters to deep sea waters has been accompanied by many changes in structure and functions. Coelacanths lack internal nostrils; there is, instead, a series of canals filled by a jelly-like substance in the snout, which is obviously a sense organ of unknown nature (possibly an electroreceptor?). The nasal sacs are quite unlike those of most other fish (they do resemble those of *Polypterus* mentioned below—which makes no sense). Lungs, which are of no use in deep marine waters, are represented only by a large sac filled with fat and connective tissue. Bone is much reduced in the skeleton, and cartilage is dominant. The heart is constructed in a simple manner; the chambers are almost linear in arrangement, with little of the folding seen in most vertebrates. The braincase is large, but the brain within is tiny. There is no pineal foramen. The branchial septa are more highly developed than in typical bony fish. The intestine retains a spiral valve. Overconcentration of salts internally is prevented as in sharks by retention of urea in the blood (cf. Chap. 13). There is a cloaca in the male. The embryological picture is poorly known, but the eggs are large, almost 10 cm in diameter, and develop within the maternal oviduct. The young reach a length of over 30 cm but appear to lack any "placental" connection with the oviduct; thus *Latimeria* is presumably ovoviviparous. Coelacanths share with the rhipidistians the possession of a transverse joint across the center of the skull, through both the dermal roof and the braincase—a trait known otherwise only in the very earliest of the amphibians.

As is so common with extinct or largely extinct groups, the classification here is much debated. Some investigators deny that the Rhipidistia are a natural group distinct from coelacanths, whereas others deny that the two are closely related at all; as usual, we are following a conservative line.

Lungfishes. The Dipnoi, or lungfishes (Figs. 30, 31), are represented today by three genera, living, one each, in tropical regions of Australia, Africa, and South America. In many anatomic features and in their mode of development, the lungfishes closely resemble the amphibians, and they were once thought by many to be actual amphibian ancestors, a view now becoming more popular again. However, most investigators believe that these features were present as well in their relatives, the ancestral crossopterygians, and that the lungfishes are to be regarded as "uncles" rather than the actual progenitors of terrestrial vertebrates. The skull structure of lungfishes, living and fossil, is of a peculiar type quite unlike that expected in an amphibian ancestor; the transverse joint seen in crossopterygians is never present. During the course of lungfish evolution, ossification is much reduced in the skeleton as a whole. In connection with a diet of invertebrates and plant materials, almost all lungfishes possess specialized fan-shaped toothplates. It is of interest that the lungfishes have survived only in regions where today we find conditions of seasonal drought similar to those that we believe to have been present in the Devonian. The Australian form can survive in stagnant water by air breathing; the other two forms are able to withstand even the complete drying up of the water by digging a burrow in the mud in which they "hole up" until the wet season of the year comes round. So dependent is the African lungfish on air that it will "drown" if kept under water. However, although almost all fossil lungfish were, like the living forms, dwellers in fresh water, the earliest and most primitive were, it now seems, marine—a point that casts much doubt on some of the standard ideas given in this and other books.

Both crossopterygians and lungfish are sarcopterygians; they share such features as fleshy fins and cosmoid scales. However, as just noted, they also differ in many ways, and no connecting links are known. The differences are great enough that many

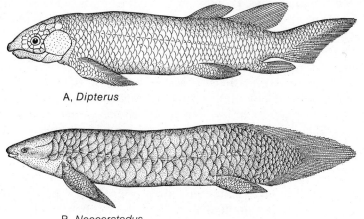

A, *Dipterus*

B, *Neoceratodus*

Figure 30. Lungfishes. *A,* An ancient Devonian fossil type; *B, Neoceratodus* of Australia. The median fins have changed greatly during the history of the group. *(A* after Traquair; *B* after Dean.)

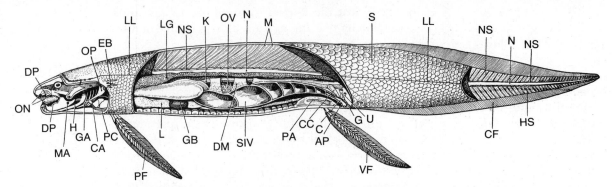

Figure 31. Diagrammatic dissection of a lungfish. *AP*, Abdominal pore; *C*, cloaca; *CA*, conus arteriosus; *CC*, rectal gland; *CF*, caudal fin; *DM*, dorsal mesentery; *DP*, tooth plates; *EB*, external gills (normally present only in larvae); *G*, genital duct; *GA*, gill arches; *GB*, gallbladder; *H*, ceratohyal; *HS*, hemal spine; *K*, kidney; *L*, liver; *LG*, lung; *LL*, lateral line; *M*, muscle segments; *MA*, lower jaw; *N*, notochord; *NS*, neural spine; *ON*, external and internal openings of nostril; *OP*, operculum; *OV*, ovary; *PA*, pelvic girdle; *PC*, pericardium; *PF*, pectoral fin; *S*, scales; *SIV*, spiral valve of intestine; *U*, urinary duct; *VF*, pelvic (ventral) fin. (From Dean.)

workers deny a close relationship and consider them to be separate subclasses (or even classes).

Actinopterygii. As types ancestral to higher vertebrates, the Sarcopterygii are of major interest; but as successful fishes, the Actinopterygii, or ray-finned fishes, are vastly more important. From Carboniferous times on, these have been the dominant fishes. In contrast with many sarcopterygians, internal nostrils are absent; the scales were primitively of quite another type; and except in a few primitive forms, there is never a fleshy lobe to the fins. Instead, as the name implies, the paired fins are webs of skin supported by horny rays. Primitively, but not in most living or extinct forms, the skull was divided by a transverse joint as in crossopterygians, though located more posteriorly and probably never capable of movement.

The actinopterygians have long been divided into three groups—here considered as superorders—which are, in ascending order: the **Chondrostei, Holostei**, and **Teleostei**. The names are not particularly appropriate from the point of view of our present knowledge of the evolution of ray-finned fishes but may be retained for convenience. All these now appear to be "unnatural" groups—"grades" or "levels of organization" rather than "clades." They are, in normal words, primitive, intermediate, and advanced forms.

Chondrostei. In the Paleozoic the ray-finned fishes were represented by abundant genera of Chondrostei known as **palaeoniscoids** (Fig. 32 *A*). These were generally fishes of small size, with rather uptilted sharklike tails (the heterocercal type, cf. Fig. 136) and with scales covered by a shiny material known as ganoine (cf. p. 171); the term "ganoid" is sometimes applied to fishes of this group but is to be avoided, because it was applied indiscriminately and variably to fish with shiny scales. In the earliest Devonian days of bony fish history, primitive ray-finned forms were outnum-

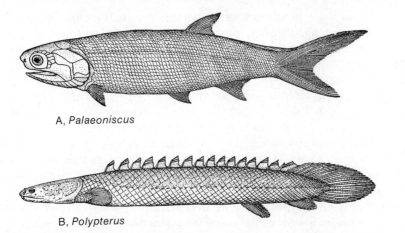

A, *Palaeoniscus*

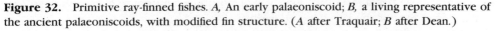

B, *Polypterus*

Figure 32. Primitive ray-finned fishes. *A,* An early palaeoniscoid; *B,* a living representative of the ancient palaeoniscoids, with modified fin structure. (*A* after Traquair; *B* after Dean.)

bered by crossopterygians and lungfishes, but in the late Paleozoic they became far more numerous than their early rivals and swarmed in ancient lakes, streams, and seas in immense numbers and variety, ranging from long, thin, almost eel-like forms to high, laterally compressed forms shaped like modern angelfish. In the Triassic these ancient actinopterygians were still abundant but were mainly represented by advanced types transitional to the Holostei. The palaeoniscoids then rapidly declined and became extinct before the end of the Mesozoic.

This primitive ray-finned group still survives in the form of three aberrant types. Two, the sturgeons and paddlefishes (both represented in North America), are rather "degenerate" (Fig. 33). They have lost the ganoid scales of their ancestors. Scales may still be present on the tail, but the paddlefish has otherwise only a naked skin and the sturgeon a partial armor of rows of plain bony plates. The internal skeleton, highly ossified in their ancestors, is modified similarly to that of the sharks; it is mainly

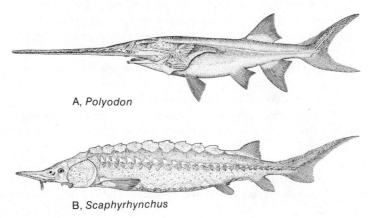

A, *Polyodon*

B, *Scaphyrhynchus*

Figure 33. Chondrosteans. *A,* The paddlefish or "spoon-billed cat" of the Mississippi; *B,* a sturgeon. (After Goode.)

cartilaginous and little bone remains. Different, too, is their method of feeding. In both sturgeons and paddlefishes, the jaws are feeble. In advance of the jaws is a sensitive rostrum that explores for food ahead of them; sturgeons and paddlefishes are bottom-dwelling scavengers or food-strainers. Only their persistently sharklike tail recalls the older palaeoniscoids.

The third type of chondrostean survivor is *Polypterus* (Fig. 32 *B*), the bichir of Central Africa, which lives in much the same environment as the lungfish of that continent.* In its fins, *Polypterus* is much modified from the ancestral type. Its caudal fin has become essentially symmetric; its dorsal fin is split up into a series of small sail-like structures (to which its scientific name refers), and its paired fins, unlike those of any proper actinopterygian, have short fleshy lobes. Again unique among living actinopterygians is the fact that *Polypterus* has typical (if simple) lungs, whereas other ray-finned fishes have in their place a structure termed the swim bladder (cf. p. 360), which seldom has respiratory functions and is, instead, a hydrostatic organ. The nose has a structure like that of no other actinopterygian or indeed any other animal except the surviving coelacanth.

Because of the presence of lungs and the fleshy paired fins, *Polypterus* was long considered to be a crossopterygian. But closer study shows this to be incorrect. Lungs were probably present in all primitive bony fishes, and the survival of *Polypterus* in peculiar drought conditions appears to be due (as with the lungfish) to their retention. Although the fins are somewhat fleshy, they differ markedly in pattern from those of crossopterygians. The anatomy of the animal generally agrees with that of actinopterygians rather than that of the sarcopterygians, and the scales are of the true ganoid type, contrasting strongly with those of the sarcopterygians. *Polypterus* is thus best interpreted as a somewhat modified descendant of the ancient palaeoniscoids, although not all investigators agree; some even interpret it as the type of a separate subclass or class, the Brachyopterygii.

Holostei. Succeeding the chondrosteans as dominant fishes in the middle Mesozoic were the holosteans. In them, the old, long, upturned sharklike tail had become shortened; the jaws tended to have a shorter gape; and the scales, in many cases, tended to lose their shiny ganoid covering. Another trend, too, was apparent at this time: the ray-finned fishes were invading the seas in large numbers. The major center of actinopterygian evolution from the Jurassic period onward appears to have been the ocean. The oceanic holosteans, however, are extinct (the group became rare in the Cretaceous), and the only two survivors are North American freshwater forms. The gar pikes, *Lepisosteus* (Fig. 34 *A*), are fast-swimming fishes fairly representative of the ancestral holosteans in many ways but specialized in their elongated jaws associated with predaceous habits. A more advanced type is *Amia* (Fig. 34 *B*), a lake and river fish of the Midwest and South, popularly termed the "dogfish," "mudfish," or bowfin. In these holosteans, the internal skeleton is well ossified, but in *Amia,* the scales have lost their ganoine covering, and the tail is much like that of the teleosts.

Teleostei. The teleosts, as the name suggests, form the end group of the ray-finned fishes and are the fishes dominant in the world today. They appear to have originated

---

Calamoichthys is a closely related form, but with a more elongate eel-like shape, from the same region; comments on *Polypterus* usually apply to *Calamoichthys* as well.

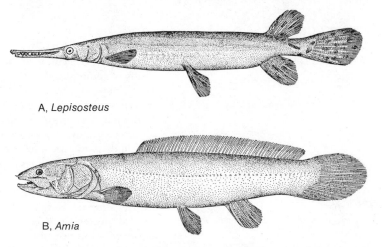

A, *Lepisosteus*

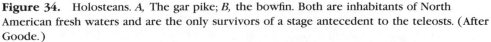

B, *Amia*

Figure 34. Holosteans. *A,* The gar pike; *B,* the bowfin. Both are inhabitants of North American fresh waters and are the only survivors of a stage antecedent to the teleosts. (After Goode.)

from the holosteans in the Mesozoic oceans, and before the close of the Cretaceous had replaced the older group as the most flourishing of fish types. In teleosts, the originally sharklike tail has been reduced, and the tail fin has a superficially symmetric appearance, although the vertebral axis still turns dorsally (cf. Fig. 123). The paired fins are small; the pectorals are usually well up the sides of the body and may function as effective brakes; and the pelvic fins frequently are found well forward. The scales have lost all trace of the original shiny ganoid covering and are generally thin, flexible, bony structures; they may be lost completely. The basic modifications of actinopterygians and the ones that most probably are responsible for their great success are in the locomotor and feeding mechanisms. We have already mentioned the changes in the tail; those in the jaws are more complex (Fig. 35). During the course of actinopterygian evolution, the jaws tend to shorten as the hyomandibular and other elements involved in their suspension from the braincase rotate, so that in typical chondrosteans they project posteroventrally, in holosteans ventrally, and in teleosts anteroventrally from the otic region of the braincase. As this occurs, the maxilla gradually becomes free from the other bones of the skull roof, is excluded from the gape, and forms a lever rotating about its anterior attachment. In most teleosts, the premaxilla also becomes independently mobile, so that the upper jaw is protruded as the mouth opens (watch a goldfish—the upper jaw actually slides forward). These bony changes are, of course, correlated with important changes in the muscles, which become larger and very complex. The net result is a system that has become adapted to almost every diet possible (and even some that seem impossible).

In the oceans, the teleosts constitute (despite the presence of sharks and skates) the vast majority of all piscine inhabitants. They have invaded every possible marine habitat from the strand line to the abyssal depths. Further, in fresh waters they constitute almost the entire fish population. Teleosts are unquestionably the most numerous of all vertebrates. It is estimated that there are about 20,000 different species, and the number of members of one species alone—the common herring—is probably on the order of a billion billion! The prosperity of the teleosts is certainly due in part

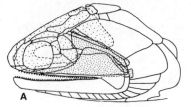

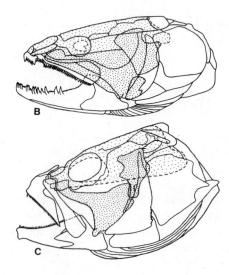

Figure 35. Diagrams of the skulls of actinopterygians in lateral view. *A,* The palaeoniscoid *Pteronisculus; B,* the holostean *Amia; C,* the advanced teleost *Epinephelus.* The lines show the bones of the skull roof; the braincase is lightly stippled; the jaw suspension and palate are heavily stippled. Note how the jaw suspension keeps its dorsal position near the back of the braincase, but how its ventral end swings anteriorly as the jaw shortens. Note also the freeing of elements at the anterior end of the skull. (After Schaeffer and Rosen.)

to an efficient body organization but is also at least partially due to extraordinary fecundity. Existing bony fishes of other groups lay only a modest number of eggs; among teleosts, the herring, for example, may lay 30,000 eggs in a single season, and a female cod is estimated to produce up to 9,000,000 eggs. Individual survival is not that important; only two eggs need to grow to maturity to keep up the numbers of the race. Because of their abundance, the teleosts are a major source of human food, making available to us in highly edible form the organic materials of the ocean— materials particularly plentiful on the relatively shallow "banks" of the continental shelves where most major fisheries are located.

The salmon and trout and the herrings and their relatives (Fig. 36 *A*) represent primitive groups of teleosts. The carps and catfishes are characteristic of a major freshwater division of the teleosts. More progressive and numerous, but almost all marine, are the spiny-finned forms—the perch (Figs. 36 *B,* 37) is typical—in which parts, at least, of the fins are supported by stout spines rather than softer rays.

Teleosts are the most versatile of vertebrates. Within both lower and higher divisions of the teleosts, there has evolved a great variety of body shapes, a few of which are shown in Figure 38. Equally diverse are their habits. In food, they range from eaters of microscopic plant particles to predaceous forms that attack other fishes. And although they have not successfully invaded the land or air, a surprising number of teleosts, such as the "climbing perch," can clamber about on land, and "flying fishes" can glide above the water.

In any comparable study of vertebrate anatomy of physiology in which the teleosts are involved, their ecologic history must be kept in mind. Because terrestrial

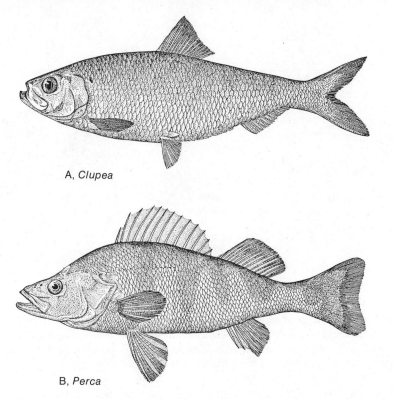

A, *Clupea*

B, *Perca*

Figure 36. Teleosts. *A,* Primitive type, the herring; *B,* an advanced, spiny teleost, the yellow perch. (After Goode.)

vertebrates are usually thought to come from freshwater fishes, one tends to assume that structures or functions seen in freshwater teleosts may be representative of those once present in the ancestors of the tetrapods. But we must remember that our common fishes belong to a different branch of the fish family tree from that which gave rise to tetrapods. It must further be kept in mind that modern freshwater teleosts have not been, in all probability, continuous residents in that environment since the early days of fish history; between that time and the present, there almost certainly intervened a long marine phase.

Amphibians

Greatest, perhaps, of all ventures made by the vertebrates during their long history was the development of tetrapods and the invasion of the land—a step that involved major changes in function and resulted in profound structural modifications. The shifts from swimming to four-footed walking and from breathing with gills to the dominance of lungs are the most obvious of the modifications necessary in this step. But analysis shows that functional and structural changes were necessitated in almost every organ or organ system of the body.

The basic group of terrestrial vertebrates is the class **Amphibia**. There are three living orders (Fig. 39): the frogs and toads (**Anura**), the newts and salamanders (**Uro-**

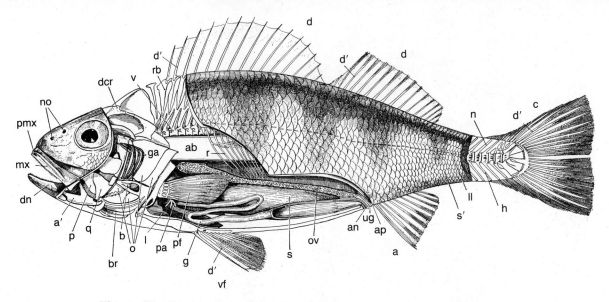

Figure 37. Diagrammatic dissection of a teleost, the perch *(Perca)*. *a,* Anal fin; *ab,* air bladder; *an,* anus; *ap,* abdominal pore; *a',* articular; *b,* bulbus arteriosus (conus arteriosus); *br,* branchiostegal rays; *c,* caudal fin; *d,* dorsal fins; *d',* dermal rays of fins; *dcr,* dorsal crest of skull; *dn,* dentary; *g,* intestine; *ga,* gill arches; *h,* hemal spines (expanded to hypurals in caudal fin); *l,* liver; *ll,* lateral line; *mx,* maxilla; *n,* neural spines; *no,* nasal openings; *o,* opercular bones; *ov,* ovary; *p,* pterygoid; *pa,* pyloric appendices; *pf,* pectoral fin; *pmx,* premaxilla; *q,* quadrate; *r,* ribs; *rb,* basal fin supports; *s,* stomach; *s',* scales; *ug,* urogenital opening; *v,* vertebral centra; *vf,* pelvic (ventral) fin. (From Dean.)

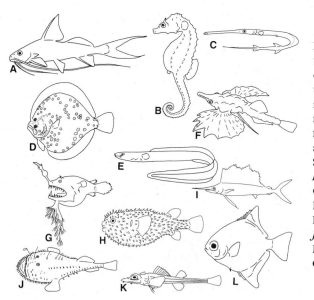

Figure 38. A variety of teleosts, showing the great diversity of form found in these fishes. For those who may wish to trace these further, the names of the families to which they belong are given. *A,* Pimelodidae; *B,* Syngnathidae; *C,* Fistulariidae; *D,* Scophthalmidae; *E,* Congridae; *F,* Pegasidae; *G,* Linophrynidae; *H,* Diodontidae; *I,* Istiophoridae; *J,* Lophiidae; *K,* Agonidae; *L,* Monodactylidae. (After Greenwood et al.)

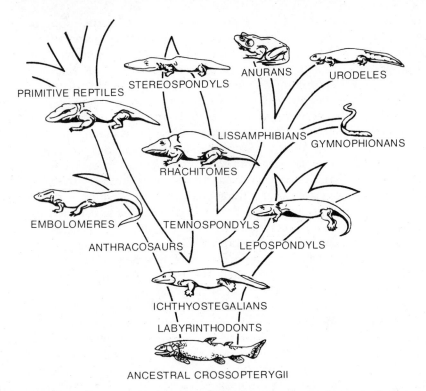

PRIMITIVE REPTILES

STEREOSPONDYLS

ANURANS

URODELES

LISSAMPHIBIANS

GYMNOPHIONANS

RHACHITOMES

EMBOLOMERES

TEMNOSPONDYLS

LEPOSPONDYLS

ANTHRACOSAURS

ICHTHYOSTEGALIANS

LABYRINTHODONTS

ANCESTRAL CROSSOPTERYGII

Figure 39. A "family tree" of the Amphibia. The surviving groups, shown in the upper right, may have a common origin, in the late Paleozoic, from primitive rhachitomes, but this is still open to debate. The arrangement here is tentative and highly debatable; for example, it is unlikely that the lepospondyls are a simple group as shown here.

dela), and some wormlike burrowers (**Gymnophiona**). The most common are the anurans, familiar to us in temperate regions and represented in great variety in the tropics. Their specialized nature is obvious in their jumping habits, which are responsible for many structural modifications, particularly in the skeletal system. The salamanders are retiring but common dwellers in moist north temperate zone habitats. In their external form, the salamanders resemble the ancestral amphibians that first sprang from ancestral fishes. There is a fairly elongate but stoutly built body, with powerful axial musculature and a well-developed tail, which is an aid in swimming. The median fins of fishes are gone, and the paired fins have developed into the typical limbs that are the trademark of the tetrapods. The Gymnophiona (or Apoda) will not be familiar to many readers of this text, because they include only a few genera of small, blind tropical burrowers that resemble earthworms.

In salamanders and in the other modern amphibians as well, we find various internal features that bridge structural gaps between the sarcopterygian fishes and the higher classes of terrestrial vertebrates. Amphibian structure and function are important in comparative studies. But due caution must be exercised. Frogs and salamanders and apodans are amphibians, and the amphibians are the most primitive of tetrapod classes. However, we must not complete a false syllogism by concluding that these modern amphibians are really primitive tetrapods. In the modern orders, the skeleton is reduced, with considerable loss of bone from the head and, particularly in salamanders, a trend for retention of embryonic cartilages. In all, vertebral structure is apparently quite far removed from that which was present in truly ancestral land forms, and the great shortening of the trunk and reduction of the tail in frogs shows a high degree of specialization. The limbs of newts are seemingly not too aberrant, but those of a frog are certainly highly modified; the apodans have abolished limbs altogether. Concerning the soft anatomy, we cannot be as certain, but there are indications that here, too, many features are likewise aberrant in these modern orders. A frog is, in many ways, as far removed structurally from the oldest terrestrial vertebrates as is a human, and even a salamander must be regarded with suspicion.

For the actual ancestors of the tetrapods as a whole, we must turn to the fossil record of the late Paleozoic, when, in the Carboniferous and early Permian, there lived numerous and varied amphibians of a more primitive nature. Unfortunately, there is no general agreement on the higher classification of these and other amphibians. Most other investigators would accept all the orders described here; however, their interrelationships are hotly debated. We are retaining three subclasses for convenience but without any confidence in them as natural groups.

One includes a series of small animals, termed **Lepospondyli**, of which a diagnostic feature is the spool shape of the central portions of the vertebrae. We divide the lepospondyls (Fig. 40) into four orders (others will have more or less). The **Aistopoda** included a few totally limbless, elongate, snakelike forms. Although the earliest known in the fossil record, they are surely highly specialized lepospondyls. The **Nectridea** were a group of rather salamander-like forms; most were normal looking, but some had bizarre "horns"—really extensions of the posterolateral corners of

Figure 40. Representations of the extinct lepospondyl amphibians. *A, Dolichosoma*, a member of the Aistopoda; *B, Cardiocephalus*, one of the Microsauria; *C, Diplocaulus*, one of the Nectridea. All were small, more or less salamander-like forms from the swamps of the upper Paleozoic. (*A* after Špinar and Burian; *B* after Gregory et al.; *C* after Colbert.)

the skull. All nectrideans had rather odd and easily recognized vertebrae. The **Micro-sauria** formed another group of variable but basically salamander-like animals; they are probably not closely related to the other orders just noted and, indeed, not even truly lepospondylous. Sometimes included in this group are the forms here termed the **Lysorophia**, another group of small semiaquatic animals with reduced limbs. It seems likely that the older lepospondyls were, despite their antiquity, a side branch, if an early one, of the basal stock of land animals.

The true base is to be sought among a second early group of amphibians, the **Labyrinthodontia** (Fig. 41). These animals of variable size were in general considerably larger than contemporary lepospondyls; some attained the size of crocodiles. Their vertebral construction was a diagnostic feature and was one from which that of reptiles and higher vertebrates could have been readily derived (cf. Figs. 119, 120). Except for the absence of median fins and the presence of short but sturdy legs developed from paired fins, many features of the earliest labyrinthodonts, the **Ichthyostegalia** from the uppermost Devonian of Greenland and Australia, are highly comparable with those of the sarcopterygians from which they came. They were the first vertebrates to walk on land.

In the late Paleozoic the labyrinthodonts flourished greatly. One major group, termed the **Temnospondyli**, was especially abundant and varied. Typical among them were semiaquatic forms resembling crocodiles as well as more terrestrial animals like large, clumsy lizards; they survived, in the shape of purely water-dwelling forms with flat bodies and small limbs, until the late Triassic. Less abundant and less long-lived but important in phylogeny were the **Anthracosauria**, among which were forms closely approaching the reptiles in skeletal structure. It is from these labyrin-

Figure 41. Labyrinthodont amphibians. *A, Ichthyostega,* the earliest known tetrapod from the Devonian of Greenland, a member of the Ichthyostegalia; *B, Eryops,* a "typical" member of the Temnospondyli; *C, Metoposaurus,* an advanced and completely aquatic temnospondyl; *D, Diplovertebron,* an early member of the Anthracosauria. The labyrinthodonts were a varied group of primitive amphibians, many of which were large and more like reptiles than the small and "degenerate" modern amphibians. (*A* and *D* after Špinar and Burian; *B* after Romer; *C* after Fenton and Fenton.)

thodonts that there arose, in the latter part of the Carboniferous, the first reptiles, and members of it are often taken as "typical primitive tetrapods" (as in Fig. 167 and the accompanying descriptions).

Whence did the three modern orders arise? This is an unsettled problem. The first prefrog is present in the Triassic and typical, if rather primitive, anurans are present in the Jurassic. In this latter period are found the first urodeles; there is almost no fossil record of the Gymnophiona (Tertiary vertebrae only). There is little in the structure of the modern orders to give a clear clue to their origin from either labyrinthodonts or lepospondyls. The adaptations seen in anurans and urodeles are so divergent that many investigators have believed them to have evolved separately from discrete Paleozoic origins. Recently, however, it has been pointed out that there are significant common features in ear apparatus, dentition, and so on. This situation suggests that despite their later divergence, the three modern types may have had a common origin, most probably though very uncertainly from temnospondyl labyrinthodonts, and may be united in a major amphibian subclass, the **Lissamphibia**, equal in rank to the Labyrinthodontia and Lepospondyli. Other investigators would derive only the anurans from temnospondyls; according to them, urodeles and gymnophionans are more likely descendants of the microsaurs.

The development of the early land vertebrates has sometimes been "explained" as the result of some "urge" toward terrestrial life among their fish ancestors. This is, of course, absurd; the evolution of the earliest amphibians capable of walking on land seems to have been essentially a happy accident. The amphibians appear to have evolved from crossopterygian ancestors toward the close of the Devonian. Lungs, already present in the ancestral bony fishes, are an excellent adaptation for use in poorly oxygenated water, whether because of seasonal drought in streams or of similar conditions in estuarine or coastal environments. But when a stream or pool dries up completely, a typical fish is rendered immobile and dies. Some further development of the fleshy fins already present in crossopterygians would give their fortunate possessor the chance of crawling (albeit with considerable pain and effort at first) and enable it to reach some surviving body of water where it could resume a normal piscine existence.

Legs, the diagnostic feature of the tetrapod, may thus have been, to begin with, only another adaptation for an aquatic life. The earliest amphibian was little more than a four-legged fish. Life on land would have been the farthest thing from its thoughts (had it had any), for early amphibians were eaters of animal food, and until insects became widespread in the later Carboniferous, there was precious little for an amphibian to eat on land. It was probably only after a long period of time that its descendants began to explore the possibilities of terrestrial existence opened out before them through their new locomotor abilities. And even today, few of its descendants that have remained amphibians have capitalized fully on these potentialities.

The term amphibian implies the double mode of life exhibited by many members of this class. Some toads spend much of their lives on good dry land, but most amphibians do not venture far from the stream banks, and some modern forms are still essentially water-dwellers like their ancestors. The typical amphibian mode of development, as exemplified by the familiar frogs and toads of northern temperate regions, is still essentially that of the ancestral fishes. The eggs are laid in the water and develop there into water-dwelling, gill-breathing tadpoles. Only when they approach adult size do lungs replace gills, do limbs develop and mature, and does terrestrial life become possible. An amphibian is chained to the water by its mode of development and the necessity of returning to that element periodically for reproductive purposes. Although numerous modern amphibians have adapted variously to avoid this complication, none has been a complete success as a fully terrestrial form. Indeed, some salamanders have, as it were, abandoned the attempt; such forms as the American mud puppy (*Necturus*) never emerge onto land at any age, retain external gills and water breathing, and reproduce, in neotenic fashion, in essentially a larval condition.

Reptiles

The reptiles are the descendants of the ancient amphibians that, happily, solved this reproductive problem and became the first fully terrestrial vertebrates. The "invention" of the amniote egg (with the associated developmental processes; see Chapter 5) is the major diagnostic feature that distinguishes reptiles from amphibians.

The reptilian egg is laid on land—thus is avoided the necessity of any adaptation for aquatic existence in either young or adult. This type of egg is the familiar one preserved in the reptiles' avian descendants. The shell offers protection. A large yolk furnishes an abundant food supply so that the reptilian young (unlike the tadpole) can hatch out, at a fairly good size, as a miniature replica of the adult and thus has no need of prematurely foraging for its food (obviously this apparent advantage can be abandoned—see most birds). Of embryonic membranes developed within the egg shell, one externally sheathes both young and yolk. A second forms a lunglike breathing mechanism for absorption of the oxygen that penetrates the porous shell. A third (the amnion, from which the egg type gets its name) encloses the developing embryo in a liquid-filled space, a miniature replica of the ancestral pond. The development of such an egg was so important an advance in the evolution of land vertebrates that, as noted earlier, the reptiles, together with the birds and mammals that descended from them, are often classified collectively as the amniotes.

Possibly the oldest reptiles were still amphibious in their habits and the amniote egg was merely an adaptation parallel to, but better than, other adaptations seen in modern amphibians—a device that removed the eggs from the dangers of drought, and of predators present in the ancestral waters. However, it seems more probable that the first reptiles were small terrestrial forms—ones that would be considered odd lizards rather than odd salamanders, to use living forms for comparison.

Modern reptiles—mainly consisting of lizards, snakes, and turtles—are moderately abundant in the tropics but unimportant in temperate zones and absent from cold climates where survival is difficult for these "cold-blooded" creatures. The modern forms, however, are but sparse remnants of the great array of reptiles that, beginning in the late Paleozoic, radiated out into a bewildering variety of forms that long ruled the earth and caused the Mesozoic Era to be popularly known as the Age of Reptiles (Fig. 42). The basic stock from which they sprang, the long-extinct order **Cotylosauria** (Fig. 43 *A*) or "stem reptiles" were, apart from reproductive improvements, still very archaic, with limbs sprawled out sideways from the body, and, in most regards, little more advanced than their amphibian forebears and cousins. The earliest and most primitive were small animals, but some of the later cotylosaurs became large, ungainly herbivores. Due, no doubt, to the breaking of the chains that

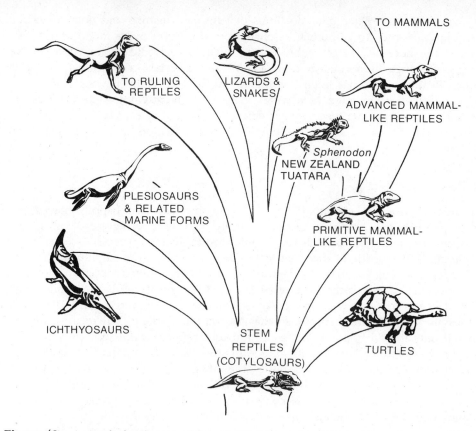

Figure 42. A simple family tree of the reptiles (the archosaurs are shown in more detail in Fig. 47; after Romer.)

bound them to the water, there presently developed from cotylosaurs the groups that became prominent in the Mesozoic.

A side branch of the stem reptiles was that of the turtles, order **Testudines** (Fig. 43 *B, C*). In their sprawling gait, they are reminiscent of their Paleozoic forebears, but turtles have made a conspicuous advance in the development of a protective shell of bone, covered by horn, guarding both back and belly. This shell involves the rib cage plus dermal elements with the shoulder and hip girdles inside it. No other vertebrate puts the girdles *inside* the ribs. Once encased in armor, turtles turned conservative and since the Triassic have advanced but little in most aspects. The only later improvement of any note made by the order as a whole was the acquisition of the ability, lacking at first, to pull the head back into the shell. In all familiar types (termed cryptodires) this is done by a straight backward pull, with the neck bent in a vertical S-curve; some odd tropical forms (the pleurodires) tuck the head in sideways against the shoulder. Like the most ancient reptiles, most modern turtles are amphibious marsh and pond dwellers. Some, however, reverted to a purely aquatic life, and several marine forms have developed, with paddle-like limbs for propulsion. At the other extreme, one group, the tortoises, has become purely terrestrial.

In both the reptilian orders mentioned thus far, the skull roof, like that of their

Figure 43. Anapsid reptiles. *A,* The primitive and very lizard-like cotylosaur *Cephalerpeton; B* and *C,* two modern turtles. *B,* The snapping turtle *Chelydra,* and *C,* the marine green turtle *Chelonia. (A* from Carroll and Baird; *B* and *C* from Young.)

amphibian ancestors, had openings only for the major sense organs, the nostrils, eyes, and pineal eye. Such a condition is known as "anapsid," and reptiles with it are placed in the subclass **Anapsida** (Fig. 44; see also p. 265). In all other reptiles, "extra" openings appear in the temporal region; these are called **temporal fenestrae** and are diagnostic features in reptilian classification. Some, members of the subclass **Euryapsida**, have a single temporal fenestra high on each side of the cheek; this is the euryapsid or parapsid pattern. Reptiles of the subclass **Synapsida** show the synapsid pattern, a single fenestra on each side but more ventrally placed than in euryapsids. Finally, in the diapsid condition, both pairs of fenestrae occur. Two great subclasses, the **Lepidosauria** and **Archosauria**, have diapsid skulls.

Euryapsid reptiles are all extinct and poorly understood (Fig. 45). Some appear to have been fairly primitive, lizard-like terrestrial forms; these are placed in the order

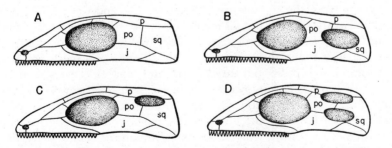

Figure 44. Diagrams to show types of temporal openings in reptiles. *A,* Anapsid type (stem reptiles, turtles); *B,* synapsid type (mammal-like reptiles); *C,* euryapsid type (extinct plesiosaurs, and so forth); *D,* diapsid type (rhynchocephalians, ruling reptiles; lizards and snakes derived by loss of one or both temporal arches). Abbreviations; *j,* jugal; *p,* parietal; *po,* postorbital; *sq,* squamosal.

Figure 45. Euryapsid reptiles, having a single pair of dorsal temporal openings. *A, Araeoscelis,* a small lizard-like member of the Araeoscelidia; *B, Placodus,* one of the mollusc-eating Placodontia; *C, Plesiosaurus,* a member of the Sauropterygia; *D, Stenopterygius,* one of the extremely fishlike Ichthyosauria. (*A, B,* and *C* after Fenton and Fenton; *D* after Špinar and Burian.)

Araeoscelidia. Unfortunately, the more these forms are studied, the fewer appear to be euryapsid; thus, many investigators deny the existence of the group completely.

Despite their newly won ability to conquer the land, a number of Mesozoic reptile groups took up (like the sea turtles) a marine existence. Prominent Mesozoic types of this sort were the plesiosaurs, the placodonts, and the ichthyosaurs. Recent work suggests that these were not all really euryapsid, but some were probably modified diapsids. The plesiosaurs (order **Sauropterygia**) have been popularly compared to "a snake strung through the body of a turtle." Plesiosaurs were not at all related to either of these two reptiles, but the description is apt. They possessed a long neck or long snout or both; the body was short, broad, and relatively flat. Reversion to a truly fishlike means of locomotion was impossible, because the trunk was inflexible and the tail was short; instead, the limbs were developed into powerful oarlike structures, with which the creature "rowed" its way through the sea; actually, the motions, as in sea turtles, may have been those of flying, not rowing. More primitive forms (nothosaurs) were less spectacularly but similarly modified.

Placodonts (order **Placodontia**) were Triassic forms that had become eaters of molluscs or other hard-shelled invertebrates. They were heavily built, rather flattened forms, some of which developed rather turtle-like armor. Their relation to the plesiosaurs, like most other things about euryapsids, is currently debated and several investigators deny any close affinities between them.

Even more unusual structurally were the **Ichthyosauria,** the "fish-reptiles." Possibly the plesiosaurs were able at least to waddle about on a beach in the manner of a marine turtle or a seal. The ichthyosaurs, however, had become as completely adapted to a marine life as a porpoise or dolphin (with which there are many analogies); there is fossil evidence indicating that egg-laying on land had been abandoned and that the young were born alive. The body shape was completely reconverted to that of a fish—the neck telescoped to give a fusiform body shape, the limbs shortened into small steering devices. Locomotion was performed in a fishlike manner by undulations of the trunk and tail; a fishlike fin was developed on the back (but, like that of whales, it lacked the skeletal supports found in the dorsal fins of fishes), and the tail became a powerful swimming organ, like that of a shark in appearance. In this last regard, however, there is a notable structural difference: whereas in a shark the end of the backbone tilts into the upper lobe of the tail fin, that of the ichthyosaur turns sharply down at the back, the fin (as we know from excellent slabs of fossils) expanding above it. Most ichthyosaurs were presumably fish eaters, but some also fed on ammonites, large molluscs related to the modern nautilus.

The **Lepidosauria** (Fig. 46), the first subclass of diapsid reptiles, include a basal ancestral group, the order **Eosuchia,** of extinct, more or less lizard-like (as usual) forms. Another group, which appeared early in the Mesozoic and has persisted to the present day although never prominent, is the order **Rhynchocephalia,** now represented by the tuatara (*Sphenodon*). This creature, lizard-like in appearance, has survived in the relative safety and isolation of New Zealand, where, once widespread, it is now preserved on a few small islands.

Descended from ancient forms related to the tuatara is the much more successful order of the **Squamata,** the lizards, amphisbaenians, and snakes. Technically, they are readily distinguished by the fact that the cheek and temple region of the skull has been reduced, so as to leave but one temporal arch (most lizards) or none at all (snakes; cf. Fig. 188). The "scaled reptiles" are not only the most flourishing but also the most modern of reptile orders; even lizards amounted to little until late in the Cretaceous and the deployment of most of the various snake types did not take place until Cenozoic times. Lizards (suborder **Lacertilia**) are widespread and highly varied

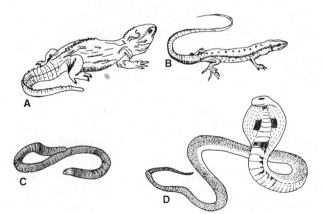

Figure 46. Lepidosaurs, the dominant group of living reptiles. *A, Sphenodon,* the extremely primitive tuatara from New Zealand, the only surviving member of the Rhynchocephalia; *B, Lacerta,* an ordinary lizard or member of the Lacertilia; *C, Amphisbaena,* a limbless member of the Amphisbaenia; *D, Naja,* the cobra, suborder Serpentes. (*A, B,* and *C* after Young; *D* after Newman.)

in the tropics. Most prominent of American lizards are the iguanas and their relatives, such as the collared lizard or "mountain boomer" of the Southwest and the little "horned toad." In the Old World, the largest forms are the monitor lizards (*Varanus*), one of which, in the East Indies, may reach a length of 4 m. Relatives of the monitors in the late Cretaceous had a temporary success as a group of giant marine lizards, the mosasaurs. The (true) chameleons of the Old World tropics, with peculiar grasping feet and highly protrusible tongue, are a specialized side branch of the lizard stock. In several lines of lizards, there have developed burrowing types, with limbs reduced or absent. Indeed, there are several families entirely composed of such limbless forms.

One such group, often considered a family of lizards, is better regarded as a separate suborder, equivalent to the snakes or lizards: the **Amphisbaenia.** They are mainly tropical and limbless (with one exception—a form with small arms but large hands and no hind limbs), characterized by having the scales arranged in distinct rings on the body, which makes them look even more like earthworms than do most other limbless forms.

Derived from lizards are the snakes (suborder **Serpentes**), which are highly modified in two major regards. As in some lizards and amphisbaenians, the limbs have been reduced and generally lost completely, and locomotion is accomplished, in most cases, by undulations of the body and tail; essentially, a snake swims on dry land. More distinctive is the fact that the skull and jaw apparatus are markedly altered in the direction of flexibility and, as a result, allow the swallowing of the prey whole. Some snakes of primitive type are burrowers, and it is possible that snake evolution began with such forms, but even among such primitive types as the constrictors—the boas and pythons, some of great size—most now live above the surface. The great majority of snakes belong to an assemblage, often called a family but probably better considered as several (unfortunately, not all investigators agree on how many or what!), of which the common harmless forms of northern temperate regions are representative. But even within this family, many tropical genera have developed poison glands. These, however, are generally small and inoffensive forms and the fangs, situated in the back of the mouth, present little danger to man and other large animals. Two further families include the major poisonous snakes, with highly developed fangs and powerful and varied venoms that attack the nervous system or cause destruction of tissues: (1) a group, mainly Old World, that consists of the cobras and their kin, including the coral snakes, and, as a separate subfamily, the venomous sea snakes, found in the Indian and Pacific Oceans; (2) the vipers, with erectile fangs, including the adders and other vipers of the Old World, and the pit vipers, mainly American, with such representatives as the rattlesnakes, copperhead, and water moccasins.

An exceedingly important reptile group with the same two-arched type of skull build as *Sphenodon* is the great subclass **Archosauria**, the ruling reptiles (Figs. 47, 48). Today they survive only in the form of the rather aberrant crocodiles and alligators, but most of the dominant land reptiles of the Mesozoic were archosaurs, and the birds are descendants of this group.

The basal stock of the archosaurs is found in the Triassic in the shape of predaceous reptiles forming the order **Thecodontia.** Some early thecodonts were four-footed; however, elongated hind legs, a modified hip structure, and other features suggest that others were becoming adapted to a bipedal mode of life. Still other thecodonts became armored or paralleled closely the later crocodilians. From these modest beginnings came the dinosaurs. These are popularly considered as constitut-

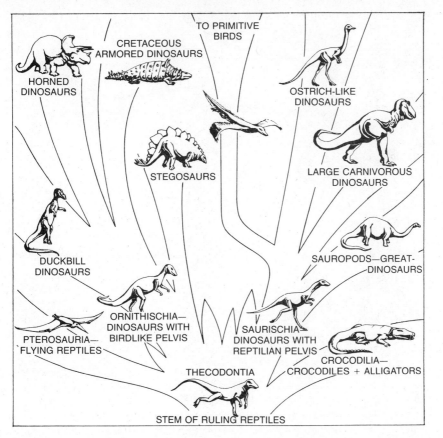

Figure 47. A simple family tree of the ruling reptiles. (After Romer, The Vertebrate Story, University of Chicago Press.)

ing a single group of gigantic reptiles. This concept is far from correct, because although many dinosaurs were large, some were small (one was no bigger than a rooster). There were two major dinosaur stocks, possibly not closely related to one another, although both were descended in common from thecodont ancestors.

In one group, termed the **Saurischia,** or reptile-like dinosaurs, a large proportion of known forms was bipedal carnivores (suborder **Theropoda**). Some of the smaller and more primitive of these bipeds can scarcely be distinguished from their thecodont ancestors. Others grew to immense size; *Tyrannosaurus* was the most ponderous flesh-eater the earth has ever seen. Forming a major side branch were the amphibious (sauropod) dinosaurs and their ancestors (suborder **Sauropodomorpha**), which changed to a herbivorous mode of life, and instead of being bipeds, they were four-footed walkers and grew to such giants as *Apatosaurus* and *Diplodocus.* These great reptiles may have spent much of their lives in lagoons where soft vegetation abounded. Their weight (one is estimated to have exceeded 45 tonnes) was such that it seems improbable that their stocky limbs could have supported them effectively on land, although this is open to question; some recent investigators suggest they were purely terrestrial. Both groups had normal reptilian pelvic girdles, the character that gives them their ordinal name.

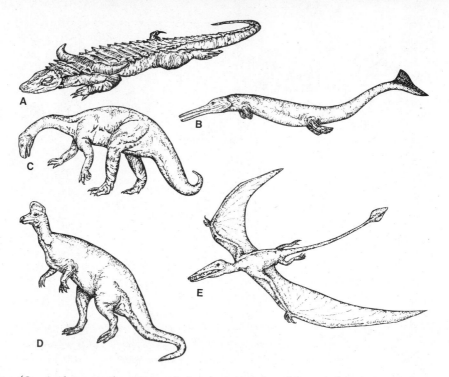

Figure 48. Archosaurs, the most varied and spectacular of the reptiles. *A, Desmatosuchus,* an armored and rather atypical member of the Thecodontia; *B, Geosaurus,* an extinct marine member of the Crocodilia; *C, Plateosaurus,* an early dinosaur of the order Saurischia; *D, Corythosaurus,* a "duck-billed" dinosaur of the order Ornithischia; *E, Rhamphorhynchus,* a flying reptile, a member of the Pterosauria. (*A* after Breed; *B* after Fenton and Fenton; *C* after Colbert; *D* and *E* after Špinar and Burian.)

A second major group was that of the **Ornithischia**, or birdlike dinosaurs, in which the hip girdles (but not other anatomic features) were comparable to those of birds. Like some, at least, of their saurischian cousins, primitive members of this group were bipeds (suborder **Ornithopoda**); but in contrast with that other dinosaur stock, all the birdlike forms were herbivores. Best known of the bipeds of this group are the duckbills (hadrosaurs), which were abundant in the closing days of the Age of Reptiles. In most of the ornithischians, there was reversion to a quadrupedal pose. There were, in fact, three distinct types of quadrupeds developed in this order, all with some sort of defense against the great carnivores of the day. These types are exemplified by such popular museum exhibits as those of *Stegosaurus* (suborder **Stegosauria**), whose backbone is capped with odd triangular plates and defensive spines; *Ankylosaurus* (suborder **Ankylosauria**), low and flat and heavily armored on back and tail; and the horned dinosaurs (suborder **Ceratopsia**), such as *Triceratops,* with horns—often a trio of them—and a great frill of bone protecting the neck. The horned dinosaurs, all from the late Cretaceous, were in a sense better "rhinoceroses" than true rhinoceroses.

Dinosaurs flourished during the Jurassic and Cretaceous, and even in the closing phases of the latter period, they were present in considerable numbers and variety. Then, within a very short space of time (geologically speaking), they disappeared completely. The reason for this abrupt end of the Age of Reptiles is far from fully understood. Geologic events are perhaps basically responsible. The Cretaceous period was one of mountain building during which great ranges, such as the American Rockies, began to emerge from formerly flat country. Many of the low-lying marshy and lagoonal areas where dinosaurs browsed on lush vegetation disappeared. Climatic conditions were radically changed, and new types of plants became dominant with which herbivorous dinosaurs were probably unable to cope. As the herbivores consequently dwindled and disappeared, their flesh-eating cousins, who preyed upon them, would of necessity follow them to extinction. However, there are many other theories, with various astronomic ones currently receiving much consideration.

Still another group of extinct archosaurs was that of the order **Pterosauria**, the winged reptiles. In them, the front limbs had one finger (the fourth) enormously elongated. From it, there was extended, in a somewhat batlike manner, a great wing membrane. Manipulation of a wing of this sort would appear to have been an awkward matter, and flight almost certainly consisted mainly of soaring rather than active beating of the wings. Further, without any intermediate fingers extending into the wing membrane to strengthen it, the dangers of a disastrous tear would seem to have been great. The hind legs of pterosaurs, quite in contrast with those of birds, were feeble structures, and it is difficult to see how these creatures could have stood upon them, much less get a running take off, as a bird of any size must do. Probably they perched in a batlike manner, although recent investigations have cast doubt on this idea. But it is hard to imagine what safe perch could be found for a pterosaur with a wing spread of up to 16 m (estimated in one Cretaceous form). All in all, it is not difficult to understand why the pterosaurs became extinct after more efficient flying forms, the birds, had evolved. However, this extinction was not rapid; pterosaurs survived and were successful for about a hundred million years!

Sole survivors today of the archosaurs are the alligators and crocodiles, comprising the order **Crocodilia**. Although many of their structural features resemble those of bipedal dinosaurs, the crocodilians, like many of their dinosaurian relatives, have a quadrupedal gait and are, further, amphibious. The crocodiles are phylogenetically remote from the base of the reptilian family tree; their anatomic features are hardly to be considered characteristic of reptiles as a whole, and, as might be expected, they show numerous features found in the birds, whose ancestors were archosaur relatives of the crocodiles.

Living reptiles, including crocodilians, are "cold-blooded"—their temperature varies, more or less, with that of their environment. People have tended to assume extinct reptiles, dinosaurs among them, to be similar. Recently, this view has been challenged and some investigators claim that all dinosaurs and pterosaurs as well as some thecodonts were "warm-blooded"—technically homoiothermic and endothermic—like birds and mammals. None of the many lines of evidence are conclusive, so the argument continues, with many investigators now taking a middle ground, that some dinosaurs were warm-blooded but others were not. In any event, sheer size would mean that deep temperatures in a large dinosaur would vary slowly. Oddly enough, pterosaurs have been less frequently discussed, though there is at least as much evidence for temperature control in them as in any dinosaur.

BIRDS

Birds have been aptly termed "glorified reptiles." We customarily treat them as a separate class, **Aves**, but in many regards they are not much farther removed from the general reptilian stock than are some of the ruling reptiles from which they sprang. Within that group, as we have noted, was included one series of flying forms, the pterosaurs. The birds are not descended from pterosaurs but are a second archosaur flying type, in which, instead of membrane, feathers—diagnostic of the class—form the wing surfaces of the modified pectoral limbs. In certain respects, notably bipedal adaptations, the birds are similar to their dinosaurian relatives, but almost every notable avian characteristic is an adaptation to flight. The maintenance of a high and constant body temperature and improvements in the circulatory system are associated with the need of a high metabolic rate for sustained flight. Lightening of the body in various ways (particularly by the development of air sacs and hollow bones) is also associated with flight, as are modifications in the brain and sense organs. In the course of their evolution, the birds necessarily "discovered" long ago many of the principles of aerodynamics that man has learned as a result of much scientific work and hard experience; they are "expert" in wing construction and in the utilization of winds and air currents in gaining distance or elevation. In general, birds with an active beating type of flight tend to have relatively short, wide wings; those that rely mainly on gliding and soaring usually have relatively larger wing surfaces with the length greatly increased in most cases. Flying birds (in contrast to ostrich-like ground-dwellers) are seldom of any great size. To maintain flight, wing area must increase, on the whole, proportionately to weight, and too great size of body would require a wingspread so great as to be difficult to manipulate.

Except in primitive Mesozoic birds, teeth were lost and reliance placed on a bill for gathering food. Numerous variations are seen in bill structure, with such extremes as the parrot beak and the effective drilling organ of the woodpeckers. Presumably, the food of primitive birds was of some relatively soft nature, and teeth were not necessary. Many modern birds, however, are eaters of grain. For this type of food, mammals need highly developed grinding teeth; birds have used as a functional substitute a grinding apparatus in the muscular gizzard containing grit or small pebbles; such a gizzard is also present in crocodilians and may well be a general character of archosaurs.

In the birds, we see a class of vertebrates that in many regards is to be considered as on as high a level of organization as the mammals but (even disregarding differences in aerial vs. terrestrial locomotion) organized in a very different manner. Birds can certainly be trained but on the whole seem relatively much less capable of learning by experience than mammals. On the other hand, they exhibit innate behavior patterns of a complexity unknown in mammals. Many of these, connected (for example) with social behavior, courtship, nestbuilding, and rearing of the young, are familiar to any bird lover. Avain "knowledge" of geography is most remarkable. The homing ability of birds is great, and the ability, for example, of young golden plovers to migrate successfully from the Arctic tundras to the Chaco region of South America—unaccompanied by older birds, over a complicated course of about 10,000 km, much of it over open ocean—is an accomplishment apparently verging on the supernatural.

A happy accident of preservation has given us knowledge of five skeletons of an ancestral bird, *Archaeopteryx*, the sole representative of the subclass **Archaeor-**

nithes, from deposits of late Jurassic age (Fig. 49 A). In this bird, teeth were still present, the wing had clawed fingers, and there persisted a long reptilian type of tail; *Archaeopteryx* so nearly splits the differences in skeletal structure between ruling reptiles and modern birds that its systematic position would be a matter of doubt if imprints of its feathers had not been preserved with the skeletons. Comparison of *Archaeopteryx* with various extinct archosaurs suggests that it, and birds generally, are descended from small theropod dinosaurs; however (as usual), there is disagreement, and thecodonts, ornithischian dinosaurs, and even basal crocodilians are also proposed as avian ancestors by some paleontologists.

Birds, mainly owing to the delicacy of their skeletons, are relatively rare in the fossil record. However, there is evidence that before the close of the Cretaceous there had evolved forms that were quite modern in almost every structure. A few intermediate forms are known—intermediate in that they possess teeth. *Hesperornis* (Fig. 49 B) and relatives were extremely specialized flightless, fish-eating forms that, together with a few other dubiously related flying forms such as *Ichthyornis,* comprise the superorder **Odontognathae**. Because they are considered closer to living birds than to *Archaeopteryx,* all birds except the latter are placed in one subclass, the **Neornithes**.

Although taxonomists divide the birds into a considerable series of orders, the structural differences between these groups are for the most part small. There is, however, one partial exception to this; for there has been thought to be a distinction between two modern groups, representing primitive and advanced stages in bird evolution. Technically, the two have been defined by details of palatal and jaw structure (with which we need not here concern ourselves), which gave rise to the terms "palaeognathous" and "neognathous" (or superorders **Palaeognathae** and **Neognathae**) to distinguish them. Most birds with which the reader is ordinarily familiar, indeed, the majority of all living birds, are members of the latter group. To the other

Figure 49. Primitive birds. *A, Archaeopteryx,* the earliest known bird and sole member of the Archaeornithes, had teeth, claws in the "hand," a long bony tail, and other reptilian characters; in fact, were it not for the preservation of feathers on the fossils it would be considered a small saurischian dinosaur. *B, Hesperornis,* a member of the Odontognathae, was primitive in retaining teeth; however, the wings were almost totally lost and it was a highly specialized swimming form, like a flightless loon. (*A* after Heilmann; *B* after Fenton and Fenton.)

assemblage (Fig. 50) belong the ostrich and other forms generally termed **ratites** (referring to the reduced condition of the breastbone common in flightless birds), such as the cassowary and emu of Australia, the rhea of the South American pampas, the extinct moas and the little kiwi of New Zealand, and the giant extinct birds of Madagascar.

Most palaeognathous birds have tiny wings and are flightless, a fact that has given rise to the claim that they represent a primitive stage in bird evolution in which flight had not been attained. But anatomic study strongly indicates that this is not the case and that the ratites are probably secondarily terrestrial descendants of once flying types. Most of them are found on islands where there are few terrestrial predators or on continents (Australia, South America) where, the fossil record tells us, the same was true at the time that the native ratites evolved. If ground-dwelling predators are absent, much of the "point" of flying has been lost. The tinamous of South America are birds that have the power of flight (although they are awkward flyers) and yet have the "old-fashioned" type of palate; they may represent an ancestral group from which the ratites have developed. It is quite probable that the various groups of ratites are not closely related; probably the Palaeognathae are not a "natural" group.

Apart from the ratites, most birds, as has been said before, are rather uniform in basic anatomic features, with differences between orders no greater than those that distinguish smaller groups, such as families, among mammals. To the student of habits, songs, and plumage, the birds have much to offer; they offer less, however, to those interested in anatomic structure and function. Hence, the remaining bird groups may be dismissed briefly. In Figure 51 are thumb-nail sketches of representatives of the major groups of birds other than the ratites, pictured in the sequence in which the orders are listed by many ornithologists. A majority of all birds, including the song birds, are members of the final order, the Passeriformes or perching birds. These highly evolved bird types are all relatively small; the crows and ravens are giants of the order. Several orders of aquatic birds and oceanic types are customarily placed at the beginning of the series, although there is little evidence that they are actually

Figure 50. Palaeognathous birds, including the ratites. *A,* The kiwi, *Apteryx,* of New Zealand. *B,* The cassowary of Australia. *C,* The ostrich. *D,* The tinamou of South America. *E,* The emu of Australia. *F,* The rhea of South America.

Loon
GAVIIFORMES

Grebe
PODICIPITIFORMES

Albatross
PROCELLARIIFORMES

Penguin
SPHENISCIFORMES

Pelican
PELECANIFORMES

Stork
CICONIIFORMES

Goose
ANSERIFORMES

Hawk
FALCONIFORMES

Pheasant
GALLIFORMES

Rail
GRUIFORMES

Gull
CHARADRIIFORMES

Dove
COLUMBIFORMES

Parrot
PSITTACIFORMES

Cuckoo
CUCULIFORMES

Owl
STRIGIFORMES

Goatsucker
CAPRIMULGIFORMES

Hummingbird
APODIFORMES

Hornbill
CORACIIFORMES

Woodpecker
PICIFORMES

Swallow
PASSERIFORMES

Figure 51. Thumb-nail sketches of representative members of the major orders of birds above the ratite level. The ordinal name and the popular name of the bird representing it are given in each case.

primitive. Of special interest here are the penguins, Southern Hemisphere forms that are flightless but nevertheless have powerful wings that have been transformed into swimming flippers. Here, as in the case of the ratites, it has been argued that absence of flight is primitive. It is, however, more probable that they have descended from flying oceanic birds, probably petrels, that also used their wings in swimming.*

Mammals

Mammal-like Reptiles (Fig. 52).   The mammals are descended from reptiles; but the fossil record shows that the reptilian line leading to them, the subclass **Synapsida**, diverged almost at the base of the family tree of that class. Their relationship to the existing reptilian orders is thus exceedingly remote.

Oldest of the reptiles antecedent to mammals were members of the order **Pelycosauria** (Fig. 52 *A*), a group that appeared in the Carboniferous and flourished in the early Permian. They were in most regards exceedingly primitive reptiles, but certain characters in skull structure (such as the presence of but a single, rather ventrally placed opening in the temporal region of the skull) indicate that they represent a first stage in the evolution toward mammalian status. Many pelycosaurs, such as *Dimetrodon,* had greatly elongated dorsal spines of the vertebrae to support a large "sail." Succeeding them in late Permian and early Triassic days were the **Therapsida**, progressive mammal-like forms that were among the most common animals of their day (Fig. 52 *B*). The characteristic therapsids were flesh-eaters, active four-footed runners in which, as in their mammalian descendants (and, independently, many archosaurs), elbow and knee had been swung in toward the body, resulting in better support and greater speed—this in contrast to the sprawled limb pose of primitive land animals. In advanced Triassic members of the group, many features of skull, jaw, dentition, and limbs approach closely the mammalian pattern. Other therapsids were, for a time, very successful herbivores.

The evolution of mammal-like reptiles was a major feature of early reptilian evolution. However, during the Triassic period, other reptile groups had become prominent—notably the dinosaurs. It appears that therapsids, for the most part, could not compete successfully with them and rapidly dwindled and disappeared from the scene. There survived, however, small therapsids from which evolved the oldest mammals, sparse remains of which are found in Mesozoic deposits from the end of the Triassic onwards. Living as they did for tens of millions of years as contemporaries of the dinosaur dynasties, our small Mesozoic mammal ancestors were seemingly insignificant in the life of their times.

Intelligent activity may be reasonably regarded as the hallmark of mammalian progress. With activity may be correlated not only the efficient locomotor apparatus characteristic of mammals but also (as in birds) circulatory improvements and constant body temperature (with which the development of hair is related). In enterprise and ingenuity, even the most stupid of mammals is an intellectual giant compared with any reptile. The habit of bearing the young alive, characteristic of all except the most primitive forms, and the development of the nursing habit, with concomitant

---

*For those of literary tastes, it is mentioned that the penguins of which Anatole France wrote are the auks of English-speaking peoples—Northern Hemisphere birds that paralleled the penguins in taking up a swimming mode of life.

Figure 52. Synapsid reptiles. *A,* The primitive pelycosaur *Ophiacodon* from the early Permian of Texas; *B,* the therapsid *Lycaenops* from the later Permian of South Africa. Note the more mammal-like posture and projecting canine tooth in *Lycaenops.* If hair were drawn on it in place of scales, it would appear to be a mammal; this would not be true of *Ophiacodon.* *(B after Colbert.)*

care and training of the young, are mammalian innovations that have resulted in giving a long period for the development and elaboration of delicate nervous and other mechanisms before the young are sent out into the world. Most of these advanced characters came into existence during the long period of time when mammals lived under dinosaurian dominance. Alertness was necessary for survival, and mammals owe the dinosaurs a vote of thanks for having been unwittingly responsible for the eventual success of higher mammals.

Mammals are properly defined by the presence of mammary glands and, therefore, the habit of suckling the young. Other characters that always apply to living forms are the presence of hair and of a muscular diaphragm. However, all these are "soft" characters—none will be preserved in a fossil. Another character that appears to work and *can* be used is the nature of the jaw articulation; in reptiles (and other submammalian forms), the joint is located between the quadrate and the articular, two parts of the original visceral skeleton, but in mammals these have become ear ossicles, and the joint lies between two dermal bones, the squamosal and dentary. Unfortunately, intermediates are known (unfortunate, that is, for a nice simple system—having a good fossil record is not really unfortunate!), but this character is still the most widely used criterion for distinguishing mammals from their therapsid ancestors.

The fairly numerous Mesozoic mammals (Fig. 53) are almost all very poorly known, often only from isolated teeth. Their classification has long been debated, and

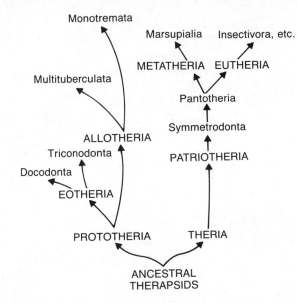

Figure 53. Diagrammatic family tree, to show the evolution and relationships of the primitive groups of mammals. Pictures are omitted because many of the forms are known only from partial specimens and other insufficient evidence.

there is still no general agreement. We here adopt the idea that all mammals may be divided into two separate lines. One line, the subclass **Prototheria** (a word of warning: this term is frequently used in a much more restricted sense), consists of forms in which much of the lateral wall of the braincase is formed by the periotic rather than the alisphenoid as in therians (see pp. 256–261). Three of the four orders, the **Multituberculata**, **Triconodonta**, and **Docodonta**, are extinct; only one of these, the rather rodent-like **Multituberculata**, survived the Mesozoic and even they disappeared early in the Tertiary. Triconodonts and docodonts are poorly known small animals, comparable in a rough way to some of the eutherian insectivores—small, long-snouted, rather unpleasant carnivores. The prototherians may be divided into two infraclasses, but the evidence for relationships is poor, so these are probably not worth worrying about. Whether prototherians and eutherians form a single "natural" group is debatable. For a long time it was assumed that they did, but then opinion gradually shifted and mammals were thought to have arisen independently several times from therapsids; now more investigators seem to be returning to the original idea.

Monotremes. The monotremes, order **Monotremata**, are the only surviving prototherians; as such they are extremely different from any other living mammals. They include only the duckbill and spiny anteaters of the Australian region. These curious animals possess many diagnostic mammalian characters but retain primitive features in that, alone of mammals, they still lay shelled eggs and possess a cloaca as did their reptilian ancestors. The duckbill is a semiaquatic, web-footed, fur-covered frequenter of streams in which it finds a food supply of snails and mussels. The spiny anteater, protected from enemies by a coat of spiny hair, subsists on termites; its powerful clawed feet give it phenomenal digging ability. Both types make nests in burrows where the young are nursed after hatching. These two types are so specialized in many ways that they cannot be regarded as in themselves ancestral types. Their survival in Australia may be due to the relative isolation of that area. Unfortunately, we know nothing of their history. Much of our knowledge of the relationships of extinct mammals is based on diagnostic characters of molar teeth (discussed in a later chapter). The monotremes are, unfortunately, toothless (as adults), having instead horny bills—flat and ducklike in the one type, slender in the other. It is not known whether other prototheres resembled monotremes in their soft anatomy, but recently it was suggested that the pelvis of multituberculates would allow the birth of very small young (as in marsupials) but not the laying of an amniote egg.

Primitive Therians and Marsupials. All normal modern mammals belong to the other subclass, the **Theria** (Fig. 53). Aside from a large alisphenoid on the lateral wall of the braincase, they are characterized by molar teeth that start as triangles and become more complex (a story treated in some detail in Chapter 11). Three infraclasses are recognized. The first, **Patriotheria**, consists of two more orders of small, obscure, insectivorous, Mesozoic forms, the **Symmetrodonta** and **Pantotheria**. They differ in dental patterns, and the first could be ancestral to the second. From the pantotheres, there appear to have arisen sometime in the Cretaceous the two groups of higher mammals, the infraclasses Metatheria and Eutheria.

The pouched mammals or marsupials—technically termed the infraclass **Metatheria** (Fig. 54)—owe their popular name to the fact that, although the young are born alive, they are born at a tiny and immature stage; typically, the female marsupial carries on her belly a pouch in which the newborn young are kept and nourished for a further period after birth. The common opossum is characteristic of the group and is a primitive mammal in many regards. In most regions of the world, the marsupials have not been able to compete successfully with more progressive mammals, and even the hardy opossum failed to survive except in the Americas. South America proved a haven for many marsupials during Tertiary times, when that continent was long isolated; a variety of marsupials, the mainly carnivorous orders **Polyprotodonta** (or Marsupicarnivora), and **Caenolestoidea**, developed there, almost all of which became extinct when the isthmian link was re-established and a host of more advanced mammals invaded that continent. Australia is the one region where marsupials have flourished greatly. The geologic evidence suggests that this latter continent was separated from the rest of the world by Cretaceous times and has since remained isolated. No placental mammals, it appears, had reached Australia at the time of separation, and few have been able to reach it since (bats and whales obviously had no problems, and rats did succeed in getting there). There the marsupials had little opposition, and expanded and diversified to fill almost every type of adaptive niche that placental mammals have occupied in other regions. Fairly directly from the ancient opossum-like, polyprotodont ancestors came such pouched carnivorous forms as the native "cats," the Tasmanian devil, the native "wolf," and pouched parallels to anteaters and even moles of other continents. A second branch of the Australian marsupials (order **Diprotodonta**) acquired chisel-like front teeth and developed types comparable to the rodents among placentals. There are varied native squirrel-like types, even a marsupial flying "squirrel." The wombat is analogous to the woodchucks or marmots of other regions; the native "bear," the koala, is a leaf-feeder belonging

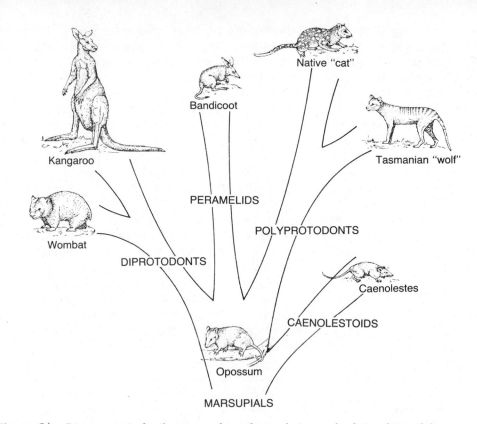

Figure 54. Diagrammatic family tree to show the evolution and relationships of the marsupials. The opossum and *Caenolestes* are basically South American; all the rest are members of the great Australian radiation of this group.

to this group. The marsupials apparently failed to parallel the placentals in one respect: there has been no development of hoofed forms comparable in structure to horses, cattle, and antelopes. But the kangaroos—speedy grass-eating plains-dwellers—fill much the same place in nature. Somewhat intermediate between the polyprotodonts and diprotodonts are the Australian bandicoots (order **Peramelida**). Sometimes all the marsupials are placed in a single order Marsupialia, with the orders here recognized considered suborders; however, their diversity would seem to support a division into separate orders.

Placental Mammals. The major progressive group of mammals includes the host of living forms properly termed the **Eutheria**, the "true mammals," but usually called placentals. The latter name is due to the fact that, in contrast with most marsupials, there is an efficient nutritive connection, the placenta, between mother and embryo; as a result the young can develop to a much more advanced stage before birth. On the extinction of the dinosaurs, these highly developed mammals were already in existence (they are known, albeit rather poorly, from the Upper Cretaceous); they rapidly expanded into a host of types, many of which have continued down to modern times. In some other groups of vertebrates, the family "tree" is actually treelike, with a main trunk or at least major branching limbs. That of the placental mammals, however, is comparable to a great bush: the various orders are difficult to assemble into groups. Various attempts have been made, but none are generally accepted or presented in formal taxonomic terms here. We may briefly note some of the main components of the placental assemblage (Fig. 55). Not all the extinct groups known (and listed in Appendix 1) are discussed here, but those omitted are very obscure and unlikely to be encountered unless one studies paleontology.

The ancestral placentals, indeed, the early mammals as a whole, seem to have been small shy animals that were potential flesh-eaters but were forced to live, owing to their size, on small prey such as insects, grubs, and worms, and presumably on some of the softer vegetable materials. This phase of mammalian existence lasted for many millions of years before the dinosaurs became extinct and the mammals were set free. This time was not, however, entirely wasted. It was a time, it seems, of training and preparation, during which advances were made in mental development and in reproductive processes. As a result, there had evolved by the end of the Age of Reptiles highly developed if small placental (as well as marsupial) mammals, ready to take over the surface of the earth.

At the beginning of the Cenozoic, there rapidly developed a great radiation of mammals into a variety of orders. Some few forms, however, have remained not too distantly removed in structure and habits from their small insectivorous ancestors; these constitute the order **Insectivora** (Fig. 56). The little shrews closely resemble,

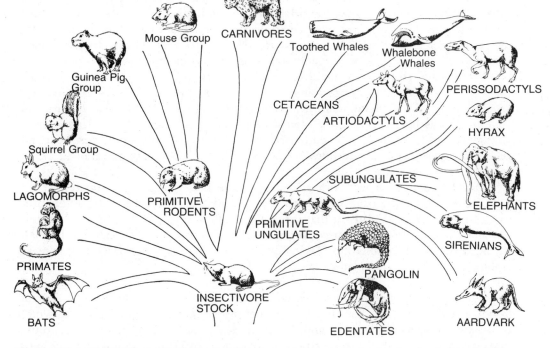

Figure 55. Diagrammatic family tree of the major orders (and some suborders) of eutherian (placental) mammals. Separate diagrams (Figs. 57–61) give in more detail the evolution of primates, carnivores, and odd- and even-toed ungulates.

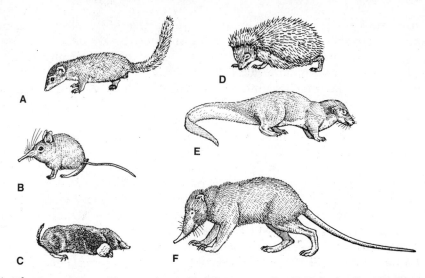

Figure 56. Insectivores. These rather varied forms are all relatively small and more or less shrew-like. *A, Tupaia,* a tree shrew; *B, Elephantulus,* an elephant shrew; *C, Talpa,* a mole; *D, Erinaceus,* a hedgehog; *E, Potamogale,* an otter shrew; *F, Solenodon,* a large Cuban form. (From Thenius.)

in habits at least, their remote ancestors; in many regions, shrews are exceedingly abundant in woods and meadows but are so shy that they are almost never seen. Other familiar insectivores are the prickly European hedgehog and the moles, which, with powerful digging limbs, have taken up the pursuit of grubs and worms underground. Other related forms include the tree shrews (*Tupaia*) of the Oriental region, long thought to be close to the ancestry of the primates; however, this now seems less likely, although they may be close to the ancestry of higher mammals as a whole. Odd groups of insectivores are found in the tropics; one that is often considered a separate order is the "flying lemur," a gliding form not closely related to lemurs. The **Insectivora** as here described is almost certainly an unnatural assemblage—almost any small, relatively primitive eutherian may be placed there—but, despite much recent progress, the separate lines are not yet clear so they may be retained, provisionally, as a single order.

Developed from insectivore ancestors is the one group of mammals that has attained true flight, the bats, forming the order **Chiroptera**. The bat wing differs from that of pterosaurs and birds in that it is a web stiffened by four of the five "fingers." The majority of bats (**Microchiroptera**) have tended to remain insectivorous in habits; however, one major group (**Megachiroptera**), abundant in the tropics, consists mainly of relatively large fruit-eaters.

Primates. This order, to which we ourselves belong, was an early offshoot of the insectivores; indeed, so close are the ties that it is an unsettled question as to which of the two orders certain fossil and living forms should be assigned. Many primitive mammals were to some degree arboreal; this mode of life was emphasized in early primates and appears to have been responsible for the development of many features of importance: general body agility and coordination; the ability to climb by grasping a limb, which has resulted in giving us that most useful of "tools," the hand; and the high development of vision, so necessary for arboreal life. Most important of all, the high degree of development of the brain, which is the outstanding character of higher primates, may have been closely correlated with the needs and opportunities found in arboreal life.

Lowest of acknowledged primates alive today are the lemurs (suborder **Lemuroidea**; Fig. 57), still flourishing in the isolation of Madagascar and little changed from early Cenozoic ancestors. These are four-footed arboreal forms, with thick fur, relatively poor eyesight, and a nose forming a typical mammalian muzzle. Certain extinct forms, put in a separate suborder **Plesiadapoidea**, were equally primitive but had certain rodent-like characters such as gnawing incisors.

A step in advance is represented today by the curious little creature named *Tarsius* (suborder **Tarsioidea**) from the East Indies. The living form is somewhat specialized in such characters as an elongate ankle region, a hopping adaptation to which it owes its name. But although still lemur-like in many ways, *Tarsius* shows advances in such features as excellent eyesight and reduction of the nose to a mere button.

The higher general level of primate evolution is that represented by the monkeys, apes, and man. Although man may pride himself on mental attainments, anatomic differences between the various members of the group are few and are mostly

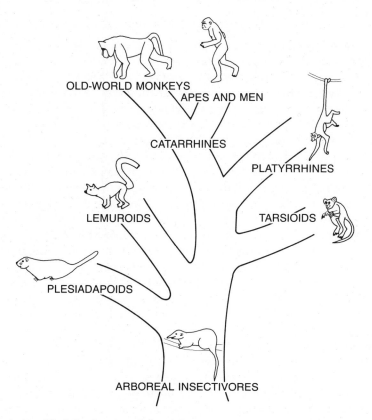

Figure 57. A simplified family tree of the primates.

matters of proportions of various structures or obvious adaptations, such as the distinctive human features associated with upright posture. In all these higher primates, the cerebral hemispheres are relatively large, eyesight is highly developed, the nose is reduced, and the hands are useful grasping organs. Two distinct groups of monkeys appear to have arisen independently from a *Tarsius*-like stock. One, termed the platyrrhines (suborder **Platyrrhini**), is found in South America, where it is represented by a variety of forms such as the common organ-grinder's monkey—the Capuchin— and the little marmosets. Incidentally, it is only among South American monkeys that a prehensile tail is to be found. A second group of higher primates, the suborder **Catarrhini**, developed in the Old World. Primitive members are the familiar monkeys and baboons of Africa and Asia. More advanced members are the great apes, including the gibbon, orangutan, chimpanzee, and gorilla. All these apes are of relatively large size; in all the tail has been lost; and in the last two forms mentioned, the anatomic similarities to man are close indeed, although neither is a human ancestor. The gibbon is an agile arboreal acrobat, and the orangutan is a good traveller in the trees; arboreal specialization is less marked in the chimpanzee, and some gorillas have become almost completely terrestrial, although as quadrupeds. Man is essentially a fifth member of this great ape series; he has become a terrestrial biped but has, stamped deeply into his structure, features probably acquired during a long sojourn in the trees. We do not fully know our own pedigree, but the australopithecines and other fossils from South and East Africa structurally bridge the gap between man and his simian relatives.

Carnivores.   The insectivores were potential flesh-eaters. With the development of numerous mammals of the more harmless varieties, there soon arose from the primitive placental stock varied predaceous types (Fig. 58). All the living carnivores are included in the order **Carnivora**. However, in the early Tertiary the common flesh eaters were members of a different stock, which are often considered to form a distinct order **Creodonta**. These forms appear, however, to have been relatively slow, clumsy, inefficient, and, it would seem, stupid and soon dwindled and eventually disappeared, giving place to members of the proper stock of Carnivora. The modern terrestrial carnivores, the suborder **Fissipedia**, may be essentially divided into two great infraorders, of which dog and cat are familiar examples, and the weasel tribe and civet group are, respectively, more primitive members. A third group, the infraorder **Miacoidea**, is recognized for various extinct ancestral forms.

The weasels and their close kin, small and short-legged in build and often purely carnivorous in habits, are seemingly primitive members of the general "dog" group (infraorder **Arctoidea**). But within the weasel family, there developed a considerable series of forms variant in habits and diet—badgers, skunks, otters, even a marine member, the sea otter of the Pacific. The dog family developed into terrestrial types adapted to running down their prey and is represented today by a series of wolves, jackals, foxes, and other doglike forms. The raccoon is related to the dog stock but is a persistently arboreal animal with an omnivorous taste in food; there are several American relatives, and the lesser panda of Asia is allied. The bears are the members of the general dog group that have departed most widely from ancestral conditions; these clumsy fellows have (except for the polar bear) swung far away from flesh-eating to take up a mixed, but mainly herbivorous, diet. The giant panda, basically an odd bear, has become completely herbivorous.

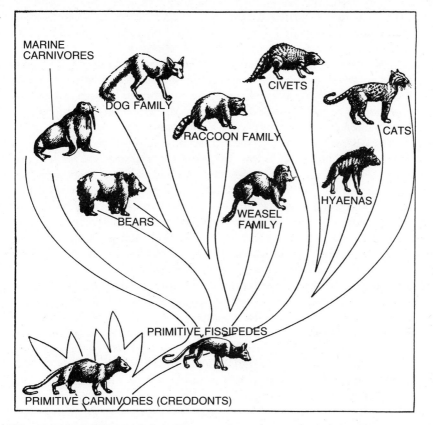

Figure 58. A simplified family tree of the carnivores. (From Romer, The Vertebrate Story, University of Chicago Press.)

Of the "cat" group of carnivores (infraorder **Aeluroidea**), the civets and their relatives appear to occupy a primitive position comparable to that of the weasel family in the "dog" group. These varied forms are mainly Old World tropical forest dwellers; the mongoose is the most familiar representative. The unattractive hyenas are an overgrown offshoot of the civets, which have taken up a life of scavenging, although some are highly efficient predators, and one, the aardwolf, has become an eater of termites. The felids form the most highly specialized development in this subgroup of carnivores. The cats are adapted for stalking their prey and making an agile jump on to the victim, rather than running it down as do the dogs; the teeth are highly specialized for stabbing and shearing, and, in nature, the cats are purely flesh-eating. Most living cats, such as the lion, tiger, panther, and so forth, are very similar to one another in structure; a variety of extinct sabre-tooths were notable for an exaggerated development of stabbing teeth (cf. Fig. 243 A).

Last of the carnivores to be mentioned are the marine **Pinnipedia**—the various seals, which are fish-eaters, and the grotesque walrus, with its digging tusks and blunt molars for crushing the mussels upon which it feeds. These appear to have evolved, possibly as two or three separate lines, from terrestrial carnivores in mid-Tertiary times; in pinnipeds, among other adaptations, the limbs have been transformed into "flippers"; the hind limbs, turned backward, replace the reduced tail as a swimming organ.

Ungulates.   Notable in Tertiary history has been the development of great series of forms, often of relatively large size, that have assumed an herbivorous mode of existence and developed dentitions with grinding molars for chewing vegetable foods. The more advanced forms of this sort have tended to become good running types, with their limbs elongated by lengthening the bones of the "palm" and "sole" region of the feet. They have tended to walk on the tips of the toes, which are generally reduced in number. The claws borne by primitive mammals have generally given place to hooves—hence the term "ungulates" applied to these herbivores.

Although the varied ungulate types have developed some common characters, it is far from certain that all ungulate stocks spring from a common source; there has surely been considerable parallelism in their development. The statement, all too often seen in print, that "the ungulate" has such-and-such physiologic or structural features is valueless unless it is more specifically stated just what type of ungulate is meant; a cow may be as closely related to a lion as to a horse.

Early in the Age of Mammals, there rapidly sprang into existence a host of ungulates of varied but archaic types; most passed rapidly out of existence and need not concern us here. One order of these archaic forms, the **Condylarthra**, appears to have been very close to the ancestry of most other ungulates (and even of the carnivores—the two are very difficult to differentiate at the start of their history). Some of these extinct orders produced large, clumsy, more or less rhinoceros-like forms. One group of orders developed in South America, which until very recently lacked representatives of the "higher" groups of ungulates. Figure 59 shows a variety of these odd forms.

The dominant ungulates of later Cenozoic and recent times belong to two very distinct orders, characterized by the horse and cow: the orders Perissodactyla and Artiodactyla, respectively, the odd- and even-toed types.

In the **Perissodactyla** (Fig. 60), the key character has been the early reduction of the toes from five to three, and in the case of the later horses, further reduction to a single-toed or monodactyl condition (cf. pp. 223, 228). Primitive forms such as little *Hyracotherium* (*Eohippus* is probably a synonym), which was not only a "dawn horse" but also close to the ancestry of the entire order, had already reached a three-toed condition in the hind foot but in front had lost merely the thumb. Early perissodactyls were browsers, living on relatively soft food materials in forests and glades. In the progressive series of the horses, mid-Tertiary types occupied the spreading grasslands, developed high-crowned teeth to cope with a diet of grasses and grains, and reduced the toes to three on each foot. With further approach to the modern genus *Equus,* there was attained a single-toed pattern; the middle toe was the sole survivor. The tapirs of Old and New World tropics remain browsers that, although larger, have departed little from the mode of life of early perissodactyls. More divergent among extinct odd-toed ungulates were the large and ungainly horned titanotheres and the grotesque chalicotheres, which combined a somewhat horselike body with feet armed with powerful claws (perhaps used for digging tubers). More successful, despite large size, were the rhinoceroses, which during their history developed hornlike defensive weapons; once common and widespread, they are still represented by a few species in the Old World tropics. Although rhinos are often said to

Figure 59. Various archaic ungulates (to very different scales). *A,* The condylarth *Phenacodus; B,* the pantodont *Coryphodon; C,* the dinocerate *Uintatherium; D,* the notoungulate *Toxodon; E,* the pyrothere *Pyrotherium.* (From Kurtén.)

be clumsy and may appear so in a zoo, people who have been chased by them or watched them in the wild disagree.

The perissodactyls were very successful in the earlier part of the Age of Mammals but are now reduced to a relatively few species of three types: horses, tapirs, and rhinoceroses. Quite in contrast has been the story of the order **Artiodactyla** (Fig. 61). Rare in the early Tertiary, these even-toed ungulates became increasingly abundant in later times and are widespread today. In these forms, toe reduction began with the loss of "thumb" and "big toe," giving a four-toed pattern. Among the four, the two side toes tended to reduction or loss, leaving toes three and four to form the so-called "cloven hoof" diagnostic of the group. The pigs of the Old World and the related peccaries of the New World are relatively primitive types, omnivorous in habits, as were certain extinct forms that looked like pigs but had skulls up to a meter long. The hippopotamus is a lumbering amphibious cousin of the pigs, a vegetarian in diet. The more successful artiodactyls became purely herbivorous, developed a series of grinding cheek teeth with characteristic crescentic cusps (cf. p. 345), and perfected a multichambered stomach associated with the cud-chewing habit (rumi-

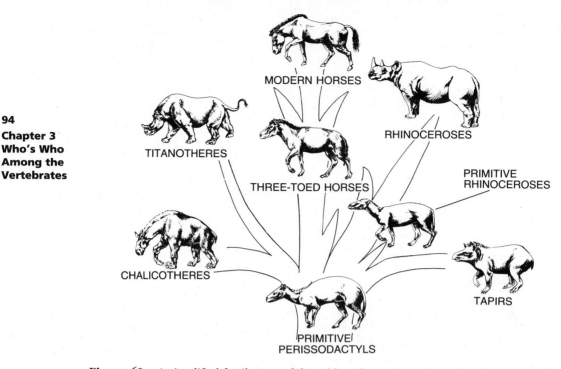

Figure 60. A simplified family tree of the odd-toed ungulates, the perissodactyls.

nation, cf. p. 383) for dealing with vegetable foods. The camels (Tylopoda), originating in North America, but now surviving in the Old World and (as the llamas) in South America, are relatively primitive cud-chewers. The most advanced artiodactyls are the Pecora—agile, swift-running ungulates, with highly developed ruminating stomachs and with heads generally armed with some kind of horns or antlers. The deer and giraffe are relatively primitive browsers. Much more abundant are the "cow-like" forms, the bovids, most of which have become (parallel to the horses) grass-eating plains-dwellers. Cattle, sheep, and goats are familiar domesticated bovids; for a host of others, mainly inhabitants of the Old World tropics, we have no familiar specific names and tend to lump them as "antelopes." Finally, in this ruminant assemblage we may mention the prong-buck of the western American plains, survivor of a New World group paralleling the true antelopes.

Subungulates. Frequently grouped as the subungulates are a series of orders, probably originating in Africa, that are perhaps best regarded as aberrant offshoots of a primitive ungulate stock (cf. Fig. 55). The little hyraces—order **Hyracoidea**, the "conies" of Scripture—are animals of rather rabbit-like size and habits (though some are larger and others live in trees) but are definitely hoofed ungulates whose pedigree goes back to the fossil records of the early African Tertiary. They have, however, never progressed structurally to any degree and have never ranged beyond Africa and the Mediterranean region. Strange as it appears, these little forms appear to be related to two other quite divergent groups (plus, possibly, a couple more extinct orders of uncertain affinities): the elephants and the sea cows.

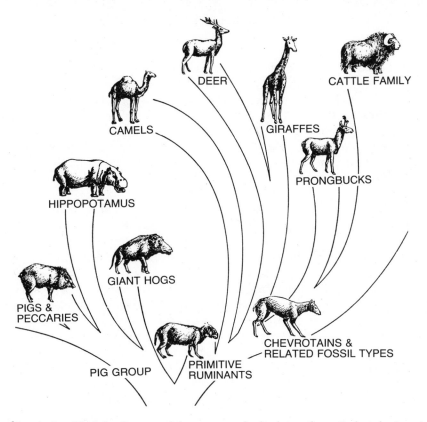

Figure 61. A simplified family tree of the even-toed ungulates, the artiodactyls. A major cleavage is into the pig group, to which the hippopotamus and extinct giant hogs belong, and the cud-chewing ruminants. The little chevrotains of tropical Africa and southern Asia are close to the ancestry of advanced ruminant types. (From Romer, The Vertebrate Story, University of Chicago Press.)

The **Proboscidea** (whose name derives from their trunk) are represented today only by the two elephant types proper to Africa and to southern Asia. Their history, however, has been long and varied. The older types are termed mastodonts. The most primitive known form, from the Eocene of Egypt, was already large for its day—much the size of a large pig—with chisel-like front teeth and a good set of grinders in the cheek. Later mastodonts increased rapidly in size and developed long jaws, with short tusks both above and below. Later the jaws diminished in length, but the upper tusks, in seeming compensation, tended to elongate and eventually there developed the characteristic head pattern of tusks and trunk seen in the modern elephants. Meanwhile, from mid-Tertiary times on, the mastodonts had spread widely over Eurasia and eventually over the Americas; in the Ice Age a variety of true elephants, most of them termed mammoths, roamed every continent except Australia and South America. Toward the end of the Pleistocene, almost all of the proboscideans vanished. The sudden reduction of this seemingly flourishing order of mammals and the nearly simultaneous extinction of a variety of other large mammals may be the result of overhunting by man.

To class the sea cows—the manatees and dugongs of the order **Sirenia**—with the ungulates would appear superficially to be a gross misuse of terms. These animals, browsers in shallow waters of the tropical Atlantic and Indian Ocean regions, are purely aquatic beasts, with front limbs transformed into paddles, hind legs reduced to concealed vestiges, and the tail redeveloped to form a horizontal swimming fluke. Now relatively rare, they ranged widely over the world for much of the Tertiary. The oldest fossil remains, also from the Eocene of Egypt, show a number of resemblances, especially in the teeth and their replacement, to the most primitive conies and mastodonts and add strength to the belief that all three of these curious subungulate groups are varied descendants of some common archaic ungulate ancestor that existed in Africa at the dawn of the Age of Mammals.

Whales. Although both carnivores and subungulates have developed aquatic types, none of these is as highly specialized for marine life as the whales and porpoises that constitute the order **Cetacea**. Much as in the sirenians, the front legs have been transformed into flippers, the hind legs have vanished, and the tail has become a highly developed swimming organ, with horizontal flukes. As in ichthyosaurs among the reptiles, a fishlike dorsal fin may redevelop and the neck is shortened, so that a streamlined fusiform fishlike body is re-attained. The skull is peculiarly altered, for the external nostrils have moved upward to become the "blowhole" atop the head. Most mammals cannot survive long under water, but the physiology of whales has been so modified that some can remain submerged for the better part of an hour. The greater part of the order, including the porpoises and dolphins and a few larger whales (suborder **Odontoceti**), are toothed forms that subsist on animal food—fishes, octopodes, and squids. However, the very largest of the "noble cetaceans," the whalebone whales (suborder **Mysticeti**), live on much smaller food materials—the small marine organisms that constitute the ocean's plankton. Teeth are absent, and instead, the "whalebone" is used to strain the plankton from the water (cf. p. 328). The oldest known whales, from the Eocene, were already aquatic types but were less specialized in body and skull and were suggestive of a descent from some primitive type of terrestrial carnivore; however, the relationships of the sorts of modern whales are debated.

Edentates. Extremely aberrant in many ways are the members of the order **Edentata**, mainly South American in distribution and history. Characteristics of this group include extra articulations between the vertebrae, a trait responsible for the alternate name of Xenarthra. The living representatives are the tree sloths, dull, sluggish arboreal leaf-eaters; the long-snouted anteaters; and the armadillos, omnivorous feeders that have well-developed dorsal armor. Extinct are two other types that attained large size: the glyptodonts, relatives of the armadillos, with a dome-shaped protective "shell" and an armored tail, and the great ground sloths, elephant-sized and massive herbivores. The anteaters are actually edentulous, but the ordinal name is a misnomer for the other forms, which have at least a good series of molars (although the enamel covering of the teeth is reduced). The group attained its development in South America, but during the Ice Age, ground sloths and glyptodonts invaded North America with momentary success. However, only the armadillo has survived in the north; like the mammoths, the larger edentates may have been exterminated by man.

In the past, two odd types of mammals were often grouped with the South American edentates. The slender-snouted pangolins (*Manis;* order **Pholidota**) of the Old World tropics are unique among mammals in that they are completely covered by overlapping horny scales, which give them somewhat the appearance of animated pine cones. Recent work indicates that these may have some, though distant, relationship to the edentates. The aardvark (*Orycteropus;* order **Tubulidentata**) is a grotesque long-snouted African beast; it is thought to have some remote relationship with archaic ungulates. Both make a living by invading termite nests and are armed with powerful claws. The aardvark has retained a few peglike cheek-teeth; the pangolin, like its South American analogues, has lost its teeth, useless in this mode of existence.

Rodents.   Most impressive of all mammals from the point of view of both numbers of genera and species and numbers of individuals are the gnawing animals, the order **Rodentia**. The key character of the group lies in the development of an enlarged pair of front teeth in both upper and lower jaws into an effective chisel-like gnawing apparatus. Rodents have never developed into flying forms, nor into marine types; but in almost every known terrestrial habitat, rodents are the most flourishing group. Rodents have spread out into a great variety of forms that are difficult to classify into major subgroups—indeed, their classification is probably as hotly debated as that of any group of vertebrates. Three "clusters" of living forms, however, stand out rather prominently: (1) A relatively small but familiar group (suborder **Sciuromorpha**) is that of the squirrels, prairie dogs, marmots, and relatives. (2) The guinea pig is representative of a great group (suborder **Caviomorpha**), which includes most of the rodents of South America. The North American porcupine appears to be of South American origin; the Eurasian porcupine and a few other Old World rodents are often assigned to this group. Although this relationship is now frequently thought to be due to parallelism, opinion once again seems to be swinging toward the view that they are a "natural" group (to which a variety of names may be given). (3) Most flourishing of rodents are the rats and mice (suborder **Myomorpha**), ubiquitous in distribution; an indication of their versatility is the fact that they include the only terrestrial placentals that were able to reach Australia before the coming of man. Still other rodents, such as beavers, cannot be placed in any of these groups.

Lagomorphs.   The order **Lagomorpha** is a small one, including almost exclusively the hares and rabbits. They were at one time classed with the rodents because, as in that group, there are chisel-like front teeth. But there are few other resemblances; furthermore, the lagomorphs have two pairs of upper chisels rather than one. For decades, students of living and fossil mammals have recognized the lack of relationship, but despite this, instances constantly recur in which biologists ignorant of the animal world have described structures or functions found in the rabbits or hares as characteristic of "the rodent."

In our brief survey of the world of vertebrates, we have seen the rise of many groups to prominence, followed by decline to extinction or to a point where there remain today only a few relict types, often aberrant or "degenerate." Major success lies currently with the teleosts, the birds, and the placental mammals.

Chapter 4

Cells and Tissues

Although the anatomic and physiologic studies of the vertebrate body treated in this book generally deal with gross structures—organs and organ systems—it must never be forgotten that these structures are composed of tissues, and these in turn of cells, and that the cells are the basic living units from which and by which the entire complex body is built. A knowledge of the structure and function of cells and tissues is basic in the study of biology; our purpose here is to review briefly these familiar topics as a background for a better understanding of those structures of greater complexity with which this work is primarily concerned. The functioning of organs depends basically upon the activities of the cells of which they are composed; conversely, almost all the varied organ systems are engaged, directly or indirectly, in furnishing the cells with the materials required for their vital processes and in maintaining them in a suitable environment. Each cell lives its own life; however, each is dependent upon other cells for its continued existence, and each in turn makes its contribution to the welfare of the total organism.

Chemical Materials. Much of vertebrate structure and function is related to the collection, transformation, and transportation to the cells of the basic chemicals needed to form and maintain the protoplasm and enable it to play its proper part in the work of the body and, again, to the disposal of wastes formed by its activity. In consequence, it is important to know the basic chemistry of the various molecules of which the body is composed. However, this is an immense topic in itself and is beyond the scope of this volume.

Cellular Structure and Function. The cells of the body are, of course, highly varied in their structures and functions; a liver cell and a nerve cell, for example, appear very different and have very different activities. But despite such differences, all cells, at least of higher animals such as vertebrates, have in common basic structural features and basic functional activities.

The "gross" features of cellular anatomy are readily visible under the ordinary compound microscope, although the electron microscope is needed to see the finer details (Fig. 62). Often centrally situated is the **nucleus**. Within it are amounts of readily stainable—i.e., chromatic—material, to be identified chemically with nucleic acid. Before cell division, the chromatin is compacted into a series of paired rodlike structures, the **chromosomes**; each chromosome splits lengthwise, so that when the cell divides, each daughter cell receives a full complement of chromosomes and,

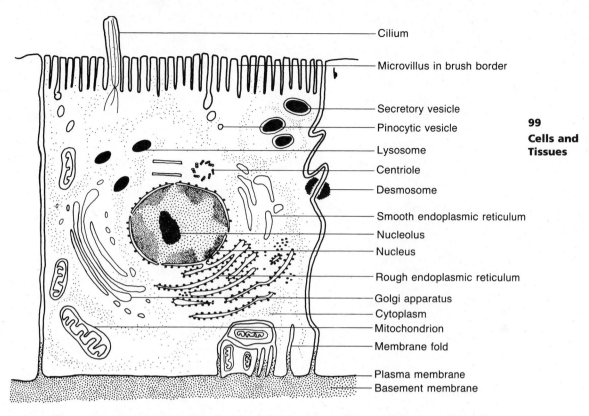

The following labels appear on the diagram:

- Cilium
- Microvillus in brush border
- Secretory vesicle
- Pinocytic vesicle
- Lysosome
- Centriole
- Desmosome
- Smooth endoplasmic reticulum
- Nucleolus
- Nucleus
- Rough endoplasmic reticulum
- Golgi apparatus
- Cytoplasm
- Mitochondrion
- Membrane fold
- Plasma membrane
- Basement membrane

Figure 62. Diagram of a "typical" cell. Actually, no cell is really typical; because this cell, has cilia, a brush border, desmosomes, and a basement membrane, it must be an epithelial cell. Even then, it is highly schematic—only a sensory cell would be expected to have a single cilium, and most of them would lack one or more of the other structures shown here.

hence, of the contained genetic material. Other chromatic material (mainly RNA) may be more or less permanently consolidated into a mass termed a **nucleolus**. Except at times of division, the nucleus is generally seen to be separated by a membrane from the general cell body. This membrane is a much flattened sac, broken or penetrated by "pores," which here are solid connections rather than holes as might be imagined from their name. The cell in turn is bounded externally by a definite **cell membrane,** which, to a considerable degree, regulates the traffic in substances entering or leaving. The membrane, which under the electron microscope appears to be composed of three layers, contains lipids and proteins. To the cell materials in general, the term **protoplasm** is applied; **cytoplasm** is a more restricted term, designating the contents of the cell body external to the nucleus. With the ordinary microscope, certain formed structures can generally be made out in the cytoplasm. Near the nucleus, there are often visible small particles, the **centrioles**, which play a part in cell division. Scattered through the cytoplasm are numerous little structures, the **mitochondria**. With the aid of the electron microscope, the mitochondria, which appear as mere dots or dashes through the light microscope, are seen to have a definite structure as various elongate spheroids with a series of internal cross partitions or cristae. The mitochondria are centers of enzyme concentration and activity in which much of the chemical work that supplies energy for the cell takes place. It is in them that sugars and fatty acids are oxidized to yield adenosine triphosphate (ATP), the universal energy-carrier of the cell. Mitochondria, unlike other cytoplasmic organelles, also contain some DNA and are partially independent of the nucleus. Special stains reveal in many cells a **Golgi apparatus** (named for its discoverer), often called "a reticular apparatus," a series of flattened and smaller spherical vesicles whose function appears to be concentration of proteins, which, combined with large carbohydrates manufactured here, are given off from the cell as secretory granules. They also, we think, give off lysosomes, small packets containing hydrolytic enzymes.

It was long thought that, apart from such structures as those just mentioned, cytoplasm was essentially an amorphous jelly-like colloid, with organic materials in solution in a fluid, watery medium. The electron microscope shows that, on the contrary, the cytoplasm is finely organized, with a network, or **endoplasmic reticulum**, of delicate membranes or tubules with more liquid material between them. The cytoplasm contains numerous tiny particles termed **ribosomes**, invisible to the ordinary light microscope, embedded in these membranes; the ribosomes are major centers of protein synthesis. When ribosomes are present, the reticulum is a **rough endoplasmic reticulum**; parts of its membrane may contribute to the formation of the Golgi apparatus. The endoplasmic reticulum may lack ribosomes and be **smooth**; this latter sort of reticulum is involved in the synthesis of lipids (including steroid hormones) and in the regulation of calcium ions (notably as the sarcoplasmic reticulum of muscle cells).

A cell may also contain various other inclusions or bodies. Secretory granules are an obvious type, but there are various others such as vacuoles with engulfed matter, pigment granules, and lipid droplets.

The surface of the cell may be fully as complex as its interior. Adjacent cells are frequently "attached" by specialized areas. **Desmosomes** merely hold two cells together, but gap junctions are implicated in the transfer of materials between adjacent cells. Free surfaces may bear a brush border composed of large numbers of microvilli (the villi of your intestine are each composed of many cells, which may themselves have microvilli). Larger and more complex are cilia, threadlike motile structures that are discussed in more detail in the section entitled Epithelia (p. 103).

Such complexity of cell structure gives the impression of a static condition. This is, of course, the reverse of the actual situation, for the cytoplasm is the seat of constant chemical activities of an exceedingly complex nature. Every cell must produce energy for its particular function in the body (whether secretion, nervous conduction, muscular movement, and so on), but, in addition, it constantly undergoes chemical change in the maintenance of its internal economy. A cell, like a body, has, so to speak, its own "basal metabolism," necessary for the support of its vital activities.

Cell Environment; Tissue Fluid.   An organism must provide the environment, physical and chemical, required by its cellular units. These requirements are rather rigidly fixed for vertebrate cells.

Although the organism as a whole may be able to exist over a fairly wide range of external temperatures, the temperature to which the cells in the interior of the body can be subjected and still survive is limited, although the optimum temperature and permissible departures from it vary considerably from form to form (cf. p. 155). In general, lower primitive water–dwelling forms have a temperature range pitched lower in the scale than do terrestrial vertebrates. With few exceptions, even fishes are unable to stand internal temperatures much below the freezing point. Thirty degrees Celsius is about the upper limit for fish cells, 45° C or so for the higher, terrestrial vertebrate classes. Beyond such a point, "heat death" occurs as some proteins undergo irreparable denaturation.

A cell can live and avoid desiccation only if bathed in a watery liquid medium. Such a material, the **tissue fluid** (also called interstitial or intercellular fluid), pervades the body. Water itself is not enough. This liquid must contain a considerable amount of material in solution; otherwise, through osmotic pressure, swelling and disruption of the cells that it bathes may take place. Further, cells flourish only if the materials in solution, mainly inorganic salts, adhere rather closely to the formula actually present in tissue fluid normally found in the body of vertebrates. This liquid contains considerable amounts of sodium and chloride ions, lesser quantities of potassium, calcium, and magnesium ions, and small amounts of other elements.

Reasons for the requirement of this salt formula (different from that in the watery constituents of the cells themselves) are poorly understood. It is of interest, however, that with two exceptions (the absence of sulfates and the lesser amounts of magnesium), the "formula" of the tissue fluid of vertebrates is similar to that of sea water, although it is generally much more dilute. Possibly, this is no mere accident; the ancestors of the vertebrates were simple animals bathed in and permeated by the waters of the early Paleozoic ocean; their cellular physiology was evolved with this type of environment as a basic feature, and when, as complex organisms, they developed an independent internal environment, this salty interstitial liquid persisted as, so to speak, a remnant of the ancient seas. The tissue fluid is in relatively free communication with the similar liquid constituting the bulk of the blood plasma, and it is therefore through the blood circulation that the even distribution of the fluid materials through the body is accomplished.

It is, of course, through the tissue fluid bathing the cell that the latter is supplied with most of the materials that it needs for its existence. Through this same liquid, it is relieved of its waste materials. The tissue fluid must be in intimate contact with the circulatory system, so that food materials may be constantly supplied and waste constantly taken away.

The nature of the cell membrane and the necessity of there being a sufficiency of materials in solution in the tissue fluid to balance the concentration within the cell bring to our attention the basic physical-biologic phenomenon of **osmosis**, a phenomenon of importance in a great variety of activities and hence influential in the molding of structure and function in many parts of the body.

If two bodies of liquid—such as, of water with materials in solution—are separated by a membrane, manmade or natural, the membrane may be so impermeable that little or no exchange of materials may take place or so tenuous that it may be readily penetrated by all the materials of the liquids present, which may thus exchange freely. Intermediate conditions, however, may exist in which large molecules or even large ions cannot pass through the membrane, although small ones may pass readily. Such membranes are termed **semipermeable**.

Cellular membranes of variable degrees of permeability, noted in later chapters, are common in the organs of vertebrates. Obviously, osmosis is a phenomenon that enters notably into the determination of the architecture of the body. It is to be kept in mind as a process quite distinct from the active transport of specific materials across cell membranes—another phenomenon often encountered in vertebrate tissues.

Epithelia (Fig. 63).   The body cells are not isolated entities; they are part of **tissues**—organized associations of cells of similar origins and functions. In some cases, notably the connective tissues in a general sense, the aggregation of cells may be relatively diffuse and sometimes ill defined or amorphous. Indeed, the concept that the cells in the blood stream form a tissue is somewhat of a strain on one's mentality. In many instances, however, the cell association is that termed an **epithelium**—a regular and compact arrangement of cells in a sheet, which, usually, borders on one of its aspects the surface of the body or one of its cavities.

In the embryo, most epithelia are simple and diagrammatic in appearance. Later, however, by thickening and modification, they may lose much of the early epithelial appearance, as in a mass of liver tissue. This differentiation of epithelia during development roughly parallels their probable phyletic differentiation. The earliest and simplest multicellular organisms presumably had but a single epithelium, covering the surface and necessarily serving a variety of functions in protection, sensory reception, and exchange of food and waste materials with the outside world. Later, with the development of internal cavities and internal structures, specialization could and did occur. We may note, however, that in much of the gut and in other internal structures, the epithelium, protected from external harm, remains persistently thin and delicate in nature, and its cells may have important properties of absorption or secretion. Such epithelia are generally termed **mucous membranes**, since some or all of the cells generally secrete mucus (cf. p. 105); as a result the epithelium is covered and moistened by a thin film of liquid.

For convenience, various names are given to different shapes of epithelial cells. **Squamous cells** are very low relative to their width, appearing in transverse section as thin lines. **Cuboidal cells** are those in which height and width are about equal, so

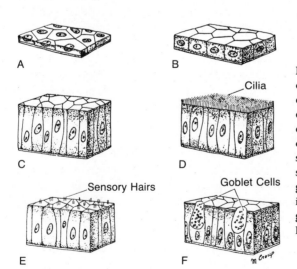

Figure 63. Types of simple epithelial tissue. *A,* Squamous epithelium; *B,* cuboidal epithelium; *C,* columnar epithelium; *D,* ciliated columnar epithelium; *E,* sensory epithelium, showing special sensory processes; *F,* glandular epithelium, including mucus-producing goblet cells. (From Villee, Biology.)

that in section they have a square outline. In **columnar cells**, the height is in excess of the width.

Embryologically, epithelia often consist at first of a single layer of cells, and this arrangement is retained in various adult cases as a **simple epithelium**, frequently of the mucous membrane type; such epithelia tend to be present when there is little wear and tear on a surface or where there is absorption or filtration. In such an epithelium, any one of the three shapes of cells noted may be present. Embryonic cells tend to be rounded, but when closely packed, they usually assume a cuboidal shape in a simple epithelium; this type is not uncommon in the adult. A thin squamous type is characteristic, for example, of the lining of blood vessels and other areas where osmotic exchange through the membrane is of importance. Superficial cells with secretory or other important functions tend to be of a deeper, columnar type.

When an epithelium shows two or more layers of cells, it is termed a **stratified epithelium**. The stratified types are named from the nature of the cells on their surface; these may be squamous, cuboidal, or columnar, but both the latter two are rare. The deeper cells of a stratified squamous epithelium may be cuboidal or columnar, and the underlying cells of a columnar type may be much flattened.

On surface view or in section parallel to the surface, the cell boundaries of an epithelium may sometimes be irregular, but they are typically polygonal in arrangement, frequently with the hexagonal outline, which (as the bee knows) gives the best compromise between the ideal circular individual shape and the necessities of a compact arrangement. The adjacent cells in an epithelium may be tightly packed together, but are separated by small spaces filled by the tissue fluid. These spaces may be bridged by "tight junctions" between adjacent cells; these junctions appear to restrict the movement of substances within the tissue fluid. Because they are usually relatively impermeable, they do not serve to allow substances to be transferred between cells, an action that probably occurs, as noted earlier, at gap junctions—exceedingly fine pores through the neighboring cell membranes. As a rule, blood vessels are absent in epithelia, and nutrient material reaches the cells by passing upward in the interstitial fluid from vessels beneath the base of the epithelium. In most cases, epithelia are bounded below by connective tissue; between the two there is typically a **basement membrane** formed largely or entirely by the epithelial cells (the basal lamina) and sometimes also by the connective tissue (the reticular lamina).

In a stratified epithelium, there is frequently a constant loss or destruction of cells at the surface; these are replaced from below, and the lower cells of the epithelium form a **germinative layer** capable of making replacements by cell division (cf. Fig. 90). The major activities of an epithelium are in general associated with its outer, free surface, so it is natural to find that, in many stratified epithelia, the cells are progressively more specialized as the surface is approached. We also find that in almost any epithelium, the individual cells are more or less polarized, the proximal (deeper) and distal (superficial) ends being different in nature.

The free surface of an epithelium frequently shows marked specializations. The cells may secrete a **cuticle**, a more or less solid layer of material covering the epithelium. More frequently, however, the surface of the cells themselves may be modified. Among the most specialized of surface structures are **cilia**, found on the surface of a variety of embryonic and adult columnar epithelia of vertebrates as well as invertebrates, and remarkably uniform in structure in all organisms. They are slender, mobile, hairlike "organs," containing 11 microtubules, two in the middle and nine around them, connected with a complex basal structure. Cilia beat in but one direction; the beat (independent of nervous control) in a ciliated surface usually progresses in a wavelike manner so as to carry mucus or other materials along in a constant unidirectional stream. However, in some instances, all the cilia may beat in unison.

In certain cases, a simple epithelium may present a seemingly stratified appearance, a **pseudostratified epithelium**. If, for example, a simple but tall epithelium contains cells of two sorts, with nuclei at different levels, or with cell bodies expanded at different levels, one gets the impression that two layers are present. The term is better justified when two types of cells are present at the base of a simple epithelium but when only one type reaches the surface. Incidentally, it will be realized that if a simple epithelium is cut at an angle, several cell layers may appear, deceptively, to be present in microscopic section. The poorly named **transitional epithelium** found in the amniote bladder or urinary tubes is really a special type of pseudostratified epithelium. When the cavity it surrounds is empty or not greatly expanded, this type of epithelium appears to be highly stratified; with distention of the cavity, the epithelium expands in area and consequently thins. Other epithelia, such as those lining blood vessels, must also stretch and contract with the surrounding tissues.

In a broad sense, the term epithelium may be applied to any tissues of the sort described here (that is, sheets of adherent cells), regardless of where they are found in the body. Many authors, however, restrict the term to layers of tissue on the outer surface of the body on the one hand, and the gut and its derivatives on the other, or to cavities obviously originating from these outer or inner surfaces. The theoretic consideration behind this is the fact that (as will be seen in the next chapter) such epithelia are derived during individual development from the surfaces, ectodermal and endodermal, of the early embryo. If such a restriction is made, one or more alternative terms are used for tissues that line cavities formed secondarily in the deeper layers of the body. The name **endothelium** is generally used for the epithelial material lining the vessels of the circulatory system. Sometimes this term is also applied to the lining of the body cavities; however, **mesothelium** is a preferable name.

Glands. Secretory activity—the production and discharge of fluid materials—is characteristic of a variety of cells; cells in which this is a dominant activity are termed **gland cells**.

Glands are almost always epithelial in nature, although nerve and other tissues may also secrete substances. Frequently, their nature is obvious in microscopic preparations, owing to the presence of granules of material being secreted or vacuoles filled by it in the cytoplasm. Glands and gland cells may be classified with regard to the effect that the giving off of secreted materials has on the cell. Commonly, secretory activity may be carried on indefinitely without apparent harm to the cell; this is the condition in the **merocrine** type of gland. But if the secretion is thick or viscous, the cell is usually damaged in discharging it. In the **apocrine** type, the superficial part of the cell breaks down as the secretion is released, but no permanent harm is done; this part of the cell is reconstituted and secretion recurs. But in some instances (as in sebaceous glands of mammalian skin), the entire cell is destroyed with the discharge of its contents—the **holocrine** type of secretion.

In many instances, gland cells may occur individually or as small groups in restricted areas of an epithelium that also serves other functions. Mucus, a thick, slimy lubricating material, moistening and protecting various membranes, may be produced by formed glands but is frequently produced by individual cells scattered about the surface of an epithelium; this is common, for example, in the skin of various fishes. The secretion frequently accumulates in the superficial parts of the cells in the form of large droplets, causing the cells thus distended to be termed **goblet cells** (Fig. 63 *F*).

Although gland cells or a glandular area may be simply part of the surface of an epithelium, we frequently find that, in the course of embryonic development, glandular tissues withdraw from the general epithelial surfaces to form discrete glands; the trend for the formation of formed glands is greater among the higher vertebrate groups. In most cases, these glands discharge by ducts into some cavity of the body or to the body surface, and are hence termed **exocrine glands**. Withdrawal from the surface serves a variety of purposes. The folding of the glandular epithelium allows a great increase in secretory area; the gland cells are sheltered from the vicissitudes that might be encountered on the general epithelial surface; the presence of a narrow duct (in which the cells generally are nonsecretory) may allow regulation of discharge of the secretion; and provision may be made for storage of considerable quantities of the secretion.

Some of the varied forms that exocrine glands may assume are shown in Figure 64. The secretory areas may take the shape of **tubules** or of rounded pockets (**alveoli** or **acini**) and may be simple or subdivided. Intermediate forms (tubuloacinar, and so on) also occur. Further, in larger glands, a compound structure may be brought about by branching of the ducts as well, tending to divide the gland into lobes and lobules. The shape of glands is often quite irregular, for appropriate volume is the desideratum, and form is immaterial. The larger glands are usually surrounded by a sheath of connective tissue; such a sheath also extends into the gland between the lobes. Generally there is an ample supply of capillaries and nerve endings.

In contrast to these "normal" glands of exocrine nature, in a variety of instances, we find glands that do not discharge to any surface but, instead, pour their secretions into the blood stream. These are the **endocrine glands** (cf. Chap. 18). In some instances, these can be seen to arise embryologically from epithelia; in such cases they may have been originally exocrine in nature (in the case of the thyroid gland, for example, this seems certain; cf. p. 609). In other cases, however, there is no evidence of such a pedigree and the phylogenetic origin of many of the endocrines is quite obscure. Indeed, in some cases the "gland" is clearly a neural structure. Because the secretions, the hormones, given off by endocrine glands may act, by way of the blood stream, in any part of the body, their position need have no anatomic relation to the organs that they influence, and (as described later) they are found in a variety of situations. As with exocrine glands, their shape is irrelevant to their function and is thus variable.

Nonepithelial Tissues.   Although a great series of body tissues, particularly those of the lining of the gut and derivatives and of the body surface, are clearly epithelial in nature in the adult, many others that are derived from embryonic epithelia are so greatly modified in the adult that their epithelial origin is obscured. Liver tissue, for example, is derived from the epithelium of an outpocketing of the gut, which devel-

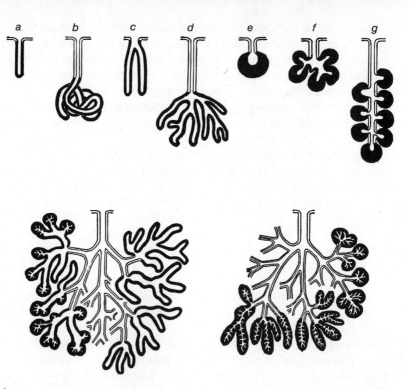

Figure 64. Diagrams of various exocrine gland types. *Above (a–g),* simple glands; *a,* tubular; *b,* coiled tubular; *c, d,* branched tubular; *e,* alveolar; *f, g,* branched alveolar. *Below,* examples of compound glands. In all diagrams, ducts are in double lines, secretory portions in solid black. (From Maximow and Bloom, Histology.)

ops into the adult bile duct. But although the cells of the adult liver retain a connection with branches of the bile duct, the arrangement of the masses of liver tissue has little regard for the original epithelial arrangement (cf. Fig. 286). Much of the nervous system begins its history as an epithelium surrounding a neural tube; although cavities representing the tube (and epithelial cells around them) persist in the adult, there is little trace of typical epithelial arrangement in the nervous tissues. The striated body musculature of amphioxus and the vertebrates can be regarded as of epithelial origin, because in amphioxus, it arises as epithelia surrounding a series of paired pouches. But in the adult vertebrate, no trace of the ancestral epithelial structure persists, and, even in the embryo, this musculature shows at the most only fleeting traces of an epithelial origin.

We can still, in a broad sense, regard these tissues as epithelial in nature, because they are ontogenetically or phylogenetically derived from true epithelia. However, it is customary to recognize at least nervous and muscular tissue (see Chapters 16 and 9, respectively) as separate "basic" sorts, specialized for conduction of impulses and contraction, respectively. Other epithelial derivatives are not as easy to define and so continue to be termed epithelial despite their modifications.

It is not so with certain other tissues of the vertebrate body—the connective and skeletal tissues and those of the circulatory system.* These arise, as will be noted in the next chapter, from a type of embryonic material known as **mesenchyme**—cells that may be derived from the under surfaces of epithelia, but which themselves are never arranged in a compact epithelial type of tissue. They are, instead, diffusely arranged within a ground substance, a **matrix**, that they have created. In the case of the skeletal elements, the matrix becomes the hard material of cartilages or bones; in the case of connective tissue proper, it is gelatinous in nature; in the case of the circulatory system, the matrix is a liquid—the blood plasma. All may be considered connective tissues in a broad sense, this being traditionally the fourth and last of the basic tissues (the others being epithelium, nerve, and muscle). Connective tissue is described in more detail in Chapter 7.

*However, parts of the walls of blood vessels may become epithelial (the endothelium) or muscular (smooth muscles).

Chapter 5

The Early Development of Vertebrates

The adult phase of an organism is merely a last major stage in a long series of stages through which it passes during its lifetime. In later chapters, the development of the various specific organs and tissues will be noted. Here we shall discuss briefly the early developmental history of vertebrates from the egg to the point where the major organ systems have differentiated and the basic ground plan of the body has been established. In doing so, we shall simplify the story in a diagrammatic manner and will omit many features of interest to the embryologist.

Egg Types

The vertebrate egg, exceedingly variable in size, is typically a spherical cell, containing, in addition to a nucleus and some clear cytoplasm, yolk to furnish material for the developing embryo. Although embryonic development, as we customarily think of it, begins with fertilization, the egg has already made various preparations for this event. The nucleus of the egg must be put in proper condition for union with that of the sperm, and part of the chromosomal materials are cast out from the egg in small polar bodies (a process generally incomplete but temporarily arrested at the time of fertilization). Further, the cytoplasmic materials of the egg cell undergo a considerable degree of organization before fertilization, and much of the future pattern of symmetry appears to have already been established by that time, although it may be modified by later events.

The sperm supplies nuclear material that is of importance in heredity and in the later phases of development but that has little effect on the initial phases. The egg has within itself all the potentialities needed for the complete development of an adult. When ripe, the egg is "set" and ready, waiting only a proper stimulus to begin the cleavage into cellular units, which is the necessary first stage toward the development of the tissues and organs of the complex adult. Physical or chemical stimulation can in many cases set this process in motion. Under normal conditions, however, the trigger is the entrance of the sperm.

The quantity of yolk present in the egg varies greatly; its amount essentially determines the size of the egg and is crucial in regard to the pattern of cleavage. In some eggs, particularly those of amphioxus and mammals, there is little yolk. Such an egg may be termed **microlecithal** (alecithal is also sometimes used but should mean "without any yolk," which is not true of these forms). A second egg type is that which we will call **mesolecithal**. The egg is somewhat larger, containing a moderate amount of yolk. The frog has a characteristic mesolecithal egg, and the same type is found in urodeles, lungfishes, lower actinopterygians, and lampreys; it is so widespread in primitive aquatic forms that it is reasonable to conclude that the mesolecithal egg was characteristic of the ancestral vertebrates. In the sharks and skates, on the one hand, and reptiles and birds, on the other, we find eggs of large size, the **macrolecithal** type, with yolk constituting most of the volume of the cell, and with the relatively small amount of cytoplasm concentrated at one pole. So overwhelming is the yolk mass that in the kitchen, the cell body of the hen's egg is simply termed the "yolk," to the neglect of the tiny amount of clear cytoplasm that it contains.

Eggs can also be divided into types based on the distribution of the yolk within them. In some eggs, mainly microlecithal, the yolk is fairly evenly distributed throughout the cell; this condition is termed **oligolecithal** (or, linguistically more logically, **isolecithal**). In most mesolecithal and macrolecithal eggs, the yolk is more concentrated in one hemisphere, the lower hemisphere in an egg floating in water. This arrangement is termed **telolecithal**. In the common modern bony fishes, the teleosts, the egg is likewise heavily loaded with yolk, but is variable in size. The complexities and peculiarities of its development will not be discussed here.

The concentration of the yolk in one hemisphere brings to light clear evidence of organization in the egg—a polarity, with an **animal pole** in the relatively clear cytoplasmic area above, and a **vegetal pole** in the yolky region below. In many invertebrates, the axis connecting these two poles becomes the anteroposterior axis of the body, the vegetal pole becoming the posterior end. In the vertebrates and lower chordates, this is not the case. Related to the greater complexity of development, the adult axis in amphioxus, for example, lies about 45 degrees off the egg axis, so that the animal pole slants down beneath the prospective chin of the adult, and the vegetal pole slants upward and posteriorly toward the back of the animal. To make a bilaterally symmetric rather than a radial animal, we need, in addition to an anteroposterior axis, a medial plane separating right and left halves of the future body. In some cases, at least, this plane is known to have been established in the unfertilized egg; in others the point of entry of the sperm may influence if not determine it.

To illustrate the varied patterns seen in the early development of vertebrates, we shall select eggs of three contrasting types. For an almost diagrammatically microlecithal and oligolecithal egg, we shall, in fact, leave the true vertebrates and resort to amphioxus, a lower chordate relative. The frog or urodele egg is a characteristic mesolecithal type. That of the shark or skate is illustrative of an extremely heavy-yolked macrolecithal type, and the bird's egg is similar in nature in early stages. We will, further, note the peculiar early development seen in the mammalian egg. This is tiny and almost devoid of yolk; but mammals have descended from reptiles with large-yolked eggs, and their developmental pattern is of a type with many "reminiscences" of that of macrolecithal forms. We shall follow each of these types through three successive major processes: (1) cleavage and the formation of a blastula, (2) gastrulation, with laying down of the main body layers, and (3) formation of neural tube and mesodermal structures.

Cleavage and Formation of the Blastula

Amphioxus. The process of **cleavage**, leading to the stage known as the **blastula**, is almost diagrammatically seen in amphioxus.

In this form (Fig. 65), the first cleavage of the egg extends along a "meridian" from pole to pole (much as one cuts an apple into two halves) and results in the formation of two cells of equal size. In the amphioxus egg, bilateral symmetry is already established before fertilization, and the two cells resulting from the first cleavage are destined, in normal development, to become the right and left halves of the body. The two adhere to one another, but, nevertheless, each tends to round up and approach a spherical shape, a tendency repeated in later divisions. A second cleavage is, like the first, from pole to pole, but at right angles to the first, and the process is similar to that of cutting an apple into quarters.

The third division is at right angles to both those preceding and is essentially a cut around the equator of the egg, dividing each of the four cells then present into upper and lower components and resulting in an eight-celled stage. By this time, however, it is manifestly difficult for each of these cells to assume a spherical shape and yet remain in contact with all former neighbors. In consequence, a central cavity tends to develop inside the sphere of cells, a cavity that becomes increasingly large as division proceeds and is known as the **segmentation cavity** or **blastocoele**.

Although there is but little yolk in the amphioxus egg, its distribution is not absolutely uniform; as said previously, there is a somewhat greater concentration toward the vegetal pole. A cell tends to divide, not through the center of its total mass, but through the center of its living protoplasm, without regard for such relatively inert materials as yolk. In consequence, the equatorial division that results in the

110
Chapter 5
The Early
Develop-
ment of
Vertebrates

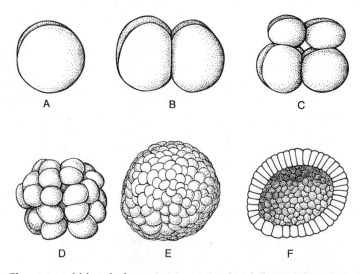

Figure 65. Cleavage and blastula formation in a microlecithal egg—that of amphioxus (cf. Figs. 66 to 68). *A,* First cleavage; animal pole of uncleaved egg is at top of figure. *B,* Second cleavage, to four-celled stage. *C,* Third cleavage; cells of animal hemisphere are somewhat smaller. *D,* After about two further cleavages. *E,* Blastula. *F,* Hemisected blastula, to show segmentation cavity in interior, and single-layered surface. (After Cerfontaine, Conklin.)

eight-celled stage is not exactly through the equator, but slightly above this plane. Hence, we find that the upper quartet of cells is slightly smaller than that below, and the concentration of yolk is slightly greater in the lower quartet.

The next set of cleavages is, as would be expected, longitudinal; each cell of both quartets is subdivided into two, to give a 16-celled stage. Then follows a horizontal set of cleavages in both animal and vegetal hemispheres to give a 32-celled stage with four (rather irregular) rings of cells around the circumference, and with distinct (if small) differences in size among these rings as we travel from pole to pole. Beyond the 32-celled stage, since the process is a geometric progression, the number of cells increases rapidly, to about 64, 128, and so on; now, however, there is much less regularity in the divisions and a less synchronous cleavage. Within a short time, the original egg cell has given rise to several hundred cells, essentially arranged as a single-layered shell about a central cavity. This sphere is termed a **blastula.** Within it is a large cavity, the blastocoele, the initiation of which was noted above. For the most part, the cells of the blastula are not too dissimilar from one another, but observation reveals the presence of smaller cells toward the original animal pole, larger and somewhat more yolky cells toward the vegetal pole, and still a third type occupying part of the vegetal hemisphere. In consequence, there is a differentiation of types of cells which, as will be seen later, have varied destinies in the embryo.

Mesolecithal Eggs. The amphibian egg (Fig. 66) as seen in a typical frog or salamander is of the mesolecithal type, with a certain amount of granular yolk present throughout, but with a heavy concentration of it in the vegetal hemisphere. In such an egg, the first two divisions are comparable to those seen in amphioxus—longitudinal and cutting from pole to pole at right angles to one another. In many amphibians, the distribution of pigments in the egg indicates that the first cleavage here, as in amphioxus, is a median sagittal one, separating right and left halves of the future body. There is, however, one difference between the first two cleavages here and those of amphioxus. The great amount of yolk in the vegetal hemisphere has the effect of slowing up cleavage of this part of the egg. The first cleavage begins promptly enough in the animal hemisphere. But as the furrow progresses downward, it encounters difficulty in cleaving the inert yolk and is slowed up. In consequence, the second division may begin in the animal hemisphere before the first has been completed. This tendency for retardation of division in the lower portion of the egg persists through the period of cleavage.

The third division of an egg of this type is equatorial, as in amphioxus. This cleavage tends, as there, to divide the protoplasm in each of the four cells into equal halves; but because there is considerably more yolk and less protoplasm in the vegetal hemisphere, the line of cleavage is well above the equator (almost as far north, one might say, as the Tropic of Cancer). In consequence, we find in the frog or urodele blastula a marked difference between the small cells of the animal hemisphere and the large, yolky cells of the vegetal hemisphere. As the end result of the process of cleavage in an egg of this sort, we arrive at a blastula that in certain regards resembles that of amphioxus but in which the blastocoele is not as extensive as in amphioxus. The yolk-filled protoplasm toward the vegetal pole fails to divide promptly in the frog or urodele, and a mass of large yolky cells partially fills the internal region of the sphere; even in the animal hemisphere, the surface layer of the blastula is a number of cells thick.

112
Chapter 5
The Early
Develop-
ment of
Vertebrates

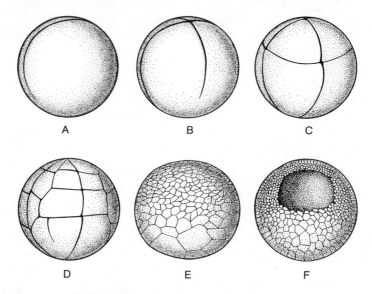

Figure 66. Cleavage and blastula formation in a mesolecithal type of egg found in amphibians. These six figures are comparable to the six in Figure 65 (cf. also Figs. 67 and 68). *A,* First cleavage. *B,* Second cleavage. *C,* Third—meridional—cleavage, with smaller cells in animal hemisphere. *D,* About 36-celled stage; cleavages irregular, but slower and with larger cells in vegetal hemisphere. *E,* Blastula, with strong contrast between cells at two original poles. *F,* Section of blastula, showing segmentation cavity of restricted size; blastula is a number of cells thick. Yolky mass at vegetal pole has cleaved, but slowly and into a mass of large cells.

Macrolecithal Eggs. Although the amphibian egg contains a considerable amount of yolk, it is, nevertheless, as is that of amphioxus, **holoblastic;** that is, there is a complete cleavage of the entire egg in blastula formation. The **meroblastic** type of cleavage (Fig. 67) is found in such forms as the elasmobranchs, reptiles, and birds, in which most of the egg consists of a great, inert mass of yolk. The protoplasm is confined to a small area capping the yolk at the animal pole; cleavage and blastula formation are confined to this clear protoplasmic area. The yolk mass cleaves hardly at all; nuclei invade its margins adjacent to the area of cleavage but fail to make much impression on it.

As in eggs of other types, the first and second cleavages pass through the center of the animal pole. They are not, however, complete cleavages of the egg; they affect only the materials of the protoplasmic cap and do not extend into the yolk. There follows an "equatorial" division, actually well above the Arctic Circle, separating a cluster of cells at the pole from more peripheral members. Further subdivisions, generally rather irregular, occur, until the entire protoplasmic area at the region of the animal pole is divided into a considerable number of cells, to form a plate-like structure, which we must regard as a blastula, and may be termed a **blastoderm.** In amphioxus, the blastula was only one cell thick. In a mesolecithal egg, however, the surface of the blastula is generally a thicker structure several cells deep; and in macrolecithal eggs, the entire blastula region of the egg may be a thick mass of cells that tends to be slightly lifted off the underlying yolk. The "blastula" is here essentially a

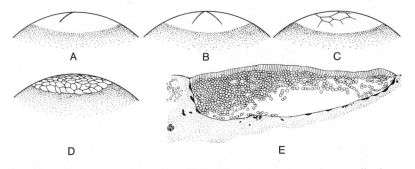

A   B   C

D   E

113
**Cleavage
and
Formation
of the
Blastula**

Figure 67. Diagrams to show cleavage and blastula formation in a large-yolked egg, as that of a shark, reptile or bird (cf. Figs. 65 and 66). In *A* to *D,* only the animal pole of the egg, containing an area of clear protoplasm on top of the large, inert yolk mass is shown. *A* to *D* show cleavage stages comparable to those in *A* to *D* of the two preceding figures; the result of cleavage is not a sphere, but a flattened plate of cells. In *E,* at higher magnification, a section through the formed blastula of a shark is shown. The blastula is a flat plate, a number of cells in thickness, with an irregular segmentation cavity lying below it, but above the unsegmented yolk mass. (*E* after von Kupffer.)

flat sheet, not a sphere. Its margins, bounded all about by the yolk, consist of cells that in less yolky eggs would lie in the vegetal hemisphere but are here unable to attain such a position. In cartographic terms, the blastula is a sphere flattened down into a two dimensional "map" on a north-polar projection.

  Despite their small size, the eggs of teleosts are macrolecithal in structure, with most of the clear cytoplasm forming a cap over a solid yolk mass. The meroblastic cleavage and general pattern of development parallel those of larger macrolecithal types.

Mammals. The early stages in mammalian development (Fig. 68) are quite specialized and unlike those of any other vertebrates. Except for the monotremes (whose eggs are essentially like those of reptiles), mammals carry the developing young within their bodies, and nutritive materials for growth of the embryo are derived, to begin with, from uterine secretions and, later, from the blood stream of the mother through the **placenta.** As noted later (p. 133), this is formed by a union of maternal uterine tissues and of fetal tissues, which are modifications of embryonic membranes found in other amniotes. The necessity for the rapid development of the placenta is responsible for the unique early embryonic histories.

  Because food materials are soon to be obtainable from uterine fluids and later from the maternal blood stream, the mammalian egg can be (and is) a tiny object practically devoid of yolk. In consequence, total cleavage, on the mode of amphioxus, is possible. This cleavage results presently in the production of a score or two of cells. With further division, there appear marked variations among different mammalian types. We shall follow the pattern seen in primates, in which there differentiate (1) a compact cluster of cells termed the **inner cell mass** or blastoderm, and (2) an expanded sphere of cells external to it, the **trophoblast.** The trophoblast is to become a part of the placenta, and its early development is a functionally vital embryonic activity. The egg is fertilized far up the oviduct but reaches the uterus, in the walls of which it is to become embedded, within a few days, and the trophoblast

114
Chapter 5
The Early
Develop-
ment of
Vertebrates

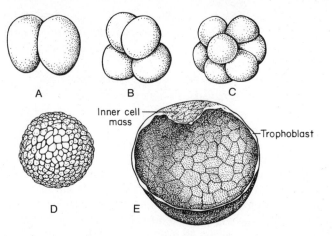

Figure 68. Cleavage and blastula formation in a mammal. The small, almost yolkless egg cleaves (*A* to *D*) in a fashion similar to that of amphioxus. The formed blastula (seen in section at *E*) has a deceptive resemblance to that of amphioxus. Actually, however, the thin external sphere is the trophoblast, which forms a connection with the uterine wall, and the true blastula (or blastoderm) is merely the inner cell mass. This mass is a sheet of cells placed above the internal cavity, much as the blastula of a macrolecithal egg (Fig. 67 *E*) is situated atop the yolk mass. (After Streeter.)

develops rapidly so that contact may be made with the maternal uterine tissues when conditions are appropriate. This is an excellent example of an embryonic adaptation, the development of a structure never present in either adult or embryo of "lower" vertebrates.

Gastrulation and Formation of the Germ Layers

We have seen as the result of cleavage and blastula formation the development of the egg into an early embryonic stage, which, in most types, consists of a single body layer in the form of a sphere or sheet of cells. In some instances (amphioxus, for example), differences in size, pigmentation, or amount of yolk present in different parts of the blastula indicate the differentiation of specific areas destined to form one or another major tissue of the later embryo and adult. In other instances, a real differentiation is not readily visible, but the future fate of any given area can often be discovered by applying a stain to cells in the blastula stage and following the stained area into later stages of embryonic development. As a result of such observations and experiments, it has been found in a considerable number of different chordate types that the future fate of various groups of cells is already established in the blastula stage; "fate maps" of the blastula regions can be drawn, and several of them are given here (Fig. 70 *A–H*).

There now begins a series of movements of specific areas of cells toward assumption of the position that they will eventually occupy in the later embryo and adult. A major step is the process of **gastrulation**—the transformation of the surface pattern of the sphere or disc of the blastula stage into a folded, double-layered early embryo, of which part of the outer layer corresponds to the skin surface of the adult, and part of the inner layer will form the adult gut lining.

Amphioxus. In some primitive metazoans, various coelenterates, for example, gastrulation is a simple process, merely the folding of the spherical shell of cells of the blastula stage into a double-layered hemisphere, with skin outside, lining of the gut within. All the cells of the original animal hemisphere, which form the outer layer of the gastrula, constitute the **ectoderm,** or outer germ layer, of the late embryo and adult. The inner cells are the **endoderm,** or inner germ layer, forming the gut, reasonably termed the **archenteron** in the embryo. The single opening into the cavity, less reasonably, is the **blastopore.**

The process of gastrulation in amphioxus (Fig. 69) appears to be equally simple but is actually more complex. A special area of the superficial ectodermal layer is needed for the formation of the complicated nervous system, a **neurectoderm.** There must be formed the materials of the third major germ layer, the **mesoderm,** which constitutes the greater part of the bulk of the chordate body. A distinctive mid-dorsal area of mesoderm, the **chordamesoderm,** forms the notochord. It is of particular importance in bringing about by induction the development of the nervous system from the overlying neurectodermal tissue.

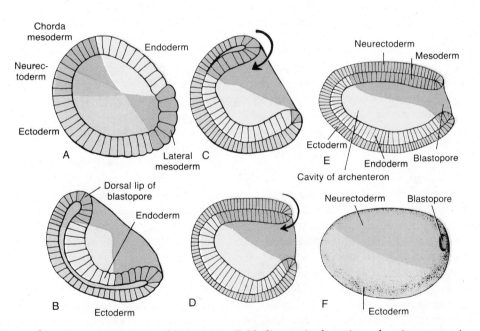

Figure 69. Gastrulation in amphioxus. *A* to *E,* Median sagittal sections showing successive stages. The embryo has been rotated from the original egg position to that of the future adult, with head end at left, the former vegetal pole region posterodorsal. *A,* The endoderm cells are a flattened plate. In *B,* the endoderm has invaginated and the lateral mesoderm cells, originally widely separated from the notochordal materials, are moving upward to join them. In *C,* this movement has been mainly accomplished, and the inturning dorsally of chorda materials (*arrow*) continues. *D* and *E,* Gastrula formation has been completed, and the embryo (particularly notochord and overlying neurectoderm) is elongating. *F,* Surface view of late gastrula, seen from the left. (After Hatschek, Cerfontaine, Conklin.)

In this and subsequent figures in this chapter, the following colors are used to distinguish germ layers: skin ectoderm, blue; neurectoderm, green; mesoderm, red; endoderm, yellow.

All these areas are already laid out in the blastula of amphioxus (Fig. 70 *A, B*) and are involved in the process of gastrulation. As gastrulation begins, the large yolky cells at the vegetal pole, which are to become the endoderm, form a flat plate and then bend inward at the ventral rim of the forming blastopore. A crescent of meso-dermal cells destined to form the notochord rolls inward over the dorsal lip of the blastopore and pushes forward internally, lengthening in the process and tending to lengthen somewhat the gastrula as a whole. On either margin of the blastopore, be-tween the ectoderm above and endoderm below, cell masses of the mesoderm proper stream in, push forward and upward, in the inner layer, and align themselves in a strip on either side of the forming notochord. With the infolding of this meso-dermal tissue, the future neural ectoderm comes to occupy a large area on the dorsal surface anterior to the blastopore.

116

Chapter 5
The Early
Develop-
ment of
Vertebrates

When gastrulation has been completed, we find the embryo forming a somewhat elongated spheroid, with the only entrance to its interior the posteriorly placed blas-topore, now reduced in size. On the outer surface, the future skin ectoderm forms the ventral and anterior epithelial covering; the neurectoderm, the more posterodor-sal area. Internally, the endoderm occupies an area essentially comparable to that of the skin ectoderm on the outer surface, with the chordamesoderm occupying a mid-dorsal position along the roof of the archenteron between bands of typical mesoder-mal materials.

Although not properly a part of gastrulation, a further stage in the development of the amphioxus mesoderm will be mentioned at this point (Figs. 78, 79). The cells round up to form a lengthwise cylindrical structure as the definitive notochord. At either side, the lateral mesoderm folds outward into a pair of longitudinal grooves. At the anterior end of each groove, a pouch, more or less cubical in shape, pinches off as a **mesodermal somite.** This contains a cavity, originally continuous with that of the archenteron, which is destined to become part of the coelom; its walls later differentiate into mesodermal tissues. Posteriorly, successive pairs of somites are bud-ded off along the length of the trunk. As these pouches are formed, the endoderm, originally lateral and ventral in position, grows upward ventromedial to them and across beneath the notochord to form a continuous lining to the definitive gut. This pouch type of mesoderm formation is highly comparable to the process seen in acorn worms and echinoderms and is a prime basis for the belief that chordates are related in origin to the echinoderm phylum.

Mesolecithal Types (Figs. 70 *C, D,* 71, 80 *A, B*).   In such types as the urodeles and frogs, with mesolecithal eggs, the potential germ layers are laid out over the surface of the blastula in a manner comparable to that of amphioxus (except for the relative position of chorda materials and lateral mesoderm), and gastrulation takes place in a manner basically similar. However, the presence of a large mass of yolk cells filling much of the interior of the blastula causes modification of the process. It is a physical impossibility for all this yolky material of the vegetal hemisphere—the potential en-doderm—to fold itself neatly within the mantle of cells formed from the animal hemi-sphere of the egg, as is the case in amphioxus. A furrow forms on the surface at a point corresponding to the dorsal lip of the amphioxus blastopore. Cells from the exterior—at first, endodermal cells, mainly from the region below the furrow—stream into this pit. Then, rolling inward from above and turning forward, cells flow that are destined to lie anterior to the notochord, followed by chorda tissue that

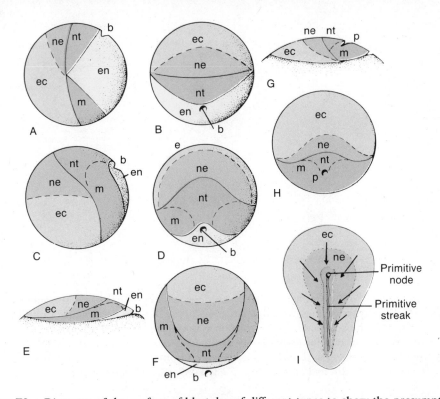

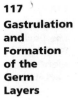
Figure 70. Diagrams of the surface of blastulae of different types to show the presumptive fate of various regions in normal development. *A, B,* Left side and dorsal views of the blastula of amphioxus (cf. Fig. 65 *E*). *C, D,* Similar views of an amphibian egg (cf. Fig. 66 *E*). In these figures, the embryo has been rotated from the position in which the egg originally floated to that assumed by the gastrula; the shift is such that the vegetal pole region, originally ventral, has rotated posteriorly and upward to essentially the position from which the endoderm develops. The position at which the blastopore develops is arbitrarily indicated by an indentation. In *E, F, G,* and *H,* similar lateral and dorsal views are shown for the flattened blastula of macrolecithal eggs of shark and bird. Note that in all forms, the pattern of potential germ layer areas is similar; however, in the bird (an amniote), the blastopore is replaced, except for endoderm formation, by the primitive streak. *I,* Stage in bird development beyond *H;* the embryo is elongating, and mesoderm and neurectoderm are moving medially (as indicated by arrows) to the primitive streak. *b,* Position at which blastopore develops; *ec,* future skin ectoderm; *en,* endoderm; *m,* mesoderm; *ne,* neurectoderm; *nt,* notochordal region of mesoderm; *p,* position in which primitive node and streak make their appearance. (Data from Conklin, Vogt, Vandebroek, Pasteels.)

comes to lie in the dorsal midline below the surface anterior to the blastopore. The shape of the blastopore continually changes throughout gastrulation, as the lip continues to be the site of active involution of superficial materials. At first mid-dorsal, it is added to on either side as mesodermal cells lateral to the notochord join in the inturning movement; finally, midventral mesodermal material is invaginated, and a circular blastopore is completed. Correlated with these movements, there is a downward growth of the dorsal lip over the yolk, until, on the surface, only a small plug of yolk within a reduced blastopore is visible.

118
Chapter 5
The Early
Develop-
ment of
Vertebrates

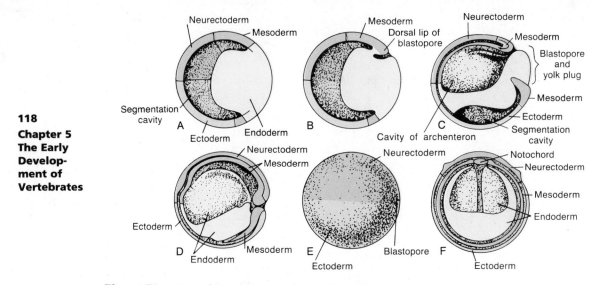

Figure 71. Gastrulation in an amphibian type of egg. *A* to *D* are views comparable to *A* to *E* of Figure 69. The presence, however, of a large mass of yolk restricts invagination to the extent shown at *B;* the remainder of gastrulation is performed by further growth of the blastopore lips, as indicated for dorsal lip by arrow in *C. E,* Gastrula from left side. *F,* Transverse section, looking anteriorly. As can be seen from *C, D,* and *F,* the mesoderm folds inward between ectoderm and endoderm (cf. Fig. 80 *A, B*). (After Hamburger.)

With the exception that in some amphibians completion of infolding of meso-derm may be delayed beyond this point, we have now attained externally a gastrula comparable to that of amphioxus. Externally, skin ectoderm lies ventrally and anteri-orly, and neural ectoderm occupies a posterodorsal oval area above the dorsal lip of the blastopore. Internally, much of the endoderm is still a yolky mass of cells, but this has begun to outline an archenteron. This is of good size anteriorly but posteriorly is a narrow tube. As in amphioxus, the archenteron is at first roofed by the chorda and lateral mesodermal tissues. Soon the notochord, essential for induction of the neural tube above it, develops its distinctive structure along the mid-dorsal line; the meso-derm turns outward laterally, and the endoderm grows upward to complete the de-finitive gut roof. Mesoderm formation here, and, indeed, in all true vertebrates, differs markedly from the pattern seen in amphioxus. Never is there any pouch formation. Instead, the mesoderm pushes outward as a continuous sheet on either side of the notochord, to grow laterally and ventrally between the ectoderm and endoderm. Only later do coelomic cavities appear in the mesoderm, and (except in cyclostomes) only the dorsal part of the mesodermal sheets develop a segmental pattern.

Elasmobranchs. Obviously, the geometry of movements of tissues during gastrula-tion must be of a different pattern in a macrolecithal egg from those thus far dis-cussed, for we are dealing with a flattened blastoderm, not a spherical blastula; the processes are, nevertheless, basically the same. Major events in more primitive modes of gastrulation are the placing in position of endoderm and mesoderm, notably chorda tissues, by inturning at the lips of the blastopore. But where are these lips to form on a flat disc? The most reasonable answer is that they should be at the margins of the plate and that the equivalent of the dorsal lip should be the posterior end of the prospective embryo.

The surface of the blastoderm is found to be laid out in a pattern of potential germ layers arranged in a manner similar to that on the surface of an amphibian egg (Fig. 70 E, F). At one margin is an area of prospective endoderm and, adjacent to it, the mesodermal area that is to form the notochord. Here then is the region corresponding to the dorsal blastopore lip. At this point (Fig. 72), endodermal tissue rolls inward and then forward over the surface of the yolk to underlie the surface of the central disc, to be followed immediately by the chorda and lateral mesoderm; these latter tissues separate from the endoderm to assume their definitive position between ectodermal and endodermal layers.

The early elasmobranch embryo resulting from the processes just described has attained the essential structure of an amphibian gastrula: an outer surface composed of skin ectoderm and neurectoderm, an inner surface of endoderm, and, between, mesodermal tissues with a median band of chorda tissue. But because of the presence of the great mass of yolk beneath, the topography is radically different. Not until much later in development do the embryo's germ layers completely surround the

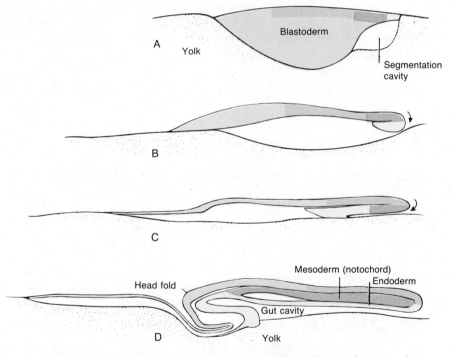

Figure 72. Longitudinal sections of successive stages in gastrulation of an egg of macrolecithal type, as seen in an elasmobranch. Only the disc of the blastula and the neighboring part of the yolk are shown. Anterior end at left. A, Blastula (cf. Fig. 68 E); B, involution of endoderm at posterior end of disc, corresponding to blastopore; C, continued process of inturning of mesoderm; D, mesoderm separated from endoderm; gut cavity formed, open below, roofed by endoderm. (After Vandebroek.)

yolk. Instead, they are at first merely spread out over its upper surface. The embryo is, so to speak, unbuttoned ventrally.

This infolding of tissues posteriorly corresponds only to the activities seen at the dorsal lip of the blastopore of such a form as an amphibian. What are the elasmobranch equivalents of the lateral and ventral lips of the blastopore of more simple types, which function in the involution of further mesodermal and endodermal materials and, finally, cover over the yolk of the vegetal pole? Their equivalents are the margins of the blastoderm, curving forward on either side from the dorsal lip homologue. The disc gradually expands over the yolk and eventually covers it (Fig. 84 *D*). As its ectodermal surface progresses to form a yolk sac, endoderm continually differentiates at the disc margins to form an internal covering over the yolk and, between inner and outer layers, mesodermal components in which blood vessels are formed to carry food materials inward from yolk to embryo.

Reptiles and Birds (Figs. 73–75, 85). The ancestral amniotes, like the sharks, developed a large-yolked egg and met similar problems regarding gastrulation. But the development of a macrolecithal egg took place independently in the two groups, and the solutions reached were different. As seen today in the reptiles and birds, the amniote solution differs more radically from the primitive type than does that of elasmobranchs. The cartilaginous fishes infold endodermal and mesodermal tissues at the margin of the blastoderm, utilizing this marginal area in the manner of the blastopore lip. The amniotes use the disc margin to form ectoderm and, probably, the extra-embryonic endoderm, but formation of the mesoderm and, probably, much of the endoderm (although this is still open to question—different experimental techniques give different results) takes place only in the central part of the blastoderm. Further, in the pattern of distribution of presumptive areas spread over the surface of

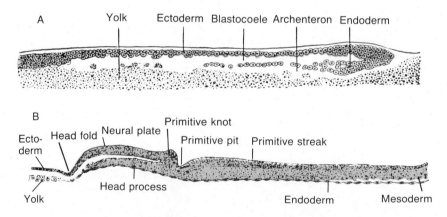

Figure 73. Two successive longitudinal sections of bird embryos at successive stages to show gastrulation. *A,* Figure comparable to the shark of Figure 72 *B,* but the endoderm is delaminating rather than involuting posteriorly to form a roof to the archenteron and, despite the seeming similarity, there is no development of a blastopore posteriorly. *B,* Later stage, comparable to Figure 70 *I;* from the primitive pit backward, cells are turning down inward from the surface (in the plane of the paper) and rolling laterally to form typical mesoderm and are moving forward from the pit to form the notochord ("head process").

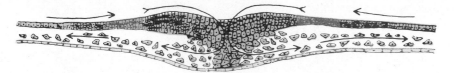

Figure 74. A cross section of the primitive streak at the stage shown in Figures 70 *I* and 73 *B*. Mesoderm, as indicated by arrows, is rolling medially, downward in the primitive streak, and then outward laterally on either side; above, the presumptive neural ectoderm is moving inward toward the midline. (After R. Bellairs, in Marshall, Biology and Comparative Physiology of Birds, Academic Press.)

121
**Gastrulation
and
Formation
of the
Germ
Layers**

the blastoderm (cf. Fig. 70 *G, H*), no specific endodermal area can be demonstrated experimentally. Some endoderm may form by delamination (as indicated in Fig. 73 *A*), but, as already noted, there is not yet general agreement on how this layer forms in amniotes.

An infolding of mesodermal (and possibly endodermal as well) tissues occurs at a peculiarly modified representative of the original blastopore, the **primitive streak** (cf. Fig. 70 *I*). This appears on the posterior part of the clear area of the expanding blastoderm and grows forward to form, when completed, a longitudinal groove bounded by parallel ridges; at its anterior end, there is a raised node of tissue, with a pit extending downward and forward beneath it. Observation and experiment show that the primitive streak is not a static topographic feature; it is a region of crucial activity in the formation of the embryo, possessed of part of the functions of its predecessor, the blastopore. The primitive streak is formed within the area of the blastoderm where the potential mesodermal tissues are situated. Medially into the ridges bounding the primitive streak, there is a continued movement of mesodermal cells from either side; these cells next migrate downward along the central groove and then fan out laterally to interpose themselves in proper mesodermal position between outer and endodermal layers. The chordal tissue lies, to begin with, on the surface anterior to the primitive streak. It moves backward, turns down into the anterior end of the streak, and then pushes forward in the midline, beneath its original external position. Its cells here develop into a definitive notochord, which expands in length. It grows backward beneath the surface (where neural ectoderm has come into position). As it lengthens, the notochord incorporates into its substance material from the front end of the area at first occupied by the primitive streak, while on either side of the streak differentiation of mesodermal tissues occurs. The primitive streak thus tends to be more and more restricted anteriorly. However, with continued lengthening of the blastoderm, it continues to be active posteriorly as further mesodermal tissues roll down into place on either side to form the posterior part of the body (cf. Fig. 85 *B, C*).

Mammals. Gastrulation in mammals is unique. In later stages, the mammal embryo comes to be identical in almost all major respects to its amniote relatives, but until gastrulation is completed, it is still quite atypical; it has not yet recovered, so to speak, from its initial vagaries. Methods of gastrulation vary among mammalian groups; described here (as in the case of cleavage) is that characteristic of higher primates.

The blastula, as we have seen, consists of an external trophoblast, which makes

122

**Chapter 5
The Early
Develop-
ment of
Vertebrates**

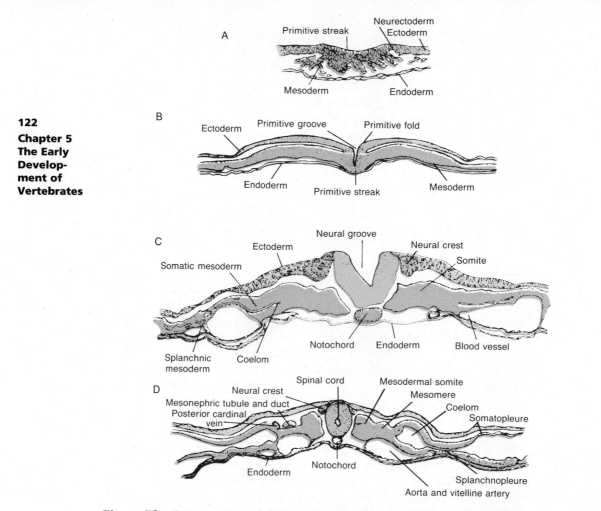

Figure 75. Cross sections of chick embryos to show successive stages in development of mesoderm and neural tube. *A,* Stage in which the inturning of mesoderm in the primitive streak has begun. *B,* The mesoderm has spread widely on either side between ectoderm and endoderm, but has not differentiated further (cf. Fig. 74). *C,* The coelom has begun to appear, splitting the lateral part of the mesoderm into outer (somatic) and inner (splanchnic) parts. Centrally, the notochord has separated from the remainder of the mesoderm, and neural folds and crests are appearing on either side of a neural groove. *D,* The neural folds have closed to form a tube, the spinal cord. The mesoderm has divided into somites, mesomeres, and a lateral plate in which a coelom separates inner and outer layers of the body-splanchnopleure and somatopleure. (From Arey.)

contact with the uterine tissues, and an inner cell mass. The first stage in further development is the appearance in the inner cell mass, above and below, of cavities that expand to leave between them a flat, two-layered plate of cells (Fig. 76 *C*). The upper cavity, lined with ectoderm, is that of the amnion; the lower is a yolk sac with an endodermal lining. The cavities and the materials lining them are parts of the amniote membrane system, to be described later; the flat two-layered plate between them is a blastoderm, in which the embryo is to arise. This appears to have the same pattern of potential germ layers as the blastoderm of a reptile or bird. The cells of the upper surface are those of the future skin ectoderm, neurectoderm, and mesoderm, both chordal and lateral. Its under surface, facing the yolk sac, is endoderm, formed here, as in reptiles and birds, without benefit of blastopore or inrolling of cells. As in other amniotes, a primitive streak now arises on the blastoderm (cf. Fig. 86 *A, B*); chordamesoderm and lateral mesoderm roll inward and downward along

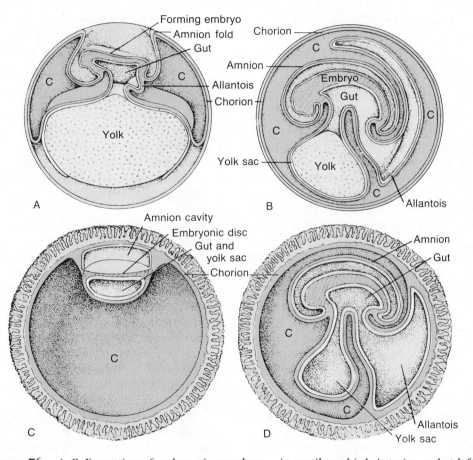

Figure 76. *A, B,* Formation of embryonic membranes in reptile or bird. Anterior end at left. *A,* An early stage. The embryo has been lifted somewhat off the yolk, but gut cavity proper and yolk sac are broadly connected. Yolk sac incompletely formed; amnion folds and chorion incomplete; allantois barely indicated. *B,* Later stage; embryonic membranes formed and yolk already partially reduced. *C, D,* Comparable views of mammalian type of development as seen in primates. *C,* Stage beyond the blastula seen in Figure 68 *E.* The inner cell mass has split ventrally to produce a gut cavity—constituting the major act of gastrulation—and split dorsally to produce an amniotic cavity. Between the two cavities, the embryo forms a disc in which primitive streak formation occurs much as in a reptile or bird. Mesoderm has already appeared, and chorionic villi are establishing connections with the surrounding uterine wall. *D,* Later stage in mammalian development, corresponding to *B. c,* coelomic cavity.

the course of the primitive streak to push anteriorly and laterally between ectoderm and endoderm, just as in birds and reptiles.

Development of the Neural Tube and Mesoderm

Formation of the Neural Tube. When gastrulation has been completed, with movements that place in proper position the various major tissues of the body—neurectoderm, chordamesoderm, and lateral mesoderm in addition to skin ectoderm and endoderm—processes of organ formation begin. Early stages in structural development bring the embryo to a stage termed the **neurula**.

Prominent on the outer surface is the formation of a neural tube; stimulus for its formation is the presence of chordamesodermal tissues beneath this area of the ectoderm. In amphioxus, the potential neurectoderm occupies a large oval area on the dorsal and posterior surface of the gastrula (cf. Fig. 69 *E, F*). On either side is found a folding upward of tissue at the junction of the future body ectodermal and neurectodermal areas (cf. Fig. 78). In amphioxus (not in true vertebrates), the two tissues separate as this fold forms. The ectodermal margins from right and left sides grow over the neural region, beginning posteriorly, and finally meet to form a complete layer of "skin" over the top of the body. Beneath this layer, the lateral margins of the neurectodermal sheet gradually roll dorsally, meet, and form a circular tube—the neural tube (cf. Figs. 78 *F*, 79). In amphioxus, the tube first closes midway along its length; the process gradually progresses both forward and backward. For some time the anterior end still opens to the surface as a **neuropore**. More curious is the situation at the posterior end of the canal. Here, the folds cover over the blastopore (cf. Fig. 79). This remains open; however, in consequence of neural tube folding, it now opens not onto the surface but into the posterior end of the neural tube. The primitive gut thus remains in connection with the surface, but the connection is indirect, via the neural canal to the neuropore. The connecting piece between gut and neural tube is termed the **neurenteric canal**. In later development, as the tail sprouts out, this canal closes, and neural tube and gut become discrete structures.

Most vertebrates show a type of neurectodermal differentiation differing somewhat from that seen in amphioxus (cf. Figs. 75 *C, D*, 77, 83 *B*, 84 *B*, 85 *C*). The ectoderm folds upward on either side as a **neural fold**. Eventually, the two folds meet one another mid-dorsally; ectodermal and neural tissues in each fold separate and fuse with their "opposite numbers" from the other side, and skin and neural tube are jointly completed. The end result is much the same as in amphioxus, although attained in a different manner. During the folding process, high crests are formed on either side, from which masses of cells are pinched off into the interior. Most of these **neural crest cells** are destined to become part of the nervous tissues; but others, as noted in later chapters, have an exceedingly varied history. In the head, still other future neural elements and sensory structures may arise as **placodes**—thickenings of the embryonic ectoderm lateral to the neural tube region, which detach themselves from the under surface of the future skin.

Mesodermal Development. The mesoderm forms the greater part of the body. Except for the brain and spinal cord, the ectoderm forms little but the superficial portion of the skin. Except for a mass of liver and pancreatic tissue, the endoderm forms little but a thin film of epithelium lining the gut and its outgrowths (lungs are endo-

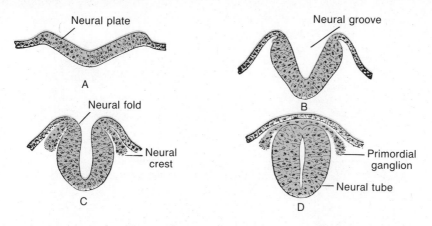

Figure 77. Formation of the neural tube and crest as seen in a typical vertebrate (mammal); a series of transverse sections at successive embryonic stages. (From Arey.)

125
Development
of the
Neural
Tube and
Mesoderm

derm—but mostly air). Practically all of the remainder of the body—muscles, connective tissues, most of the skeleton, and circulatory, urinary, and genital tissues—is derived from the mesoderm. If comparison is made with a house, the ectoderm corresponds to the paint on the outside and the wiring system; the endoderm to the floor varnish, wall paper, and perhaps the kitchen stove and stovepipes. All the rest— frame, plumbing, sheathing, flooring, even the floor boards, lath, and plaster—is comparable to the mesodermal derivatives.

We have here considered the notochord as part of the mesoderm, in a broad sense of that term. Many, however, would regard the chordamesoderm as a quite separate tissue; it certainly becomes distinct at a very early stage, as a long band of cells lying lengthwise along the roof of the archenteron. This rapidly separates from the sheets of mesoderm on either side and rounds up in section to gain its characteristic cylindric shape (Figs. 75 C, D, 78 C–F, 79, 82). It is much reduced in most adult vertebrates but forms a center about which vertebral formation takes place. As noted earlier, presence of the underlying notochordal tissue is necessary for the induction of the neural tube.

Apart from the notochord, in amphioxus, the mesoderm forms a paired series of segmentally arranged somites, which from the beginning contain a coelomic cavity (Fig. 78). In true vertebrates, as we have seen, there is a marked modification of this pattern. At first, there is no segmentation in the mesoderm, which pushes out on each side as a longitudinally continuous sheet between ectoderm and endoderm; nor does the mesoderm at first contain any cavities of coelomic type. In mesolecithal eggs, the mesoderm of either side grows as a hemicylinder, following the curve of the formed body wall ventrally and then medially to the midventral line of the belly (Figs. 71, 80, 82). In macrolecithal eggs, where the body is at first spread out flat above the yolk mass (and in mammals where the same pattern is followed despite the absence of yolk), the mesoderm spreads out laterally as a flat plate (cf. Fig. 75) and continues beyond the region of the body proper to contribute to the development of the extraembryonic membranes. In these latter types, we noted, the body is at first "unbuttoned" ventrally, so that it is only at a late stage that the mesodermal sheets of the

126
Chapter 5
The Early
Develop-
ment of
Vertebrates

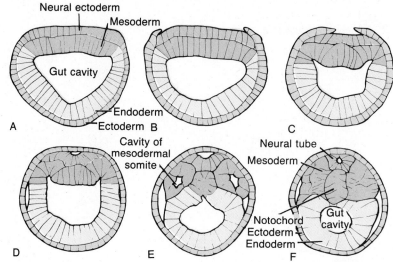

Figure 78. A series of cross sections to show formation of mesodermal pouches and neural tube in amphioxus. (Sections *E, F* are somewhat diagrammatic, since the somites of the two sides are alternating in position.) (After Cerfontaine.)

two sides gain contact ventrally and the contours of the body (and those of the mesoderm within) come to resemble those of mesolecithal forms.

There presently appears, in all vertebrates, a differentiation, from the dorsal midline outward, of three divisions of the mesoderm, each extending the length of the trunk (in head and tail, mesoderm is more restricted in its development). Next to the neural tube and notochord, the mesoderm thickens and subdivides on either side into a longitudinal row of cubical blocklike structures, the **mesodermal somites** (Figs. 75 *C, D,* 80, 85 *C, D,* 86 *C*), comparable to the somites of amphioxus. These are the first indications of true segmentation in the vertebrate body, and (apart from the independently derived serial arrangement of the branchial structures) the segmentation seen in other vertebrate organs appears to be largely due to the influence of the segmental arrangement of the mesodermal somites.

Within each somite, there is generally, for a short time, the development of a small coelomic cavity comparable to that which exists from the very beginning in

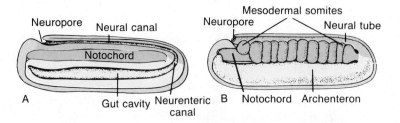

Figure 79. Amphioxus embryos at a stage in which the neural tube has formed and mesoderm is differentiating. *A,* Sagittal section. *B,* Longitudinal view with skin ectoderm sectioned medially, but internal structures preserved intact. (After Cerfontaine and Conklin.)

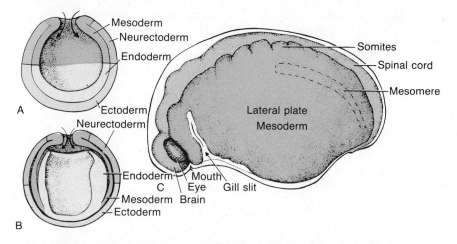

127
Development
of the
Neural
Tube and
Mesoderm

Figure 80. Mesoderm formation in an amphibian. *A,* Section of a urodele gastrula, cut transversely through the blastopore, showing involution of mesoderm into the lateral walls of the archenteron. This is essentially similar to the situation in amphioxus at the stage of Figure 69 *C* or *D. B,* Later stage; the mesoderm, instead of forming hollow pockets, as in amphioxus (Fig. 78), attains its intermediate position by pushing downward and forward between ectoderm and endoderm. *C,* A later embryo of an amphibian, after the neural tube is formed, seen in side view; the skin has been removed. The mesoderm forms a long, continuous sheet on either side of the body. The dorsal part is beginning to subdivide into somites. The part of the mesoderm that will later form kidney tissue is indicated by broken lines. The lateral plate is being broken up anteriorly by the formation of gill clefts. (*A* and *B* after Hamburger; *C* after Adelman.)

amphioxus. Soon differentiation appears within the somite, with a loss of its epithelial nature (Fig. 81). There is a great proliferation of cells from its ventromedial corner. These cells form an area of loose embryonic tissue that expands around the nerve cord and notochord and forms much of the axial skeletal structures, particularly the vertebrae and at least the proximal parts of the ribs; in relation to this, the medial part of the somite is termed the **sclerotome**. The epithelial layer on the outer side of the somite disintegrates and proliferates mesenchymal cells that appear to form part of the connective tissues of the skin; hence, the name **dermatome** is given to that part of the somite. After the loss of these two areas, medial and lateral, the remaining portion of the somite, termed the **myotome**, gradually differentiates and expands to form striated axial musculature, the history of which we shall follow in a later chapter.

Lateral or ventral to the somites, a relatively small intermediate region of the mesoderm develops in the trunk into materials, the **intermediate mesoderm**, from which are derived the kidney tubules and their ducts (hence formerly called **nephrotome**), and the deeper tissues of the gonads as well. This region may exhibit a segmentation comparable to that of the somites adjacent to them and form **mesomeres**, but frequently the kidney-forming tissue develops for most or all the length of the trunk as a continuous band (Fig. 80 *C*).

Beyond the mesomeric region—curving ventrally in mesolecithal forms, but at first extending straight laterally in other types—is a great sheet of mesoderm termed

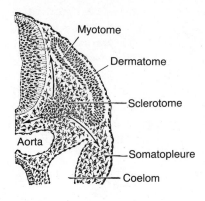

Figure 81. Hemisection through a mammalian embryo to show the subdivision of the somite into myotome, dermatome, and sclerotome. Arrows show directions in which mesenchyme grows from sclerotome to form vertebra and rib. The small notochord is present above the aorta, and part of the wall of the gut is shown below that large vessel. (From Arey.)

the **lateral plate** (Figs. 80 *C,* 82). Apart from the cyclostomes, there is never any segmentation in this plate. At first it is a solid sheet; later, however, it cleaves, and a **coelomic cavity**, the cavity that in adult life surrounds most of the viscera, develops within it (Figs. 75 *C, D*). The mesoderm external to the coelom, plus the adjacent ectoderm, is termed the **somatopleure**; the inner mesodermal layer plus endoderm is the **splanchnopleure**.

During much of embryonic development, there exist, between the epithelia and tissue masses of the major organs, relatively empty spaces filled with fluid. Scattered through these spaces, however, we find a diffused network of star-shaped cells that are believed to have, in many cases, the properties of ameboid movement. These cells compose the **mesenchyme**, the embryonic connective tissue. Much of it is formed by proliferation from the somites, but, in addition, the lateral plate is a source of mesenchyme. These areas of origin are both mesoderm, and certainly most mesen-

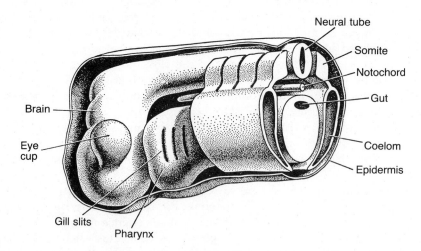

Figure 82. Stereogram of the front part of an embryo after partial differentiation of mesodermal components and nervous system. There is relatively little yolk here; in a macrolecithal form, the pattern would be similar, but the embryo would be open ventrally where it rested on the mass of yolk. (After Waddington, Principles of Embryology, George Allen and Unwin.)

chyme is of mesodermal origin. But in certain instances, the ectoderm gives rise to materials of a mesenchyme-like nature, as does the neural crest; the endoderm, too, it seems, may produce tissues of a sort normally derived from mesenchyme; its production is not confined to one region or one germ layer alone.

Although the mesenchyme acts as an embryonic connective tissue, its activities are not confined to such an essentially passive and transient function. It is a most versatile material. It gives rise to the connective tissues (including the skeletal tissues—cartilage and bone) of the adult. The mesenchyme gives rise to most of the circulatory system—to the blood vessels and to the blood corpuscles as well. Much of the musculature of the body is of mesenchymal derivation, including all smooth musculature, the specialized cardiac musculature, and even considerable striated musculature.

Body Form and Embryonic Membranes

In earlier sections, we have seen that development from the egg through cleavage and gastrulation takes place in a variety of ways. But once gastrulation has occurred and mesodermal and neural components have been sorted out, we find that, much as the means of attaining it may vary, the end result is the same—the laying out of a rather uniform pattern of body regions and potential organ systems (Fig. 82). The further history of these systems will be considered in later chapters. In consequence, we shall not discuss in any detail the further development of the embryo. We may, however, describe briefly the gradual assumption of definitive body shape and the nature of the embryonic membranes that are important in the development of many large-yolked eggs.

Primitive Types. In amphioxus, we left the embryo at the neurula stage as a rather short cylinder. The remainder of the developmental story is, superficially, mainly one of bodily elongation. Anteriorly, there is a development of a mouth opening and of complex pharyngeal gills; posteriorly, the neurenteric canal closes, and there buds out a tail in which develops a continuation of the spinal cord and notochord and numerous mesodermal somites. The mouth and gills of cephalochordates develop in a unique asymmetric manner.

In vertebrates with a moderate amount of yolk, such as amphibians, lampreys, and less specialized bony fishes (Fig. 83), the neurula is likewise essentially a spheroid. Dorsally, there is a rapidly growing nervous system; below this, a bulging belly contains the gut cavity and a mass of endodermal cells distended with yolk. With growth of the brain, the head expands anteriorly; posteriorly, there is a major development of a tail beyond the anal region, much as in amphioxus. Soon a body shape is attained that is not far removed from that of the later fish or amphibian larva except for a distended gut, which is gradually reduced as the yolk is absorbed.

Cartilaginous Fishes. Body form is more slowly attained in forms with large-yolked eggs. In the elasmobranchs (Fig. 84), the neurula is little more than a flat sheet overlying a thick mass of yolk with the neural tube, above the notochord, marking out the "main line" of body organization. Anteriorly, the growing brain comes to extend forward over the yolk surface, with the ectoderm folded about it, and the anterior end of the digestive tract assumes a tubular form separate from the yolk cavity. At

130

**Chapter 5
The Early
Develop-
ment of
Vertebrates**

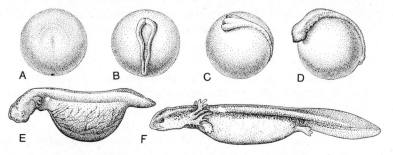

Figure 83. Development of body shape in a form with a mesolecithal egg—the urodele *Necturus* (the mud puppy). *A,* Late gastrula, seen from above, head end at top. *B,* Neural folds forming. *C,* View from left side, neural tube formed, brain bulging upward above sac partly filled with yolk. *D,* Head and trunk taking shape dorsally. *E, F,* Steps in reduction of yolk-filled belly sac and assumption of normal form. External gills and eye appear in *E,* limbs in *F.* (After Keibel.)

the posterior end, too, the body lifts off the yolk as the tail buds out; only near the midlength of the body does the gut long remain connected with the mass of yolk. Meanwhile the margins of the blastoderm continue to expand over the yolk and eventually completely enclose it in a **yolk sac,** the cavity of which is essentially an extraembryonic part of the gut. As the blastodermal rim advances over the yolk, it continually differentiates to extend the ectodermal covering, the endodermal sheet facing

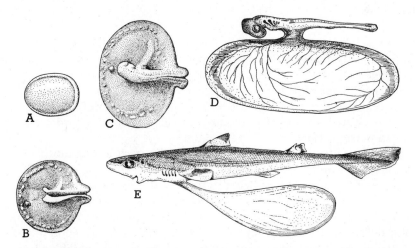

Figure 84. Development of body form in a shark. *A* to *C* are dorsal views of the blastodisc from which the embryo forms; the underlying yolk is omitted in these figures. *A,* The embryonic disc at gastrulation; the endoderm is rolling under at the thickened posterior and lateral margins (cf. Fig. 72 *B*). *B,* The disc is enlarging, and the neural folds are developing on the upper surface. *C,* The neural folds are closed except at the growing posterior end; the body of the embryo is lifting off the yolk and head region and somites are visible. *D,* The yolk sac is completely formed and the embryo connnected with it by a stalk; eyes and gill slits are visible. *E,* Almost normal shape has developed except for retention of a relatively small yolk sac. (After Ziegler, Dean.)

the yolk, and, between the two, mesoderm with forming blood vessels. The yolk is gradually digested and carried to the embryo by these vessels; the sac dwindles and is eventually resorbed. In addition, there may be, in viviparous elasmobranchs, provisions for supplying the uterine young with food materials from the mother.

The embryos of many teleosts have a yolk sac that appears similar to that of elasmobranchs. It is not, however, formed in the same manner, being completely separate from the embryonic gut and not lined by endoderm.

Reptiles and Birds (Fig. 85).   The shelled amniote egg has evolved as a structure in which, in contrast to the eggs of lower vertebrates, development can, indeed must, take place on land rather than in the water. Apart from the protection of the shell, we find that the embryo becomes surrounded by a series of membranes that provide it protection and aid its metabolic activities. These membranes include a yolk sac, amnion, chorion, and allantois (cf. Fig. 76 *A, B*).

In reptiles and birds, the development of a yolk sac is somewhat as in sharks. The endoderm, to begin with, is a flat sheet of cells capping the yolk; it gradually grows downward to enclose the yolk in an almost complete sac, followed in its growth by mesoderm, spreading outward and developing vessels for circulation in the yolk sac. The connection of the sac with the growing embryo is eventually constricted to a stalk.

In sharks, the ectoderm lying outside the region of the embryo grows downward around the yolk sac, the only structure external to the embryo proper. In amniotes, the situation is more complicated; this extra-embryonic ectoderm (with accompanying mesoderm) does not directly envelop the yolk sac but gives rise to two of the other membranes. Folds grow upward over the developing embryo to form a closed

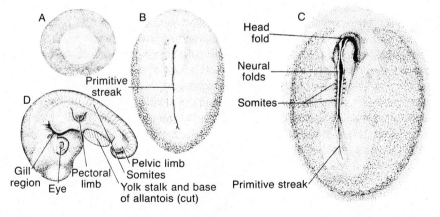

Figure 85.   Some stages in amniote development as seen in reptile or bird. *A,* Small germinal disc situated on the upper surface of the yolk. *B,* Formation of primitive streak and elongation of germinal disc (cf. Figs. 70 *I,* 73 *B,* 74, and 75). *C,* The embryo is enlarging to cover more of the yolk; the head is lifting off the yolk surface; neural folds and somites are appearing; the primitive streak, now relatively small, is still active in formation of posterior part of body. *D,* Side view of a considerably later stage, comparable to Figure 76 *B.* The embryo is separated from the yolk except by a stalk (cut). Many structures of the head and body are formed, and limb buds are appearing. (*B* and *C* after Huettner.)

sac, the **amnion**; its liquid-filled cavity furnishes a miniature replica of the former aquatic environment for development of the embryo. Externally, the ectodermal sheet, with mesodermal reinforcement, expands to enclose almost the entire set of embryonic structures in a protective membrane, the **chorion**.

With the constriction of the connection to the yolk sac to a stalk, the remaining part of the gut assumes a tubular form. However, soon a stalk of endoderm plus mesoderm grows out ventrally from the posterior end of the gut. This rapidly expands into a large sac that underlies and is attached to the chorion over much of its extent. This is the **allantois**. The kidneys of the embryo begin to function at an early stage, and the expanding allantoic cavity becomes an embryonic bladder; indeed, the allantois seems to have arisen as a "precocious" embryonic development of a bladder such as is present in the adult amphibian but not notably developed in the embryo in that class. Equally important, however, in an embryo reptile or bird is the function of the allantois as a breathing organ. The combined chorionic and allantoic membranes operate as a respiratory surface for absorbing the oxygen that enters through the porous shell; blood vessels in the allantoic walls carry oxygen into the embryo and carry carbon dioxide out in exchange. With these membranes formed, the embryo takes shape and grows within the expanding cavity of the amnion (Fig. 76 *B*).

Mammals (Figs. 76 *C, D,* 86, 87).   We have previously noted that there are many variants among mammals in early developmental processes. In primates, which we chose as a "sample," there appear, as a preliminary to gastrulation, cavities lying above and below the embryonic disc within the inner cell mass of the developing egg. These cavities are comparable to those of the amnion and yolk sac of reptiles

132
Chapter 5
The Early
Develop-
ment of
Vertebrates

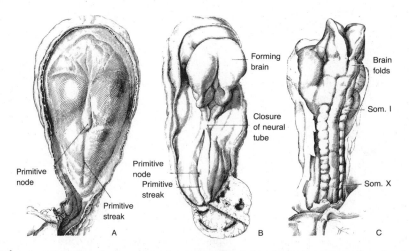

Figure 86.   A series of early human embryos, to illustrate stages in mammalian development. All three are dorsal views of the embryo, with the embryonic membranes cut away. *A,* Primitive streak stage, comparable to Figure 85 *B,* for a bird or reptile. *B,* Later stage in which the primitive streak is still active posteriorly, but the neural tube is forming more anteriorly. This stage is comparable to Figure 84 *C,* for the shark, not quite so advanced as Figure 85 *C,* for the bird. *C,* More advanced stage, with neural tube almost completely closed and somite formation well advanced. (After Heuser, West, Corner.)

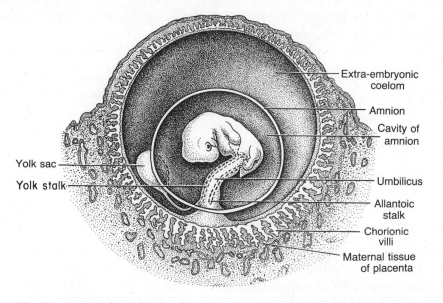

Extra-embryonic
coelom

Amnion

Cavity of
amnion

Umbilicus

Allantoic
stalk

Chorionic
villi

Maternal tissue
of placenta

Yolk sac

Yolk stalk

Figure 87. Diagram to show the development of a primate embryo inside its membranes. The stage represented is one at which the embryo, though well formed, is still of small size.

and birds, but because of the precocious development of the trophoblast, the order of events is quite different here. The blastoderm, present between amniotic and yolk sac cavities, is connected posteriorly with the superficially placed trophoblast by a stalk of mesodermal cells. Along this stalk, there pushes an outgrowth of the gut that expands beneath the trophoblast into an allantoic sac. Mesodermal tissues, in which blood vessels develop for the embryonic circulation, come to reinforce the endoderm of the yolk sac and allantois; mesoderm reinforces the ectoderm in formation of a stout amnion and sheathes the trophoblast internally to form a chorion comparable to that of reptiles. As an end result, we find that the mammal has, by a different route, arrived at the formation of a series of embryonic membranes highly comparable to that of its reptilian forebears.

There are, of course, notable differences; the yolk sac contains no yolk in any but the most primitive of mammals, and a placenta, as noted below, is formed by the external membranes. As in reptiles and birds, the growing embryo gradually takes form and in this process separates its body from the yolk sac (Fig. 87). Eventually, its only connection with other structures is by a narrow ventral area, the **umbilical cord**. This cord is surrounded by the ectoderm of the amnion and contains stalks connecting embryo with yolk sac and allantois, together with mesodermal tissues and blood vessels associated with those structures.

A major difference between a typical mammal and its amniote relatives is the development of a **placenta**, through which the embryo receives food and oxygen from the mother and sends back waste materials in return. This characteristic mammalian structure is formed by an intimate union of the embryonic membranes with the surrounding maternal uterine tissues. A primary embryonic component of the placenta is the chorion, formed from the trophoblast plus the mesodermal tissues that come to reinforce it. On its outer surface, there develop finger-like processes, which project into the maternal tissues. These are highly variable in size, disposition, and the manner in which they interlock with the uterine walls. A number of different placental types, which need not concern us here, are found in the various mammalian groups.

134
Chapter 5
The Early
Develop-
ment of
Vertebrates

The placenta increases in size and complexity during the growth of the embryo. It is richly supplied with blood vessels from both mother and embryo, and, in it, there is a constant exchange of materials between the two organisms. It is, however, important to note that the two blood streams are never in contact; they are everywhere separated by a membrane through which small molecules may pass readily, but not larger bodies such as blood proteins or blood corpuscles. Occasionally there may be "leakage" between the blood streams, but this is a pathologic condition and may cause severe problems.

The chorion, however, is only part of the placental structure supplied by the embryo (and, in fact, may disappear during late stages of development). This membrane is not in direct communication with the embryo within; some means must be supplied through which the blood may freely flow from embryo to placenta and return. In most marsupials, the yolk sac adheres to the inner surface of the chorion, and its mesoderm develops blood vessels that serve this purpose; to some extent a yolk sac type of placenta persists in some higher mammal groups such as rodents. In almost all higher forms, however, the dominant structure is an allantoic placenta. In reptiles and birds, the allantois extends far around the outer surface beneath the chorion and becomes closely attached to that structure, and allantoic blood vessels are important in those groups in carrying oxygen and carbon dioxide from surface to embryo, and vice versa. In most mammals, the allantois is similarly developed beneath the chorion as an integral part of the placenta; its vessels here are vital as carriers of food materials as well as oxygen to the embryo and of carbon dioxide and wastes back to the mother's blood stream. Although such a placenta is characteristic of higher mammals, more or less similar adaptations have arisen in other forms. Those in elasmobranchs are basically different, but some lizards and snakes have chorioallantoic placentae, reduced yolk, and other features normally thought of as purely mammalian.

Larvae. In amniotes and the sharklike fishes with large-yolked eggs, in which an abundant food supply is available to the embryo, development proceeds rather directly toward adult structure; the young, at birth, is essentially a sturdy little replica of the adult, soon capable, in a modest way, of setting about its business of making a livelihood in the manner of its elders. This is not so in many water-dwelling lower vertebrates such as lampreys, many bony fishes, and amphibians. In these forms, the supply of yolk available for embryonic growth is limited; the young, when hatched, is of tiny size. At such a stage it is in many cases unable to take up adult habits and is liable, because of its size, to dangers to which the adult is not subjected. Under such conditions, it is to be expected that frequently the young of these forms assumes a mode of life quite different from that of the adult, often lives in a different environment, eats a different type of food, and develops, temporarily, structures suited to its livelihood that may differ markedly from those of the adult. A **larval stage** is thus interjected into the life history. With growth, larval features are eventually lost and adult structures and habits assumed—the process of **metamorphosis** (in which activity of the thyroid gland often plays an important part). The tadpole stage of the frog or toad is a familiar larval form. Aquatic water-breathing larvae are found in the

Urodela; most metamorphose into a somewhat more terrestrial gill-less adult stage, but some fail, partially or entirely, to metamorphose—the common mud puppy, *Necturus,* of Midwest streams and the axolotl of Mexico are examples. Many teleosts have larval stages; an extreme type is the translucent, leaf-shaped larva of the eel, found in midocean, whose relationship to the adult was long unsuspected. The adult marine lamprey is a large predaceous form; the larva is a tiny food strainer, sedentary in the mud of stream bottoms.

Regeneration

In this description of development, it has been more or less tacitly assumed that organs and tissues, when finally differentiated, are permanent, that they are made once and for all. But this is, of course, far from true in many instances. In even the most normal of existences, blood cells are constantly replaced; hairs, feathers, skin cells, teeth, and other structures may be discarded and renewed. Further, accident and disease may cause destruction of essential tissues or organs. In such cases, regeneration and reformation of lost parts often occur; development is thus not only characteristic of embryonic and youthful stages but is also to some degree a lifelong process.

Mammals and birds are not notable for regenerative powers, but even so large epidermal areas may regenerate after damage; injured blood vessels and nerves (but not whole nerve cells) may redevelop, and destroyed liver tissue may be regrown. Lower tetrapods tend to have greater regenerative powers. In many lizards, much of the tail is readily shed and as readily regrown (although never fully reattaining its former structure). Most labile of all vertebrates are the urodele amphibians, which exhibit remarkable regenerative powers; a complete limb may be redeveloped following amputation.

Developmental Mechanics

In the earlier sections of this chapter, we have given a description of the orderly series of events that take place during the development of a vertebrate but have said little as to the "why" of these occurrences.

The answer to this question is the major interest of embryologists today. The development of the individual from seemingly simple egg to complex adult is a miracle so common that we regard it as commonplace. When some accident occurs in the normally well-regulated process, we tend to be puzzled or disturbed about the abnormality that results. Rather, we should marvel that the processes of development normally proceed so effectively; that development takes place at all. Most of the major events in vertebrate embryology are well known, but the mechanisms of development are in many regards still a mystery. However, new tools and approaches are making possible the examination of the roles played by the genes and their modes of action in the differentiation of the many and varied types of cells during embryonic development. Also, studies of the movements and other behavior of cells are providing a clearer understanding of how the different cells become organized and arranged to form the tissues and organs of the animal. Work of this sort is considered in any recent embryology text. Although this work is interesting and important for anatomic studies, space does not allow its review here.

Ontogeny and Phylogeny

136
Chapter 5
The Early
Develop-
ment of
Vertebrates

Even in the early days of the scientific study of embryology, over a century and a half ago, it was noticed that animals vastly different as adults are similar in structure and appearance as embryos, and that the embryos of "higher" vertebrates often exhibit conditions quite different from those of the adult, but similar to those seen in members of "lower" groups. From such observations came the idea of a biogenetic "law," that individual development—**ontogeny**—repeats the history of the race—**phylogeny**; in other words, an animal in its development climbs its own family tree, successive embryonic stages representing the adult stages of ancestral types.

The main proponent of this extreme version of the biogenetic "law" was the German biologist Haeckel, who believed that it would provide answers to almost all evolutionary problems. His ideas go back to the earlier work of von Baer; the latter, however, made no such exaggerated claims. Von Baer's "laws" noted (1) that more general characters appear before more specific ones during development, (2) that more specific characters develop from the more general, (3) that animals progressively diverge more from related forms during development, and (4) that the early stages of young animals resemble those early stages (*not* adult stages) of more primitive forms.

For decades, the biogenetic "law" was an important factor in the stimulation of embryologic work and in the study of homology. But further consideration has shown that it is only a half truth. A mammalian embryo at an early stage is fishlike in many regards; it has, for example, prominent "gills" that are, of course, later reduced or modified (Fig. 88); but there is actually little resemblance to an adult fish. The "gills" are the branchial pouches seen in the fish embryo; they fail to open to the surface as do the branchial slits in the mature fish, and branchial membranes never develop. Actually, as von Baer had noted just over 150 years ago, it is the fish embryo, not the adult fish, that the mammalian embryo resembles. During evolution, there may sometimes have been progression ontogenetically by piling one adult stage on top of another. However, new forms, it would seem, have far more often arisen by diverging at some point from old developmental sequences and thus arriving at new goals in the adult.

Development tends to be a conservative process, for departure from the old, tried and true methods will usually result in failure and death. In consequence, much of the developmental pattern in forms only remotely related to each other may be similar, and the development may follow a devious course, with indications of structures that were once present in the adult of ancestors, but that now never reach maturity. Ontogeny repeats many important steps in the developmental pattern of ancestral forms; it is especially likely to repeat them when they are structurally or functionally useful in the derived type's own development.

But embryos as well as adults are often, even usually, modified in relation to their environment, and such modifications may result in the creation of embryonic structures never seen in adults at any lower level. Larval types, as noted earlier, typically possess many features of this sort. Again, early amniotes developed a shelled **egg** with a large yolk mass as food and with embryonic membranes as devices for improved development; no amniote ancestor ever had, of course, an adult stage with a pendant yolk sac, nor was there ever an adult amniote ancestor enclosed in a membrane—the amnion—or in a relatively hard shell.

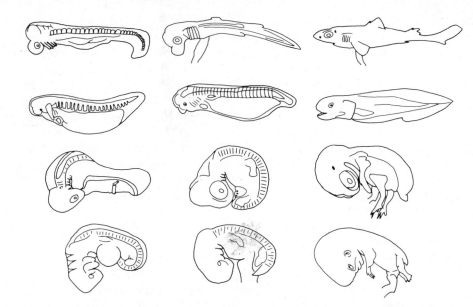

Figure 88. Embryos of various vertebrates, early stages on the left, intermediate in the center, and advanced on the right. *Top row,* a dogfish (*Squalus*); *second row,* a lungfish (*Neoceratodus*); *third row,* the chick (*Gallus*); *bottom row,* human (*Homo*). Note that, as described by von Baer, the early stages are all very similar, but later stages are less similar. The early stages of advanced forms do *not,* as claimed by Haeckel, look like adults of more primitive animals.

Although, in general, ancestral developmental patterns are conservatively followed or are at the most subject to gradual modifications, there may be drastic modifications of the pattern, and later embryonic stages may be reached by a different route from that followed by ancestral types—presumably in relation to powerful adaptive requirements. We have seen that large-yolked amniotes have radically modified the older methods of gastrulation and endoderm formation, although reaching the same goal by another path. A further striking example is seen in the early development of the mammal. Other amniotes have a neat, "logical" method of forming the embryonic membranes. The mammal departs radically from its reptilian ancestors in the process of membrane formation but arrives at the same end result. However, even here we have retention of odd characters—our yolk sacs do nothing but still appear and then disappear.

The Germ Layers

In our account of development we have placed emphasis on the early differentiation and segregation of the major structural elements of which the adult body has been built. The theory of the germ layers was an early concept in embryology and a most useful and fruitful one. In the early embryo, two layers, ectoderm and endoderm, already potentially determined in the egg, may be distinguished as comparable in general to the outer and inner layers that alone constitute the entire body of a coelenterate. Soon there becomes distinct a third, intermediate germ layer, the mesoderm, from which in vertebrates, as in all invertebrates above the coelenterate level, the greater part of the body is derived. In this chapter, we have adhered to the germ layer concept, although emphasizing the early separation between superficial and neural portions of the ectoderm.

In the adult vertebrate, tissue components of organs and organ systems can, for the most part, be sorted out readily in regard to their derivation from these germ layers. Details and exceptions will be found in later chapters; major derivatives are listed here.

From the body ectoderm: the superficial portion (epidermis) of the skin of the body surface together with its extensions into both ends of the digestive canal (mouth, cloacal region); epidermal derivatives, such as hairs, feathers, skin glands (mammary, sweat, and so on); sensory epithelia of nose and internal ear; eye lens; enamel of the teeth.

From the neural ectoderm: almost the entire nervous system; the retina of the eye; and from the neural crest, sensory ganglia of the nervous system and certain non-neural structures (much of the cranial skeleton, pigment cells).

From the mesoderm: connective and most skeletal tissues; the musculature; the vascular system; most of the urinary and genital systems; the lining of the coelomic cavities; the mesenteries; the notochord.

From the endoderm: the lining of the major part of the digestive tract and the substance of organs (liver, pancreas) connected with it; much of certain endocrine glands (thymus, parathyroids, thyroid); much of the branchial membranes of lower vertebrates and all the lining of the lung-breathing apparatus of higher forms; terminal parts of the urinary and reproductive tracts.

So generally do the patterns of origin of various adult tissues adhere to tabulations of this sort that beliefs as to the absolute specificity or origin of such tissues completely dominated the minds of many embryologists for many decades. In more recent times, however, further observation and experimental work have shattered such a faith. It is found, for example, that although most skeletal material arises in orthodox fashion from mesodermal mesenchyme, certain cartilages and bones in the head and throat come from cells of the neural crest, otherwise mainly devoted to the development of neural structures; that the lining of the gills may come from either ectoderm or endoderm; that cell areas that normally produce nerve cord will, if transplanted, produce skin, and vice versa. There has been, in consequence, a tendency on the part of some investigators to abandon the germ layer concept as meaningless. This, however, is a counsel of despair. As we have noted, what a cell becomes depends upon the potentialities present in it at any given stage of development and on the influences to which it is subjected. But, in general, we find that in normal development, the embryonic cells and tissues do follow a consistent pattern of regional movement and arrangement of components. If nothing more, the germ layer terminology is useful as a description of the topography of development. It is, however, more than this. Experimental work has shown that, although in early stages there may be little differentiation among various regions of the embryonic germ layers, there is increasingly, in later stages, a limitation of capacities in different regions. The prospective fate in normal development of the germ layers and subsidiary areas of these layers is in general accord with the experimentally deduced story of their prospective potencies.

Chapter 6

The Skin

Skin Functions. Forming a covering for the entire body, the **integument**—the skin with its accessory structures—is an organ system performing varied and important functions, many of which are protective. The tough "hide" of many vertebrates, often reinforced with dermal scales or bones, is a protection against mechanical injury and against the attack of predators; the latter function may involve scales for armor plating, color patterns for concealment, horns and claws for active defense, or other such modification. The skin is a continuous unbroken line of defense against the invasion of bacteria and other microorganisms. As a sheath of tissue isolating the internal structures from the exterior, the skin may ward off physical or chemical influences disturbing the inner economy. It may aid in regulation of the water content of the body, preventing too great a loss of water in marine or terrestrial vertebrates, too great an influx of water in freshwater dwellers. Pigment in the skin prevents the intake of injurious amounts of light. The skin itself, in addition to auxiliary structures such as hair or feathers, insulates the body against too great a loss (or gain) of heat and plays a major part in temperature regulation in mammals and birds.

The skin may further play an active physiologic role in the absorption or elimination of materials through moist surfaces or glandular structures. Breathing, the absorption of oxygen and release of carbon dioxide, is a function of the skin in many forms, and, in some vertebrates, the skin is a major respiratory organ. The skin may, through its glands, function as an accessory to the kidney in the elimination of wastes. Whether or not such obviously skeletal elements as bones are present in it, the skin may function as a skeleton—a structure to which muscles may attach and on which they may act.

As the region of the body in immediate contact with the outer world, the skin would appear to have been the area of origin of sensory and nervous structures in ancestral metazoans. In the vertebrates, the nervous tissues have become a separate organ system withdrawn from the surface; but embryologically, we have seen, the nervous system still arises in continuity with the skin, and the skin is still the seat of abundant sensory structures.

The skin is not a single structural entity but consists of two parts: **epidermis** and **dermis**, closely united but differing greatly in their nature and origins. The epidermis is superficial in position and is an essentially cellular material, an epithelium derived from the ectoderm of the embryo; the deep-lying dermis has primarily a fibrous structure, with relatively few cells, and is derived from embryonic mesenchyme, mainly of mesodermal origin. The epidermis is relatively thin; the dermis is usually much thicker. The epidermis gives rise to a host of differentiated structures, such as hair, feathers, various glands; the dermis, on the other hand, has (except for bone and scale formation) a relatively simple and uniform composition.

Epidermis.   The epidermis, the superficial body covering, is usually much the thinner of the two skin layers. Derived directly from the embryonic ectoderm, it forms a continuous epithelium over the entire body. It tends to give rise in higher groups to varied special skin structures (described in later sections).

In amphioxus, the epidermis is in the simplest possible condition; it consists merely of a single layer of columnar cells, without glandular elements; it is ciliated in the young, but in the adult is covered by a thin film of cuticle, which the cells have secreted. A cuticle is absent in adult vertebrates except cyclostomes, and, in every case, the epidermis of true vertebrates is a stratified epithelium.

In fishes and aquatic amphibians (Fig. 89), the epidermis remains in general a simple structure, apart from the presence of glandular elements. Pigment may be present in the vertebrate epidermis in the form of **melanin**, an organic compound, which in various concentrations gives shades of brown and black. However, in lower vertebrates, coloration is mainly due to color-bearing cells or chromatophores (described later). The melanin of the epidermis is derived, by cellular transfer, from pigment cells situated in the dermis.

In fishes and amphibians, generally the entire thickness of the epidermis consists of "live" cells, containing normal protoplasm. The superficial cells, however, include a certain amount—usually small—of **keratin**, a waterproof protein abundant in the horn sheaths of cattle, fingernails, and similar superficial structures.* These outer cells tend to be lost by wear or injury, and there is a constant replacement from below. The basal layer of cells, usually more or less columnar, is the vitally important part of the epidermis in any vertebrate. This is the "matrix" of the epidermis, the layer from which successive generations of cells are budded off to form the outer portions of the epithelium. Superficial damage to the epithelium is readily repaired. But if, through major injury such as serious burns, a large area of this basal matrix is destroyed, a re-covering of the flesh by skin becomes difficult, if not impossible. The skin of most fishes and aquatic amphibians is permeable to some degree, and in most modern amphibians, in default of any backing-up of the epidermis by bone or dense connective tissue, this permeability is availed of to make the skin a major respiratory organ, supplied by special blood vessels beneath the epidermis. Indeed, in some lissamphibians, there are infra-epidermal capillaries, structures rarely found in other vertebrates. In hibernating amphibians, it appears, all breathing is done through the skin.

With the assumption of a definitely terrestrial life by certain of the amphibians and the amniotes, the nature of the epidermis is changed. In land dwellers, in which water loss through the skin is a serious matter, the outer part of the epidermis becomes differentiated from that beneath. As the surface is approached, the layers of cells are successively flatter, the protoplasm becomes increasingly modified and lifeless, and the cells are mainly filled with keratin. The surface of the skin is dry and covered by thin, "dead" cells. These may be rubbed off and lost piecemeal (dandruff is a familiar example) or shed seasonally in some reptiles and amphibians.

Keratos and *cornu* are, respectively, the Greek and Latin words for horn, whence are derived English words concerned with this substance.

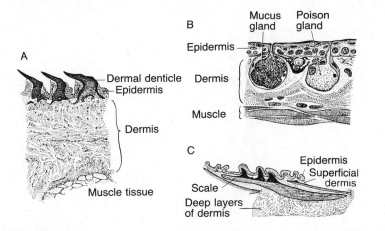

Figure 89. Sections of the skin of *A,* a shark; *B,* a salamander; *C,* a teleost. These are shown at quite different magnifications; actually the teleost skin would be as thick as or thicker than the others and that of the salamander the thinnest. (After Rabl.)

The transition between the inner and outer portions may be gradual, as in amphibians, reptiles, and birds (in the last group, incidentally, keratin is mainly concentrated in the feather covering), but in mammals (Fig. 90), there may be a sharp contrast between the lower zone of "live" cells, the **stratum germinativum**, and the flattened, deadened, and cornified cells of the **stratum corneum**. Intermediate layers may be recognized, but they are not very distinctive or always present.

Keratin Skin Structures. Throughout the higher vertebrates, keratin-filled epithelium develops into a variety of special structures. Simplest perhaps are thickenings or swellings of the stratum corneum, such as in the "warts" of toads. Such thickenings are frequently present on surfaces subject to wear, as the under surface of the feet; such **calluses** often involve thickening of the dermis as well as the epidermis. In mammals, generally we find a characteristic arrangement of **foot pads** (Fig. 91), in-

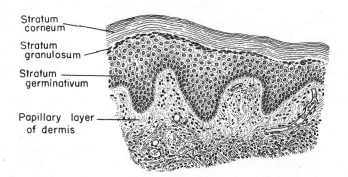

Figure 90. Section of the skin of the human shoulder, × 125. In addition to the germinative layer and stratum corneum, there is (as here) an intermediate granular layer in the epidermis in many areas of mammalian skin; in some situations, there is, further, a transparent layer (stratum lucidum) between horny and granular levels. (After Maximow and Bloom.)

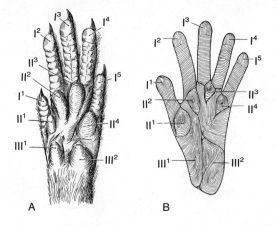

Figure 91. Palmar surface of the manus of *A,* an insectivore; and *B,* a monkey (macaque). The insectivore shows a presumably primitive mammalian structure, with thick pads on either side at the proximal end of the palm (III^1, III^2), pads between the bases of successive digits (II^1 to II^4), and pads at the tip of each toe (I^1 to I^5). In higher primates, these pads are replaced by patterns of friction ridges. (After Whipple.)

cluding, in five-toed types, a pair of pads on the proximal part of palm and sole, one on the base of "thumb" or big toe, a series between the bases of the other toes, and one on each toe tip. In higher primates, where the hand and foot grasp limbs of trees in locomotion, we find palm and sole covered instead by a pattern of **friction ridges** (Fig. 91 *B*), which aid in obtaining a firm grip. These epidermal ridges, in the position of the pads of other groups, form complex series of loops and whorls. As is well known, the pattern of the fingertips is almost infinitely varied and affords a ready means of individual identification in man.

In reptiles, thickening and hardening of the cornified epidermis results in the formation of **horny scales** or **scutes**. In crocodilians and most turtles, they form flat plates; in the latter (as the tortoise "shell" of commerce) these overlie the bony dermal skeleton of the back and belly. In lizards and snakes (Fig. 92), overlapping scales are commonly present, and in snakes, these are highly developed as aids to locomotion. Though such structures are termed scales, it cannot be too strongly emphasized that these horny epidermal scales of reptiles (and other amniotes) are not homologous with the largely bony scales most characteristically developed in fishes.* A reptilian horny scale develops embryologically as an outpushing of the epidermis containing a papilla of mesodermal tissue; the broad upper surface of the papilla becomes the intensely cornified scale.

In mammals and birds, the horny scales once present in their ancestors have for the most part disappeared. They persist, however, on the legs of birds, and on the legs and tails of a variety of mammals, notably certain rodents, insectivores, and marsupials. The pangolin of the Old World tropics is notable as a mammal that has redeveloped a complete body covering of large horny scales (cf. Fig. 55).

In many amniotes in which teeth are reduced or absent, the skin on the jaw

---

*In many lizards, bony scales may be present as well as horny scales, the latter, of course, being more superficial in position (Fig. 92). Bony underlying plates are also present to a variable degree in crocodilians.

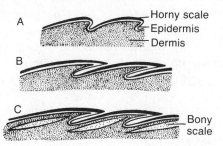

Figure 92. Diagrammatic sections of reptile skin to show scale types. *A,* Lizard skin with simple, horny epidermal scales, gently overlapping; *B,* deeply overlapping horny scales of snake type; *C,* type of scale present in many lizards, with bony scale underlying horny element. (After Boas.)

margins may cornify as a substitute, producing a **bill** or **beak**. Such structures are, of course, characteristic of the birds as a group, are well developed in the turtles and some extinct reptiles, and are found in a few mammals, such as the monotremes.

Claws, nails, and hoofs are keratinized epidermal structures tipping the digits of amniotes (Fig. 93). The claw is the basal type; nails and hoofs are mammalian modifications. A typical **claw** forms a protection for the top, sides, and tip of the terminal joint of a digit; an inverted V in cross section, the claw becomes increasingly narrow distally and curves downward beyond the tip of the toe. Beneath the claw (except for its projecting tip), there is a germinative layer, protected at its base by a fold of skin; from this matrix the keratinized epithelium continually grows outward over the dermis to be as continually worn away at the tip of the claw. On the under surface, distally, is a pad of softer, less cornified tissue, a **subunguis**, which effects a transition between claw and normal epidermis. A **nail**, as developed in the arboreal, grasping hands and feet of the primates, is essentially a broadened and flattened claw restricted to the upper surface of the finger or toe. In a mammal claw, the germinative "nail bed" is confined to the proximal end of the structure (seen as a white area in the

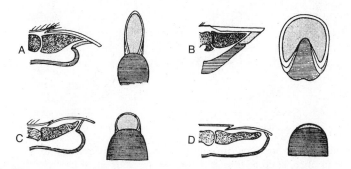

Figure 93. Longitudinal sections and ventral views of terminal phalanges of mammals to show the build of claw, nail, and hoof. Toe phalanges, stippled; subunguis, fine stipple; epidermis of ventral surface of foot, hatched; epidermis of upper surface and horny material of claw, clear. *A,* Claw of a carnivore; *B,* a horse's hoof; *C,* a nail of a typical primate; *D,* a human nail. (After Boas.)

translucent human nail, for example). **Hoofs** are characteristic of the various ungulate mammals, which have reduced the number of toes and walk on the tips of the remaining digits. The original claw has shortened and broadened to become essentially a hemicylinder sheathing the tip of the toe; it is the curved or V-shaped distal end of the hoof that rests on the ground, the subunguis forming a pad within the curve of the hoof.

Horns and hornlike structures are widely distributed, particularly among mammals. Though actual horn, i.e., keratin, is not always present in these structures, the general picture may be discussed here (Fig. 94).

A true **horn** is seen in cattle and is present in other members of the cattle family—sheep, goats, and antelopes. The core of the horn is a spike of bone arising from a dermal bone of the skull. Sheathing and extending this, however, is a hollow cone of true horn substance, formed by keratinization of the epidermis. Neither core nor sheath is ever shed, and these typical horns, although variously curved, are never branched.

Although often called a horn, the **antler** of the deer, almost always confined to males, is quite a different structure. When matured, it consists solely of bone; only during growth is it covered by skin in the form of "velvet"; no actual horn substance is present. As further points of contrast, we may note that an antler is shed annually and tends to be branched, increasingly so in older animals.

Still other types of "horns" are found among mammals. The horn of the prongbuck, like that of cattle, has a horn-sheathed bony core; but in contrast, the horny sheath is branched and is shed annually. The giraffe has short "horns," which are bony projections permanently covered by skin and hair. Again, the "horn" of a rhinoceros

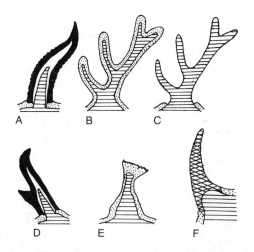

Figure 94. Diagrammatic sections through various horns and related structures. Anterior is to the left. *A,* A true horn, such as that of a cow or sheep; *B,* an antler of a deer in the velvet (the velvet is simply skin over the developing antler—it dies and is rubbed off when the antler is full grown); *C,* the same antler after the velvet is removed; *D,* the "horn" of the prong-buck; *E,* the "horn" of a giraffe; *F,* the "horn" of a rhinoceros. In all cases, the bone of the skull is shown by lines, skin by stippling, horn in solid black, and the odd rhino horn by cross-hatching.

is formed entirely of keratinized epidermis, but this is a fused mass of long modified hairlike epidermal papillae, rather than a typical horn. Comparable structures are seen, although less commonly, in reptiles.

Feathers. The possession of feathers is the distinguishing mark of a bird. They are primarily horny epidermal structures evolved, it is believed, from reptilian scales and perform two major functions in the avian economy. As a body covering, they form an effective insulation, aiding in the maintenance of the high body temperature that is as characteristic of birds as of mammals. Secondly, bird flight is rendered possible through the development of large feathers that form most of the surface of the wing and tails.

Three main types of feathers may be distinguished (Fig. 95): the down feather, the contour feather, and the filoplume. The most familiar (if the most complicated) are **contour feathers**, large feathers that sheathe the body surface and (as the term will be used here) also include the feathers of the wing and tail. A mature feather is formed entirely from greatly modified and cornified epithelial cells. The basal portion of the feather is the **quill** or **calamus**, a hollow cylinder more or less filled by a pithy material—the remains of mesodermal tissue that was present here during the development of the feather. At the base of the feather is a small opening into this cavity, an **inferior umbilicus**, and a similar **superior umbilicus** is present at the distal end of the quill. The quill lies in a **follicle**, a deep cylindric pit surrounded by a sheath of epidermal tissue, sunken into the dermis of the skin.

Beyond the quill lies the exposed and expanded portion of the feather, the **vane**. The axis is continued by the **shaft** or **rachis** (some feathers have a smaller, secondary axis at the plume base, an **aftershaft**). Unlike the quill, the shaft is solid, its interior being filled with a sponge of horny, air-filled cells. Major branches, or **barbs**, extend obliquely out from either side of the shaft. These are so closely connected with one another as to appear to the naked eye to form a continuous sheet. Closer inspection, however, shows that the continuity of the feather surface is effected by tertiary elements, **barbules**, that arise in rows on either side of each barb. By a complicated system of hooks and notches, each barbule interlocks with neighbors on the next barb. Disruption of this arrangement can be repaired by the bird's preening action. In nonflying ostrich-like birds, in which smooth feather contours are unnecessary for

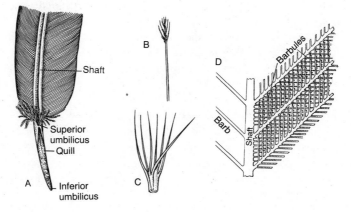

Figure 95. Feathers. *A,* Proximal part of a contour feather; *B,* filoplume; *C,* down feather; *D,* diagram of part of a contour feather, to show interlocking arrangement of barbules. (After Gadow, Bütschli.)

wing surfaces or for streamlining of the body, there is little development of such hooks, and the contour feathers may be fluffy plumes.

Basically similar, but simpler in build, are **down feathers**, or **plumules**. These cover the entire body of the chick and underlie the contour feathers over much of the body of the adult, forming the main insulation. A proximal quill is present in these small feathers as in the contour type. Distally, however, beyond the termination of the quill, there is no shaft, but, instead, a spray of slender simple branches.

Filoplumes are still simpler in appearance. These are small, hairlike feathers with a slender shaft continuing out from the body; they may terminate in a tiny tuft of barbs.

Much of the beauty of birds lies in their varied and attractive coloration. In part this is due to pigments present in the feathers. Dark colors may be due here, as in the epidermis, to pigment transferred from special color-bearing cells in the dermis (described later in this chapter); brighter hues are in part brought about by pigments obtained from the blood stream by the cells of the developing feather. However, much of the color and the iridescent sheen of feathers are not attributable to pigments but to the refraction of light from the horny feather substances; all blues are produced this way.

Feathers are not, of course, uniformly distributed over the bird's body. Contour feathers occur in definite feather tracts, **pterylae**. Most prominent is a long row of large feathers along the back of the "forearm" and "hand," the **remiges**, which form the wing surface; another group of large feathers, the **retrices**, forms the tail. Over a number of body areas, no contour feathers are present, but the arrangement of the pterylae is such that the entire body is completely and smoothly ensheathed. In ratites, neither plumules nor filoplumes are present.

In its initial stages, the development of a feather is comparable to that of a reptilian scale (Fig. 96). Mesodermal tissues, including small blood vessels, gather beneath the site of the prospective feather to form a feather papilla. Above this, the epidermis extends outward into a conical structure with an epithelial surface and a pulp-filled center.

From this point onward, development diverges from that of the horny reptilian scale. The papilla does not remain on the surface; instead, the epidermis surrounding it sinks inward to form the feather follicle, out of which the papilla extends in further development.

In the development of a down feather, we find that the papilla becomes a greatly elongated tubular structure. The living mesodermal tissues of the papilla, a nutritive pulp, extend the length of the tube during development. The epidermal surface gradually takes on a horny character. Proximally, in the future quill region, the epidermal tube is a simple cylinder. Distally, however, a thin outer layer separates as a conical sheath over the inner regions of the epithelium. The inner mass develops as a series of thickened longitudinal ridges. When feather growth is completed, the mesodermal tissue of the distal part of the feather is resorbed, the thin sheath breaks down, and the ridges are freed to become the spreading distal filaments of the down feather. Proximally, however, the tube remains intact as the quill. The mesodermal tissue filling it dries to a pith; the two umbilici of a mature feather are the original proximal and distal openings of this tubular segment.

More complicated, but basically similar, is the development of a contour feather. The feather rudiment still forms an epidermal cone, mesoderm-filled, in the distal part

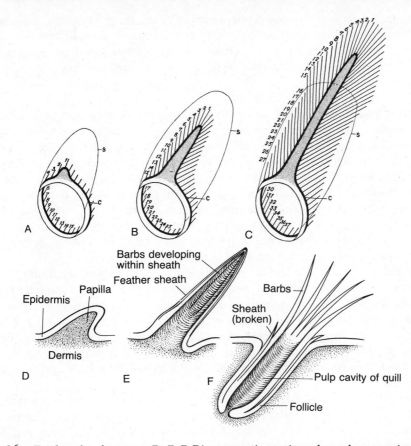

Figure 96. Feather development. *D, E, F,* Diagrammatic sections through successive stages in the development of a down feather. Development begins in the form of a mesodermal papilla; later the structure sinks into a follicle. An outer layer of the ectodermal covering separates as a thin feather sheath. The remainder of the ectoderm forms basally a hollow tube that becomes the quill. More distally, it divides into a number of parallel columns. With rupture of the sheath, these are freed as the barbs. *A, B, C,* Diagrams to show the development of a replacement contour feather; the basic pattern is comparable to that of a down feather. Feather growth begins at a basal ectodermal collar (*c*), from which develop, as in the down feather, parallel columns of tissue within the feather sheath (*s*). One exceptionally strong upgrowth (stippled) becomes the shaft; the parallel columns of tissue migrate successively (as shown by the numbering) onto this to become the barbs. (*A* to *C* from Lillie and Juhn.)

of which a surface layer separates as a thin sheath. One might expect the complicated branching structure of the mature feather to grow in a treelike manner. This is not at all the case, because the entire process of differentiation takes place within the feather sheath as shown in Figure 96 *A–C*. At a later stage, the barbules form from the barbs. When the feather is completely formed within the sheath, this ruptures, and the feather has simply to unroll to attain its mature state.

Feather replacement continues throughout life. At the base of each follicle, the papilla and its epidermal covering persist as a feather matrix; from these materials, new feathers form. In many birds, particularly those of temperate and arctic regions, feather replacement is the seasonal phenomenon of **molting**; in others, there may be a gradual replacement throughout the year. At the base of contour feathers, muscle fibers may be present. Sensory nerve endings at the feather base may register touch, and in night-flying birds, such as the owls, hairlike sensory feathers may be present on the face.

Hair. As an insulating device formed of keratinized epidermis, hair is a mammalian analogue to the avian feather. In other regards, however, the two structures contrast. As will be seen, they develop in a different manner, and, unlike a feather, a hair is a purely epidermal structure, without the mesodermal component seen within the feather during its development. Hairs, unlike feathers, are not modifications of horny scales but are new structural elements of the skin. Hairs probably developed before scales were lost by our reptilian forebears, possibly as special sensory projections. In certain instances in which horny scales are retained in mammals, hairs are found growing in definite patterns between the scales; even when (as usual) scales are absent, the same arrangement of hairs may persist (Fig. 97).

A typical hair (Fig. 98) includes two portions, the projecting **shaft** and the **root**, which lies in a pit sunk, usually at a slant, within the dermis, termed the **hair follicle**; several follicles may, as in dogs, join to produce a common opening. Both shaft and root consist (except at the very base) of essentially dead and heavily keratinized epidermal cells. At its base, however, the root expands into a hollow bulb. Beneath is a dermal **papilla** containing connective tissue and blood vessels. Through the latter come nutrient materials for the sustenance of the cells in the bulb and thus for the growth of the hair.

In the bulb alone, the hair includes typical "living" cells, with a normal proto-plasmic content—the hair **matrix**. Growth and division of these cells results in the development of the hair. As cells are budded off distally in the bulb region and pushed outward by the development of further cells beneath, they become part of the root, and eventually of the shaft section of the hair, and gradually become kera-tinized to form typical hair cells.

Within the hair, two or three layers may be distinguished in section. Externally, a thin **cuticle** consists of a single layer of transparent cells that often overlap in scalelike fashion. Beneath the cuticle is the main substance of the hair, a dense horny material of modified cells containing a variable amount of pigment and air vacuoles. In stout hairs, a thin central area of shrunken cells and large air spaces may be distin-

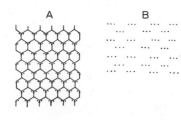

Figure 97. Hair patterns of mammals to show presumed derivation from structures developed in the interstices between scales. *A,* Part of the scaly tail of a tree shrew, with the hair (represented by dots) in this position; *B,* skin of a marmoset, with the hairs arranged in a similar pattern despite the absence of scales. (After de Meijere.)

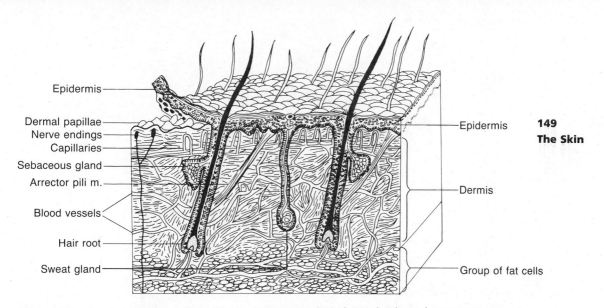

Epidermis

Dermal papillae
Nerve endings
Capillaries
Sebaceous gland
Arrector pili m.

Blood vessels

Hair root

Sweat gland

Epidermis

Dermis

Group of fat cells

Figure 98. Section of mammalian skin, to show particularly hair, glands, and accessory structures.

guished; when present, this is considered a **medulla**, and the main mass of the hair substance is termed a **cortex**. Lining the follicle is the hair sheath, which includes several layers.

Adjacent to the follicle and emptying its oily lubricating material into it may be a sebaceous gland (cf. p. 152). Each hair is further provided with a small **arrector pili muscle**, composed generally of smooth muscle fibers. Attached at one end to the superficial part of the dermis, the muscle slants inward to insert into the hair follicle deep down at the side away from which the hair inclines. Contraction of this muscle (under the control of the autonomic nervous system) erects the hair (and by its pull on the skin causes "gooseflesh"). Of little avail in such a thinly haired form as man, this potentiality of raising and depressing the hair can bring about, in a thickly furred animal, marked changes in the insulating powers of the coat.

In the development of a hair (Fig. 99), there is no formation of a mesoderm-filled papilla such as we find initially in feather development. Instead, a solid column of epidermal cells grows downward into the dermis. At the bottom of this column, a thick mass of these epithelial cells develops as a hair germ; a hollowing of its base allows the development of a vascular papilla. Above, the column of epithelial cells that had descended from the surface hollows out to form the epithelial walls of the follicle, and growth of the hair upward from its germinal base takes place in this tube.

Once developed, a hair is not (any more than a feather) a permanent structure; throughout life, most hairs are cast off and replaced, either as a gradual process or as a seasonal shedding of much of or the entire coat. Resorption takes place at the bulb, and a new hair develops in the same follicle from the matrix cells surrounding the papilla.

Although hair may serve a variety of special functions, its main use is as an insulating and thermal regulating material. It is perhaps difficult for us, as members of a

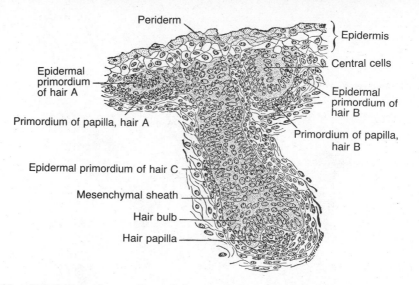

Figure 99. Hair follicles from a 3-month human embryo, showing successive stages of development at *A, B,* and *C.* (From Arey.)

mammal group in which hair is much reduced, to appreciate this fact fully; but it must be remembered that, in mammals, generally the hair forms a thick covering over almost the entire body. It is a furry coat, containing numerous "dead" air spaces that give it many of the attributes of commercial rockwool insulation. In connection with its insulating qualities, we may note the seasonal shedding of hair in mammals of temperate and arctic regions, whereby a thicker coat is grown for winter, a thinner one for milder summer weather.

Most hairs contain pigment to some degree, and hair color may be independent of that of the skin. Melanin, taken over from dermal pigment cells, is the common substance, producing in different concentrations various shades of brown and black; a related pigment is responsible for reddish tinges. Air bubbles present in the hair will lighten the intensity of coloration; when abundant, air spaces will turn darker colors to gray and in the absence of pigment will produce a silvery white.

Loss of hair color with age in an animal is due to reduced pigmentation and to increase in the air content of the hair. Coat color may change notably with the seasons when the hair is shed and replaced, as in the white winter coats of various arctic animals, admirably adapted to concealment in the snowy landscape and contrasting sharply with the dark summer pelage of the same species.

Mammalian hair is, of course, highly variable in a number of ways. It varies from form to form, between one area of the body and another, from one season to another, from youth to old age. The hairy covering in many mammals is a thick fur. In others, such as the higher primates that are primarily tropical forms, it is a much thinner and uneven coat. In whales, which depend on fat rather than hair for insulation against the cold of sea water, it may be entirely absent in the adult, and the sea cows are almost hairless. An animal may have a double pelage, consisting of a sparse outer set of long coarse hairs and an insulating undercoat of much more numerous finer hairs. Hair is generally arranged in tracts, in which all hairs have a slope in the same direction; these directions appear to be related in various cases to the manner in which the animal combs or scratches its pelage or to the better shedding of water.

Specialized types of hair are found in many instances. Each hair has a nerve plexus at its base, but various mammals may develop **vibrissae**, special tactile organs such as the cat's "whiskers" on the face and, particularly in arboreal types, near the carpus and tarsus. There is some evidence for such vibrissae, even in mammal-like reptiles, and they may still carry on the original function of hairs. Hairs that are rounded in cross section tend to be straight and erect, and, if the hairs are stoutly built, may develop as bristles or protective spines as in the hedgehog or porcupine. Hairs that have an oval or flattened section are more readily bent; their presence may result in the production of a curly or woolly covering.

Skin Glands. Glandular structures are developed in the epidermis in every vertebrate class. In fishes, they occur mainly as discrete, mucus-secreting cells widely distributed over the surface of the body, but serous cells, secreting a watery material, are common in elasmobranchs and are found in a few teleosts. The mucous cells are responsible for the slimy feel of fish skin and aid in insulation and in a sealing off of the body surface; it may also function to decrease turbulence in the flow of water over the surface. In the hagfishes, mucus-forming glands are abundant, and it is said that a single one of these animals can turn a whole bucket of water into a jelly of slime.

In some cases, formed glands are present in the fish skin, generally serving some special function. **Poison glands** are found associated with spines on the fins, tail, or gill cover in several sharks and chimaeras and a number of teleosts.

A most unusual development in various deep sea fishes is that of luminous organs, **photophores**, formed by modified mucous glands (Fig. 100). In some instances, the light is produced by the symbiotic phosphorescent bacteria present in the organ; in others, complicated chemical processes of oxidation in the glandular material appear to result in the production of a "cold light." In many teleosts, these organs develop accessories, including a pigment-backed reflector behind the light

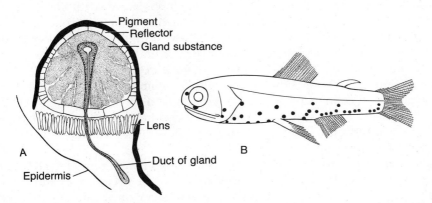

Figure 100. *A,* Section through a light organ of a teleost (*Cyclothone*). Part of the duct leading in from the body surface is seen. *B,* Light organs of a small teleost (*Myctophum*). The organs are in black. (After Brauer.)

organ and a lens in front, giving the photophore much the build of an automobile headlight. They are not, however, designed to illuminate the fish's pathway. Recognition, particularly at mating time, may be important. Frightening enemies, on the one hand, and luring food on the other, may enter the picture (one deep sea angler fish dangles a phosphorescent lure, formed from the dorsal fin, above its gaping mouth).

In water-dwelling amphibians, mucous production persists. Here, however, we are dealing not with individual mucous cells, as in most fishes, but with formed **mucous glands**, although glands of a simple alveolar nature (Fig. 89 *B*). A second type of gland is also frequently present in amphibians: the **granular glands**. These glands owe their name to the granulation in the protoplasm of the secreting cells. The varied products are poisonous, ranging from mildly irritating substances to alkaloids that may be highly toxic. The common toad, for example, has a large gland of this type on either shoulder.

When the reptilian stage is reached, the primitive mucous cells have disappeared, and glands of any kind are rare in the hard dry skin, although there are various scent glands used in intraspecific behavior. In birds, too, skin glands are rare; the only conspicuous gland is the **uropygial** or **preen gland**, a large, compound holocrine alveolar structure present in most birds on the back above the base of the tail. Its function appears to be the supplying of an oily "water-proofing" material used by the bird in preening its plumage.

In mammals, other types of gland make their appearance, seemingly *de novo*. Wax-producing glands are developed within the mammalian ear pinna (and also in the ear canal of a few birds—the turkey, for example). Alveolar **sebaceous glands** are typically associated with hair follicles (cf. Fig. 98), but may persist in regions in which hair is absent; their oily secretion, produced through cellular breakdown, is a protection and lubricant for hair and skin. **Sweat glands** have a simple tubular structure; a much elongated tube coils in a complicated manner in the dermis (Oliver Wendell Holmes, Sr., who was an anatomy professor as well as an author, compared them to fairies' intestines). The exocrine secretion of the sweat glands contains salt, urea, and other waste products. These structures thus act as accessory kidneys, and we are familiar with the fact that excessive sweating results in a considerable loss of salt from the body. Two types of sweat glands may be found. The smaller type, predominant in human skin, gives off a watery secretion; larger sweat glands, common in many mammals and typically connected with hair follicles, produce a thicker, milky, and odoriferous fluid. The evaporation of sweat on the skin is a cooling device important in the temperature regulation of many mammals. The excretion of sweat is aided by the contraction about the gland of small cells that function like smooth muscle but appear to be of ectodermal origin. In climbing forms, high development of sweat glands on palm and sole aids the grip. In carnivores, generally, sweat glands are much reduced in number; the panting of a dog utilizes salivary evaporation on the tongue as a substitute. Skin glands are reduced in many aquatic animals and are absent in whales and sirenians.

Many mammals possess special scent glands, important in reproductive and other behavior. They may be derived from either sebaceous or sweat glands and occur in a variety of places: on the face (some antelope and deer), on the feet (sheep and various other ungulates), around the anus (most carnivores, such as skunks), on the back (hyraces), and elsewhere.

Still another type of mammalian gland—one to which, in fact, the class owes its name—is the **mammary gland**, inactive in males but in females secreting a nourishing milk, of casein, lactose, fats, and salts, for the young. No antecedents for these are found in lower classes; it is thought, from their construction, that they are specialized derivatives of the sweat glands.

There is considerable variation in the structure of mammary glands. In monotremes, there are simply two bundles of discrete glands in the abdominal wall that discharge a sticky secretion into depressions on the belly surface, to be licked off by the young; they are functional, exceptionally, in both sexes. In all higher mammalian groups, there are well-formed mammae, with the gland opening through projecting nipples or teats from which the fluid is sucked by the young. Active milk glands typically have the structure of clusters of alveoli, from each one of which a duct leads toward the surface. Properly, the term **nipple** should be applied to types in which the ducts lead directly to the tip; in a teat (as in the familiar cow), the ducts open into a common storage cavity, from which a large duct leads to the surface. Embryonic development of these glands begins with the appearance of a pair of longitudinal swellings, **mammary ridges** (Fig. 101), running along the length of the trunk on either side. Concentration of tissue at specific points on the lines results in the eventual development of the mammary glands. They develop further in the female when sexual maturity is reached and become functional at the time of parturition, the development being under hormonal control. The number is generally correlated with the number of young produced at a birth; they range from a single pair up to half a dozen pairs or more. Their position is likewise variable. In marsupials, they are generally enclosed within an abdominal pouch in which the young are carried after birth, each one tightly clamped to a nipple. In many ungulates, the mammae are abdominal in position; in higher primates, on the other hand, they are in the pectoral region; forms with large litters (pigs, many carnivores) usually develop two long rows of nipples.

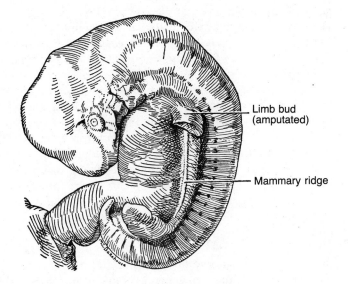

Figure 101. Mammalian embryo, to show mammary ridge or "milk line." (From Arey.)

Limb bud (amputated)

Mammary ridge

Dermis. Thicker but less varied in structure than the epidermis is the deeper layer of skin, the dermis. Basically this consists, in most vertebrate groups, of a tightly felted fibrous mass of connective tissue, derived from embryonic mesenchyme (cf. Fig. 98). It is this part of the skin of various animals that, after treatment, becomes commercial leather (a fact that the alternative name of **corium** for this layer implies). The dermis is an effective insulating material, and its flexible yet unyielding nature makes it a major defense against injury. The deeper part of the dermis is generally loose in texture, frequently containing fatty deposits; there is often a practically imperceptible transition to the connective tissue covering the muscles or other deeper organs. The skin may thus be closely bound to the underlying structures, whereas, in other cases, the skin may be quite loose; compare, for example, the skin on the palm and that on the back of the human hand.

In amphioxus, the dermis is, however, of a different (and perhaps more primitive) nature; it consists mainly of a layer of gelatinous material containing few fibers except on its inner and outer borders. In most bony fishes, too, we find a different kind of dermis from that described above. Some loose connective tissue is present, but in considerable measure it is replaced by more specialized mesenchyme derivatives: bony dermal scales or plates (cf. Fig. 89). These structures, forming a defensive armor of a harder and stiffer type than that offered by the normal connective tissues, are part of the external skeleton, and as such are described in Chapter 7. Except for parts of the skull, this dermal armor is much reduced or absent in most terrestrial vertebrates (turtles and armadillos form conspicuous exceptions), and it is likewise absent in the cyclostomes and sharklike fishes (except for the small, toothlike denticles embedded in the skin of the latter).

Offhand, one would assume that the more common condition of the dermis in modern vertebrates—that in which it is fibrous—is primitive and that the presence of bone in the dermis is secondary. The history of vertebrates suggests, however, that the reverse is the case. The evidence is definite in the terrestrial vertebrates; tetrapods are descended from fishes in which the dermis consisted mainly of thick bony scales, and stages in the reduction of the bony armor to the modern fibrous condition of the skin are observed in the paleontologic record of early amphibians and reptiles. For lower fish groups, the evidence, although not absolutely conclusive, indicates that here too the ancestors had bony dermal scales, for the oldest known vertebrates had an armored skin. A bone-filled dermis was, it would seem, a primitive vertebrate condition; the leathery type prevalent today was arrived at as the result of bony reduction.

Although collagenous fibers form the main bulk of the typical dermis, other materials are present here as well. A certain (though small) percentage of the fibers are of the elastic type. Cellular components are present among the fibers, notably fibroblasts (p. 160), by which the tissue is formed. The deeper layers of the dermis are, in mammals, a major locus for the development of fatty tissues, termed the **panniculus adiposus**. Fat is an excellent insulating material; in whales, as the thick "blubber," it substitutes for the absent hair. Down into the dermis project, in higher vertebrates, the basal portions of glands and of feather or hair follicles. Smooth muscle fibers may develop in the dermis, and striated muscle tissues, derived from underlying body muscles, may attach to the under surface of the skin as a **panniculus carnosus**. The sensitivity of the skin is due to the presence in the dermis (less commonly in the epidermis) of nerve fibers; some end freely, but most terminate in sensory corpuscles, often numerous, generally situated in the outer part of the dermal layer.

Circulatory vessels are, in general, abundant—lymphatics, small arteries and veins, and capillary networks of complicated pattern. The blood vessels supply nutriment not merely to the dermis, but also, by diffusion of materials through the tissues, to the overlying epidermal cells. In vertebrates with a thick epidermis, its under surface is irregular and may be deeply indented by an out-pushing of **dermal papillae** (cf. Figs. 90 and 98), which bring blood vessels in closer relation to the epidermis as well as affecting a closer union of dermis and epidermis. In forms with moist skins, the presence of this rich vascular supply provides the possibility of exchange of materials with the surrounding medium. Thus, the skin may be used as an auxiliary breathing organ in many amphibians and some bony fishes; certain salamanders and eels, indeed, are able to supply their entire oxygen need in this way.

Temperature Regulation. Proper functioning of the body in vertebrates can take place only at internal temperatures ranging from close to freezing to about 48°C (118°F). In fishes, amphibians, and reptiles, internal temperature tends to vary with the external temperature, with consequent climatic restriction of situations in which many of these forms can live. Because water temperatures generally lie well within the limits mentioned above, internal regulation is of relatively little concern to aquatic vertebrates. However, in fast-swimming fishes, deep body temperatures may be considerably higher than superficial ones because of circulatory arrangements such as retia mirabilia, discussed on page 455. Also, in many reptiles there is a tendency for regulation, partly perhaps by internal mechanisms but mainly by the animals' taking advantage of environmental variations—such as the way in which reptiles seek shade and sunlight, for example. Some extinct reptiles probably had regulatory mechanisms as effective as our own. Among mammals, the internal temperature is highly irregular in the archaic monotremes, and somewhat variable in marsupials, edentates, and bats, but in mammals generally body temperatures are so regulated that the internal temperature varies little from a norm, which is usually somewhat below 38°C (100°F); in birds, temperatures range up to 43°C (109°F).

Mammals and birds are said to have a **homoiothermous** condition, contrasted with the varying **poikilothermous** state of lower vertebrate groups. Regulation is under the control of a center in the hypothalamic region of the brain stem that is sensitive to heat and acts as a neural "thermostat." At least one reptile also has neurons sensitive to temperature in the hypothalamus. Internal temperature depends upon the amount of heat created by metabolic processes and upon the amount lost (or gained) by the body. Most of the heat loss in mammals (some 70 percent in man) is through the skin, which is thus of the greatest importance in regulation. It may be noted that in a small animal, the skin surface is relatively larger in proportion to body bulk than in a larger form; size, therefore, is a factor in heat conservation. In general, the larger members of any mammalian species are those that live in the colder climates. Hair tends to be reduced, because it is not as necessary for insulation in such large tropical mammals as elephants and rhinoceroses. On the other hand, such forms as shrews and small bats find it difficult, even with the best of fur, to maintain body temperatures without great expenditures of energy. In mammalian hibernation, the hypothalamic thermostat is, so to speak, readjusted, and body temperatures drop sharply; however, control is normally maintained.

The skin is to some extent a static insulating material in every vertebrate, but in birds and mammals, the heat loss can be regulated by various devices. Feathers and hair are adjustable insulators produced by the epidermis, and evaporation of liquid from the sweat glands produces a cooling effect. The dermis supplies an important factor in regulation in its abundant blood vessels. Under control of the autonomic nervous system, the dermal arterioles can be dilated, the skin flushed, and heat lost rapidly; with constriction of the arterioles (and a blanched skin), a major saving in heat is effected. Color changes between light and dark shades aid in the limited heat control available to lower vertebrate classes.

Chromatophores. The functional "reasons" for the varied colors and color patterns of animals and the relative importance in their development of camouflage, warning, mating, attraction, or even pure evolutionary "accident" have been warmly debated and need not be discussed here. In birds and mammals, feathers and fur are mainly responsible for surface coloration. In lower vertebrate groups, however, skin color is due almost entirely to special color-bearing cells, the **chromatophores**, located in the outer part of the dermis (Fig. 102). The chromatophores contain numerous granular elements. They are typically stellate, with elongate cell processes, and are capable of amoeboid movement. Common types include (1) **melanophores**, with the dark brownish pigment melanin, (2) **lipophores**, with red to yellow carotenoid pigments, and (3) **guanophores** (or **iridocytes**). The last contain not pigment, but crystals of an organic substance, guanine, which by reflection may alter the effect of the pigmented materials present. Almost all the varied colorations seen in fishes, amphibians, and reptiles are due to chromatophores of these three types, present in varied numbers and in varied arrangements. Although they lie in the dermis, chromatophores are, oddly enough, derived as migrant cells from the neural crest.

The chromatophores are capable of producing striking color changes in a variety of fishes, amphibians, and reptiles (particularly lizards). The chameleon is proverbial in this regard, and the flounders are equally remarkable in the variety of color and color patterns that they are able to display. In part, the color changes are attributable to shifts in the relative position of the chromatophores in the skin, one or another type becoming more broadly exposed at the dermal surface or more fully masked by neighboring cells. Mainly, however, the color changes are due to changes in the distribution of the color granules within the individual cell. If the granules are dispersed

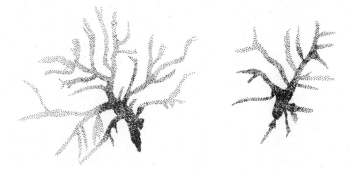

Figure 102. Melanophores from the skin of a larval urodele. In these, the pigment is dispersed throughout the cell, so the animal would appear dark.

throughout the cell, the effect is maximum; if they are concentrated in a tight cluster, little color appears.

In addition to their role in pigmentation of the dermis, the melanophores are responsible for dark pigments in the epidermis of birds and mammals and in feathers. The melanophore cell bodies remain in the dermis but produce long slender processes that gain contact with cells in the epidermis or in developing feathers and transfer pigment particles to them.

Chapter 7

Supporting Tissues—
The Skeleton

Most of the functionally "active" tissues of the body are epithelia or (as in the case of the central nervous system, liver, or muscles) tissues derived from epithelia. But if the vertebrate body were composed solely of epithelia and epithelial derivatives, it would be a flabby and amorphous mass. Materials are needed to back up and reinforce epithelia, to weld them together into a formed body, to protect this body, and, particularly in nonaquatic forms, to give strength and support. Such materials are described in this chapter. The notochord is a distinctive structure of this sort. Connective tissues are widespread and important. Most prominent in vertebrates are the cartilages and bones that compose the skeleton.

Notochord

The notochord is an ancient structure, present even in lower chordates such as amphioxus and the tunicates. As has been seen, it is derived from cells that fold in at the dorsal lips of the blastopore or at the corresponding position at the anterior end of the primitive streak, and, pushing forward, come to lie beneath the neurectoderm. It soon differentiates, along the median axis, from adjacent regions of the mesoderm.

The notochord is always well developed in the embryo, and, in many lower vertebrates, it continues little changed throughout life. It is prominent in adult cyclostomes (in which vertebrae are little developed) and even in many bony fishes. There it performs, as in amphioxus, the function of preserving body shape during locomotion, acting as an antitelescoping device (Figs. 17, 121). In all vertebrate embryos, it extends anteriorly to the region just behind the hypophysis and infundibulum of the brain; posteriorly, it extends to the terminus of the fleshy part of the tail. Its cells are soft and gelatinous; the notochord, however, is surrounded and made firm by a cylindric sheath and an external membrane, apparently of its own making. In lower vertebrates with a large and persistent notochord, this sheath is of considerable thickness; it renders the whole a relatively strong yet flexible structure, resembling a slender, elongate sausage.

In more advanced vertebrate groups the notochord is progressively replaced functionally by the central elements of the vertebrae, which develop around it and give greater strength to the back. As this replacement occurs, we find a correlated reduction of the adult notochord. In many fishes and in ancient fossil amphibians and a few reptiles, the notochord is present in the adult as a continuous, unbroken structure running the length of the body. It is, however, generally greatly restricted by the vertebrae; it may expand between successive vertebral centra but constrict within each segment, so that its contours resemble those of a series of hourglasses set end to end. In most tetrapods, even this tenuous connection between successive segments fails, and the notochord is represented in the adult only by gelatinous materials that may persist in intervertebral discs between successive centra of the vertebral column.

Connective Tissues

Even in such lowly invertebrates as the coelenterates, which lack a true mesoderm, there is generally interposed between inner and outer layers an intermediate zone of more or less gelatinous material, sometimes fibrous and containing sparsely distributed cells; this material aids in filling out the form of the body. Such materials are comparable in a sense to the mesenchyme of the vertebrate embryo's mesoderm. In the adult vertebrate, the most direct products of the mesenchyme are the connective tissues, which form the "stuffing" of the body and reinforce the epithelia of many of the body organs. The simplest form of connective tissue is of a loose, **areolar** type (Fig. 103). Much of it is a gelatinous protein ground substance. It contains a loose network of small branching **reticular fibers** and, more prominently, long and slender

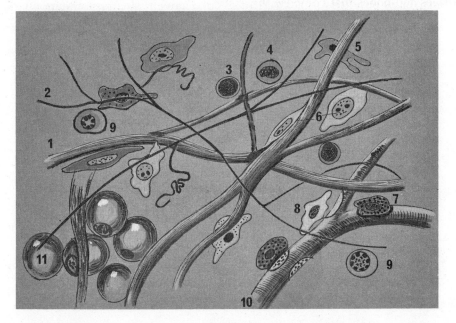

Figure 103. Diagram showing the types of cells present in loose (areolar) connective tissue. (1) Collagenous fiber, (2) elastic fiber, (3) lymphocyte, (4) monocyte, (5) macrophage, (6) fibroblasts, (7) mast cell, (8) undifferentiated mesenchymal cell, (9) plasma cell, (10) capillary, (11) fat cells. (After Leeson and Leeson, Histology, W. B. Saunders Co., 1976.)

white **collagenous fibers**, flexible but inelastic structures that are formed by the spindle-shaped or stellate connective tissue cells, the **fibroblasts**; each fiber is made up of a bundle of very small fibrils that are formed of long molecules of the protein **collagen**.

Another abundant form of connective tissue is the **compact** type, such as the dermis, in which there is a densely packed mass of interlacing fibers, producing a feltlike structure. Most connective tissues contain a small percentage of coarse, yellow **elastic fibers**; in some instances, the elastic type of fiber is dominant, producing **elastic tissues**. **Tendons**, forming the attachment of many muscles, consist of parallel bundles of collagenous fibers, which at one end are tightly bound to the connective tissue of the muscle and at the other may be closely attached to cartilage or bone, into which the fibers may penetrate. Some actively moving tendons are, further, surrounded by a fibrous connective tissue sheath. Where tendons play over the surface of a bone a **bursa**, a liquid-filled sac, may be formed beneath the tendon by the connective tissues. **Ligaments** are comparable to tendons but unite skeletal elements; **fasciae** are sheets of connective tissue investing muscles or other objects.

We may further note the presence in connective tissues of **mast cells**, irregularly oval in shape with deeply staining granules. They contain supplies of the anticoagulant heparin, the basic amine histamine, and, in some but apparently not in all animals, the vasoconstrictor serotonin, but their function is still not completely understood. Still further, macrophages (histiocytes), important in body defense, may be found. These will be discussed in connection with the circulatory system (p. 447). Blood cells may also be present, notably lymphocytes (p. 445).

Fatty, or adipose, **tissue** is a variant of connective tissue most commonly devel-

160
Chapter 7
Supporting
Tissues—
The
Skeleton

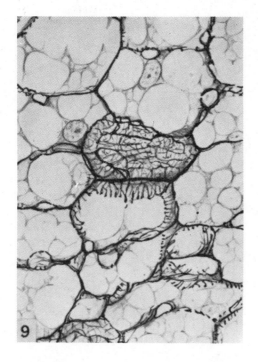

Figure 104. Brown fatty tissue from a rat. Fat-filled cells, with the fat represented by spaces in this preparation, are separated by fibers in the connective tissue. (From Fawcett.)

oped beneath the skin or in mesenteric folds among the abdominal organs. Fat droplets form and join together within modified mesenchyme cells, which become greatly distended. Most fat is whitish, but some is darker. This brown fat is used by the animal to produce heat and is most prominent in newborn or hibernating mammals, in which it forms distinct masses, often called **hibernating glands**, underneath the skin or in other localized areas (Fig. 104).

Skeletal Tissues

From a physiologic point of view, the skeleton is sometimes thought of as a relatively inert system. From a broad functional viewpoint, however, it is of the greatest importance. Evolved, in ontogeny and in phylogeny, from the connective tissues, or the mesenchyme that precedes them in the embryo, the hard skeletal structures are vital in supporting and protecting the softer organs and helping to maintain proper body form. Almost all the striated musculature attaches to the skeleton, which is hence the agent through which bodily movement is accomplished. Still further, the seemingly inert nature of bones and cartilages as viewed in a skeletal preparation is deceptive, because not only the cellular materials but also the apparently "dead" matrix surrounding them are centers of continual biochemical action; and throughout life bones undergo constant remodeling, both internally and externally, in correlation with mechanical strains and stresses.

Cartilage. Two skeletal tissues are characteristic of vertebrates: cartilage ("gristle") and bone. Although both are specialized derivatives of the connective tissues and arise from mesenchyme, they differ markedly in nature, in mode of origin, and (frequently) in position.

Typical **hyaline cartilage** (Fig. 105) is a flexible, rather elastic material, with a translucent, glasslike appearance. Its ground substance, or matrix, is mainly a sulfated polysaccharide (chondromucoprotein) which forms a firm gel containing a network of connective tissue fibers. Scattered spaces, lacunae, contain cartilage cells, **chondrocytes**; except in some lower vertebrates or in early stages in formation in higher forms, these are typically rounded and without the branching processes characteristic of bone cells. They are isolated within the matrix that they have secreted. Blood

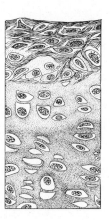

Figure 105. Section through part of a cartilage (from the sternum of a rat). The surface layers (at the top) show a fibrous condition transitional to the perichondrium. (After Maximow and Bloom.)

vessels are absent except in very large cartilages; the nutriment supplied to the cells must thus reach them by diffusing through the ground substance. The outer surface of a cartilage is covered by a layer of dense, cell-containing connective tissue, the **perichondrium**, the inner cells of which may transform into chondrocytes and thus add to the diameter of the cartilage.

There are numerous variants from ordinary, hyaline cartilage. Some fishes have a type of cartilage in which there is little matrix between the cells. At the opposite extreme, in animals such as sharks in which the adult skeleton is largely cartilaginous, there is a strong tendency for the development of **calcified cartilage**. A deposition of calcium salts in the matrix produces a relatively hard, brittle, and opaque substance superficially somewhat similar to bone. This gives a stouter structure; however, calcification prevents nutriment from reaching the enclosed cells, resulting in their death and the danger of a breaking down of the tissue. Calcification precedes the replacement of cartilage by bone during the ossification of deep-lying bones. **Elastic cartilage** (seen characteristically in our external ears) gains flexibility through the presence of many elastic fibers in the ground substance. **Fibrocartilage** is a form of material transitional between a dense connective tissue and cartilage; it is often present in the neighborhood of joints and the attachments of ligaments and tendons and forms the thin **intervertebral discs** between the vertebral centra of mammals. **Chondroid tissue**, found in larval lampreys and to some degree in sharks, is a relatively soft and gelatinous cartilage-like tissue, with branching cells.

In cartilage formation, the irregularly shaped mesenchyme cells become more spherical and develop between themselves the interstitial ground substance and fibers. At first closely packed, the cells tend to separate as more material is laid down around each cell. Frequent cell divisions are seen in growing cartilage; we may find cells grouped in pairs or quartets, indicating their origin by division from a single parent cell. Cartilage may grow by the addition of new materials to its outer surface, formed by cells derived from the perichondrium. But it is also capable of growth, particularly when newly formed, by internal expansion—interstitial growth—a very important character of cartilage as contrasted with bone.

Cartilage is essentially an *internal* skeletal structure. With few exceptions (e.g., the mammalian ear pinna), it is never present in the skin or near the surface of the body but is, rather, characteristic of deep-lying parts of the skeleton. It is abundant in the embryo and young of vertebrates. In the higher groups of living vertebrates and in many of the older fossil types of lower vertebrates, however, the adult skeleton is in great measure bony, and cartilage is relatively reduced. Only in living lower groups—cyclostomes, Chondrichthyes, and a few Osteichthyes—is cartilage the major skeletal material in the adult.

Cartilage is a deep tissue, an embryonic tissue, and a relatively soft and pliable and readily expandable tissue.

Bone (Fig. 106). Bone is the dominant skeletal material of the adult in most vertebrate groups. Like cartilage, it is a derivative of the mesenchyme and consists of transformed mesenchymal cells enclosed in a ground substance, containing connective tissue fibers. The two materials differ, however, in numerous ways. The bone matrix rapidly becomes a hard, opaque, calcified material. Laid down in it are salts in which phosphate and carbonate are combined with calcium in complex fashion, forming the mineral hydroxyapatite, the principal substance being calcium phosphate.

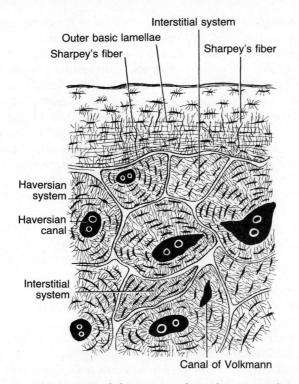

Interstitial system

Outer basic lamellae

Sharpey's fiber

Sharpey's fiber

Haversian system

Haversian canal

Interstitial system

Canal of Volkmann

Figure 106.   Bone structure. A ground thin section through a mammalian metacarpal. Toward the top (outer) margin are parallel lamellae of bone formed from the periosteum; within are a number of haversian systems cut at various angles. The "interstitial system" includes remains of earlier formed bone layers not destroyed when the present haversian systems were created. A cementing substance binds the different areas of bone together. "Volkmann's canals" carry blood vessels from the surface or bone marrow cavity to haversian systems. Sharpey's fibers are connective tissue fibers, which run from the periosteum inward through the bone substance. (After Maximow and Bloom.)

Enclosed within the matrix are the bone cells, the **osteocytes**, which have secreted it. They differ markedly from cartilage cells, in that they (and the lacunae in the bone in which they are situated) are irregular, star-shaped structures.* In early stages, the bone cells have long branching processes that continue outward from the lacunae in all directions, in tiny **canaliculi** that reach neighboring lacunae; later, the protoplasmic processes may be retracted, but the canaliculi persist. In contrast to cartilage, bones are penetrated by blood vessels. The solid matrix of the bone is impervious to nutritive materials, and the cells receive sustenance by the transfer of food materials from the blood vessels via the canaliculi. The external surface of bone (like that of a cartilage) is covered by a layer of dense connective tissue, the **periosteum**, the deeper portion of which may, when required, furnish cells for additional bone formation. In a mature bone, the periosteum is bound to the bone by bundles of fibers (**Sharpey's fibers**), which penetrate the bone surface.

*In many teleosts, particularly the advanced spiny-finned forms, cells may be lacking in the mature bony tissue; a similar condition was present in some ancient ostracoderms.

Whereas a cartilage is generally rather uniform in appearance when seen in section, bones have a complicated internal structure. Many areas (particularly the outer portions of bony elements) consist of **compact bone**, a hard mass in which openings of microscopic size only are present. **Spongy (cancellous) bone**, generally present in the interior of large skeletal elements, contains a framework of bars (trabeculae) of bony material, enough to preserve the shape and rigidity of the structure; but much of its area is occupied by vascular, fatty, or other tissues that may form a **bone marrow**. The development of this type of bone aids in lightening the skeleton, and the marrow itself has positive functions (fat storage, blood cell formation). The trabeculae tend to line up with the lines of stress in the bone and presumably supply an optimal combination of strength and elasticity for the amount of material present; however, the subject is complex and we lack the space to review it here.

Much of the substance of any bone is laid down during the development in the form of layers—**lamellae**—and a lamellar arrangement is common in much adult bone, particularly surface layers. However, bones are continually reworked throughout life by resorption of material and redeposition of new bone, so that the histologic pattern is often highly complicated. Large multinucleate cells, **osteoclasts**, are found in areas of resorption. They apparently arise from monocytes (see p. 445) and are involved in resorption of bone, although the exact mechanism is not understood. Characteristic in bone remodeling of higher vertebrates is the development of **haversian systems**, consisting of small **haversian canals** containing blood vessels and nerves, which branch through the bone. Surrounding them are concentric cylinders of bone, the cells of which obtain their nutriment from the canal. The formation of haversian systems is preceded by the destruction of the bone materials present in a given region; tubular channels are formed, and bone is redeposited in concentric layers inside their walls.

One customarily thinks of bone as a static, inactive material. This is far from true. As just noted, bone is constantly being reworked morphologically. Further, the bones are highly vascular and serve as major storehouses of calcium and phosphate—substances most important for body metabolism.

Bone Development. Bone arises through the activity of mesenchyme cells that become **osteoblasts** and then form about themselves a thick matrix containing numerous fibers. The rapid deposition of calcium salts in this intercellular material completes the formation of bone, and the osteoblasts, their main activity being over, become the osteocytes.

Two radically different modes of bone formation—i.e., **ossification**—are to be seen in the embryos and young of vertebrates. The simpler of the two is that in which the bone forms directly from mesenchyme. This occurs in the dermal bones formed in the skin (Fig. 107). A group of mesenchyme cells in the dermis takes on the properties of osteoblasts and lays down between them thin, irregular membranes or plates of dense fibrous interstitial substance in which bone salts are presently deposited. Such plates gradually expand at their margins and become thickened by the deposition of successive layers of additional bone on inner and outer surfaces. In bony fishes, dermal bones are usually formed over almost the entire body (including the oral cavity), in the form of large plates anteriorly, and of bony scales over most of the trunk. In tetrapods, the extent of dermal bone formation is much restricted, because bony scales tend to disappear; in birds and mammals, dermal bones are normally found only in the skull, jaws, and shoulder girdle. In adult form, platelike der-

164
Chapter 7
Supporting
Tissues—
The
Skeleton

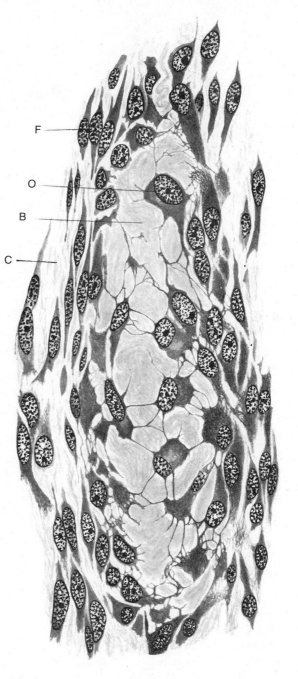

Figure 107. An early stage in the development of a dermal bone in the cat skull. *B,* Homogeneous thickened collagenous fibers, which become the interstitial bone substance; *C,* collagenous interstitial substance; *F,* fibroblasts; *O,* connective tissue cells with processes. These last cells become osteoblasts and later osteocytes. (From Bloom and Fawcett.)

mal bones usually consist of inner and outer layers of compact bone with an intervening reworked layer of spongy bone. Dermal bones, with rare and specialized exceptions in higher vertebrates (mammalian jaw articulation for one), lack cartilage completely.

A similar mode of ossification is seen in **membrane bones,** deeper or endo-skeletal elements that are entirely bony throughout their development. Such elements

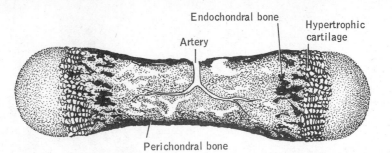

Endocondral bone

Artery

Hypertrophic cartilage

Perichondral bone

Figure 108. Section of a mammalian embryonic metapodial, in which ossification is taking place in the shaft (bone in black). Perichondral ossification is taking place superficially. At either end is normal cartilage; toward the center of the shaft, the cartilage becomes "hypertrophic"; the cells are swollen and arranged in rows; calcification occurs, followed by replacement of the cartilage with dermal bone.

166
Chapter 7
Supporting
Tissues—
The
Skeleton

may represent a secondary simplification with loss of the embryonic cartilaginous stages of elements such as the ribs in some teleosts or be new "inventions" such as various heterotopic bones. Dermal bone and membrane bone are often used synonymously, but such usage confuses two quite distinct parts of the skeleton.

Quite different and much more complicated is the formation of **endochondral bone** (Figs. 108, 109). This is primarily the replacement of an embryonic cartilage by an adult bony structure. However, even here, much of the bone is really laid down directly in membranous fashion external to the embryonic cartilage.

In such typical internal structures as the long bones of the limbs of tetrapods, the cartilage tends to assume the definitive form of the adult bone at an early stage and tiny size. Presently the cartilage begins to undergo modification and degeneration near the middle of its length; the chondrocytes swell and arrange themselves in longitudinal columns; the material between these columns calcifies; the cartilage cells die. Blood vessels break into the cartilage from the surface, and destruction of most of the cartilage of this central area takes place. Osteoblasts that enter with the blood vessels lay down bone in place of the destroyed cartilage. Ossification proceeds from the center toward either end of the element.

If the cartilage failed to grow, its replacement by bone would be accomplished in a short time, and a completely ossified bone of tiny size would result. But this does not occur, because the cartilage grows at either end about as fast as it is destroyed centrally. The cartilage, so to speak, leads the osteoblasts a long "stern chase"; indeed the process of ossification may never catch up with the cartilages at either end of the element. With complete ossification, growth stops; because, at their ends, internal skeletal elements are usually articulated with their neighbors, and bone cannot be readily added to the articular surfaces without damage to the joints.

In lower vertebrates, generally, internal bones ossify from a single center—the shaft region of a long bone, for example—and even in the adult, the articular ends are often largely cartilaginous. But in mammals (and to a much more limited extent in certain reptiles), accessory ossifications—**epiphyses**—are found (Fig. 109).*
These are characteristically developed at either end of a long bone and on prominent

---

*The term "epiphysis" is sometimes applied in a broader sense, to include terminal cartilages without accessory ossifications.

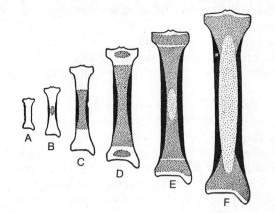

Figure 109. Ossification and growth in a long bone of a mammal. *A,* Cartilaginous stage. *B,*
C, Deposit of spongy, endochondral bone (stippled) and compact, perichondral bone (black).
D, Appearance of an epiphysis at either end. *E,* Appearance of the marrow cavity (sparse
stipple) owing to resorption of endochondral bone. The lengthwise growth of the bone is
confined to the thin strips of actively growing cartilage between shaft and epiphyses. *F,* Union
of epiphysis with shaft, leaving articular cartilages on end surfaces; enlargement of marrow
cavity by resorption, centrally, as deposition continues peripherally. (From Arey.)

processes for muscular attachment. In epiphyses, there develop centers of ossification
similar to those of the shaft but more limited in extent. These centers may ossify the
articular regions of the bone long before the growth of the shaft, the **diaphysis**, is
completed and thus allow the element to function despite incomplete ossification.
Between epiphysis and shaft, there is a long-persistent band of cartilage. This would
seem at first sight to be a relatively inert and functionless region. Actually, of course,
it is highly important. This is a zone of growth; the cartilage here is continually growing and continually being replaced by bone from shaft and epiphysis. Once ossification eliminates this band of cartilage, shaft and epiphysis unite; growth of the bone is
over, and definitive adult size has been attained.

The process of endochondral bone formation in higher vertebrates was long believed to be an example of the ontogenetic recapitulation of phylogenetic history.
The replacement of cartilage by bone in embryonic development was thought to
repeat the phylogenetic story of the vertebrate skeleton, under the assumption that

Though much of the growth of an internal "cartilage bone" takes place by the
replacement of cartilage, this is not, as one will readily see upon reflection, the whole
story. When the center of the shaft of a long bone is first ossified, the bone has a
small diameter. As the cartilage grows at either end, these ends gradually expand in
width toward adult size; but the bone, if formed solely by internal replacement,
would be shaped rather like an hourglass, with a constricted middle portion. To rec-
tify this imperfection, we find that, in addition to internal replacement, cartilage
bones have a major increment produced by the direct formation of bone on the
surface of the cartilage as **perichondral bone**, formed in successive concentric layers rather after the fashion of membrane bone. These layers are thickest in the originally thin middle part of the shaft. This process of superficial addition of bony material may continue after the underlying shaft has been ossified; at this stage the term
periosteal bone rather than perichondral is more appropriate.

the cartilaginous condition found in the adult cyclostome or shark was a truly primitive one. However, the fossil evidence strongly suggests that this is not the case; that, on the contrary, the living lower vertebrates have reduced skeletons and that cartilage in the adult is a retention of an embryonic condition.

Why this roundabout way of forming bones if there is no phylogenetic story concerned?

168
Chapter 7
Supporting
Tissues—
The
Skeleton

The answer may be deduced from the fact that not all bones are preformed in cartilage. It is significant that the dermal elements, which lack that stage, are usually simple and platelike and are capable of free growth at every surface until adult size is reached, whereas most of the deeper elements, which are preformed in cartilage, have complicated articulations with their neighbors and important muscular attachments. The vertebrate skeleton at very early stages is an almost perfect if tiny model of that of the adult; it must perform all its normal functions in larval or young animals. If the internal elements, with their complicated articular relations, were formed directly in bone, growth to adult size would be impossible; because bone can grow only at its surface, and surface growth would disrupt the relations of the bone with surrounding structures. Some type of moldable material capable of growth without disturbance of surface relations is needed for the growth of deep-lying elements. Cartilage, with its capability of growth by internal expansion, is an ideal embryonic adaptation for this purpose. Except where reduction has occurred—in various fishes and amphibians—bone is the normal adult skeletal material in a vertebrate; cartilage is its indispensable embryonic auxiliary.

Joints. Bones and cartilages are joined to one another by structures of varied types. In such cases as the bones of the skull, where one abuts directly against another, there is usually little possibility, or desirability, of movement. The connection between such elements, termed a **synarthrosis**, may be formed by thin intervening sheets of cartilage or connective tissue, the visible line of apposition being termed a **suture.** If the union is completely immovable because of actual fusion of bone with bone, we speak of **ankylosis**.

A freely movable joint between adjacent elements is termed a **diarthrosis** (Fig.

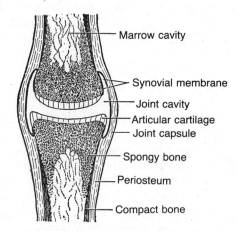

Marrow cavity

Synovial membrane

Joint cavity
Articular cartilage
Joint capsule

Spongy bone

Periosteum

Compact bone

Figure 110. A typical diarthrodial joint. (From Arey.)

110); the adjacent surfaces are typically covered by a film of cartilage even if the articulating elements are bony and are bound together by stout connective tissues; frequently there is a liquid-filled **joint cavity**, surrounded by a **synovial membrane** formed of dense connective tissue. Complex thickenings of this membrane form ligaments. In some cases there may be a division of the joint cavity into two portions by a thin plate of fibrous cartilage, a **meniscus**.

Classification of Skeletal Elements. The skeleton includes a wide variety of elements of varied form, structure, function, position, and embryonic origin, associated in variable combinations and showing major modifications from group to group. Classification of skeletal elements is difficult, and no classification is absolutely satisfactory. We here adopt the following scheme of major subdivision:

Dermal

Endoskeleton $\begin{cases} \text{Somatic} \\ \text{Visceral} \end{cases}$ $\begin{cases} \text{Axial} \\ \text{Appendicular} \end{cases}$

A primary distinction may be made between the **dermal skeleton**, consisting of bony structures—plates and scales—developed in the skin (mainly the dermis) and never preformed in cartilage, and the more deeply placed skeletal parts, the **endoskeleton**, in which the elements are almost always preformed in cartilage and, particularly in lower vertebrate groups, may remain cartilaginous throughout life.

Within the class of endoskeletal elements, a distinction may be made between two categories of unequal size, the somatic and visceral groups (as noted in Chapter 9, the associated muscular system may be similarly subdivided). As the **visceral skeleton**, we may classify the elements that lie between and operate the branchial openings and structures (such as the jaw cartilages and ear ossicles) derived from them. The **somatic skeleton** includes all the remaining internal structures. As further discussed in other sections, the branchial region is distinctive in regard to its musculature and nerve supply. Its skeleton is less distinctive. The visceral skeleton forms from the neural crest, and this was long considered to differentiate it from the rest of the skeleton that formed, it was thought, from normal mesoderm. However, now most of the cranial bones, whether dermal, visceral, or somatic, have been shown to be derivatives of neural crest, and there is no sharp distinction between somatic and visceral elements. The visceral skeleton is associated with the theoretic "visceral animal" (see page 35), whereas remaining skeletal structures are those of the theoretic "somatic animal." This somatic skeleton includes in its more primitive, axial portions the vertebrae, the ribs, and other related elements of the trunk and tail, and, anteriorly, most of the braincase, which forms a major element of the skull. These structures constitute the **axial skeleton**. The limb girdles and the elements of the free appendages, forming the **appendicular skeleton**, may be considered derivatives of the axial skeleton.

In certain instances, we find that structural units or even single bones of the adult skeleton contain elements derived from two or more categories. Thus, for example, the shoulder girdle of many vertebrates contains both dermal and endoskeletal components; the lower jaw of several vertebrate classes includes both visceral and dermal elements. Most complex of all is the skull, which in bony fishes and land vertebrates includes dermal, somatic, and visceral structures in its formation.

Dermal Skeleton

Fishes. The skin over most of the body of many living vertebrates contains no hard skeletal parts; however, dermal bony structures are usually present in the head, at least, and the fossil evidence leads to the conclusion that ancestral vertebrates were almost completely ensheathed in a bony armor. Although there may be a superficial covering of other hard skeletal materials, the major component of such plates and scales is dermal bone. Body armor of this sort was present in the most ancient vertebrates, the jawless ostracoderms, was present over part or all of the body of the extinct placoderms, most primitive of jawed vertebrates, and is present in varied and sometimes reduced form in almost all the great group of Osteichthyes. Armor is absent in the living cyclostomes and present only as denticles in the skin of sharklike fishes. This condition, once thought to be primitive, now seems quite certain to be one of secondary simplification.

In many ostracoderms and placoderms, a pattern of microscopic structure in scales and plates was present that, with variations, persisted into the primitive bony fish stage (Fig. 111 *A*). The inner layer consisted of lamellae of compact bone. Above this was a layer of spongy bone, with numerous spaces presumably containing blood vessels. Toward the outside was a further compact layer. The surface of the plates and scales was frequently ornamented with tubercles or ridges. The surface of each "denticle" or **odontode** and, indeed, of the whole plate or scale, was covered with a hard shiny material resembling the enamel of a tooth.* Beneath was a "pulp" of blood

170
Chapter 7
Supporting
Tissues—
The
Skeleton

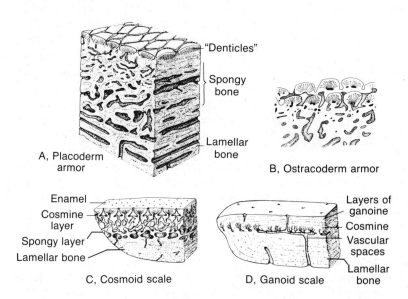

Figure 111. The structure of dermal plates and scales in primitive vertebrates. *A,* Devonian placoderm. *B,* Devonian ostracoderm, showing superimposed layers of odontodes. *C,* Primitive crossopterygian with a cosmoid scale. *D,* Paleozoic ray-finned fish with a ganoid scale. (After Kiaer, Goodrich, Ørvig.)

*There is considerable dispute about the nature of the enamel-like substances in anamniotes; the problem is discussed on page 331. In general, the tissues of armor and teeth are quite variable in anamniotes and the definitions, based on mammalian material, are often inappropriate.

vessels. Between enamel and pulp cavity, the substance of the tubercle was formed of a bonelike material. In some presumably very primitive cases, this appears to have been typical bone, with star-shaped lacunae for osteocytes. In more advanced types, however, the cells had retreated to the pulp cavity, but tiny canals, presumably filled in life by cell processes, penetrated the bony material—the structure now resembling the dentine of a tooth. It seems clear that this primitive armor was of a type from which the scales of more modern fishes have been derived and that the superficial tubercles were structures from which were evolved shark dermal denticles.

The odontodes are very similar to teeth, and it is plausible that they developed in a manner comparable to that seen in the teeth of living forms (described on pages 332–333). Certainly both the epidermis and dermis were involved and, we assume, also neural crest material. Such involvement of all these tissues occurs in the development of teeth and scales of fish whether or not any of the hard tissue is actually formed from each of them. The formation of this armor was clearly complex because there are often several superimposed layers of odontodes that must have formed one after the other (Fig. 111 *B*). Often, during the formation of a younger generation of odontodes, the older ones were partly resorbed, making the resultant structure even more complicated; the living coelacanth, although having reduced armor, still shows this pattern.

In the Osteichthyes, the body is generally covered with an overlapping series of scales arranged in diagonal rows and either rectangular in shape or with a curved external border. Among early bony fishes, two structural types of scales were present. In primitive sarcopterygians, the **cosmoid scale** (Fig. 111 *C*) was characteristic. The general arrangement of the three layers was that just described, but the main substance of the tubercles was not typical dentine but a variant termed **cosmine**, differing in that the tiny canals penetrating it from the pulp cavity were branching structures rather than simple tubes. Scales of this sort were characteristic of the typical crossopterygians and the very earliest lungfish. In coelacanths, however, the scales became thinner, though retaining complex odontodes, and in the lungfishes, they are reduced to a rather fibrous and leathery type of bony structure—true cosmoid scales are nonexistent in living animals.

Primitive actinopterygians possessed quite a different sort of scale, the **ganoid** type (Fig. 111 *D*). This differs from the cosmoid type in that, instead of a single layer of enamel-like material, above the cosmine, there is laid down during scale growth layer after layer of a comparable material termed **ganoine**, and, correspondingly, successive layers of compact bone are formed on the lower surface of the scale. Hosts of Paleozoic and early Mesozoic ray-finned fishes, the palaeoniscoids, had scales of this type. However, actinopterygians, like the sarcopterygians, have tended to simplify their scales. Ganoid scales are present today only in *Polypterus* and the garpikes, and even in the latter, they are in somewhat simplified form, in that the cosmine is lost; in most Mesozoic actinopterygians, the ganoine was reduced or lost, and the cosmine layer as well.

By the time the teleosts were developed, there remained only relatively thin, pliable scales, formed of a thin sheet of bonelike material (but usually lacking cells) and an underlying fibrous layer (cf. Fig. 89 *C*); even these reduced structures may be lost, leaving (as in catfish) a naked skin over the trunk and tail. These scales represent the basal, bony layer of the original armor; possibly the outer layers of the armor are present as an extremely thin sheet, but there is no general agreement on this point.

Teleost scales are typically translucent and overlapping—imbricating. The "free" margins may be smoothly rounded, the "cycloid" type, or this margin may be frayed out, the "ctenoid" scale. The distinction between these two types of scales, was once thought to be important for classification but is now seen to be of little significance. Teleost scales frequently show annual growth rings, which are of practical value in determining the age of a fish in life history studies.

172
Chapter 7
Supporting
Tissues—
The
Skeleton

Dermal scales and bones are completely lost in the lampreys and hagfishes. In the Chondrichthyes, there are, in some instances, fin-spines of a dentine-like material, but the skin is otherwise naked except for the **dermal denticles**, or **placoid** "scales" (Figs. 89 A, 112). These are small, isolated, often conical structures that resemble simple teeth (cf. Fig. 233). The base is bonelike in substance and contains a pulp cavity; a mass of dentine forms the bulk of the tubercle; there is a shiny film of enamel-like material on the surface. This may be partly true enamel, but most of it appears to be a hard external layer of dentine, **vitrodentine**. It was formerly assumed that the presence of these dermal denticles in sharks marked the first phylogenetic appearance of dermal armor and that bony scales and plates appeared later in world history as the result of the fusion of such denticles. It now seems, however, that the reverse is the case. Dermal denticles are the last remnants of an ancestral armor; the bone proper has, so to speak, melted away, leaving only its superficial tubercles.

The anterior portion of the body in fishes with dermal armor is generally covered, not with discrete scales, but with bony plates, sheathing the head and gills and extending down over the "shoulder" region. In many ostracoderms, we find these regions covered by armor plates of varied patterns. In placoderms, we can distinguish plates in the pectoral region that roughly correspond to elements of the shoulder girdle of higher fishes, and anterior to this a "head" shield, the elements of which are (as with ostracoderms) difficult to compare with those of "higher" vertebrates. In bony fishes, we find, at last, a more familiar pattern in the arrangement of head plates, although individual bones are highly variable. There is a well-defined cranial shield that forms part of a typical skull; dermal bony plates are further associated with the lower jaw and the inner surfaces of the mouth, and there is a dermal shoulder girdle. These structures will be discussed later.

We may here, however, note briefly the presence, between head and shoulder, of a series of **opercular elements** lying in the flap of skin over the branchial region (cf. Figs. 170, 172, 173). There is typically a major plate, the **operculum**, on either side behind the cheek region of the skull, and other, smaller elements may be present

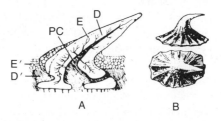

Figure 112.   Shark dermal denticles. *A,* Section through a denticle; *B,* side and surface views of denticle. Abbreviations: *D,* dentine; *D ',* dermis; *E,* hard, enamel-like surface of denticle— vitrodentine; *E ',* epidermis; *PC,* pulp cavity. (From Dean.)

adjacent to the main operculum (preoperculum, suboperculum). The opercular structures follow the contours of the branchial chamber downward and forward beneath the jaws. Here there was primitively a series of plates termed **gulars**; in teleosts, these have been reduced to a series of parallel rods, the **branchiostegal rays** (Figs. 37, 123, 172).

We may here (parenthetically) discuss the nature of the rays that stiffen the peripheral portions of fish fins, median or paired (Fig. 113). In some more primitive fishes, scales like those of the body sheathe the fins but become smaller and more attenuate distally. In higher bony fishes, these fin scales tend to be modified into elongated rays, termed **lepidotrichia**, each of which represents a row of slender scales. The very tip of the fin in bony fishes may be additionally stiffened by tiny horny rods developed in the dermis of the skin, **actinotrichia**. Larger rays of this sort, termed **ceratotrichia**, are the sole support of the web in the distal part of a shark fin.

Tetrapods. Terrestrial vertebrates are descended from fishes with a complete bony covering of plates and scales. The dermal bones of the head are retained as part of the skull and jaw structures, and a dermal shoulder girdle generally persists. The remainder of the dermal armor, however, has tended strongly to reduction and loss.

With the reduction of the gills in tetrapods, the series of opercular bones vanishes. The bony body scales may persist but only in a much reduced or modified manner. In the older fossil amphibians, part of the original fish scaly covering was retained in the form of V-shaped rows on the belly and flanks, perhaps as a protection to the low-slung body; more rarely are bony scales found over the back. In modern amphibians, scales are completely lacking except for functionless vestiges buried in the skin of the Gymnophiona.

In some reptiles, the ventral bony scales persist as a series of jointed V-shaped rods beneath the skin of the belly—abdominal "ribs" or **gastralia**. These structures are dermal and should not be confused with ventral extensions of the true ribs, which may lie deep to them but are, of course, part of the endoskeleton. Abdominal ribs persist today in some lizards, crocodilians, and *Sphenodon;* but in birds and mammals, these last representatives of the old bony scales of fish have vanished.

The skin, however, retains its potentialities of forming dermal bone, and in many reptiles and a few mammals, we find a redevelopment of bony nodules, scales, or

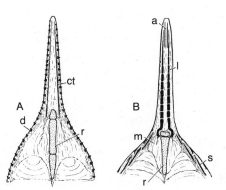

Figure 113. Sections through the dorsal fin of *A,* a shark; *B,* a ray-finned fish; to show the nature of the fin supports. *a,* Actinotrichia; *ct,* ceratotrichia; *d,* dermal denticle; *l,* lepidotrichia; *m,* muscle tissue at fin base; *r,* skeletal supports of fin; *s,* bony scales. (From Goodrich.)

plates as protective armor. Examples of this sort are common in extinct reptiles. In lizards, there is frequently a development, probably secondary, of bony scales underlying the horny scales in the epidermis (Fig. 92 C); in some crocodilians, there is an armor of subquadratic bony plates covered by horny skin. Bony plates are formed in the skin of some whales, but most mammals that have armor belong to the edentate stocks native to South America. Extinct ground sloths reinforced their tough hide with lumps of dermal bone buried in the dermis; in the armadillos and their extinct relatives, the glyptodonts, bony plates form a complete carapace covering the back; the top of the head and the tail are also frequently protected.

Turtle Armor (Fig. 114). Although there are many interesting examples of armor in extinct reptiles, that of the living turtles is unequalled for completeness and efficiency. We have noted earlier the presence of horny scutes over the surface of the turtle body; beneath these scutes most have a stout and effective bony shell. This consists of a convex oval dorsal shield, the **carapace**, and a flat belly slab, the **plastron**. At the sides, the two parts are connected by a bony bridge; front and back, the shell is open for the head, limbs, and tail.

The carapace consists essentially of three sets of plates: (1) Circling the margin is a set of **peripheral** (marginal) plates. (2) From front to back in the midline is a series of subquadratic plates, most of which are fused to the vertebrae beneath them and termed **neurals** (anterior and posterior members of the median series have special names). (3) Transversely placed between peripherals and neurals are eight paired **pleural** (costal) plates, bound to underlying ribs. The elements of the carapace are firmly fused in many cases, with obliteration of sutures; the lines of separation of the horny surface scutes, however, are often impressed on the surface of the bony carapace. For the most part, the boundaries between the horny scutes of the carapace (and of the plastron as well) alternate with the sutures between the bony plates, a condition that lends strength to the shell structure as a whole.

The plastron of typical turtles consists of four paired plates and an anterior median element; their nomenclature is given in Figure 114. In primitive types, additional paired plates may be present. The three plates at the anterior margin are actually the clavicles and interclavicle of the dermal shoulder girdle described later in this chapter; the remaining plastral elements and all the carapacial plates appear to be new developments in turtles, although there is some evidence suggesting that dermal abdominal ribs have been incorporated in the plastron.

Axial Skeleton

Vertebrae—Amniote. The host of skeletal structures remaining to be described are endoskeletal, lying (in contrast to those of the dermal skeleton) in the interior of the body; most are first formed in the embryo as cartilages, the cartilage usually being replaced and supplemented by bone. The great majority of endoskeletal structures belong to the system here termed **somatic**; postcranial elements are formed from mesenchyme of mesodermal origin, whereas cranial ones may form from the neural crest. Apart from the limb supports, the somatic skeletal elements are classed as axial.

The major axial structure is the vertebral column, which replaces the notochord as the main longitudinal girder of the vertebrate body. The greater flexibility given by the notochord is sacrificed for greater strength; need for the strength given by

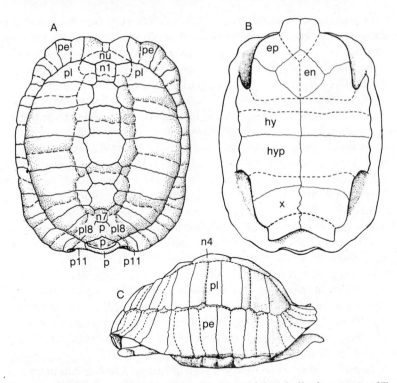

Figure 114. *A,* Dorsal, *B,* ventral, and *C,* lateral views of the shell of a tortoise (*Testudo*). Sutures between bony plates in solid line, outlines of horny scutes in broken line. Abbreviations: *en,* entoplastron; *ep,* epiplastron; *hy,* hyoplastron; *hyp,* hypoplastron; *nu,* nuchal; *n1–n7,* neurals; *p,* pygal plates (postneurals); *pe,* peripheral plates; *p11,* most posterior (eleventh) peripheral; *pl,* pleural plates; *x,* xiphiplastron.

bony elements is especially great in tetrapods in which the back must suspend the weight of the body. As an additional function, the vertebral elements extend upward from the notochordal region to surround and protect the spinal cord.

As an introduction to vertebral structure, we may consider the relatively simple situation seen in typical vertebrae of amniotes (Fig. 115). Each vertebra, formed of bone in these forms, includes two principal portions, a neural arch and a centrum.

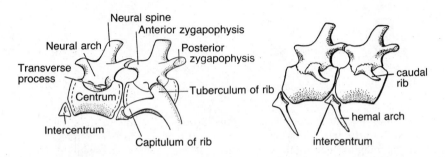

Figure 115. *Left,* two trunk vertebrae of an early generalized reptile (anterior end to the left). *Right,* two caudal vertebrae.

176
Chapter 7
Supporting
Tissues—
The
Skeleton

Ventrally placed is the **centrum**, which functionally replaces the notochord. Typically this is a spool-shaped cylinder articulated at either end with the adjacent centra. In some reptiles, the centrum may be hollowed at both ends, a condition termed **amphicelous** (Fig. 116 A). The cavity thus present between successive centra is filled with soft material that may be, in part at least, derived from the embryonic notochord. In early fossil reptiles and in *Sphenodon* and some lizards (geckos) today, there is a tiny hole, piercing the centrum lengthwise, through which the notochord extends as a continuous (though much constricted) structure.

In most living amniotes, however, the notochord has disappeared except for, at the most, thin pads of tissue lying between successive centra, and the centra have become solid structures, whose end faces are closely apposed to those of their neighbors. Those vertebrae that have flat terminal faces, as is common in mammals, are termed **acelous** (Fig. 116 D). In many cases, one end of the centrum expands in spherical fashion, to be received in a hollowed socket in the neighboring centrum. If the concave socket is at the anterior end of the centrum (a condition found in most living reptiles), the vertebra is termed **procelous** (Fig. 116 C); if posteriorly placed, **opisthocelous** (Fig. 116 B). Because the nature of the articular faces may change regionally along the column, various irregularities and combinations of structure may occur, and there are numerous variants from the orthodox types. Notable is the development in the neck of birds of **heterocelous** vertebrae, with saddle-shaped articular faces between the centra, and the development of complicated central connections in the retractable neck of turtles.

In reptiles and in the tail of mammals, a series of small elements, crescentic in end view, may be present and wedged in ventrally between successive centra. These are **intercentra** (or hypocentra), remnants of elements important, as we shall see, in lower tetrapods and their presumed crossopterygian ancestors. Fused to the intercentra in the tail are the **hemal arches** or chevrons, which extend downward in the septum between the muscles of either side. The base of each hemal arch consists of a pair of processes that descend from the intercentrum to enclose a space in which the major blood vessels of the tail lie; usually the two branches join below the vessels to give a V or Y shape as seen in end view. As will be seen later, the hemal arches are represented in fishes not only in the tail but also in the trunk as a type of rib, the ventral rib, not present in terrestrial vertebrates.

In most cases, a rib attachment is present in the central region. Primitively, this was with the intercentrum, but in some amniotes, this point of attachment shifts to

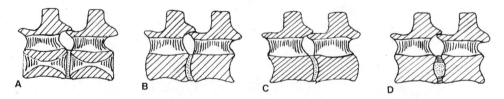

Figure 116. Diagrammatic longitudinal sections through vertebrae to show various types of centra. Anterior is to the left. *A,* Primitive amphicelous type, in this case with the centrum pierced by an opening for a continuous notochord; *B,* the opisthocelous type; *C,* the procelous type; *D,* an essentially acelous type but with the centra slightly biconcave to allow room for an intervertebral disc. (After Romer.)

the centrum, where a special process, the **parapophysis**, may develop, or it may come to lie, in the trunk of mammals, at the point of junction between two successive centra.

The upper surface of the centrum forms the floor of the canal carrying the spinal cord. At either side rises the broad base of a **neural arch**, firmly bound to the centrum in most adults. Above the spinal cord, the two arches fuse to form a **neural spine**, which extends upward between the dorsal muscles of the two sides; successive spines are usually bound together by an elastic ligament. At the base of the arch on either side, there is frequently a stout **transverse process (diapophysis)** as a rib attachment.

In tetrapods (and in a few bony fishes), strength is added to the column by the presence of articulations, paired "yoking" processes, or **zygapophyses**, between successive neural arches; these allow a degree of vertical and horizontal movement of the column but help prevent undue torsion. The anterior zygapophyses terminate in rounded surfaces that generally face upward and inward; to them are opposed surfaces facing downward and outward on the posterior zygapophyses of the next anterior arch (for mnemonic purposes, one may recall that in any procession, those at the front of the parade are up and in, those at the rear are down and out). There are sometimes additional articulating processes between neural arches for firmer vertebral union. Below the zygapophyses, there are gaps between successive neural arches, **intervertebral foramina**, through which the spinal nerves emerge.

The structures described are those normally found in the higher vertebrate classes; in amphibians and the various fish groups, there is a great variety of vertebral types. Before discussing them, however, we should briefly consider the embryologic development of vertebrae as seen in amniotes.

In the discussion of mesodermal differentiation, it was noted that a considerable amount of mesenchyme proliferates from the medial side of each somite—the area known as the sclerotome (cf. Fig. 81). Its major function is the formation of vertebrae; the sclerotome cells spread out in a sheet applied to the sides of both notochord and spinal cord opposite each somite.

The number of vertebrae in the adult corresponds to the number of somites from which their materials are derived, so one might assume that each vertebra would correspond in position to the muscle segment produced by the same somite. Actually this is not the case; in amniotes, at any rate, the vertebrae alternate with the muscle segments, and the mesenchyme from which a vertebra is composed is derived, half and half, from each of two adjacent somites. In many amniotes, there early appears a distinction between the anterior and posterior portions of the mesenchyme sheets produced by each segment. As development proceeds, it is found that the front half of a vertebra is formed from the posterior part of one segmental mesenchyme sheet, the back half from the anterior part of the mesenchyme of the next segment (Fig. 117).

In fishes and many lower tetrapods, the segmentally arranged axial muscles are powerful and are important in locomotion. To exert their force, they depend in great measure upon attachment to the vertebrae and ribs. But these attachments, to be effective, must be to two successive vertebrae; in consequence, vertebral structures, at least the neural and hemal arches, must alternate with muscular ones. The primary segmentation, that of the somites, is retained by the axial muscles; the vertebral segmentation, developing later in both phylogeny and ontogeny, is subsequent to that of

178

**Chapter 7
Supporting
Tissues—
The
Skeleton**

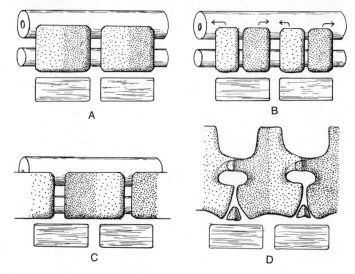

Figure 117. Diagrams to illustrate the manner of utilization of mesenchyme in the formation of amniote vertebrae. Anterior is to the left. *A* is a diagrammatic lateral view of the sclerotomes of two somites; nerve cord and notochord are shown behind them, and below is arbitrarily figured a portion of the musculature belonging to these somites. In *B*, the sclerotomes have split into anterior and posterior halves, which move apart and in *C* have fused with the adjacent halves of neighboring sclerotomes. These newly formed sclerotome masses are the materials from which successive vertebrae are derived, and in *D* is diagrammatically represented (by differences in stippling) the dual origin—from parts of two segments—of an adult vertebra. As a result of this seemingly odd developmental process, the vertebrae (as is seen by reference to the blocks of segmental origin) are intersegmental rather than segmental in position.

the muscles and alternating with it. In fish, most of the axial musculature does not insert directly on vertebrae but on sheets of connective tissue (myocommata; see page 282), but the myocommata attach to the neural and hemal arches so the principle is similar. The centra of lower vertebrates vary greatly in position relative to the segments. The complex developmental sequence in amniotes is, apparently, concerned more with the developmental mechanics than with the adult function.

Anamniote Vertebrae (Figs. 118, 119). In considering the highly variable structure of the vertebrae in fishes and amphibians, we may first discuss the fairly straightforward history of the neural arches. Such arches are present in every case. In the hagfish tail, tiny cartilages are present in the neural arch region. In lampreys, a small arch cartilage is present on either side in the more anterior trunk segments; posteriorly, two small paired cartilages may be present in each segment. In the Chondrichthyes (Fig. 118), there are, in addition to the neural arches proper, accessory neural elements that fill out the gaps between them and enclose the spinal cord in a continuous cartilaginous sheath. The presence of "extra" arches (interneurals or dorsal intercalary plates) is secondary, because in all older fossil groups in which the column is preserved (including the primitive placoderms and older sharks), there is never more

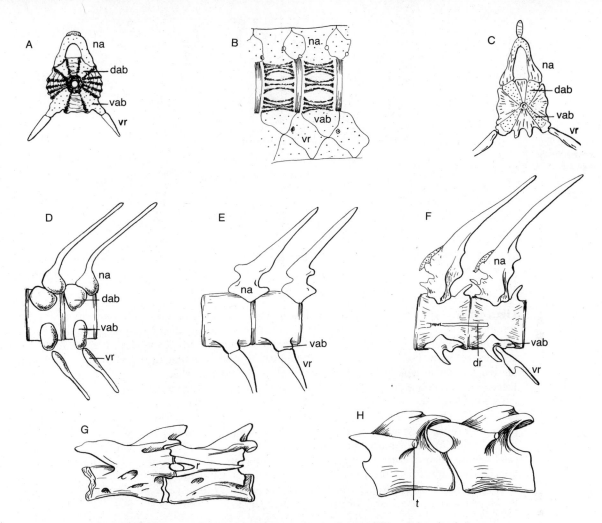

Figure 118. *A–F,* Fish trunk vertebrae. *A,* Cross section of a vertebra of the shark *Lamna,* showing arch bases wedged into centrum. The dark areas are calcified. *B,* Lateral view of two segments of the column of the same shark. Elements between the neural arches complete a closed tube for the spinal cord; a complex of small elements represents short ventral ribs (= hemal arches). The dorsal arch bases are covered by the base of the neural arches. *C,* A cross section of a vertebra of a teleost, *Esox;* the structure is comparable to that of a shark. *D,* Lateral view of two vertebrae of a young *Amia;* both dorsal and ventral arch bases are visible and distinct from the centrum proper. *E,* Adult vertebrae of the same; the arch bases are incorporated in the centrum. *F,* Side view of vertebrae of *Esox* (cf. *C*). The ventral arch bases are distinct; as in the shark, the dorsal arch bases are concealed by the neural arches. *G,* Vertebrae of the urodele *Necturus,* in which short ribs are present. *H,* Two vertebrae of a frog. *dab,* Dorsal arch base; *dr,* dorsal rib; *na,* neural arch; *r,* rib; *t,* transverse process; *vab,* ventral arch base; *vr,* ventral rib (= hemal arch). (Mainly after Goodrich.)

180

Chapter 7
Supporting
Tissues—
The
Skeleton

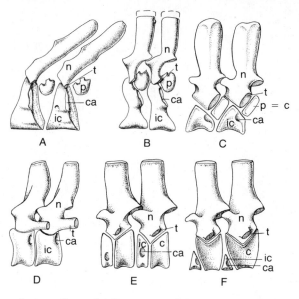

Figure 119. The evolution of vertebral structures from crossopterygians to primitive tetrapods and reptiles. Anterior is to the left. *A,* Two vertebrae of a crossopterygian; the principal central element is a large intercentrum, wedge-shaped in side view, crescentic if seen in end view; there are small paired pleurocentra. *B,* Vertebrae of the most primitive known amphibian type, of similar construction. *C,* The typical rhachitomous type, common in Paleozoic labyrinthodonts. *D,* The stereospondylous type of vertebra, found in late labyrinthodonts; the pleurocentra have disappeared and the intercentrum forms the entire centrum. *E,* The embolomerous type, found in extinct forms related to reptile ancestors; the pleurocentra have expanded and fused to form a complete ring-shaped centrum, and the intercentrum forms a similar but thinner structure. *F,* The primitive reptilian type in which (as in *E*) the pleurocentra have expanded to form a large true centrum, but the intercentra are reduced. Abbreviations: *c,* centrum; *ca,* capitular attachment of rib to intercentrum; *ic,* intercentrum; *n,* neural arch; *p,* pleurocentrum (paired); *t,* transverse process on neural arch for attachment of tuberculum of rib. (*A, B,* after Jarvik.)

than one arch per segment. In bony fishes and amphibians, characteristic single neural arches are present throughout.

The story of hemal arches in lower vertebrates is likewise a simple one. They are essentially comparable to those of amniotes except that their attachment to central structures is variable. They are not developed in cyclostomes but are present in the caudal region of all higher types; in elasmobranchs, there are here (as in the case of neural arches) accessory cartilages.

In the history of the centrum, great diversity is to be found.* No such structure is present in the cyclostomes (which hence can be called vertebrates only by courtesy). In sharks (Fig. 118 *A, B*), the centra are short amphicelous cylinders. Their structure is far from simple. In part, they are formed of concentric layers of cartilage

---

*It was long widely believed that a vertebra was formed basically from four pairs of "arcualia," variously arranged or combined. Although still frequently mentioned in texts and manuals, this attractive theory has (alas!) proved to be untenable.

around the notochord and even within the notochordal sheath; in these layers, calcification usually occurs, often in complicated and characteristic patterns. But this continuity of structure is interrupted, above and below on either side, by plugs of cartilage of a different texture. These lie beneath the areas of attachment of neural and hemal arches (the latter represented by ventral ribs), function as bases for the arches, and are frequently in continuity with them. In the bony fishes, similar **arch bases** are present as distinct structures during development. In most lungfishes, *Latimeria,* sturgeons, and paddlefishes, cartilaginous arch bases are the only skeletal elements in the adult centra, the notochord persisting as a large unconstricted structure. But in most actinopterygians—*Polypterus, Amia,* the garpikes, and all teleosts—the whole region of the centrum is ossified and the arch bases are fused with or buried within the remainder of the centrum at maturity (Fig. 118 *C–F*). A few fishes, such as the bowfin, *Amia,* have two centra per segment in the back part of the body (paralleling the condition seen in some fossil amphibians).

The fossil record gives us a series of stages apparently leading from the vertebrae of crossopterygian fishes to those present in varied form in the great extinct group of amphibians, the labyrinthodonts, and on to the amniote structure (Figs. 119, 120).

In the crossopterygian vertebrae, two sets of ossifications are present in the central region. The notochord is, of course, not preserved in the fossils but appears to have been persistently large. On either side, just below the neural arch, is a small plate of bone. This we believe to be the element called a **pleurocentrum** in fossil Amphibia; it appears to have expanded and developed into the true centrum of amniotes. Below the curve of the notochord is a single median element, wedge-shaped in side view, crescentic if seen from the end. This, although much larger than in reptiles, is certainly the **intercentrum**, as is shown by the attachment to it (as in reptiles) of the hemal arches of the tail.

If we wish to compare these crossopterygian structures with those of other fishes (Fig. 120), a reasonable assumption is that the pleurocentra correspond to the basal elements beneath the neural arch, the intercentra to a fusion of the pair of ventral arch bases. But in contrast to most actinopterygians, there is no further ossification in the crossopterygian centrum. The arch bases, it would seem, have (so to speak) absorbed all the potentialities for ossification present in the centrum as a whole.

This pattern of centrum is basically known as the **rhachitomous** type; it is seen not only in the crossopterygians but also in many of the oldest and most primitive amphibians and has long been known in a great percentage of later labyrinthodonts. Among these amphibians and the reptiles descended from them, however, there is a trend toward a more solid and better ossified condition of the centrum, brought about by an expansion of the pleurocentrum, the intercentrum, or both. Among the labyrinthodonts themselves, late members tend to reduce the pleurocentra and to expand the intercentrum until it alone forms a completely cylindric central structure, the **stereospondylous** condition. In forms closer to the ancestory of reptiles, the two pleurocentra, on the contrary, tend to expand and form a complete ring. In one group, we find an **embolomerous** structure in which both pleurocentrum and intercentrum are rings, giving a double centrum in each segment.* However, in the true

---

*In some fishes, notably the holostean *Amia,* "duplex" centra are developed in the posterior part of the column, and in chimaeras, there may be several discs representing one centrum; however, these are formed in ways that do not appear to be closely comparable to that seen in embolomere vertebrae.

182

**Chapter 7
Supporting
Tissues—
The
Skeleton**

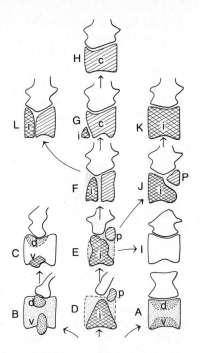

Figure 120. Diagrams to show possible evolutionary sequence in the structure of vertebral centra. *d* and *v* are the dorsal and ventral arch bases found in many fishes, parts of the centrum structure supporting neural arches and hemal arches or ventral ribs, respectively. In crossopterygians and many tetrapods, the arch bases appear to be represented by pleurocentra (*p*) and intercentra (*i*). Cartilage stippled; arch base cartilages, heavy stipple; pleurocentra hatched; intercentra crosshatched; parts of central ossifications not included in arch bases, plain white. Part or all of the neural arch is shown as well as the central structures. Front end of vertebrae to left.

A, Shark condition; arch bases incorporated in cartilaginous centrum. B, Embryonic *Amia* vertebra, showing presumed basic actinopterygian condition, with arch bases forming separately from rest of cartilaginous centrum. C, Teleost condition; centrum ossified, and incorporates dorsal arch base and in some cases ventral arch base as well. D, Crossopterygian condition, probably basic for tetrapods; much of the centrum was persistently cartilaginous, but small paired dorsal arch bases ossify as pleurocentra, and ventral elements fuse as a wedge-shaped structure. E, Oldest known amphibians, similar in structure to crossopterygians. F, Type leading toward reptiles among fossil labyrinthodonts; the pleurocentra have grown and fused to form a ring-shaped "true" centrum. G, Primitive reptiles; the centrum has grown at the expense of the intercentrum, which is much reduced. H, Advanced reptiles, birds, mammals. The intercentrum disappears, and the whole centrum corresponds to the expanded pleurocentra. I, Modern amphibians; the centrum ossifies as a simple spool-shaped structure, with no evidence of arch bases. J, Structure in rhachitomous labyrinthodonts; the intercentrum is a large wedge-shaped structure, the pleurocentrum persistently small. K, Late, stereospondylous labyrinthodonts; the pleurocentra vanish, and the intercentrum constitutes the entire centrum. L, Extinct embolomerous labyrinthodonts; both pleurocentra and intercentrum form complete rings.

reptilian ancestors, the intercentra are, in contrast, reduced to the small ventral crescents described earlier, whereas the pleurocentra are enlarged and fused to produce the true centrum. There is considerable variation in labyrinthodont vertebrae, and the story is not as simple as it often seems in brief descriptions such as this.

In this description, we have said nothing as yet of the centrum in the modern amphibian orders (Fig. 118 *G, H*) because of the embarrassing fact that we cannot fit them into the seemingly clear story just given of the evolution of other tetrapod vertebrae. In urodeles and apodans, the centra are essentially simple cylinders (despite irregularities in outlines) in which there is no evidence of structures corresponding to separate pleurocentra and intercentra; they ossify directly from mesenchyme with little or no cartilaginous preformation. In most frogs and toads, there develops a simple cylinder of cartilage that ossifies in continuity with the neural arch; in others, the centrum is represented in the adult only by ossification in the upper half of the cylinder. Unless (quite improbably) the modern amphibians have developed independently of all other land vertebrates, it is reasonable to believe that these structures of modern amphibians have developed from the conditions seen in crossopterygians and early labyrinthodonts. However, there must have been considerable secondary simplification, because at present there is little evidence to connect these modern amphibian vertebral types (or those of most extinct lepospondyls) with those of any other group.

Regional Variations in Vertebrae. It is obvious that there may be considerable variation along the length of the vertebral column. In higher vertebrates, vertebral regions may be defined according to the variations in the nature and presence or absence of ribs; but in many fishes, early amphibians, and various reptiles, ribs may be present without a break on every vertebra from neck to tail base. In fishes, a **caudal** series of vertebrae can be distinguished from that of the **trunk** by the presence of hemal arches in the tail; in tetrapods, specialized ribs, as noted later, develop for attachment of the pelvic girdle, and this distinguishes a **sacral** region separating more distinctly **presacral** and caudal segments. As a neck forms in terrestrial forms (as noted later), the ribs tend to become short, fused or absent; hence, a **cervical** region can be distinguished from the **dorsal** region, that of the trunk. Even in primitive tetrapods, the posterior ribs of the trunk tend to be relatively short, and in mam-

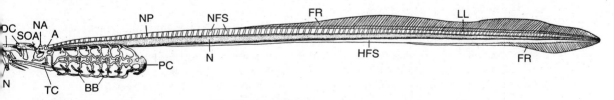

Figure 121. Skeleton of the lamprey, *Petromyzon. A,* Otic capsule; *AN,* ring cartilage surrounding mouth; *BB,* cartilages of branchial basket; *DC,* dorsal cartilages of mouth region; *FR,* dermal fin rays; *HFS,* fibrous sheath around dorsal aorta; *LL,* longitudinal ligament connecting tips of neural processes; *N,* notochord; *NA,* openings of nasal capsule; *NFS,* fibrous sheath of spinal cord; *NP,* neural processes; *PC,* pericardial cartilage; *SOA,* cartilaginous arch around orbit; *TC,* tongue (or lingual) cartilage. The openings in the branchial basket indicate the openings of the gill chambers. (From Dean.)

184

**Chapter 7
Supporting
Tissues—
The
Skeleton**

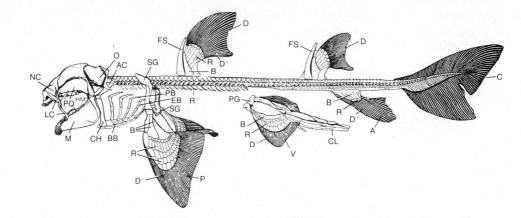

Figure 122. Skeleton of a shark (*Cestracion*). *A*, Anal fin; *AC*, auditory capsule; *B*, basal elements of fin; *BB*, basibranchial; *C*, caudal fin; *CH*, ceratohyal; *CL*, claspers of male; *D*, dorsal fins; *D'*, dermal rays of fin; *EB*, epibranchial; *FS*, fin spines; *HM*, hyomandibular; *LC*, labial cartilages; *M*, mandible; *NC*, nasal capsule; *O*, orbit; *P*, pectoral fin; *PB*, pharyngobranchial; *PG*, pelvic girdle; *PQ*, palatoquadrate; *R*, radial elements of fin; *R'*, ribs; *SG*, pectoral girdle; *V*, pelvic (ventral) fin. (From Dean.)

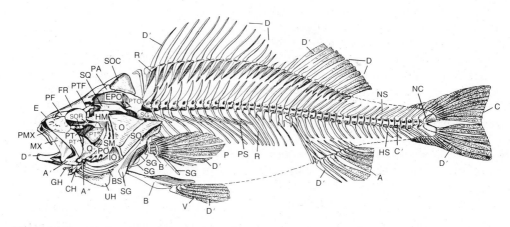

Figure 123. Skeleton of a teleost (*Perca*). The interpretation of certain skull elements differs from that used in Figure 173 but is one commonly used in works on teleosts. *A*, Anal fin; *A'*, articular; *A''*, angular; *B*, pelvic girdle; *B'*, skeleton of pectoral fin; *BS*, branchiostegal rays; *C*, caudal fin; *C'*, centrum of vertebra; *CH*, ceratohyal; *D*, dorsal fins (anterior one supported by stout dermal spines); *D'*, dermal rays of fins; *D''*, dentary; *E*, ethmoid; *EPO*, epiotic; *FR*, frontal; *GH*, hypobranchial (glossohyal); *HM*, hyomandibular; *HS*, hemal spine; *IO*, interopercular; *MX*, maxilla; *NC*, tip of notochord in tail; *NS*, neural spine; *O*, opercular; *P*, pectoral fin; *PA*, parietal; *PF*, prefrontal; *PMX*, premaxilla; *PO*, preopercular; *PS*, supporting processes of ribs; *PT, PT', PT''*, bones of palatal region; *PTF*, postfrontal; *PTO*, pterotic; *Q*, quadrate; *R*, ribs; *R'*, supports of dorsal fins; *SG*, various portions of shoulder girdle; *SM*, symplectic (binding hyomandibular to skull); *SO*, subopercular; *SOC*, supraoccipital; *SOR*, suborbitals; *SQ*, squamosal; *UH*, urohyal (ventral element of branchial skeleton); *V*, pelvic fin. (From Dean.)

mals, there develops here a region without free ribs, so that in the mammal trunk we may distinguish **thoracic** and **lumbar** divisions. Thus, there can be established, progressively as we "climb the tree" of vertebrates, a series of subdivisions as follows:

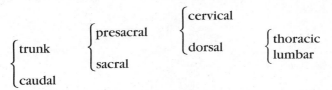

In primitive tetrapods, there is a variable vertebral count, but 30 or so presacrals, of which about 7 were cervicals, appear to have been a fairly common number in early tetrapods. In amphibians, there is normally only a single sacral (Fig. 124). Primitively, the tail was long, with up to 50 or more caudals. There is great variation in modern amphibian vertebral numbers: the elongated Gymnophiona may have 200 or so vertebrae; typical frogs, on the other hand, have reduced their backbone to 9 vertebrae plus a rodlike **urostyle** representing a number of fused caudals. The primitive reptiles appear to have had about 27 presacrals and a long tail; 2 or more sacrals are generally present. There is great variation within the reptiles in vertebral counts; snakes may have several hundred vertebrae, whereas turtles may have a very short column. In some early mammal-like reptiles, such as *Dimetrodon,* the neural spines are greatly elongated; connected by a web of skin, they formed a "sail" above the body that was probably a primitive essay in temperature regulation.

A curious condition is seen in the tail of *Sphenodon,* some lizards, and certain ancient reptiles. Halfway along each caudal centrum, there is an almost complete unossified septum separating front and back halves of the segment. The whole tail may be shed at such a "breaking point" and then regenerated (in somewhat imperfect fashion). The appearance is like that of some fishes with double caudal centra or the embolomerous amphibians, but the structure here is certainly a specialized, not a primitive, one. A parallel phenomenon of tail-breaking and regeneration is present in two families of salamanders, but here the break occurs between two vertebrae at the base of the tail.

In birds, the distinct cervical region has a highly variable number of vertebrae. Most of the thoracic vertebrae tend to be fused together, as an aid to the effective bracing of wings in flight. The posterior thoracic, lumbar, and proximal caudal vertebrae are generally fused with the original sacrals into an elongated **synsacrum** supporting the pelvic girdle; thus, there is almost no motility in the avian column back

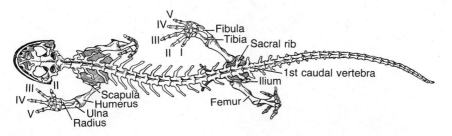

Figure 124. The skeleton of a salamander, as seen from above. (From Schaeffer.)

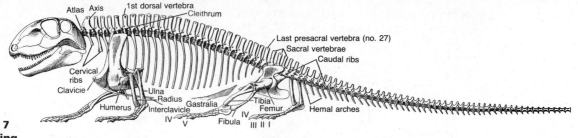

186

**Chapter 7
Supporting
Tissues—
The
Skeleton**

Figure 125. The skeleton of a generalized primitive reptile (the Permian pelycosaur *Haptodus*).

of the neck. The bony tail is much shortened and is capped by a **pygostyle**, formed of fused vertebrae, to which the tail feathers are bound.

Ancestral mammals appear to have had, much like early reptiles, about 27 presacral vertebrae and two or three sacrals. The tail, however, tended to become relatively short and weak. With remarkable consistency, the number of cervicals, which lack free ribs, remains at 7. Only in a few instances (tree sloths, manatees) is this rule violated, and a neckless whale and a giraffe both have 7 cervicals, appropriately abbreviated or elongated. The number of dorsal vertebrae is usually in the 20's and tends to remain rather constant within the limits of many mammalian families or orders. However, the number of ribs, and hence the proportion of thoracics to lumbars, is variable. Thus, in the cow-antelope family of ungulates, the number of dorsal vertebrae is almost always 19, but thoracics may vary from 12 up to 14, and the lumbars, correspondingly, from 7 down to 5. Mammalian sacrals are generally 3 to 5 in number. Mammalian tails vary greatly in length and number of vertebrae, from the practical absence of this structure in man and the great apes, to a few forms with as many as 50 caudal vertebrae.

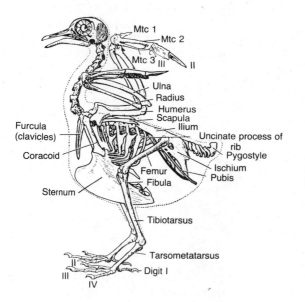

Figure 126. The skeleton of a bird (the pigeon). *Mtc*, Metacarpal; Roman numerals indicate digits. (After Heilmann.)

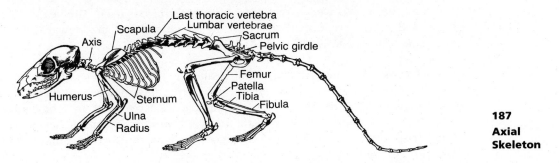

Figure 127. The skeleton of a generalized mammal, the tree-shrew *Tupaia*. (From Gregory.)

Atlas-Axis Complex (Fig. 128). In fishes, head and trunk move as a unit. In tetrapods, independent movement of the head is important, and modifications of the most anterior vertebrae and their articulation with the skull facilitate this. The articular surface at the back of the skull, the condyle, is in bony fishes a single rounded structure, comparable to the end of a centrum, and the oldest amphibians still retained this type. In most amphibians, living or fossil, the single condyle has divided into a pair, one at either side, and the articular surface of the first vertebra has become correspondingly divided. This allows the head to swing up and down in a hinge-like manner, but gives little scope for lateral movement. In most reptiles and in birds, the condyle remains single, but the first two vertebrae, termed the **atlas** and **axis**, have been modified to permit considerable freedom of movement.

In the atlas, the intercentrum and neural arch may form a ringlike structure (in many reptiles there may be, incidentally, a small, extra neural arch in advance of that of the atlas—a **proatlas**). The centrum of the first vertebra tends to remain independent of the atlas proper and to associate itself with the second vertebra, the axis, which typically has a large neural spine for the attachment of ligaments. In mammals (in which, as in amphibians, the condylar articulations with the skull are double), the axis bears anteriorly a process projecting from its centrum into the ventral part of the atlas ring; the embryologic story shows that this **odontoid process** includes the

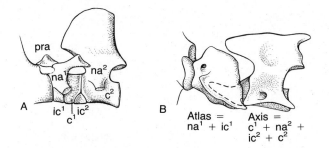

Figure 128. The atlas-axis complex. *A*, The occipital condyle and first two vertebrae in a primitive reptile (*Ophiacodon*). *B*, The same region in a typical mammal, showing fusion of elements. The proatlas in *A* is the neural arch of a "lost" vertebra of which the centrum has fused with the occiput. The broken line in *B* shows the position of the odontoid process (c^1), which runs forward inside the ring of the atlas. c^1, c^2, Centra of first and second vertebrae; ic^1, ic^2, intercentra; na^1, na^2, neural arches; *pra*, proatlas.

centrum of the atlas as well as the intercentrum of the axis itself. The head may move upward and downward on the atlas; the major component in sideways and rotary movements of the head is movement of atlas on axis.

Ribs in Fishes.   Although the vertebral column is the main skeletal structure upon which the powerful axial muscles of fish play, little musculature is directly attached to the vertebrae. For the most part, muscular force is exerted, in fishes, upon the connective tissue septa, **myocommata**, between successive muscle segments; ribs, formed at strategic points in these septa, connect with the vertebrae and render the muscular effort more effective.

In most fishes, each muscle segment in both embryo and adult is divided into dorsal and ventral parts by a longitudinal septum running the length of the flank (Figs. 2, 129). A logical place for rib formation is at the intersection of this septum with the successive transverse septa, or myocommata; in many fishes, **dorsal ribs** are present at these points. A second position in which ribs may develop is that in which the myocommata reach the walls of the coelomic cavity ventrally and internally. Ribs in this position, common in fishes, are termed pleural or **ventral ribs** and are serially identical with the structures termed hemal arches in the tail of both fishes and tetrapods; at the back end of the body cavity, the ventral ribs of the two sides approach one another and then, a bit farther back, fuse to form a V-shaped chevron. In addition to these two types of characteristic ribs, riblike intermuscular bones may be found in teleosts (the shad is all too good an example) at other points in the myocommata, making for greater muscular efficiency. The ribs, like the connective tissues in which they are found, are derived from mesenchyme and are usually preformed in cartilage even if later ossified. Proximally, close to the vertebrae, rib mesenchyme may be derived from the sclerotomes, more distally from the lateral plate of the mesoderm.

188
Chapter 7
Supporting
Tissues—
The
Skeleton

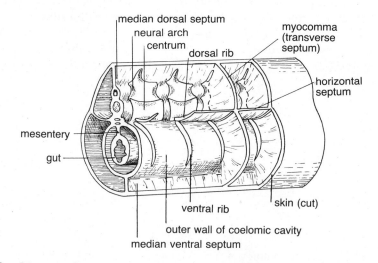

Figure 129.   Diagram of a section of the trunk of a vertebrate to show the connective tissue system and the axial skeletal elements. A view from the left side, as if partially dissected out and the muscles removed from between the septa. Vertebral elements develop in the tissue sheath surrounding the spinal cord and notochord; ribs, dorsal or ventral, develop where the transverse septum intersects the horizontal septum or coelomic wall. (After Goodrich.)

Cyclostomes lack ribs. In sharks, short ribs that appear to be members of the ventral series may be present. In many bony fishes, only ventral ribs are present, but in *Polypterus* and numerous teleosts, both dorsal and ventral ribs are to be found.

Tetrapod Ribs. In tetrapods, only one type of rib is present in the much reduced muscular wall of the body. Although doubts have been expressed, it is probable that these ribs correspond to the dorsal ribs of fishes. As in fishes, ribs were borne in ancestral tetrapods by every vertebra from the atlas to the base of the tail (cf. Fig. 125). The more anterior—cervical—ribs are short, in relation to the development of a flexible neck. The thoracic region is typically characterized by stout and elongated ribs, which are, through muscular connections, concerned in supporting the trunk on the pectoral girdles. These thoracic ribs are, in most groups of tetrapods, bound to a median ventral structure, the sternum. In primitive tetrapods, the vertebrae of the lumbar region may bear ribs, but these are relatively short. One or more highly developed sacral ribs, not differentiated in fish, connect the backbone with the pelvic girdle. Beyond the pelvis, ribs, diminishing in size posteriorly, may be present in the lateral muscle septum of the tail.

In ancestral tetrapods, the ribs were characteristically double-headed structures on the anterior vertebrae at least, although the two heads may be connected by a thin web of bone (Fig. 115). The **capitulum**, the head proper, was attached, in early forms, to the intercentrum. With reduction of the intercentrum in advanced tetrapods, this attachment tends to shift to the anterior edge of the adjacent centrum or to the intercentral space. The second head, the **tuberculum**, is essentially a short process from the curved proximal part of the rib and primitively attached to the transverse process of the neural arch. In the posterior part of the column, the two heads tend to come close to one another and may fuse, the capitular attachment shifting upward and backward. The powerful sacral rib or ribs have a broad head, a short shaft, and a distal expansion applied to the inner surface of the dorsal element of the pelvic girdle. The sacral ribs are immovably attached to the vertebrae, and the lumbar and caudal ribs, where present, tend likewise to be incapable of movement.

In modern amphibians, the ribs (like the vertebrae) have departed more radically than those of early reptiles from the primitive tetrapod type. They are always much reduced and never reach the sternum. In most frogs and toads, they are entirely absent except for a sacral. In urodeles, they are present but short and are sometimes forked distally. Proximally, each attaches by two heads to a transverse process of a peculiar type, not comparable to that in any other group. Ribs are well developed in the Gymnophiona.

Within the reptiles, there is much variation in rib structure. Particularly notable is the variation in vertebral attachment, with either a single or double head; either or both heads may attach to transverse process or centrum. In the Squamata, the single-headed ribs articulate with the centrum. Ribs, in connection with the trunk musculature, are very important in snake locomotion; they are highly developed the length of the long trunk. In the crocodilians, the ribs are two-headed; in the cervical region, the capitulum arises from the centrum, but in the trunk, both heads attach to the transverse process. Each thoracic rib in reptiles is typically formed in two segments: a proximal, ossified, rib proper, and a distal **sternal rib** (Fig. 130), which almost always remains cartilaginous. A joint between the two segments permits the flexibility often used in the expansion and contraction of the chest and consequent filling and

190

Chapter 7
Supporting
Tissues—
The
Skeleton

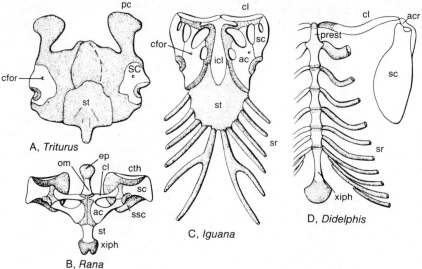

Figure 130. Ventral views of the shoulder girdle and sternal apparatus in various tetrapods. For lateral views, cf. Figures 138 and 139. *A,* Salamander; *B,* a frog; *C,* a lizard; *D,* a mammal (opossum). Anterior end at the top of the figures. In *A* and *C,* the dorsally turned scapulae are invisible. In *A,* the two coracoid cartilages overlap, as indicated by the broken line. *ac,* Anterior coracoid element; *acr,* acromion; *cfor,* coracoid foramen; *cl,* clavicle; *cth,* cleithrum; *ep,* episternum; *icl,* interclavicle; *om,* omosternum; *pc,* precoracoid region of coracoid plate; *prest,* presternum; *sc,* scapula (in salamanders, this ossification extends down into the place of the absent coracoid); *sr,* sternal ribs; *ssc,* suprascapula; *st,* sternum; *xiph,* xiphisternum. Cartilage stippled. (*A, C,* and *D* after Parker.)

emptying of the lungs. In turtles, the ribs are reduced in number; eight pairs of them are firmly fused to the overlying costal plates of the carapace.

In birds, the cervical ribs are fused to the vertebrae (cf. Fig. 126); free ribs, with ossified sternal segments, are confined to the compact thoracic region. These carry, as do those of some living reptiles and early fossil tetrapods, posteriorly directed **uncinate processes** for the attachment of muscles supporting the shoulder blade. In mammals, generally cervical ribs appear to be absent, but the embryologic story shows that the so-called transverse processes of the cervical vertebrae actually include short, fused, two-headed ribs; a foramen piercing the process represents the gap between capitulum and tuberculum of the rib. Distinct although immovable cervical ribs are present in monotremes. Ribs are present on all the thoracic vertebrae in mammals and are, indeed, diagnostic of that region; most curve forward ventrally to reach the sternum through cartilaginous sternal rib segments (**costal cartilages**), but the shorter posterior "floating ribs" are bound to that structure only via ligamentous connections with their longer anterior neighbors (cf. Fig. 127). The lumbar vertebrae have long transverse processes; as with the cervical vertebrae, these processes include a fused rib and are hence termed **pleurapophyses.** Caudal ribs are never developed in mammals.

Braincase. The braincase, or **chondrocranium**, of cartilage or replacement bone, forms the anterior end of the axial skeletal system, here greatly modified in relation to the brain and specialized sense organs of the head. In most vertebrates, the brain-case is fused with dermal and visceral skeletal materials to form a definitive and complex skull. In cyclostomes and Chondrichthyes, however, the braincase, owing to the absence of dermal bones—presumably a secondary condition—is a discrete skeletal element. The shark braincase shows a structure that, often with considerable modification, is repeated in all jawed vertebrates. In sharks, cyclostomes, and a few bony fishes, it retains its embryonic cartilaginous nature; in most other forms, it is more or less replaced by bony elements.

As seen in sharks, the braincase (Fig. 131) is a troughlike structure that forms a floor, side walls, and a partial roof for the brain. The roof, however, is never complete; one or more dorsal gaps, **fontanelles**, may be present. This lack of a complete roof

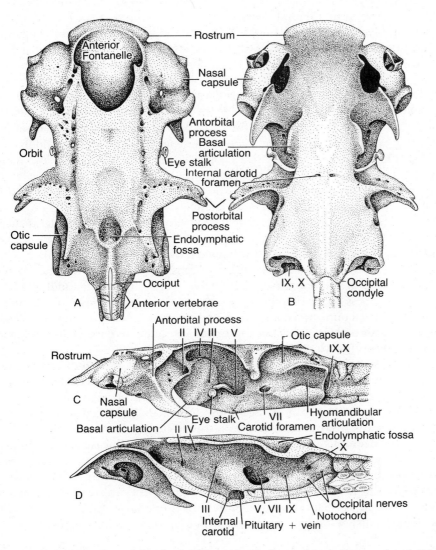

Figure 131. The braincase of the shark *Chlamydoselache; A,* dorsal, *B,* ventral and *C,* lateral views, and *D,* sagittal section. Nerve exits in Roman numerals. (After Allis.)

is repeated in most other vertebrates; in forms with an ossified dermal skeleton (such as placoderms), a bony roof was present overhead, and a complete braincase roof was unnecessary and only rarely present.

In adult sharks and in at least the embryos of many other forms, four major braincase regions may be distinguished, from back to front: occipital, otic, orbital, and ethmoid. The most posterior part of the braincase, the **occipital region**, is relatively narrow. Posteriorly, there is an opening, the **foramen magnum**, through which passes the spinal cord. Below this, a rounded **occipital condyle** abuts against the first vertebra. The notochord in the embryo, and sometimes in the adult, extends forward in the floor of the braincase from the condyle to a point just back of the pituitary body pendant from the brain. Anterior to the occipital region, the braincase expands laterally in the otic region; there is here incorporated on either side an **otic capsule** enclosing the sacs and canals of the internal ear. Still farther forward, the braincase narrows in the **orbital region** to aid in forming sockets, the orbits, for the eyeballs and their musculature. Here a median depression in the braincase floor may mark the position of the pituitary. Still farther forward, the braincase expands in an **ethmoid region** to terminate in a rostrum; on either side, in front of an antorbital process, are the **nasal capsules**, which enclose the olfactory organs.

Numerous openings, **foramina**, are present in the braincase for the cranial nerves and blood vessels. Their arrangement differs in various forms, but that shown in Figure 131 is reasonably typical. Nerves I and VIII, connecting with nasal pocket and internal ear, do not, of course, show superficially. In sharks, the endolymphatic ducts from the internal ear (cf. Chap. 15) open to the surface, typically into a median depression, or fossa, in the braincase roof. Ventrally, in the pituitary region, an opening is present for the carotid arteries, which supply blood to the brain, and other arterial and venous openings are present laterally.

A specialized element of the branchial skeleton, the hyomandibular, articulates rather loosely with the outer surface of the otic capsule and props the jaws on the braincase. The cartilages that form the upper jaws of sharks also articulate with the braincase (Fig. 166). In most sharks, the cartilages are loosely joined anteriorly with the under surface of the braincase; in primitive sharks, there is an additional articulation with a process of the braincase posterior to the orbit. In higher fishes and early tetrapods, this last connection tends to be reduced or absent, and there is instead a strong **basal articulation** between the upper jaw and the base of the braincase in the orbital region (Figs. 132 *E,* 167, 168).

In the embryo vertebrate (Fig. 132), the brain and the notochord extending forward beneath it to the pituitary region are already far advanced in development before skeletal structures appear. Braincase development begins ventrally, then continues laterally; because of the "precocious" growth of the brain to large size, it is difficult, so to speak, for the braincase to surround it, and the roof, as we have noted, often remains incomplete. The two basic embryologic braincase structures are the parachordals and trabeculae (Fig. 133). The **parachordals** are a pair of cartilages that develop beneath the posterior part of the brain on either side of the notochord; farther forward in lower vertebrates is a similar pair, the **trabeculae** (in mammals, however, a single median structure represents the trabeculae).* Presently, these paired elements fuse with their mates and with each other to form a nearly solid floor

---

*As noted later, the trabeculae originally may have been part of the branchial skeleton but have been "borrowed" by the axial skeleton to complete enclosure of the brain.

192
Chapter 7
Supporting
Tissues—
The
Skeleton

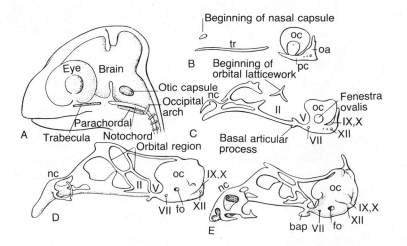

Figure 132. Stages in the embryonic development of the braincase of a lizard. In *A,* the outlines of the head, brain, and notochord are given for orientation. The stage shown in *A* is the earliest, that in *E* the latest. The main elements of the braincase structure are appearing—trabeculae, parachordals, otic capsule, and occipital arches (the nasal capsule appears later); further development consists in great measure of the growth and fusion of these elements. In the lizard, the orbital region grows as a complicated latticework, rather than a plate. Nerve positions are indicated by Roman numerals. *bap,* Basal articular process; *fo,* fenestra ovalis (for stapes); *nc,* nasal capsule; *oa,* occipital arch; *oc,* otic capsule; *pc,* parachordal; *tr,* trabecula. (After DeBeer.)

to the braincase. There may, however, be a long-persistent median opening for the hypophyseal pouch, which grows upward from the roof of the mouth to form part of the pituitary body; there are, as well, openings for the carotid arteries.

Alongside the posterior part of the brain, the sacs and canals of the internal ear develop at an early stage. A shell of cartilage, the **otic capsule**, forms around the outer surface of these structures and tends to enclose them almost completely. Farther forward, the eye, of course, is not enclosed (although there may be considerable

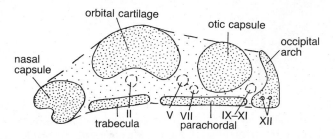

Figure 133. Diagram to illustrate the main embryonic components in the formation of a braincase. Trabeculae and parachordals are the main ventral elements to which an occipital arch (or arches) is added posteriorly. Usually later in development are more dorsal elements: otic capsule, nasal capsule, and orbital plate (the last, as in Fig. 132, often develops merely as a latticework). These primary elements are later bound together by a further growth of cartilage, leaving, however, gaps for nerves (indicated by the Roman numerals) and blood vessels.

cartilage in the wall of the eyeball—in a sense an optic capsule), but there soon appears, in one manner or another, a plate or latticework of cartilage internal to the eye, which often forms an **orbital plate**. The nasal region is relatively slow to develop, but there eventually appears a **nasal capsule** of complicated structure. The posterior boundary of the skull varies somewhat from group to group. The occipital region appears to be formed from one or several modified vertebrae; thus a variable number of **occipital arches**, somewhat resembling the vertebrae, appear behind the otic capsule.

194
Chapter 7
Supporting
Tissues—
The
Skeleton

As development proceeds, the ventral elements of the braincase fuse with the lateral structures and with the occipital arches to form side walls as well as a floor for the brain (cf. Fig. 163 *F, H*). Most of the openings for nerves and vessels lie in spaces between the major components. Thus, nerves V and VII have their exits in openings in front of the otic capsule; nerves IX, X, and XI emerge between the otic capsule and occipital region. Although a roof is slow to form, there is always a union of the two otic capsules above the foramen magnum.

The above description of the adult braincase was mainly based on that of sharks. In the other lower vertebrates in which the braincase is a separate entity—cyclostomes and chimaeras—its structure is highly modified and need not be considered in detail. In cyclostomes (Figs. 121, 163 *A*), it consists of little more than a rather simple trough, a pair of otic capsules, and a capsule for the single nasal sac. The chimaera braincase (Fig. 134) is short, high, and specialized in various features, of which the most noticeable is the fusion of the short upper jaw to the braincase. This peculiarity (repeated in lungfishes and in modified fashion in various terrestrial ver-

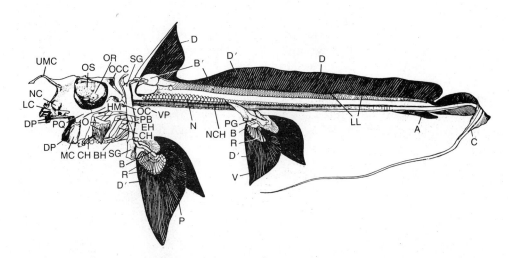

Figure 134. The skeleton of a female chimaera. *A,* Anal fin; *B,* fin basals; *B',* dorsal fin supports; *BH,* basibranchial; *C,* caudal fin; *CH,* ceratohyal, ceratobranchial; *D,* dorsal fin, with spine; *D',* dermal fin rays; *DP,* dental plates; *EH,* epibranchial; *HM,* hyomandibular; *LC,* labial cartilages; *LL,* ligament connecting fin supports; *MC,* lower jaw (Meckel's cartilage); *N,* cartilages of neural arch regions; *NC,* nasal capsule; *NCH,* notochord, with ring calcifications of sheath; *O,* cartilages in operculum; *OC,* occipital condyle; *OCC,* crest on occiput; *OR,* orbit; *OS,* interorbital septum; *P,* pectoral fin; *PB,* pharyngobranchial; *PG,* pelvic girdle; *PQ,* palatoquadrate, fused with skull; *R,* radial fin supports; *SG,* shoulder girdle; *UMC,* upper median cartilage of snout (this is not the clasper of the male); *V,* pelvic fin; *VP,* plate formed of fused anterior vertebrae. (From Dean.)

tebrates) is here associated, apparently, with the stronger support necessary for the shell-crushing activity of these mollusc-eating fishes.

The braincase of more highly developed fishes and tetrapods, described later as part of the skull, is usually well ossified in the adult but always develops in cartilage in a manner similar to that of sharks. In elasmobranchiomorphs and modern amphibians, the floor of the braincase is flat; in most other vertebrates, however, the braincase tends to be relatively high and narrower ventrally in its anterior portion.

Median Fins. Primitive vertebrates were aquatic. As the builder of ships is aware, the most efficient propulsive shape, that offering the least resistance to the water, is a fusiform one, in which the surfaces are as smoothly contoured as possible—"streamlined"—and in which, following a pointed anterior tip, the greatest width is not far from the anterior end of the body. Most fishes approximate this shape, although there are conspicuous variations. In general, the posterior part of the body tends to be flattened from side to side. This flattening is associated with the nature of the propulsive force; forward movement is due to lateral muscular movements of the body, which effect a backward pressure of trunk and tail upon the surrounding water (Fig. 135 *A*). Tension of the muscles on one side of the body produces lateral curvature

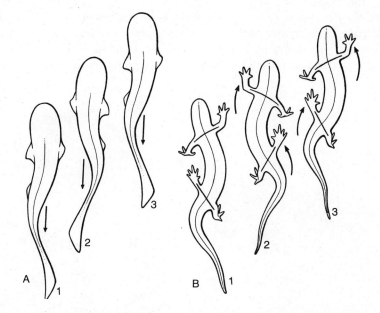

Figure 135. *A,* Dorsal views of a fish swimming, to show the essential method of progression by the backward thrust of the body on the water, resulting from successive waves of curvature traveling backward along trunk and tail. The curve giving the thrust (indicated by arrow) in *1* has passed down the tail in *2* and is replaced by a succeeding wave of curvature; the thrust of this wave carries the fish forward to the position seen in *3,* and so on. *B,* Dorsal views of locomotion in a salamander. Although limbs are present, much of the forward progress is still accomplished by throwing the body into successive waves of curvature. In position *1,* the right front and left hind feet are kept on the ground, the opposite feet raised; a swing of the body in *2* carries the free feet forward, as indicated by arrows. If these feet are now planted and the other two raised, a following reversed swing of the body will carry them forward another step, as seen in *3.*

in a given region; such curves, successively produced on opposite sides, travel back along the tail and often the trunk as well, pushing the body forward as a result of their backward thrust. The "snap" of this whiplike motion is accentuated at the posterior tip of the body, and an expanded caudal fin gives the maximum effect from the push at this point.

These propulsive movements would tend to be poorly regulated (as are those of a tadpole) in the absence of fins as aids in stabilization and steering. Rotary, rolling motions are checked in part by the development of median fins in addition to the caudal: dorsal fins—usually one or two of them—above, and an anal fin below, posterior to the anus. Originally, all the median fins may have formed a continuous fold down the back, around the tail, and forward to the anal region. Certain fish larvae do exhibit a continuous median fin, and dorsals and anals are often more broadly based in the embryo than in the adult. These facts tend to support such a theory; however, the oldest known fossil vertebrates, with a few exceptions, show median fins as discrete as any modern fish.

The skeleton of median fins is formed in the median dorsal septum above the neural spines and in the tail also in the ventral septum in which the hemal arches occur. In the tail, neural and hemal spines may themselves contribute directly to the support of the caudal fin (Fig. 136 A). This, however, is never the case with the other median fins—the dorsals and anal (cf. Figs. 122, 123, 134). Here, the supports are commonly rods of cartilage or bone, the **radials** or **radial pterygiophores** (sometimes in two rows), which may articulate with neural or hemal elements of the column but are often separated by a gap from these structures. Probably the radials originally corresponded in number to the body segments, but frequently in teleosts

196
Chapter 7
Supporting
Tissues—
The
Skeleton

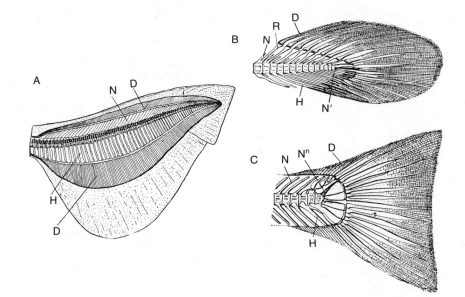

Figure 136. Caudal fins. *A*, Heterocercal type as seen in a shark (a similar structure occurs in sturgeons and paddlefish); *B*, diphycercal type, as seen in *Polypterus; C*, homocercal type of a teleost. Abbreviations: *D*, Dermal fin rays; *H*, hemal spines; *N*, neural arches; *N'*, tip of notochord; *R*, fin radials. In *C*, enlarged elements beyond *H* are hypural bones. (From Dean.)

double the number is present. Primitively, it appears, the radials extended well into the fin, as is the case today in sharks; in most actinopterygians, however, they hardly extend at all into the free fin, which is supported mainly by fin rays (lepidotrichia) articulating with the ends of the bony radials (cf. p. 173). There may be, as in many sharks, a concentration of elements at the fin base. With such concentration there may appear larger elements termed **basals** (Fig. 122), presumably representing a fusion of radials.

In many sharks and in the chimaeras, the median fins carry, anteriorly, spines that act as cutwaters, and spines were present on all the fins (except the caudals) in the fossil acanthodians (Fig. 27). In some ostracoderms, we find rows of spines as stabilizers in the absence of fins. Possibly, spines were the basic structures from which dorsal and anal fins evolved; a web of skin may have developed behind projecting spines, with a later strengthening of the structure by skeletal elements and fin rays.

Dorsal fins are normally one or two in number. Ostracoderms, placoderms, and Chondrichthyes all show both conditions in various cases; among bony fishes, the actinopterygians primitively have a single dorsal (Fig. 32 A), and early sarcopterygians have two (Fig. 29). Dorsal fins may be reduced or fused with the caudal as in modern lungfishes; on the other hand, there may be a secondary increase in numbers, as in the "many-winged" *Polypterus* (Fig. 32 B). An **anal fin** is almost always present ventrally behind the anus but may be lost in such bottom-dwelling forms as the skates and rays, and in lungfishes it fuses with the caudal.

Caudal Fins. In the caudal fin (Fig. 136), the major skeletal supports are the arches, neural and hemal, of the distal elements of the vertebral column. Additional radial elements are sometimes present, but these are seldom well developed; the fin proper consists of a stout web of skin supported by powerful fin rays.

Three main types of caudal fins are to be found in fishes: heterocercal, diphycercal, and homocercal. The **heterocercal fin** type is that familiar in sharks (Figs. 24 C, 122); the tip of the vertebral column turns upward distally, and the greater part of the fin membrane is developed below this axis. In the **diphycercal fin**, the vertebral column extends straight back to the tip of the body, with the fin developed symmetrically above and below it; the living lungfishes and *Polypterus* are examples of fishes with fins of this type (Figs. 30 B, 32 B). The **homocercal fin** is characteristic of teleosts (Figs. 36, 123). It is superficially symmetric, but dissection shows that the backbone, which terminates at the base of the fin, tilts strongly upward at its tip; the fin expanse is purely a ventral structure, despite its appearance, and is supported by enlarged hemal spines known as **hypurals**.

It seems logical to suppose that the symmetric diphycercal caudal fin is the primitive type, from which the others have been derived. This is not the case. The ancestral vertebrate, like amphioxus, may have had a simple straight tapering tail, but in almost every ancient fish known, the tail is strongly asymmetric. In most instances, diphycercal tails are shown by fossils to have arisen from heterocercal ones, and the homocercal type is likewise of heterocercal origin. Some of the oldest vertebrates, the ostracoderms, have tails of a "reversed heterocercal" type (Fig. 18 B, C), in which the backbone tilts downward rather than upward posteriorly. The modern cyclostomes have as adults a diphycercal tail, but the larval lamprey shows some development of the reversed heterocercal condition. In the jawed vertebrates of the Devonian period in which the tail structure is known, the heterocercal type is dominant

(Figs. 22, 24 A, C, 29, 30 A, 32 A). The tails of all well-known placoderms had this structure; the heterocercal type is found in the oldest Chondrichthyes and in the oldest bony fishes of all types as well; it is clearly the ancestral form in all jaw-bearing fish groups.

In later Chondrichthyes, the heterocercal tail remains dominant and is seen in typical form in modern sharks. However, skates and rays, with depressed bodies, and chimaeras (Figs. 25 B, 134) tend to reduce the tail to a rather whiplike structure, although the pattern is basically that of an attenuated heterocercal form; swimming in skates and rays is mainly accomplished by undulations of the enormous pectoral fins, and a caudal fin is redundant.

198
Chapter 7
Supporting
Tissues—
The
Skeleton

Among the actinopterygians, the typical Paleozoic forms, the palaeoniscoids (Fig. 32 A), had a good heterocercal tail. The sturgeons and paddlefishes have retained it (Fig. 33) but *Polypterus*, although primitive in other ways, has modified the tail to a diphycercal type (Fig. 32 B). In Mesozoic days, the holosteans, the middle group of ray-finned fishes, exhibited a tail technically heterocercal, but an abbreviate modification; the distal extension of the column into the fin is much shortened. This condition is retained in the garpikes and *Amia*, the modern holostean representatives (Fig. 34). Their tails are superficially symmetric and appear much like those of teleosts, but it is not until this last group appears that further shortening of the tip of the column and development of large hypural bones give us the true homocercal type with its deceptively simple and seemingly primitive appearance (Figs. 36, 123).

In the Sarcopterygii, the earliest lungfishes had typical heterocercal tails (Fig. 30 A). Soon, however, in geologic history, the tip of the tail tended to straighten and the dorsal and anal fins to lengthen in the direction of the caudal. In modern genera (Fig. 30 B), the tail fin is diphycercal; it reaches far anteriorly both dorsally and ventrally, owing to the incorporation in it of dorsal and anal fins. The most primitive of Devonian crossopterygians likewise had a heterocercal tail (Fig. 29 A). In this group, however, there was a rapid trend toward straightening of the tip of the vertebral column, and most crossopterygians show a characteristic type of three-lobed, diphycercal caudal fin (Fig. 29 B).

The development of symmetric tails, of both homocercal and diphycercal types, has taken place mainly among bony fishes with either lungs or an air bladder. The heterocercal tail may facilitate swimming with the body in a tilted position and thus counteract the effect of gravity; this is not as important in fishes that have air sacs capable of functioning as hydrostatic organs. However, even in forms with heterocercal tails, a direct anterior thrust can be achieved by twisting the caudal fin in the proper manner.

In tetrapods (except the earliest known fossil amphibian), the original fish median structures have been completely abandoned; a tadpole or salamander exhibits a tail expanded dorsoventrally as a good swimming organ, but completely lacks the internal structures characteristic of fish fins. Many amniotes have returned to an aquatic mode of life. Some, such as the turtles, the extinct plesiosaurs, the penguins, and the seals, never redeveloped the tail as a propulsive organ and rely instead upon limbs for locomotion. The extinct ichthyosaurs among reptiles and the whales and sea cows among mammals have redeveloped caudal fins, but these are never the same as those of their fish ancestors, long since lost. The caudal fins of whales and sirenians are expanded transversely, not vertically, and have fibrous rather than true skeletal supports. The ichthyosaurs have come closest to the redevelopment of a fish type of caudal fin supported by the axial skeleton; the backbone extends into the sharklike fin but extends into its lower, not into its upper lobe (Fig. 45 *D*). Ichthyosaurs and cetaceans also redeveloped dorsal fins, but skeletal supports are absent.

Heterotopic Elements. In addition to the normal cartilages and bones that are welded into the comprehensive pattern of the vertebrate skeleton, there may be mentioned here a variety of bones and cartilages that are not part of the proper skeleton but that arise as accessories to body organs in one vertebrate group or another. Generally these elements take the place of fibrous connective tissues otherwise present in the same situation; bone may form directly from such tissues or pass through an intermediate cartilaginous stage. Common, particularly in mammals, are small bones or cartilages formed in connection with muscle tendons at points of potential friction where, for example, a tendon passes over a bony ridge or joint. Such structures (frequently found in the manus or pes) are termed **sesamoids** (Fig. 137 *A*); the kneecap of mammals, the patella, is an "overgrown" sesamoid structure. Other bones that are found here and there in one vertebrate type or another and seem to be skeletal "afterthoughts," late in development, are termed **heterotopic bones**.

A **baculum** (os priapi, os penis, Fig. 137 *B*) is a heterotopic bone, a skeleton of the penis, found in all insectivores, bats, rodents, and carnivores, and in most primates except man; its widespread distribution suggests that it developed early in the evolution of mammals. It forms in the fibrous septum between the corpora cavernosa and above the urethra (cf. p. 441). In shape it is the most varied of all bones; it may be a plate or a rod of varied form—straight, bent or doubly curved; round, triangular, square, or flat in section; simple, pronged, spoon-shaped, or perforated; long or short. A similar but smaller os clitoridis is present in the females of many of these groups.

Other heterotopic elements have a more limited distribution in special groups. We may mention, as examples, the **os palpebrae** of crocodilians (and some dinosaurs), a plate embedded in the eyelid; the **os cordis** formed in the heart of deer and

A.

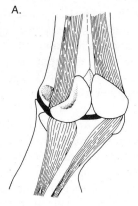

Figure 137. Heterotopic bones. *A,* Two sesamoid bones at the posterior side of the articulation between the metacarpal (*top*) and proximal phalanx (*bottom*) of a horse; the dashed lines show the tendons leading to and from the sesamoids. *B,* The baculum of an otter; the proximal end is to the left.

B.

bovids; **rostral bones** found in the muzzles of some mammals (and some dinosaurs); the **os falciforme**, a sickle-shaped element in the digging "hand" of moles, functioning as the skeleton of a supernumerary digit.

Appendicular Skeleton

200
Chapter 7
Supporting
Tissues—
The
Skeleton

The girdle and limb elements constituting the appendicular skeleton definitely belong to the general somatic system of endoskeletal structures; their history, however, has been distinctive. Normally two pairs of appendages are present, as fins in fishes and as the limbs of tetrapods. The anterior **pectoral appendages** are situated just behind the gills in fishes; in tetrapods, in an equivalent position at the border of neck and chest. The posterior **pelvic appendages** are typically placed at the back end of the trunk, just anterior to anus or cloaca.

Origin of Paired Fins.   The origin of paired fins has been much debated. A century ago, a famous anatomist of the day suggested, ingeniously, that these fins are modified gills. The limb girdles were compared with branchial bars; the fin was regarded as developed from the flap of skin, which, in a shark, forms the outer margin of the branchial pouch; its skeleton was supposed to have been derived from a concentration and extension of the small cartilages that stiffen the shark branchial septum. A host of embryologic and morphologic facts show that this idea is purely fanciful, but a reminiscence of this theory remains in the name archipterygium (p. 210) given to a leaf-shaped type of fin (Figs. 144 *H,* 145 *D*) that, under this theory, was supposedly primitive (but apparently is not).

In reaction against this theory arose a rival **fin fold theory** of the origin of paired fins. Its advocates pointed out that paired fins are similar to median fins in essential structure. In both, there is a fold of skin extending outward from the body, a centrally placed set of skeletal supports, and, on each surface of the skeletal structures, a sheet of musculature that moves the fin. If, as we believe, median fins have evolved in a simple manner as keels growing outward dorsally or ventrally from the trunk, may not the paired fins have developed from the flanks in a comparable way?*

The functional aspect of paired appendages may be considered in relation to this problem. As tetrapods, we tend to think of paired limbs as active propulsive organs. This is, however, not a function of the paired fins in most fish. In many modern fishes, they have relatively narrow bases and are capable of a free movement. This is of aid in steering but generally has little to do with propulsion. In fins that seem to be of relatively ancient patterns, there is generally, in contrast, a broad base that makes the fin relatively immobile. Primitively, it would seem, the paired fins were little more than horizontal stabilizing keels that aided the median fins in the prevention of rolling and had the further function of preventing fore and aft pitching. Both structure and function suggest that, like the median fins, the paired fins were developed as stabilizing aids in the aquatic locomotion of early vertebrates.

Among living fishes, all jawed types have developed typical paired fins; in contrast, they are completely lacking in cyclostomes. If we turn to the oldest and more

---

*In an extreme form of this theory, it was suggested that the paired fins developed out of a pair of long folds running the length of either flank and that these folds were continuous with a median fold running along the back and tail. There is little evidence to support this.

primitive fossil fishes with the expectation of finding a clear and simple answer to the problem of paired fin origins, we meet with disappointment. Among the ancient jawless ostracoderms, one major group (Heterostraci) shows no trace of paired fins. In the Anaspida, some forms, at least, had stabilizing lateral folds (cf. Fig. 18 C), and in some of the Osteostraci, there are paddle-like structures in the "shoulder" region; although these two groups may be related to the ancestry of lampreys, the relation of these paired fins to those of jaw-bearing forms is obscure. The fossil placoderms developed pectoral and pelvic paired fins (cf. Fig. 22); but in the most primitive members of the group, the pectoral fins seem to consist only of a long bony spine, and it is probable that fin development in placoderms took place independently of that in bony fishes. The Acanthodii, the problematic little "spiny sharks" (Figs. 27, 144 A), have paired fins supported by strong spines. However, even though pectoral and pelvic fins are present, further pairs may be added; in one form there are as many as seven pairs.

Pectoral Girdle—Dermal Elements (Figs. 138, 139). Each vertebrate limb includes not only skeletal elements lying within the free portion of the appendage but also a basal supporting structure, the limb girdle, within the substance of the trunk. This girdle acts as a stable base for the motions of the fin, and, in terrestrial vertebrates, it forms an intermediary through which the weight of the body is transferred to the limb.

The pectoral girdle is, in all major groups except the Chondrichthyes, a duplex structure, including both dermal and endoskeletal elements. The latter form the fin supports; the dermal bones, however, give added strength and help to tie the endoskeletal girdle to the body.

In early fishes, the anterior part of the body was armored by solid plates of bone, in contrast to the flexible scales found more posteriorly. In varied placoderms, particularly the arthrodires, this armor extended backward to enclose part of the trunk (Figs. 22, 138 A). The side walls of the thoracic "chest" armor, behind the gill opening, can be considered a **dermal shoulder girdle**, although differing greatly from that in higher fishes; an endoskeletal girdle was present within and behind the dermal girdle. The Chondrichthyes lack dermal elements, in correlation with the general reduction of the bony skeleton in that group.

In the more primitive bony fishes (Fig. 138 C), we find a dermal pectoral girdle with a pattern basic to that of all remaining vertebrate types. On either side, a vertical band of superficially placed dermal bone extends dorsoventrally along the back edge of the gill opening; the endoskeletal girdle, usually of modest dimensions in fishes, is bound to its inner and posterior surfaces. The main element in the fish dermal girdle is a vertically placed paired **cleithrum**; below it is the smaller **clavicle**, which curves downward and forward beneath the branchial chamber and is expanded ventrally, where the two clavicles have a union—a symphysis—with one another. Above each cleithrum, additional elements curve upward and forward above the gills and anchor the dermal girdle to the skull. This general type of structure is found in all known crossopterygians and lungfishes and all chondrosteans, living and extinct. In the holosteans and teleosts (Fig. 138 D), however the clavicle is lost.

In early fossil amphibians (Fig. 138 E), the dermal girdle is retained, but with notable changes. Dorsally, the connection with the skull is lost, a condition that allows the head to move more freely on the trunk. For part of their length, cleithrum

202

Chapter 7
Supporting
Tissues—
The
Skeleton

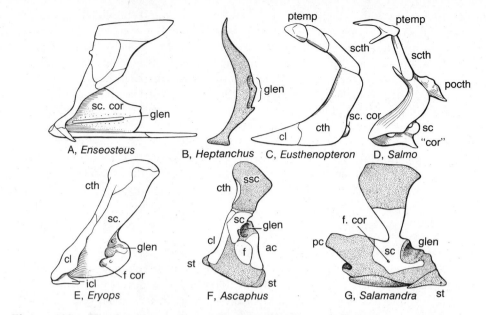

Figure 138. The shoulder girdle in fishes and amphibians. *A,* Devonian placoderm; *B,* a shark; *C,* a Devonian crossopterygian; *D,* a teleost (trout); *E,* a Paleozoic primitive amphibian; *F,* a frog; *G,* a salamander. Cartilage stippled. In all except *B* and *G,* a dermal girdle is present; in fishes *A, C, D,* this is the most prominent part of the girdle, including all parts except that labeled scapula and coracoid, and in the placoderm (*A*), the dermal girdle is the lateral part of an extensive thoracic armor. In amphibians, the dermal girdle is reduced or absent. Except in the shark, the endoskeletal girdle is relatively small in fishes and partially hidden beneath the dermal elements. In amphibians, the endoskeletal girdle is expanded, but generally ossifies from a single center, comparable to the scapula of amniotes; the frog, however, has a coracoid element. Much of the endoskeletal girdle is cartilaginous in living amphibians. *ac,* Anterior coracoid element; *cl,* clavicle; *"cor",* coracoid of teleost (homology with that of tetrapods doubtful); *cth,* cleithrum; *f,* foramen in coracoid plate of frog; *f. cor,* coracoid foramen for nerve and blood vessels; *glen,* glenoid cavity, point of fin attachment in fishes; *icl,* interclavicle; *pc,* precoracoid process of coracoid plate; *pocth,* postcleithrum; *ptemp,* posttemporal; *sc,* scapula; *sc. cor,* single scapulocoracoid ossification of fishes; *scth,* supracleithrum; *ssc,* suprascapula; *st,* sternum. Cartilage stippled. Anterior is to the left. (*A* after Stensiö, *C* after Jarvik; *D* and *G* after Parker.)

and clavicle are rather slender dermal rods lying along the front margin of the much enlarged endoskeletal girdle. The upper end of the cleithrum may expand to cap the shoulder blade. Ventrally, the clavicles typically bear triangular expansions; between them there is a diamond-shaped median **interclavicle**—a new element, although foreshadowed in some crossopterygians by an enlarged scale in this position.

Among modern amphibians (Fig. 138 *F, G*), the dermal girdle has disappeared completely in the urodeles. In the frogs, it is much modified. There is no interclavicle, but a cleithral rudiment may be present in the scapular region. The clavicle forms a bar connecting scapula and sternum. In some frogs, the clavicle is, quite unorthodoxly, developed in connection with a bar of cartilage; this cartilage may be an intrusion from the underlying endoskeletal girdle.

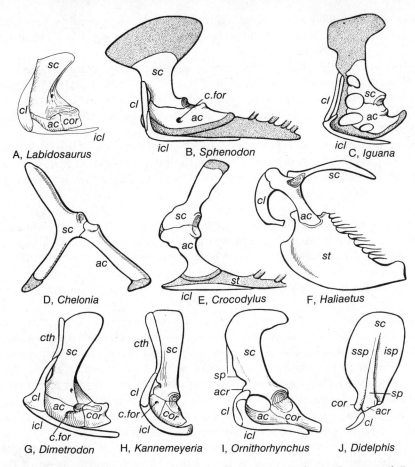

Figure 139. The shoulder girdle in reptiles, birds, and mammals. *A,* "Stem reptile" (cotylosaur); *B,* Sphenodon; *C,* a lizard; *D,* a turtle; *E,* a crocodile; *F,* a bird; *G,* a pelycosaur (primitive mammal-like reptile); *H,* a therapsid; *I,* a monotreme (duckbill); *J,* opossum. In *B, E,* and *F,* the sternum is shown. In most reptiles and birds, only one (anterior) coracoid element is present; in mammal-like forms, the true coracoid appears and persists, despite the loss of the coracoid plate area. The borders of the scapula and coracoid are often cartilaginous in reptiles. In lizards, scapula and coracoid are often fenestrated at points of muscular origins. The cleithrum has vanished in all living amniotes, but persisted long in mammal-like forms (*G, H*); its position is represented by the scapular spine, which lies at the front edge of the scapula in monotremes, but back of the new supraspinous fossa in higher mammals. *ac,* Anterior coracoid bone of reptiles, birds, monotremes; *acr,* acromion; *c. for,* coracoid foramen; *cl,* clavicle; *cor,* true coracoid; *cth,* cleithrum; *icl,* interclavicle; *isp,* infraspinous fossa; *sc,* scapula; *sp,* spine of scapula; *ssp,* supraspinous fossa; *st,* sternum. (*A* after Romer; *C* and *I* partly after Parker; *H* after Pearson.)

In primitive fossil reptiles (Fig. 139 *A*), the dermal girdle was like that of early amphibians; the cleithrum, however, disappeared at an early reptilian stage and is not present in any living amniote. Clavicles and interclavicle are little changed today in *Sphenodon* and lizards (Fig. 139 *B, C*) but have been subject to change and loss in other groups. In turtles (Fig. 114) they are incorporated in the plastron; in crocodilians, the clavicles have disappeared, but the interclavicle persists ventrally (Fig. 139 *E*). In birds, the clavicles and interclavicle are present (Fig. 139 *F*) in fused form as the **furcula**, or wishbone.

As witnessed by the living monotremes, some primitive mammals retained the general reptilian pattern of interclavicle and paired clavicles (Fig. 139 *I*), but above the monotreme level, the former has vanished (Fig. 139 *J*). The clavicles remain in many of the more generalized marsupials and placentals, articulating ventrally with the sternum. They are, however, frequently lost or reduced (as in the cat), particularly in running or bounding types in which complete freedom of the shoulder blade from the axial skeleton is desirable to relieve the jolts transmitted from front legs to body and to increase the functional length of the limb by allowing the scapula to rotate.

204
Chapter 7
Supporting
Tissues—
The
Skeleton

Endoskeletal Shoulder Girdle (Figs. 138, 139).   Functionally, the endoskeletal girdle always carries the limb articulation and is used as a major base of attachment of limb muscles. It is cartilaginous in the Chondrichthyes and partly or entirely cartilaginous in some of the less well-ossified bony fishes; otherwise it is a bony structure, although cartilage may persist in peripheral areas in many fishes, amphibians, reptiles, and even mammals.

Centrally situated on the endoskeletal girdle on either side is the limb articulation. In tetrapods with but a single proximal appendicular element, this is an articular socket termed the **glenoid fossa**; in most fishes, however, a number of skeletal elements articulate with the girdle. An area above the limb articulation, affording attachment to limb muscles, may be termed the **scapular blade**; a similar area below it is the **coracoid plate**. These regional names are derived from those of bony elements found in these areas in tetrapods, but we shall use the terms here merely as descriptive of regions of the girdle, without reference to ossifications.

In the Chondrichthyes, the dermal girdle is not developed, and in compensation, apparently, the endoskeletal cartilages (Fig. 138 *B*) are enlarged, those of the two sides fusing ventrally to produce a U-shaped structure. In bony fishes, the endoskeletal girdle is relatively small; it is generally ossified, with, in teleosts, as many as three small bony elements present.

In tetrapods, in which the dermal girdle tends to undergo reduction, the endoskeletal girdle is, in contrast with bony fishes, much expanded. This expansion is obviously related to the larger size of the land limb, compared with the fish fin, and the more extensive limb musculature attaching to the girdle. Above the glenoid fossa, in primitive tetrapods (Fig. 138 *E*), there is an elongate yet broad scapular blade; below, there is an expanded coracoid plate with a characteristic **coracoid foramen** for a nerve and blood vessels. In primitive amphibians, the entire endochondral girdle may be bony, but each half ossifies as a single element that corresponds to the scapular ossification of amniotes. The modern urodeles (Fig 138 *G*) also have a single ossification, but much of the girdle remains unossified, and there is an elongate cartilaginous anterior process from the coracoid plate region. In urodeles and primitive frogs, the lower margins of the two coracoid plates may overlap ventrally. In some frogs, a **firmisternal** condition develops (Fig. 130 *B*) in which the margins of the two coracoid plates are butted together ventrally, bracing the animal against the jar of landing. In frogs (Fig. 138 *F*) and in all early reptiles, there are two ossifications on either side. One, the **scapula**, centers in the scapular blade. The other, a ventral ossification, is sometimes called the **coracoid**, but, because it is not homologous with the mammalian bone of that name, it is better termed **procoracoid** or **anterior coracoid**.

In many reptiles (Fig. 139 *A–C*), there persists a girdle highly comparable to that of primitive tetrapods. However, turtles (Fig. 139 *D*) show a triradiate girdle, with two ventral prongs. These are not both coracoid bones, but actually the anterior ventral process is a downward extension of the scapula, which thus retains its connection with the clavicle (here a plate situated ventrally in the plastron). In crocodilians (Fig. 139 *E*) and dinosaurs, both scapula and coracoid tend to be elongate elements that meet at an angle at the glenoid fossa. This construction is in general retained in birds (Fig. 139 *F*); the coracoids brace the wing on the sternum.

The earliest mammal-like reptiles, the pelycosaurs (Fig. 139 *G*), show a third ossification not present in most other reptiles (although it does occur in some cotylosaurs, including that shown in Fig. 139 *A*). Most of the coracoid plate was still ossified as an anterior coracoid, but behind it there appeared a new element, equivalent to the true **coracoid** of mammals. This bone was small in pelycosaurs; in therapsids (Fig. 139 *H*) it grew forward at the expense of the anterior coracoid to occupy the greater part of this ventral plate. Meanwhile, excavation of the front edge of the coracoid plate produced a distinct process, the **acromion**, at the point of attachment of the clavicle to the scapula.

In the most primitive of living mammals, the monotremes, the endoskeletal girdle (Fig. 139 *I*) greatly resembles that of the therapsids. With the step upward from monotremes to marsupials and placentals, however, there is a major change (Fig. 139 *J*). The entire coracoid plate has vanished; the anterior coracoid has disappeared; and the coracoid has dwindled to a small process like a crow's beak (as its name implies), attached to the lower margin of the scapula. The girdle is thus practically reduced to a dorsally situated scapular blade, with the glenoid cavity on its ventral surface. The scapula has also changed radically, for, instead of being a simple plate, it has a ridge, the **scapular spine**, running down its length and separating **supraspinous** and **infraspinous fossae**; the acromion is situated not at the front end of the bone, but at the ventral end of the spine.

This major change in the structure of the girdle is related to the shift in limb posture in mammals as compared to reptiles, and consequent major changes in limb musculature (cf. p. 296). Much of the musculature that once arose from the coracoid plate has shifted upward to the scapula, rendering the former useless. The spine and acromion actually represent the original anterior margin of the scapula; the supraspinous part of the bone is a new shelf built out in front of the erstwhile front margin to accommodate part of the musculature that has migrated upward.

Sternum (Figs. 130, 139).   In many reptiles, the sternum is a ventrally placed, shield-shaped element, usually cartilaginous, that articulates anteriorly with the shoulder girdle and that posterolaterally connects on either side with the ventral ends of thoracic ribs to form a complete enclosure of the chest region, or thorax. Although there is little fossil evidence, the sternum was probably developed in early tetrapods; no such structure is found in fishes.

In modern amphibians, most urodeles have a typical sternal plate of cartilage; in frogs, there is a sternal plate or rod posterior to the girdle ventrally and, in some forms, a second element anterior to the girdle. Among reptiles, the sternum has been lost in turtles and in snakes and certain snakelike lizards. In birds (Fig. 139 *F*), the sternum is an enormous ossified structure, its size being due to the attachment to it of most of the mass of the chest muscles, important in flight; except in the ostrich-like birds, in which flight has been lost, the bone bears a huge ventral keel for additional muscle attachment. In typical mammals, the sternum, usually ossified, has a different type of structure from that of either reptiles or birds. It forms an elongated, jointed rod with which the ribs articulate at the "nodes."

Although the sternum is associated topographically with the shoulder girdle, it has been generally considered part of the axial skeleton because of its close association with the rib system. Its embryologic development, however, shows that it arises independently of the ribs and is, thus, to be regarded as an "appendix" to the shoulder girdle.

Pelvic Girdle (Figs. 140–143).   The pelvic girdle, a purely endochondral structure, differs markedly from the pectoral throughout the vertebrate series—even more markedly, perhaps, in lower than in higher types. Theories that pectoral and pelvic girdles are comparable, bone for bone, are purely fanciful. In fishes (Fig. 145), each half of the girdle is a small and simple ventral plate, often triangular, embedded in the muscles and connective tissues of the abdomen, primitively just anterior to the cloaca; it may shift anteriorly in actinopterygians. Never in fishes is there any connection with the vertebral column; the ventral tissues seem to be sufficient to anchor girdle and limb in place. The half-girdles of the two sides are usually in contact anteromedially. This contact, a primitive **pelvic symphysis**, affords further mutual support, and in the sharks and lungfishes, the two half-girdles are fused to form a single element. Along its posterolateral margin, the girdle affords a point of attachment for the fin. Although the half-girdle of either side is usually a single, compact structure, the sturgeons and the early shark *Cladodus* show evidence of subdivision, suggesting that the pelvis arose from a fusion of originally independent basal fin elements. In the Chondrichthyes, lungfishes, and a few ray-finned forms, the girdle remains cartilaginous in the adult; in other fish groups, each half of the girdle ossifies as a single element.

With the presence, in tetrapods, of large limbs that support much of the weight of the body, the pelvic girdle changes radically. Greater areas are needed on the girdle, particularly ventrally, for the attachment of the limb musculature; the problems of support and providing thrust in locomotion necessitate that the girdle be tied more closely into the body (Fig. 140).

In tetrapods, the original ventral part of the girdle on either side becomes a large plate of bone lying in a tilted position in the flank. This ossifies from two centers, the **pubis** anteriorly and the **ischium** posteriorly, and may be termed the **pubo-ischiadic plate**. Externally, it offers an area of origin for limb muscles; it is pierced by the **obturator foramen**, carrying the nerve supply to part of these muscles. Above the pubo-ischiadic plate, there develops the **acetabulum**, a large rounded socket that receives the head of the proximal limb bone, the femur. Pubis and ischium enter into the formation of the acetabulum; its upper margin is occupied by the **ilium**, which forms the dorsal third of the girdle. In fishes, there is sometimes a small dorsal process on the girdle; in even the most primitive of known amphibians, this process, as the ilium, had grown to be a large rodlike structure, and there early developed an

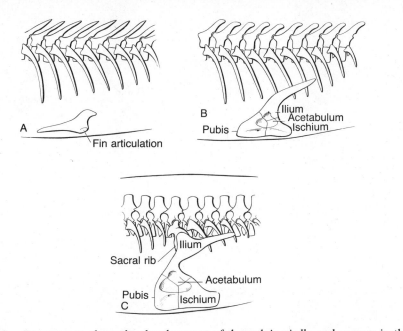

Figure 140. Diagrams to show the development of the pelvic girdle and sacrum in the evolution of amphibians from fishes. *A,* Left lateral view of the pelvic region of a fish, with the vertebral column and ribs above, the small pelvic girdle placed ventrally. *B,* Primitive tetrapod stage, found in some early fossil amphibians. The girdle has expanded, with the three typical bony elements. The ilium extends upward, but was presumably connected with the column only through ligaments binding it to the neighboring ribs. *C,* The girdle has grown further, and the ilium is firmly attached to an enlarged sacral rib. Anterior is to the left.

iliac blade in addition to the primitive rod. Beneath this blade lies an expanded sacral rib, which ties girdle and body together and is a major factor in the support of the body on the limbs.

The two halves of the pelvic girdle are, in most cases, broadly apposed ventrally in a pelvic symphysis. The cloaca lies posterior to the pelvic girdle; hence the girdle and the ribs and vertebrae with which it joins form a ring of bone bounding a **pelvic outlet**, through which materials in genital, urinary, and digestive systems must pass before leaving the body. The size of the outlet is an important factor in forms laying large eggs or bearing young alive.

There are marked variations among amphibians in the structure of the pelvic girdle (Fig. 141). The posterior projection of the ilium was usually reduced even in ancient times, and this bone became only a narrow vertical blade apposed to the sacral rib. In many extinct amphibians, the pubis failed to ossify, and it is never ossified in modern forms. The urodele pelvis is basically similar to that of older types except for the development anteriorly of a median cartilaginous prepubic process (not figured), which serves as an added support for the belly wall. In the anurans, in correlation with the hopping gait and shortening of the vertebral column, the ilium is a long slender bone extending far forward as well as upward from the acetabulum.

208
Chapter 7
Supporting
Tissues—
The
Skeleton

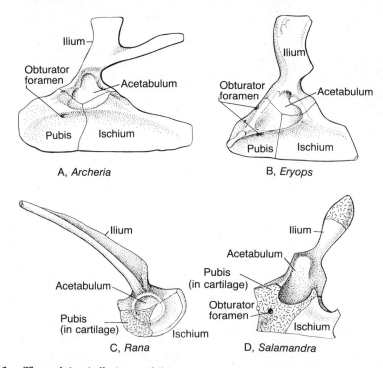

Ilium

Obturator
foramen

Acetabulum

Pubis   Ischium

A, *Archeria*

Ilium

Obturator
foramen

Acetabulum

Pubis   Ischium

B, *Eryops*

Ilium

Acetabulum

Pubis
(in cartilage)

Ischium

C, *Rana*

Ilium

Acetabulum

Pubis
(in cartilage)

Obturator
foramen

Ischium

D, *Salamandra*

Figure 141. The pelvic girdle in amphibians. *A,* A primitive fossil labyrinthodont; *B,* a typical later labyrinthodont; *C,* a frog; *D,* a urodele. A posterior process of the ilium, present primitively, is retained in many reptiles (cf. Figs. 142 and 143) but is lost in most Amphibia (represented by a prong in *B*). In the anurans, the ilium is a specialized and elongate rod. The pubis was primitively well ossified but remains cartilaginous in many fossil forms and in all modern amphibians. Anterior is to the left.

In reptiles (Fig. 142 *A*), the blade of the ilium tends to be expanded for the attachment of more extensive limb muscles on the outer surface and for the attachment medially of two or more sacral ribs rather than the single rib usual in amphibians. Within this class there occurs great variation in pelvic structure. Much of the outer surface of the pubo-ischiadic plate was primitively occupied by the broad, fleshy origin of a muscle (obturator externus; cf. Fig. 210) running to the under side of the femur. An opening, or fenestra, covered by a membrane, tends to develop beneath the origin of the muscle, separating pubis and ischium and turning the pelvic girdle into a tripartite structure; this **pubo-ischiadic** (or **thyroid**) **fenestra** is present in turtles, *Sphenodon,* and lizards (Fig. 142 *B*).

In the archosaurs (Fig. 142 *C–E*), the girdle becomes highly modified in relation to the development of bipedal habits. In these reptiles and their avian descendants, the acetabulum usually becomes open at its base for the better reception of the head of the femur. The ventral plate tends to be reduced, and the attachments of the limb muscles move fore or aft toward the ends of pubis or ischium, gaining a more advantageous leverage. Both these bones curve downward at their ends, and in all except primitive fossil members of the Archosauria, the symphysis becomes restricted to their tips. In bipedal archosaurs and in birds, the ilium is elongate, in correlation with

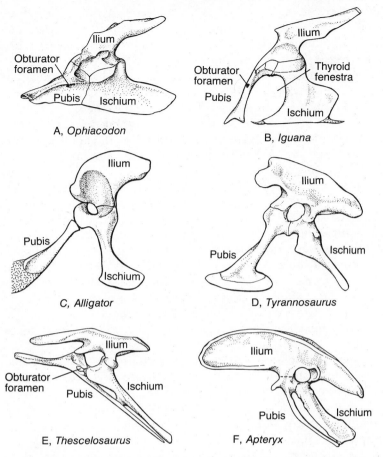

Figure 142. The pelvic girdle in reptiles and birds. *A,* Primitive reptile; *B,* a lizard; *C,* a crocodilian; *D,* a reptile-like dinosaur (Saurischia); *E,* a birdlike dinosaur (Ornithischia); *F,* a bird (the kiwi). In *A,* the ilium is a low blade, and the same is true in lizards; in other forms, this structure is more expanded. In the dinosaurs and birds, with a bipedal type of locomotion, the ilium has grown forward somewhat as in mammals (cf. Fig. 143). In the archosaurs *(C, D, E)* and the birds descended from them, there is typically an open bottom to the acetabulum for the better reception of the head of the femur. In the primitive reptile, the pubo-ischium is a solid plate. In lizards, there is a large thyroid fenestra developed between pubis and ischium, from which a large muscle to the femur takes origin (*Sphenodon* and turtles are similar). This fenestra is comparable to one seen in mammals, but there the obturator foramen is concerned in the development of the fenestra. In such archosaurs as the alligators and saurischians, there appears at first sight to be a similar structure. Actually, however, this is not the case; the pubis and ischium are twisted downward, and the true ventral margin of the girdle is the curving lower margin of pubis and ischium. In *C* and *D,* the pelvis is triradiate, with a simply built pubis; in *E* the pubis is two-pronged, and the pelvis is tetraradiate. In the bird, the anterior process of the pubis is reduced or absent. In the alligator, the pubis is excluded from the acetabulum by the ischium; the pubic region of the girdle extends forward along the belly as a fibrous cartilage. Anterior is to the left.

a considerable increase in the number of sacral vertebrae that must transfer the entire body weight to the limbs via the ilium. In the so-called reptile-like dinosaurs, the saurischians, the pubis and ischium tend to become rodlike, giving a triradiate type of pelvis (Fig. 142 *D*). In the crocodiles (Fig. 142 *C*), the pelvic structure is basically similar to that of saurischians except that the ischium has expanded and excluded the pubis from the acetabulum. The ornithischians (Fig. 142 *E*), the so-called birdlike dinosaurs, get their name from the fact that (as in birds) the slender main shaft of the pubis has swung far back to parallel the ischium; this pattern may allow better balancing of the large digestive organs of herbivores over the hindlegs of bipedal forms. The pubis may send out, as a support for the belly, a stout process in much the position of its original shaft.

In birds (Fig. 142 *F*), as in ornithischians, the pubis has swung back to a posterior position. There is, however, little development of an anterior pubic process; the greatly developed sternum forms a bony sheath over most of the ventral surface of the body, and further support by such a process is little needed. Except in the ostrich, the pelvic symphysis has disappeared, and the two halves of the girdle are widely separated ventrally, a situation correlated with the relatively large size of the bird egg and the consequent necessity for a large pelvic outlet. Some strengthening of the ventral parts of the girdle is, however, afforded by a fusion between the posterior ends of the three pelvic elements of either side in many birds.

In the mammal-like reptiles and mammals (Fig. 143), marked changes associated with changed limb posture and changed musculature have occurred. The muscles and the bones from which they originate undergo a marked rotation, counter-clockwise as seen from the left side. The ilium, which primitively grew mainly backward dorsally, comes to extend anteriorly as it reaches upward to the sacrum. The pubis and ischium, on the contrary, have moved posteriorly so that hardly any of the ventral plate of the girdle extends anterior to the acetabulum.

The mammalian ilium is primitively a rather slender rod, triangular in section; however, in heavy-bodied ungulates (e.g., horses, cattle, elephants) or bipeds (e.g., man), in which there are powerful gluteal muscles running from ilium to femur, this bone may be much expanded. As in many reptiles, the pubo-ischiadic plate is fenestrated. Here, however, in contrast to reptiles, the foramen for the obturator nerve is included in this opening, termed the **obturator fenestra**. In connection with viviparity, the pelvic opening is often broader in the female, and in some forms the symphysis (mainly between the pubes) is loosened under hormonal control at the time of parturition. A few mammals, such as shrews, have no pelvic symphysis. In monotremes and marsupials, a pair of "marsupial bones" (**prepubes**), extends forward from the pubes and supports the body wall. They were not present in the ancestral reptiles and are absent in almost all placentals;* the reason for their presence in these two groups of primitive mammals is not clear (although they help to support the abdominal muscles and the pouch for the young).

Paired Fins In Fishes (Figs. 144, 145).   Apart from various peculiar early patterns, we see in most fishes, living or fossil, two basic types of paired fin skeletons, together with various intermediates. One type is the **archipterygium**, typically developed in

*Prepubes are reported in at least one very primitive eutherian, a fossil from the Cretaceous of Mongolia.

210
Chapter 7
Supporting
Tissues—
The
Skeleton

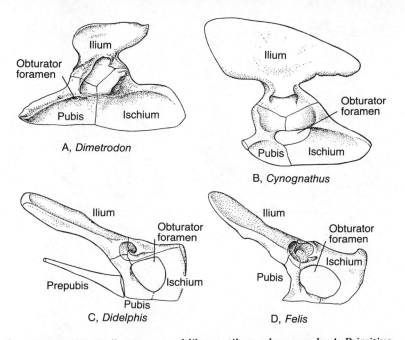

Figure 143. The pelvic girdle in mammal-like reptiles and mammals. *A,* Primitive pelycosaur; *B,* a therapsid; *C,* opossum; *D,* cat. The *Dimetrodon* pelvis is of a primitive reptilian type. In the therapsid, the ilium has grown forward dorsally, pubis and ischium have swung back ventrally, and the obturator foramen has expanded (comparable to the situation in many modern reptiles) into an obturator (or thyroid) fenestra. The opossum and cat show a typical mammalian structure, with a large obturator fenestra, a shortened ischium and a slender ilium (secondarily broadened in many heavy mammals, however). The opossum, like other marsupials and the monotremes, has a pair of "marsupial bones" (prepubes) rarely found in other groups. In the cat, as in certain other mammals, an accessory element is seen in the acetabulum.

the lungfish *Neoceratodus* (Figs. 144 *H,* 145 *D*). The skeleton of the leaf-shaped, narrow-based fin consists of a jointed central axis and side branches that are usually better developed on the anterior (preaxial) margin of the fin. Typical archipterygia are found in fossil lungfishes and in some early crossopterygians; hence, the archipterygium may be a basic type in the sarcopterygians. But except for an aberrant family of extinct sharks, it is unknown in other fish groups, ancient or modern, and hence is unlikely to have been antecedent to other types of fins. In most crossopterygians, we find an abbreviated type of archipterygium (Figs. 144 *I;* 145 *E*) in which the axis is short and the branches are generally confined to the anterior margin—a fin pattern of great interest as one seemingly antecedent to the skeletal plan of the tetrapod limb.

In strong contrast to the archipterygium is the finfold type of fin seen in ancient Paleozoic sharks such as *Cladoselache* (Figs. 144 *B;* 145 *A, Cladodus*). Here the fin has a broad base and hence was little more than a fixed horizontal stabilizer; there is no projecting fin axis, and the skeleton consisted of parallel bars of cartilage. This type of fin persists with modification in modern cartilaginous fishes (Figs. 144 *C, D;* 145 *B, C, F*). The major change is that the base of the fin has generally become much narrowed, making it more flexible and a more effective steering device. With this

212

**Chapter 7
Supporting
Tissues—
The
Skeleton**

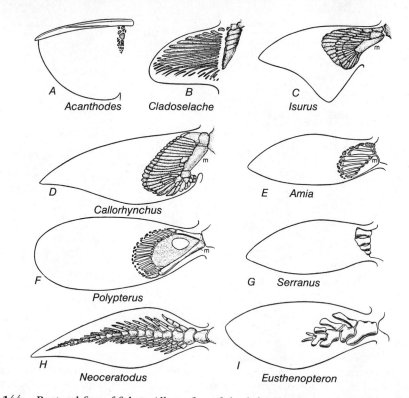

Figure 144. Pectoral fins of fishes. All are fins of the left side, viewed from the upper surface, so oriented that the long axis of the body is vertical on the page, the anterior end above. The outline of the complete fin is indicated, and (except for *A*) the articular region of the girdle is included at the right. Cartilage is stippled and bone shown as white. *A,* Fossil acanthodian, with the fin skeleton little developed, and a spine forming a cutwater and main fin support. *B,* Primitive fossil shark, with a parallel-bar type of fin. *C,* Modern shark with a narrow-based flexible fin and a basal concentration of bars with formation of a metapterygial axis (the metapterygium is the elongate posterior basal element, *m*). *D,* Comparable type found in chimaeras. *E,* Primitive actinopterygian type, with parallel-bar construction, but metapterygial axis present. *F,* An aberrant modification of the last, found in the archaic actinopterygian *Polypterus. G,* Teleost (sea bass) with a much reduced skeleton. *H,* The typical archipterygium of the Australian lungfish. *I,* The abbreviate archipterygium of a fossil crossopterygian. (*A* after Watson; *B* after Dean; *C* and *D* after Mivart.)

constriction of the base, the basal cartilages have, of necessity, become crowded together; one of the most posterior cartilages tends to become particularly prominent and acts as an axis on to which many of the other bars articulate. This axial element is frequently termed the **metapterygium**.* It is obvious that the modern type of shark fin shows a condition intermediate between the finfold type and the archipterygium; there is here a main fin axis with side branches, although the branches are almost all on the exterior or anterior margin of the axis. In the elasmobranchs, at

---

*Some sharks have three elements attaching basally, termed the pro-, meso-, and metapterygium, but there is great variation in elements other than the metapterygium.

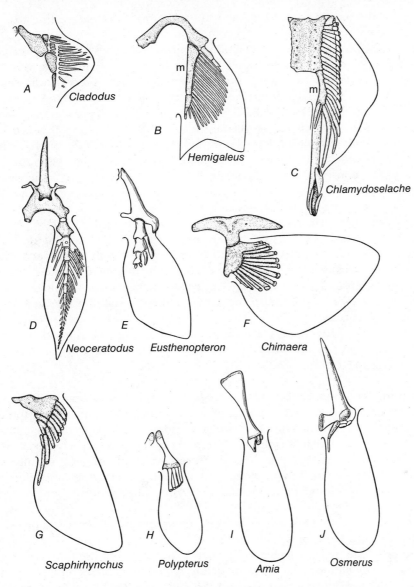

Figure 145. Pelvic fin of fishes. All are of the left side, viewed from below, the midline of the body is at the left, the anterior end above. The fin outline is indicated in each case. The left half of the pelvic girdle is included, or the whole girdle if the halves are fused. Cartilage is stippled and bone shown as white. *A,* Primitive Carboniferous shark, with a broad-based fin and parallel bars as fin support. *B,* Modern female shark, with similar construction. *C,* Male shark, with additional cartilages supporting the clasper. *D,* The archipterygium of the Australian lungfish. *E,* The abbreviate archipterygium of an ancient crossopterygian. *F,* Female chimaera fin, essentially similar to that of sharks. *G,* A sturgeon, also with a sharklike construction. *H, Polypterus; I,* the bow-fin; *J,* a teleost, showing nearly complete reduction of the bony fin skeleton. *m,* Metapterygium. (*A* after Jaekel; *B* after Garman; *C* after Goodey; *D, H, I* after Davidoff; *E* after Gregory.)

least, the parallel-barred finfold type appears to be primitive, suggesting that the archipterygial type is a secondary, not a primitive, type of fin.

The actinopterygians are by definition a group of fishes in which, in general, skeleton and flesh extend hardly at all into the fin, which is mainly a web of skin supported by dermal rays. Consideration of a series of actinopterygian fins indicates that they have been derived from the finfold type and have evolved in a manner comparable to that of sharks. The basic structure is a series of parallel bars; in such a primitive ray-finned form as the sturgeon, the pelvic fin (Fig. 145 *G*) is broadbased, with numerous bars extending into the fin much as in a primitive shark. As in later sharks, in ray-finned forms, there is a trend toward a narrow fin base and greater flexibility, with a consequent development of a metapterygial axis in some cases (Fig. 144 *E*).* In teleosts (Figs. 144 *G*, 145 *J*), the bony elements are much reduced; only a few short parallel bars typically remain in the pectoral fin, and in the pelvic fin, these may be further reduced to nubbins of bone or cartilage.

In fishes, generally the pectoral fins are more important as steering devices than are the pelvics; they are thus larger and have a more highly developed skeleton. Pectorals are practically universal in jawed fishes, but the pelvics may be reduced or may, in many teleosts, move forward to a position beneath the shoulder region (cf. Fig. 36 *B*) or even forward beneath the "chin," aiding in stabilization; such displacement is usually correlated with a dorsal migration of the pectorals. In the Chondrichthyes and in at least one placoderm, associated with internal fertilization (cf. Chap. 13), the finger-like **claspers** are supported by extra posterior cartilages on the pelvic fins (Fig. 145 *C*).

The Primitive Tetrapod Limb. The limbs of tetrapods first appear strikingly different from fish fins, but, although the greater size and greater complexity of the former tend to obscure the relationship, the two are readily comparable in basic structure. The numerous muscles that sheathe the tetrapod appendage can be resolved on comparative and embryologic grounds into two series comparable to the simpler muscle masses on the upper and lower surfaces of the fish fin, and the limb elements of terrestrial forms are comparable in essence to the fin supports of crossopterygian fishes with abbreviate archipterygia.

The limb of a primitive land vertebrate is composed of three major segments (Fig. 146). In both front and hind limbs—corresponding, respectively, to pectoral and pelvic fins—the proximal segment projects laterally from the body and includes only a single bony element, humerus or femur, which primitively moves fore and aft in an essentially horizontal plane. Beyond the elbow or knee a second segment lifts the body off the ground; ancient tetrapods did not crawl on their bellies. Here movement in a vertical plane parallel to the long axis of the body is capable of increasing the progression caused by the fore and aft movement of the proximal elements. Two bones are present in this segment: radius and ulna in the forelimb, tibia and fibula in the hind, the first named in each case being the preaxial (anterior or medial) element of the pair.

The third segment in either pectoral or pelvic limb is that of the foot, termed the manus in the pectoral limb, pes in the pelvic limb. This includes proximally the

214
Chapter 7
Supporting
Tissues—
The
Skeleton

Polypterus (Fig. 144 *F*) shows a peculiar development of the pectoral fin, but the primary structure is still one of parallel bars.

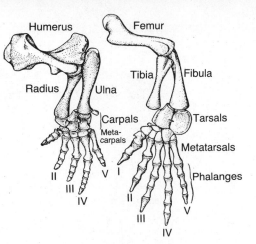

Figure 146. Left front and hind limbs of a primitive reptile *(Ophiacodon),* to show the general pattern of limb construction in early tetrapods. Roman numerals indicate the digits.

wrist or ankle region, carpus or tarsus, which forms a more or less flexible adjustment between the foot and the more proximal part of the limb and, distally, the toes, or digits, placed upon the ground. The proximal element of a toe (contained in palm or sole) is a metapodial—a metacarpal or metatarsal; the remaining elements are the phalanges.

This pattern of skeletal construction is, as mentioned, basically comparable to that found in crossopterygian relatives of the tetrapods (Fig. 147): a single proximal element articulating with the shoulder girdle and equivalent to humerus or femur, and, in a second segment, two elements comparable to the radius and ulna of the pectoral limb or tibia and fibula of the hind leg. Beyond this point, the fin skeleton of known crossopterygians shows an irregular and variable branching arrangement; detailed homologies of these distal elements with particular bones of the manus or pes are obscure.

From the first, there were notable differences between the front and hind legs of tetrapods despite the basic similarity of their bony patterns, and many of these differences are retained in later tetrapods. We note particularly the differences in the major

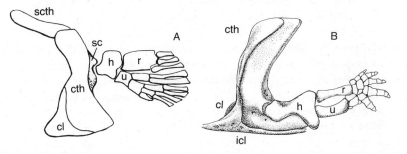

Figure 147. The shoulder girdle and pectoral fin of a crossopterygian, *A,* and the same structures in an ancient fossil amphibian, *B,* placed in a comparable pose to show the basic similarity in limb pattern. *h, r,* and *u,* Humerus, radius, and ulna of the tetrapod and obvious homologues in the fish fin. *cl,* Clavicle; *cth,* cleithrum; *icl,* interclavicle; *sc,* scapula; **scth,** supracleithrum. Anterior is to the left. (*A* after Gregory.)

216

**Chapter 7
Supporting
Tissues—
The
Skeleton**

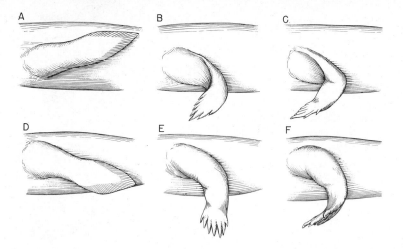

Figure 148. Diagrams to show the postural shift in the paired limbs in the transition from fish to amphibian. *A* to *C,* Pectoral limb; *D* to *F,* pelvic limb; *A, D,* fish position; *B, E,* transitional stage; *C, F,* amphibian position (cf. text).

joints between limb segments. Below the elbow, radius and ulna rotate rather freely on the humerus, but the comparable knee joint is invariably a simple untwistable hinge. Wrist articulations are essentially hingelike; although the ankle is a hinge in mammals, the foot could be readily rotated on the shin in reptiles and early amphibians. This contrast in articular structure can, it seems, be traced back to a contrast in the position of the pectoral and pelvic fins in the ancestors of the tetrapods (Fig. 148). The changes involved are still open to some question, but the theory outlined here seems most probable. As in living lungfishes, the fins of the fish forebears of tetrapods extended outward and backward from the body. For the pectoral fin, a major forward twist occurred at the future elbow; this brought the forearm forward, with the "hand" directed anteriorly, and only a simple hinge was needed in the future wrist region. In the pelvic fin, the future limb, projecting laterally, developed a simple hinge at the knee that brought the lower leg into vertical position; this segment, however, faced more or less outward and hence sharp rotation at the future ankle was needed to bring the foot into a fore and aft position.

Limb Function and Posture. The development of terrestrial habits brought about a revolution in vertebrate locomotion. In fishes, propulsion is normally accomplished by undulatory movements of the trunk and tail. Undulatory movement long remained important in early tetrapods and is still important in salamanders (Fig. 135). Here the feeble limbs play only a minor role as positive propulsive agents; they are in great measure stationary organs through which the push of body undulations may be exerted on the ground. In tetrapods generally, however, the limbs have taken over a positive and dominant role in progression.

Primitive land vertebrates, both amphibians and reptiles, had a sprawling posture, with the limbs extending far out from the side of the body. This pose is still preserved in living urodeles and with little modification in turtles, lizards, and *Sphenodon.* But it wastes energy in that much muscular effort is used up in merely keeping the body off the ground, and most terrestrial vertebrates have modified this primitive structure in various ways. The frogs have become specialized for a hopping gait, with highly modified hind legs. Archosaurs show a strong trend toward bipedalism, and the Mesozoic saw a variety of bipedal dinosaurs (cf. Fig. 47). The front legs were little modified, but hind limbs were strengthened and lengthened, and the knees rotated anteriorly to a position essentially beneath the body, thus transferring much of the burden of weight directly to the limb bones. Many of these giant reptiles reverted to a quadrupedal gait, and the living archosaurs, the crocodilians, are likewise quadrupedal. With the front limbs freed from locomotor function, the development of wings and flight became possible among archosaurs. Pterosaurs evolved a wing membrane supported by a single elongate finger; birds, with feathered wings, developed from another group of bipedal archosaurs.

A second type of major improvement in terrestrial locomotion, one in contrast with bipedalism, was the development of the efficient four-footed gait characteristic of mammals. The knee is turned forward as in bipedal reptiles and birds, but in addition, the elbow was turned backward and brought in close to the body. The entire limb movement thus becomes a fore and aft swing, allowing a longer stride and greater speed; further efficiency is gained through the fact that the body is supported directly by the bones of the four limbs, without the constant muscular effort needed in the old-fashioned sprawled position. This pattern of locomotion was developed by the therapsids and remains relatively unchanged in most modern mammals. There are, however, many variations, such as the flying mechanisms of the bats, the digging specializations of moles, the bipedalism of man, and so forth.

Many amniotes, despite their fitness for terrestrial life, have returned to the water, with consequent modification of limb structure. In general, the limbs of aquatic tetrapods tend to reassume a finlike shape, with shortened proximal segments and an expanded "foot." In less specialized aquatic types, such as seals, the proximal elements may remain little modified, but in advanced types, such as the whales and the extinct ichthyosaurs and plesiosaurs, the whole limb may be modified into a flipperlike structure (Fig. 149). Aquatic birds usually use webbed feet for swimming, but penguins and auks use their powerful wings as swimming organs.

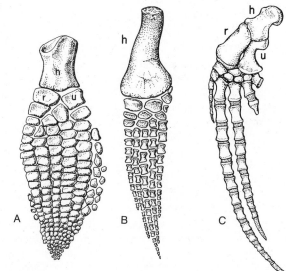

Figure 149. Examples of tetrapod appendages retransformed into fishlike paddles in marine forms. Pectoral limbs of *A,* an ichthyosaur; *B,* a plesiosaur; *C,* a whale. In all, there has been a shortening and broadening of the proximal bones and a multiplication of the phalanges. In *A,* there has also been an increase in the number of digits. *h,* Humerus; *r,* radius; *u,* ulna. (*A* after Huene; *B* after Williston; *C* after Flower.)

The gymnophionans, snakes, amphisbaenians, and certain lizards have abandoned limbs and reverted to an essentially eel-like piscine type of locomotion based on body undulation; gymnophionans, amphisbaenians, and most limbless lizards are burrowers, as were probably the ancestral snakes. In whales and sirenians, the pelvic limbs are reduced or absent, and among bipedal forms, the "arm" is reduced in certain dinosaurs and nonflying birds.

218
Chapter 7
Supporting
Tissues—
The
Skeleton

Humerus (Fig. 150).   In many primitive tetrapods, the two ends of the humerus are much expanded and seemingly "twisted" on one another; the proximal end of the bone is essentially horizontal, the distal end tilted so that its lower surface faces the forward-slanting forearm. Proximally, stout processes are developed on the humerus for the pectoralis, deltoid, and subscapular muscles, the insertion of the last represented in mammals by the **lesser tuberosity**. In the mammalian humerus, the **greater tuberosity** affords attachment for other muscles from the scapula. In primitive tetrapods, the proximal articular surface is an elongate oval capping the end of the bone; in later types, there tends to develop a spherical head distinct from the shaft. Primitively little if any distinct shaft was present between proximal and distal expansions, but in many small reptiles and in most birds and mammals, the bone is more slenderly built and elongate.

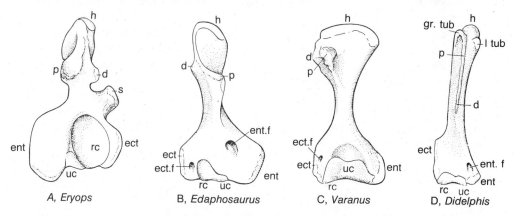

Figure 150.   Humeri of *A,* a Paleozoic amphibian; *B,* a primitive fossil reptile; *C,* a lizard; *D,* the opossum, all from the ventral side. *A* shows a left humerus; the others are right humeri. Primitive humeri were short, practically without a shaft, and much expanded at both ends (in *A* and *B,* the proximal end is twisted about 90 degrees with the distal and hence appears thin). In all, a prominent crest is present to which is attached the pectoralis and deltoid muscles. In later types, the bone became relatively long and slender, particularly in small animals. In primitive reptiles, foramina developed distally on the inner or posterior side (entepicondylar foramen, *ent. f*) and on the outer or anterior margin (ectepicondylar foramen, *ect. f*). The former foramen persists in various mammals and in *Sphenodon,* the latter in many reptiles. *d,* Deltoid crest; *ect,* ectepicondyle, for attachment of extensor muscles of forearm; *ent,* entepicondyle, for attachment of flexor muscles of forearm; *gr. tub,* greater tuberosity, for attachment of supraspinatus and infraspinatus muscles; *h,* head; *l tub,* lesser tuberosity, for attachment of subscapular muscle; *p,* pectoral crest; *rc,* radial condyle; *s,* process (supinator), which in reptiles aids in formation of ectepicondylar foramen; *uc,* ulnar condyle, or trochlea.

Distally, the humerus bears on its anteroventral surface a rounded condyle for the radius; posterior to this, the end of the bone is notched for a pulley-like ulnar articulation. Expansions on either side distally, for the attachment of forearm muscles, are the **ectepicondyle** on the front (or outer) side and the larger **entepicondyle** on the back (or inner) margin. Universally in early reptiles (seldom in amphibians), there is a large **entepicondylar foramen** for a nerve and blood vessels. Proximal to the ectepicondyle, there is primitively a separate projecting process that in some reptiles fuses with the ectepicondyle, bridging a foramen. This **ectepicondylar foramen** persists in many turtles and lizards and in *Sphenodon,* and the entepicondylar opening is present in the last form and in many of the more primitive mammals; but in many reptiles and mammals, and in birds as a whole, both foramina have disappeared.

Radius and Ulna (Fig. 151).   The stout columnar radius supports the body on the front foot and articulates at its upper and lower ends with humerus and carpus, respectively. Except in extreme aquatic specialization, there is little variation in its structure.

The ulna lies lateral to the radius in the forearm. It seldom bears any notable part of the weight but is important for muscular attachment. Below, it articulates with the carpus; above, at the elbow, it articulates by a notch with the distal edge of the humerus. Above this notch, the ulna projects as the **olecranon**, the "funny bone"; to this attaches the triceps muscle, and the pull of this muscle on the olecranon is the major force extending (i.e., opening out) the forearm. In some mammals, the shaft of the ulna may fuse with the radius or may be entirely lost, but the olecranon is persistently present.

Manus (Figs. 152–156).   The proximal part of the hand, or manus, consists of a series of small elements that form the **carpus**, a region of flexible adjustment be-

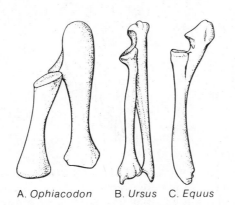

A. *Ophiacodon*   B. *Ursus*   C. *Equus*

Figure 151.   Left radius and ulna, seen from the extensor or anterior lateral surface in *A,* a primitive reptile; *B,* a bear, representing a typical mammalian condition; and *C,* a horse. The humerus articulates with the curved surface of the notch in the ulna and the adjacent head of the radius; above, the projecting olecranon of the ulna serves for the attachment of the powerful triceps muscle, which extends the forearm. In many mammals, as in the horse, the lower part of the ulna is reduced and fused with the radius.

tween the forearm and the digits. Primitively, as seen in early tetrapods (Fig. 152), there appear to have been 12 elements in the carpus, arranged in three series:

1. Three proximal elements, beneath the two forearm bones, are reasonably termed **radiale**, **intermedium**, and **ulnare**.
2. A series of central carpals or **centralia**, wedged in between this proximal set and a distal one; four were primitively present.
3. A row of **distal carpals**, one typically lying opposite the head of each toe, or digit.

There are numerous reductions and fusions of carpals in various tetrapods. The three proximal elements persist in most cases. The centralia, on the other hand, are almost always reduced. Even the most primitive reptiles never have more than two; a single centrale is a common reptilian and mammalian condition. Distally, there is a general trend for the loss of a fifth distal carpal, even when the fifth digit is retained, and with reduction of toes, there is generally further loss. In reptiles and mammals, there is an additional small bone, the **pisiform**, attached to the outer margin of the carpus and forming a point of attachment for the tendons of muscles along this aspect of the limb.

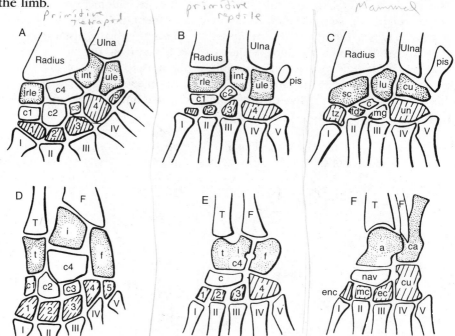

Figure 152. Diagram of carpus (*A* to *C*) and tarsus (*D* to *F*) to show essential homologies between primitive tetrapod (*A, D*), primitive reptile (*B, E*), and mammal (*C, F*). Proximal row of elements stippled; central row (and pisiform) unshaded; distal row hatched. Digits indicated by Roman numerals; distal carpals and tarsals by Arabic numerals. *a,* Astragalus; *c, c1* to *c4,* centralia; *ca,* calcaneum; *cu,* cuneiform in carpus, cuboid in tarsus; *ec,* external cuneiform (ectocuneiform); *enc,* internal cuneiform (entocuneiform); *f,* fibulare; *F,* fibula; *i, int,* intermedium; *lu,* lunar; *mc,* middle cuneiform (mesocuneiform); *mg,* magnum; *nav,* navicular; *pis,* pisiform; *rle,* radiale; *sc,* scaphoid; *t,* tibiale; *T,* tibia; *td,* trapezoid; *tz,* trapezium; *ule,* ulnare; *un,* unciform.

220
Chapter 7
Supporting
Tissues—
The
Skeleton

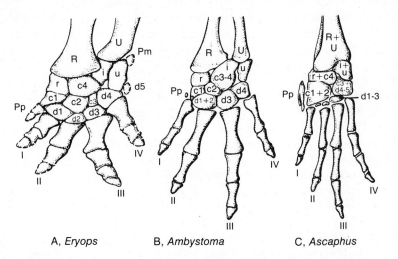

A, *Eryops*   B, *Ambystoma*   C, *Ascaphus*

Figure 153.   The left manus of amphibians, including an early labyrinthodont, a urodele, and an anuran. Restored elements in *A* in broken line. All twelve elements thought to have been present in the primitive carpus are found in *Eryops;* in modern amphibians, fusions of various sorts have occurred. As in most amphibians, four toes are present in the forms figured, although all have some development of a digit medial to the pollex, and an element in *Eryops* may have been the stub of an extra toe beyond the reduced fifth digit. The phalangeal count of two or three is usual in amphibians. *c1* to *c4,* Centralia; *cu,* cuneiform; *d1* to *d5,* distal carpals; *i,* inter-medium; *I,* lunare; *m,* magnum; *m1, 3, 5,* metacarpals; *p,* pisiform; *Pm,* postminimum digit; *Pp,* prepollex; *R,* radius; *r,* radiale; *s,* scaphoid; *td,* trapezoid; *tm,* trapezium; *U,* ulna; *u,* ulnare; *un,* unciform; *I* to *V,* digits. (After Gregory, Miner, and Noble.)

The terminology of the carpals in lower tetrapods is relatively simple. In mammalian anatomy, unfortunately, this is not the case; each was early given a distinct name, still used in comparative studies (Fig. 152); and (to make matters worse) still other names, not useful elsewhere, have been applied to some of these elements in the medical school dissecting room. We list this plethora of terms (and there are still other minor variant systems) without comment:

General Terminology	*Mammalian Anatomy*	*Human Anatomy*
Radiale	Scaphoid	Os scaphoideum (naviculare)
Intermedium	Lunar (or semilunar)	Os lunatum
Ulnare	Cuneiform	Os triquetrum (pyramidale)
Pisiform	Pisiforme	Os pisiforme
Centrale	Centrale	(Absent)
Distal carpal 1	Trapezium	Os trapezium (multangulum majus, cubiforme)
Distal carpal 2	Trapezoid	Os trapezoideum (multangulum minus)
Distal carpal 3	Magnum	Os capitatum
Distal carpal 4	Unciform	Os hamatum (uncinatum)

Beyond the carpus lie the toes, or **digits**. For each toe there is a proximal segment, or **metapodial**, contained in the flesh of the "palm"; in the forelimb these are the **metacarpals**. In mammals with speedy locomotion, particularly the ungulates, there may be a considerable increase in the length of the metapodials of such toes as are present, thus adding an extra functional segment to the limb. Distal to the metapodials are the elements of the free toes, the **phalanges**; distal **ungual phalanges** may be much modified to bear claws, nails, or hoofs.

Because the complicated foot pattern had no fixed antecedents in the ancestral fish fin, it is probable that considerable variation occurred in foot structure in the earliest tetrapods. Unfortunately, however, complete feet are seldom preserved in fossil forms. The presence of five toes in both front and hind feet is a condition primitive and common in amniotes and is found as well in some of the oldest amphibians. However, there are never more than four toes in the front foot of living amphibians, and the same number is found in many fossil forms.

In ichthyosaurs (Fig. 149 *A*), there is great variation in the number of digits present in the "fin"; there may be as few as three or as many as eight longitudinal rows of disc-shaped phalanges. In the bipedal archosaurs, the "arms" may be freed from duty as locomotor organs, and there may be much modification of the hand (Fig. 154 *C*), with reduction of the outer elements; in some dinosaurs only the two or three inner fingers remain. The flying reptiles, the pterosaurs, lost the fifth digit; the three inner ones are feeble, clawed structures, but the fourth is enormously elongated as a support for the wing membrane. The birds appear to have descended from archosaurs in which only the three inner fingers remained; these three digits are preserved, in a modified manner, in the bird wing (Fig. 154 *D*).

Five toes were present in the manus in the ancestral mammals and are found in many forms today (Fig. 155). Early mammals, many believe, were more or less arboreal, and the first digit, as a thumb or **pollex**, became more or less opposable to the others to give a grasping power useful in locomotion in the trees. Most mammals, however, soon abandoned arboreal life; this divergent digit was of little use in terrestrial locomotion and is frequently reduced or lost.

222
Chapter 7
Supporting
Tissues—
The
Skeleton

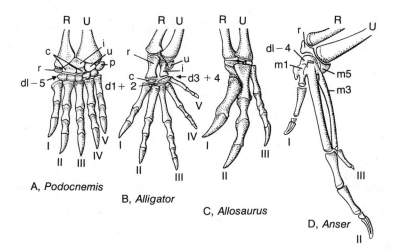

Figure 154. The manus in *A*, a turtle; *B*, the alligator; *C*, a carnivorous dinosaur; *D*, the goose. Abbreviations as in Figure 153. The bird manus includes the fused elements of the first three digits only; a somewhat comparable structure is seen in certain dinosaurs. (*A* after Williston; *C* after Gilmore; *D* after Steiner.)

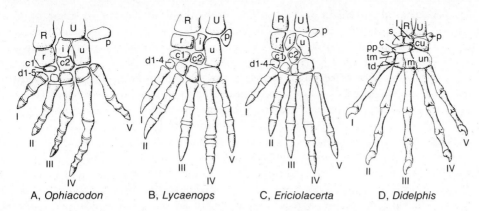

A, *Ophiacodon*   B, *Lycaenops*   C, *Ericiolacerta*   D, *Didelphis*

Figure 155. Evolution of the mammalian manus. *A,* Primitive reptile; *B,* a primitive therapsid; *C,* an advanced therapsid; *D,* a primitive mammal, the opossum. Abbreviations as in Figure 153. In the carpus, there is a loss of the fifth distal element, a reduction from two centralia to one; distally, "supernumerary" phalanges are lost from digits III and IV. (*B* after Broom; *C* after Watson.)

Further digital reduction or modification occurs in a great variety of mammals. For example, bats develop a wing membrane that, in contrast to that of pterosaurs, is supported by all the digits except the "thumb," and the whale manus becomes a structure somewhat comparable to that of ichthyosaurs (Fig. 149 *C*).

The development of the ungulate foot is accompanied by a reduction and elimination of digits. In most ungulates, the "thumb" was early lost, and a four-toed manus was characteristic of forms as dissimilar as ancestral horses and cows. Further reduction, however, followed two different paths, as diagrammed in Figure 156. Several extinct orders of mammals paralleled these changes to varying degrees.

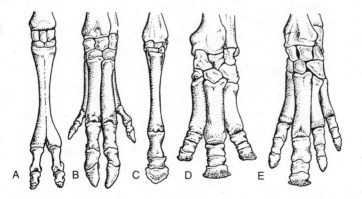

Figure 156. Left front feet of ungulates—in order, camel, pig, horse, rhinoceros, and tapir. The first two are artiodactyls, in which the axis of symmetry lies between the third and fourth toes. In the pig (*B*), lateral toes, 2 and 5, are complete but small. In most artiodactyls, as in the camel (*A*), the two main metapodials are fused into a cannon bone. The three remaining forms are perissodactyls, in which the axis runs through the third toe. In the tapir (*E*), the pollex is lost, but the other four toes remain; in modern rhinoceroses (*D*), the fifth toe has disappeared; in modern horses (*C*), the second and fourth are reduced to splints. (After Flower.)

A **phalangeal formula** indicates the number of phalanges in each digit in succession from inner to outer toes. In amphibians, the number of phalanges is in general low, usually two or three per toe. In primitive reptiles, however, there was established a formula of 2.3.4.5.3 for the manus. This persists with little change through a great variety of reptilian groups. Thus, the number of phalanges (and the length of the toes) increases steadily from pollex to fourth toe; the fifth toe is reduced in size and phalangeal count and tends to diverge laterally from the other four, perhaps because in the sprawling limb posture of primitive tetrapods, the foot, although turned somewhat forward, strikes the ground at an angle. The toes toward the higher, outer side must be longer to get a proper "footing"; the fifth toe merely extends out sideways as a lateral prop. In turtles, there is frequently a reduction in phalangeal formula (Fig. 154 A), often to that seen in mammals (see below). On the other hand, the extinct marine plesiosaurs and ichthyosaurs greatly increased the number of phalanges in each toe; each individual phalanx was, however, much abbreviated (Fig. 149 A, B).

In the mammal-like reptiles, the earlier types (Fig. 155 A, B) retained the reptilian formula. Advanced forms, however, reduced it to 2.3.3.3.3, which is the generalized formula for primitive mammals (Fig. 155 C, D). One joint has been lost from the third toe, two joints from the fourth; some therapsid genera show a transitional stage in which the joints to be lost are present but vestigial. This change in formula may be related to the changed limb posture; with the limb in a straight fore and aft plane, all the toes strike the ground equally well and hence tend to be subequal in length. Mammals frequently show a loss of digits but seldom a reduction in phalangeal count in any persisting toe. In the Cetacea, alone among mammals, we find an increase in the phalangeal count to as high as 13 or 14 in a digit; the whales here parallel the extinct marine reptiles (Fig. 149 C).

Femur (Fig. 157). The thighbone in primitive tetrapods is essentially a cylindric structure with expanded ends. Proximally, there is a terminal head fitting into the acetabulum. Beneath the head is a cavity, or fossa, into which some of the short ventral muscles insert, and at the anterior (or inner) margin of this fossa is seen ventrally in primitive types a strong **internal trochanter** for the attachment of further ventral muscles. Along the length of the shaft ventrally is an **adductor crest** for the attachment of strong adductor muscles. Near the proximal end of this ridge is frequently seen the so-called **fourth trochanter**, to which attach powerful muscles running onto the thigh from the tail. Distally, the femur expands, with rugosities on either side for the origins of shank muscles; there is a deep groove along the dorsal midline distally for the tendon of the extensor muscle of the thigh, which extends to the tibia. The distal end bears a broad, double articular surface for the head of the tibia, and along the posterior (or lateral) margin of the femur, an adjacent area for the fibular articulation.

The femur remains little modified in a majority of tetrapod groups except for the head of the bone. This usually develops into a more or less spherical structure set off from the shaft. In forms in which the hind limbs are rotated forward (archosaurs, birds, mammals), the femoral head faces inward for better reception in the acetabulum, and a **greater trochanter** for iliac muscle attachment may develop on the upper end of the shaft. In mammals, the internal trochanter disappears, and a new **lesser trochanter** develops in much the same region but for different muscular attach-

224
Chapter 7
Supporting
Tissues—
The
Skeleton

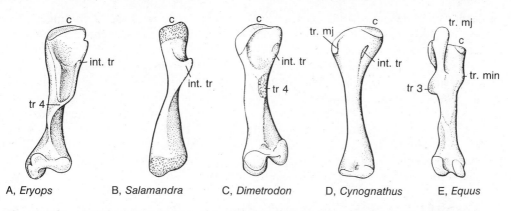

A, *Eryops*   B, *Salamandra*   C, *Dimetrodon*   D, *Cynognathus*   E, *Equus*

Figure 157.   Left femora seen from the ventral surface. *A,* Primitive fossil amphibian; *B,* a urodele; *C,* a primitive reptile; *D,* a mammal-like reptile; *E,* the horse. The proximal end above; distally, the articular surfaces for the tibia. *c,* Head (capitulum); *int. tr.,* internal trochanter of primitive forms, for attachment of obturator externus muscle or equivalent; *tr. 3,* third trochanter of perissodactyls for part of gluteal muscles; *tr. 4,* fourth trochanter, to which are attached tail muscles pulling the femur backward in many amphibians and reptiles; *tr. min.,* lesser (minor) trochanter for iliopsoas muscles of mammal; *tr. mj.,* greater (major) trochanter of mammals, for gluteal muscles.

ments. The fourth trochanter is absent in mammals, in which the muscles once attaching to it have dwindled or disappeared. The adductor crest may persist or be represented by rugose lines, the **linea aspera**.

Tibia and Fibula (Fig. 158).   The tibia, corresponding to the radius in the front limb, is the main supporting element of the lower segment of the leg; it is always stoutly developed, articulating broadly below with the inner side of the tarsus, and above by a much expanded head, triangular in section, with almost the entire under surface of the distal end of the femur. The front margin of the head bears a strong ridge, the **cnemial crest**, to which attaches the terminal tendon of the main extensor muscle of the thigh. Primitively the attachment of the tendon was a direct one. However, the sharp curve that this tendon must make down over the end of the femur is disadvantageous. In mammals, there generally develops the **patella**, a large sesamoid bone

Figure 158.   The left tibia and fibula, seen from the extensor (dorsal) surface, of *A,* a primitive reptile (a Permian pelycosaur) with a fibula of good size; *B,* the pig, showing a primitive mammalian condition, with the fibula complete, although slender; *C,* the horse, exemplifying a type with reduced fibula. *cn,* Cnemial crest.

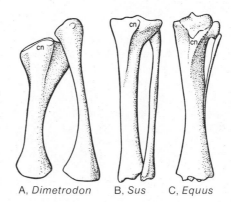

A, *Dimetrodon*   B, *Sus*   C, *Equus*

that rides over the knee joint, with functions comparable to those of the ulnar olecranon in the front limb; the extensor tendon inserts into this ossification which is, in turn, tied to the cnemial crest.

The fibula is in some respects comparable to the ulna in the forearm; it bears little weight and, in consequence, tends to undergo reduction. Because in the hind leg the extensor tendon attaches to the tibia, the fibula has no development comparable to the olecranon. The fibula is articulated proximally with the posterior (or outer) aspect of the end of the femur, distally with the outer side of the tarsus. In many living groups, the fibula has lost its femoral articulation, and the tarsal attachment may be much reduced or lost. It usually persists as a slender structure, but its ends may fuse with the tibia, or the shaft may be lost.

226

**Chapter 7
Supporting
Tissues—
The
Skeleton**

Pes (Figs. 152, 159–161). The structure of the ankle region, the **tarsus**, was primitively quite similar to that of the carpus; the same number of bones were present in much the same arrangement, and a terminology similar to that of the carpus may be used. A proximal series of three elements are termed **tibiale**, **intermedium**, and **fibulare**. There were, as in the manus, four **centralia** in primitive types and five **distal tarsals**. As in the carpus, free central elements are reduced to two in primitive reptiles and to one in mammals, and the fifth distal element tends early to be reduced and to disappear.

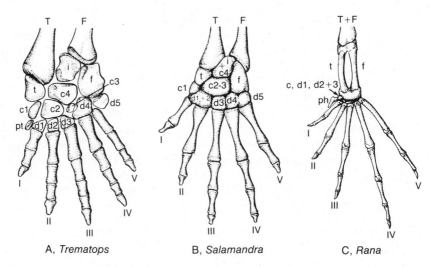

A, *Trematops*   B, *Salamandra*   C, *Rana*

Figure 159. The pes of amphibians, including *A,* an early labyrinthodont; *B,* a salamander; and *C,* a frog. The tarsus of *Trematops* includes all elements presumed to be present in ancestral tetrapods, as well as an additional pretarsal bone. In the urodele some fusion of tarsal elements has occurred. In the frog, tibiale and fibulare are so elongated as to constitute an additional limb segment; except for a few small distal elements, the fate of the other tarsal bones is not clear. Five toes are typically present, as contrasted with four in the manus, and the count of phalanges is generally higher in the pes than in the amphibian manus (cf. Fig. 153). The frog has a developed prehallux. *c1* to *c4,* Centralia; *d1* to *d5,* distal tarsals; *F,* fibula; *f,* fibulare; *i,* intermedium; *ph,* prehallux; *pt,* pretarsal element; *T,* tibia; *t,* tibiale. (*A* after Schaeffer; *B* after Schmalhausen; *C* after Gaupp.)

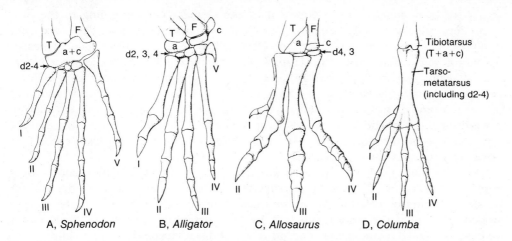

A, *Sphenodon*   B, *Alligator*   C, *Allosaurus*   D, *Columba*

Figure 160. The pes in reptiles and birds. *A, Sphenodon; B,* the alligator; *C,* a carnivorous dinosaur; *D,* the pigeon. In all forms there have been various types of reduction and fusion of the tarsal elements and a trend toward the development of a main joint within the tarsus between proximal and distal elements. In the birds all the tarsals are fused with the tibia proximally or with the fused metatarsals distally. In archosaurs, such as the alligator and dinosaurs, there is a strong trend toward the loss of the fifth toe, and it has disappeared in the dinosaur figured and in the bird. Except for the lack of fusion of the metatarsals, the *Allosaurus* foot is quite similar to that of birds. *a,* Astragalus; *c,* calcaneum; other abbreviations as in Figure 159. (*B* after Williston; *C* after Gilmore).

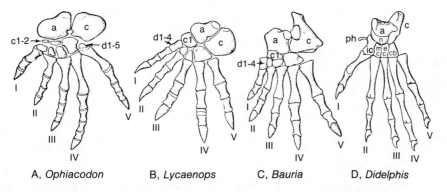

A, *Ophiacodon*   B, *Lycaenops*   C, *Bauria*   D, *Didelphis*

Figure 161. Evolution of the mammalian type of pes. *A,* Primitive reptile (of the early Permian); *B,* a primitive therapsid (of the late Permian); *C,* an advanced Triassic mammal-like reptile; *D,* the opossum. Principal changes include development of a pulley surface on the mammalian astragalus, development of a heel on the calcaneum (in *C* and *D*), loss of two tarsal elements, and reduction in phalangeal count (a transitional stage in *B*). *a,* Astragalus; *c,* calcaneum; *c1* to *c2,* centralia; *cb,* cuboid; *d1* to *d5,* distal tarsals; *ec,* external cuneiform; *ic,* internal cuneiform; *mc,* middle cuneiform; *n,* navicular; *ph,* prehallux (exceptional in mammals). (*B* and *C* after Schaeffer.)

There is, however, no development of any element corresponding to the pisiform. More important is the change that occurs in the proximal row. There are here in reptiles only two elements that are often called by their mammalian names of **astragalus** and **calcaneum**. The latter is simply the fibulare under another name. The astragalus, however, appears to be a fusion of a reduced tibiale with the intermedium and a central element. We have noted that in primitive tetrapods, the hind foot would tend to project out laterally from the body if it were not twisted sharply forward on the leg. The fusion of elements to form the astragalus appears to give a firmer support to the tibia in the twisted position; in primitive reptiles there is free movement between tibia and astragalus.

Reptiles have tended to develop an "intratarsal" joint where the major movement of lower leg on foot occurs. In crocodilians, the main movement of the foot on the shank occurs between astragalus and calcaneum, the former bone being closely applied to the tibia, the latter to the foot. In dinosaurs, both astragalus and calcaneum are closely applied to the lower leg, and reduced distal tarsals are functionally joined to metatarsals. In birds, actual fusion of bones has taken place, so that we speak of a **tibiotarsus** as a single element, succeeded distally by a **tarsometatarsus**; there are no free tarsals at all.

In mammals, astragalus and calcaneum are much modified. The astragalus develops a head as a keeled rolling surface that fits into a corresponding set of grooves in the distal end of the tibia. The joint is a hinge, allowing almost no twisting but a great amount of extension and flexion of the foot. Characteristic of artiodactyls is an astragalus in which the lower end also has a keeled rolling surface, so that an extreme amount of backward and forward movement of the foot in running is possible. The calcaneum has little or no articulation with the more proximal limb elements in mammals but develops a powerful heel projection to which attach the calf muscles through the tendon of Achilles.

In the tarsus, as in the carpus, we find (unfortunately) a complicated series of names applied to the mammalian elements as shown in the table below.

General Terminology	Mammalian Anatomy	Human Anatomy
Tibiale + Intermedium.....Astragalus		Talus
Fibulare...................Calcaneum		Calcaneus (Os calcis)
Centrale..................Navicular		Os naviculare pedis
Distal tarsal 1 ....Internal cuneiform		Os cuneiforme mediale (primum)
Distal tarsal 2 ....Middle cuneiform		Os cuneiforme intermedium (secundum)
Distal tarsal 3 ....External cuneiform		Os cuneiforme laterale (tertium)
Distal tarsal 4 ....Cuboid		Os cuboideum

The description given for the digits of the manus applies in general to the pes, but the proximal elements are, of course, termed **metatarsals** rather than metacarpals. The phalangeal formula of the pes in amphibians and reptiles frequently corresponds closely to that for the manus except that in reptiles there is usually a fourth phalanx in the last toe. In bipedal archosaurs, there was often, in relation to speedier locomotion, an elongation of the metapodials comparable to that noted in ungulate mammals. There was, further, a reduction in the number of digits in these reptiles; the divergent fifth toe often disappeared, and the first toe, or **hallux**, was turned back as a prop for the foot. There remained, in the more advanced bipedal types, three

228
Chapter 7
Supporting
Tissues—
The
Skeleton

forwardly directed toes, symmetrically disposed, the center one the longest; these toes were, of course, the second, third, and fourth. Despite the modification of their comparative lengths, the primitive phalangeal count had been retained; the second and fourth toes, although subequal in length, had three and five phalanges, respectively; the middle (third) toe, four. The birds, as descendants of bipedal archosaurs, have exactly the same toe structure; it is thus natural that dinosaur footprints when first discovered were thought to be those of gigantic birds.*

In mammals, the history of the hind foot is comparable to that of the manus. In the horses and their relatives, digital reduction proceeded more rapidly in the hind foot than in the forefoot; the earliest known fossil horses had already attained a functionally three-toed stage.

Visceral Skeleton

Between the branchial openings of lower vertebrates is a series of cartilaginous or bony bars—arches that aid in the support and movements of the gills. These arches are the basic components of a set of structures termed the **visceral skeleton**. Except in agnathous forms, the elements of the visceral system never form more than a small proportion of the skeletal materials of the body; they are, however, notable for their versatility, the degree of modification of form and function that they have undergone in the course of vertebrate history. Primitively, to be sure, they merely supported the gills. But early in fish history, they played a leading role in the development of the jaws; and although gills as such have disappeared in amniotes, persistently surviving visceral skeletal elements are to be found even in mammals in such varied areas as the skull, the auditory ossicles, and the larynx.

It may seem difficult to justify classifying these usually modest structures in a special category contrasting with all other endoskeletal elements. Consideration, however, shows that such distinction is well warranted. The branchial region, or pharynx, although relatively insignificant in terrestrial vertebrates, is, if traced downward in the chordate series, increasingly prominent, important, and in every way distinctive. In the nature of its muscles and nerve supply, the branchial region is in strong contrast, primitively, with any other body region, and its skeleton is likewise unique.

Other endoskeletal structures form in the "outer tube" of the body, the "somatic animal"; those of the branchial region are formed in the walls of the gut (at least in all jawed vertebrates—the situation in the Agnatha is confusing) and are truly visceral in position. In addition, they are unusual in their embryologic origin. The visceral cartilages are derived from mesenchyme, but not mesenchyme of mesodermal origin—it is, instead, derived from the ectoderm (Fig. 162). Some of the neural crest cells of the head (p. 124) migrate downward and inward, form a mesenchyme material similar to that elsewhere derived from the mesoderm, and ultimately form skeletal structures. This mesenchyme, sometimes termed **mesectoderm**, is concerned with the formation of the gill bars and their derivatives.

Although such an origin from neural crest material was long thought to differentiate the visceral from the somatic skeleton, more recent work has shown that this

*Unfortunately, however, those dinosaurs with the most birdlike feet are included in suborder Theropoda—meaning *mammal*-footed.

230
Chapter 7
Supporting
Tissues—
The
Skeleton

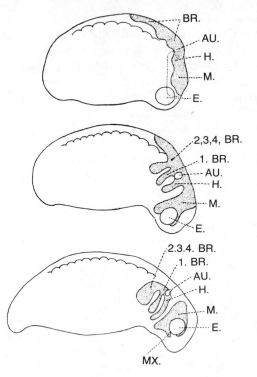

Figure 162. Side views of salamander embryos to show the downward growth of the neural crest (stippled) to form visceral skeletal structures in the head and throat region. *E,* eye; *AU,* auditory capsule; *MX, M, H, BR,* neural crest downgrowths in maxillary, mandibular, hyoid, and branchial arch regions. (From Stone, L. S., J. Exp. Zool., *44:*95–131, 1926.)

mesectoderm is also involved in the formation of most, possibly all, dermal bones and some of the axial skeleton within the head. Unfortunately, few forms have been studied at all, and the older ideas are based largely on amphibian and the newer ones on avian material.*

The Skeleton of the Gills. In the lampreys (Figs. 121, 163 *A*) certain of the gill structures are highly specialized, serving to stiffen the rasping "tongue" that has evolved in these predaceous forms as a jaw substitute. The remainder of the visceral skeleton forms a continuous latticework surrounding the branchial region and connecting with the braincase. Thus here and, as far as we can tell, in the ostracoderms, the visceral skeleton lies *lateral* to the actual gills and immediately beneath the skin, in marked contrast to its deeper position in jawed vertebrates. In the cephalaspid ostracoderms, the skeleton of the branchial pouches, as far as preserved, is fused with the rest of the "cranial" skeleton (cf. Fig. 20 *A*). This suggests that in the ancestral vertebrates, the gill supports were probably not discrete elements but were fused with one another and with the braincase. In other ostracoderms (certain anaspids; see Fig. 163 *B*), the visceral skeleton looks more like that in the modern lampreys. Our knowledge is, thus far, too imperfect to be certain of the true ancestral situation (we know nothing of the visceral skeleton in the heterostracans, by far the oldest known vertebrates), and in all gnathostomes, at any rate, the branchial skeleton consists (except for cases of secondary fusion) of numerous separate movable structures.

*Even in amphibians, part of the braincase, the trabeculae, has long been known to develop from the neural crest; it was, logically enough, assumed that these were part of the visceral skeleton that were, early in vertebrate history, incorporated into the skull.

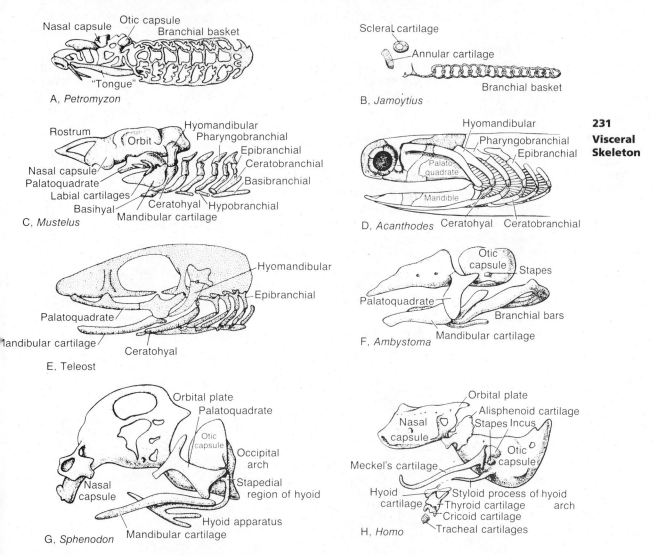

A, *Petromyzon*
Nasal capsule — Otic capsule — Branchial basket
"Tongue"

B, *Jamoytius*
Scleral cartilage — Annular cartilage — Branchial basket

C, *Mustelus*
Rostrum — Orbit — Hyomandibular — Pharyngobranchial — Epibranchial — Ceratobranchial — Basibranchial — Nasal capsule — Palatoquadrate — Labial cartilages — Basihyal — Ceratohyal — Hypobranchial — Mandibular cartilage

D, *Acanthodes*
Hyomandibular — Pharyngobranchial — Epibranchial — Palato-quadrate — Mandible — Ceratohyal — Ceratobranchial

E, Teleost
Hyomandibular — Epibranchial — Palatoquadrate — Mandibular cartilage — Ceratohyal

F, *Ambystoma*
Otic capsule — Stapes — Palatoquadrate — Branchial bars — Mandibular cartilage

G, *Sphenodon*
Orbital plate — Palatoquadrate — Otic capsule — Occipital arch — Stapedial region of hyoid — Nasal capsule — Hyoid apparatus — Mandibular cartilage

H, *Homo*
Orbital plate — Alisphenoid cartilage — Nasal capsule — Stapes — Incus — Otic capsule — Meckel's cartilage — Hyoid cartilage — Styloid process of hyoid arch — Thyroid cartilage — Cricoid cartilage — Tracheal cartilages

Figure 163. Visceral skeleton and braincase in representative vertebrates. *A* to *D,* Adult forms; *E* to *H,* embryos; cartilage stippled, bone unstippled. *A,* The lamprey, with a peculiar braincase and associated cartilages not readily comparable to those of other groups; the branchial basket is fused anteriorly to the braincase. *B,* An anaspid, one of the extinct ostracoderms, showing a branchial basket somewhat like that of the lamprey; the braincase is not preserved. *C,* A modern shark, with hyostylic jaw suspension. *D,* A Paleozoic acanthodian, a very primitive bony fish; the sclerotic plates of the orbit are included as is a slender dermal bone beneath the mandible. *E,* A rather diagrammatic teleost embryo; in the adult, almost the entire visceral skeleton, as well as the braincase, would be encased in dermal bones. *F,* A urodele; the palatoquadrate is reduced, and the branchial arches are reduced even in the embryo or larva. *G,* The reptile *Sphenodon;* the braincase is incompletely formed (cf. lizard in Fig. 132 *D*), and hyoid and branchial bars are reduced to a hyoid apparatus and stapedial cartilage. *H,* Human foetus; the braincase develops only ventrally and anteriorly around the brain; the palatoquadrate is reduced to alisphenoid and incus (= quadrate); the lower jaw (Meckel's cartilage) is reduced, the proximal part becoming the malleus in the adult; other visceral elements include the hyoid and its styloid process, stapes, laryngeal, and tracheal cartilages. (*A* and *C* after Goodrich; *B* after Ritchie; *D* after Watson; *E* after Daget; *F* after Hörstadius and Sellman; *G* after Howes and Swinnerton; *H* after Gaupp, Macklin).

232

Chapter 7
Supporting
Tissues—
The
Skeleton

The branchial skeleton in jaw-bearing fishes consists of jointed bars along the walls of the pharynx that form arches between successive gill slits. These **visceral arches** are serially arranged structures, as are the gill openings themselves and the muscles and nerves of the gill region. However, no proof has been found, despite many attempts, that this segmentation is in any way related to the segmentation present in other parts of the body, which is founded on the segmentation of the mesodermal somites rather than on that of the endodermal gill pouches (see pages 562–565).

Each typical gill bar of a jawed fish (Figs. 163 *C–E;* 164 *A;* 213) includes on either side a major dorsal and a major ventral element, the two bent somewhat forward on each other. The dorsal elements are the **epibranchials**; the ventral ones, **ceratobranchials**. There are generally **pharyngobranchials**, turned inward over the pharynx, at the upper end of each normal arch. Below, the successive arches are tied to one another and with their mates of the opposite side by elements that extend downward, forward, and inward to the midline; there is a great variation here, but frequently there are short paired **hypobranchials** below the ceratobranchials, and median ventral structures—**basibranchials** or **copulae**. The gill bars generally bear on their inner margins **gill rakers**, and on their outer surfaces a row of **gill rays**, which stiffen the gill septum or gills (Fig. 251). The elements of the branchial arches of bony fish typically bear dermal tooth-bearing bones on the side toward the pharynx. In most jawed fishes, five sets of typical branchial bars are present, in addition

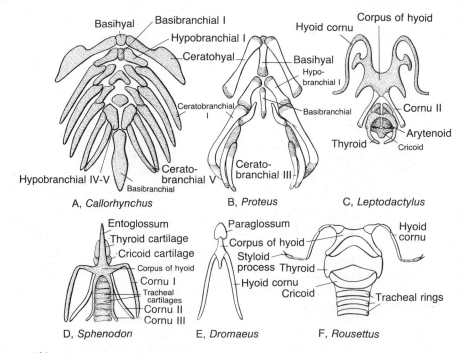

Figure 164. Gill bars and their derivatives in tetrapods, in ventral view, in *A,* a chimaera; *B,* a water-dwelling salamander; *C,* a frog; *D,* a reptile; *E,* the cassowary; and *F,* a bat. In *A,* the dorsal arch elements are not included. In *C, D,* and *F,* laryngeal cartilages are included. The entoglossum of *D* and paraglossum of *E* are anterior, tongue-supporting developments from the body of the hyoid.

to the specialized jaw and hyoid arches now to be described, but a few sharks have six or even seven such bars.

Development of Jaws. Perhaps the greatest of all advances in vertebrate history was the development of jaws and the consequent revolution in the mode of life of early fishes. In this process, the visceral skeleton played a leading role, because transformed branchial bars are basic structures in jaw formation and, indeed, form the entire skeleton of the jaws in the Chondrichthyes.

In the sharks (where there is no complication of investing dermal bones), both upper and lower jaws are formed solely of paired cartilages that seem to represent transformed branchial elements (Figs. 163 *C*, 166). The shark upper jaw, running fore and aft below and beside the braincase, is the **palatoquadrate** (or pterygoquadrate) **cartilage**; it is comparable to the epibranchial of a normal branchial arch. Posteriorly, it articulates with a lower jaw or **mandibular cartilage**, presumably derived from the ceratobranchial of the same arch as the upper jaw. It appears that, in the development of jaws, a pair of branchial bars lying adjacent to the expanding oral cavity became armed with teeth and enlarged to function in a new capacity as biting jaws (Fig. 165). That the visceral arch, which became modified into jaws in the ancestral gnathostome, was the most anterior one present in the jawless ancestor is doubtful; it is more than probable that one or two anterior arches were destroyed in the expansion of the mouth. Although it has been suggested that the small **labial cartilages** present in the angle of the mouth in sharks (Fig. 166) represent such arches, this is doubtful; more probably the trabeculae may represent an arch otherwise lost. Unfortunately, we have nothing resembling a transitional stage showing the development of jaws; indeed, no known agnathan has visceral arches lying in the pharyngeal wall like those of all jawed vertebrates.

In forms other than sharklike fishes, dermal elements are also present in the complex upper and lower jaws, and the part played in jaw formation by visceral elements is much reduced. The jaw articulation, however, remains persistently that between upper and lower visceral bars in every group except the mammals, in which (as discussed later) the entire structure of the jaws and palate is taken over by dermal elements. Even in mammals, however, relics of the visceral jaw structures persist, although curiously transformed into ear ossicles and a component of the braincase.

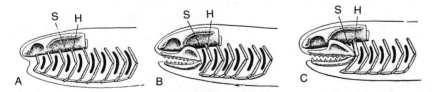

Figure 165.   Diagrams to show evolution of the jaws and hyoid arch. Gill openings in black. *H,* Hyomandibular; *S,* spiracular gill slit. In *A,* a primitive jawless condition. *B,* Jaws formed from a pair of gill arches (two anterior arches and slits may have been lost in the process); spiracular gill slit unreduced and hyomandibular not specialized. *C,* Condition seen in primitive jawed fishes; the hyomandibular has become a jaw support, and the intervening gill slit reduced to a spiracle. *A* and *B* both represent essentially hypothetical conditions.

234
Chapter 7
Supporting
Tissues—
The
Skeleton

Jaw Suspension. In most fishes (Figs. 165 *C,* 166), the second or **hyoid arch**, lying behind the first branchial opening (the spiracle), is specialized as an aid to jaw support. The ventral part of this arch is normal; on either side there is a major element termed the **ceratohyal** and usually a smaller **basihyal**. In the upper part of this arch, however, the epibranchial segment is an element, often of considerable size, termed the **hyomandibular**. This is braced dorsally against the otic region of the braincase; below, it runs to the region of the jaw articulation and is bound by ligaments to the jaws. Thus it forms an effective support for the jaws, propping them on the braincase. In tetrapods, in which the addition of dermal bones produces a compact skull structure, the upper jaw and the palatal structures are satisfactorily bound to the remainder of the skull without the aid of the hyomandibular support. In them, the hyomandibular has been transformed into an auditory bone, the stapes; its history is discussed in connection with the ear.

The nature of the hyomandibular serves to introduce the general problem of the suspension of the jaws on the skull in fishes. If, as is the case in most modern fishes, the upper jaw has no direct connection with the braincase (except far anteriorly) and the jaw joint is braced entirely by the hyomandibular (with the effect of widening the gape), the condition is said to be **hyostylic**. In a few sharks, however, there is support both from the hyomandibular and from a direct articulation of upper jaw and braincase; this is an **amphistylic** condition. A condition in which the hyomandibular is nonfunctional as a supporting structure and the upper jaw region is entirely responsible for its own suspension is termed **autostylic** (Fig. 165 *B*). In theory, we may believe that the ancestral gnathostome was autostylic, with a palatoquadrate articulating movably on the braincase. This theoretical ancestor, however, is as yet missing in the fossil record (as mentioned earlier, see p. 47), and the only known autostylic fishes are the specialized chimaeras (cf. Fig. 134) and lungfishes, in which the autostylic condition is associated with the fact that the upper jaw itself is fused tightly to the braincase. Tetrapods are all autostylic, of course, in a broad sense of this term, because the hyomandibular never functions as a jaw support.

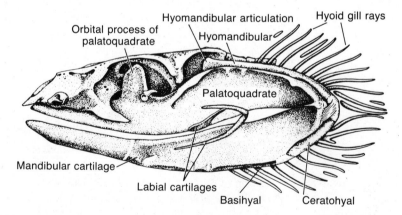

Figure 166. The cranial skeleton of a shark, *Chlamydoselache:* braincase, jaws, hyoid arch. (After Allis.)

Tetrapod Gill Arch Derivatives (Figs. 163 *F–H,* 164 *B–F*). In larval amphibians, gills are still functional, and thus the branchial bars remain prominently developed. There is already, however, some tendency toward reduction; there never are more than four arches present behind the hyoid arch, instead of the five customary in fishes, and these are often continuous bars, without clear-cut divisions into the various dorsal and ventral elements typical of their fish ancestors.

In metamorphosed amphibians and in amniotes, functional gills are totally lost; visceral arches nevertheless persist (apart from the jaw and ear ossicles) as two sets of structures formed in connection with new tetrapod developments in the former gill region. *(a)* A tongue, never characteristically present in fishes, arises in the floor of the mouth where the more anterior and ventral parts of the visceral bars once lay; reduced and transformed remnants of the anterior branchial bars serve, as the hyoid apparatus, to support the tongue. *(b)* With increased importance of the lungs, skeletal structures derived from or related to the more posterior part of the visceral skeleton are developed about the larynx and trachea that form the entrance to the lungs. These structures are the laryngeal skeleton, the "Adam's apple" of man, and ring-shaped tracheal cartilages.

The tetrapod **hyoid apparatus** typically includes a main body, the **corpus**, formed from one or more of the median ventral arch elements (copulae or basals), and remnants of the hyoid arch and of one to three branchial arches extending outward and upward as "horns" or **cornua** (singular, cornu). The body of the hyoid is situated in the floor of the throat at the base of the tongue, ventral to the anterior end of the windpipe. In some reptiles and in birds, the corpus of the hyoid develops a long anterior process for better support of the tongue. The horns extend upward on either side of the windpipe, and, although the most anterior—"principal"—horns represent only the lower half of the original hyoid arch (the upper part forms the stapes), they may nevertheless gain a slender secondary connection with the otic region of the skull, either directly or by ligaments. Mature amphibians and reptiles often have several pairs of cornua; birds generally have a single pair; in mammals, two pairs typically represent the hyoid and first branchial arch. In mammals, additional ossifications may occur in the slender process of the hyoid horn, reaching up to the otic region, and in some mammals, including man, the distal end of this horn may fuse with the skull as a **styloid process**.

But the hyoid apparatus is only one tetrapod derivative of the fish branchial skeleton; skeletal structures supporting the windpipe itself have developed. Just beyond the entrance to the windpipe, the glottis, there typically develops in tetrapods an enlargement in the tube as the **larynx** (cf. Chap. 11). The basal part of the hyoid apparatus generally lies close to the front end of the floor of this expansion; additional cartilages, at least part of which appear to be modified skeletal structures, tend to form a complex **laryngeal skeleton**. When fully developed, these include unpaired **thyroid*** and **cricoid cartilages** in line ventrally behind the body of the hyoid and, dorsally, paired **arytenoids**; these cartilages may be ossified. In salamanders, only the arytenoids are developed; in anurans, a cricoid is present as well. The thyroid is present only in mammals. The **epiglottis**, found in the flap of skin covering the open-

*The term **thyroid** means simply shield-shaped; the thyroid gland, also shield-shaped, happens to lie in this general region of the throat, but the topographic juxtaposition is pure coincidence.

ing to the windpipe (glottis), is present in mammals but may be represented in some reptiles by a cartilage in this position. In most tetrapods, the windpipe (trachea), running down from the larynx, tends to be strengthened by ring-shaped **tracheal cartilages**; these may also ossify. These last structures are not readily comparable to any specific elements seen in fish but are reasonably regarded as new developments of the visceral skeletal system.

236
Chapter 7
Supporting
Tissues—
The
Skeleton

Chapter 8

The Skull

The term "skull" is used in a somewhat variable way. In a broad way, it refers to any type of skeletal structure found in the head. In this sense, one may consider the shark as having a skull consisting of braincase and isolated visceral bars. But in common parlance, and as we shall use it here, the term has a somewhat different meaning. The familiar skull of every form from a bony fish to a mammal is a unit structure in which braincase and endoskeletal upper jaws are welded together by a series of dermal bones; the lower jaw is not included.

The older Agnatha and the extinct placoderms appear to have possessed a fused cranial skeleton that included dermal as well as endoskeletal materials and is hence to be considered a skull in the present sense. But in most of those archaic fishes, the structures seem to have been aberrant, and they are as yet too incompletely known to warrant their being considered in detail here; further, in these ancient forms, the branchial region is frequently included as well as proper head structures. In the higher bony fishes and all tetrapods, there is a well-constructed skull, with many common features present throughout these groups. Even so, it is difficult to give a generalized description that will hold in all cases. The task is further complicated by the fact that in most living groups considerable modification and specialization have taken place, obscuring the true phylogenetic story.

In the study of soft anatomy, we suffer under the handicap of knowing nothing of the actual ancestral types from which the organs of modern animals have been derived; we can, at best, merely make guesses (which we hope are intelligent) at the structural conditions in such ancestors, extinct these many millions—or hundreds of millions—of years. In the case of skeletal history, however, we frequently have preserved as fossils the bones of actual ancestors. In regard to the skull, we have a fairly complete knowledge of the structure in the ancient labyrinthodont amphibians, whose skulls are of the exact type from which those of all later tetrapods have been derived. Further, although bony fish history is still obscure in some respects, the skulls of the higher fishes, particularly fossil crossopterygians, are not very different from those of the ancient amphibians. We shall, therefore, at this point give a rather full account of this central type of skull seen in the early amphibians and then discuss the major modifications seen in the various fish and tetrapod groups. The general pattern of the skull figured (Fig. 167) is that of Carboniferous embolomerous labyrinthodonts.

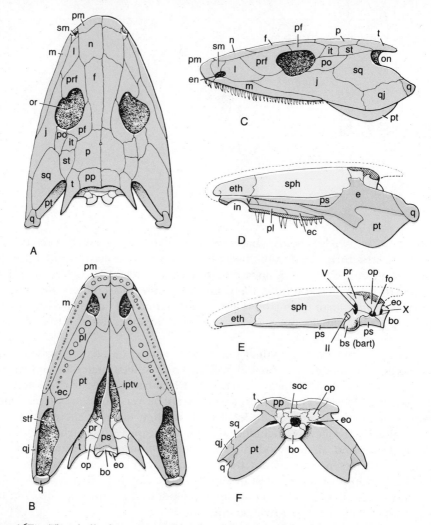

Figure 167. The skull of an ancestral land vertebrate, based primarily on the Carboniferous labyrinthodont *Palaeoherpeton*. *A,* Dorsal view of skull; *B,* palate; *C,* lateral view; *D,* lateral view with dermal skull roof removed (outline in broken line); palatal bones—dermal and endoskeletal—of left side are shown; deep to them, the braincase. The ventral hatched area is the sutural surface of palatal bones and maxilla. *E,* Lateral view of braincase; *F,* posterior view. Dermal bones, red; endochondral bones of palate, orange; endochondral bones of braincase, yellow; cartilage, blue. (After Watson and Panchen.)

Abbreviations: *bart,* basal articulation of braincase and palate; *bo,* basioccipital; *bs,* basisphenoid; *e,* epipterygoid; *ec,* ectopterygoid; *en,* external naris; *eo,* exoccipital; *eth,* ethmoid region; *f,* frontal; *fo,* fenestra ovalis; *in,* internal naris; *iptv,* interpterygoid vacuity; *it,* intertemporal; *j,* jugal; *l,* lacrimal; *m,* maxilla; *n,* nasal; *on,* otic notch; *op,* opisthotic; *or,* orbit; *p,* parietal; *pf,* postfrontal; *pl,* palatine; *pm,* premaxilla; *po,* postorbital; *pp,* postparietal; *pr,* prootic; *prf,* prefrontal; *ps,* parasphenoid; *pt,* pterygoid; *q,* quadrate; *qj,* quadratojugal; *sm,* septomaxilla; *soc,* supraoccipital; *sph,* sphenethmoid; *sq,* squamosal; *st,* supratemporal; *stf, subtemporal fossa; t,* tabular; *v,* vomer. Roman numerals, foramina for cranial nerves.

Skull Components. To avoid (as much as possible) mental indigestion in treating of this complex structure, it is best that the skull be resolved into components (Fig. 168). It includes dermal bones and cartilages or their bony replacements of both somatic and visceral nature (cf. p. 169). One possible approach would be to consider the elements derived from each of these three embryologic sources as units. It is, however, preferable to recognize that a primitive skull seems to be composed of three major functional units: one purely dermal, the other two composites, as follows:

A. **Dermal skull roof.** This shield of membrane bone covers thoroughly the top and sides of the head and extends down to the jaw rims, where elements of the shield bear the marginal row of teeth. The roof is unbroken except for openings for the nostrils (**external nares**), eyes (**orbits**), and a small **parietal foramen** for a third, median eye. On either side, the shield is primitively notched behind the orbit for the opening housing the eardrum, which replaces the spiracle present here in bony fishes.

B. **Palatal complex.** This includes ossifications in a palatoquadrate of visceral origin (which in sharks forms the entire upper jaw) and, in addition, a series of membrane bones formed in the roof of the mouth below this cartilage and in great measure replacing it. The anterior portion of this complex forms a broad palatal plate with, anteriorly, lateral gaps for the **internal nares**, or **choanae**. Posteriorly, the palatal complex on either side is separated from the edge of the shield by the **subtemporal fossae**; through these descend the adductor muscles, which close the jaws.

C. **Braincase.** This is formed in cartilage, mainly of somatic origin but usually ossified to a considerable degree; to its lower surface, in early tetrapods and bony fishes, is applied a sheet of dermal bone formed in the central area of the mouth vault.

The Primitive Amphibian Skull. *Dermal Roof.* In a primitive tetrapod, the dermal skull roof includes a considerable number of elements, suturally connected to form a practically solid shield. In general, these bones are paired; occasionally we find as

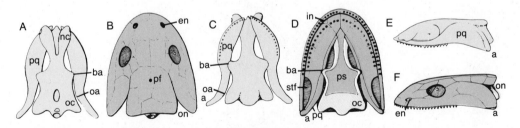

Figure 168. Diagrams to show the components of the skull. *A, C, E,* Dorsal, ventral, and lateral views of endoskeletal structures of braincase and palate (or upper jaw) as present in a shark or an embryo of higher jawed vertebrates; *B, D, F,* similar views with dermal elements added. In *B* and *F,* a dermal roof shield binds together and covers braincase and endoskeletal jaw structures; in *D,* dermal palatal structures are seen reinforcing the endoskeletal palate, and a further dermal element underlies the braincase. Dermal bones, red; endochondral bones of palate, orange; endochondral bones of braincase, yellow; cartilage, blue. Abbreviations: *a,* articulation with lower jaw; *ba,* basal articulation of braincase with palatoquadrate; *en,* external naris; *in,* internal naris; *nc,* nasal capsule; *o,* orbit; *oa,* otic articulation of palatoquadrate; *oc,* otic capsule; *on,* otic notch; *pf,* parietal foramen; *pq,* palatoquadrate; *ps,* parasphenoid; *stf,* subtemporal fossa for jaw muscle.

variants unpaired median ossifications of little importance. It is a strain to commit to memory the names of the numerous elements present, so, as an aid to memory, we may arbitrarily group them into several series (Fig. 169):

(a) Tooth-bearing marginal bones. At the tip of the skull there is on either side a small **premaxilla** bearing the most anterior teeth. The remaining marginal teeth are carried on the elongate **maxilla**. The suture between premaxilla and maxilla lies below the external naris.

(b) A longitudinal series along the dorsal midline. From the premaxilla back to the occipital end of the skull, we find in order the following bones: nasal, frontal, parietal, and postparietal. The **nasals** lie medial and posteromedial to the external nares. The **frontals** lie between the orbits but primitively do not enter into their margins. The **parietals** are generally large and occupy the central part of the skull table—the flattened area of the roof behind and between the orbits; the parietal foramen for a median eye lies between them. The **postparietals** (dermal supraoccipitals) are at the back rim of the roof and may descend as flanges onto the posterior surface of the braincase.

(c) A circumorbital series of five bones—prefrontal, postfrontal, postorbital, jugal, and lacrimal—which in many early tetrapods form a complete ring around each orbit. **Prefrontal** and **postfrontal** form the upper rim of the orbit; the **postorbital** occupies a place at the posterodorsal orbital margin. The **jugal** is usually a large bone centered beneath the orbit. The **lacrimal** lies between the orbit and the external

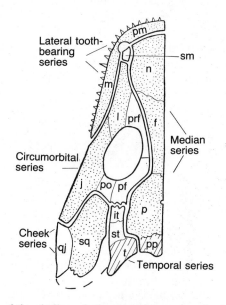

Figure 169. A diagram of the skull roof of a primitive tetrapod with the elements grouped (rather arbitrarily) into regional series. The little septomaxilla does not readily fit into any series. The stippled elements are retained as such in the skull of typical mammals; hatched elements at back appear to be fused into the mammalian occipital bone; elements unshaded are lost in mammals. Abbreviations: *f,* frontal; *it,* intertemporal; *j,* jugal; *l,* lacrimal; *m,* maxilla; *n,* nasal; *p,* parietal; *pf,* postfrontal; *pm* premaxilla; *po,* postorbital; *pp,* postparietal; *prf,* prefrontal; *qj,* quadratojugal; *sm,* septomaxilla; *sq,* squamosal; *st,* supratemporal; *t,* tabular.

naris and may enter the margin of either or both of these openings; the lacrimal (tear) duct lies in this area in tetrapods, and the bone derives its name from the fact that it usually contains a canal for this duct. There is also a small and obscure dermal element, the **septomaxilla**, which may be present at the back margin of the external naris adjacent to the lacrimal or may appear as a small shell of bone within the nasal cavity; it is found in numerous tetrapods of one group or another, but its history is poorly known.

(d) Temporal series. The notch in the skull behind the orbit is known as the **otic notch** in early tetrapods, because the eardrum is believed to have been placed here. Extending back above the notch at the margin of the skull table is a row of three small bones—**intertemporal**, **supratemporal**, and **tabular**—which tend to be reduced and lost in most later tetrapods. The names are not especially descriptive, but happily, they are in alphabetical order from front to back.

(e) Cheek bones. Most of the cheek, behind the orbit and below the otic notch, is occupied by a broad, platelike bone, the **squamosal**; below it, there is primitively a smaller marginal element, the **quadratojugal**.

In primitive tetrapods, the skull roof usually had a sculptured surface bearing ridges or pits, as in modern crocodilians, to which the dermis was strongly attached. This pattern is repeated in the skulls of primitive bony fish but is generally replaced by a smooth surface in living types. Fishes and water-dwelling amphibians bear sensory lateral line organs (cf. Chap. 15) in a pattern of canals or grooves on the head. In fishes, the canals are usually sunken into the substance of the roofing bones as closed tubes with occasional pores opening to the surface; in amphibians, they are more superficial, and, in the roofing bones, are indicated, at the most, by grooves. Similar canals or grooves are found in the lower jaws and in the operculars of fishes.

Palatal Complex (Fig. 167 *B, D*).   In sharks, which lack dermal bones, the only part of the "skull" lateral to the braincase is the cartilage, which functions as an upper jaw. This, however, is not comparable to the jaw margins of bony fishes or tetrapods, because, in them, the homologue of the shark jaw is combined with dermal bones to form a complex palatal structure (Fig. 168 *D*), and the margins of the jaws are formed by bones of the skull roof.

The major part of the palatal complex consists of four pairs of dermal bones developed in the skin lining the roof of the mouth. The largest and most medially situated is the **pterygoid**, which primitively runs the greater part of the length of the skull; it is applied to the superficial, that is, lower, surface of the palatal cartilage or the ossifications formed in it and in part replaces that structure as an element of strength in palatal construction. Lateral to the pterygoid is a row of three further dermal elements: the **vomer**, which meets its partner anteriorly and extends back medial to the internal naris; the **palatine**, extending back from that opening along the palatal margin; and the **ectopterygoid**, terminating at the subtemporal fossa. Teeth may be present on any or all of these four dermal elements in bony fishes and early tetrapods, and the three lateral bones sometimes carry a series of teeth even larger than those of the jaw margins (shark teeth are perhaps comparable to this set, rather than to the marginal teeth of bony types).

These dermal palatal bones serve many of the functions performed by the cartilaginous upper jaw of sharks, but that structure persists, although in somewhat reduced form, in the embryonic skull of bony fishes and tetrapods as the **palatoquad-**

rate (or pterygoquadrate) **cartilage**. In bony fishes, a number of ossifications, variously named, may be found in the adult along the length of this bar; in tetrapods, only two ossifications are found. Posteriorly, the **quadrate bone** forms the articular surface for the lower jaw. More anteriorly, a second ossification, the **epipterygoid**, lies, as the name implies, above the pterygoid. This bone connects at the basal articulation with the sphenoid region of the braincase; a movable articulation at this point between palate and braincase was characteristic, it would seem, of early bony fishes and tetrapods. A vertical bar of the epipterygoid extends upward toward the skull roof; more posteriorly, the epipterygoid sometimes extends backward to meet the quadrate. Anteriorly, the palatoquadrate cartilage originally extended toward the nasal region, but this area is reduced and never ossifies in tetrapods.

The anterior part of the palatal complex is an essentially horizontal plate of dermal bone, notched laterally by the internal nares, but otherwise tightly apposed to the marginal, tooth-bearing, dermal bones. The palatal plates of the two sides meet anteriorly; posteriorly, they separate, leaving an **interpterygoid vacuity** on either side of the ventral surface of the braincase. The posterior half of the palatal complex is, by contrast with the anterior part, a vertical sheet of bone, bounding medially the subtemporal fossa for the jaw muscles. At the junction of anterior and posterior parts of the palatal complex is the basal articulation with the braincase, in which the pterygoid as well as the epipterygoid may take part. Behind this point, there is an open cleft between the vertical palatal plate and the lateral walls of the braincase. Through this cleft pass important blood vessels and nerves; through it also passes, in primitive bony fishes, the spiracular gill slit and, in amphibians, the ear passage that replaces the spiracle.

The Braincase (Fig. 167 *B, E, F*).   In most bony fishes and lower tetrapods, a median dermal element, the **parasphenoid**, forms in the skin of the roof of the mouth beneath the braincase and is applied closely, sometimes indistinguishably, to its lower surface. Anteriorly, it is a slender and sometimes rodlike structure in tetrapods; more posteriorly, it spreads broadly over the expanded ventral surface of the braincase and may gain contact with the pterygoids of either side at the region of the basal articulation.

The braincase is well ossified in most bony fishes and tetrapods. However, the region of the nasal capsules never ossifies in any but the very oldest of tetrapods, and in the skeletally reduced living amphibians, ossification of the braincase is greatly reduced. In the better ossified forms, the bony elements are frequently fused in the adult, so that it is difficult to delimit the areas of the individual bones.

In the occipital region, a ring of four bones—median **basioccipital** and **supraoccipital** elements, and paired **exoccipitals**—surrounds the foramen magnum. In bony fishes and primitive tetrapods, the occipital condyle is a single concave* circular structure, formed mainly by the basioccipital, but with the exoccipitals entering into its upper and lateral borders. Basioccipitals and exoccipitals are formed, embryologically, from fused occipital arch elements and hence tend to resemble the centrum and neural arches of a vertebra. The supraoccipital, on the other hand, arises in a cartilage connecting the otic capsules of the two sides; it is unossified in some bony fishes and amphibians. The basioccipital may extend well forward beneath the floor

*Because it is concave, it should properly be called a cotyle, not a condyle, but it is not.

of the cranial cavity. The exoccipitals, however, are relatively small. Except in the modern amphibians, they are pierced in all tetrapods—and in crossopterygian fishes as well—by openings for the twelfth (hypoglossal) nerve. The vagus and accessory nerves (X and XI) and usually a vein draining part of the brain pass through a large foramen (jugular) lying anterior to the exoccipital and representing the embryologic gap between occipital arch and otic capsule; nerve IX may use this exit or pierce, more anteriorly, the otic capsule.

Except where bridged posteriorly by the supraoccipital, the roof of the braincase is open between the two otic capsules, but the region is, of course, protected by overlying dermal elements. Two ossifications are present in the otic region, the **prootic** and **opisthotic**, but the two may be firmly fused and difficult to distinguish in the adult. The sacs and canals of the internal ear are buried within these ossifications of the capsule and may extend into the supraoccipital as well. In fishes, there is no external opening from the otic region, but the head of the hyomandibular bone abuts against its outer surface; in tetrapods, the **fenestra ovalis** is present at this point to receive the footplate of the sound-transmitting stapes (evolved from the hyomandibular). Nerve VIII, for the reception of sensory stimuli, enters the auditory capsule from within; nerve VII usually pierces the prootic near the anterior end, and the large trigeminal nerve (V) emerges through one or more openings at or near the anterior margin of the prootic. There is frequently an opening, the **post-temporal fenestra**, through the posterior surface of the skull above the otic capsule and below the dermal bones of the skull roof.

Forward from the otic region, the braincase rapidly narrows to the interorbital or sphenoid region. A major element here is the **basisphenoid**, an unpaired bone with its center of ossification in the floor of the braincase; it is almost completely sheathed ventrally by the parasphenoid. At either side, the basisphenoid sends out a **basipterygoid process**, which forms the basal articulation with the palatal complex. A pocket, the **pituitary fossa** or **sella turcica** (not seen in Fig. 167), extends down into this element from the floor of the cranial cavity and contains the pituitary body. Nearby there is, ventrally, an opening from below on either side through which the internal carotid artery enters the braincase, and just behind the pituitary, there is primitively a canal containing a vein that crosses from one side to the other. The lateral walls of the braincase in the sphenoid region are unossified in many early tetrapods. If cartilage or bone is present in this region between the orbits, there will be tiny foramina for the nerves to the eye muscles (III, IV, VI). The optic nerve (II) may pass out here or (as in the form figured) emerge through the back part of the sphenethmoid wall.

A median ossification, the **sphenethmoid**, sheathed below by the anterior process of the parasphenoid, is present in the narrow ethmoid region. Within it, the two olfactory nerves diverge toward the nasal capsules. The more anterior part of the ethmoid area and the nasal capsules remain unossified in almost all tetrapods; they are ossified in various fishes, but even when bone is present, there is also extensive cartilage.

Cranial Kinesis. It will have been noted that in the primitive amphibian skull just described, there are three main functional components and that these components are *movably* attached to each other. Obviously, no such motion, or **cranial kinesis** to use the proper term, is allowed in our or other mammalian skulls. It is also absent in many other skulls: those of chimeras, lungfishes, frogs, turtles, and crocodiles for example; however, most vertebrates, including sharks, teleosts, snakes, and birds, do exhibit such kinesis, which thus seems to be the primitive and "normal" condition.

Unfortunately, however, the situation is anything but simple. Many groups have kinetic skulls, but the number and positions of the functional components vary greatly; thus, kinesis has clearly evolved (or re-evolved) numerous times. As would be expected from the structural variation, the function of kinesis also varies from group to group. Most commonly, kinesis allows the upper jaw, in a broad sense, to work in partial independence of other parts of the skull. Thus, in a shark, the palato-quadrate may slide fore and aft on the braincase; in many teleosts, the upper jaw is protruded (that is, slides forward) as the mouth opens; in snakes, the two sides of the upper jaw may even function independently; in birds, the upper jaw or dorsal half of the beak usually bends slightly on the rest of the skull. Such movements may, in various cases, allow their possessor to open its mouth more widely, to close it more rapidly, to aim its bite more accurately, or to ingest large prey more readily; these are not mutually exclusive, and kinesis may serve more than one function in a single animal.

The Skull Roof In Bony Fishes. Paleontologic work makes it increasingly evident that the ancestors of the tetrapods are to be sought among the Paleozoic sarcopterygians and probably the crossopterygians. We have considerable knowledge of their cranial structure; this shows a resemblance to the early tetrapods in many features.

The dermal shield of these forms (Fig. 170 *A, C*) is generally comparable to that of tetrapods, and, despite considerable confusion in past interpretations, many of the individual dermal elements can be directly homologized. It is, however, interesting to note the marked evolutionary change in proportions between front and back regions of the skull. In crossopterygians, the paired eyes and parietal foramen are far forward on the head; there is a long skull table and cheek region and a short "face." In early tetrapods, the facial region is much more developed and the posterior part of the head is relatively short. This suggests that the ancestors of the bony fishes were very short-faced and that the facial region is in the process of expansion in the crossopterygians. This idea appears to be borne out when we examine the dermal roof pattern. In the back part of the head, most of the bony elements are directly comparable to those of tetrapods. Anteriorly, however, the small elements present are variable, forming a mosaic of scales; thus, they cannot be readily homologized with the facial bones of tetrapods. The face, it would seem, was a new, growing development in which stability of the bone pattern had not yet been attained.

The dipnoans are regarded as an aberrant offshoot of the primitive stock of the Sarcopterygii. The skull roof is certainly aberrant. In modern lungfishes, it is incomplete and includes only a few large dermal elements not readily comparable to those of crossopterygians or amphibians (Fig. 171). In the oldest fossil lungfishes, the roof shield consisted of numerous small polygonal bones, which varied not merely from form to form but even from one individual to another. In later genera there was a trend toward reduction in the number of these elements and for the preservation of a relatively few large plates such as those present today.

The presence of a variable pattern of numerous small bony plates over the entire roofing shield in early lungfishes and on the snout of crossopterygians suggests that the earliest Osteichthyes had numerous, small, and variable bony centers in the head

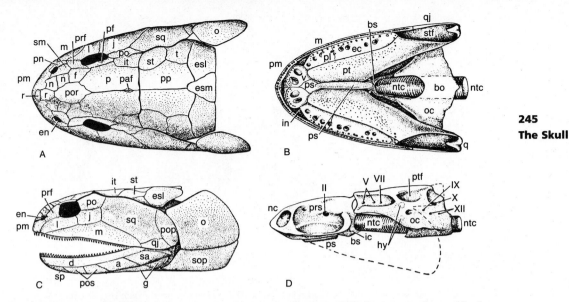

Figure 170. The skull of a Paleozoic crossopterygian (composite), for comparison with that of a primitive tetrapod (cf. Fig. 167). *A,* Dorsal view; *B,* palatal view; *C,* lateral view; *D,* lateral view of braincase. From the crossopterygian dermal covering, that of a labyrinthodont differs primarily in loss of opercular elements, relative reduction of length of posterior part of skull and elongation of "facial" region, and reduction of small elements in rostral and nasal region; the "extrascapulars" at the back of the skull are not true skull bones but enlarged cervical scales. The palate is similar in the two. The labyrinthodont braincase is less completely ossified, formed in one piece instead of the two present in crossopterygians, and lacks the greatly expanded notochord of the latter.

Abbreviations: *a,* angular; *bo,* basioccipital; *bs,* basisphenoid; *d,* dentary; *ec,* ectopterygoid; *en,* external naris; *esl, esm,* lateral and medial extrascapulars; *f,* frontal; *g,* gulars; *hy,* hyomandibular articulation; *ic,* foramen for internal carotid; *in,* internal naris; *it,* intertemporal; *j,* jugal; *l,* lacrimal; *m,* maxilla; *n,* nasal; *nc,* nasal capsule; *ntc,* notochord (restored); *o,* opercular; *oc,* otic capsule; *p,* parietal; *paf,* parietal foramen; *pf,* postfrontal; *pl,* palatine; *pm,* premaxilla; *pn,* postnasal; *po,* postorbital; *pop,* preopercular; *por,* postrostral; *pos,* postsplenial; *pp,* postparietal; *prf,* prefrontal; *prs,* presphenoid (sphenethmoid); *ps,* parasphenoid; *pt,* pterygoid; *ptf,* post-temporal fenestra; *q,* quadrate; *qj,* quadratojugal; *r,* rostrals; *sa,* surangular; *sm,* septomaxilla; *sop,* subopercular; *sp,* splenial; *sq,* squamosal; *st,* supratemporal; *stf,* subtemporal fossa; *t,* tabular; *v,* vomer. Roman numerals, foramina for cranial nerves. (*A* and *B* based on *Eusthenopteron; C* based on *Osteolepis; D* based on *Ectosteorhachis;* data from Jarvik, Romer, Säve-Söderbergh, Stensiö.)

shield and that only later was there a tendency at various times for a reduction in the number of elements present and for the stabilization of the pattern. Because it is improbable that this process of elimination and stabilization would have been the same in all cases, it is presumably too much to expect that homologies can be firmly established between individual roofing bones in the various groups of bony fishes.

The actinopterygians (except the paddlefishes) have a well-developed dermal roof in which many of the elements are quite stable and placed in positions comparable to those of certain of the bones in crossopterygians and tetrapods (Figs. 123,

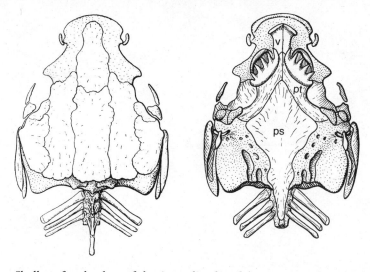

Figure 171. Skull roof and palate of the Australian lungfish, *Neoceratodus*. The braincase (stippled) is cartilaginous but is underlain by a large parasphenoid (*ps*); the dermal roof includes only a small series of large plates that cannot be readily homologized with those of other forms. The upper jaws are fused to the braincase, and the only ossifications in the upper jaw or palate are large pterygoids (*pt*) and vomers (*v*). Large fan-shaped toothplates are borne by the pterygoids, and a small cutting tooth is present on each vomer. (After Goodrich.)

172, 173). In consequence, these elements are customarily given familiar names, such as parietals, frontals, and so on. We have, however, little guarantee of true homology. A notable difference from crossopterygians and tetrapods is that in ray-finned fishes, there is seldom, apart from primitive fossil types, much development of a cheek region behind the orbit and no development, normally, of a bone there corresponding to the squamosal. Except in a few early forms, there is no parietal foramen. In actinopterygians, the orbits are generally large, and (apart from specialized types such as pike and sturgeons) the facial region anterior to the orbits is short and undeveloped. The jaws primitively extended the full length of the head and the maxillae were well developed. In higher actinopterygians, however, the gape is reduced and the jaw articulation is moved forward; the marginal dentition tends to disappear in teleosts, and the maxillae may be eliminated from the borders of the mouth (see Fig. 35).

The Palate In Bony Fishes. In crossopterygians, the palatal structure (Fig. 170 *B*) is quite comparable to that of their tetrapod descendants; the dermal elements are almost identical in every regard.

In lungfishes, the marginal teeth of the mouth are lost (as are the marginal bones that bore them), and the remaining upper dentition includes only a pair of large, fan-shaped tooth plates (Fig. 171) fused to the pterygoids and a much smaller anterior pair on the vomers. In correlation, we find that the dermal pterygoid bone is well developed and that there is a small vomer; the other palatal bones, however, have been lost. There is no ossification of any sort in the palatoquadrate cartilage, which is firmly fused to the braincase in autostylic fashion.

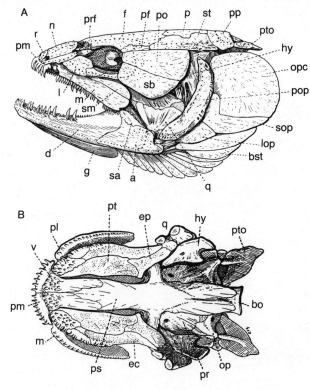

Figure 172. *A,* Lateral and *B,* palatal views of the skull of *Amia,* least specialized of living actinopterygians. Abbreviations: *a,* angular; *bo,* basioccipital; *bst,* branchiostegal rays; *d,* dentary; *ec,* ectopterygoid; *ep,* epipterygoid; *f,* frontal; *g,* gular; *hy,* hyomandibular; *iop,* interopercular; *l,* lacrimal; *m,* maxilla; *n,* nasal; *op,* opisthotic; *opc,* opercular; *p,* parietal; *pf,* postfrontal; *pl,* palatine; *pm,* premaxilla; *po,* postorbital; *pop,* preopercular; *pp,* postparietal; *pr,* prootic; *prf,* prefrontal; *ps,* parasphenoid; *pt,* pterygoid; *pto,* pterotic; *q,* quadrate; *r,* rostral; *sa,* surangular; *sb,* suborbital; *sm,* supramaxillary; *sop,* subopercular; *st,* supra-temporal; *v,* vomer. The identity of certain of these elements with similarly named bones in crossopterygians and tetrapods is doubtful. (After Goodrich.)

In actinopterygians, the palatal construction appears primitively to have been similar to that of crossopterygians, with two or more replacement bones forming in the palatoquadrate cartilage. In most living actinopterygians, the basal articulation of palate and braincase is absent, and, in compensation, the hyomandibular is large and well ossified (Figs. 123, 172 *B,* 173 *A*); an extra bone, the **symplectic**, appears in this area to bind further the hyomandibular to the upper jaw.

The Braincase In Bony Fishes. In lungfishes, the braincase, like other parts of the skeleton, is reduced, persistently cartilaginous except for a pair of exoccipitals; in the sturgeons and paddlefishes, much of the braincase is cartilaginous (and presumably neotenic). In the ancient crossopterygians and older fossil members of the actino-pterygians, the whole structure was thoroughly ossified, but most sutures were obli-terated during growth, and hence little can be said as to the individual bones present.

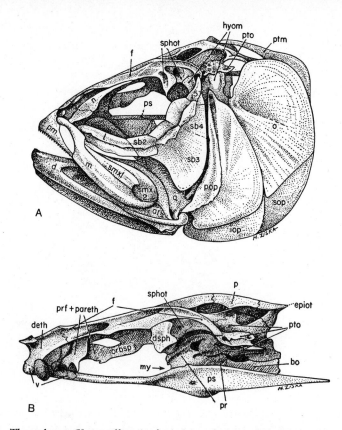

Figure 173. The teleost *Clupea* (herring). *A,* Lateral view of skeleton of head, including the operculum; *B,* lateral view of braincase. The structure of teleost skulls is highly specialized but derived from the general pattern seen in *Amia.* Abbreviations: *ar,* articular; *bo,* basioccipital; *d,* dentary; *deth,* dermethmoid; *dsph,* dermosphenotic; *epiot,* epiotic; *f,* frontal; *hyom,* hyomandibular; *iop,* interopercular; *l,* lacrimal; *m,* maxilla; *my,* opening of myodome (arrow); *n,* nasal; *o,* opercular; *orbsp,* orbitosphenoid; *p,* parietal; *pareth,* parethmoid; *pm,* premaxilla; *pop,* preopercular; *pr,* prootic; *prf,* prefrontal; *ps,* parasphenoid; *ptm,* posttemporal; *pto,* pterotic; *q,* quadrate; *sb 1–5,* suborbitals; *smx 1, 2,* supramaxilla; *sop,* subopercular; *sphot,* sphenotic; *v,* vomer. (After Gregory.)

In modern higher ray-finned forms, including the teleosts, the well-ossified braincase generally shows discrete bones; many appear to be the same as in tetrapods, but the otic region often includes as many as five bones that are not directly comparable to those of land animals (Fig. 173 *B*).

In many regards, even in small details, the crossopterygian braincase (Fig. 170 *B, D*) is quite similar to that of typical early tetrapods. There are, however, striking differences. Whereas the braincase of most vertebrates is built, in the adult, as a single compact unit, that of the crossopterygians is built in two quite separate pieces.* The anterior segment is rather firmly bound to the upper jaw and palatal structures; it

*In primitive fossil actinopterygians, the braincase may also consist of two separate parts; however, the division is farther posterior, with the ear lying in the anterior part, leaving only the occipital region posterior to the division between "halves."

includes the nasal, ethmoid, and sphenoid regions, back to the area of the pituitary fossa and the basal articulation. The posterior segment includes the otic and occipital parts of the braincase. The two parts are movably articulated with one another, and a further specialization tends to give them a flexible union: the notochord runs forward, without constriction, through a canal in the floor of the posterior moiety to terminate in a socket on the posterior surface of the anterior segment.

This unusual type of braincase was long believed to be an insuperable barrier to the derivation of tetrapods from crossopterygians. It is now known that the most ancient of all known amphibians (from the late Devonian) had a braincase in which there is at least a partial separation into two moieties and in which a large notochordal canal persists. It thus appears that even in their peculiar braincase construction, the crossopterygians are antecedent to the tetrapods.

In actinopterygians, the eyes and orbits are usually large, and in relation to this fact, we find that, between the orbits, the braincase may be reduced to a membranous septum. In most actinopterygians, there develops behind the pituitary region a large excavation in the floor of the braincase, a **myodome**, in which the rectus muscles of the eyeball take their origin.

History of the Mammalian Skull—the Skull Roof (Figs. 174–177). Having finished an account of the fish skull, an orthodox procedure would be to discuss, seriatim, the

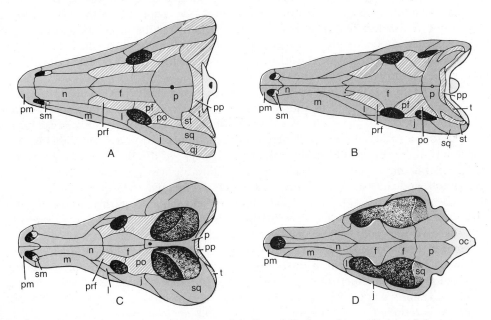

Figure 174. Diagrammatic dorsal views of skull roof to show the evolution of the mammalian condition. The principal changes are (1), loss of many primitive elements; (2), development of a temporal opening, and reduction of the area between the openings to a sagittal crest, followed by an expansion of the braincase beneath. Abbreviations: *oc*, occipital. Colors and other abbreviations as in Figure 167; however, dermal skull roof elements lost or fused with others in mammals are shown as hatched here and in Figures 175 and 183. *A*, a stem reptile (cotylosaur); *B*, a primitive mammal-like reptile (pelycosaur); *C*, an advanced mammal-like reptile (therapsid); *D*, a placental mammal. The forms shown in Figure 177 are representative of stages *B–D*.

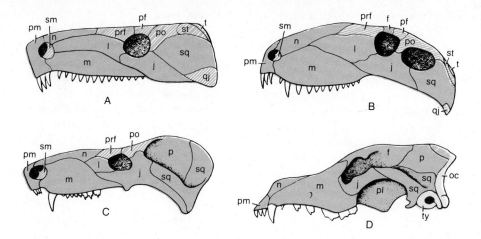

Figure 175. Diagrammatic views of the stages of the mammalian skull roof shown in Figure 174, seen in lateral view. Abbreviations: *oc,* occipital; *ty,* tympanic. Colors and other abbreviations as in Figures 167 and 178. The forms shown in Figure 178 are representative of stages *B–D.*

cranial changes found in the various groups of amphibians, reptiles, and birds, and only at the end come to the mammals. However, we suspect that the major interest of most of the readers of this book lies in the evolution of the interesting but complex skull of mammals from primitive beginnings. In consequence, we shall now deal directly with the evolution of the mammalian skull, and only after that return to describe the variations seen in other tetrapod groups.

First, let us consider the evolution of the dermal skull roof. In the progression from ancestral amphibians to the basal reptile stock (the captorhinomorph cotylosaurs, to be precise), some changes had already occurred. The primitive reptile skull tended to be rather higher and narrower than in the ancestral amphibians. More important, however, was a solidification, so to speak, of the back part of the roof. In the ancestral amphibian, the cheek and the skull table were partially separated by an otic notch, and the two areas were but loosely connected with one another. In reptiles, the otic notch has been eliminated; cheek and table are firmly bound together. The back margin of the skull is essentially a straight line from one side to the other, and the eardrum, which presumably lay in the primitive notch, was forced back to lie behind the cheek. Further, in the fusion of cheek and table, there begins the reduction of the row of three little bones—intertemporal, supratemporal, and tabular—which earlier formed the margin of the skull table above the notch. The intertemporal is absent in all reptiles, and except in very primitive cotylosaurs, the supratemporal is reduced. In the "lining up" of the back edge of the roof, the tabulars and, medially, the postparietals, have been, so to speak, pushed off the skull roof and survive for the time as elements forming the upper margin of the occiput (cf. Fig. 183).

Very early in reptile history, the line leading toward mammals split off the cotylosaur stock as the Pelycosauria of the late Carboniferous and early Permian. The pelycosaurs were still very primitive reptiles but in a number of respects show indications of a drift in a mammalian direction. All the roofing elements present in stem reptiles are still present, although the little supratemporal is reduced to a splinter.

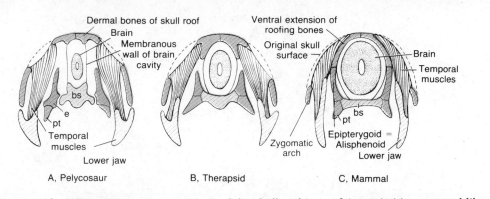

Dermal bones of skull roof
Brain
Membranous wall of brain cavity
bs
e
pt
Temporal muscles
Lower jaw

Ventral extension of roofing bones
Original skull surface
Zygomatic arch
bs
pt
Epipterygoid = Alisphenoid
Lower jaw

Brain
Temporal muscles
bs
pt

A, Pelycosaur   B, Therapsid   C, Mammal

Figure 176. Diagrammatic cross sections of the skull and jaws of *A,* a primitive mammal-like reptile; *B,* an advanced mammal-like form; and *C,* a mammal; to show features in the development of the skull roof and braincase. (1) Part of the lateral walls enclosing the expanding brain were originally membranous. This lateral area comes to be ensheathed by ventral extensions of the roofing bones and by the incorporation of the epipterygoid of the palatal structure as the alisphenoid. (2) This extension of the roofing bones downward around the brain gives the appearance in mammals of being the original skull surface; the original surface, however, lay external to the temporal muscles, as indicated by the broken lines. (3) The upper jaw elements (*e,* epipterygoid; *pt,* pterygoid) originally articulated movably with the braincase (*bs,* basisphenoid); the two structures are fused in *B* and *C.* Dermal bones of skull roof and palate, red; endochondral bone of braincase, yellow; endochondral bone of upper jaw (epipterygoid = alisphenoid), orange.

The most notable change is the development of an opening, a temporal fenestra, in the dermal bones of the cheek region.

In primitive amphibians, the powerful mandibular adductor muscles that close the jaw are shut inside the solid dermal roof of the cheek or temporal region of the skull. In most reptiles, openings develop in the primitively solid dermal shield (cf. Figs. 44, 176). As noted later, such openings are variable in reptiles; there may be two openings, fenestrae, one above the other, or a single fenestra, high up on the temple or lower down. The type characteristic of the synapsids is a **lateral temporal fenestra**, primitively well down on the side of the cheek, with the postorbital and squamosal bones meeting above it. The function of these fenestrae is not clear. Bone appears to be lost near the center of the muscular attachments, where the pull of the muscle would be minimal, and retained around its edges. The muscle fibers originating from the area that becomes the fenestra then attach to sheets of heavy connective tissue closing the opening. Fenestration does, of course, lighten the skull, but presumably this would be important only in large forms with well-developed fenestrae.

But in the therapsids, mammal-like reptiles of more advanced types that appeared during the Permian and continued to flourish in the Triassic, the approach to mammalian conditions is much closer. The temporal fenestra usually enlarged greatly toward the top of the skull; the openings of the two sides in some cases almost meet dorsally, leaving only a narrow strip of the original parietal surface between them. Ventrally, only a bar, often narrow, formed by the jugal and squamosal, remains below the temporal opening as a **zygomatic arch** (Fig. 176). The bar between orbit and temporal opening disappeared in some advanced therapsids and is absent in primitive mammals; its upper end may remain as a process of the frontal. Farther forward, the

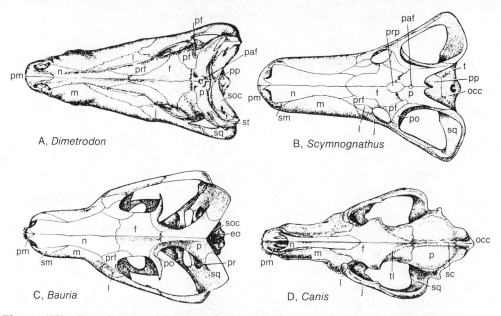

A, *Dimetrodon*

B, *Scymnognathus*

C, *Bauria*

D, *Canis*

Figure 177. Dorsal views of skulls to show the evolution of the mammalian skull roof. *A,* Primitive early Permian mammal ancestor (pelycosaur); *B,* a later Permian therapsid; *C,* a progressive Triassic therapsid; *D,* the dog. Abbreviations: *eo,* exoccipital; *f,* frontal; *j,* jugal; *l,* lacrimal; *m,* maxilla; *n,* nasal; *occ,* occipital bone; *p,* parietal; *paf,* parietal foramen; *pf,* postfrontal; *pm,* premaxilla; *po,* postorbital; *pp,* postparietal; *pr,* prootic; *prf,* prefrontal; *prp,* preparietal; *q,* quadrate; *qj,* quadratojugal; *sc,* sagittal crest; *sm,* septomaxilla; *soc,*supraoccipital; *sq,* squamosal; *st,* supratemporal; *t,* tabular; *tl,* temporal lines. (*B* after Watson; *C* after Boonstra.)

maxilla (bearing the large canine teeth) expands upward and reduces the size of the lacrimal. The two external nares crowd toward the midline, where they form (as also in turtles) a single bony orifice. Various elements tended toward reduction and were lost by the time the definitive mammalian stage was reached; these include the septomaxilla, prefrontal, postfrontal, postorbital, supratemporal, and quadratojugal. The originally paired postparietals form a small median element in therapsids; this persists in mammals but in most adults is fused with the occipital elements (as apparently are the tabular bones, cf. Fig. 183).

The primitive pattern of the mammalian skull roof is mainly preserved in such a form as the dog (Figs. 177 *D,* 178 *D*). There are, however, numerous variations in proportions and disposition of skull parts in the different groups of mammals, most of which are apparent in the arrangement of the dermal bones. The bar behind the orbits, lost in the ancestral mammals, has been rebuilt in a number of different groups, and, in higher primates, a complete, deep, bony wall between orbit and temporal fossa has developed. In some forms (notably among insectivores and edentates), the zygomatic arch may be incomplete. The face is much lengthened in insectivores and is especially elongate in anteaters of several orders; on the other hand, it is shortened in higher primates and withdrawn, so to speak, beneath the expanded brain vault. The external narial opening may move backward over the roof of the face

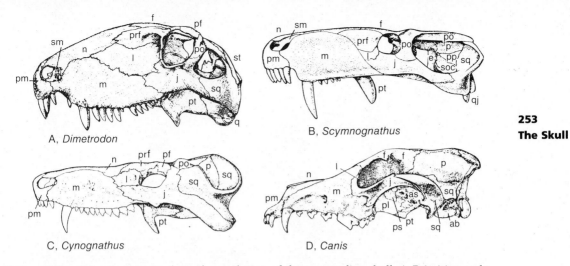

A, *Dimetrodon*   B, *Scymnognathus*

C, *Cynognathus*   D, *Canis*

Figure 178. Lateral views to show the evolution of the mammalian skull. *A,* Primitive early Permian mammal ancestor (pelycosaur); *B,* a late Permian therapsid; *C,* a progressive Triassic therapsid; *D,* the dog. Abbreviations: *ab,* auditory bulla; *as,* alisphenoid; *e,* epipterygoid; *f,* frontal; *j,* jugal; *l,* lacrimal; *m,* maxilla; *n,* nasal; *p,* parietal; *pf,* postfrontal; *pl,* palatine; *pm,* premaxilla; *po,* postorbital; *pp,* postparietal; *prf,* prefrontal; *ps,* presphenoid; *pt,* pterygoid; *q,* quadrate; *qj,* quadratojugal; *sm,* septomaxilla; *soc,* supraoccipital; *sq,* squamosal; *st,* supratemporal. (*B* after Watson; *C* after Broili and Schroeder.)

region in forms that develop a flexible snout (as the tapirs) or trunk (elephants) and in the aquatic sirenians and whales. In the cetaceans, the nasal opening, the "blow-hole," is at the top of the skull; there has been a backward movement of the nasal region and a forward push of the occipital bones as well, so that the original skull roof is practically eliminated from the skull surface; in some whales, this is further complicated by being asymmetrical! In rodents, the masseter muscles of the jaw (cf. p. 312) are highly developed; the side of the face may be deeply excavated for their origin, and a canal leading forward from the orbit onto the snout may be opened out into a large fenestra for their accommodation. In small animals within any group, the brain is relatively large, and, in consequence, the dorsal surface of the skull appears swollen; larger species or races of domestic animals, with relatively small brains but heavy temporal muscles, are more likely to develop a median longitudinal **sagittal crest** for muscular attachment (the dogs show typical variations here). A transverse **nuchal crest** may form across the back margin of the roof for muscles and ligaments supporting the head. The tall, domed shape of the skull roof of the elephant is not due primarily to brain expansion; largely filled with air sinuses, it provides a broad surface for attachment of the cervical muscles and ligaments that support the heavy head and tusks.

History of the Mammalian Skull—the Palatal Complex (Figs. 179–181). The palatal construction seen in primitive tetrapods persists relatively little changed in primitive reptiles such as the cotylosaurs and even the pelycosaurs. The four dermal components persist, the vomers narrowing between the choanae, the ectopterygoid remaining small, and the pterygoids still running most of the length of the palate. A

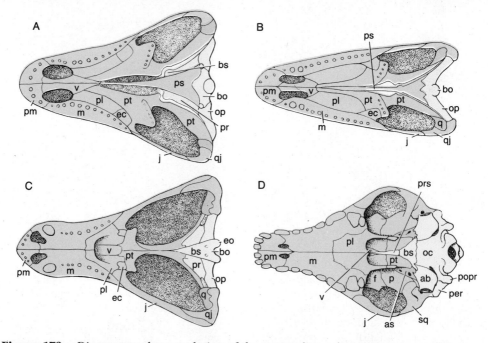

Figure 179. Diagrams to show evolution of the mammalian palate as seen in ventral view. *A,* a stem reptile (cotylosaur); *B,* a primitive mammal-like reptile (pelycosaur); *C,* an advanced mammal-like reptile (therapsid); *D,* a placental mammal. The principal changes are: (1) development of a secondary palate; (2) fusion of palatal structures with braincase at midlength of skull; (3) reduction of pterygoid and quadrate, and loss of latter from skull structure. Abbreviations: *as,* alisphenoid; *per,* periotic; *prs,* presphenoid. Colors as in Figure 167; other abbreviations as in Figures 167 and 181. The forms shown in Figure 181 are representative of stages *B–D.*

prominent lateral flange, primitively tooth-bearing, is developed on the pterygoid. The two cartilage replacement bones of the palatal complex persist; the epipterygoid still forms a movable basal articulation with the braincase and sends a rodlike process upward toward the skull roof. The quadrate continues to be a prominent bone, which, in addition to forming an articulation for the lower jaw, extends far forward along the inner surface of the pterygoid.

The therapsids, however, show a series of modifications leading to the mammalian type of palate. Anteriorly, a major change is the gradual development, within the group, of a **secondary palate**. The premaxillae, maxillae, and, farther back, the palatines fold ventrally and then turn medially to produce an elongate shelf lying below the primary roof of the mouth; channels are thus created through which the air passes far back from the original position of the internal nares before entering the mouth. The vomers, primitively paired, fuse into a single element (they maintain their position in the original roof of the mouth and hence are almost concealed in ventral view). They aid in forming a floor beneath the presphenoid and ethmoid and, further, form a partition between the two nasal cavities. The bony secondary palate is continued backward by a fold of skin as a "soft palate" (Fig. 231). The origin of the second-

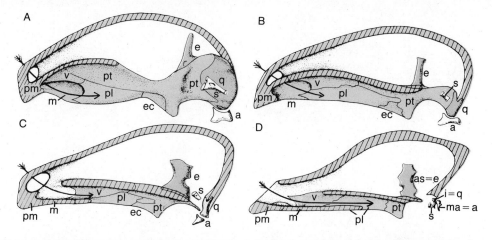

Figure 180. Diagrams to show the evolution of palatal structures and auditory ossicles from primitive reptiles to mammals. Longitudinal sections of skulls, with bones of skull roof and palate represented as if cut vertically just to the right of the midline (hatched surfaces). The left half of the skull and the entire braincase are removed, so that the palatal structures of the right side are seen in medial view. In addition, the stapes and the articular element of the lower jaw are figured to show the evolution of the auditory ossicles. *A,* Primitive pelycosaur (*Dimetrodon*); *B,* a primitive therapsid; *C,* an advanced mammal-like reptile; *D,* a mammal. In *A* and *B,* incoming air (*arrow*) enters the mouth cavity directly through anteriorly placed internal nostrils; in *C* and *D,* a secondary palate is developed, and there is some degeneration of the primary palate. In *A,* the palate articulates movably with the braincase at a socket on the epipterygoid; in *B,* the palate becomes fixed to the braincase; the epipterygoid loses its original function but survives as the alisphenoid bone of mammals. In *B–D,* the pterygoid is reduced; the quadrate and articular, forming the articulation between upper and lower jaws, are reduced in size and lose their original function but survive as auditory ossicles. Abbreviations: *a,* articular; *as,* alisphenoid; *e,* epipterygoid; *ec,* ectopterygoid; *i,* incus; *m,* maxilla; *ma,* malleus; *pl,* palatine; *pm,* premaxilla; *pt,* pterygoid; *q,* quadrate; *s,* stapes; *v,* vomer. Colors as in Figure 167.

ary palate in mammal-like reptiles and mammals may be reasonably associated with the development of the constant body temperature characteristic of mammals. Continuous breathing is practically a necessity, and the secondary palate is an aid in the maintenance of breathing while the mouth is functioning in chewing food; birds, of course, also have a constant body temperature and must breathe continuously—but they do not chew their food.

In the mammalian line, as in many lower tetrapods, movable articulation of palate and braincase has been lost (Fig. 176); the pterygoid and epipterygoid bones fuse on either side to the sphenoid region of the skull. In this process, the pterygoids, once large elements, become much reduced in size. The lateral flange on the palate disappears (and with it the little ectopterygoid bone), and the posterior arm of the pterygoid, once running backward and outward to the quadrate, gradually dwindles. In a mammal, the pterygoids are little more than small wings of bone projecting ventrally on either side below the sphenoid region of the braincase with which they are usually fused. They form the walls of the posterior end of the air passage.

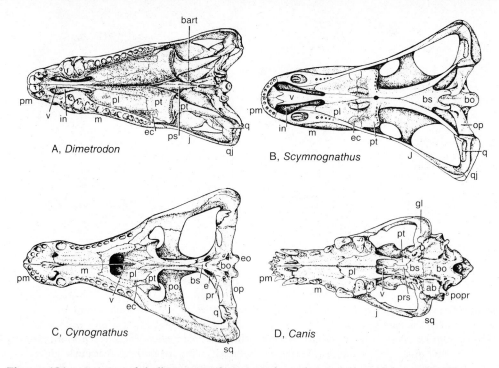

A, *Dimetrodon*

B, *Scymnognathus*

C, *Cynognathus*

D, *Canis*

Figure 181. A series of skulls in ventral view to show the evolution of the mammalian palate. *A,* Primitive early Permian pelycosaur; *B,* a late Permian therapsid; *C,* an advanced Triassic therapsid; *D,* the dog. Principal changes include development of secondary palate in *C* and *D;* loss of movable basal articulation and fusion of braincase and palate in *B–D;* reduction of pterygoid; loss of quadrate from skull structure and development of new jaw joint in *D;* addition of auditory bulla in *D.* Abbreviations: *ab,* auditory bulla; *bart,* basal articulation of palate and braincase; *bo,* basioccipital; *bs,* basisphenoid; *e,* epipterygoid; *ec,* ectopterygoid; *eo,* exoccipital; *gl,* glenoid cavity; *in,* internal nares; *j,* jugal; *m,* maxilla; *op,* opisthotic; *pl,* palatine; *pm,* premaxilla; *po,* postorbital; *popr,* paroccipital process; *pr,* prootic; *prs,* presphenoid; *ps,* parasphenoid; *pt,* pterygoid; *q,* quadrate; *qj,* quadratojugal; *sq,* squamosal; *v,* vomer.

The original endochondral elements of the palate have changed markedly in mammals and no longer form part of the palatal structure. The mammals have acquired a new jaw joint; the cranial portion of it, the **glenoid fossa,** is formed by the squamosal, and the quadrate has thus lost its major function. The quadrate was already small in therapsids and rather loosely connected with other bones. In mammals, it has disappeared from the skull; it is, however, present as a tiny auditory ossicle. After the fusion of palate with braincase, the epipterygoid lost much of its original function. The epipterygoid, in addition to forming a socket for the basal articular process of the braincase, primitively had a vertical process extending up to the skull roof lateral to the braincase. In mammals, this has fused into the braincase walls as the alisphenoid bone (cf. p. 261).

History of the Mammalian Skull—The Braincase (Figs. 182, 183). In early reptiles, the braincase retains much of the structure already described for primitive am-

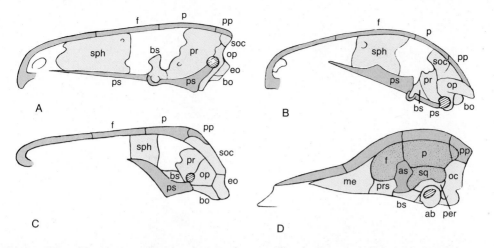

Figure 182. Diagrams to show evolution of mammalian braincase. *A,* Primitive reptile (cotylosaur); *B,* primitive mammal-like reptile (pelycosaur); *C,* advanced mammal-like reptile (therapsid); *D,* placental mammal. Colors as in Figure 167. The principal changes are (1) development anteriorly of presphenoid from sphenethmoid and in some mammals an added mesethmoid; (2) disappearance of parasphenoid; (3) addition of alisphenoid to braincase by transformation of epipterygoid of upper jaw; (4) fusion of otic elements into a periotic; (5) addition of tympanic element(s); (6) fusion of occipital elements plus postparietals into a single occipital bone. Except in mammals, the brain is confined to the posterior part of the cranial cavity (essentially between the otic and occipital elements), and the sphenethmoid serves merely for passage of the most anterior nerves. In the expanded mammalian brain, not only are sphenethmoid and epipterygoid (as presphenoid and alisphenoid) pressed into service to aid in brain coverage, but also flanges (stippled) from frontal, parietal, postparietal, and squamosal form parts of the lateral wall. Abbreviations: *ab,* auditory bulla; *as,* alisphenoid; *bo,* basioccipital; *bs,* basisphenoid; *eo,* exoccipital; *f,* frontal; *me,* mesethmoid; *oc,* occipital bone; *op,* opisthotic; *per,* periotic; *pr,* prootic; *prs,* presphenoid; *ps,* parasphenoid; *soc,* supraoccipital; *sph,* sphenethmoid. Other abbreviations as in Figure 183.

phibians, with a ring of occipital bones, pro- and opisthotics, a basisphenoid, and, forward, a tubular sphenethmoid—the whole underlain by the dermal parasphenoid. On the whole, however, the early reptilian braincase tends to be higher and narrower than that of amphibians. The space between the orbits is narrower, and the walls here (often incomplete in early amphibians) are rarely ossified. For much of the height in the orbital region, only a membranous septum is present between the otic region posteriorly and the sphenethmoid in front; almost the entire brain lay posterior to this point.

In the mammal-like reptiles, the braincase elements are those seen in early amphibians and reptiles. When the mammalian condition is attained, notable changes, partly associated with the increase in the size of the brain, have occurred (Figs. 184, 185). The occipital region is basically the same as that of early reptiles, but there is a division of the originally single condyle into a paired structure. In the adult mammal, the elements of the occipital region are usually fused into a single **occipital** bone, although ossification centers of all four primitive elements are seen in the embryo. Usually added to the occipital, dorsally, are dermal bone centers that appear to

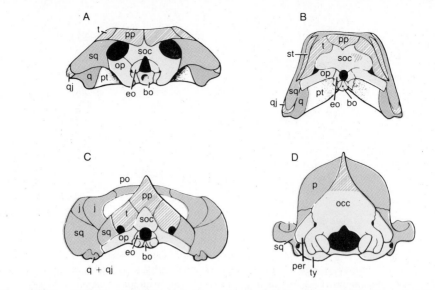

Figure 183. Diagrammatic views to show the evolution of the occiput from reptile to mammal. Colors as in Figures 167 and 174. *A,* primitive reptile, similar to primitive tetrapod. *B,* primitive mammal-like reptile (pelycosaur). *C,* advanced mammal-like reptile (therapsid); postparietal and tabulars on occipital surface; condyle double. *D,* mammal; occipital elements, postparietals, and (?) tabulars fused into a single occipital element. Abbreviations: *bo,* basioccipital; *eo,* exoccipital; *j,* jugal; *occ,* occipital bone; *op,* opisthotic; *p,* parietal; *per,* periotic (including mastoid); *po,* postorbital; *pp,* postparietal; *pt,* pterygoid; *q,* quadrate; *qj,* quadratojugal; *soc,* supraoccipital; *sq,* squamosal; *st,* supratemporal; *t,* tabular; *ty,* tympanic bulla.

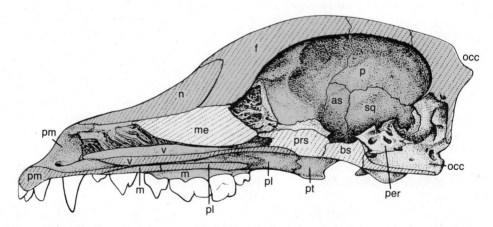

Figure 184. Median sagittal section of the dog skull. Colors as in Figure 167. Diagonal lines indicate sectioned bones. It will be seen that most of the bone enclosing the brain cavity is derived from dermal elements (*f, p, sq*), and even part of the occipital is embryologically of dermal origin. Abbreviations: *as,* alisphenoid; *bs,* basisphenoid; *f,* frontal; *m,* maxilla; *me,* mesethmoid; *n,* nasal; *occ,* occipital bone; *p,* parietal; *per,* periotic; *pl,* palatine; *pm,* premaxilla; *prs,* presphenoid; *pt,* pterygoid; *sq,* squamosal; *v,* vomer.

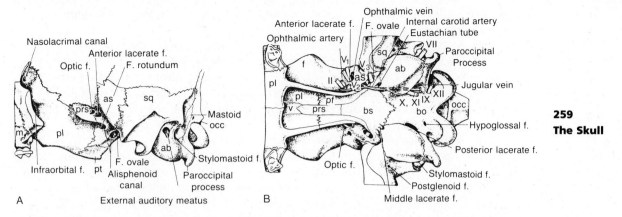

Figure 185. Braincase region of a dog in *A,* lateral view; *B,* ventral view, to show foramina. In *B* the main nerves, the course of the internal carotid artery and its palatine branch, and the jugular vein are indicated. Abbreviations: *ab,* auditory bulla; *as,* alisphenoid; *bo,* basioccipital; *bs,* basisphenoid; *f,* frontal; *j,* jugal; *m,* maxilla; *occ,* occipital bone; *pl,* palatine; *prs,* presphenoid; *pt,* pterygoid; *sq,* squamosal; *v,* vomer.

be remnants of the postparietal and (questionably) tabular bones of the ancestral skull roof.

Primitively, the otic capsule was, we noted, formed from two centers (prootic and opisthotic). In mammals, however, a variable number of centers fuse in the adult into a compact mass, the **periotic**. In reptiles, the otic capsule occupies a relatively large area on the side wall of the braincase; in mammals, in contrast, it is relatively small and is situated low down at the lateral margins of the braincase. For the most part, the capsule is buried deeply beneath the surface of the skull, but its posterior end may be exposed as the **mastoid process**; this is frequently adjoined, posteriorly and medially, by a **paroccipital process** of the occipital bone.

In the otic region, in most mammals, we find a new addition to the skull in the shape of the **auditory bulla** (Figs. 181 *D,* 185). Just outside the periotic region lies the middle ear cavity, containing in mammals the three small and delicate auditory ossicles and, at its outer margin, the eardrum (cf. p. 531). In the earliest mammals, as in most marsupials today, this region had no bony protection against injury from without; in most placentals, this protection is supplied by the development of the bulla as an ensheathing capsule. Frequently present here is a **tympanic** bone; primitively, it was merely a bony ring surrounding the eardrum; in other cases, this bone expands inward to form part or all of the bulla. The tympanic is a dermal element, and embryologic evidence suggests (unlikely as it seems at first) that it is a modification of the angular bone of the reptilian lower jaw, which lies close to the eardrum in reptiles (cf. Fig. 193). In addition to the tympanic, there forms in many mammals a second bone, an **entotympanic**, which may surround the middle ear cavity and may fuse in the adult with the tympanic to form a compound bulla. The entotympanic is formed in cartilage. This cartilage, however, is not part of the original tetrapod skull; we have here, most exceptionally in the skull, the addition of a new element. The otic elements, periotic and new bulla, remain discrete in some forms; in others, they fuse with one another and with the adjacent squamosal bone of the skull roof to form a compound **temporal** bone.

The basisphenoid forms the braincase floor in the region of the pituitary. In many cases, it fuses with neighboring elements—alisphenoid, pterygoid, presphenoid, and orbitosphenoid—to form a complex (if small) structure termed the **sphenoid**. The parasphenoid, which sheathed the braincase floor in lower forms, has disappeared as a separate ossification in mammals; whether or not it is represented in the complex sphenoid bone is debatable.

The sphenethmoid of lower tetrapods persists in mammals, anterior to the basisphenoid, as the **presphenoid**; its lateral portions, in the orbit, are sometimes distinct and termed **orbitosphenoids**, but the two form essentially a unit structure. In some mammals, including various primitive types and both major orders of ungulates, the presphenoid is the most anterior part of the braincase; but in other orders, including primates, rodents, and carnivores, a new ossification, the **mesethmoid**, is present still farther anteriorly. When both are present, the presphenoid is restricted to the anterior part of the braincase floor, appearing externally in the roof of the choanae ventrally and (as the orbitosphenoid portion of the bone) in the orbital walls anterior to the alisphenoid. In mammals, the nasal pouches have extended backward so that they are separated from the expanded cranial cavity only by a transverse plate of bone, the **cribriform plate**, perforated for the passage of bundles of olfactory nerve fibers. The mesethmoid, when present, ossifies in this area and extends forward as a median septum between the two nasal cavities. In these cavities, variable scrolls of cartilage or bone, **turbinals**, which represent the original nasal capsule, are developed in connection with the neighboring bones (maxilla, nasal, and mesethmoid). Covered with nasal mucosa, they increase the surface area of the nasal passages (cf. p. 504).

The bones named thus far include the entire roster of elements formed in the braincase proper. But if we examine the actual "braincase" of a typical mammal (Fig. 184), it will be seen that these original braincase elements form little but the floor and the back wall of the cranial cavity. The mammalian brain has expanded to such proportions that the braincase proper has been unable, so to speak, to keep up with its growth; most of the walls of bone surrounding the brain in the adult placental mammal have been derived from other parts of the skull.

The parietals and frontals in early reptiles formed a broad roof over the brain region. With the development of the temporal fenestrae, their surface area was much reduced (Fig. 176 A). They tended, however, to send flanges downward, deep to the jaw muscles, along the walls of the cranial cavity beneath. With the great expansion of the brain in mammals (Fig. 176 C), these flanges became much enlarged and spread out laterally over the swollen brain surface to form much of the roof and side walls of the brain box; the squamosal bone from the side of the dermal roof also helps to form a secondary covering for the brain. These newly expanded areas of the dermal roofing elements give the appearance of being part of the original surface of the skull. They are not, however. They lie beneath the temporal muscles, which are those closing the jaws; the original skull roof was external to these muscles. The upper limits of temporal muscle attachment on either side are marked by the **temporal lines**, which may become confluent posteriorly as the sagittal crest (Fig. 177 D). Only the area of the frontals and parietals that lies between these lines, or forms the sagittal crest represents the original cranial surface; the remainder of the expanse of these bones is a new, deeper, internal growth. One tends, unthinkingly, to regard the zygomatic arches as superficial "handles" external to the skull proper much as the handle or handles of a jug projecting outward from the vessel itself. This is, as we have seen, the reverse of the case. The arches are a last remnant of the former lateral walls of the skull; the solid "brain box" is mainly a new formation deep to the muscles of the jaw.

Still another addition has been made to fill out the walls of the expanding braincase: the incorporation of the epipterygoid bone of lower vertebrates as the **alisphenoid** (Figs. 180 *D*, 182, 184). We have noted the lack of ossification in the reptilian braincase wall between the orbits. The epipterygoid in reptiles is mainly a vertical rod, which, as part of the upper jaw, lies external to this unossified interorbital region. In therapsids, this bar was expanded to a plate but still lay external to the braincase. In mammals, however, it has been incorporated into the braincase wall, filling the gap above the basisphenoid and so firmly fused with this last bone that (as its new, mammalian name implies) it appears to form a wing of the combined sphenoid element.

In sum, the bony capsule surrounding the mammalian brain is a composite affair. The original braincase is able to do little more than form its floor; dermal elements plus the alisphenoid are called into play to form the walls and roof of this expanded structure.

Foramina in the Mammalian Braincase (Fig. 185).   We have mentioned the various openings for nerves and vessels found in the braincase of primitive tetrapods. The situation may be reviewed for the mammals, because, although many of the openings are the same, the nomenclature (unfortunately) differs; further, the incorporation of the alisphenoid and auditory bulla into the braincase has modified the structure considerably.

The olfactory nerves in mammals consist of numerous small filaments that pierce the ethmoid bone in the sievelike cribriform plate. The optic nerve usually enters the cranial cavity through an **optic foramen** in the orbitosphenoid.

In early reptiles, the area posterior to the ethmoid region, on the side of the braincase as far back as the otic capsule, was open for the exit of nerves III to VI. The filling in of the braincase wall by the alisphenoid has changed this situation. We find a large opening in front of the alisphenoid, the **anterior lacerate foramen** (sphenoidal or orbital fissure); two openings—**foramen rotundum** and **foramen ovale**—usually piercing the alisphenoid; and a **middle lacerate foramen**, lying at the back edge of the alisphenoid (with which the ovale may fuse). Typically, the nerves to the eye muscles and the most anterior division of the trigeminal nerve pass through the anterior lacerate foramen; maxillary and mandibular branches of the trigeminal use the round and oval foramina; the middle lacerate foramen contains no major nerves, but the internal carotid artery enters the braincase here or close by.

Nerves VII and VIII enter the otic capsule from the cranial cavity in an **internal auditory meatus**. The former nerve primitively emerged at the surface toward the front end of the otic capsule; in mammals, however, in relation to the complicated middle ear structure, it remains buried in the otic bones until it emerges back of the bulla at the **stylomastoid foramen**. Posterior to the bulla, in a gap between the otic and occipital regions, is a **posterior lacerate foramen** (jugular foramen) through which nerves IX, X, and XI and the internal jugular vein draining the skull make their exit. Nerve XII usually occupies a separate **hypoglossal foramen**, sometimes multiple, in the floor of the occiput.

The auditory bulla has, of course, an external opening, the **external auditory meatus**, for access to the eardrum and, at its front end, an opening for the auditory or eustachian tube from the throat. The internal carotid artery originally ran forward on the under surface of the skull past the middle ear. This area is covered in most mammals by the bulla, beneath which the artery runs forward in a **carotid canal** to a point near that at which it enters the braincase. Near this latter point an **alisphenoid canal** may transmit a branch of the artery, which continues forward toward the palate.

Other openings in the skull and ducts not connected with the braincase may include the **incisive foramen**, which is near the front of the palate, and connects the mouth with the vomeronasal organ in many mammals (cf. p. 504); the **infraorbital foramen**, sometimes enlarged to a canal, carrying blood vessels and nerves forward from the orbit to the surface of the snout; the **nasolacrimal canal** from orbit to nasal cavity, for the tear duct; and variable canals in the palate for the forward passage of nerves and small blood vessels. In some rodents, the infraorbital canal is greatly enlarged to accommodate the expanded masseter muscle of the jaws.

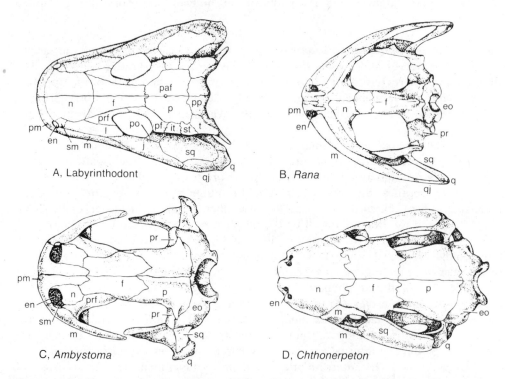

Figure 186.   Dorsal views of amphibian skulls. *A,* Primitive labyrinthodont; *B,* a frog; *C,* a salamander; *D,* a gymnophionan. Abbreviations: *en,* external naris; *eo,* exoccipital; *f,* frontal (fused with parietal in frogs); *it,* intertemporal; *j,* jugal; *l,* lacrimal; *m,* maxilla; *n,* nasal; *p,* parietal; *paf,* parietal foramen; *pf,* postfrontal; *pm,* premaxilla; *po,* postorbital; *pp,* postparietal; *pr,* prootic; *prf,* prefrontal; *q,* quadrate; *qj,* quadratojugal; *sm,* septomaxilla; *sq,* squamosal; *st,* supratemporal; *t,* tabular. (*A* after Watson; *D* after Marcus.)

The Skull Roof in Lower Tetrapods (Figs. 186–188).   Having followed through the story of mammalian skull evolution, we will return to give an outline of the evolution of the skull in other tetrapod types. Rather than discuss the changes rung on the primitive pattern as a whole in group after group of nonmammalian tetrapods, we shall, as in the case of mammal evolution, follow through separately the history of the major components.

The later history of the dermal skull roof is almost exclusively one of loss and reduction. There is almost never development of a new element; always there is, in time, a greater or lesser degree of reduction. No living tetrapod has retained in full the pattern of its early ancestors, and few have preserved a solid dermal roof.

Although a great variety of Paleozoic and early Mesozoic amphibians retained the primitive pattern with little change, the modern amphibians show a greater reduction of the skull roof than do most amniotes. In the broad, flat skull of a modern frog or salamander (Fig. 186 *B, C*) only a small proportion of the ancient dermal roof elements is preserved. Premaxilla, maxilla, nasal, frontal, parietal, and squamosal form the usual complement of bones; in urodeles, a prefrontal is also present, but, in anurans, the parietals (as seen from embryos) appear to be fused with the frontals, and, in some urodeles, maxillae and nasals are absent. All other elements have vanished, and much of the upper surface of the head is bare of dermal bone. In the Gymno-

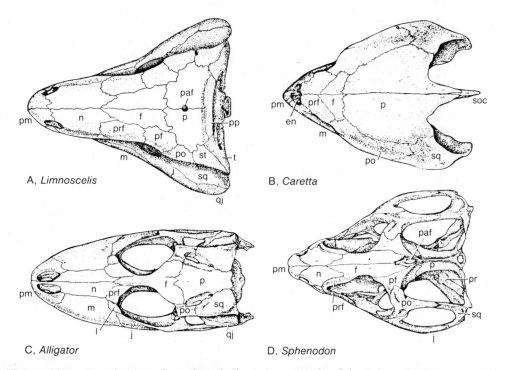

Figure 187.   Dorsal views of reptilian skulls. *A*, Stem reptile of the Paleozoic; *B*, a sea turtle; *C*, a young alligator; *D*, *Sphenodon*. Abbreviations: *en*, external naris; *f*, frontal; *j*, jugal; *l*, lacrimal; *m*, maxilla; *n*, nasal; *p*, parietal; *paf*, parietal foramen; *pf*, postfrontal; *pm*, premaxilla; *po*, postorbital; *pp*, postparietal; *pr*, prootic; *prf*, prefrontal; *qj*, quadratojugal; *soc*, supraoccipital; *sq*, squamosal; *st*, supratemporal; *t*, tabular.

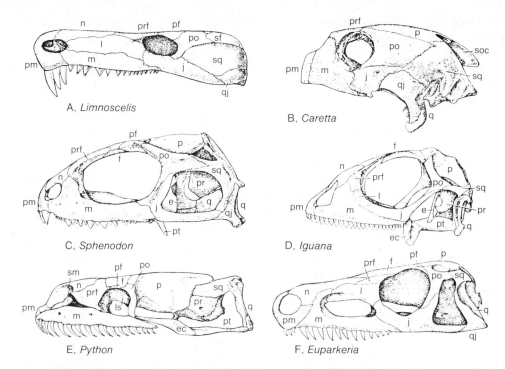

Figure 188. Side views of reptilian skulls. *A,* Stem reptile of the Paleozoic; *C,* *Sphenodon; D,* a lizard; *E,* a python; *F,* a primitive ruling reptile of a type from which birds, dinosaurs, and crocodilians have descended. Abbreviations: *e,* epipterygoid; *ec,* ectopterygoid; *f,* frontal; *j,* jugal; *l,* lacrimal; *ls,* laterosphenoid; *m,* maxilla; *n,* nasal; *p,* parietal; *pf,* postfrontal; *pm,* premaxilla; *po,* postorbital; *pr,* prootic; *prf,* prefrontal; *pt,* pterygoid; *q,* quadrate; *qj,* quadratojugal; *sm,* septomaxilla; *soc,* supraoccipital; *sq,* squamosal; *st,* supratemporal. (*F* after Broom.)

phiona (Fig. 186 *D*), the skull roof is a solid structure, but there is a similar reduction in the number of bones present; this solidity is probably a secondary condition associated with the burrowing nature of those wormlike forms.

In the stem reptiles (cotylosaurs*; Figs. 187 *A,* 188 *A*), the primitive pattern of the skull roof, we have noted, was in general retained in regard to the number of elements present, with no loss except the intertemporal. In later reptiles (Figs. 187 *B–D,* 188 *B–F*), however, there are notable losses. The elements that most generally disappear are those in the temporal region and the back margin of the skull: supratemporal, tabular, and postparietal; no one of the three is positively identifiable in any modern reptile. The parietal foramen became unfashionable, it would seem, in many reptiles; *Sphenodon* and some lizards retain it, but it has vanished in turtles, snakes, and crocodiles (and it is absent, too, in birds and mammals). We noted earlier that the primitive otic notch had disappeared in stem reptiles, apparently leaving the eardrum "free" behind the cheek region. But in many modern reptiles, notably chelonians, lizards, and primitive archosaurs, there was redevelopment of a notch for support

*The form figured, *Limnoscelis,* is now thought by some to be a very cotylosaur-like labyrinthodont rather than a labyrinthodont-like cotylosaur; in any event, it shows the primitive reptilian condition.

of the anterior rim of the eardrum in a concavity, bounded by an enlarged quadrate bone, at the back of the cheek region.

Many major changes in the skull roof of reptiles are associated with the development of **temporal fenestrae** (Fig. 44). The stem reptiles and the turtles retained the primitive, solidly roofed—**anapsid***—condition. In most reptiles, however, one or two openings on each side have been made in the cheek or temple. As already noted, these openings are important features in the classification of reptiles, defining in many cases the subclasses. Two openings are developed in the **diapsid** condition: one high up toward the skull roof, a second well down the side of the cheek; a horizontal arch of bone between the two openings connects the postorbital and squamosal elements. The Squamata are derived from early diapsids; the bar closing the lower fossa has, however, been lost in lizards, leaving but a single arch and single closed opening, and, in snakes, the upper bar has been lost as well, leaving the whole cheek clear of dermal bone. This reduction leaves the quadrate as a relatively free element, particularly so in snakes, giving a high degree of motility (useful in swallowing large prey) to the jaws. In the extinct euryapsids, the single opening was dorsally situated, the postorbital and squamosal bones meeting below the fenestra in a (logically enough) **euryapsid** condition; possession of an opening lower down on the cheek, already discussed in relation to mammal evolution, is termed the **synapsid** structure.

The turtles (Figs. 187 *B,* 188 *B*) are technically anapsids, because they have no true temporal opening. In many turtles, however, the skull roof has been, so to speak, eaten away from behind and is often strongly scalloped out on either side of the parietals; in other cases, emargination takes place at the lower margin of the cheek. In extreme cases, the entire temporal region has been excavated, and there is no dermal roof over the cheek region from the orbits to the back edge of the skull. Posterolaterally, the turtle skull is uniquely specialized for the eardrum and middle ear structures.

In Aves (Fig. 189 *A*), the lightly built skull has such a close fusion of elements that most sutures are obliterated. Birds have evolved from archosaurian reptiles, and the dermal roofing pattern is essentially that found in generalized members of that group. However, the expanding brain and large orbit have, so to speak, crowded the cheek region, originally diapsid in structure; the bars behind the orbit and between the two fenestrae have largely disappeared.

The Palatal Complex in Lower Tetrapods.   The palatal elements present in the earliest tetrapods persist for the most part, but the structural arrangement became considerably modified in the palate of the broadened and flattened skull of the modern urodeles and anurans (Fig. 190). The basal articulation has lost its motility, and palate and braincase are broadly fused here (and in many extinct amphibians as well). Anterior to this point of union, the gap between the dermal bones of the palate and the base of the braincase is widened out into broad interpterygoid vacuities. The ectopterygoid is absent, and, in urodeles, the palatine is also small or absent, so that the palate is poorly developed laterally. Further, ossification is reduced in the embryonic palatal cartilage. The epipterygoid is never ossified, and, in urodeles, ossification of

---

*Early students of the subject named the various conditions according to the arches (apsid = arched) that bound the openings, although they are normally defined in terms of the openings.

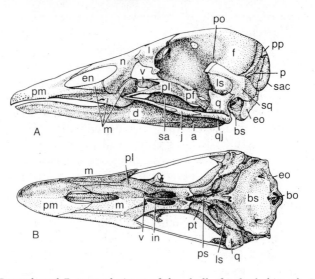

Figure 189. *A,* Lateral, and *B,* ventral views of the skull of a duck (*Anas*). Abbreviations: *a,* angular; *bo,* basioccipital; *bs,* basisphenoid; *d,* dentary; *en,* external naris; *eo,* exoccipital; *f,* frontal; *in,* internal naris; *j,* jugal; *l,* lacrimal; *ls,* laterosphenoid; *m,* maxilla; *n,* nasal; *p,* parietal; *pl,* palatine; *pm,* premaxilla; *po,* postorbital; *pp,* postparietal; *ps,* parasphenoid; *pt,* pterygoid; *q,* quadrate; *qj,* quadratojugal; *sa,* surangular; *soc,* supraoccipital; *sq,* squamosal; *v,* vomer. (After Heilmann.)

the quadrate is incomplete; here, as elsewhere, the modern amphibian skeleton is definitely reduced. The gymnophionan palate has a more compact structure but is essentially comparable to that of other modern amphibians, and the ossifications present are the same as in the frog. In the oldest tetrapods, the jaw gape was long, and the quadrate articulation was consequently well back of a plane drawn transversely through the occiput. In living amphibians, particularly the urodeles and apodans, the jaws are much shortened and the quadrates level with or in advance of the occiput.

The reptiles (Fig. 191) tended to hold to the primitive type of palate more persistently than did the amphibians; the primitive construction developed in the first reptiles is still seen little changed in *Sphenodon* and lizards. A prominent lateral flange, often tooth-bearing, developed on the pterygoid in early reptiles and is preserved in *Sphenodon* and lizards. As noted previously, the quadrate has become a relatively free and movable element in lizards. In snakes, the palate is still more flexible; indeed, the entire skull, except for the compact braincase, is pliable.

In the turtles, the palate has become fused to the braincase; the epipterygoid may disappear in the process. The pterygoids have fused with the lateral margins of the braincase floor, closing the interpterygoid vacuities; the quadrate is massive and is broadly in contact with the otic region of the braincase, thus increasing the solidity of the skull. The ectopterygoid is lost. The internal nostrils have usually moved some distance back in the roof of the mouth; they are generally situated in a deep dorsal pocket, which may be covered ventrally by bone to a varying degree. We see here the beginning of a secondary palate, paralleling the development already seen in mammals; the form figured here (Fig. 191 *B*) shows extreme development of this secondary palate.

In crocodilians, palatal mobility is likewise lost. The two pterygoids have gained

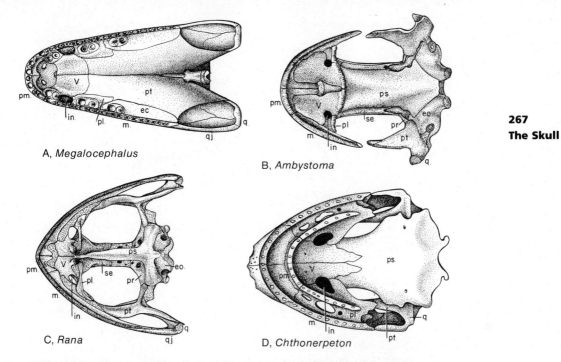

A, *Megalocephalus*

B, *Ambystoma*

C, *Rana*

D, *Chthonerpeton*

Figure 190. The palate of amphibians. *A,* Paleozoic labyrinthodont; *B,* a salamander; *C,* a frog; *D,* a gymnophione amphibian. Abbreviations: *ec,* ectopterygoid; *eo,* exoccipital; *in,* internal naris; *m,* maxilla; *pl,* palatine; *pm,* premaxilla; *pr,* prootic; *ps,* parasphenoid; *pt,* pterygoid; *q,* quadrate; *qj,* quadratojugal; *se,* sphenethmoid; *v,* vomer. (*A* after Watson; *D* after Marcus.)

a firm median union beneath the floor of the sphenoid region and have welded themselves onto the braincase. A further notable change from that of primitive forms is the presence of a very highly developed secondary palate. Beneath the original palatal roof, the maxillary, palatine, and pterygoid bones have built a secondary shelf of bone; between the braincase and this underlying shelf, canals run back from the external nares to open into the mouth far back, above the opening to the windpipe. A flap of skin just in front of this point can close off the pharynx from the mouth so that breathing may be accomplished under water, even with the mouth open, so long as the tip of the nose is above the surface—a useful adaptation in a predaceous water-dweller.

In birds (Fig. 189 *B*), the palatal structures are lightly built and flexible. There is a fair amount of variation, but, in most forms, the vomers are small bones, centrally placed between the large choanae; the ectopterygoids and epipterygoids are lost. The palatines run lengthwise lateral to the choanae to terminate at a movable articulation with the braincase. The pterygoids are short bars that extend back from this point to the quadrates, which are movably socketed onto the side of the braincase. In large flightless birds, the Paleognathae, the palate is more rigidly constructed.

The Braincase in Lower Tetrapods. In modern amphibians, the braincase tends to become relatively broad and flat and to become greatly reduced in its degree of ossification. On its under surface, it is covered by a parasphenoid, very broad in uro-

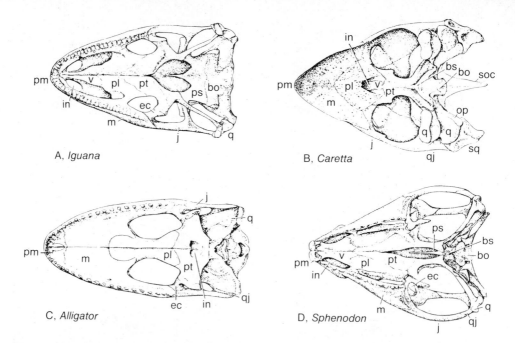

A, *Iguana*

B, *Caretta*

C, *Alligator*

D, *Sphenodon*

Figure 191. The palate of reptiles. *A,* Lizard; *B,* a sea turtle; *C,* a young alligator; *D,* *Sphenodon.* Abbreviations: *bo,* basioccipital; *bs,* basisphenoid; *ec,* ectopterygoid; *in,* internal naris; *j,* jugal; *m,* maxilla; *op,* opisthotic; *pl,* palatine; *pm,* premaxilla; *ps,* parasphenoid; *pt,* pterygoid; *q,* quadrate; *qj,* quadratojugal; *soc,* supraoccipital; *sq,* squamosal; *v,* vomer.

deles and gymnophionans (Fig. 190). In the braincase itself, little ossification remains. There are paired exoccipitals that may extend into the otic region; often there is a prootic bone, and anteriorly, a sphenethmoid. Other ossifications have disappeared. In most fossil amphibians and in all modern types, the condyle, primitively single, is paired. The skull appears to have been shortened in the occipital region in modern amphibians, and no foramen for a twelfth nerve is present.

In reptiles, the braincase tends to remain better ossified than in amphibians. There is almost never any of the flattening tendency seen in amphibians; the occipital region remains well developed; there is always a foramen for the twelfth nerve; and the condyle, a convex structure, remains single except in the line leading to the mammals. The ossifications present are generally those seen in early tetrapods. Fusion of the palatal structures with the braincase in turtles, crocodiles, and various fossil reptiles results in a loss of the basipterygoid process of the basisphenoid, a union of the otic region with the quadrate, and so forth; but usually the basic structure is readily discernible.

There is, however, one notable difference between the structure of the reptilian braincase and that of modern amphibians. In the latter, the part of the braincase between the orbits, in front of the otic region, is relatively thick and has in many cases complete side walls. In primitive reptiles, however, and in many early amphibians, the walls are here incomplete. For much of the height, as noted earlier, only a membranous septum is present between the orbits; the brain primitively lay entirely back of this point. However, in certain reptiles—the snakes and the crocodilians—

there appears to have arisen a functional need for a better enclosure of the cranial cavity here. This is accomplished by the development of a new paired ossification of the braincase, the **laterosphenoid** (Fig. 188 *E*).

In birds, the brain is much expanded, and the braincase is a swollen and completely ossified structure (Fig. 189 *A*). There has occurred here a process somewhat similar to that described earlier for mammals. Much of the expanded brain is sheathed by new extensions of the old roofing elements—frontals, parietals, and squamosals. In addition, however, a laterosphenoid element is present.

Lower Jaw (Figs. 192, 193).   The lower jaw, or **mandible**, although not part of the skull as we have used the term, may be described here. It is in most vertebrates a complex of both dermal and endoskeletal structures. Presumably, dermal elements were present in the jaw of the ancestral jawed vertebrates; in the living Chondrichthyes, however, they are absent, and the lower jaw of sharklike fishes is, like the upper, simply a tooth-bearing bar of cartilage derived from a branchial element (Figs. 163 *C*, 165, 166). This **mandibular cartilage**, also termed **Meckel's cartilage** after its discoverer, is present in the jaw of the embryo in bony fishes and all tetrapods

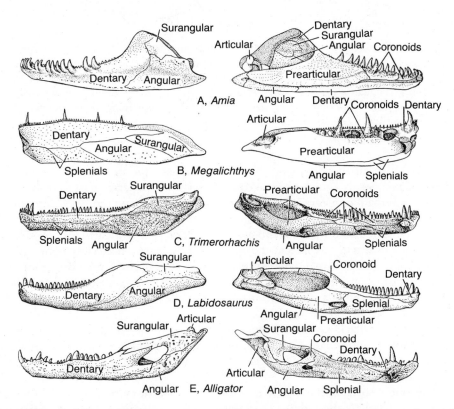

Figure 192.   Left lower jaws, outer views at left, inner views at right, of *A*, a ray-finned fish, the bow fin; *B*, a primitive crossopterygian; *C*, a primitive labyrinthodont; *D*, a primitive reptile; E, an alligator. The jaws of modern teleosts, amphibians, and reptiles, in which the number of elements is reduced, have been derived from the types shown in *A*, *C*, and *D*, respectively.

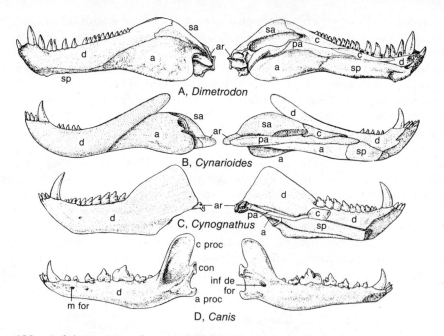

Figure 193. Left lower jaws of mammal-like reptiles and mammals, illustrating the reduction of jaw elements. Outer views (*left*) and inner views (*right*). *A,* Primitive mammal-like reptile (pelycosaur); *B,* a primitive therapsid; *C,* an advanced therapsid; *D,* a typical mammal (dog). Abbreviations: *a,* angular; *a proc,* angular process; *ar,* articular; *c,* coronoid; *con,* condyle; *c proc,* coronoid process; *d,* dentary; *inf de for,* inferior dental foramen; *m for,* mental foramen; *pa,* prearticular; *sa,* surangular; *sp,* splenial.

(Fig. 163 *E–H*). In all these types except the mammals, the cartilage, or a bone derived from it, persists as a minor part of the adult jaw. Its anterior part may disappear in the adult or remain as a cartilage buried deep within the jaw, rarely (as in common frogs) an anterior part of the cartilage may ossify as a **Meckelian bone**. The posterior portion, which articulates by a socket with the upper jaw, is a cartilage in skeletally reduced forms but usually ossifies as the appropriately named **articular** bone, moving on the quadrate of the upper jaw. Anterior to it is a **prearticular fossa** opening downward into the jaw from above, which affords an area of insertion for muscles closing the jaw and offers a place of entrance into the interior for nerves and blood vessels.

Except in the sharklike fishes, the lower jaw consists in great part of dermal bones. Early fossil amphibians and bony fishes had a complex series of such bones; their descendants have simplified the pattern in various ways. The primitive pattern will be described first.

On the outer surface, the major element is the **dentary**, which bears the marginal teeth and parallels the maxilla (and premaxilla) of the upper jaw. Below and to the back of the dentary, there is primitively a row of dermal bones that includes anteriorly two small **splenials**, followed by a larger **angular** and a **surangular**. The symphysis between the rami of the jaw is mainly formed by interlocking rugosities on the inner surface of the dentaries, but the splenials often enter into the symphysis as well. All these external elements may curve inward ventrally to appear also on the lower part of the inner surface of the jaw.

The internal sheathing elements of the primitive jaw were four in number. A major dermal bone is the elongate **prearticular**, which runs forward from the articular. Toward the upper margin, there is a series of three slender bones, the **coronoids**. They bear, in early types, teeth of variable size, and teeth may be present on the prearticular as well. It is of interest that the inner aspect of the lower jaw is primitively very much a mirror image of the upper jaw and palate (cf. Figs. 190 *A*, 192 *C*). The prearticular, running forward from the articular, internal to the prearticular fossa, is comparable to the pterygoid, running forward from the quadrate internal to the subtemporal fossa; the three coronoids are comparable to the three lateral palatal elements.

The structure described above is found in both early crossopterygians (Fig. 192 *B*) and early amphibians (Fig. 192 *C*). In other groups of vertebrates, the changes from this primitive condition are almost entirely a matter of fusion or, more frequently, loss of elements.

In ray-finned fishes, various losses and fusions occur, leading to a condition in which, in teleosts, only three bones are recognized, termed **dentary**, **articular**, and **angular**. In sturgeons, paddlefishes, and lungfishes, skeletal reduction leads to the retention in the adult of a large unossified Meckel's cartilage and little dermal ossification.

In modern amphibians, too, ossification is much reduced. Typical urodeles have three dermal bones, the dentary and two bones on the inner surface, which are probably the prearticular and a coronoid. In frogs, there is, apart from the dentary, only a single other element, presumably the prearticular. In reptiles (Fig. 192 *D*), there is never more than one splenial and almost never more than a single coronoid, but a general retention of other elements—articular, dentary, angular, surangular, and prearticular. In crocodilians (Fig. 192 *E*) and other archosaurs, a large lateral fenestra is usually developed in the angular region of the jaw. In birds, the characteristic reptilian elements (except, apparently, the coronoids) are retained in a fused condition, and the archosaurian lateral fenestra is present.

In mammal-like reptiles (Fig. 193 *A–C*) stages can be seen in a notable transformation of the jaw structure that is completed in mammals (Fig. 193 *D*). The dentary increases steadily in importance; posteriorly, it extends to a position beneath the squamosal close to the old jaw articulation, and, more anteriorly, an ascending process grows upward beneath the temporal region for the attachment of temporal muscles. Concomitantly, the other jaw ossifications are reduced in size. In mammals, we find that these other structures have been lost from the jaw and, insofar as they are still present, are to be looked for in the auditory apparatus (cf. Fig. 385). The transitional stages in advanced therapsids are surprisingly well known; indeed, we now have numerous forms in which both the old reptilian joint (articular-quadrate) and the new mammalian one (dentary-squamosal) are simultaneously developed and functional (Fig. 194). The system will work as long as all the joints are on, or very close to, one transverse line—a door can have as many hinges as desired along its side as long as they are all in a straight line. This new jaw joint is only part of a major reworking of the entire masticating apparatus involved in the origin of mammals; the

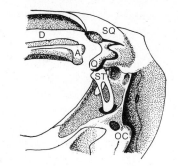

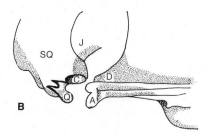

Figure 194. The jaw articulation of an advanced therapsid, the cynodont *Probainognathus,* in (*A*) ventral and (*B*) medial views. In this form, both the old, reptilian articular-quadrate joint and the new, mammalian dentary-squamosal joint are functional. Note how the articular (soon to be the malleus), quadrate (incus), and stapes form a chain leading from the region of the tympanic membrane (not shown; posteroventral to the jaw articulation) to the otic capsule of the braincase. Abbreviations: *A,* articular; *C,* condyle in the squamosal with which dentary articulates; *D,* dentary; *J,* jugal; *OC,* occipital condyle; *Q,* quadrate; *SQ,* squamosal; *ST,* stapes. (After Romer.)

muscles (cf. p. 312) are also quite unlike those of more primitive forms. The need for more food in relation to temperature control and the development of chewing are further aspects of this complex change.

The mammalian jaw consists of the dentary bone alone. This includes a **ramus**, the original tooth-bearing part of the bone; an ascending **coronoid process** to which temporal muscles attach; a new backward projection that may bear ventrally an **angular process** (often prominent); and posteriorly a transversely elongate rounded **condyle** that fits into the glenoid fossa of the squamosal bone of the skull and has thus functionally replaced the articular.

In mammals, alone of all vertebrates, the ancestral type of jaw articulation between endochondral visceral jaw elements has been abandoned for a new type of articulation between dermal bones both above and below.

Chapter 9

Muscular System

The muscular system, judged quantitatively, should loom large in any study of the present sort, because muscle tissue constitutes from one third to half the bulk of the average vertebrate. Functionally, too, the musculature is of highest importance. The activity of the nervous system, even the highest functioning of a human brain, has little mode of expression other than the contraction of muscle fibers. From locomotion to the circulation of the blood, the major functions of the body are caused by or associated with muscular activity. Although movement of the trunk, limbs, or jaws or of an organ or organ part is the main product of muscular activity, subsidiary functions should also be kept in mind. Often muscular work is expended in a negative manner in maintaining stability of the body or its parts, and muscular activity is of primary importance in the production of body heat. Muscular tissue may also function by producing electric fields.

Smooth muscle. Histologically, two major types of muscular tissue are distinguished: smooth and striated fibers. Smooth muscle fibers are the simpler and smaller of the two. These fibers are derived from the embryonic mesenchyme, primarily mesodermal in origin and hence developed in association with connective tissue. The main seat of smooth muscle is in the lining of the digestive tract. Certain other sites, such as the ducts of the glands associated with the gut, the bladder, and the trachea and bronchi of the lungs, are outgrowths of this tract. Still other loci for smooth muscle fibers, however, are independent of the gut. These include circulatory vessels, genital organs, and the connective tissue of the skin and other areas.

A typical smooth muscle fiber (Fig. 195) is a slender, spindle-shaped cell averaging a few tenths of a millimeter in length. There is a single, centrally situated nucleus; the cytoplasm appears homogeneous, but, with special chemical treatment, tiny fibrils can be seen running the length of the cell. In smooth muscle, as the name implies, the fibrils appear to be simple; there is no cross banding of the fiber such as is seen in striated muscle cells. Smooth muscle fibers are sometimes scattered. More generally, however, they are arranged in bands or bundles, with interspersed connective tissue fibers uniting them into an effective common mass (cf., for example, Figs. 273, 282) and, in some cases at least, with protoplasmic processes affording direct contact between neighboring cells. In great measure, smooth muscle cells are stimulated by autonomic nerve fibers; but apparently they may be stimulated by the contraction of their neighbors as well, and, particularly in the digestive tract, one may find a wave of contraction passing along a band of smooth muscle tissue.

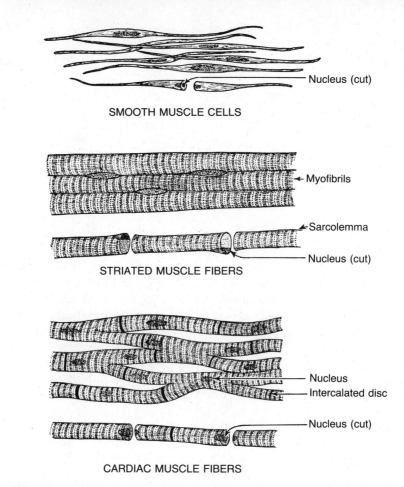

SMOOTH MUSCLE CELLS

Nucleus (cut)

Myofibrils

Sarcolemma

Nucleus (cut)

STRIATED MUSCLE FIBERS

Nucleus

Intercalated disc

Nucleus (cut)

CARDIAC MUSCLE FIBERS

Figure 195. Major types of muscle fibers. (From Hunter and Hunter.)

Cardiac Muscle. In the heart is found a special type of musculature not present elsewhere (Fig. 195). This tissue shows a cross banding similar to that seen in striated fibers, but, in other regards, cardiac muscle differs markedly from striated muscle. The whole heart musculature is a continuous network of dividing and recombining strands. Prominent cross bands, termed **intercalated discs**, more or less separate this network into short units. It was once thought that the heart musculature was a syncytium, without boundaries between individual cells; recent work, however, indicates that the intercalated discs are actual boundaries between cellular units, each of which has a centrally situated nucleus. Embryologically, heart muscle, like smooth muscle, is derived from mesenchyme; its histologic characteristics are presumably associated with the important functions of this constantly active tissue. The heart beat may be influenced by autonomic nerves, but, as discussed in a later chapter, the heart musculature is essentially autonomous.

Striated Muscle. Fibers of this type form the "flesh" of the body—the voluntary muscles, derived in great part from the myotomes of the embryo; in general, they are arrayed as formed muscles that attach to and move skeletal structures. Striated muscle fibers (Fig. 195) are large cells, elongate and cylindric, with lengths that vary from about 1 mm up to several centimeters (lengths of over 30 cm have been reported). The fibers are arranged in parallel fashion to form muscles. Connective tissues run between the fibers and bind them together, forming sheaths for fiber bundles—fascicles—and, further, forming an external sheath, the epimysium, for the muscle as a whole. The toughness of meat depends upon the amount of connective tissue present; for example, the psoas major muscle of mammals has little connective tissue and is the source of "filet mignon."

Striated muscle cells are multinucleate, with numerous nuclei scattered along the entire length. In most vertebrate groups, the nuclei are situated in the interior of the fiber; in mammals and to some degree in birds, the nuclei are, on the contrary, superficial in position. The fiber surface is covered by a thin membrane, the **sarcolemma**, or cell membrane plus a basal lamina (only the former is present in smooth muscle). As in smooth and cardiac muscle cells, the striated muscle fiber contains a large number of closely packed longitudinal fibrils. The striated appearance of the fiber is due to the fact that each fibril consists of alternating light and dark portions; these occur at the same point on each fibril and hence appear as cross striations on the fiber as a whole. Cardiac muscle shows similar banding but in a less clear-cut way.

Each nerve fiber usually divides terminally to supply a number of adjacent muscle fibers at **myoneural junctions** (motor end-plates). In a resting condition, the sarcolemma is electrically polarized. A nerve impulse results in the giving off from the nerve fibril tips a minute quantity of the chemical acetylcholine. This appears to cause a momentary depolarization of the sarcolemma and thereby sets up contraction.

Muscle Fiber Function. The force of a muscle is exerted through a shortening of its fibers and hence a pull on the bones or other structures to which the two ends of the muscle are attached. There is, of course, no reduction in total bulk; in consequence, the shortened fibers, and the muscle as a whole, become thicker with contraction. In smooth musculature, the contraction is relatively slow and the amount of contraction is less than in a striated muscle, but contraction may be long sustained; in contrast, striated muscle may be more rapidly stimulated and contract more vigorously but is more readily fatigued. In addition to the vigorous contraction of muscles, there is the phenomenon of **tonus**. Seldom is a muscle completely relaxed; both in smooth and in striated muscles, there is usually a condition of constant slight contraction, as a result of which organs may maintain their proper shape and position or, in the case of striated muscles, proper posture of the body or its parts can be maintained.

A striated muscle as a whole may contract slightly or strongly, briefly or for a considerable period of time; the result varies according to the number of fibers stimulated and the number and rapidity of the nerve stimuli. Individual muscle fibers, however, work on an "all or none" basis; each fiber either contracts as fully as possible or fails to contract at all. A single sharp contraction of a striated fiber and the more gradual relaxation that follows occupy altogether but a tenth of a second or so. Normally, however, muscle contraction is not caused by a single stimulation but by a continuous rapid tattoo of nerve impulses. A second impulse given before the effect of the first has worn off will increase the contraction, a third will further increase it, and a long series may bring the muscle fibers to a state of maximum and almost continuous contraction or tension (tetanus).

Classification of Muscle Tissues. Attempts to range the muscular tissues of the body in major categories have been based on a variety of criteria; some contradictions are encountered, no matter what criterion is considered fundamental. One obvious suggestion is a classification according to histologic structure, with two main divisions consisting, on the one hand, of smooth muscles (including cardiac muscles) and on the other hand of the striated muscles. This seems at first sight reasonable. The striated muscles are for the most part under voluntary control, whereas the smooth muscles are under the influence of the involuntary or autonomic nervous system; the striated muscles are in general large, well-formed structures, whereas the smooth muscles are often diffuse and may be incorporated in the substance of other organs; most of the striated muscles are formed in the "outer tube" of the body, whereas the smooth musculature is in great measure associated with the gut tube; much of the striated muscle is derived from the mesodermal somites, whereas the smooth musculature arises from mesenchyme.

One prominent group of muscles, however, is in many ways anomalous and ruins the seeming simplicity of such a classification. This is the **branchiomeric system** of striated muscles, which in primitive vertebrates is associated with the branchial bars (the visceral skeletal system) and which remains prominent in the head and neck in all vertebrates forming, among other things, the jaw muscles. The branchial muscles and their derivatives are striated but exhibit strong contrasts with all other striated muscles and, curiously, show affinities in important diagnostic characters with the smooth musculature. Muscles of this group are not derived from the somites as are most other striated muscles but arise from mesenchyme of the splanchnopleure like smooth muscle fibers (Fig. 196). Other striated muscles are innervated from brain or spinal cord by the somatic motor system (cf. p. 546). In contrast, the branchiomeric muscles are innervated, through special cranial nerves, by a different category of neurons, the visceral motor system; the smooth muscles are likewise innervated by visceral nerves. Finally, whereas "ordinary" striated muscles lie in the "outer tube" of the body, the muscles of the visceral arches, although sometimes becoming superficial in higher vertebrates, are, like the smooth muscles, primarily associated with the digestive tube, that is, its anterior, pharyngeal region.

All this suggests that the primitive vertebrate had two discrete sets of musculature. The first, which we may term the **somatic musculature**, is that forming the muscles of the "outer tube" of the body, the "somatic animal"; this musculature is derived, directly or indirectly, from myotomes, is universally striated, is innervated by somatic motor neurons, and is associated with the adjustment of the organism to its external environment. The second is the **visceral musculature**, connected mainly with the gut or "inner tube" of the "visceral animal"; muscles of this group are all derived from the mesenchyme of the splanchnopleure (never from myotomes), are innervated by visceral motor fibers, and are mainly associated with nutrition, digestion, secretion, and circulation—the internal economy, that is, of the animal. In this second group of muscles, those of the more posterior segments of the gut retained a simple structure as smooth muscle; but, in the pharyngeal region, important in early vertebrates for food gathering and breathing, the striated condition is assumed, and, in some instances, striated musculature rather than smooth continues

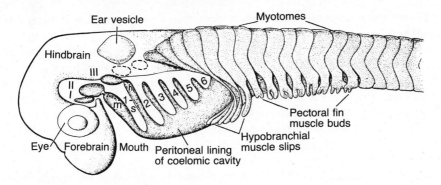

Figure 196. A diagrammatic view of a shark embryo to show the development of the muscles. Skin and gut tube removed; the brain and eye and ear vesicles are included as landmarks. Posteriorly, the myotomes have extended downward to form myomeres; in the region of the pectoral fin, paired buds are forming from neighboring myotomes as potential fin muscles. Anterior to this, buds from anterior myotomes extend ventrally to form hypobranchial muscles. In the ear region, myotomes (broken lines) are rudimentary or absent, but farther forward, three myotomes (I to III) persist to form eye muscles. The position of the spiracular slit (s) and normal gill slits (2 to 6) is indicated. These interrupt the continuity of the coelom and its peritoneal epithelium. Buds of this project upward between the branchial slits; from the splanchnopleure, there arise the visceral muscles of the mandibular arch (m), the hyoid arch (h), and the more posterior gill arches. (In part after Braus.)

farther back along the gut to the esophagus or (in teleosts) even to the anterior end of the stomach.

It appears, then, that despite the seeming impropriety of associating in one major group all smooth and some striated elements, the most natural classification of muscles is as follows:

$$\text{Somatic} \ldots \ldots \begin{cases} \text{Axial} \\ \text{Appendicular} \end{cases}$$

$$\text{Visceral} \ldots \ldots \begin{cases} \text{Branchiomeric} \\ \text{Smooth (gut, and the like)} \end{cases}$$

The smooth muscles are, in general, component parts of various organs and require no separate consideration here. In this chapter, we shall discuss only the formed muscles of striated type—the various somatic muscle groups and the branchial muscles of the visceral system.

Muscle Terminology. Several terms are frequently used to describe muscles, particularly limb muscles, according to the type of action they perform. An **extensor** muscle is one that acts to open out a joint; a **flexor** closes it. An **adductor** draws a segment toward the midline of the body (or, rarely, of a part like the hand); an **abductor** does the reverse. A **levator** raises a structure, in contrast to a **depressor**. A **rotator** twists a limb segment; a **pronator** or **supinator** rotates the distal part of a limb toward a prone or supine position of the foot (i.e., with palm or sole down, or vice versa). **Protractors** draw parts forward or out; **retractors**, naturally, retract them. **Constrictor** or **sphincter** muscles are those that surround orifices (as gills, anus) and tend to close them when contracted; they may be opposed by **dilators**.

Muscles most often attach to skeletal elements at either end. One attachment is usually the more stable and is considered the area of **origin**, the other, the **insertion**. By convention, in limb muscles, the proximal end is always considered to be the point of origin; this does *not* mean it moves less (in locomotion it frequently does not). The fleshy mass of a muscle is its **belly**. Muscles with two bellies in sequence, with an intervening tendon (as in one of the mammalian jaw muscles), are, reasonably, termed **digastric**. A muscle with multiple heads may be termed bicipital, tricipital and so forth.

Muscle fibers never attach directly to a bone or cartilage; the attachment is always mediated by collagenous fibers. The muscle as a whole is surrounded by a thin sheet of connective tissue, the **epimysium**; within the muscle, further sheets, **endomysium** and **perimysium**, surround and separate bundles and other subdivisions (Fig. 197 *A*). Fibers from these sheets are continuous with those around skeletal elements, that is, periosteum or perichondrium, and form the actual attachment. In many cases, the muscle fibers approach the bone closely, and hence we speak of a "fleshy" attachment; in others, the muscle terminates in a **tendon**, a ropelike structure formed almost entirely of connective tissue fibers. Some tendons form thin, flat sheets; these are commonly termed **aponeuroses** and are comparable to the fasciae that surround groups of muscles and generally bind parts of the body together. Tendons and aponeuroses may lie along the surface or even penetrate the substance of a muscle; when this occurs, there may develop complex arrangements of muscle fibers, producing **pinnate muscles** (Fig. 197 *B*). Such muscles, consisting of many but short fibers, may produce considerable power but have only a small distance of contraction; however, their function is too complex to be summarized neatly in one short sentence.

Muscle Homologies. The comparative study of musculature is difficult because of the variability of muscles and the apparent ease with which their function may alter. A mass of muscle tissue that is a unit in one animal may be split into two or more distinct muscles in another animal, and, in other cases, originally separate muscles have fused. The general pattern of their arrangement may be of aid in an attempt to sort out individual muscles, but the pattern may be obscured by the fact that a muscle's origin or insertion may shift in relation to differing functional needs.

As with other organ systems, muscles are given, as far as possible, names used in human anatomy. Unfortunately, the comparative study of musculature is still in its infancy, and we are in doubt in many cases as to the homologues of human muscles in lower vertebrates. Elementary students in zoology, for example, frequently dissect the thigh of a frog and find applied to its muscles the names of those found in the human thigh; but it is doubtful that many of the muscles given like names are really homologous. Another, and better, procedure often followed is to give muscles of lower vertebrates of which the homologies are, or have been, in doubt names that are simply descriptive of their general position or attachments. Thus, reptiles have an iliofemoralis muscle running from ilium to femur; it may be more or less homologous with some of the gluteal muscles lying in this region in man and other mammals (cf. p. 297), but since exact homologies are doubtful, it is safer to give the reptile muscle a name that avoids a definite commitment.

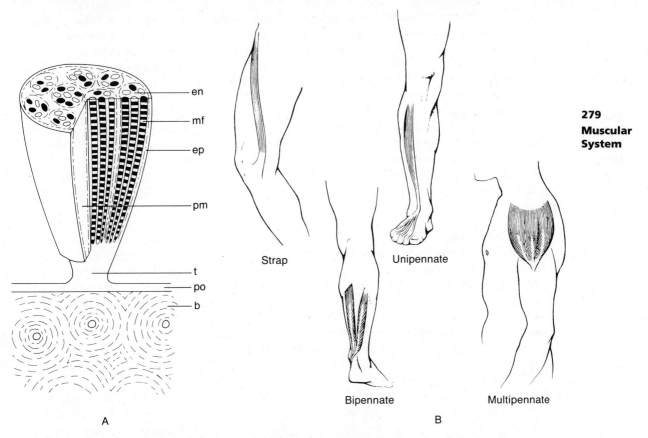

Strap

Unipennate

Bipennate

Multipennate

A

B

Figure 197. *A,* Diagram of the end of a muscle, showing the various layers of connective tissue. Although they are shown as partially separate, they are in fact continuous and indistinguishable except by position. The connective tissue would be fibrous, and the muscle fibers would occupy more volume than they appear to in the drawing. Abbreviations: *en,* endomysium; *ep,* epimysium; *pm,* perimysium; *mf,* muscle fiber; *t,* tendon; *po,* periosteum; *b,* bone.

B, Diagrams showing various arrangements of muscle fibers, based on human examples. The strap or parallel fibered muscle is the sartorius; the unipennate muscle is one of the extensor digitorum longus muscles; the bipennate muscles are the flexor digitorum longus and flexor hallucis longus; and the multipennate muscle is the deltoid. (*B* from Young.)

Embryologic origin is here, as ever, an important criterion for identification of homologies. In many cases, groups of individual muscles in the adult can be traced back to larger aggregations of muscular or premuscular tissue in the embryo (cf. Fig. 205 *B*); the mode of breaking up these mother masses gives valuable evidence of homologies. Unfortunately, only a limited amount of work has been done on muscle embryology, and the information usually indicates homologies of groups of muscles more than of individual ones.

The motor innervation of muscles provides valuable clues. Limb muscles, for example, are generally innervated by a series of trunks and branches arising out of a nerve plexus associated with the limb (Fig. 396). The pattern of the plexus and the nerves extending into and from it tend to remain rather uniform, and muscles known to be homologous usually receive their nerve supply from like branches of the plexus. In consequence, there arose in the minds of many investigators a belief that there is an unalterable phylogenetic relation between a given nerve and muscle—a proposition known as the Fürbringer hypothesis, after the anatomist who first explicitly stated it. Embryology, however, gives no indication that there is any mysterious affinity between specific nerve fibers and the specific muscle fibers which form a given muscle, and, in some cases, it seems quite certain that the innervation of a muscle *is* different in different animals. Nevertheless, actual practice indicates that, as the embryologic picture suggests, the nerve supply to a particular mass of muscle does tend to remain constant and that innervation affords an important—indeed, except for a nearly complete series of intermediate forms, the best—clue to the identity of muscles.

Muscle Function. The action of muscles has, of course, always been of great interest to anatomists. Indeed, in many cases the names of muscles—flexor carpi radialis, adductor femoris, and so on—reflect our ideas on their function. Often investigators have tried to deduce the actions of muscles simply from dissections of preserved animals, probably an impossible task. Sometimes this has required simplified assumptions, such as the absence of other muscles or the absolute immobility of one end of the system. Such assumptions are reflected in the familiar diagrams illustrating actions, such as Figure 198. Here one muscle, the biceps, is shown as the flexor of the elbow, while another, the triceps, is the extensor. Even in this simplified system, flexion does not depend only upon the contraction of the biceps; note that the triceps must relax at the same time or no movement will be produced. Also contraction of one or the other of these muscles may not bend the elbow at all. Both cross the shoulder joint as well as the elbow; they may rotate the humerus on the scapula instead. All the other muscles, conveniently omitted here, will complicate the picture greatly.

This problem of oversimplification also arises when we try to show muscle action as a system of simple levers. In Figure 198, contraction of the biceps always works as a third class lever system, because the force is between the load and the

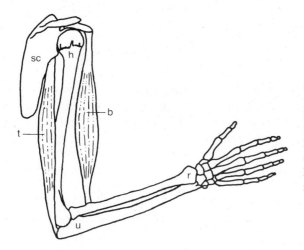

Figure 198. Diagram showing muscles acting on the human elbow. The long head of the biceps (*b*) flexes the elbow while the long head of the triceps (*t*) extends it. Abbreviations: *b,* humerus; *r,* radius; *sc,* scapula; *u* ulna.

fulcrum; however, when a book is being lifted, the hand is the load and the elbow joint is the fulcrum; the reverse is true if the subject is chinning himself on a bar. Use of the triceps will involve either a first or a second order lever system, depending upon which end moves. The basic physical principles apply, of course, but not necessarily in diagrammatically simple or unchanging ways; it is often impossible to say that a given muscle acts in one way.

Thus to describe accurately the actions or functions of muscles, much more than casual inspection is needed; knowing how an animal uses its muscles requires considerable analysis. First, behavioral studies are important to see what the animal actually does. Then, dissection may show what muscles might be involved in the activity being studied. Finally, more physiologic techniques, such as electromyography, may indicate which muscles are actually contracting when a given action is being performed. Naturally, all these stages may overlap and there will almost certainly be complexities not even hinted at here. If the motion is fast, the exact timing of all steps will be crucial and may be difficult to determine. If there are many moving parts, determination of exactly which are moving may require X-ray cinematography. Many techniques are now being developed and used, and functional analysis of muscular systems has become an important aspect of anatomy.

Axial Muscles

Trunk Musculature In Fishes. The major part of the somatic muscle division in fishes is the axial musculature, which forms much of the bulk of the body (Figs. 199, 200 *A, B*). Arranged for the most part in segmental masses along either flank, it forms the major locomotor organ of a fish. By rhythmic, alternate contractions of the muscles of the two sides, the fish's body is thrown into undulations which, traveling posteriorly and gaining a cumulative effect at the tail, thrust backward against the resistance of the water and propel the fish forward (cf. Fig. 135).

Dorsally, the axial muscles form a thick mass extending outward and upward on either side from the vertebrae to the skin. In the tail, there is a corresponding ventral development of the musculature; in the trunk, of course, the body cavity occupies much of the ventral region of the body, and the ventral trunk musculature takes the form of an enveloping sheath.

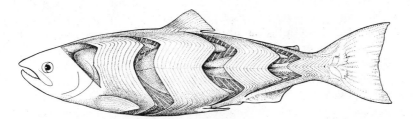

Figure 199. Dissection of a salmon to show axial musculature. In four places, a series of myomeres has been removed to show the complicated internal folding of these segmental structures. Within the body, each V projects farther anteriorly or posteriorly than it does at the surface. The horizontal septum is visible, cutting the main, anterior-pointing V. (After Greene and Greene.)

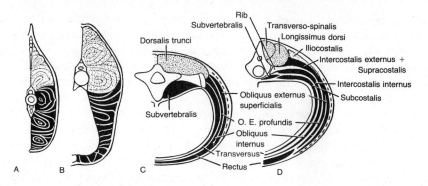

Figure 200. Diagrammatic sections to show the divisions of the trunk musculature in *A,* a shark tail; *B,* the shark trunk; *C,* a urodele; *D,* a lizard. The epaxial muscles are stippled, hypaxial muscles are in black. In *D,* a rib is assumed to be present dorsally, and the adjacent parts of the hypaxial muscles are labeled as in the rib-bearing region; more ventrally, the names are those of the corresponding abdominal muscles. (Mainly after Nishi.)

In fishes, this axial musculature is of direct myotomic origin. In the embryo of a shark, for example, one can observe parallel strips of musculature growing down the body on either side as extensions of the dorsally placed myotomes (Fig. 196). The segmental arrangement is for the most part preserved in the adult fish, so that the major part of the trunk musculature consists of successive segments, the **myomeres,** running along each flank, corresponding in number to the vertebrae. In tetrapods, the vertebrae and myomeres alternate, so that each myomere lies opposite the back half of one vertebra and the front half of the next; fish are basically similar, but the extensive folding of the myomeres makes their relation to the vertebrae less obvious. The muscle fibers are oriented in an anteroposterior position in each myomere. Few reach skeletal parts directly; between successive myomeres are stout sheets of connective tissue, the **myocommata,** into which the muscle tissue is bound. The myocommata reach inward to tie into the vertebral column, and it is in these segmentally arranged septa that ribs and the extra intermuscular bones of many teleosts develop to give further support (cf. Fig. 129).

The myomeres develop initially as simple vertical bands but soon become folded in a zigzag manner that appears to promote muscular efficiency. In amphioxus, the folds form relatively simple V's, with the point turned forward along the flank. In cyclostomes, the myocommata are essentially vertical in position. In jawed fishes, there is a complicated folding, generally greater dorsally than toward the belly (Fig. 199). Seen on the surface, each myomere is generally in the form of a W with the upper edge turned forward. There is thus a forwardly projecting V halfway down the flank and backwardly projecting V's above and below; there may also be one or two small extra jogs above. Deep to the surface, the angles of the zigzag become increasingly sharp, so that each V of a myomere has the shape of a plow. Each myomere folds forward or backward beneath its neighbors, so that any vertical cross section through trunk or tail cuts a considerable number of myomeres; their tissues are often seen in section to be grouped in a number of complex folded masses, each of which corresponds to one of the V's seen on the surface (Fig. 200 *A, B*).

In fishes above the cyclostome level, there develops on either flank, usually just below the tip of the anterior-pointing V, a **horizontal septum** of connective tissue dividing every myomere into dorsal and ventral portions. It is in this septum, at the point where it intersects successive myocommata, that dorsal ribs are developed (cf. Figs. 2 *D*, 129). As a result of the formation of the septum, the axial muscles of gnathostomes may be divided into two major groups: the dorsal or **epaxial musculature**, lying above the septum and external to the dorsal ribs (when developed) and the **hypaxial musculature** of the flanks and belly, lying below the septum and in the main internal to the dorsal ribs (Fig. 200).

Although most of the trunk musculature of fish is involved in the major structural organization of myomeres, there may nevertheless be minor units separated from the embryonic myotomes as buds or cell masses that form discrete muscles. Most important of these are the muscles of the median fins—small, symmetric muscle slips placed on either side of the radials supporting them (m on Fig. 113). Muscles of the paired fins have a similar phylogenetic origin, we believe, but because of their importance in tetrapods, the muscles of the paired appendages will receive consideration in a special section.

The general description of fish axial musculature just given applies primarily to that found in the main part of the trunk. Anteriorly and posteriorly, specialized regions may be found. We shall first follow the history of the two major groups of trunk muscles—epaxial and hypaxial—from fishes upward through the tetrapods and then return to pick up the story of more specialized anterior and posterior regions.

Epaxial Trunk Muscles. The dorsal musculature has led a relatively uneventful phylogenetic career. In fishes (Figs. 199, 200 *A, B*), generally it is a massive column of segmented muscle, lying lateral to the vertebrae and the median septum, which runs along the back of the body, from the skull or braincase to the end of the fleshy tail, without, in most instances, any great degree of regional differentiation. We may simply consider the whole mass as a single dorsolateral trunk muscle, the **dorsalis trunci**.

In tetrapods (Figs. 200 *C, D,* 201, 202), this dorsal musculature—indeed, the trunk musculature as a whole—is much reduced in relative volume. This reduction is, of course, associated with the fact that in terrestrial vertebrates, the limbs tend to take over the propulsive duties of the axial muscles. The axial muscles still function, however, in such lateral movements as continue, and, in urodeles, in which the limbs are weak and much of the work of progression is still accomplished by body undulation, they are still highly developed. In addition, the epaxial muscles are responsible for dorsoventral bending of the column, a type of movement practically unknown in fishes.

In most tetrapods, there is little height to the back above the level of the neural arches, and the dorsal muscles are restricted to a channel, of limited dimensions, running the length of the body between the neural spines medially and the transverse processes laterally and ventrally.* In amphibians, the dorsal musculature may still be considered a single muscle, the dorsalis trunci, which is persistently segmental and shows little in the way of distinct subdivision.

In reptiles, the dorsal trunk muscles remain in part segmentally divided but tend to have a complex build. Laterally, a thin sheet of dorsal musculature, the **iliocostalis**, extends downward onto the flank, external to the ventral muscles, and attaches laterally to the ribs. A **longissimus dorsi**, the main member of the dorsal series, lies

*However, neural spines are often taller than you realize (cf. Fig. 203).

above the transverse processes of the vertebrae. Most medially, between the longissimus and the vertebral spines, we find a complicated crisscross of little muscles that tie together successive vertebral spines and transverse processes. Although they pass under a variety of names, they may be termed collectively the **transversospinalis** system. In snakes, in which the axial muscles have resumed their propulsive function, the dorsal muscles are highly developed and extremely complicated. In turtles, with the development of a solid shell, the dorsal trunk muscles (and the ventral ones, as well) are, of course, much reduced, and in birds, in which the trunk vertebrae are in great measure fused, they are also weakly developed.

In mammals, the three reptilian divisions of the epaxial musculature are still present, although little of the original metameric arrangement is preserved. These muscles form a complex series of small slips to which a plethora of names has been given. Along the lumbar region, the two more lateral divisions may unite into a strong **sacrospinalis**, which helps preserve the arch of the backbone.

Hypaxial Trunk Muscles. In fishes, the hypaxial musculature lies below the horizontal septum and extends downward around the outer body wall to the ventral midline. It is essentially a unit in most cases, but there is often a degree of subdivision into lateral and ventral bundles.

Although phylogenetically the hypaxial muscles are derived from the myotomes, they tend in at least some tetrapods to arise in the embryo not from these muscle blocks but from mesenchyme in the body wall lateral to them. These muscles, like their epaxial colleagues, are considerably reduced in volume and for the most part form but thin muscle sheets around the belly and flanks (Figs. 200 *C, D,* 201, 202). They have, however, taken on a new function in land life in the support of the abdominal cavity and the enclosed trunk viscera, a function in which they are aided by the development in all tetrapods (except the modern amphibians) of a powerful system of ribs. The presence of ribs adds to the complexity of the hypaxial musculature of tetrapods through a tendency for the development of individual slips for each rib from various sheets of muscle. Somewhat arbitrarily we may divide the hypaxial trunk muscles of tetrapods into three groups arranged dorsoventrally as follows:

1. Subvertebral muscles dorsally and medially
2. A lateral series of superimposed muscle sheets along the flanks
3. A rectus group ventrally

Subvertebral musculature is little developed in fishes. In tetrapods, a deep longitudinal band of hypaxial muscle tissue forms at the sides of the vertebrae below the costal articulations, acting as an opponent of the dorsal musculature in dorsoventral movements of the spinal column. The subvertebral musculature is often continuous with the deepest layer of the flank muscles.

Complicated and varied are the **flank muscles**, which follow the curve of the trunk over the general area extending from the transverse processes of the vertebrae to the ventral territory held by the rectus system. Three superimposed major sheets of muscle may be generally found; each of these, however, may be subdivided in various regions and areas. Although primitively segmentally arranged, these are continuous unbroken sheets in most tetrapods.

In urodeles, the reduction of ribs gives us a relatively simple picture of these muscles (that the reduction of ribs and the resulting simplicity is a secondary condition need not cause us undue concern). The most external sheet is the **external**

oblique muscle, the fibers of which run in general anteroposteriorly but tend to slant upward anteriorly in each segment; a thin, external subcutaneous sheet is generally present in addition to the main muscle. A middle layer is that of the **internal oblique**; its fibers run from anteroventral to posterodorsal. The third and deepest layer is that of the **transverse** muscle, whose fibers are oriented, in contrast to the obliques, in a transverse, dorsoventral direction. None of the three layers is of any great thickness, but the diverse orientation of the three muscle sheets makes for a body sheath of considerable strength, as in a piece of plywood. In anurans, there is reduction, with loss of the internal oblique.

In amniotes, the presence of ribs on some or all of the trunk vertebrae creates a more complicated situation except in the lumbar region, where, in the absence or reduction of ribs, abdominal sections of the muscle sheets may be present in a relatively primitive manner. In the abdomen, the external oblique is split into two layers, superficial and deep (in addition to a subcutaneous sheet in reptiles), and the transversus may be split into two layers as well. In the major rib-bearing region of the trunk, the outer layer of the external oblique lies in general superficial to the ribs and produces **supracostal** muscles, and the transversus is for the most part transformed into a **subcostal** series. The intermediate muscle layers, however, form **intercostal** muscles—external intercostals arising from the deep part of the external oblique, internal intercostals from the internal oblique. In most amniotes, the transversus tends to persist internally as a continuous if thin sheet; the outer muscle layers, however, tend to become discontinuous.

Ventrally, the **rectus abdominis** runs anteroposteriorly from the shoulder region to the pelvis; the portions from each side of the trunk are frequently separated by an intervening tendinous area, the **linea alba**. There is a partial differentiation of a rectus abdominis in some fishes. In tetrapods, a superficial portion may be more or less continuous with the external oblique; for the most part, however, the rectus appears to be more closely associated with the internal oblique group. With the growth of the sternum, the rectus may be restricted in length, and in some mammals, such as ourselves, it is a mainly abdominal muscle. Thin transverse bands of connective tissue, termed **inscriptions**, are sometimes present in the rectus and may reflect to some degree the original segmental condition of the muscle.

The components of the axial musculature of the tetrapod trunk, as described above, may be classified, in somewhat simplified manner, as shown.

Epaxial . . . . {
Transversospinalis system
Longissimus system
Iliocostalis system

Hypaxial . . . . {
Dorsomedial . . . Subvertebralis

Lateral {
Obliquus abdominis externus {
(Subcutaneous)
Superficialis = Supracostals
Profundus = External intercostals

Obliquus abdominis internus = Internal intercostals
Transversus abdominis = Subcostals

Ventral . . . Rectus abdominis

Trunk Muscles of the Shoulder Region (Figs. 201, 202).   In many urodeles, the ventral trunk musculature extends forward past the pectoral region almost without break to continue on beneath the throat. This continuity, however, is probably not primitive, but due, rather, to a secondary loss of the dermal shoulder girdle. Even in sharks, there is a partial interruption of the hypaxial muscles here, and in bony fishes and tetrapods other than urodeles, there is a complete break at the shoulder; the rectus system and the lower margins of the flank muscles typically terminate anteriorly in muscle slips attaching to sternum and coracoid, and more anterior ventral trunk muscles reappear in front of the girdle as the hypobranchial muscles, described below.

Whereas the pelvic limb bears body weights by a solid sacral connection with the backbone, there is no such connection at the shoulder. In almost all tetrapods,* in contrast with fishes, the pectoral girdle and limb have no skeletal connection with the backbone other than that gained indirectly via the sternum. The body is, instead, suspended between the two scapular blades in slings formed by muscles, which, on either side, run from the top of the scapular blade downward to insert mainly on the ribs (Fig. 203). The utilization of muscles for this supporting system furnishes an elasticity that makes up in the shoulder region for the lack of the bony strength present in the pelvis; body and limb movements can be pliably adjusted to one an-

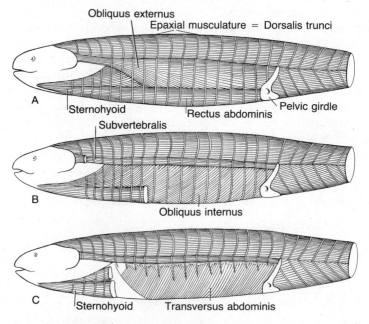

Figure 201.   Lateral views of the axial musculature of a urodele. *A,* Surface view (a thin superficial sheet of the external oblique, however, has been removed). *B,* The external oblique and rectus have been cut to show the internal oblique and subvertebral muscles. *C,* The internal oblique has been removed to show the transversus. (Modified after Maurer.)

*Some pterosaurs are, for our purposes, a very minor exception, and, in turtles, there is an indirect connection via the shell.

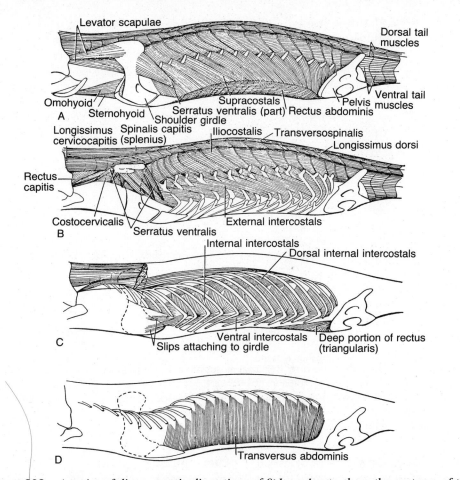

Levator scapulae

Dorsal tail muscles

Omohyoid
Sternohyoid
Shoulder girdle
Serratus ventralis (part)
Supracostals
Rectus abdominis
Pelvis
Ventral tail muscles

A

Longissimus
cervicocapitis
Spinalis capitis
(splenius)
Iliocostalis
Transversospinalis
Longissimus dorsi

Rectus
capitis

Costocervicalis
Serratus ventralis
External intercostals

B

Internal intercostals
Dorsal internal intercostals

Ventral intercostals
Slips attaching to girdle
Deep portion of rectus
(triangularis)

C

Transversus abdominis

D

Figure 202. A series of diagrammatic dissections of *Sphenodon* to show the anatomy of the axial muscles. In *A,* a thin superficial sheet of the external oblique has been removed. In *B,* the supracostals, rectus, throat muscles, and more superficial muscles to the scapula have been removed. In *C,* the epaxial muscles are cut posteriorly, and the internal intercostals and triangularis (not shown in the last figure) are indicated. In *D,* the ribs are cut, and all other muscles are removed to show the transversus. (After Maurer and Fürbringer.)

other, and (analogous to the use of springs in automobile construction) the body is eased of much of the jolts and jars of locomotion.

The muscles that form those slings are derivatives of the external oblique muscle that have been pressed into service for this special purpose. The major elements of this system spread out fanwise fore and aft, mainly from the under surface of the scapula, to attach posteriorly to the thoracic ribs, anteriorly to cervical ribs or transverse processes (Fig. 202 *A, B*). Because of their jagged appearance, elements of this group are known as the **serratus** muscles (serratus ventralis or anterior); the most anterior part is termed the **levator scapulae.** In addition, we find in mammals a more dorsally placed sheet, the **rhomboideus,** which tends to keep the upper end of the scapular blade in place by a pull toward the midline. Ventrally, an appendicular muscle, the **pectoralis,** holds the base of the forelimb in place.

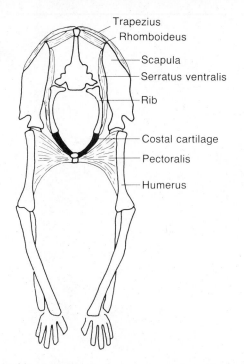

Trapezius
Rhomboideus
Scapula
Serratus ventralis
Rib
Costal cartilage
Pectoralis
Humerus

Figure 203. Diagram showing how the body is supported on the forelimbs of a mammal by muscles. The main support is by the serratus ventralis, the two of which form a sling suspending the trunk between the scapulae. The trapezius and rhomboideus dorsally and the pectoralis ventrally hold the scapula and leg in the proper position and prevent the system from collapsing. (After Walker.)

Dorsal Muscles of the Neck (Figs. 201, 202*B*). In fishes, the gills interrupt anteriorly the course of the axial muscles that continue forward, however, above and below the gills, as **epibranchial** and **hypobranchial** groups. In tetrapods, the same situation persists (the effect is somewhat masked by the presence, superficially, of visceral striated muscles of the trapezius group; cf. p. 309, Fig. 214). Dorsally, the axial muscles run forward as neck muscles, which serve for the support and movement of the head. The epaxial series extends forward here in an unbroken manner, although with a special arrangement of small muscle slips connecting the skull to the cervical vertebrae. A minor dorsal contribution to the neck musculature is made by the hypaxial muscles; small muscles of the subvertebral series continue forward beneath the column to the skull base.

Throat Muscles (Figs. 201, 202*A*). The anterior ventral part of the fish axial muscles, the **hypobranchial musculature**, runs forward along the "throat" beneath the gills. The main elements of this musculature are known collectively as the **coracoarcuales**, because they originate posteriorly from the coracoid region of the shoulder girdle and in their forward course to the jaw area send off slips that attach to the lower segments of the adjacent gill arches. Note that these muscles are *not* part of the special group of striated visceral muscles situated in the branchial region. In tetrapods, this musculature persists in its original location, in reduced form, as a series

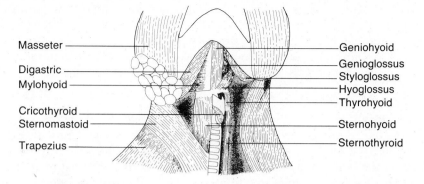

Masseter

Digastric

Mylohyoid

Cricothyroid

Sternomastoid

Trapezius

Geniohyoid

Genioglossus

Styloglossus

Hyoglossus

Thyrohyoid

Sternohyoid

Sternothyroid

Figure 204. Throat and neck muscles of the cat in ventral view. The more superficial muscles and other structures are removed on the right side of the figure to show deeper layers. The muscles labeled on the right side of the figure are all hypobranchial; those labeled on the left are some of the branchiomeric ones.

of ventral slips running from sternum and shoulder girdle to the thyroid cartilage of the larynx and to the hyoid, and from these elements to the symphysis of the jaw (**sternohyoid, omohyoid** muscles, and so forth; Fig. 204). One remarkable development of the hypobranchial musculature in tetrapods, the tongue, develops in the floor of the mouth from the region of the base of the hyoid arch. As it develops, it carries with it a mass of the hypobranchial muscle fibers present in this region; these constitute the flesh of the tongue (cf. p. 328).

The embryology and nerve supply of the hypobranchial muscles are not without interest. As axial muscles, they are of myotomic derivation; however, the development of the gills separates the throat region from direct connection dorsally with the myotomes of the occipital and cervical regions from which we would expect them to come. In some embryos, slips from these myotomes are seen in the process of migrating circuitously backward above the gills, down behind the gill chamber, and then forward in the throat to form the hypobranchial (and tongue) muscles (Fig. 196). We have noted earlier that there tends to be a constant nerve supply to a given mass of muscle, even if it has migrated far from its original position. Thus, the hypobranchial muscles in fishes are innervated by nerves, from the occipital region of the skull and the anterior part of the cervical region, which follow the same path of migration as did the muscle tissue, around the back of the branchial chamber and forward along the throat. In amniotes, comparable nerves form the hypoglossal nerve and cervical plexus; in the embryo, these nerves follow the ancestral route back and down behind the embryonic gill pouches, and, even in the adult, they pursue a roundabout course to the throat and tongue (cf. p. 562).

Diaphragm Muscles. Mammals are notable for the development of the **diaphragm**, a partition separating thoracic and abdominal cavities and of functional importance in the expansion (and contraction) of the lungs (cf. pp. 321–323). Most of the diaphragm is formed by a series of striated muscles that converge from all sides toward its center—from the sternum ventrally, the ribs laterally, and the lumbar vertebrae dorsally. These muscles appear to be derivatives of the rectus abdominis system of axial muscles. They are innervated by a special (phrenic) nerve that arises in the cervical region. This strongly suggests that the muscles of the diaphragm originated from an anterior part of the rectus musculature, perhaps in the shoulder region, and migrated posteriorly in an early stage of pre-mammalian history in conformity with a backward expansion of the lungs. Various submammalian forms have muscles or septa sometimes called diaphragms, but these are not very similar or, apparently, homologous.

Caudal Muscles. In fishes, the axial musculature continues into the tail without interruption, except for the cloacal, or anal, opening. The epaxial musculature of the tail is simply a continuation of that of the trunk. Ventrally, however, in the absence, in the tail, of the body cavity and its contained viscera, the hypaxial muscles change from a series of sheetlike structures to a compact pair of ventral bundles similar to the epaxial muscles above (Fig. 200 *A*).

In tetrapods, the great development of the pelvic girdles and the limb muscles arising from them has tended to break the continuity of the axial muscles between trunk and tail. The epaxial muscles are relatively little affected; in many amphibians and reptiles, they extend backward over the sacral ribs onto the tail with little or no interruption. The hypaxial muscles, however, are completely, or almost completely, interrupted at the pelvis. In the more distal regions of the tail, they form a relatively uniform cylindric muscle bundle occupying a position on either side below the transverse processes of the caudal vertebrae similar to that of the dorsal muscle above. Anteriorly, approaching the pelvic region, the hypaxial caudal muscle mass becomes constricted and sends terminal fibers downward and forward to insert onto the ischium, or upward and forward to the ilium. In urodeles and in reptiles, the basal region of the tail is expanded to contain, ventrally, beside the caudal muscles proper, large muscles that run forward and outward to the femur; these, however, are parts of the limb musculature, not axial muscles. Needless to say, there is a reduction or loss of caudal musculature in forms in which the tail is reduced. Remnants of this musculature may still be found attached to the bony stump of the tail of birds, frogs, and the like. In mammals, generally the tail is still moderately elongate but slender, and the caudal muscles are correspondingly reduced in volume. From the ventral muscles just behind the pelvic girdle, there usually develops in tetrapods a sphincter muscle that can close the cloaca; in mammals, there is a corresponding anal sphincter.

Eye Muscles. The muscles that move the eyeball form a far-flung anterior outpost of the axial musculature. In embryonic cyclostomes, there is present, as in amphioxus, a series of mesodermal somites continuous from the front part of the head back to the trunk. In most vertebrates, the series is interrupted, because, with the expansion of the braincase around the ear, the somites there tend to be vestigial or absent. More anteriorly, however, in the region of the developing eye, small somites persist in every vertebrate class (Fig. 196). When well developed, three pairs are to be found. They are in great measure separated at an early stage from the lateral mesoderm by the development of the gill slits and play little if any part in the formation of skeletal or connective tissues; they appear in the embryo as small **head cavities** from the walls of which arise the muscles of the eyeball. Associated with these three somites and innervating the muscles that they form are three small cranial nerves: III, IV, and VI of the numbered series (p. 561).

In the vast majority of vertebrates, six typical straplike muscles develop from

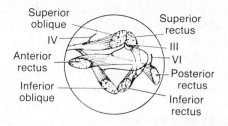

Figure 205. Eye muscles. A lateral view, with the eyeball (in outline) removed; the ovals are the insertions on the eyeball. The three nerves to the eye muscles are shown (III, IV, VI). Anterior is to the left. (After Goodrich.)

these somites (Fig. 205). They take origin from the surface of the braincase and fan outward to attach to the eyeball somewhat internal to its "equator"; in varied combinations, their pull will rotate the eye in any desired direction. Four of them, the **rectus muscles**, arise posteriorly, often close to the eye stalk or optic nerve; in actinopterygians, they usually arise from within a deep pit in the skull, the **myodome**. The other two, the **oblique muscles**, spring generally from the anterior part of the orbit. Typically, the superior rectus and the superior oblique attach to the upper margin of the eyeball, a pair of inferior muscles to its lower edge; the two remaining rectus muscles insert on the anterior (or internal) and posterior (or external) margins. In mammals, the superior oblique has shifted its origin dorsally and passes to its insertion through a pulley—trochlea—to which the little nerve innervating it owes its name.

Four of the six muscles are innervated by nerve III, the superior oblique by nerve IV (trochlear nerve), the posterior rectus by nerve VI. As this would lead us to suspect, we find that, in the embryo, four of these muscles arise from the first of the three somites concerned, and one each from the other two.

Variations are, of course, found on this basal pattern of six eye muscles. In hagfishes and other vertebrates with degenerate eyes, the muscles are correspondingly degenerate or absent, and the homology of certain lamprey muscles is uncertain. Accessory muscles are frequently present, particularly in tetrapods. In a majority of tetrapods (birds and primates are exceptions), a **retractor bulbi** muscle tends to pull the eyeball deeper in its socket; in most amniotes, a **levator palpebrae superioris** derived from the eyeball muscles raises the upper lid, and rather variable slips move the nictitating membrane of the eye.

Limb Muscles

The musculature of the paired appendages is derived, historically, from the general myotomic musculature of the trunk and hence is part of the somatic system. The limb muscles, however, are so distinct in position and nature and so important in higher vertebrates that they deserve special treatment.

We have previously noted the decline of axial musculature in tetrapods; limb muscles, modest in size and simple in composition in fishes, grow, on the other hand, to relatively enormous bulk and to great complexity in terrestrial vertebrates, as they take over the major duties of locomotion.

As derivatives of the somatic system, the limb muscles should, in theory at least, originate in the embryo from the myotomes. In some lower vertebrates—specifically, sharks—this origin appears to be demonstrable. Paired finger-like processes extend out from the ventral ends of a considerable number of myotomes developing near the base of the paired fins; these form masses of premuscular tissue above and below the skeleton of the fin (Fig. 196). In tetrapods, however, the myotomic origin of appendicular muscles has not been demonstrated, and these muscle masses appear to develop as condensations in the mesenchyme of the region of the budding limb. It is, of course, possible that this mesenchyme is derived from the myotomes, but proof is lacking.

292
Chapter 9
Muscular
System

Paired Fins. The fin musculature of fishes is simple: two opposed little masses of muscle are usually readily discernible, running outward from the girdle to the base of the fin (Fig. 206 *B*). A dorsal muscle mass serves primarily to elevate or extend the fin, a ventral muscle mass to depress or adduct it. There may be, in addition, small slips developed from either group that serve to give rotary or other special movements to the fin.

Tetrapod Limbs. In tetrapods, we meet with a different situation. Not only is the limb musculature much more bulky, but it is also much more complex. The mode of development affords, however, a clue to a natural classification of the muscles present.

Early in the ontogeny of a land limb, while the arm or leg is still a relatively short bud from the body, a mass of premuscular tissue (Fig. 206*A*) is formed on both the upper and lower surfaces of the developing skeleton. Clearly, these two masses are equivalent to the opposed dorsal and ventral muscle masses that in fishes raise and lower the paired fins. From them arise, by growth and cleavage, all the complicated muscles of the mature limb. From the dorsal mass develop the muscles on the exten-

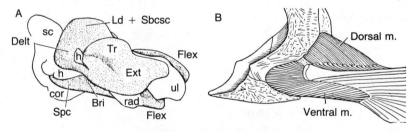

Figure 206. *A*, External view of the left pectoral girdle and limb and its musculature in a lizard embryo (the skeleton unshaded, the muscle tissue stippled); *B*, the pectoral girdle and fin in a fish (sturgeon). In the fish fin, the musculature consists simply of opposed dorsal and ventral muscle masses. In the adult tetrapod, the limb has a large number of discrete muscles, but, in the embryo, these are arranged in two opposed masses comparable to those of the fish fin. At the stage figured, the two masses are barely beginning their differentiation into the muscles of the adult (cf. Fig. 207 *A*). The dorsal mass is well shown; the ventral mass is mostly concealed beneath the limb (in which the foot is not yet developed). Abbreviations: *Bri*, brachialis inferior; *cor*, coracoid region; *Delt*, deltoid; *Ext*, extensor muscles; *Flex*, flexor muscles; *h*, humerus; *Ld*, latissimus dorsi; *rad*, radius; *Sbcsc*, subcoracoscapularis; *sc*, scapula; *Spc*, supracoracoideus; *Tr*, triceps; *ul*, ulna.

sor surface of the limb and related muscles of the girdle; from the ventral mass arise the flexor muscles of the opposite limb surface.

In older texts, a distinction was frequently made between "extrinsic" and "intrinsic" limb muscles, the former name being given to muscles that arise from the body and run either to girdle or limb, the latter to muscles that do not leave the confines of the limb. This distinction is misleading and should be avoided. Embryologic study shows that, on the one hand, muscles that arise from the trunk and run only to the limb girdle are not limb muscles at all but are modified axial muscles, which arise directly from somites. On the other hand, various true limb muscles may take origin from the trunk and run outward onto the limb. Their real nature is shown by the fact that they arise in the embryo as part of the limb muscle mass and only during the course of development spread inward onto the trunk. Where a muscle inserts, not where it arises, indicates whether or not it is a true limb muscle.

How to treat such a complex series of structures as limb muscles in a brief work of this sort is a problem, because there is an immense amount of variation from group to group in the arrangement of the musculature (and also, sad to say, in the nomenclature). On the one hand, one might describe, seriatim, the limb muscles of the various groups, in a manner that would be not merely exhaustive, but exhausting. On the other hand, one might (as is sometimes done) make a few hasty generalizations about extensors, flexors, and so on, and pass hurriedly on. We shall take a middle course (which probably includes the disadvantages of both extremes). The various muscle groups of the limbs will be considered and compared in (1) lizards, representing a rather primitive and generalized tetrapod condition,* and (2) a primitive mammal, the opossum (*Didelphis*), representing a mammalian type of specialization. A few remarks will follow on the musculature of other forms, and a table of homologies be presented.

The musculature of each limb can be naturally divided into dorsal and ventral groups, proximally none too readily distinguishable in the adult, but distally forming clear extensor and flexor series. We shall follow this logical order in considering successively the muscles of pectoral and then of pelvic limbs. Little will be mentioned of the functions of individual muscles. It must be emphasized that rarely is a movement performed by the contraction of a single muscular element. In general, several muscles cooperate; the movement is a resultant of forces. The function of a given muscle varies according to the company it keeps. In only a few cases has this been well studied.

Pectoral Limb. *Dorsal Muscles* (Figs. 207, 208). A number of dorsal muscles attach to the humerus near its head and are responsible for much of the movement of that bone on the shoulder girdle. Superficially, both reptile and mammal have prominent fan-shaped dorsal muscles of this sort, the **latissimus dorsi** and the **deltoideus**, the former arising from the surface of the back and flank, the latter from the scapula and the clavicle, often in two parts. In mammals, the scapular part of the deltoid, which in reptiles took origin from the front edge of the scapula, now arises from the centrally situated spine of that bone. A slip of the mammalian latissimus has gained contact with the scapula and separated as the **teres major**. A deep muscle associated

---

*Other reptiles (except *Sphenodon*) and the frog are more specialized, and the urodeles show secondary reduction of their musculature.

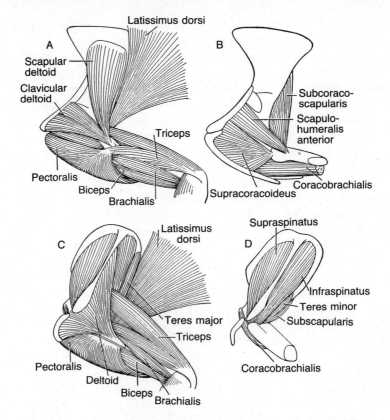

Figure 207. Shoulder and upper arm muscles in the lizard *(A, B)* and opossum *(C, D)*, lateral views; the right hand figures in each case are comparable deep dissections with latissimus, deltoid, pectoralis, and long muscles (triceps, biceps, brachialis) removed. Notable is the upward migration of the supracoracoideus to become the two "spinatus" muscles.

with the latissimus at its insertion—the **subcoracoscapularis** of reptiles, **subscapularis** in mammals—arises from the inner surface of the girdle. In reptiles, a small deep muscle arises from the outer surface of the scapula as the **scapulohumeralis anterior**; in mammals, this has been pushed to the back margin of the scapular blade as the **teres minor**.

The dorsal surface of the humerus is covered by a stout muscle, the **triceps**, which arises from the humerus and by one or more heads from adjacent parts of the girdle; the muscle attaches distally to the olecranon of the ulna—this attachment is in fact the reason for existence of that process—and serves to extend the forearm. The dorsal arm musculature is continued beyond the elbow by muscles of the extensor series. Lizard and opossum show basically similar arrangements (c.f. Fig. 208 *A, D*).

Ventral Muscles (Figs. 207, 208). On the underside of the shoulder, a superficial and important muscle, the **pectoralis**, spreads far back over the sternum and ribs and inserts on a powerful process beneath the proximal end of the humerus. The muscle gives a strong pull on the humerus downward and backward, and thus is of great

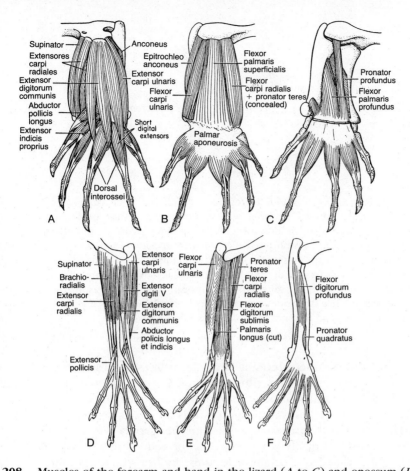

Figure 208. Muscles of the forearm and hand in the lizard (*A* to *C*) and opossum (*D* to *F*), somewhat diagrammatic and simplified. *A* and *D* are views of the extensor surface; *B* and *E* superficial, *C* and *F* deeper dissections of the flexor aspect. On the extensor surface, the most prominent change from reptile to mammal is the reduction of short muscles on the manus and the development from the common extensor of tendons to the toes. Long special muscles have developed for movement of the "thumb" and fifth digit. On the flexor aspect, a prominent feature in reptiles is the presence of a stout and complex aponeurosis on the "palm" with which connect the long flexors proximally and a variety of tendons and short muscles to the toes. In mammals, this is broken up; the palmaris longus inserts into a superficial aponeurosis over the wrist, which is cut away in the figure, and the two deeper flexors present here have each developed a broad palmar tendon. Deeply placed in the hand are various short muscles of the digits, which are not shown.

importance in locomotion. The pectoralis also gives rise to a major dermal muscle in mammals, the **panniculus carnosus** (cf. p. 313). From the back end of the coracoid, the **coracobrachialis** runs along the underside of the humerus. Two muscles flex the radius and ulna: the **brachialis**, running down the outer and anterior margin of the humerus, and the **biceps**, originating from the coracoid.

These four muscles are present in a similar fashion in lower tetrapods and mammals. But a fifth reptilian muscle appears, at first sight, to have no homologue in a

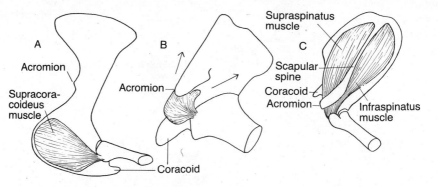

Figure 209. Diagrams of the shoulder region of *A,* a lizard, *B,* an embryonic opossum, and *C,* an adult opossum, to show a major shift in shoulder musculature between reptiles and mammals and the consequent modification of the shoulder girdle. In the lizard, the supracoracoideus is a large ventral muscle running from coracoid plate to humerus. In the embryo opossum, a comparable muscle is found, but this is, at the stage figured, tending to split and grow upward (*arrows*) on either side of the acromion. In the adult mammal, this muscle mass has become the dorsally situated supraspinatus and infraspinatus muscles; a new portion of the scapula has developed for the reception of the supraspinatus muscle, while the coracoid has been reduced to a nubbin. (*B* after Cheng.)

mammal. The **supracoracoideus**, a large fleshy muscle, arises from the coracoid plate and runs to the underside of the humerus. In primitive tetrapods, with the legs sprawled at the sides, this muscle is very important in keeping the body from sagging downward between the limbs. In mammals this muscle is present in the form of the **infraspinatus** and **supraspinatus** muscles on the scapula (Fig. 209); this major muscular migration is presumably responsible for the reduction of the coracoid region of the girdle, which afforded it origin in reptiles, and for the development of the spine and supraspinous fossa of the mammalian scapula. The reptilian muscle (as can be seen in the mammal embryo) has pushed its way upward beneath the deltoid and has *(a)* taken over the old scapular blade as the infraspinatus, and *(b)* occupied a newly formed shelf, the supraspinous fossa. The scapular deltoid muscle is then restricted to the old front margin, now the spine of the scapula. With the changed limb posture of mammals, the supracoracoid muscle has retained its supporting function as the two homologous mammalian muscles insert at the very tip of the humerus anterior to the glenoid; the resulting lever action tends to swing the limb downward and forward or, conversely, to pull the body up and back on the arm.

In the distal part of the limb, the main propulsive effort is a backward push of the forearm and digits, accomplished by the muscles of the ventral, flexor surface, which are, in consequence, more powerful than the extensors. Long flexors arise from the entepicondyle of the humerus and fan out to the lower end of the forearm and to the hand. Flexion of the digits by a muscle from the elbow is rendered difficult by the fact that it must pass around the curve on the underside of the wrist. This difficulty is avoided in reptiles by the development of an aponeurosis beneath the carpus to which long flexors attach proximally and certain of the short toe muscles distally. In mammals, this pad of tissue is subdivided into several superimposed tendinous sheets.

Pelvic Limb. Dorsal Muscles (Figs. 210, 211). As in the pectoral appendage, the muscles of the distal part of the limb are readily separable into dorsal and ventral (flexor and extensor) series, and the proximal muscles are equally divisible on an embryologic basis.

Certain of the muscles or muscle groups in the hip and thigh are readily comparable in lower tetrapods and mammals. Reptiles have a powerful fleshy **puboischiofemoralis internus** (what names these muscles have!), which arises from the lumbar region and the inner surface of the girdle and runs back to insert onto the femur near its head. In mammals, this is split into the **psoas** from the lumbar region, the **iliacus** from the ilium, and part of the **pectineus** from the pubis. Frequently grouped as the **quadriceps femoris** is a series of extensor muscles (**rectus femoris** from the ilium and the **vasti** from the femur) running to the head of the tibia; the insertion is by a stout tendon in which the patella of mammals is placed. The **sartorius**, the "tailor's muscle," is sometimes grouped with the quadriceps. In reptiles, **iliotibialis** and **femorotibialis** are clearly homologous with the first two, and the **ambiens** of reptiles probably with the sartorius.

Two other muscles arising from the reptilian ilium are not, at first, readily compared with mammalian muscles. We have noted that in mammals there has been a major shift in posture of the thigh, and, in consequence, shifts in its musculature are to be expected. In reptiles, the **iliofemoralis** runs from the iliac blade directly laterally to insert on the upper surface of the femur. In mammals, the powerful **gluteal muscles** (including the pyriformis), running from ilium to femur, appear to correspond to this muscle. They run backward and downward (not outward) from the ilium to the femur and attach to its outer (back) margin—particularly strongly to the proximal end of the bone, where they exert a powerful leverage in pulling the knee back, or, conversely, pushing the body upward and forward on the leg. In reptiles, there is, further, a long **iliofibularis** muscle running to the fibula from the ilium; its homologue in mammals is the deeper and often lost **tenuissimus**.

Beyond the knee, the extensor muscles of the pelvic limb in both lizards and mammals show patterns essentially comparable to those of the pectoral appendage (cf. Figs. 208, 211).

Ventral Muscles (Figs. 210, 211). The muscles of the hip and thigh region mainly adduct the femur and flex the knee; in locomotion, that is, they raise the body off the ground and push it forward. Hence, they are large, important, and complex. They are disposed in three main groups:

1. The muscles running from pelvis to femur are for the most part deeply buried. They include three major elements: *(a)* A large muscle arises fleshily from much of the outer surface of the pubis and ischium (the fenestration of these elements is related to this muscular attachment) and inserts on the underside of the femur near its head; it pulls the femur downward. In reptiles, this is termed the **puboischiofemoralis externus**; in mammals, the **obturator externus** is essentially homologous, but the **quadratus femoris** is a separate fraction of this same muscle group. *(b)* A smaller muscle emerging from the inner side of the ischium and running to the head of the femur is called in lower tetrapods the **ischiotrochantericus**, in mammals, the **obturator internus** (plus smaller **gemelli**). *(c)* In reptiles, an **adductor fe-**

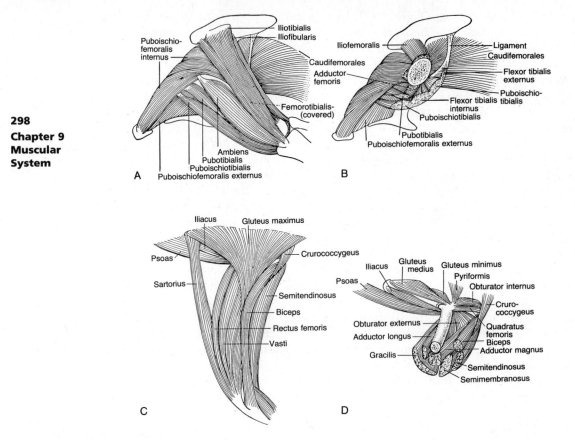

Figure 210. Limb muscles of the pelvis and thigh in a lizard (*A, B*), and opossum (*C, D*), lateral views. *A, C,* Superficial views; *B, D,* dissections to show deeper layers of musculature. Anterior is to the left.

moris, arising from the puboischium, runs far down the underside of the femur; most of the variable femoral adductors of mammals are homologous.

2. Covering the under surface of the thigh is a large and complicated group of long muscles that flex the tibia. A superficial sheet is the **gracilis** (generally called the **puboischiotibialis** in reptiles); variable deeper bellies termed the **flexor tibialis externus** and **flexor tibialis internus** in reptiles appear to be homologous in mammals with the **semimembranosus, semitendinosus**, and **biceps**. A separate deep reptile slip, the **pubotibialis**, is homologous to the **adductor longus** of mammals.

3. Powerful ventral limb muscles, the two **caudifemorales** (long and short), arise in typical reptiles from the underside of the caudal vertebrae. Running forward, these two muscles narrow to tendons that insert midway along the femur; they act powerfully in a backward pull of the femur and hence contribute greatly to forward locomotion in a reptile. In mammals, however, this group of muscles has been reduced, its insertion has been moved distally, and its name has been spelled differently (caudofemoralis).

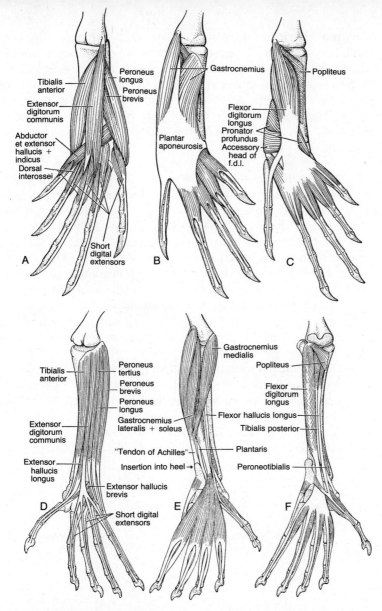

Figure 211. Muscles of the lower, leg and foot in a lizard *(A to C)* and opossum *(D to F)*, somewhat diagrammatic and simplified. *A* and *D* are views of the extensor surface; *B* and *E* superficial, *C* and *F* deeper dissections of the flexor aspect. The extensor surface of the lizard hind leg is comparable to that of the forearm and manus except for a lesser development of individual muscles on the inner (tibial = radial) side. In the change to mammals the modifications are similar to those seen in the front leg, including the development of separate tendons from the common extensor and the development of long muscles working on the first and fifth toes. The flexor aspect of the lizard hind leg resembles that of the front in many regards, including the development in reptiles of a stout plantar (sole) aponeurosis, into which much of the musculature attaches both proximally and distally. It differs, however, in the absence of the flexors running down either side and the development of a powerful two-headed "calf" muscle, the gastrocnemius. In mammals the calf musculature has (except for the plantaris) changed to a new attachment on the "heel bone." The long flexor of the digits, however, runs on to the reduced plantar aponeurosis. From this extend, as in reptiles, distal tendons and toe muscles (in *F,* the superficial muscles of this set have been removed to show the deeper tendons and muscles). In both reptile and mammal, there are, on the flexor surface, deep, short toe muscles not shown in the figures.

The long ventral muscles of the lower leg are in all tetrapods a powerful series; most of their bulk is centered in the **gastrocnemius**, whose muscle bellies constitute the "calf" of the leg. "Rounding the turn" of the tarsus, to the underside of the foot is a major structural problem. Reptiles generally have solved this problem by placing an aponeurosis beneath the ankle, with muscles and tendons attaching at its proximal and distal ends. In mammals, however, a new type of foot-raising device has been evolved, with the calf muscles inserting on the heel of the calcaneum by the "tendon of Achilles" (Fig. 212).

Limb Muscles in Other Tetrapod Groups. We shall not consider all the variations found in other nonmammalian tetrapods. Among reptiles, *Sphenodon* is similar in almost every regard to lizards. In turtles, the presence of the shell has tended to make locomotion a specialized affair. Limb muscles as well as limb skeleton have been modified accordingly, but the basic pattern is comparable to that of lizards; the most notable differences are those caused by the peculiar orientation of the shoulder girdle (cf. p. 205). In crocodiles, most of the muscles are comparable to those of lizards, but certain modifications are suggestive of the avian relations of that group. The modified pelvic structure is associated with marked changes in the arrangement of the ventral muscles of the thigh, and these changes are, in turn, due to a modified position of the hind leg, which moves fore and aft rather than assuming the sprawled position of typical reptiles.

In the wing of birds, the pectoral muscles that give the main down and back pull to the wing in flight are enormous; they constitute most of the "white meat" of the fowl. Deep to them, the smaller **supracoracoideus** (often incorrectly called the pectoralis minor) sends a tendon to the dorsal surface of the humerus and thus acts to elevate the wing; its ventral position helps keep the center of gravity well ventral, which is important for stability during flight. The dorsal wing muscles are relatively

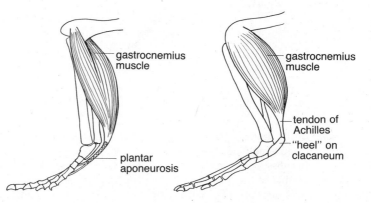

Figure 212. Side views of the hind leg of a lizard (*left*) and a typical mammal (*right*) to show the contrast in action of the main calf muscle, the gastrocnemius, in extending the foot. In lower tetrapods, it rounds the ankle region to act on the under surface of the foot by attaching to a sheet of connective tissue, the plantar aponeurosis, which in turn makes connections with the toes (cf. Fig. 211 *B*). In mammals, its action is simplified by the development of a heel on the calcaneum (cf. Fig. 211 *E*); attachment of the muscle tendon here raises the foot off the ground.

slender. Most of the short musculature of the reptilian "fingers" has disappeared, but special dorsal slips to the thick dermis of the wing membrane are developed. Even more than in the crocodilians, the pelvic muscles are specialized in relation to the forward rotation of the hind limbs and the associated modifications of the pelvic girdle.

The urodeles show a simply built type of limb musculature, but this is probably more a secondarily simplified than a truly primitive condition. The loss of the primitive dermal shoulder girdle has resulted in specialization of the shoulder muscles. The clavicular deltoid arises from a long anterior process of the coracoid as a **procoracohumeralis** muscle. The short dorsal extensors of the toes are not developed. There is no biceps in the arm in amphibians. Probably this amniote muscle has arisen by a fusion of fibers of the supracoracoideus with those of the adjacent brachialis; in some urodeles and in frogs, part of the former muscle separates as a **coracoradialis**, which sends out a long tendon to the radius and thus serves somewhat the functions of a biceps in flexing the elbow.

There are numerous differences from reptiles in the urodele hind leg. The quadriceps femoris equivalent lacks the strong femoral heads present in all amniotes. On the ventral surface of the thigh, the long tibial flexors are much simplified, and the caudifemoralis longus does not reach the femur but simply ties into the side of the gracilis. There is no development of a gastrocnemius of reptilian or mammalian type.

In the frog, the musculature is highly developed, but much modified in relation to the jumping habits of the animal and the peculiar skeletal structure of the girdle and limb. In Table 2, we have given the usual names accorded many of the major muscles; the chart is far from complete and does not attempt to list all the muscles of any specific animal. The nomenclature is patterned on that of mammals, but, as may be seen, not always accurately.

As might be expected from the wide variety of skeletal structures of the limbs, there are many variations in the different mammalian groups from the generalized muscle pattern. The loss of the clavicle in some mammals tends to bring about confusion in the deltoid and pectoral regions; the great modifications in the feet have, of course, resulted in wide differences in the build of the distal musculature of the limbs.

Branchiomeric Musculature

Markedly different from the striated musculature thus far considered is the musculature of the branchial region, highly developed in fishes and persistently prominent, although in much modified form, in even the highest groups. Like the skeleton and nerves of the pharyngeal region of primitive vertebrates, its musculature—musculature of the "visceral animal"—is distinctive and noteworthy. In contrast to other, somatic, striated musculature, it arises, not from the myotomes, but from mesenchyme derived from peritoneum of the lateral plate (cf. Figs. 80 C, 196). The smooth musculature of the gut proper, posterior to the pharynx, arises in a similar manner, and branchial and gut muscles, striated or smooth, are only anterior and posterior parts of a single great visceral system of muscles whose primary locus is in the walls of the digestive tract. This contrast between the parts of the digestive tract may be associated with the contrast in functions between anterior and posterior regions. For food gathering and breathing, the primitive functions of the mouth and pharyngeal

Table 2. Muscle Homologies in Various Tetrapod Groups*

Pectoral Limb, Dorsal Musculature

Mammal	Reptile	Bird	Frog	Urodele
Latissimus dorsi Teres major } Subscapularis	Latissimus dorsi Subcoracoscapularis } Scapulohumeralis posterior	Latissimus dorsi Coracobrachialis posterior Subcoracoscapulares } Scapulohumeralis anterior	Latissimus dorsi —	Latissimus dorsi Subcoracoscapularis
Deltoideus Teres minor Triceps	Dorsalis scapulae Deltoideus clavicularis } Scapulohumeralis anterior Triceps	Deltoideus and tensor propatagialis Scapulohumeralis anterior Triceps	Dorsalis scapulae Deltoideus } Scapulohumeralis brevis Anconeus	Deltoideus scapularis Procoracohumeralis longus Procoracohumeralis brevis Triceps
Supinator	Supinator	Extensor antibrachii radialis	Extensor antibrachii radialis	Supinator longus
Brachioradialis } Extensores carpi radiales	Extensores carpi radiales	Extensor metacarpi radialis	Extensor carpi radialis	Extensores carpi radiales
Extensor digitorum communis Extensor digiti quinti	Extensor digitorum communis	Extensor digitorum communis	Extensor digitorum communis	Extensor digitorum communis
Extensor carpi ulnaris Anconeus	Extensor carpi ulnaris Anconeus	Extensor carpi ulnaris Ectepicondyloulnaris	Extensor carpi ulnaris } Epicondylocubitalis	Extensor carpi ulnaris
Abductores pollicis Extensores digitorum 1–3	Abductor pollicis longus Extensores digitorum breves	Abductor alulae Extensores digitorum breves	Abductor indicis longus Extensores digitorum breves	Supinator manus Extensores digitorum breves
Dorsal interossei	Dorsal interossei	Dorsal interossei	Dorsal interossei	Dorsal interossei

Pectoral Limb, Ventral Musculature

Mammal	Reptile	Bird	Frog	Urodele
Pectoralis	Pectoralis	Pectoralis	Pectoralis	Pectoralis
Supraspinatus ⎱	Supracoracoideus	Coracobrachialis anterior ⎱ Supracoracoideus	Coracohumeralis	Supracoracoideus
Infraspinatus ⎰				
Biceps brachii	Biceps brachii	Biceps brachii	Coracoradialis	Coracoradialis
Coracobrachiales	Coracobrachiales	Coracobrachiales	Coracobrachiales	Coracobrachiales
Brachialis	Brachialis inferior	Brachialis		Brachialis
Pronator teres	Pronator teres	Pronator superficialis	Flexor antibrachii medialis ⎱ Flexores carpi radiales ⎰	Flexor carpi radialis
Flexor carpi radialis	Flexor carpi radialis			
Flexor digitorum sublimis ⎱	Flexor palmaris superficialis	Flexor digitorum superficialis	Palmaris longus	Flexor palmaris superficialis
Palmaris longus ⎰				
Epitrochleoanconeus ⎱	Epitrochleoanconeus		Flexor antibrachii laterales ⎱ Flexor carpi ulnaris Epitrochleocubitalis ⎰	Flexor antibrachii ulnaris
Flexor carpi ulnaris ⎰	Flexor carpi ulnaris	Flexor carpi ulnaris		Flexor carpi ulnaris
			Ulnocarpalis	Ulnocarpalis
Flexor digitorum profundus	Flexor digitorum profundus	Flexor digitorum profundus	Flexor palmaris profundus	Flexor palmaris profundus
Pronator quadratus	Pronator profundus	Pronator profundus	Pronator profundus	Pronator profundus

Superficial short digital flexors: palmaris brevis, contrahentes, lumbricales, etc.

Deep short digital flexors, digital interossei, etc.

Table continued on following page

Table 2. Muscle Homologies in Various Tetrapod Groups* *(Continued)*

Hind Leg, Dorsal Musculature

Mammal	Reptile	Bird	Frog	Urodele
Sartorius	Ambiens	Ambiens	?Tensor fasciae latae	Iliotibialis
Rectus femoris	Anterior iliotibialis	Iliotibiales	Crureus glutaeus	Ilioextensorius
Vasti	Femorotibialis	Femorotibiales		Puboischiofemoralis externus
Gluteus maximus Femorococcygeus }	Posterior iliotibialis	Posterior iliotibialis	?Iliofibularis	Iliofibularis
Iliacus Psoas Pectineus }	Puboischiofemoralis internus	Iliofemoralis internus	Iliacus Pectineus Adductor longus }	Puboischiofemoralis internus
Tensor fasciae latae Gluteus minimus Gluteus medius Pyriformis }	Iliofemoralis	Iliofemoralis externus Iliotrochanterici }	Iliofemoralis	Iliofemoralis
Tibialis anterior	Tibialis anterior	Tibialis cranialis	Extensor cruris brevis	Tibialis anterior
Extensor digitorum longus Extensor hallucis longus Peroneus tertius }	Extensor digitorum communis	Extensor digitorum communis	Tibialis anticus longus	Extensor digitorum communis
Peroneus longus	Peroneus longus	Fibularis [peroneus] longus	Peroneus	Peroneus longus Peroneus brevis }
Peroneus brevis	Peroneus brevis	Fibularis [peroneus] brevis		
Extensores digitorum breves	Extensores digitorum breves	Extensores digitorum breves	Tibialis anticus brevis	Extensores digitorum breves
Dorsal interossei	Dorsal interossei			Dorsal interossei

Hind Leg, Ventral Musculature

Mammal	Reptile	Bird	Frog	Urodele
Obturator externus; Quadratus femoris	Puboischiofemoralis externus	Obturator lateralis	Adductor magnus (pt.)	Puboischiofemoralis externus
Obturator internus; Gemelli	Ischiotrochantericus	Ischiofemoralis	Gemelli; Obturator internus	Ischiofemoralis
Adductor femoris brevis; Adductor femoris longus	Adductor femoris; Pubotibialis	Puboischiofemoralis	Obturator externus; Quadratus femoris	Adductor femoris; Pubotibialis
Adductor femoris magnus	Adductor femoris			
Caudofemoralis; Dorsal semitendinosus	Caudifemorales; Flexor tibialis externus	Caudoiliofemoralis	Caudalipuboischiotibialis	Caudofemoralis; Ischioflexorius
Ventral semitendinosus	Flexor tibialis internus II	Flexor cruris lateralis	Semimembranosus; Gracilis	Ischioflexorius
Biceps femoris; Semimembranosus	Flexor tibialis internus I	Flexor cruris medialis	Semitendinosus; Sartorius	
Gracilis	Puboischiotibialis			Puboischiotibialis
Gastrocnemius medialis; Flexor hallucis longus	Gastrocnemius internus	Gastrocnemius pars medialis; Flexor hallucis longus; Flexor digitorum longus	Plantaris longus	Flexor digitorum sublimis; Flexor digitorum longus
Flexor digitorum longus; Tibialis posterior	Flexor digitorum longus; Pronator profundus			Pronator profundus
Popliteus	Popliteus	Plantaris; Popliteus		Popliteus
Gastrocnemius lateralis; Soleus	Gastrocnemius externus	Gastrocnemius pars intermedia	Tibialis posticus	Fibulotarsalis
Plantaris				
Interosseus	Interosseus			Interosseus

Superficial short digital flexors: flexor digitorum brevis, contrahentes, lumbricales, etc.

Deep short digital flexors: digital interossei, etc.

*No attempt has been made to include all the variants present in different members of the groups tabulated, and many details are omitted. For each group, only one common name is given for each muscle; often there are synonyms. In many instances, homologies are doubtful.

region, the vigorous movement characteristic of striated muscles is appropriate; for the slower movement fitting for digestive processes, smooth musculature suffices.

In higher vertebrates, much of the striated visceral musculature has abandoned the wall of the gut and assumed a variety of natures as facial and jaw muscles and even a part of the shoulder musculature. Primitively, however, the pharyngeal walls appear to have been its proper, truly visceral locale. In many instances, the branchial muscles are associated with skeletal structures, primarily the branchial bars, whereas the more posterior smooth visceral muscles lie exclusively in soft tissues of the gut. But, in cyclostomes, the skeletal connections are of little importance. Note that the striated muscles of the gut are not confined to the pharyngeal region of the digestive tube, where these skeletal structures lie. In fishes, generally the striated muscles may extend backward into the esophagus, and the same condition is repeated in mammals (although here their presence is perhaps secondary). There is thus no sharp, definite line of demarcation between regions with branchial and smooth muscles, a fact that emphasizes their relationship.

In the gnathostome fishes and all higher types, the most anterior set of branchial muscles have been much modified in function to serve the jaws, and, in tetrapods, the surviving branchial elements serve a further series of aberrant functions, as facial, neck, and shoulder muscles, and even as tiny elements accessory to the auditory apparatus.

Branchial musculature is well developed in the cyclostomes, both as sheets of muscles constricting the gill pouches and as specialized muscles operating the peculiar "tongue." The construction of the lamprey musculature is, however, quite unlike that of other vertebrate groups, and it will not be considered further here.

In the sharks (Fig. 214 A), we find the branchial muscles well developed, with a pattern that in many regards may be considered basal to that in all other gnathostomes. In the jawless ancestors of the gnathostomes, the musculature of the mandibular and hyoid arches was presumably similar to that of the more posterior visceral arches; it is, however, already highly specialized in sharks. We shall therefore use an indirect method of attack and consider first the simple and presumably more primitive arrangement seen in the typical gills farther back in the shark pharynx. We shall follow the fate of the muscles of these arches upward through the higher vertebrates, then return to consider the muscles farther forward in the region of the hyoid and mandibular arches.

The branchial musculature is, of course, primarily a pumping device to force water through the gills, replacing the ciliary action seen in lower chordates. Although the branchial muscles are those most intimately connected with the visceral skeletal elements, somatic trunk muscles may also become associated with the gill bars. We have already mentioned that the hypobranchial muscles of the floor of the throat may attach to the arches from below, and, in sharks, the epibranchial muscles of the neck may also gain contact with the upper margins of the branchial arches.

Behind the hyoid arch there are typically, in fishes, five gill slits with four intervening arches, each arch with its own proper musculature as well as its own skeletal bars; a final arch lies behind the last branchial opening. Even when, in tetrapods, the gills themselves have disappeared as landmarks, muscles derived from various parts of the branchial system can be readily traced because of their innervation by a special series of cranial nerves, numbers V, VII, IX, and X (cf. pp. 558–561, and Fig. 405). The mandibular arch is supplied by nerve V; the hyoid by nerve VII; the first typical

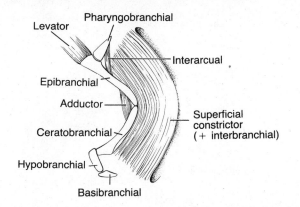

Figure 213. A single gill arch of a shark and its musculature. In many sharks, including the familiar *Squalus* or spiny dogfish, the levators are fused to form a cucullaris associated with the scapula rather than the epibranchials, and the fibers of the superficial constrictors are not parallel to those of the interbranchial but form a shallow V with the apex anterior (to the left here).

branchial bar is the territory of nerve IX; the more posterior gill bars are innervated by special branches of nerve X (which continues onward far down the gut).

Muscles of the Typical Branchial Bars and Their Derivatives (Figs. 213, 214 A).

Although there is frequent fusion of muscle tissues above and below the gills, each typical branchial arch of a shark has a characteristic series of muscle slips proper to it, which are distinguishable by innervation from those belonging to neighbors in the series. The most prominent element in a shark gill is the **superficial constrictor**; this is a broad, if usually thin, sheet of muscle whose fibers generally run dorsoventrally in the flap of skin extending outward in the branchial septum. In many cases, separate dorsal and ventral constrictors may be recognized. Dorsally and ventrally most of the constrictor fibers terminate in sheets of fascia on the back and throat. The deeper fibers, however, may attach to the outer surface of the skeletal bar and form separate **interbranchial** muscles.

In addition, deeper muscles develop. The **adductors** of the visceral arches run from epibranchial to ceratobranchial and tend to bend the two together; dorsal **interarcual** muscles connect the epibranchials with the pharyngobranchials of the same or neighboring arches. Dorsal to the constrictor, muscle fibers run diagonally downward and backward from the dorsal fascia and may insert on the dorsal region of successive branchial bars as **levators** of the arches. In many sharks, however, most or all of these fibers slant back to attach to the shoulder girdle; this is characteristic of the tetrapod muscle derived from this muscle series, the trapezius, and this name or **cucullaris** is frequently applied to the shark muscle.

In bony fishes, the development of the musculature of the gills is more restricted. Because the branchial septa are reduced (in the presence of an operculum), the superficial constrictors are absent, but, ventrally, slips of this system persist as varied **subarcual** muscles; in teleosts, the levators are lost, and the adductors and interbranchial muscles are reduced or absent.

Among tetrapods, gill-breathing larval amphibians retain a series of branchial

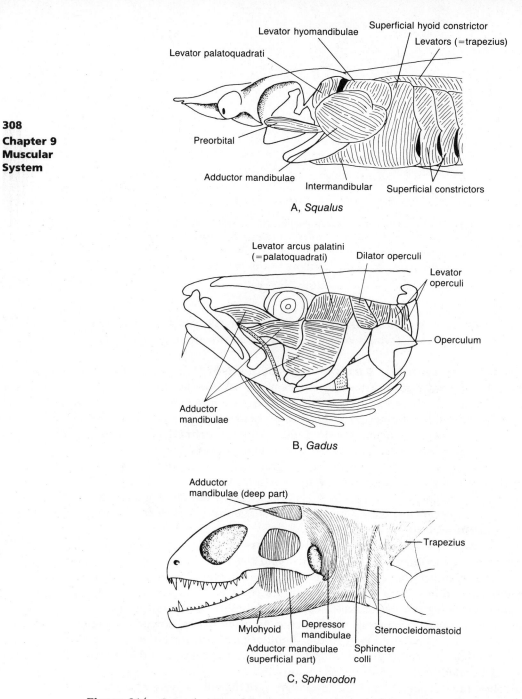

A, *Squalus*

B, *Gadus*

C, *Sphenodon*

Figure 214. Lateral views of the branchial arch musculature and its derivatives in a shark, a teleost, and the reptile *Sphenodon.* In the teleost, the muscles of the more posterior arches are hidden beneath the operculum. *B* after Dietz; *C* after Adams and Fübringer.)

muscles that resemble those of bony fishes; but in true terrestrial forms, the muscles moving the typical branchial bars have disappeared, except for small slips associated with the hyoid apparatus and larynx (though small, these can be important—they control, among other things, your vocal cords). There is, however, one conspicuous if aberrant group of muscles of this series that survives in land vertebrates—the **trapezius** musculature (Fig. 214 *C*), derived, as mentioned, from the levators of fishes.* The trapezius is almost always present and frequently greatly expanded in land vertebrates, and its origin may extend far down the back. The thin trapezius sheet converges downward from the occiput and dorsal fascia to insert along the anterior margin of the shoulder girdle. Primitively, the attachment was to the dermal elements— cleithrum and clavicle; but with their reduction or loss, the insertion lies along the primitive front margin of the scapula (the scapular spine in mammals) and may reach the sternum ventrally. More anterior and ventral slips of the trapezius group may separate as individual muscles—**sternomastoid**, **cleidomastoid**, and so on—and, with reduction of the clavicle, may (in many mammals) fuse with slips from the deltoid to form long, slender compound muscles extending directly from head to front limb.

Muscles of the Hyoid Arch. Presumably in the ancestral jawless fishes, the muscles of the hyoid arch were comparable to those of the typical branchial arches. But in all living jawed vertebrates, the hyoid arch (and its gill slit) has become highly modified, and the musculature is correspondingly modified. The muscles of this arch are always identifiable through their innervation by nerve VII, the facial nerve. In the lamprey, hyoid muscles work the odd "tongue"; as usual, lampreys are simply different and tell us nothing about the ancestors of jawed forms.

Related, presumably, to the loss of independence of the hyoid (which is modified for jaw support), all the deeper hyoid muscles are absent in sharks, and only the superficial constrictor (and a small levator) remains (Fig. 214 *A*). This muscle, however, may be variously subdivided in many fishes. Some deep slips may separate to connect parts of the hyoid arch with one another and with the jaw joint. Certain of these slips may persist in tetrapods in the hyoid and ear region; some are insignificant, but two of them, as noted later, achieve importance in tetrapods in connection with mouth opening. A ventral part of the constrictor may form, in fishes, part of the muscle (intermandibular) connecting the jaws ventrally, but in tetrapods, the hyoid contribution to this muscle sheet vanishes.

The most important part of the hyoid musculature is the dorsal part of the constrictor sheet developed in the hyoid septum. In bony fishes, this septum is greatly expanded to form the bony operculum covering the gills; the hyoid constrictor is highly and complexly developed to control the movements of this cover (Fig. 214 *B*).

In tetrapods, the operculum disappears; the constrictor of the hyoid, in the absence of the more posterior constrictors (they had disappeared in bony fishes), is the only superficial muscle remaining in the developing neck. It expands over this territory as a thin sheet, the **sphincter colli**, circling the neck ventrally and laterally (Fig. 214 *C*) and tending to adhere to the skin. In mammals, it expands in a spectacular

*The trapezius is mainly of branchial origin; however, its innervation in most mammals suggests that the axial musculature of the neck also contributes to it.

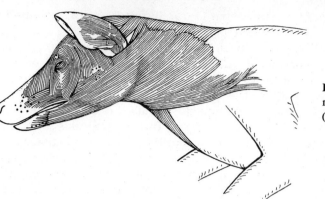

Figure 215. Facial musculature of an opossum. (After Huber.)

fashion over the surface of the head to form the **facial muscles** or muscles of expression (Fig. 215). Although widely distributed over the scalp and face, these muscle bands tend to be especially concentrated about the orbits, the outer ear, vibrissae (when developed), and lips (which last, hence, acquire a mobility and usefulness unknown in lower vertebrates). Lacking facial muscles, nonmammalian vertebrates generally can neither smile nor frown; lacking an oral sphincter, they can neither kiss nor suck; lacking ear muscles, movement of the ear pinna (as seen in most mammals, including especially gifted human beings) would be impossible. Presumably, the facial muscles appeared early in mammalian evolution, but not in mammal-like reptiles; those of monotremes show basic differences from those of marsupials and eutherians.

Mechanisms for mouth opening are not well developed in vertebrates (mouths tend, rather, to open by themselves), and various makeshift devices for this purpose are seen in different groups. Ventrally, the primitive series of hypobranchial axial muscles run forward, as we have noted earlier, from shoulder to jaw, and a backward pull of these muscles is sometimes used for depressing the jaw in fishes. In most tetrapods except mammals, there is substituted a **depressor mandibulae** (Fig. 214 *C*), an anterior slip of the hyoid constrictor that arises laterally from the posterior part of the skull and passes down behind the one-time spiracle (now the region of the eardrum) to attach to the back end of the lower jaw (there is frequently a retroarticular process of the jaw that gives the muscle good leverage).

In mammals, the jaw is refashioned; the elements about the former region of attachment of the depressor are either lost or taken into the ear. The depressor muscle, its function lost, vanishes. Another slip of hyoid musculature, however, emerges to take the place of the depressor and contributes to the formation of the **digastric** (Fig. 216 *B*). This mammalian jaw-opener is, as its name implies, a compound muscle with two bellies. An anterior belly is formed by fibers of mandibular origin running back beneath the ramus of the jaw. Posteriorly, this connects with a slip of hyoid muscle that runs upward behind the jaw to attach to the skull near the ear. The two bellies lie at an angle to one another, but between them, they successfully perform the none too arduous task of depressing the jaw.

Jaw Muscles (Figs. 214, 216). With the modification in gnathostomes of the elements of an anterior branchial arch to form basic jaw structures, the muscles of this arch have become highly modified to serve special functions. Because no typical gill

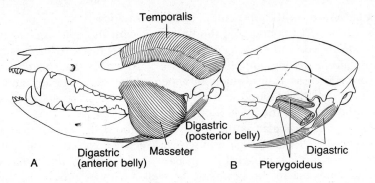

Temporalis

Digastric
(posterior belly)

Digastric
(anterior belly) Masseter

A

Digastric

Pterygoideus

B

Figure 216. Jaw musculature of the opossum. *A,* Superficial view; *B,* deeper dissection. The jaw is represented as transparent in *B* to show the pterygoid muscles, which attach to its inner surface.

opening adjoins the jaw, there is, of course, no development here of an expanded gill flap; in consequence, there is no development of a typical constrictor sheet. The jaw musculature, as seen in sharks, consists of three parts:

1. The upper jaw in sharks is only loosely attached to the braincase. A dorsal segment of the jaw musculature runs out and down from the braincase in front of the spiracle to attach to the upper jaw as an elevator of the upper jaw, a **levator palatoquadrati**, rather comparable serially to one of the levators or components of the trapezius that attach to the ordinary branchial arches.

2. The largest muscle of the mandibular segment is the **adductor mandibulae**, which performs the most important function of any single branchial muscle in gnathostomes: the pressing together of the jaws in the biting or grinding motions essential for the ingestion of food. In sharks, the main mass of the adductor muscle is arranged in a simple manner, running from the palatoquadrate cartilage, which here forms the upper jaw, down to the mandibular cartilage. A specialized portion, a **preorbital** muscle, runs forward from the main adductor to the front of the orbit, and, pulling the upper jaw forward, aids in anchoring it to the braincase. The adductor of the mandible is comparable in position to the adductors of the normal branchial arches but probably includes in its composition the material that farther back forms the main bulk of the constrictor, which is absent here.

3. Less important is a ventral element: fibers, representing the most ventral part of the constrictor, form in fishes (together with a hyoid derivative) an **intermandibular** muscle, a thin sheet of fibers between the two rami of the lower jaw.

In higher vertebrates, the more ventral and dorsal parts of the jaw musculature are unimportant. In tetrapods, the ventral fibers continue to form a thin sheet between the lower jaws as the **mylohyoid** muscle; a slip from this forms (as noted earlier) one belly of the digastric muscle, which lowers the jaw in mammals. The upper jaw and palate continue to be movably articulated with the braincase in most bony fishes and a number of tetrapod groups. In such types, the levator of the palatoquadrate persists, but in varied form. In bony fishes, it may expand to connect the hyomandibular and even the operculum with the braincase. In tetrapods, it is represented by small muscles connecting the braincase with the quadrate (as in birds) or pterygoid (as in lizards and snakes). If the upper jaw is fused to the braincase, as in chimaeras, lungfishes, living amphibians, turtles, crocodiles, and mammals, this type of musculature is useless. It has usually disappeared; in many reptiles and amphibians, however, a relic portion is found in the inner part of the orbit where its swelling or contraction tends to push the eyeball outward.

The adductor mandibulae is prominent throughout the gnathostomes. In bony fishes, it is bulky. The "cheek" region in which it lies is covered by dermal bone, and the attachments of the jaw muscles are in part on the underside of this dermal roof and may extend inward to the braincase as well. In lower tetrapods, the adductor musculature is divided into three main parts separated by the major branches of the trigeminal nerve. In many forms, the muscle is extremely complex and heavily subdivided—one recent work recognizes up to a dozen parts in some snakes and lizards. The fibers primitively arose from the under surface of the cheek region of the skull roof and from adjacent parts of the palatal complex and braincase to insert in and about the fossa in the lower jaw. Fenestration or emargination of the originally solid skull roof in all amniotes except some turtles and the extinct cotylosaurs brings the adductor musculature to a more superficial position. Associated with the modification of the jaw structure and articulation in mammals, we find that this musculature has been divided into three parts. One, the **temporalis**, has much the original position of the muscle and inserts into the coronoid process of the mandible. A second muscle, the **masseter**, is more superficial in position, lying external to the original skull roof. Its fibers run at a considerable angle to those of the temporalis and pull the jaw forward and upward; in rodents, particularly, it may be highly developed and is often stronger than the temporalis. Finally, the little **pterygoideus** muscles form a deep division of the adductor mass. They typically originate from the pterygoid region of the palate (but may extend onto the braincase) and insert on the inner or back surface of the jaw.

Dermal Musculature

In fishes, the surface of the skeletal muscles is tied in closely to the dermis. In tetrapods, particularly the amniotes, this is no longer the case, and the skin lies relatively loosely over the underlying musculature. However, in many instances, thin sheets or ribbons of muscle are found partly or completely embedded in the skin, and functioning in movement of the skin; these muscles are derived during development from the body muscles beneath them. In some cases, the distinction between the major skeletal muscle system and its superficial dermal derivatives is clear; in other cases, however, this is not so, making it difficult to decide whether a given muscle is a proper muscle of the trunk or limb or a dermal muscle.

In amphibians and reptiles, the dermal musculature is in general little developed except for the occasional presence of a dermal slip of the pectoral musculature—a condition repeated in many mammals. The Ophidia are a notable exception. The large ventral scales of snakes each have attached to them a stout dermal muscle that is important in certain rare types of locomotion; they are not used in ordinary undulatory motion. In birds, dermal elements, developed from a number of the shoulder

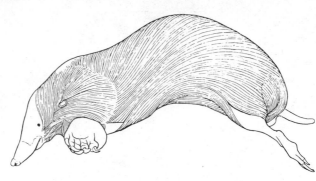

Figure 217. Dermal muscles sheathing the body of a mole. Those anterior to the forelimb are parts of the facial musculature; the large one posterior to it is the panniculus carnosus. (After Nishi.)

and wing muscles, insert distally into the thick skin of the wing and aid in flight, and dermal muscular slips can alter the slope of elements of the plumage.

We find the greatest development of dermal musculature in mammals. In most mammals (higher primates are exceptional), almost the entire trunk and neck are wrapped about by a continuous sheath of muscle termed the **panniculus carnosus** or cutaneous maximus (Fig. 217); the twitch of a horse's skin where a fly has settled is evidence of the presence and functioning of this dermal muscle sheet. Although it covers much of the trunk, the panniculus carnosus is, for the most part, derived from the pectoralis, an appendicular muscle; other muscles may also contribute fibers to this or other dermal muscles of the trunk, but they are rarely important. The ring of superficial fibers about the neck, the sphincter colli, is, as we have noted, a part of the visceral musculature of that region and is innervated by the facial nerve (VII). In mammals, this dermal musculature undergoes a striking development (to which, indeed, the facial nerve innervating it owes its name). Slips grow forward from the neck over the skull and onto the cheeks to form the facial muscles.

Electric Organs

In a restricted number of fishes—rays of the genus *Torpedo* and several tropical freshwater teleosts, such as the electric "eel" *(Gymnotus)* and the electric catfish *(Malapterurus)*—there is a development of special organs capable of producing a heavy electric shock to any animal in contact with them. In several other types of rays and teleosts, weaker electric phenomena are produced. The present chapter is the most appropriate place for their consideration, because these electric organs (Fig. 218) appear to be, in most and probably all cases, modified muscular tissues. Muscle fibers are adapted for the rapid utilization and release of energy; in the present instance, however, the energy is used for the production of electricity rather than for muscular contraction.

The elements of which these organs are composed are generally flattened plates of multinucleated protoplasm, termed **electroplaxes**, each innervated by a nerve fiber. In certain instances in which their development is known, they are seen to arise from embryonic striated muscle cells but become much altered in shape and internal structure. In other cases, it is apparently the myoneural junction rather than the muscle fiber itself that develops as the electroplax. These electric plates are arranged, one above another, in series of piles. This arrangement is immediately comparable to the

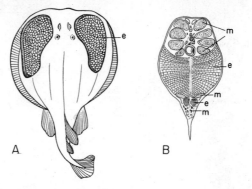

Figure 218. Electric organs. *A, Torpedo,* a ray in which musculature of the expanded pectoral fins has been transformed into **electric cells** (*e*). The skin is dissected away to show the electric organs. *B,* Section of the tail of the electric "eel" of South America (*Gymnotus*). Typical axial musculature (*m*) is present above and below, but much of the caudal muscle has been transformed into electricity-producing tissue (*e*). (*A* after Garten; *B* after duBois-Reymond.)

old-fashioned voltaic pile, famous in the history of electrical discovery; these piles of electric cells form essentially an organic battery, made effective by plus and minus differences between the two surfaces of each plate. In *Torpedo,* strengths of more than 200 volts and 2000 watts have been recorded; the electric eel can discharge more than 500 volts.

In forms with a powerful electric organ, it is obvious that this structure can be of considerable value to a fish either in warding off predators or paralyzing its prey. Those fishes with weaker electric organs have long been utilizing the principles of radar, discovered by man only in recent years. Waves emitted by electric organs in the tail of such fishes form an electric field; the presence of nearby objects disturbs the field as sensed by receptor organs modified from the lateral line system (cf. p. 522). Such organs, present in a number of primitive freshwater teleosts in South America and Africa (such as *Gymnarchus* of the Nile) and in one skate, are especially useful in dark or turbid waters. In the teleosts possessing this system, the areas of the brain associated with the lateral line system are enormously enlarged.

Despite the basic similarity of their construction, the electric organs vary greatly in position and appearance in the different types that have them; presumably such organs developed independently a number of times. In *Torpedo,* they are present as two large groups of piled-up plates, one on either side of the head in the expanded pectoral fin of this flattened elasmobranch. There are hundreds of piles and some hundreds of plates in each pile; in one species of *Torpedo,* it has been calculated that over 200,000 plates are present. In the electric "eel" *Gymnotus* of South America, in which the tail makes up four fifths of the body length, the massive electric organ is formed from the modified ventral half of the tail musculature. In the catfish *Malapterurus* from the Nile, the electric tissue forms a sheath encircling the whole body just beneath the skin from the back of the head to the base of the tail. In this case, the plates, although numerous, are not arranged in neat piles, and because the electric tissue is superficial to the body musculature, its origin from musculature is not so certain (perhaps modified glands form the source).

Chapter 10

Body Cavities

In the vertebrates, as in all the more highly organized invertebrate types, most of the body organs are not embedded in solid tissues or mesenchyme but are situated within the bounds of body cavities, more properly **coelomic cavities**. These are fluid-filled spaces in which the viscera are placed in a situation relatively free from ties and are more or less at liberty to move freely during their functional activity and to change more readily in size or shape during growth processes. We may here give a brief résumé of their arrangement before proceeding to the description of organs that are enclosed in these cavities and that border them.

Development of the Coelom.   The body cavities are developed in the mesodermal tissues, and their linings are formed by a serous mesodermal epithelium, the **peritoneum**. In amphioxus, as in echinoderms, the coelomic cavities arise in part, at least, from segmental pouches pinched off from the primitive gut as mesodermal somites (cf. Figs. 78, 79). In true vertebrates, however, this presumably primitive method of coelom formation is abandoned, and the body cavities are formed by a splitting of mesodermal sheets after that tissue has separated from other germ layers. Although small coelomic cavities may appear in the somites or the kidney-forming tissue, these are ephemeral, and the term **coelom**, in all late embryonic and adult stages, is confined to the cavities in the more lateral and ventral parts of the mesoderm, the lateral plate.

In forms with mesolecithal eggs, the lateral plate of mesoderm at an early stage extends down the flanks of the body on either side so that the two sheets presently meet, or come close to meeting, in the ventral midline (cf. Fig. 80). In large-yolked types, the lateral plates are at first directed outward laterally over the yolk and are continuous with the mesoderm of the extra-embryonic regions (cf. Fig. 75); only later, with the assumption of body form, do these sheets meet ventrally. Each of the two lateral sheets is a continuum, to begin with, from the head to the hinder end of the trunk region, and above the lamprey level never shows any indication of segmentation. At first, each plate is solid, but presently it splits in two to form a liquid-filled embryonic coelomic cavity between outer and inner walls. These walls, apart from giving rise to connective tissues, muscles, and other materials, are destined to form, respectively, the parietal and splanchnic layers of the peritoneum (Fig. 219). The **parietal** (or somatic) **peritoneum** forms the inner surface of the great external "tube" of the body, termed in the embryo the **somatopleure**, in which the somatic skeleton, somatic musculature, kidneys, and gonads presently develop. The **splanch-**

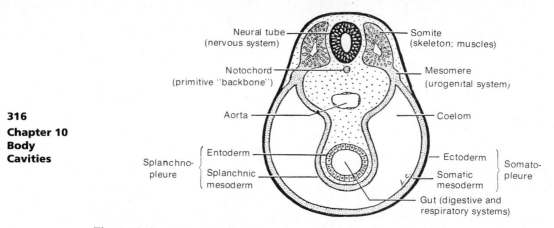

Figure 219. Diagrammatic transverse section of a mammalian embryo to show the relations of the mesoderm. (From Arey.)

nic (or visceral) **peritoneum** forms the outer covering of the digestive tract and its outgrowths, which constitute the **splanchnopleure**.

In this manner, a pair of coelomic cavities, one on either side, with a large gut cavity between them, is formed. The diameter of the gut, behind the pharyngeal region, becomes relatively small as development proceeds, and the two coelomic cavities come to lie adjacent to one another both above and below the gut (cf. Figs. 2 *B*, *D*, 222 *B*). At this stage, the right and left coeloms are separated by longitudinal thin sheets of tissue, bounded on either side by peritoneum; these are the dorsal and ventral mesenteries. The **dorsal mesentery** is usually a persistent structure, containing nerves and blood vessels supplying the gut. The **ventral mesentery**, however, is of little functional importance in the adult and usually disappears for most of its length. With this disappearance, the two coelomic cavities are merged ventrally.

In the early embryo, the sheet of lateral plate mesoderm containing the coelomic cavity is continuous along the length of the gut (cf. Fig. 80 *C*). The anal or cloacal region is the definitive posterior boundary of the trunk, and, except in a very few odd cases, neither the coelom nor the viscera associated with it extend into the tail. Anteriorly, the situation is complicated by the development of the pharyngeal gills. The developing branchial slits disrupt the lateral plate in the side walls of the pharyngeal region, so that in this part of the embryo, the coelom is confined to the ventral part of the throat, below the pharynx (cf. Fig. 196). In amphioxus, extensions of this coelom project up into the branchial bars.

The coelomic structure of an early embryo is simple and readily understood; that of the adult is not. It is complicated, owing to three factors: (1) a pushing into the coelomic space of organs other than those of the digestive tract—heart, gonads, kidneys, and lungs; (2) longitudinal subdivision into compartments—in most cases into a cavity around the heart and a general body cavity, but with further subdivision in mammals and birds; and (3) elaboration and twisting of the gut and outgrowth of its appendages—liver, pancreas—with a consequent complicated folding of mesenteries.

Pericardial Cavity. The most anterior and ventral region of the embryonic coelom becomes the pericardial cavity, in which the heart develops. This area, as noted

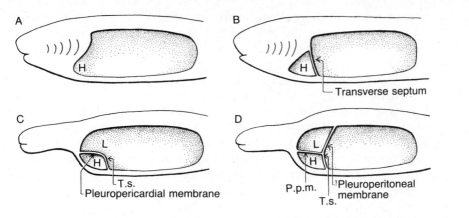

A

B
Transverse septum

C
T.s.
Pleuropericardial membrane

D
P.p.m.
Pleuroperitoneal membrane
T.s.

Figure 220. Diagrammatic longitudinal sections of the body to show the evolution of the coelomic cavity. *A,* A primitive condition, with the entire coelom forming a single cavity. *B,* Typical fish condition; the pericardial chamber is separated from the main cavity. *C,* Typical amphibian and reptilian condition; the lungs are developed, but the spaces in which they lie are in continuity with the main coelom. *D,* Mammalian condition, with a formed diaphragm (see text). Abbreviations: *H,* heart; *L,* lungs; *P.p.m.,* pleuropericardial membrane; *T.s.,* transverse septum.

above, lies primitively in the floor of the "throat," below and somewhat posterior to the pharynx. In the early embryo of all vertebrates, the pericardial cavity opens widely at the back into the general abdominal cavity. Soon, however, there forms a **transverse septum** at the level of the most posterior chamber of the heart, the sinus venosus, and of the cardinal veins that enter it (Figs. 220 *A,* 221).

The septum as thus formed blocks off the lower part of the passage between the pericardium and general coelom but leaves a gap above. In most vertebrates, this gap is closed in the adult; but in hagfishes, elasmobranchs, and a few lower ray-finned fishes, a small dorsal opening between pericardial cavity and general coelom persists into the adult stage. In the early embryo, the liver bud, growing forward below the gut, enters the transverse septum and expands within its substance, so that in lower vertebrates the main attachment of the liver is to the posterior surface of the septum.

Figure 221. Diagram of the developing coelom of a mammal showing its division, here incomplete, into various parts; dorsal view.
Abbreviations: *Dm,* dorsal mesentery; *Li,* liver; *Lu,* lung; *Pc,* pericardial cavity; *Pl,* pleural cavity; *Pm,* pleuropericardial membrane; *Pp,* pleuroperitoneal membrane; *Pt,* peritoneal cavity; *St,* stomach. (After Nelson.)

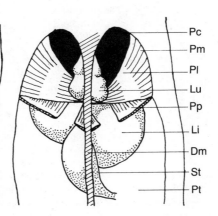

Pc
Pm
Pl
Lu
Pp
Li
Dm
St
Pt

This condition is essentially repeated in an early stage of a mammalian embryo (Fig. 221). In many higher forms, the liver tends in the adult to pull away again from the transverse septum, but remains attached to it by a relatively narrow **coronary ligament**.

In most groups of tetrapods, the disappearance of functional gills and the development of a neck are correlated with a marked change in the relative position of the pericardial cavity. It has been pushed back into the floor of the thorax, where it lies protected by the coracoids or sternum, diagonally below the anterior end of the main body cavity. Above it, in paired recesses of the coelom, lie the lungs. Separating the cavities containing heart and lung, a **pleuropericardial membrane**, bounded on either surface by coelomic epithelium, extends forward from the top of the transverse septum, of which it is an extension (Fig. 220).

General Body Cavity. With the separation of the pericardial cavity, in the majority of vertebrates there remains a single great coelomic cavity occupying, with its enclosed organs, most of the trunk. This is partially subdivided into right and left halves by the mesenteries.

The ventral mesentery, we have noted, usually disappears for most of its length, but it is persistent in dipnoans and a few other fishes and is present for most of its length in salamanders. A short posterior segment may remain as a ligament tying in the bladder with the rectum. Anteriorly, however, the mesentery persists in connection with the liver, which arises in the midst of the primitive ventral mesentery. The part of the ventral mesentery connecting stomach and liver forms the gastrohepatic ligament or **lesser omentum**; that part below the liver is the **falciform ligament** (Fig. 222).

The dorsal mesentery is a simple, continuous, longitudinal sheet of tissue in the early embryo. But, in later stages, the development of the gut and its outgrowths into a series of distinctive organs results in a parallel development of distinct mesenteric areas. The term **mesentery**, in addition to the general sense in which we have thus far been using it, is also used in a narrow sense to apply merely to that part of the sheet supporting the small intestine (where such a subdivision of the gut is developed). Special terms such as mesocolon, mesorectum, and so forth, may be used for

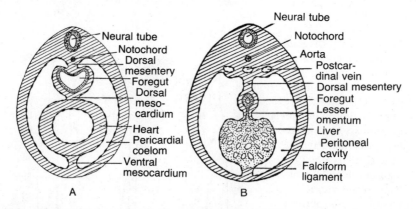

Figure 222. Diagrammatic sections through the heart and liver regions of an amniote embryo to show the relations of the mesenteries. (From Arey, after Prentiss.)

other parts. Of special importance is the mesogaster, or **greater omentum**, supporting the stomach. The dorsal mesentery system is present, though occasionally much reduced, in all vertebrates, because it is along the mesenteries that arteries, lymphatics, and nerves descend to the gut. In reptiles and mammals, the dorsal mesentery remains unbroken; in birds and amphibians, however, there may be gaps along its length; in some fish groups, it is broken up into discontinuous segments, and little of it is preserved in adult cyclostomes. In teleosts and tetrapods in which a coiled small intestine is developed, the mesentery is, of course, folded in a complicated manner, and parts of the mesentery may fuse with one another in a confusing way (cf. Figs. 223, 226).

The stomach, as developed in most groups of jawed vertebrates, swings to the left side of the body, and that part of the dorsal mesentery forming the greater omentum swings with it. In consequence, the right part of the coelomic cavity adjoining the stomach encroaches on the left side of the body. This encroachment is exaggerated, especially in mammals, by the fact that generally the greater omentum (in which the spleen is embedded) does not descend directly from the dorsal body wall to the stomach but tends to sag downward to the left of the stomach (Fig. 223). Ventrally, this folded area is, of course, closed off by the expanse of the liver, tied to the stomach by the lesser omentum. As a result this part of the right coelom becomes essentially a pouch, the **omental bursa**, opening out to the right above the liver. In many

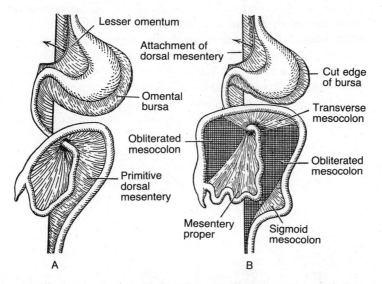

Figure 223. Diagrammatic ventral views of the gut and mesenteries of a mammal in *A*, an embryo, and *B*, essentially adult conditions. The small intestine is greatly simplified in *B*. *A* shows the general type of folding that the mesenteries must undergo because of the asymmetrical position of the stomach and twisting of the intestine. As shown in *B*, this folding may result in obliteration or fusion of parts of the mesentery. In *A*, the bursa is a structure of minor dimensions; in many mammals, the enlarged bursa would extend down, covering over much of the intestine, but has been cut off short in *B*. The opening from the bursa to the coelom of the right side (epiploic foramen) is shown by an arrow; the double line "above" the arrow (i.e., ventral to it) is the cut attachment of the lesser omentum to the liver. For lateral views of the same structures, see Figure 226. (From Arey.)

lower vertebrates, it is not especially developed and may open broadly to the right side.

Dorsally, various structures not part of the digestive system may project into the body cavity (cf. Figs. 2 *D*, 302). The testes or ovaries, formed from the genital ridges of the embryo, usually expand into the body cavity, surrounded by its peritoneum, and may be supported by special mesenteries—**mesorchium** or **mesovarium**. Farther laterally, the developing kidneys of the embryo also expand into the body cavity as longitudinal swollen ridges, and, in the adult, the kidney bulges to a varied degree into the coelom from the dorsal side. The oviducts arise in the ridges formed by the kidneys and may persist there but are often suspended by a special paired mesenteric structure, the **mesosalpinx** or **mesotubarium**.

Lung Pockets. The lungs, as developed in a few bony fishes and in tetrapods, push backward from the pharynx along either side of the esophagus and extend posteriorly into the body cavity. Here they lie above the pericardium on either side of an anterior extension of the mesentery, termed the **mediastinum**, in which the esophagus is embedded. In lower tetrapods, the lungs are usually supported by separate paired mesenteries, **pulmonary folds**, extending down to them from the dorsal wall of the body cavity and continuing downward to tie into the upper surface of the liver on either side (Fig. 224). The left fold is usually little developed. The right fold, however, is conspicuous in the embryo of every tetrapod and frequently remains so in the adult. Its persistence is associated with the fact that the posterior vena cava utilizes it to make its short circuit down into the liver and thence to the heart (cf. p. 475).

In amphibians and a majority of reptiles, the dorsal and anterior recesses that contain the lungs are still parts of the general body cavity. In some reptiles, especially turtles, the lungs may be more or less buried in the body wall dorsally, with little or no projection into the coelom. In still others, however, there may be a partial or even—in crocodilians and some lizards and snakes—complete closure of these pockets for the lungs from the remainder of the coelom, making **pleural cavities**, much as in mammals.

The Coelom in Birds. The great development of air sacs in birds (cf. p. 367) is associated with a complicated subdivision of the body cavity that appears to promote

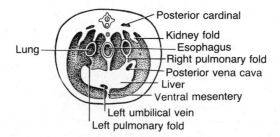

Figure 224. Diagrammatic cross section, looking forward, through the body of a lizard in the region of the lungs, to show the pulmonary folds, between which and the esophagus lie pulmonary recesses. The left pulmonary fold is little developed; it is down the right fold that the posterior vena cava makes its way from the region of the posterior cardinal vein to the liver (cf. p. 475). (After Goodrich.)

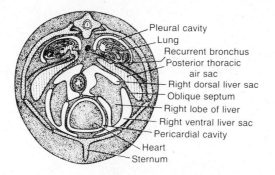

Pleural cavity
Lung
Recurrent bronchus
Posterior thoracic
air sac
Right dorsal liver sac
Oblique septum
Right lobe of liver
Right ventral liver sac
Pericardial cavity
Heart
Sternum

Figure 225. Diagram of transverse section through the thorax of a bird to show especially the subdivision of the coelomic cavity. In addition to the pericardial cavity, the pleural cavities and the sacs dorsal and ventral to the liver are shown; the main intestinal coelom is too far posterior to be included in this section. The thoracic air sacs (hatched) are also present in this section. (After Goodrich.)

the efficiency of the breathing apparatus (Fig. 225). The pericardium forms, of course, a separate anterior chamber, and, posteriorly, a transverse septum divides off a large cavity containing intestine, kidneys, and gonads. Between these two extremities lies the area of the lungs, liver, and stomach; here, three paired longitudinal septa, more or less horizontal in position, bring about the development of four pairs of cavities. The uppermost septum encloses the two lungs in individual pleural cavities; this septum somewhat parallels the development of the diaphragm of the mammals, described below, but is of independent origin. A lower septum, the oblique septum in Figure 225, causes the formation above it of a lateral pair of cavities in which are air sacs connected with the lungs (these are hence not true coelomic cavities). Farther ventrally, the liver expands on either side of the gut; here, a horizontal septum running lengthwise along either side of the liver separates coelomic cavities dorsal and ventral to that organ.

Mammalian Coelomic Cavities; Diaphragm (Fig. 227). Quite another type of coelomic development has occurred in mammals. There are, of course, no air sacs of avian type and none of the complex subdivision of the main body cavity that accompanies that development. The major features are (1) a high degree of development of the omental bursa and, especially, (2) the presence of pleural cavities closed off by the diaphragm.

We have noted the presence to a variable degree in other vertebrates of a pouch, the omental bursa, of the right coelomic cavity between stomach and liver and the development of the greater omentum associated with this condition. In mammals, this recess is much developed; the greater omentum becomes a huge fold extending down as an apron over much of the ventral surface of the abdominal cavity (Fig. 226). The opening to the bursa (on the right side) is closed except for the small **epiploic foramen** (foramen of Winslow, Fig. 223).

In mammals, the lungs are enclosed in paired **pleural cavities**, recesses in the thorax that are shut off from the major abdominal (or peritoneal) cavity (see p. 368). The partition between thoracic and abdominal parts of the coelom is the **diaphragm**; movements of this partly muscular structure, together with movements of the ribs, are responsible for the typical breathing action of mammals.

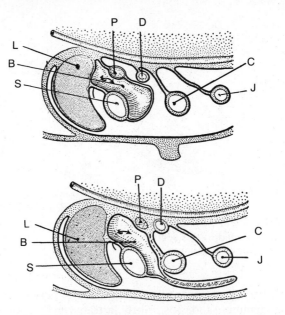

Figure 226. Diagrams of embryonic and adult conditions of the omentum and bursa in a mammal; longitudinal sections, seen from the left side, with head end at left. In the embryo (upper figure), the bursa is small; the entrance (epiploic foramen) is indicated by an arrow. In the adult of many mammals (lower figure), the dorsal mesentery, as the greater omentum, has become a long fold. The diagrams further show how the mesenteries of the gut may be fused together (as is here that of the transverse colon to the greater omentum), or obliterated (as is that of the duodenum, in the diagram). Abbreviations: *B,* omental bursa; *C,* transverse colon; *D,* duodenum; *J,* jejunum; *L,* liver; *P,* pancreas; *S,* stomach. For comparable ventral views, see Figure 223. (After Arey.)

In the adult mammal, the diaphragm has the appearance of a relatively simple partition, tendinous in its central part, but with a sheet of muscle all about its periphery; it extends across the body behind the base of the rib basket and in front of the liver and stomach to divide the thoracic region from the abdomen. Despite its apparent simplicity, the diaphragm is complicated in its mode of development and in its structure; we shall here give a very brief outline of this story.

In connection with the formation of the pericardium, we have noted the transverse septum, developed behind the heart and in front of the liver, which closes off the pericardial cavity from the main coelom. This is the principal ventral component of the diaphragm. In reptiles, the lungs are dorsal to the heart, above the top of the septum. In mammals, in contrast, the large pockets for the lungs expand downward on either side of the heart. As a result, the pleural cavities may approach one another ventrally below the heart, leaving only a median strand of tissue, a **ventral mediastinum**, connecting the pericardial sac with the ventral body wall (Fig. 266). To close off these enlarged lung pockets from the abdominal coelom as pleural cavities, adjuncts to the transverse septum are demanded (Fig. 227). The more medial parts of the opening are closed by mesodermal folds growing dorsally from the top of the transverse septum. Laterally, **pleuroperitoneal membranes** fold inward from the body wall to close much of the opening, and, as the lungs continue to expand, the

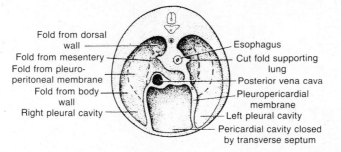

Fold from dorsal wall
Fold from mesentery
Fold from pleuro-peritoneal membrane
Fold from body wall
Right pleural cavity

Esophagus
Cut fold supporting lung
Posterior vena cava
Pleuropericardial membrane
Left pleural cavity
Pericardial cavity closed by transverse septum

Figure 227.   Anterior view of the diaphragmmatic region of an embryonic mammal, to show the various elements that comprise the diaphragm. Heart and lungs have been removed to show the posterior walls of the pleural and pericardial chambers. At this stage, the pericardial cavity still extends to the ventral wall of the chest, whereas in the later stage the pleural cavities extend below it. (After Broman, Goodrich.)

body wall itself takes part in forming the lateral areas of the diaphragm beyond the base of the original pleuroperitoneal membranes. A small opening persists for a time far dorsally on either side; here, a final closing fold develops from the dorsal mesentery or the mesentery supporting the lung itself. When in such reptiles as the crocodilians, the lung cavities are separated from the general coelom, the partition consists simply of peritoneal membranes and connective tissue. In contrast, the mammalian diaphragm is supplied with a sheet of muscle, striated and subject to voluntary control, which is obviously a derivative of the axial musculature (cf. p. 289), although its embryologic origin is not well known.

Abdominal Pores.   In the ancestral fishes, there appear to have been a pair of small **abdominal pores** connecting the coelom with the exterior in the cloacal region (cf. Fig. 316). Such openings are present in most elasmobranchs and occur in some bony fishes, although most teleosts have lost them. In tetrapods, the pores have been lost in most cases but may be seen in turtles and crocodilians. In cyclostomes, these or a pair of similar pores serve as exits for the sperm and eggs; because their relationships are not the same as those of the abdominal pores in jawed forms, their homology is questionable and they are often termed **genital**, rather than abdominal, **pores**. In all gnathostomes, special ducts have taken over the transportation of the sex cells, and the only function we can suggest for the abdominal pores is that of getting rid of surplus liquid from the coelomic cavity.

Chapter 11

Mouth and Pharynx;
Teeth, Respiratory Organs

The digestive tract, with its various outgrowths and accessory structures, looms large in both bulk and importance in body organization. The major digestive organs, including stomach and intestine, form the more posterior segments of the tract. In the present chapter, we shall consider the mouth and pharyngeal regions, "introductory" sections of the digestive tube. These play little role in alimentation beyond the reception of food but are highly important in other regards—notably as the place of origin of respiratory organs and important glandular structures.

The Mouth

In humans, the mouth appears to be a well-defined structural unit, with fixed, uniform features such as lips, teeth, tongue, and salivary glands. But a broad survey of the vertebrates shows that these structures vary widely; every one of the familiar landmarks may be absent in one group or another. Except for the fact that it is an inturned area, a **buccal cavity**, leading to the pharynx, we can make few statements about the mouth that will hold true for all vertebrates.

Development. In the embryologic story, the opening from mouth to pharynx is late in appearance. In types of vertebrates with a mesolecithal egg, the early embryonic gut may open posteriorly at the blastopore, roughly homologous with the anus of the adult, but invariably the archenteron ends blindly at its anterior end. This blind end (Figs. 228 *A,* 271) is the region of the future pharynx. Above and in front of the pharyngeal region, there early appears an expanded brain and a consequently expanded head region. The head turns downward over the surface of the yolk-swollen body or yolk sac, producing beneath it an inturned fold or pocket of ectoderm, the **stomodeum**. This forms the primitive oral cavity. At its inner end, this fold lies close to the anterior or pharyngeal end of the gut tube. A membrane separating the two structures persists a long time at this point. Eventually this membrane breaks down; mouth and pharynx are placed in continuity, and the digestive tube has acquired an anterior opening. The epithelia of the two regions concerned blend with each other and are difficult or impossible to distinguish at later stages. In a broad way, however, it is clear that the pharynx is mainly if not entirely lined with endoderm, but the oral epithelium is ectoderm, essentially a continuation of the skin epidermis.

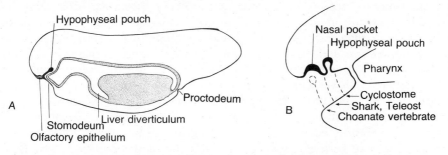

Figure 228. *A*, Diagram of a larval amphibian (at about the stage of Fig. 83 *E*), in longitudinal section, to show the extent of the endoderm (stippled) and its relation to structures of the oral region. *B*, Diagram to show the comparative position of the oral margins in various types of vertebrates (cf. text).

The embryonic gut cavity of vertebrates, the archenteron, was long ago recognized as comparable to the adult digestive cavity of coelenterates and a number of other simple invertebrate animal types. In these forms, the adult gut has only a single opening, to which the vertebrate embryonic blastopore and adult anus are frequently compared. But in the more highly organized invertebrate phyla—annelids, molluscs, arthropods, and so forth—a second opening has developed as a progressive feature; vertebrates presumably followed a similar evolutionary course, as suggested by the ontogenetic history. In echinoderms, as in vertebrates, this second opening is the mouth; in the other groups just noted, on the other hand, the blastopore becomes the mouth, and the anus is the second opening.

Mouth Boundaries. The extent of the mouth in the various vertebrate groups is highly variable (Fig. 228 *B*). One would, a priori, expect that the mouth of the adult would correspond roughly in extent with the embryonic stomodeum formed from inturned ectoderm. The posterior boundary of the mouth in the adult is by definition the boundary between the stomodeum and the embryonic gut, although this may not be recognizable and an arbitrary landmark may be used instead. However, are the anterior and outer margins of the mouth fixed in position? Does the mouth always include the same area of infolded stomodeal tissue? Not at all.

Two good landmarks are always present in the roof of the stomodeal region. Near the outer end of this funnel, beneath the swelling forebrain, is the embryonic nasal region, a pair of ectodermal pits or thickenings in most vertebrates, a single pit in cyclostomes. Farther back in the roof of the potential mouth is a median pit, the **hypophyseal pouch** (Rathke's pouch), which later closes off, its epithelium forming the major part of the pituitary body lodged beneath the brain. In an adult shark or ray-finned fish, the nasal sacs lie external to the mouth, and the area from which developed the embryonic hypophyseal pouch lies well inside it; the jaw margins bounding the oral cavity thus lie between these two landmarks. In typical crossopterygians (and possibly Dipnoi as well) and all tetrapods, the situation is different. The jaw margins form a transverse bridge beneath the nasal pockets, so that both external and internal narial openings are present. Here, then, we have a more extensive oral cavity than in most fishes.

The opposite extreme is seen in the cyclostomes. A lamprey or hagfish has a

326
Chapter 11
Mouth and
Pharynx;
Teeth,
Respiratory
Organs

well-defined mouth; but the embryonic development shows that this mouth corresponds only with the inner recesses of the oral cavity of other vertebrates. In the larval cyclostome (Fig. 229), a nasal pit and a hypophyseal pouch are found in the stomodeal depression, somewhat as in other vertebrates. But as the embryo increases in size, differential growth causes a rotation of both nostril and pit (the two are closely connected) forward and upward onto the outer surface of the head (as in the hagfish; cf. Fig. 17) and, in the lamprey, to a position high on the dorsal surface, far removed from the adult mouth (Fig. 256). Most of the outer surface of the lamprey head is thus covered by ectoderm that in gnathostomes lies within the mouth.

In a majority of vertebrates, the margins of the mouth are formed by **lips**, soft pliable structures of epidermis and connective tissue. In cyclostomes, the mouth opening is rounded (a feature to which the group owes its name); in hagfishes, it bears marginal sensory tentacles, and in the lampreys, the circular margins constitute an effective sucker by which the animal attaches itself to its prey (Fig. 256). Sensory barbels are present at the borders of the mouth in many teleosts, but in jawed fishes, amphibians, and reptiles, the lips are in general small and unimportant cutaneous folds, external to teeth and jaws. Deposits of keratin in the epithelium may transform the lips into a horny bill or beak, as in turtles, birds, and a few mammals (plus some extinct groups). In typical mammals, lips are highly developed, separated by deep clefts from the jaw margins, and rendered mobile by the presence of the facial musculature peculiar to this class. The mouth opening between upper and lower lips generally terminates posteriorly in mammals (in contrast to many lower vertebrates) well forward of the jaw articulation, and there is thus created a skin-covered **cheek** containing more facial muscles; similar muscular cheeks were, apparently, also present in many ornithischian dinosaurs. In some rodents and Old World monkeys, the potential storage spaces underneath the cheeks have been expanded into **cheek pouches**, useful in the storage of food. The pocket gophers—*Geomys* and the like— have developed enormous pouches in this region, which, however, are hair-lined external pouches, opening to the surface, not into the mouth.

Palate. In typical fishes, the roof of the mouth is a rather flattened vault, unbroken by any opening (once the embryonic hypophyseal pouch has closed). In some sarcopterygians and in amphibians, however, we find that paired openings, the **choanae** or internal nares, are present near the anterolateral margins. In amphibians (Fig. 230 *A*), the palate is almost flat; in reptiles (Fig. 230 *B*) and birds, however, the roof is well vaulted, and a pair of longitudinal **palatal folds** aid in forming a channel from the choanae back through the mouth above the tongue to the pharynx for the freer passage of air. In crocodilians and mammals (Figs. 181, 191 *C*, 230 *C*, 231), this air channel has been completely shut off from the oral cavity by the development of a

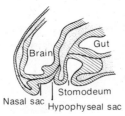

Figure 229. Section of the head of a larval lamprey. At this stage, nasal and hypophyseal sacs are still ventral in position (cf. Fig. 256).

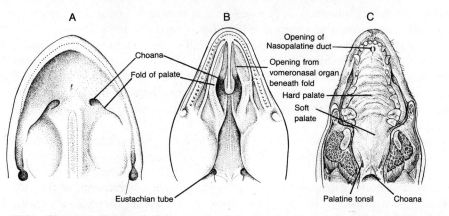

Figure 230. The roof of the mouth in *A,* a urodele; *B,* a lizard; and *C,* a mammal (dog); to show particularly the position of the choanae.

bony secondary palate, noted in our discussion of the skull; this "hard palate" is extended backward in mammals by a thick membrane, the **soft palate** (Figs. 230 *C,* 231). The vomeronasal organs (cf. p. 504) gain openings to the palate independent of the choanae in some reptiles; these organs persist in most mammals and in some open through the secondary palate into the oral cavity, through anterior palatine (or incisive) foramina.

The epithelium of the palate, and of the oral cavity in general, is usually of a stratified squamous type. In living amphibians, this epithelium is ciliated and underlain by a rich capillary network; it often functions as an important breathing organ.

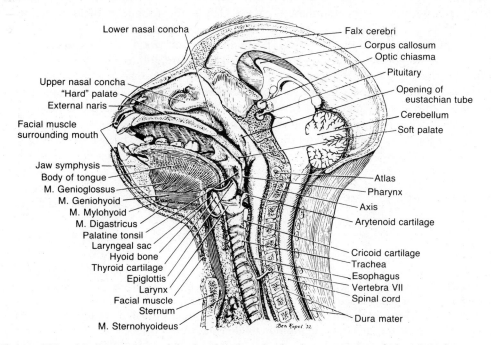

Figure 231. Median section of the head and neck of a rhesus monkey. (After Geist.)

In bony fishes, amphibians, and reptiles, palatal teeth are frequently present. In mammals, transverse ridges of cornified epithelium, which aid in the manipulation of food, are often present on the palate; they are highly developed in most ungulates and carnivores. In the toothless whalebone whales (Mysticeti), these ridges have developed into long parallel plates of horny "whalebone" hanging down into the mouth; fringes on their margins strain small marine organisms that are licked off by the tongue and form the food supply of these giant mammals (Fig. 232).

328
Chapter 11
Mouth and
Pharynx;
Teeth,
Respiratory
Organs

Tongue. In fishes, the lower ends of the gill bars, with accompanying musculature, slant forward below and between the jaws into the floor of the mouth. In many teleosts, these bars bear teeth, which work against those of the palate; in other fishes, however, the floor of the mouth is relatively smooth. Movements of the gill system may raise or depress it, but there is seldom any structure comparable to a tongue. In lampreys, this name is given to an extrusible oral structure bearing horny "teeth" wherewith to rasp the flesh of the animal's prey (Fig. 256), but this apparatus is surely a parallel development rather than one homologous with a true tongue.

In the water, the manipulation of food in the mouth is a relatively simple matter; on land, food materials are more difficult to deal with. With the reduction of the gills in tetrapods, the branchial bars and their muscles became available for other uses, and the mobile tongue characteristic of most terrestrial vertebrates developed. The musculature of the tongue, we have noted, is derived from the hypobranchial musculature, and the tongue is anchored at its base by a skeletal system—the hyoid apparatus—formed of a series of modified gill bars (*not* just the hyoid arch).

Apart from its normal use in handling food, the tongue serves various minor functions: in mammals it is the major bearer of taste buds; in man it is an aid to speech; cornified lingual papillae aid in the manipulation of food; in dogs and other carnivores with poor development of sweat glands, the tongue is of importance in temperature control. In a number of groups, the tongue may elongate and take on an active role in the gathering of food, particularly insect food. Thus, whereas some anurans are tongueless, common frog and toad types are able to flip out an elongate tongue (attached anteriorly, free at the back) to pick up an insect with its sticky tip. The chameleon's tongue can be protruded and retracted with lightning rapidity for the same purpose. In mammals, several types of termite eaters and anteaters have elongate sticky tongues; such tongues show a variety of modifications including having a sternoglossus muscle extending all the way from the breast bone into the

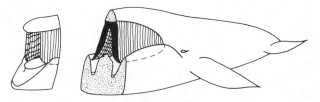

Figure 232. A whalebone whale with the front part of the snout displaced to show the whalebone, sheets of cornified tissue hanging down from the roof of the mouth. The whalebone is elastic and becomes folded back when the mouth is closed; thus, whatever the degree of opening, the sheets are in position to filter out small organisms in the water. (Based on various figures by Pivorunas.)

tongue. The tongue is well developed in snakes and some lizards and may be used (as noted later) as an accessory olfactory organ; on the other hand, it is poorly developed in turtles and crocodilians. In the woodpeckers, the tongue is extremely elongate. In some cases, a modest expansion or protrusion of the tongue may be caused by filling of blood sinuses or lymph sacs, but rapid tongue extension can be caused only by muscular action. In the woodpeckers, hummingbirds, and a few others extension is brought about through the protrusion, by hyoid muscles, of an extremely elongate but slender and flexible basihyal element; there is a special bony tube curving around the occiput and forward on the upper surface of the skull for reception of this rod when withdrawn.

Oral Glands. In fishes, there are numerous mucus-secreting cells in the mouth. Seldom, however, is there any marked development of formed oral glands. One of the few exceptions is a special gland in the lamprey, the secretion of which impedes coagulation of the blood of its prey.

In water dwellers, the water itself is a major aid in the moistening and swallowing of food. In tetrapods, in the absence of a watery medium, **salivary glands** make their appearance (they tend, however, to be secondarily reduced in groups that have returned to an aquatic existence). Such glands in general secrete mucin and a more watery material to produce the saliva.

In amphibians, mucous glands are usually to be found on the tongue, and there is normally a large median **intermaxillary gland** in the anterior part of the palate. Reptiles have a series of small mucous glands along the lips, on the palate, and on and beneath the tongue. Modified salivary glands form the poison glands of the snakes and the Gila monster (the one poisonous lizard); the opening from the gland lies at the base of the poison fang, whence the venom passes into the wound via a duct or groove in the substance of the tooth. Such glands are important in subduing the prey, and their use as defense mechanisms is surely secondary. Many birds, particularly aquatic feeders, are almost devoid of oral glands, but, in others, an abundant series of mucous glands may be present, and the mucus in some cases is an aid in nestbuilding. Highly developed salivary glands are present in mammals; most prominent are the **parotid**, **submaxillary**, and **sublingual** glands, lying in the cheek region and opening into the mouth above and below by long ducts. A few mammals (certain shrews) parallel many snakes in using their salivary glands as venom glands. For the most part, salivary glands lack chemically active materials; but in some frogs, many birds, and a number of mammals (man included), in the saliva an enzyme, **ptyalin**, initiates digestion of starch. Minor amounts of other enzymes may be present as well, but true digestion appears to have been confined exclusively to more posterior regions of the digestive tube in the lower vertebrates.

Dentition

Although the **teeth** are in reality modified parts of the dermal skeletal materials, they may be appropriately discussed here, because they are "inhabitants" of the oral cavity. Teeth are unknown in lower chordates and in jawless vertebrates, living or fossil (cyclostomes have horny toothlike structures, as do larval anurans, but these are not comparable in structure or origin). With the advent of jaws, it would seem, teeth simultaneously developed from the denticles of the dermal plates present at the margins of the mouth. Their presence made the jaws useful as biting structures; through the changed habits made possible by this new type of feeding device, the way was paved for the rise of the gnathostome vertebrates to their present estate.

330
Chapter 11
Mouth and
Pharynx;
Teeth,
Respiratory
Organs

Tooth Structure (Fig. 233).   In their simplest form, teeth are conical structures of a type such as may be seen in many fishes and reptiles and in the anterior part of the mammalian dental battery. Frequently, however, more complex forms appear, notably teeth in which the upper surface is more or less flattened to form a **crown** for crushing or chewing food materials. In the interior of a tooth is a **pulp cavity**, filled with soft materials, including small blood vessels and nerves. At the base, one or more **roots**, processes by which the tooth may be firmly implanted in the jaw, may be present. In mammals, generally the basal connection between pulp cavity and exterior is constricted to a narrow **root canal** through which pass blood vessels and nerves.

The major tooth materials are enamel (or an enamel-like substance) and dentine. **Enamel**, as found in mammalian teeth, is exceptionally hard, appears shiny, and generally forms a thin layer over the exposed surface of the tooth. It is almost entirely devoid of organic matter. Two thirds of its substance consists, as seen in mammals, of long prisms of calcium phosphate, arranged with their axes at right angles to the surface; most of the remainder is a filling of similar material between the prisms. **Dentine** forms the bulk of the tooth. It is almost identical with bone in chemical composition; about two thirds of its substance is a deposit of a form of calcium phosphate, laid down in a fibrous matrix. Structurally, however, mammalian dentine differs notably from bone. Typical dentine lacks cell bodies within its substance; the cells, **odontoblasts**, are, instead, arranged in a layer in the pulp cavity at the base of the dentine. From them, a series of long slender parallel tubules pass outward through the dentine. In mammals, the teeth are implanted in deep sockets and attached firmly to the bony jaw by a spongy, bonelike material, **cement**.

There are numerous variations from this typical mammalian tooth structure. In many mammals, the enamel, instead of forming a continuous surface layer, is penetrated by tubules reaching the tooth surface, and, in reptiles, the prismatic structure is absent. Further, although the teeth of lower vertebrates are almost always covered by a hard shiny material, this material has been thought by many investigators not to

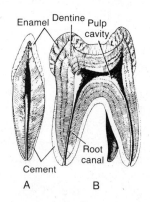

Enamel Dentine Pulp cavity

Cement

Root canal

A   B

Figure 233.   Sections through a mammalian incisor, *A,* and molar, *B.* (After Weber.)

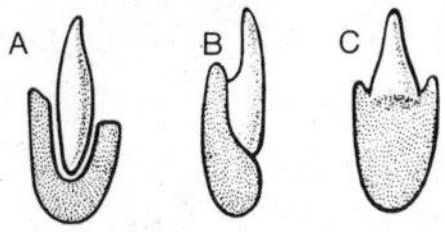

Figure 234. Diagrammatic sections through reptilian lower jaws to show the distinction between thecodont, *A*, pleurodont, *B*, and acrodont, *C*, tooth attachments.

be truly homologous with enamel. True enamel (as described later in this section) is formed by ectodermal cells. But in fishes and in some amphibians, the outer shiny layer has been claimed to be formed by the mesoderm, as is the dentine, and as a consequence is described as merely a hard outer layer of that material—**vitrodentine**. The situation, however, is uncertain. In some forms, the dentine is nourished directly by capillaries rather than by tubules from the odontoblasts in the pulp cavity, and odontoblast cells may be present in the substance of the dentine. These and other histologic variants have names such as vasodentine, osteodentine, and so on. In some lower vertebrates, the teeth may be, as in mammals, implanted in deep sockets (although almost never with the multiple roots found in the cheek teeth of mammals). This is the **thecodont** condition (Fig. 234). Usually, however, the socket is a shallow one; there is generally a cement-like material binding tooth and jaw bone. In some forms (*Sphenodon* and most teleosts), the tooth may be fused to the outer surface of the bone or the summit of the jaws—the **acrodont** type. A further variation is the **pleurodont** condition, common in lizards, in which the tooth is attached by one side to the inner surface of the jaw elements. In sharks, in the absence of bone, the teeth are attached to the cartilages of the jaws by fibrous ligaments, and ligamentous attachment is present in some teleosts and lizards and in snakes.

Tooth Position. At the top and bottom of the series of tooth-bearing vertebrates— in mammals on the one hand and sharks on the other—we find the functioning teeth limited to a single marginal series along each upper and lower jaw. Such marginal teeth are to be found in at least some representatives of every class of gnathostomes and are commonly the most important element in the vertebrate dental equipment.

It must be noted, however, that in many groups, teeth are by no means confined to the margins of the jaws. Wherever ectoderm (or even endoderm, as will be noted below) is present, teeth or toothlike dermal structures may be found.* The mouth is lined with epidermis derived from the skin ectoderm, and, in a wide variety of bony fishes, amphibians, and reptiles, teeth are found in the oral cavity internal to the marginal series. Teeth are present on the palatal elements of the oral roof in many members of these groups, on the parasphenoid bone underlying the braincase, and on the inner surface of the lower jaw. In many fishes and amphibians, certain of the teeth on the palate are large and may form a row paralleling the marginal teeth (cf. Figs. 167 *B*, 168 *D*). We have noted (cf. p. 241) that the shark jaws are not identical

---

*Teeth (or, in a sense, large, toothlike, placoid scales—such scales and teeth are virtually identical) are found outside the mouth in the sawfish (*Pristis*) and saw shark (*Pristiophorus*), elasmobranchs with a long, slender, forward-projecting rostrum that bears a long row of teeth on each margin. On the other hand, very small scales or teeth may be found over much of the inner surface of the oral cavity in elasmobranchs.

with the marginal bones of bony fishes and tetrapods, but instead appear to correspond to palatal structures. This suggests that the marginal teeth of sharks are not really comparable to the marginal row of higher vertebrates but are homologous with the palatal tooth rows of other groups. In teleosts, the maxilla is toothless, and, in some, the entire marginal dentition disappears. In compensation, there is a high degree of development of teeth in the inner parts of the mouth cavity and also of **pharyngeal teeth** situated on various bony elements of the visceral arch system.

332

Chapter 11
Mouth and
Pharynx;
Teeth,
Respiratory
Organs

Tooth Development and Replacement.   The first embryologic indication of the appearance of teeth is an infolding of the ectoderm. In marginal tooth rows (on which most developmental studies have been made—in fact almost all our information is based on pigs), this is a continuous furrow the length of the jaw, a **dental lamina**, from which individual **tooth germs** develop (Fig. 235). In later stages of development, the dental lamina is reduced and does not reach the surface of the mouth. Usually, however, a strand of tissue persists deep within the jaw connecting successive tooth germs. As will be seen presently, this longitudinal connection of tooth germs is important for tooth replacement. If, as is the case in most lower vertebrates, there is to be a continuous succession of teeth during life, the germinal material is not used directly in tooth formation. Instead, the tooth germs remain protected, well below the jaw surface, while successive teeth take their origin from bits of tissue that bud off from them and gradually move toward the surface. These ectodermal tooth buds typically form as cones or cups, hollow below, which outline the crown form of the future tooth. On the lower surfaces lies a special series of cells, **ameloblasts**, which in higher vertebrates, at least, secrete the enamel; from this fact, the tooth bud

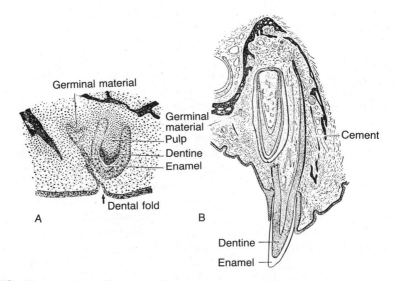

Figure 235.   Two sections of a crocodilian upper jaw to show tooth development in lower tetrapods. *A,* Embryo with a first tooth developing and germinal material in reserve at the base of the dental fold. *B,* Mature animal; a tooth is functioning but is being resorbed at the inner side of the root, where the next successional tooth is already in process of formation. Germinal material for successional teeth persists. (After Röse.)

is frequently termed the **enamel organ**. This is a rather unfortunate term, because enamel formation is at best a minor function of the ectodermal tooth bud. Within the hollow of the bud, a mass of mesenchymal cells, **odontoblasts**, which are derived from the neural crest, gather. These form an embryonic tooth pulp, a **dental papilla**, with which small blood vessels and nerves become associated. Stimulated, apparently, by the presence of the enamel organ, the odontoblasts proceed with the task of dentine deposition. They become, at the end of growth, cells of the pulp cavity. As the tooth takes shape, it works toward the surface and erupts into functional position. In forms with continuous tooth succession, each tooth eventually becomes subject to resorption at its base and is shed from the jaw. Meanwhile, one or more successional buds have already formed from the germinal material and push outward as replacements.

The prominence of the ectoderm and neural crest in early tooth development is in agreement with our general ideas of the phylogenetic origin of teeth. It has long been recognized that teeth and the dermal denticles of the shark skin (cf. Fig. 112) are essentially similar in nature and hence presumably homologous. Mouth lining and body skin are comparable in origin, both being formed from ectoderm and connective tissues gathered beneath it. They hence may be expected to develop homologous structures.

It was once believed that teeth originated directly from discrete dermal denticles that lay in the margins of the mouth overlying the forming jaws in ancient fish. Our current conceptions of the nature of primitive fish as having a bony external skeleton demand a modification of this idea. We have noted that the surfaces of the plates shielding the body of primitive fishes (Fig. 111 A) are frequently ornamented with tubercles resembling both shark denticles and primitive teeth. Shark denticles, we have said, appear to represent the last surviving parts of these plates in the skin of modern sharks; the deeper layers have, so to speak, melted away in the course of evolutionary history. When jaws evolved in ancient fishes, dermal plates were present along the jaw margins. Denticles present on these jaw plates became the teeth. Teeth and dermal denticles are thus, in modern concepts, still regarded as homologous. Teeth are not, however, derived from isolated denticles; instead, both are modified derivatives of tubercles found in the dermal bony plates of armored ancestral vertebrates.

Odontoblasts and bone cells have similar properties and form similar hard skeletal tissues, with the difference in typical cases that the bone cells are contained within the tissues they form, whereas the odontoblasts are situated peripherally. But we find that in the case of dermal tubercles of ancient and primitive fishes, the cells forming the dentine are frequently seen to be embedded in its substance (and bone in some fishes does not have included cells). Phylogenetically, it would seem that dentine arose as a specialized type of bone tissue; odontoblasts are a specialized type of bone cell which, like many osteocytes, are derivatives of the embryonic neural crest.

In mammals (as we are well aware), there is no tooth replacement apart from the succession of "permanent" teeth for a "milk dentition" in the front of the mouth. Far different is the situation in most other vertebrates. Little is known of the replacement of teeth found on the palate and inside of the jaw in many bony fishes, amphibians, and reptiles, and certain of the large tooth-plates present in various fishes and reptiles appear to undergo little if any replacement. However, in the major, marginal dentition of most fishes, amphibians, and reptiles, tooth replacement continues through life. Teeth are constantly being formed deep within the tissues from tooth germs; they grow in size, erupt, function, and presently, through resorption at their bases or loosening of their connections with the jaw elements, are shed and replaced by a new generation of teeth.

In many fishes, amphibians, and reptiles, the tooth row has a seemingly irregular appearance, with old teeth, mature teeth, and young teeth, newly erupted, scattered along the jaw in seemingly random fashion (Fig. 236). Actually, there is method in this seeming madness. Replacement is taking place in an interesting way that guarantees a continuous function of the dentition despite frequent renewal of individual teeth. The teeth and tooth germs appear to be arranged in two series, the "odds" and "evens" in each tooth row. One may find a condition in which, for example, in a given area of the jaw, the odd-numbered teeth are functional; between them, in place of the even-numbered teeth, are sockets beneath which new teeth are in the process of formation. Later, this region may show a condition in which both sets of teeth are in place at the same time, but the odd set shows evidence of age and wear. Still a bit later, the odd-numbered teeth will drop out, leaving the even series the functional set, and so on. This neat device guarantees that at least some teeth in any region will be functioning at any given time.

334
**Chapter 11
Mouth and
Pharynx;
Teeth,
Respiratory
Organs**

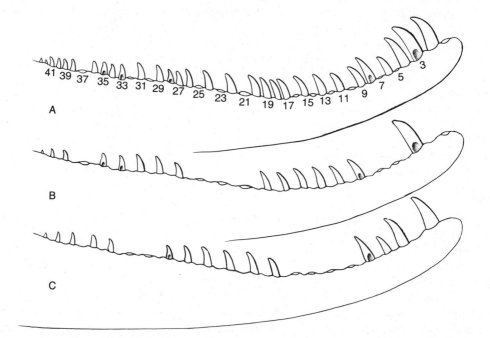

Figure 236.   Inner surface of the lower jaw of a primitive fossil reptile *(Ophiacodon)* to show the alternation in tooth replacement between odd and even tooth series. The gaps in the tooth row seem haphazard at first sight; however, if the two series are considered separately, as in *B* and *C,* it will be seen that each includes several waves of replacement arranged in a regular alternating manner. Certain teeth—3, 8, 28, 33, 35—show resorption at the root and are at the point of being lost.

This replacement appears to take place in waves that travel along the rami of the jaws; at a given time the "even" set may be functioning in some areas of the ramus, the "odd" set in others. Still further complicating and puzzling is the fact that in series of "odd" and "even" teeth over a segment of the jaw, the oldest teeth of a series are frequently at the back, the youngest at the front. This suggests that tooth formation travels from back to front. But one is reluctant to believe that this is true; in general, development of structures of any sort in a vertebrate proceeds from front to back, and this is certainly true of the teeth of mammals. How can we solve this puzzle? What is the real principle governing this rhythmic replacement? One hypothesis that gives a reasonable solution—that tooth replacement results from waves of stimulation that proceed (properly) from the front of the tooth row to the back and that the alternating appearance of "odds" and "evens" is simply due to the spacing between successive waves.

Such a situation is diagrammed in Figure 237. In *a* and *b*, the horizontal lines represent the persistent strand of tissue connecting successive tooth germs in many vertebrates, the vertical lines represent the distance from the level of the tooth germs up to the surface of the jaw. In *a1*, it is assumed that an impulse for tooth formation has begun at the most anterior germ position, and a tooth rudiment (or Anlage, represented diagrammatically by a black dot) has formed. In *a2*, the impulse has travelled to the next position and started a second rudiment; meanwhile, the first embryonic tooth in position one (represented by a small circle) has grown somewhat and moved up a bit toward the mouth surface. The stimulus continues to travel backward, initiating tooth development at successive positions until, in the diagram at *a8*, this first stimulus has produced a row of teeth of which the first (represented by a large circle) has reached maturity. But meanwhile, a second stimulus, spaced somewhat more than two tooth germs behind the first, has followed, and a third has just made its appearance.

If this sequence were to be continued, the mature jaw would obviously have the structure shown in *b,* and, as seen in side view, in *c.* Here there is, at first sight, a seemingly irregular arrangement of old, mature, and young teeth and tooth germs of various sizes; but, as we have just seen, the underlying principle is a simple and

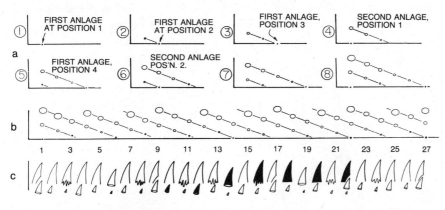

Figure 237. Diagrams (discussed in the text) to show the mode of tooth replacement in typical lower vertebrates. (After Edmund.)

336

Chapter 11
Mouth and
Pharynx;
Teeth,
Respiratory
Organs

orderly one. In the example just cited, the *apparent* sequence in "odd" and "even" series (as in one shown in black in *c*) is back to front—the reverse of the true situation. This appearance is due to the (seemingly inconsequential) fact that the spacing between impulses is assumed to be somewhat greater than two tooth germ intervals. If the spacing were exactly two intervals, there would be an exact alternation of odd and even series the length of the jaw. And if (heaven forbid!) the reader wishes to make the necessary diagrams, he will see that making the interval between impulses somewhat less than two units will result in seeming front to back waves in both series. This is a relatively simple system that many investigators believe to reflect the true situation; however, others have devised different mathematical models that account for the observed patterns in other ways.

Teeth In Fishes.   No teeth are present in lower chordates or in jawless vertebrates. The early fossil ostracoderms presumably gained their food by straining particles from the water current passing through their gills or by nibbling soft foods; teeth were unnecessary and could not function in the absence of jaws. The larval lamprey is likewise a food strainer. The adult lampreys prey on fishes and have, in the rasping "tongue," a jaw substitute. This is armed (as is the mouth cavity) with conical, pointed, toothlike structures. These, however, are not true teeth but consist of cornified epithelium. In the extinct placoderms, teeth as well as jaws made their appearance. In the arthrodires, teeth are sometimes found fused to the jaws, but in general, these ancient fishes relied for biting purposes on sharp cutting edges on the jawbones themselves.

The most generalized type of tooth is that of a simple cone. In sharks, the teeth are sometimes of this type, although usually with accessory cusps fore and aft. Frequently, however, the teeth are lengthened from front to back at the base, giving a triangular shape; anterior and posterior margins may be sharp and serrate. Flattened teeth have developed for crushing shellfish in such forms as the Port Jackson shark (Fig. 238), and many skates and rays develop a crushing battery of such teeth. In a shark (Fig. 239 *A*), the teeth develop in the soft tissues on the inner surface of the jaws, and in a dried specimen, a whole series of successional teeth can usually be seen (Fig. 238); each tooth (usually with the familiar odd-and-even alternation) gradually swings up into position on the margins of the jaw and is eventually shed. In many rays, advantage is taken of the early development of successional teeth to in-

Figure 238.   The lower jaws of a primitive living shark, *Heterodontus (Cestracion),* the Port Jackson shark of the Pacific. The teeth differ greatly between front and back parts of the jaw. Rows of successional teeth are formed down within the inner surface of the jaws. (From Dean.)

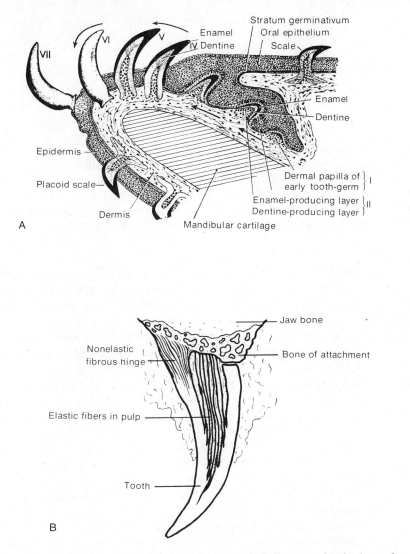

Figure 239. Attachment of teeth in fish. *A,* Section through the jaw of a shark, to show stages (I–VII) in the development of a tooth to its functional stage and to the point where it is about to be shed (VII). For comparison, placoid scales (denticles) in the adjacent epithelium are also shown. (From Rand, The Chordates, The Blakiston Company.) *B,* Section through a tooth of a type found in many teleosts, in which the tooth may be bent back into the mouth on a fibrous hinge. (From Scott and Symons, Introduction to Dental Anatomy, Williams & Wilkins.)

corporate them in an expanded crushing battery for mollusc-feeding. In chimaeras, likewise basically eaters of shellfish, the dentiton is, in contrast, reduced to a pair of large crushing plates in both upper and lower jaws and an accessory pair of small upper plates (Fig. 240 *A*). An analogous development is present in lungfishes (Fig. 240 *B*); the marginal teeth are lost, and the dentition typically consists of four fan-shaped compound dental plates (plus an accessory upper pair). In the living *Neocer-*

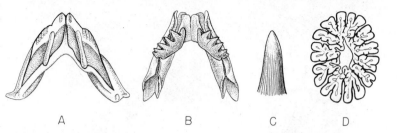

338

Chapter 11
Mouth and
Pharynx;
Teeth,
Respiratory
Organs

Figure 240. *A,* Dorsal view of the lower jaws of a chimaera, to show the pair of large tooth plates, of complex shape, which cover most of their surface. *B,* Similar view of the bony elements of the jaws of the lungfish *Neoceratodus* to show the pair of fan-shaped toothplates. *C,* External view of a grooved labyrinthine tooth characteristic of crossopterygians and ancestral tetrapods. *D,* Section through such a tooth to show the complicated folding of the enamel layer (heavy black line). (*A* after Dean; *B* after Watson; *C* and *D* after Bystrow.)

atodus of Australia, the tooth plates exhibit a series of smooth radiating ridges; the embryonic and phylogenetic stories agree in showing that each ridge is composed of a fused row of originally discrete conical teeth. Among actinopterygians, the conical tooth shape is primary. In teleosts, the cement-like material joining the tooth and jaw bone may form a massive base for the tooth as essentially a separate bone of attachment (Fig. 239 *B*). In many instances, large "fangs" are attached to the jaw by an elastic hinge, bending inward to allow the prey to enter the mouth but opposing its escape. In higher actinopterygians, the maxilla becomes toothless, and, in some teleosts, the marginal dentition disappears entirely; reliance is placed on palatal and pharyngeal teeth. Such teeth may become flattened and numerous, forming an efficient grinding mill; in one type of catfish, there are said to be 9000 such teeth. Many ancient crossopterygians were notable for the presence of longitudinal grooves on the teeth; these represent infoldings of the enamel surface layer, often of complicated pattern, giving the tooth in cross section a labyrinthine appearance (Fig. 240 *C, D*).

It was noted above that there is doubt as to whether the tooth surface in many lower vertebrates consists of true enamel (formed by skin ectoderm) or vitrodentine (of neural crest origin). The situation in sharks and skates is still a matter of controversy. In many bony fishes, both true enamel and vitrodentine may be present on the surface of a single tooth.

Teeth In Tetrapods. The crossopterygian type of tooth is repeated in many of the oldest amphibians; it is so prominent a feature that it gives the name Labyrinthodontia to the ancestral amphibian group. The marginal teeth were generally highly developed, and in addition there were powerful palatal fangs. In modern amphibians that live on small and soft materials, the teeth are small and simple in structure, frequently confined to the upper jaw in anurans, and entirely absent in some toads. A curious feature of the teeth in almost all modern forms, and at least a few labyrinthodonts, is the presence of a zone of weak calcification near the tooth bases, so that frequently in prepared skeletal specimens only the stumps of the teeth are preserved.

In a majority of reptiles, the teeth are essentially of the simple conical type, although somewhat flattened teeth are found in some lizards (Fig. 241 *A*) and crocodilians, and flat tooth-plates are present in some fossil forms (Fig. 241 *B*). Palatal teeth

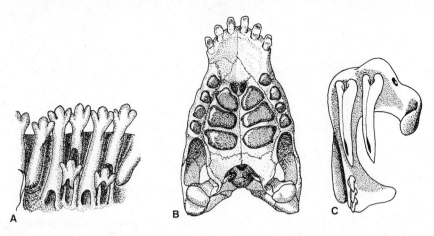

Figure 241.   Teeth of various reptiles. *A,* The lizard *Amblyrhynchus;* in many herbivorous lizards, the teeth are not simple pegs but have multiple cusps. *B,* The extinct placodont *Placodus;* the anterior teeth formed a clam rake and the more posterior ones, including palatal teeth, were flat plates for crushing hard molluscs. *C,* The maxillary fangs of a king cobra, *Ophiophagus;* the teeth are hollow hypodermic needles serving to inject the venom into the prey. (After Edmund.)

were present in ancestral reptiles and are preserved in *Sphenodon,* snakes, and lizards; such teeth are absent in crocodilians. Turtles have lost their teeth, relying instead on a horny bill. Ancestral reptiles generally had teeth inserted in shallow sockets, and typical thecodont teeth are seen in crocodiles. Most lizards, however, have a pleurodont type of tooth attachment, and some lizards (including chameleons) are acrodont, as is *Sphenodon.* Snake teeth have ligamentous attachments; in most poisonous snakes, some teeth become highly developed fangs—hypodermic needles to inject the venom (Fig. 241 *C*). The primitive bird *Archaeopteryx* was toothed, but the bill (plus gizzard stones) has replaced the dentition functionally in all post-Mesozoic birds.

In the fossil mammal-like reptiles are seen stages in the initiation of the mammalian type of dentition. The most ancient members of this series, the pelycosaurs, show the development of large upper stabbing teeth comparable to mammalian canines, separating incisors anteriorly from a series of cheek teeth (Fig. 178 *A*); various therapsids (cf. Fig. 178 *B, C*) show further stages in the evolution of a mammal-like dentition.

Tooth Differentiation In Mammals.   In mammals, teeth are reduced to a short marginal series in each quadrant of the jaw. In the more generalized members of the class, four tooth types can be distinguished in the adult, in anteroposterior order (Fig. 242 *A*). Most anteriorly are the **incisors**, nipping teeth with a simple conical or chisel-like build. Next in each series is a single **canine**, primitively a long, stout, deep-rooted tooth with a conical shape and sharp point, useful in attacking a carnivore's prey. Following the canine is a series of **premolars** (the "bicuspids" of the dentist), which frequently show a degree of grinding surface on the crown; and finally, there is a **molar** series, the members of which generally assume a chewing function and develop a complex crown pattern.

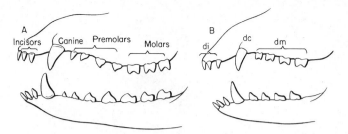

Figure 242. Left side view of the dentition of a generalized placental mammal, showing permanent dentition in *A, B,* deciduous teeth. Abbreviations: *dc,* deciduous canine; *di,* deciduous incisors; *dm,* "milk molars" (= deciduous premolars).

340

Chapter 11
Mouth and
Pharynx;
Teeth,
Respiratory
Organs

In early fossil mammals, the number of teeth in these various categories (except canines) appears to have been rather variable and the total number higher than in living forms. The primitive placentals, however, had settled down to a count in which three incisors, one canine, four premolars, and three molars were present in each quadrant. Few animals today retain precisely this number. But although there is often a reduction of teeth in one or several categories, there is seldom any increase over the primitive number (some whales and porpoises are conspicuous exceptions). Marsupials may have more incisors and can have four premolars but only three molars.

A simple "shorthand" for the nomenclature of teeth in placental mammals and a formula to express the number of teeth present in any mammal have been devised. The letters *I, C, P,* and *M* followed by a number in an upper or lower position will define any tooth in terms of the original placental formula: I^1, for example, refers to the most anterior upper incisor; M_3, the last lower molar. The number of teeth of each type present, above and below, in the dental equipment of any mammal can be formulated in succinct fashion. The **dental formula** $\dfrac{3.1.4.3}{3.1.4.3}$ indicates that in either side of both upper and lower jaws the primitive placental number of teeth is present (the four successive figures represent the number of incisors, canines, premolars, and molars present). The total number of teeth present in the mouth of such an animal is 44. The human formula is $\dfrac{2.1.2.3}{2.1.2.3}$; i.e., we have lost an incisor and two premolars of the primitive placental series from each quadrant and reduced the number of our teeth to a total count of 32.

Incisor teeth, useful in almost any mode of life, are retained in most mammalian groups. Various herbivores, however, have evolved specialized cropping devices in which the incisors are modified or lost. Ruminants, such as cows, sheep, and deer, have, for example, lost the upper incisors but retained the lower ones (Fig. 243 *B*); the elephant's tusks are greatly elongated upper incisors. Rodents have developed pairs of upper and lower incisors as gnawing chisels that grow persistently at their roots throughout life as their tips are worn away. Lagomorphs, some primitive primates, various marsupials, and other mammals parallel rodents to varying degrees in the development of a gnawing mechanism.

Canines are primitively prominent biting and piercing weapons and as such are emphasized in carnivores; they reached their peak in the now-extinct sabertooths (Fig. 243 *A*). In nonpredaceous mammals, the canines are sometimes prominent and

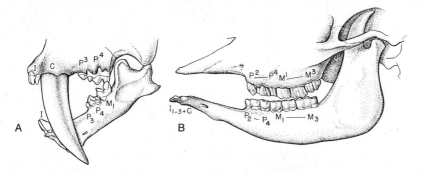

Figure 243. Two specialized mammalian dental types. *A,* Felid; *B,* the cow. For the felid, the extinct sabertooth, with extreme specialization, is shown. In the felids, the dentition is much reduced; there are modest incisors and lower canine (concealed in the figure) and stabbing upper canine. A small M^1 is present, but concealed by P^4 in this figure. In the cheek, little remains except the carnassials $\dfrac{(P^4)}{(M_1)}$. In the ruminants, the cheek teeth are expanded into a grinding battery; they are separated by a diastema from the cropping teeth—here consisting only of lower incisors and canine, working against a horny pad on the upper jaw.

are used in defense or, in males, in mating battles. More often, however, they are reduced (as in man); they are incisiform or absent in most herbivores.

The cheek teeth—the premolars and molars—whose patterns are discussed below, have a varied history. In herbivorous forms, such as the ungulate groups, these teeth are generally retained (except for frequent loss of the first premolars) to give such dental formulae as $\dfrac{3.1.3.3}{3.1.3.3}$ for the horses (at least for stallions; mares usually lack the canines), and $\dfrac{0.0.3.3}{3.1.3.3}$ for bovids. The cheek teeth here are usually elaborated to form an efficient grinding battery, which (as in a horse or cow) becomes separated, spatially as well as functionally, from the cropping teeth by a gap in the tooth row— a **diastema** (Fig. 243 *B*). In carnivores in which there is little or no chewing of food, the cheek teeth are generally reduced in number—strongly so in the sabertooth, where the dental formula is $\dfrac{3.1.2.1}{3.1.2.1}$. However, they are, as noted below, important in shearing food.

Mammalian Molar Patterns. In mammals, the cheek teeth, particularly the molars, generally develop an expanded crown with a varied and complicated pattern of projecting cusps—a structural arrangement that appears to be of major functional value in chewing and grinding food. The variations in this pattern are of diagnostic importance and are of great interest to systematists, paleontologists, and anthropologists, for a single molar tooth will often furnish precise generic or even specific identification of the animal to which it pertains. Such patterns have been much studied, but only the general features will be noted here (Figs. 244–246).

The nomenclature used is basically simple. Each cusp is termed a **cone**. The names of specific cones are formed by adding the prefixes **proto-**, **para-**, **meta-**,

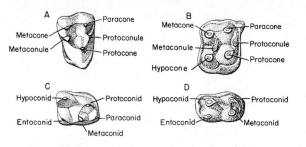

Figure 244. Diagrams of molar tooth patterns of placental mammals. In all cases, outer edge of tooth above, front edge to right. *A,* Right upper molar of a primitive form. *B,* The same of a type in which the tooth has been "squared up" by addition of a hypocone at the back inner corner. *C,* Left lower molar of a primitive form with five cusps. *D,* The same of a type in which the tooth has been "squared up" by the loss of the paraconid.

342

**Chapter 11
Mouth and
Pharynx;
Teeth,
Respiratory
Organs**

hypo-, and **ento-**; the suffix **-ule** indicates a minor cusp. The term **cingulum** is applied to accessory ridges of enamel around the margins of teeth, the term **style** to vertical marginal ridges, the term **loph** to ridges between cusps, and various other terms to minor features. The suffix **-id** indicates a structure on a lower tooth: conid, conulid, stylid, and so on.

Neglecting some variations in pattern among early fossil mammals and among marsupials, we find that the earliest and most primitive of placentals have upper molar crowns in the general shape of a right-angled triangle, with two cusps along the outer edge of the tooth row and the right angle at the antero-external margin. There are three major cusps, the **protocone** at the apex, the **paracone** and **metacone** along the external border. Between the protocone and the external cusps a pair of conules (protoconule, metaconule) is usually found. A fourth cusp, the hypocone, may be present posteromedially (see below).

The primitive lower molar has as its major portion a high, well-developed triangular area, the **trigonid**, which has a pattern similar to that of the upper molar, but reversed; there is an external apex, the **protoconid**, and an internal base, where **paraconid** and **metaconid** are present. If upper and lower teeth are fitted properly together, that is, placed in **occlusion** (Fig. 245 *B*), each lower tooth (which lies somewhat internal and to the front of its upper mate) would bite past its neighbors above without meeting them effectively if it consisted only of a trigonid. This situation is prevented through the fact that the crown of each lower molar has, posteriorly, a second portion, a **talonid**, primitively lower and less developed than the anterior part of the tooth, and bearing two additional cusps, **hypoconid** (externally) and **entoconid** (internally). The protocone of the upper tooth bites into the basin of the talonid, and proper occlusion is effected. We thus have a generalized primitive pattern of cheek teeth with three main upper cusps and five lower ones. The full pattern is present in the molars. Passing forward in the premolar series, the pattern tends to become more simplified and approaches the essential conical shape of the canine and incisors.

The origin of this pattern from the simple conical form seen in most reptiles remains a topic of considerable debate. Some early investigators suggested that mammalian molars arose by the fusion of several simple teeth, but all now agree that this is not the case; the "extra" cusps arose by differentiation within the tooth. The classic

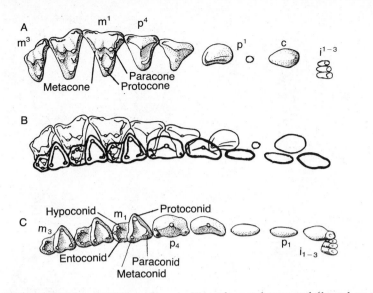

Figure 245. Diagrams of the dentition of a primitive placental mammal (based on the Eocene insectivore *Didelphodus*). *A*, Crown view of right upper teeth; *C*, crown view of left lower teeth; between, in *B*, the teeth are placed in occlusion, the outlines of the lower teeth (heavy lines) superposed on those of the uppers. (After Gregory.)

theory, proposed by E. D. Cope and elaborated by H. F. Osborn, and thus called the Cope-Osborn theory, is outlined in the lower molars shown in Figure 246. Small cusps arose anteriorly and posteriorly on the sides of the original reptilian cusp or protoconid (the names used here are based on the Cope-Osborn theory. Next the accessory cusps, the paraconid and metaconid, became large and rotated medially* to produce a triangular tooth, comparable to the trigonid of later forms. Finally the talonid was added posteriorly and elaborated. These stages can all be shown in various Mesozoic mammals and correspond, more or less, with the crown patterns seen in those animals known as triconodonts, symmetrodonts, and pantotheres. The Cope-Osborn theory postulates a similar sequence for the upper teeth except that the accessory cusps lie laterally rather than medially and a posterior talon, if present, is usually smaller than the talonid.

As more fossils were found and studied, many problems arose, and other theories or modifications of the original theory were proposed. One unhappy consequence is that, along with new hypotheses, new terminologies for all the cusps have been suggested. We will ignore almost all of these and hold to the relatively familiar names, even though, according to some views, they are far from appropriate. One other theory, often called the amphicone theory, applies to the upper molars shown in Figure 246. Here, the first stage after the simple reptilian cone is characterized by the

---

*We here use the familiar adjectives anterior, posterior, lateral, and medial, although these are ambiguous because the teeth follow a curved line around the margin of the mouth; thus, what is anterior on the most posterior teeth is medial on the most anterior. Better names are, respectively, mesial for toward the middle of the whole dental arcade; distal for the opposite, toward the back of the tooth row; buccal for toward the cheeks or outside the mouth; and lingual for the opposite, toward the tongue or inside the mouth.

344

**Chapter 11
Mouth and
Pharynx;
Teeth,
Respiratory
Organs**

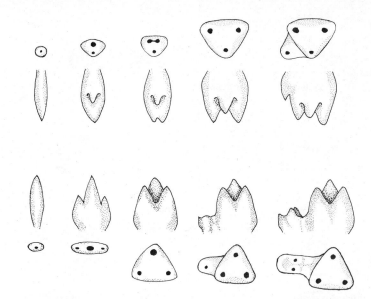

Figure 246. Diagrams to show the origin of mammalian molar structure. The top and bottom rows show occlusal views with the major cusps as black dots; in all, lateral (buccal) is to the top and anterior (mesial) is to the right. The two middle rows show medial views of the teeth, again with anterior to the right. The evolutionary sequences move from left to right. In the lower teeth, starting with the single reptilian cusp (protoconid), anterior and posterior cusps (paraconid and metaconid) are added, the tooth becomes triangular, and finally a heel (talonid) with two additional cusps (hypoconid and entoconid) is added posteriorly; this sequence is that suggested by Cope and Osborn. The upper teeth here show another theory, the amphicone theory. Here the original cusp (the eocone) divides to form the paracone and metacone; the protocone is a new development and the heel is small or absent.

addition of a small accessory cusp medial to the original cusp or **eocone**. Next, the eocone begins to divide into an anterior and a posterior cusp; the half-divided cusp is the **amphicone**, from which this theory takes its name. Continuation of this splitting and a great increase in the size of the accessory cone (the protocone of the Cope-Osborn system) produces a trigon. The amphicone theory is based on the upper teeth of various extinct forms, including some of the advanced cynodonts.

Both of these theories are oversimplified as stated here—we have ignored important conules and conulids, styles and cingula, which may have been involved in the origin of cusps, and a host of other complications. Also, as noted, these are not the only theories possible. Although the sequence is not yet completely understood, what seems to be the most probable history of the molars is a rather unexpected and, to the student, a highly unfortunate one. Lower molars appear to have evolved much as suggested by Cope and Osborn and as shown in Figure 246. The uppers probably took a different course, that postulated in the amphicone theory and also as diagrammed in Figure 246. Thus, the Cope-Osborn theory is both right and wrong; consequently, there is no really good terminology for all the cusps unless a different system for uppers and lowers is used.

In primitive placental mammals, the cusps were sharp-pointed structures, shearing along the edges of the teeth and crushing with the protocones in the talonid. Such teeth are well adapted to deal with insects and similar food materials, and some living insectivores still have teeth of this sort. Most modern mammals, however, have departed widely from this primitive pattern and have evolved a variety of tooth types that can be discussed only briefly here. Carnivores, which tend to reduce the number of cheek teeth, generally preserve a simple and primitive pattern in those that remain. Notable, however, is the development in the carnivore cheek of a pair of sharp-crested, shearing teeth that deal with tendons and bones—the **carnassial teeth**, the last premolar in either upper jaw shearing against the first lower molar (Fig. 243 *A*).

The premolars generally remain simpler than the molars in cusp pattern. In certain herbivores, however, they are "molarized," that is, assume a molar-like cusp pattern; this is exemplified in the horse. Generally, there has been a squaring up of the molars in herbivores by alterations in the cusps present (Fig. 244 *B, D*). In the originally triangular upper molars, there is often added a fourth cusp, a **hypocone**, at the back inner corner, to turn the triangular tooth into a rectangular one. Below, a four-square condition may be attained in several ways, such as losing the paraconid at the anterior edge of the trigonid.

In mammals with a mixed diet (such as men and swine), the cusps tend to become low and rounded, forming small hillocks on the crown, the **bunodont** type. In many ungulates that have taken up a herbivorous mode of life, the cusp pattern has become complicated; individual cusps may, for example, assume a crescentic outline, a **selenodont** pattern; or cusps may connect to form ridges, the **lophodont** condition (Fig. 247).

Grazing habits present a serious problem to an ungulate, because grass is a hard, gritty material that rapidly wears down the surface of a grinding tooth. If the low-crowned tooth of a primitive mammal were retained in a large grazing mammal, it would be worn to the roots in a short time. In relation to this, such hoofed mammals as the horses and cattle have developed a high-crowned and prism-shaped tooth, the **hypsodont** condition (Fig. 248), contrasting with the more primitive **brachyodont** type. One could, in imagination, develop a high-crowned tooth by elongation of the dentine-filled bulk of the tooth body, with the cusps in their original shape on the grinding surface. This type of high-crowned tooth did form in some early fossil mam-

Figure 247. Crown views of molar teeth of *A,* rhinoceros; *B,* horse; *C,* ox; to show types of molar patterns. The white areas are worn surfaces showing dentine; the black line surrounding the white is the worn edge of the enamel; stippled areas, cement covering or unworn surface. *A* is a simple lophodont type, with an external ridge (above) and two cross ridges; the development of this pattern by connecting up the cusps seen in Figure 244 *B* can be readily followed. The horse *(B)* shows a development of the same pattern into a more complex form in which the primitive lophodont pattern is obscured. *C* shows a selenodont pattern characteristic of the ruminants; each of four main cusps takes on a crescentic shape.

346

**Chapter 11
Mouth and
Pharynx;
Teeth,
Respiratory
Organs**

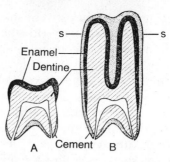

Figure 248. Diagrams to show the development of a hypsodont tooth. *A,* Normal low-crowned tooth (cf. Fig. 233). *B,* Hypsodont tooth, with cusps elevated and the whole covered by cement. As a tooth of this sort wears down to any level, such as that indicated at *s-s,* it will be seen that no less than nine successive layers of contrasting materials are present across the crown surface.

mals, but proved unsuccessful. Once the enamel surface is worn away, most of the wear comes on the relatively soft and easily abraded dentine; the resistant enamel remains only as a thin rim. Successful hypsodont forms have teeth built on quite another plan. The height is attained by a skyscraper-like growth of each cusp or ridge on the tooth; these slender peaks are fused together by a growth of cement over the entire tooth surface before eruption. As wear takes place, it grinds down through a resistant complex of layers of all the dental materials: enamel, dentine, and cement. Another modification to compensate for excessive wear is for the teeth to remain open-rooted and continually growing throughout the life of the animal; the incisors of rodents are a prime example of this device.

In eutherian mammals, tooth renewal (to the regret of all except dentists) is a much reduced process; there is only one replacing set, and that incomplete, for the most posterior—molar—teeth have no successors. We usually think of the mammalian dentition as comprised of a set of "milk" teeth—a **deciduous** or lacteal dentition—that has no molars included and a complete permanent set (cf. Fig. 242). If, however, we consider the development of the teeth we gain a different concept (Fig. 249). The "milk" teeth, in a typical mammal (man is representative), usually develop successively in anteroposterior order from incisors to deciduous premolars or "milk molars" (the canines, with long roots, may lag in eruption). After the appearance of the last "milk molar," the successive eruption of the three molar teeth begins. Meanwhile, without waiting for the completion of this first wave, the anterior teeth are replaced in order back to the premolars. But the succession proceeds no further, and the molars are not replaced. Obviously, we have one complete set of teeth and a partial second set; the molars, although permanent, belong in series to the set that, farther forward, is deciduous in nature. That is why your own first molars were in place long before you replaced either premolar—indeed, in man the first molar is the first permanent tooth to appear.

An adult mammal thus has in its mouth parts of two dental series; the molars are part of the first set, the more anterior teeth belong to the second set. This concept explains various peculiarities such as the fact (well known to paleontologists and anthropologists) that the molars in many cases resemble much more closely the short-lived "milk molars" than they do the permanent premolars, which they adjoin

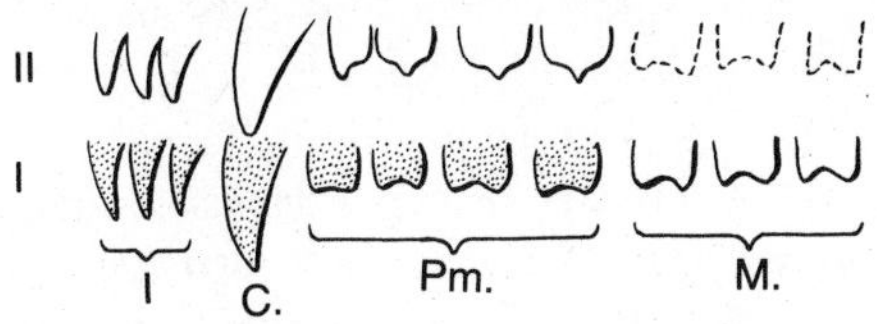

Figure 249. Diagram to show the composition of the "permanent" dentition of a mammal; left upper teeth of a generalized placental mammal. One complete set of teeth (I) develops from incisors back to molars. All these except the molars, however, are shed (as indicated by stipple). A second set of teeth develops (II) but never produces molars. Hence the "permanent" dentition includes portions of two tooth series (cf. Fig. 242).

for most of the span of life. Replacement may be modified (elephants, for example, are odd) or may be greatly reduced as in marsupials, which really have a single dentition in which only one or two teeth are replaced.

Gills

The **pharynx** is a short and unimportant segment of the digestive tract in the higher vertebrate classes, a minor connecting piece between mouth and esophagus where ventrally the **glottis** opens to the lung apparatus and, dorsally, the paired **eustachian tubes** lead to the middle ear cavities (Fig. 231). In mammals, it is merely the place where air and food channels cross one another in an awkward manner and where masses of lymphoid tissues, the **tonsils**, accumulate. But from both phylogenetic and ontogenetic viewpoints, the pharynx is an area of the utmost importance; there, the branchial pouches, basic in the construction of respiratory devices in lower vertebrate classes and persistently important in the developmental story in higher groups, develop; further, the lungs originate there.

The body cells constantly demand oxygen for their metabolic activities and as constantly must be rid of the carbon dioxide, which (with metabolic water) is the major waste product resulting from utilization of oxygen. Internal carriage of these gases to and from the cells is by means of the circulatory system. The oxygen source and the carbon dioxide destination is the surrounding medium: water or, in higher groups, air. Exchange of gases—the process of respiration or, more precisely, ventilation—occurs by diffusion through moist vascularized membranes.

In many small aquatic invertebrates, the exchange takes place through the general body surface, and this seems to be true in great measure even in amphioxus, in which the gill slits do not appear to be essential for respiration, although they may be important, especially when the animal is buried in the sand. Such surface exchange may persist among vertebrates, and cutaneous respiration is found here and there in types as far removed from one another as eels and modern amphibians; in the latter, it is often of major importance. In general, however, vertebrates and most multicellular invertebrates have developed special respiratory organs. Skins are often too thick or hard and too little vascularized to function effectively in respiration, and surface-volume relations generally come into play in animals of any considerable size, even when the body surface is competent as an organ of exchange. In water-dwellers of various phyla, **gills** of one sort or another are developed—structures with a moist, thin surface membrane, richly supplied with blood vessels and folded in a complicated manner to give a maximum amount of surface for gas exchange.

Among invertebrate groups, such gills are usually situated outside the main mass of the body as external gills. In a limited number of cases, analogous developments are seen among vertebrates—the external gills of larval fishes and amphibians (described later), projecting gill filaments of other larval fishes, and hairlike respiratory growths on the slender pelvic fins of the male South American lungfish and on the pelvic limbs of the male "hairy" frog of Africa (*Astylosternus*), for example. But the characteristic respiratory structures of the lower vertebrates—indeed, a most fundamental trademark of every chordate—are **internal gills**, forming a series of slits or pockets leading from the pharyngeal region to the outer surface of the body. As the locus of origin of the gill slits,* the pharynx is a highly developed and highly important part of the digestive tube in the lower classes of vertebrates.

348
Chapter 11
Mouth and
Pharynx;
Teeth,
Respiratory
Organs

The Gill System In Sharks.   The characteristic development of the vertebrate branchial system may be seen in the sharklike fishes, the Chondrichthyes, and the higher bony fishes. Gill construction in the two groups differs in certain important respects, but the basic pattern is the same.

In the sharks (Fig. 250 *C*), the pharynx is a long, broad duct, continuing backward from the mouth without break and terminating posteriorly at a constriction whence the short and narrow esophagus leads on to the stomach and intestine. On either side, a series of openings runs outward to the surface of the body. Anteriorly, a small and specialized paired opening, the spiracle, described later, is generally present. Back of this point, on either side, are found the typical **branchial slits**, five in most sharks, but six or seven in a few special cases. In these slits, the respiratory organs, the gills themselves, are found. Water is typically brought into the pharynx through the mouth and flows outward through the gill slits; in its passage past the gill surfaces, the respiratory exchange takes place. The term **branchial bar**† is applied to the tissues lying between successive openings. The region between mouth and spiracle is the **mandibular bar**; that between the spiracle and the first normal gill slit, the **hyoid bar**; more posterior branchial arches are generally referred to by number. As with the skeletal arches, the numbering can be a problem. Some investigators call the hyoid arch, or the mandibular arch, or even an hypothetical premandibular element number one; we will number only the branchial (posthyoid) structures and use names for the more anterior ones.

Typical shark respiratory movements are a combination of suction and force pump effects, produced by the action of the branchial muscles described previously. With the mouth open and the external gill openings closed, the pharynx is expanded and water is drawn in via the mouth; the pharynx is constricted, and the water is forced out through the slits. Negative pressure in the outer part of the slits can suck water from the pharynx so that flow is quite constant rather than in short spurts past the gills.

---

*Gill is often, as here, used as an adjective; properly it is, however, a noun—the adjective (and preferred word in situations like this) is **branchial**.

†There is a source of confusion here, because the term **arch** is often used in three different senses in connection with the branchial region. As noted in Chapter 7, it may refer to the series of bars that form the skeleton of each gill segment. The word is also used to describe the arterial blood vessel, aortic arches, that traverse each gill (cf. Chap. 14). Finally, it is used in a broader, inclusive sense, to describe the total structure lying between two successive branchial slits. To reduce this confusion, we will use **bar** rather than **arch** for the last of these three structures.

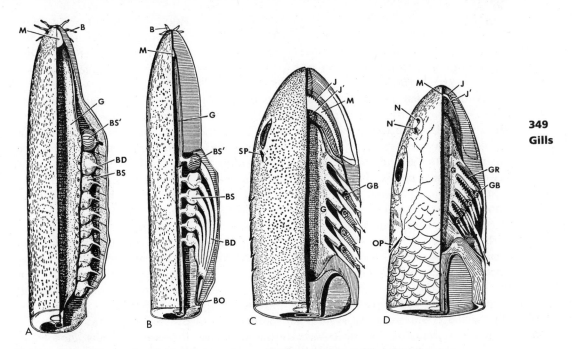

Figure 250. Heads of various fishes to show gill arrangement. *A*, The slime-hag *Bdellostoma; B*, the hagfish *Myxine; C*, a shark; *D*, a teleost. The right half of each is sectioned horizontally through the pharynx. Abbreviations: *B*, barbels around the mouth; *BD*, ducts from gill pouches; *BO*, common outer openings of gill pouches; *BS*, gill sacs; *BS'*, sacs sectioned to show internal folds of gill; *G*, gut (pharynx); *GB*, cut gill arch; *GR*, gill rakers; *JJ'*, upper and lower jaws; *M*, mouth; *N,N'*, anterior and posterior openings of nasal chamber; *OP*, operculum; *SP*, spiracle. (From Dean.)

Each of the series of branchial bars includes a number of characteristic structures (Fig. 251 *A*). Running dorsoventrally, close to the pharyngeal wall, are the skeletal supports, the branchial arch elements of the bar; skeletal **gill rakers** may extend inward from them into the cavity of the pharynx, screening food particles from the gills and diverting them backward toward the esophagus; **branchial rays** run outward toward the surface, stiffening the gills. In each bar, there is in the embryo a finger-like upward extension of the coelomic cavity (Fig. 196); this disappears in the adult, but from its mesodermal walls is derived the branchial musculature of the bar

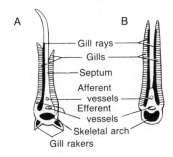

Figure 251. Horizontal sections of a gill bar of *A*, a shark; *B*, a teleost. External surface above, pharyngeal margin below; cartilage or bone in black. In the shark, there is a projecting septum, lost in the teleost.

(Fig. 213). Each branchial bar has a set of aortic arch blood vessels that bring blood from heart to gill and carry it thence, after aeration, to the dorsal aorta (cf. pp. 458–460). Down each bar passes a special cranial nerve trunk or branch (cf. Fig. 405). Extending outward in the center of each bar is a **branchial septum**, a vertical sheet of connective tissue.

350

**Chapter 11
Mouth and
Pharynx;
Teeth,
Respiratory
Organs**

The branchial passages of elasmobranchs are slits that are narrow anteroposteriorly, deep in a vertical plane; primitively, they probably opened widely both to the pharynx and to the exterior, but, in most forms, these entrances and exits are narrowed down to relatively short slits. In sharks, the external openings are lateral in position; in skates and rays, with flattened bodies, the external openings are on the ventral surface, below the expanded pectoral fins. In elasmobranchs, the branchial septum and its epithelial covering extend out to the surface of the body as a fold of skin that overlaps and protects the gill next posterior to it.

The gill itself is an elaborately folded and richly vascularized structure, covered with a thin epithelium. The branchial surface takes the form of numerous parallel **branchial lamellae**, which greatly increase the effective surface for exchange. Through them course capillaries, connected at one end with the afferent vessels from the heart and at the other with the efferent vessels leading to the dorsal aorta and thence to head and body. Blood flows anteromedially as the water goes posterolaterally; such a counter-current system allows the greatest possible extraction of oxygen from the water. For the most part, the lamellae are, in elasmobranchs, closely attached to the septum; at their outer tips, however, they may extend freely for a short distance.

Such a gill may develop as a dorsoventral sheet on either anterior or posterior side of any branchial slit, or, stated in another way, it may develop on either surface of a branchial bar (Fig. 252 A). In most cases, the bar carries a gill on both surfaces of the septum, and such a gill bar is considered to be a "complete" gill, a **holobranch**. Less commonly in fishes, a gill may be developed on only one surface of a branchial bar; in this case, the term **hemibranch** is applied. No vascular aortic arch is present behind the last gill slit; and, in consequence, there is almost never any branchial development on the posterior surface of that slit. All other typical slits in sharks, however, bear gills on both sides; in terms of gills rather than slits, there are four holobranchs. There is no development of branchial lamellae on the posterior side of the spiracle; hence, the hyoid bar behind it is a hemibranch.

The **spiracle** is the branchial opening lying between mandibular and hyoid arches. It appears to have gained its specialized nature early in the history of jawed vertebrates. We have noted that in typical jawed fishes, the hyomandibular element of the hyoid arch has become tied into the jaw joint as an effective prop for the jaws. Presumably, the ancestral vertebrates had a fully developed branchial slit between jaws and hyoid arch. When, however, the hyomandibular extended across this area, the slit was necessarily interrupted and has become restricted to a dorsal opening from throat to surface (Fig. 165 C).

Although the special nature of the spiracle may be due to skeletal developments, this opening has a positive functional history in the sharklike fishes. Among the sharks themselves, the spiracle is small in many forms and is absent in certain of the more active, fast-swimming genera. On the other hand, some of the more sluggish sharks, which spend much time on the bottom, show a considerable enlargement of the spiracle, and in the depressed, bottom-dwelling skates and rays, it is an enormous

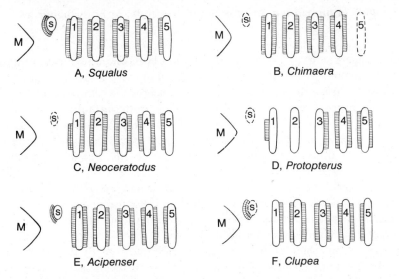

A, *Squalus*

B, *Chimaera*

C, *Neoceratodus*

D, *Protopterus*

E, *Acipenser*

F, *Clupea*

Figure 252. Diagrams to show the gill arrangement. *A,* Shark; *B,* a chimaera; *C,* the Australian lungfish; *D,* the African lungfish; *E,* sturgeon; *F,* a teleost (herring). Broken line indicates a closed slit. Hatched area adjoining slit indicates gill surface; vertical line through hatching indicates pseudobranch. The presence in *Protopterus* of a gill surface on the posterior side of the last slit is unique. Abbreviations: *M,* mouth; *s,* spiracle; postspiracular slits, are numbered.

opening on the dorsal surface behind the eye (cf. Fig. 24 *D*). In these bottom-dwelling elasmobranchs, the mouth, the usual point of entrance of the water current, may be more or less buried in mud or sand; the spiracle, dorsally placed, then affords a channel for the inflow of clear water to the gills.

Presumably, the spiracular slit, when fully developed, originally carried a typical pair of gills. Even in its reduced state, a small branchial structure is generally to be found on the anterior side of the spiracle, that is, on the posterior surface of the mandibular arch. This gill, however, is of a peculiar nature. Every proper gill receives blood for aeration directly from the heart by way of one of the aortic arches. But though such an aortic vessel runs to the mandibular arch in the shark embryo, it disappears before the adult stage is reached, and this little spiracular gill is supplied by blood that has already been purified by passage through the gills behind it (cf. Fig. 335 *B*). It is, thus, a deceptive "imitation" gill, a **pseudobranch**. It may, however, serve some useful purpose, for the blood from this gill passes, in the shark, to eye and brain, where a supply of doubly oxygenated blood may possibly be of particular importance.

The chimaeras, although related to the sharks and skates, show certain notable differences (cf. Figs. 25 *B,* 26, 252 *B*). The spiracle has been lost, and the last branchial slit is closed. A major contrast lies in the system of protecting the gills. In sharks, it was noted, a skin fold is developed from each gill bar. In chimaeras, in contrast, there is only a single large flap of skin, which grows backward from the hyoid arch and covers the entire series of gills; cutaneous folds have been reduced on the more posterior arches. This protective device is an **operculum**; with its development, we find only a single outer gill opening on either side of the body at the back end of the branchial region. Comparable developments are found in both higher and lower fish types.

Gill Development (Fig. 253).   The gill slits of elasmobranchs arise embryologically in a manner characteristic of vertebrates generally. Early in development, paired pouches push out from the endodermal lining near the anterior end of the embryonic gut. Extending toward the surface, they interrupt the continuity of the mesodermal plates of this region and come in contact with infoldings of the superficial ectoderm. Soon the intervening membranes break down, and ectoderm and endoderm join to form a continuous lining for the branchial slits. After this branchial lamellae develop along the lining of the slits.

352
Chapter 11
Mouth and
Pharynx;
Teeth,
Respiratory
Organs

In both Chondrichthyes and bony fishes, the gills appear to be formed from the ectodermal part of the epithelium of the slits. In cyclostomes, exceptionally, the evidence suggests that lamellar formation is endodermal (but the evidence for this has been questioned). Some investigators have argued that if cyclostome gills are of endodermal derivation, they cannot be truly homologous with those of other fishes. However, this point of view, which regards the germ layers as sacrosanct, need not be taken too seriously. The lining of the gill passages in any fish includes derivatives of both ectoderm and endoderm, indistinguishably fused and apparently similar in nature. The development of a gill is presumably due to the action of morphogenetic forces upon this epithelium. Apparently either ectoderm or endoderm, if lying in the area of potential branchial development, may respond.

Branchial pouches are formed in paired longitudinal series, and many investigators have tried to correlate the gill bars with the segmental arrangement seen in the myotomes and associated skeletal structures and nerves. There is, however, no conclusive evidence of any real relation between these two systems (cf. Fig. 196 and p. 562). The gill "segmentation" is based on pouches of endoderm; that of the somatic muscles and associated structures is based on somites of mesodermal origin; the two seem to be quite independent developments.

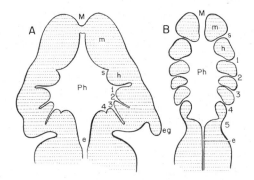

Figure 253.   Horizontal sections through the head and pharynx in embryos of *A,* a frog *(Rana);* and *B,* an elasmobranch; to show development of gill pouches. In *A,* the pouches are developing as narrow slits, which have not as yet opened to the surface. The mouth is also closed. In *B,* all the slits except the last one are open. Abbreviations: *e,* esophagus; *eg,* external gill developing; *h,* hyoid bar; *M,* mouth; *m,* mandibular bar; *Ph,* pharynx; *s,* spiracular cleft; 1 to 5, post-spiracular branchial slits.

Gills In Bony Fishes. The gills of the Osteichthyes (Fig. 250 *D*) are built on the same basic plan as those of sharklike forms; there are, however, important differences due to the universal presence of an operculum. Developing as a fold from the hyoid arch, it is basically comparable to that just mentioned for the chimaeras. The operculum of Osteichthyes, however, is a bony structure, forming a plate of armor between head and shoulder on either side (Figs. 123, 170, 172). A considerable **branchial chamber** may be present between the opercular cover and the gills. The opercular opening is primitively a great crescentic slit, convex posteriorly, lying just in front of the dermal shoulder girdle. In some specialized teleosts, however, the opening may be reduced to a small hole.

Breathing methods in most bony fishes are comparable to those of sharks except that opercular movements in great measure replace the opening and closing of individual gill openings seen in the shark. Water enters through the mouth, with a closing of the external branchial opening behind the operculum and an expansion of the branchial chamber. Valvular folds within the mouth help to close that orifice during the reverse movement of constricting the pharynx and expelling water through the opercular aperture. As in sharks, there are usually mechanisms to produce a relatively smooth, even flow of water. Some stiff-bodied bony fishes use the water current through the gills as a locomotor device, in the manner of a jet plane; on the other hand, such forms as the mackerel rely on their fast swimming movements for the production of a stream of water through the gills, simply holding the mouth open as they go. Individual gill flaps are unnecessary in bony fishes as protective devices, and we find that the branchial septum does not extend (as in sharks) outward beyond the branchial lamellae. The septum is developed to a moderate degree in lungfishes and *Polypterus*, but in teleosts, in fact, reduction has gone so far that the septum has shrunk away from the area between the pair of gills on the bar, leaving these elements projecting outward without any extraneous support as two branches of a **V**-shaped holobranch (Fig. 251 *B*). Gill rakers are frequently found, as in sharks; branchial rays are present, but, in contrast to elasmobranchs, form in two rows rather than one on each arch, stiffening the delicate branchial lamellae.

Normal branchial slits are, almost universally, five in number in bony fishes as in typical elasmobranchs (Fig. 252 *C–F*). *Polypterus*, however, has (like the chimaeras) lost the last pair of slits, and aberrant conditions are found in some teleosts. The pattern of distribution of gills along the slits is subject to reduction in various types. In actinopterygians, the hemibranch on the anterior surface of the first slit is lost except in a few primitive types (sturgeons, paddlefish, gar pikes); this is understandable, because the first slit, covered closely by the operculum, is in an unfavorable position to gain a food flow of water. The lungfish *Neoceratodus* of Australia has a full complement of gills, but that on the anterior side of the first slit is without a direct blood supply and hence is a pseudobranch. The other lungfishes exhibit further reduction of gills; these forms rely in considerable measure on lungs rather than gills for respiration and are, in fact, obligate air breathers. In *Lepidosiren* the first slit is closed, but three holobranchs remain. In *Protopterus*, gills are reduced in both size and number. All the normal branchial slits are open, and there is a true if small gill on the anterior surface of the first slit, but two branchial bars behind this slit have no gills at all.

That the spiracle was primitively present in bony fishes is attested by its retention in the surviving coelacanth and in the three living types of chondrosteans. In all other living bony fishes it is lost, although a few additional actinopterygians show a blind internal pouch in its former position. Curiously, however, we find that most actinopterygians have, despite loss of the spiracle, retained the pseudobranch that was associated with it. Tucked beneath the edge of the operculum or nearby, beneath the base of the skull, this sometimes retains its gill-like structure; sometimes it forms a small mass of tissue sunken beneath the surface. Possibly this former gill serves a sensory or endocrine function, but its nature is unknown.

An important function in addition to respiration is served by the gills in marine teleosts and lampreys. As discussed later (cf. pp. 399–402), such animals are living in an external environment containing a greater concentration of salt than is present in their own proper "internal environment." Osmotic processes constantly tend to increase the concentration of salt within the body to a lethal degree, and the kidneys are unable to counteract this entirely. In lampreys and at least some marine teleosts, cells of the branchial epithelium have the ability to take excess salt from the blood and excrete it into the water flowing through the gills (Fig. 254). Still further, these cells may actively absorb salt in freshwater teleosts. The gills of teleosts have an additional functional resemblance to the kidneys in that they have glandular elements able to excrete nitrogenous wastes such as ammonia.

Although ancestral bony fishes presumably had lungs as accessory breathing organs, these structures were later transformed by almost all actinopterygians into a hydrostatic air sac, described later, of little or no value in respiration in most cases. A variety of modern teleosts, however, live today under conditions in which air breathing may become a necessity. Some are more or less amphibious, spending part of their lives on land; others live in a muddy environment, where the water often forms too thick a gruel for satisfactory branchial respiration.

Most are in tropical or subtropical areas of seasonal drought. In the dry (and hot) season, water evaporates and ponds become smaller; this means more fish per

354
Chapter 11
Mouth and
Pharynx;
Teeth,
Respiratory
Organs

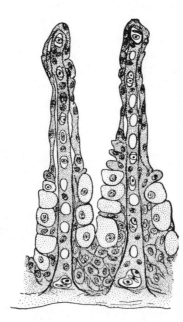

Figure 254. Two lamellae from the gill of an eel; the large cells at the base of the lamella are salt-secreting. The open (or largely open) circles in the center of the lamellae are the capillaries of the gill. (After Keys and Willmer.)

unit volume of water. However, as the water becomes warmer, it can hold less oxygen per unit volume, and each fish, as its metabolism is increased by the higher temperature, needs more oxygen.

For a number of teleosts, air gulped into the digestive tract has some respiratory value, with the epithelium of the air bladder or the gut functioning in gas exchange, and, in the electric eel (*Electrophorus*), virtually all respiratory exchange occurs through vascularized papillae in the lining of the mouth. In the nearly scaleless eels, the skin may function in aquatic or aerial "breathing." More frequently, however, we find lung substitutes in the form of structures developed in or from the branchial chambers, kept moist beneath the operculum. In various catfishes, for example, and the "tree-climbing perch" (*Anabas*) of the East Indies, familiar in popular lore, rosettes or branching filaments of vascularized epithelium are developed behind or above the gills in the branchial chambers (Fig. 255). In several cases, these chambers themselves are expanded into lung-like structures, extending backward into the trunk as a pair of elongated sacs lined with a respiratory epithelium.

Gills In Jawless Fishes. Descending the scale from the gnathostomes, we find that in the lampreys and hagfishes (Fig. 250 *A, B*), gills are equally well developed, but in a rather different way. The branchial passages are not in the form of slits, but of spherical pouches, connected by narrow openings with the pharynx on one hand and the exterior on the other. In the lamprey larva, the gills are arranged as hemibranchs on anterior and posterior surfaces of each pouch; in the adult cyclostome, however,

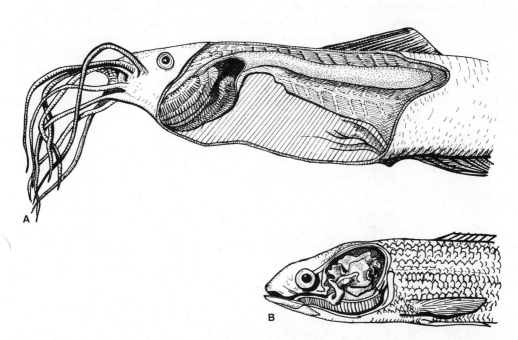

Figure 255. Accessory respiratory organs in teleosts. *A, Saccobranchus,* a tropical catfish; a rather lunglike sac extends posteriorly from the branchial chamber. *B, Anabas,* the "climbing perch"; here a complex folded structure, the labyrinth organ, projects into the branchial chamber. (After Giersberg and Rietschel.)

the branchial lamellae form a ring around the margins of each sac. The branchial skeleton forms a continuous but flexible lattice-work lateral to the pouches (Figs. 119, 163 *A*). A well-developed musculature, working upon these structures, constricts and expands the sacs. In this way, a pumping effect can be obtained; water can be pulled in through the external openings and forced out again through them, thus permitting respiration to continue even when feeding operations block anterior entrance of water to the pharynx.

Such anterior entrances vary within the group. In the larval lamprey, as in the hagfish, the mouth leads to a pharynx of normal construction which in turn connects posteriorly with the esophagus. In the metamorphosis to the adult, however, a radical change takes place in the pharynx. This splits into two separate tubes (Fig. 256). A dorsal duct leads back without interruption from mouth to what is usually called the esophagus, although it is, as we have just noted, pharyngeal in origin. Below is a large tube with which the gills connect; this ends blindly at its posterior end and is a purely respiratory part of the pharynx, hence called the **respiratory tube**. A special valve, the **velum**, guards the entrance to this tube, enabling it to be completely isolated from the digestive tract when the animal is feeding. In the hagfishes, no such subdivision of the gut is found, but an equally peculiar device is present: the combined nostril and hypophyseal sac breaks through posteriorly into the pharyngeal roof (cf. Fig. 17).

In contrast to the usual gnathostome condition of five normal slits plus a spiracle, in most cyclostomes, a higher count of branchial pouches is found. The lamprey *Petromyzon* has seven pairs of pouches; in the hagfishes, the number varies from six to 14. In addition, all members of the hagfish group have an extra unpaired gill-less duct on the left side, usually at the back of the series, leading from pharynx to exterior. The number of pouches in cyclostomes and other agnathans suggests that in primitive vertebrates, the number of gills was somewhat higher than in modern jawed fishes.

Variations are seen in the arrangement of the porelike external openings of the gills of cyclostomes. In lampreys and the slime hag *Bdellostoma*, each gill pouch opens by a separate orifice. But in another hagfish (*Paramyxine*), the tubes of each

356
**Chapter 11
Mouth and
Pharynx;
Teeth,
Respiratory
Organs**

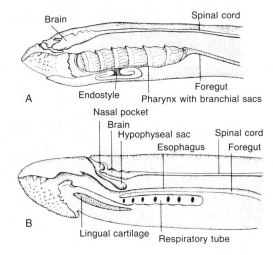

Figure 256. Sagittal sections of *A*, the ammocoete larva of a lamprey; *B*, adult lamprey; to show especially the division of the adult pharynx into two parts. The endostyle of the larva becomes the thyroid gland of the adult (not shown). (After Goodrich.)

side, although separate, are crowded close together at the surface, and, in the common hag, *Myxine,* the whole series of tubes on either side fuse externally to form a single outer opening (Fig. 250 *B*). There is thus developed here a condition somewhat similar to that of operculate jawed fishes. Whereas in most vertebrates, and even in lampreys, the gills are situated close to the cranial structures, they are removed some distance back in the hagfishes. This appears to be an adaptation to the semiparasitic life of these forms; the gill openings may remain free even if the head is completely buried in the flesh of its prey.

The ammocoete larva of *Petromyzon* spends its existence half-buried in the bottoms of streams and ponds, subsisting on microscopic food particles. The branchial system forms the food-collecting mechanism. The particles contained in the water current entering the mouth are strained in passage through the pharynx and are carried on down the digestive tract by a series of ciliated channels similar to those described for amphioxus and the tunicates in an earlier chapter (cf. pp. 20–24).

It is apparent that this food-collecting function of the gills in the ammocoete is not merely an accessory one but a function even more important here than that of respiration, which could be largely carried on by the skin of these small, soft-bodied animals. This situation leads one to the concept that in the vertebrate ancestors, food collecting was the primary *raison d'être* of the system of internal gills and that originally respiration may have been an accessory function only—the development of actual branchial lamellae along the tubes may have been an afterthought. The fact that, as we have seen, lower chordates likewise make their living by straining water through a highly developed system of pharyngeal gills adds emphasis to this concept.

Further evidence can be adduced from study of the most ancient of jawless vertebrates, the fossil ostracoderms (cf. Figs. 18, 19, 21). A considerable amount of data is available regarding the gills of these Paleozoic forms. The branchial region and pharynx are highly developed. In some, the Heterostraci, the gills opened to the exterior, rather as in a hagfish, by a common opening on either side; in others, the Osteostraci and Anaspida, a series of round holes is present on the lateral or ventral body surface, showing a condition similar to that of lampreys. The number of gills is somewhat variable, but no ostracoderm yet investigated shows more than a dozen pairs—a number above that of jawed fishes, but within the range of modern cyclostomes. The internal structure is often unknown or poorly known but in some cases is so well preserved that a fairly accurate reconstruction of the branchial region can be made (cf. Fig. 20 *A*). The gills here were probably in part respiratory—the hard armor would render the skin of little value in this regard—but the relatively enormous size of the apparatus in such a form as that figured strongly suggests that food straining was the major function in ostracoderms.

The ostracoderm story, added to that described earlier for the lower chordates, leads with certainty to the conclusion that the history of the pharynx and branchial slits begins with their development as a feeding apparatus—a function proper to their inclusion (otherwise incongruous) in an introductory part of the digestive tract. The passage through this food-screening system of a constant current of water from which oxygen could be readily obtained presumably facilitated the development here of respiratory structures; in ancestral vertebrates, this pharyngeal system served a double function, as it does today in the larval lamprey. With the development of jaws (or cyclostome substitutes for them), the feeding functions of the branchial system were, for the most part, lost, although a few fishes have secondarily readopted gill-straining as a means of feeding. However, in typical fishes the pharynx and branchial apparatus are structures that, although somewhat reduced in relative size, are still highly developed and generally indispensable for respiratory purposes.

358
Chapter 11
Mouth and
Pharynx;
Teeth,
Respiratory
Organs

Pharynx and Gills In Tetrapods.   With the appearance of lungs and the subsequent shift of tetrapods to a terrestrial life, there came about a further decrease in the size and importance of the pharynx and gills. In amphibians, we see a transitional stage. In water-dwelling larvae of this group, pharyngeal gill slits are still present, but are much reduced. The bony operculum of the fish ancestors is absent; the anurans develop a saclike opercular structure, but this is probably a new, larval adaptation. The erstwhile spiracle seldom opens and is generally incorporated into the auditory apparatus. In urodeles, three, and in anurans, four of the five typical branchial slits may open in the larvae. But typical internal gills are never developed in the walls of these slits. Further, the gill openings disappear with the end of the larval period; only in urodeles that fail to metamorphose completely do certain of the slits (from one to three of them) persist. The function of the slits in amphibians seems to be simply to ensure that a water current is set up near the external gills (described below) upon which larval amphibians mainly depend for aeration of the blood.

In amniotes, branchial pouches always push outward from the pharynx in every embryo, and surface furrows push inward to meet them; but slits open only in a transitory manner in the embryos of reptiles and birds, and seldom if at all in mammals. In no case is there, in amniotes, the slightest development of respiratory functions in the gills. The amniote embryo, in conservative fashion, repeats the early embryologic processes of countless ancestral generations, but never is there any true ontogenetic repetition of the development of the full-grown branchial system found in the fish ancestor.

In the adult amniote, the pharynx retains hardly a vestige of its former importance. It has greatly shortened, permitting the development behind it of the constricted neck, traversed by an elongated esophagus. The lungs developed from the pharyngeal region in fishes, and their entrance is still to be found in the pharyngeal floor. Of branchial pouches, the spiracle persists in modified form in connection with the ear; in the adult, the others have disappeared completely.

External Gills.   In the larvae of a few bony fishes and of amphibians, accessory respiratory structures are present in the form of external gills. Among fishes, they are found only in *Polypterus* and in the African and South American lungfishes. In the lungfishes, they consist of four feathery processes growing out on either side of the "neck" above the branchial openings. In life, they have a reddish color, due to the presence of an abundant supply of blood—the blood coming from the vascular aortic arches. *Polypterus* (Fig. 257) shows similar outgrowths, but they are limited to a single pair. Fish external gills appear at an early embryonic stage, before the operculum is formed, and arise from the dorsal ends of successive gill bars. Probably they are ancient specializations from the internal gill system.

The climatic conditions that (we have noted elsewhere) tend to favor the presence of lungs in these same animals obviously are significant in relation to the presence of external gills. A paucity of oxygen in stagnating waters provides a situation especially unfavorable for young and growing fishes, which have a high rate of metabolism. External gills increase respiratory surfaces considerably and enable the larva to

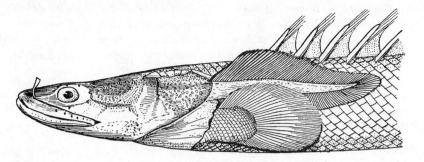

Figure 257. Larval form of the primitive African ray-finned fish, *Polypterus;* the large external gill arises from the top of the hyoid arch and extends back above the pectoral fin. (From Dean.)

make better use of such poor oxygen supplies as are available. As is the case with lungs, external gills are of little importance in the ray-finned fishes. Since, however, we find them present in some members of both major groups of Osteichthyes, we can reasonably assume that (like lungs) they were present in the ancestral bony fishes, to be lost later among the more advanced actinopterygians and some sarcopterygians. It would be most interesting to know whether acanthodians had any such structures; unfortunately, we have no evidence at all.

To be distinguished from true external gills are filaments protruding from the internal gills of various fishes. Such external filaments are found in the early stages of all elasmobranchs. They float in the albuminous fluid lying in the egg case and appear to serve for the absorption of food as well as for respiration; they disappear at the time of hatching. Somewhat similar, but purely respiratory, filaments are found in sturgeons and paddlefishes and, as noted earlier, in a few teleosts.

External gills are present in the urodele amphibians in a form similar to that seen in their lungfish cousins—feathery, coral-colored outgrowths, usually three pairs of them, above the gill slits of the larva (cf. Fig. 83 *E, F*), developed from the third, fourth, and fifth arches. Presumably they were also present in the larvae of typical crossopterygians. Throughout life, certain urodeles tend to remain aquatic larvae; retention of external gills is a diagnostic feature of these "perennibranchiate" forms. The Gymnophiona also develop external gills similar to those of fishes and urodeles. The Anura have a specialized larval gill system. Four branchial slits break through in the early tadpole stage of typical frogs and toads. Their external openings, however, are sheathed by the peculiar opercular fold, not exactly homologous to the fish operculum (there may be a modest opercular development in urodeles). This is so extensive that it envelops the under surface of the neck and even the developing forelimbs; in most anurans, it has only a single opening to the exterior, usually on the left side. There is enclosed a considerable chamber filled by a mass of vascularized filaments. Part, at least, of this branchial tissue is derived from external gills that bud out above the branchial slits at an early stage, before the operculum develops. There are, however, additional filaments, formed later, that arise more ventrally near the slits. These are apparently modified portions of the internal gills; indeed, it is highly probable that the entire series of external filaments, as seen not only in anurans but also in the external gills of other amphibians and fishes, are ultimately derived, phylogenetically, from structures of the internal gills.

Many amphibians, including representatives of all three orders, develop away from streams and ponds in a variety of situations, such that they must breathe air from an early developmental stage. A number of different types of respiratory devices have been evolved in these cases; frequently, elaborate external gills function as embryonic or larval lungs.

External gills have disappeared, without trace, in amniotes.

360
Chapter 11
Mouth and
Pharynx;
Teeth,
Respiratory
Organs

The Swim Bladder

Characteristic of most actinopterygian fishes is the presence of a **swim** (or air) **bladder**, an elongate sac arising as a dorsal outgrowth from the anterior part of the digestive tube (Figs. 258, 259 A–C, 261 C). The swim bladder is distensible and filled with air or other gases; its major function is that of a hydrostatic organ. The specific gravity of a fish (or any vertebrate) is slightly greater than that of sea water. Filling or emptying the swim bladder will change the specific gravity of the fish as a whole and facilitate its keeping to a depth, high or low in the water, proper to its habits. It is thus obvious that this structure is of greatest use to a form living in the ocean or in other deep-water bodies.

Among actinopterygians, the most primitive form, *Polypterus*, has no swim bladder as such but has, instead, paired ventral lungs. A typical dorsal swim bladder is, however, present in the other living ray-finned types below the teleost level and is retained in most teleosts. It is absent, for example, in the bottom-dwelling flounders, in which a hydrostatic organ would be functionally useless, but even there it is present in the embryo. It seems certain that the dorsal swim bladder was developed by the actinopterygians early in their history, possibly indicating a marine stage (lungs being useless in the sea). It is not found in any other fish group whatsoever.

A **pneumatic duct**, leading from the digestive tract upward along the dorsal mesentery to the bladder, is present in the more primitive teleosts and in lower ray-finned forms. Primitively, the duct is short and broad and may open from the back end of the pharynx in lower actinopterygians. In teleosts, however, it becomes a narrow and often elongate tube, opening frequently from the esophagus, but in some cases, from even farther back in the gut. Finally, in a large series of advanced teleosts, the connection with the digestive tract is entirely lost, a condition termed physoclistous.

The swim bladder is typically an elongate oval sac, lying above the coelomic lining of the peritoneal cavity, below the dorsal aorta and vertebral column. Its shape is variable; frequently it is constricted into anterior and posterior subdivisions. An

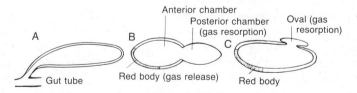

Figure 258. Diagrammatic longitudinal sections of the swim bladder in various teleosts. *A*, Primitive type, open to the gut; *B, C,* closed types, with gas-producing red body and other areas for gas resorption.

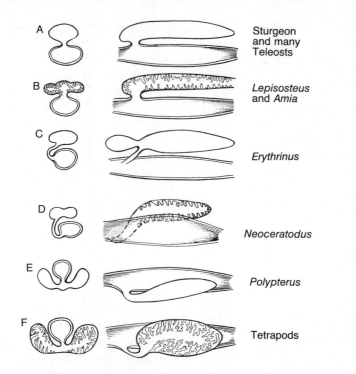

A, Sturgeon and many Teleosts

B, Lepisosteus and Amia

C, Erythrinus

D, Neoceratodus

E, Polypterus

F, Tetrapods

Figure 259. Diagrammatic cross sections and longitudinal sections of the swim bladder or lung in various fishes and in tetrapods. *A,* Typical dorsal swim bladder of actinopterygians; *B,* a bladder type found in holosteans with a folded inner surface capable of some respiratory function; *C,* an unusual teleost type with a lateral opening suggestive of a transition from lung to air bladder; *D,* the Australian lungfish, with the bilobed lung rotated dorsally, although the opening remains ventral in position; *E,* the archaic actinopterygian, *Polypterus,* with a ventral lung, probably primitive and antecedent to all other lung or air bladder types; *F,* the type of lung developed in terrestrial vertebrates, with complex internal structure. Transverse sections viewed from the front and longitudinal ones from the left. (After Dean.)

elastic connective tissue surrounding it allows for expansion, and both smooth and striated muscle fibers function in contraction.

In some primitive surface-dwelling fishes, air is taken into the sac through the pneumatic duct. But even in many forms in which the duct is present, this appears to serve merely as a release valve, and, in most cases, the swim bladder depends upon its own internal mechanisms for supply and absorption of its gaseous content. In the anterior part of the bladder of most teleosts is found the **red body**, which owes its name to the marvelous network of capillaries—a rete mirabile—contained in it. This is the gas-producing organ, which must often operate against considerable pressure; the capillaries of the rete form a counter-current exchange to allow it to do so. The gas includes, as in air, nitrogen, oxygen, and carbon dioxide, but in varying proportions. A gas-absorbing tissue is present in the posterior part of the bladder—in the posterior chamber when two are present. Primitively, this is a broad area of thin epithelium lying in the general surface of the bladder. In advanced forms, however, absorption is restricted to the **oval**, a sunken pocket that can be closed off by a sphincter.

Although the major use of the teleost swim bladder is as a hydrostatic organ, other attributes can be ascribed in some cases. As noted later (cf. p. 000), it is a hearing aid in some teleost groups. In other cases, it has the reverse function—sound is produced by vibrations set up by muscles attaching to the swim bladder. Still further, it is sometimes an accessory air-breathing organ. This function is well developed in the gar pike and *Amia*, primitive freshwater actinopterygians, and is found as well in a few freshwater teleosts of relatively primitive nature. In these fishes, the inner surface of the bladder is rather cellular and lunglike in texture (Fig. 259 *B*), quite in contrast to the smooth lining found in most teleosts.

In its dorsal position and single rather than paired nature, the swim bladder contrasts strongly with the lungs. However, most anatomists believe that the two are in some way and to some degree homologous structures. We shall discuss this further after a consideration of the early history of the lungs.

362
Chapter 11
Mouth and
Pharynx;
Teeth,
Respiratory
Organs

Lungs

Structurally distinct from the gills, although similar in basic function, and, like them, pharyngeal derivatives, are the lungs, which in typical air-breathing vertebrates replace the gills as the structure through which oxygen is brought to the blood and tissues. In most tetrapods, the air-breathing apparatus has a characteristic pattern, which may be briefly sketched at this point (Fig. 260). Entrance to the system is gained by a median ventral opening in the pharynx, the **glottis**. Immediately beyond the glottis the duct enlarges to a chamber, the **larynx**; beyond this, a ventral median tube, the **trachea**, extends backward to divide into primary **bronchi**, one leading to each lung. The paired **lungs**, primitively ventral, may expand to occupy a lateral or even dorsal position in the anterior part of the coelomic cavities.

Embryologically, the lungs first appear in the form of a ventral outpocketing of the floor of the throat, median in position, but in many cases distinctly bilobed at an early stage. Growing backward, this endodermal pouch divides into paired outgrowths from which the lungs develop; associated with it are tissues of mesenchymal origin that strengthen its walls and produce skeletal elements related to the larynx, trachea, and bronchi.

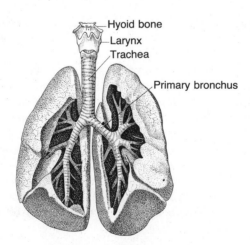

Hyoid bone
Larynx
Trachea
Primary bronchus

Figure 260. Ventral view of the respiratory system of man. The lungs are dissected to show the bronchi. (After Toldt.)

Structure of the Lung.   The inner surface of the lung and its compartments is lined by an epithelium of endodermal origin; this can be extremely thin over the respiratory capillaries. The external surface, facing the coelom or a subdivision of it, is covered by the epithelium of that cavity. Within the substance of the lung are variable amounts of connective tissue, smooth muscle fibers, and an abundance of blood vessels. In many reptiles and in mammals, the lung is a spongy structure with air passages—bronchi, **bronchioles** radiating from the bronchi, and small terminal pouches, **alveoli**, through the walls of which there is an exchange of oxygen for carbon dioxide (Figs. 260, 262). The lung is richly supplied with blood capillaries, which branch from a special pulmonary artery coming from the aortic arch of the last gill or, in higher tetrapods, from the heart itself (cf. p. 463). From the lung, the blood returns by a pulmonary vein to the heart. Between blood and pulmonary alveoli, there lie only thin membranes, through which diffusion of gases can readily take place.

In fishes that have lungs and in amphibians, the lungs have a relatively simple internal construction. The efficiency of a lung depends, of course, mainly upon the area of the surface present for gas exchange. Advanced tetrapods, with greater activity and greater needs for oxygen, increase the respiratory area, not so much by increasing the size of the lungs as by increasing the complexity of internal subdivision. Here, as usual, we must keep in mind the problem of surface-volume relationships; in large animals, the lungs must be disproportionately increased in size or attain greater complexity of subdivision, so that their surface area may keep pace with the volumetric growth of tissue demanding oxygen.

Although lungs are the structures normally present in tetrapods for air-breathing, they are not universally present in amphibians, and in many terrestrial members of that class, much of their "breathing" is accomplished by other means. Exchange of oxygen and carbon dioxide with the air can take place at the surface of any moist mucous membrane properly supplied with blood vessels. Even though lungs are present in frogs and toads, they are, in many forms, of relatively little importance. If a frog is observed, it will be noticed that the floor of the mouth is periodically lowered and raised. This movement expands and contracts the oral cavity, drawing in and expelling air through the nostrils; the moist epithelium of the mouth acts as a substitute lung. Even more important than the mouth as a breathing organ is the moist skin of many modern amphibians. This is richly vascular and appears to be so effective that in some salamanders breathing is accomplished satisfactorily by the skin despite a total absence of lungs.

Lungs in Fishes.   Although lungs are most characteristically developed in tetrapods, they are phylogenetically older structures. Lungs are present in the dipnoans, which gain their popular name of lungfishes because of this fact, and presumably were present in their close relatives, the ancient crossopterygians, from which tetrapods are probably descended. Still further, lungs are present in *Polypterus,* the most primitive living member of the ray-finned group of bony fishes.

Fish lungs are simple in construction. In *Polypterus* (Figs. 259 *E*, 261 *A*), an opening (with a sphincter) in the floor of the pharynx leads to a bilobed sac; the lobes pass back on either side of the esophagus, and the right lobe, much the longer, continues backward and upward into the mesentery above the gut. The lungs are of simple construction, with only a few internal folds. Except for the elongation of the

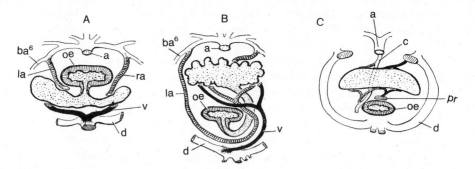

364

Chapter 11
Mouth and
Pharynx;
Teeth,
Respiratory
Organs

Figure 261. Diagrammatic cross sections of bony fishes (seen from behind) to show the position of the lung or swim bladder and the blood vessels connected with it. *A, Polypterus,* a primitive actinopterygian, with a ventral paired lung; *B, Neoceratodus,* a lungfish, with a single lung dorsal in position, but with a ventral opening from the gut; *C,* a teleost with swim bladder (duct lost). In the lung, the arterial supply is from the last aortic arch; in *B,* the curve of the vessel from the left arch below the gut indicates the path of dorsal migration of the lung. In the teleost, the arterial supply to the swim bladder is from the dorsal aorta (by way of the celiac artery). In the lung, the venous return is directly to the heart. The *Neoceratodus* veins show an asymmetrical condition comparable to that of the arteries. From the swim bladder, the blood returns to the heart via the normal venous system. Abbreviations: *a,* dorsal aorta; *ba*[6], sixth aortic arch; *c,* celiac artery, whence branches run to teleost air bladder; *d,* common cardinal vein (duct of Cuvier); *la,* left pulmonary artery; *oe,* esophagus; *pr,* hepatic portal vein, draining part of teleost swim bladder; *ra,* right pulmonary artery; *v,* pulmonary veins. (From Goodrich).

right lobe, this would appear to be a primitive type of lung, resembling the simple lung of amphibians (Fig. 259 *F*).

In the Dipnoi, this generalized pattern has been modified (Figs. 259 *D*, 261 *B*). The entrance, as always, lies in the floor of the pharynx. Instead, however, of a symmetrical development, we find that the duct to the lungs, long and unbranched, curves upward around the esophagus to a lung or lungs dorsal in position. The lung is more efficient than that of *Polypterus,* because there is some degree of internal subdivision into pockets and alveoli. In the Australian lungfish, the lung is a single structure, but in the African and South American genera, the lungs are definitely paired. The course of the duct suggests that the dorsal position of the dipnoan lungs is a secondary condition, the entire structure having twisted upward around the right side of the body. Strong evidence for this assumption is afforded by the blood supply (Fig. 261 *B*). If the dorsal position were the original one, we would expect the arterial blood supply to come from adjacent vessels in some simple manner. We find, however, that an artery from the left side of the body reaches the lungs by looping ventrally around the esophagus and then up on the right side, showing by its course the route of phylogenetic migration of the lungs.

Origin of Lungs. Although gills are present in lung-bearing fishes, the lungs are important breathing structures; indeed, the African lungfish will suffocate and drown if kept from air. We have earlier commented (cf. p. 54) on the climatic conditions of seasonal drought that appear to provide lungs with a considerable survival value in those few modern tropical fishes that have them and on the fact that such conditions were probably widespread in the Devonian period when the bony fishes arose. The probability that lungs were present as adjunct respiratory organs in early bony fishes of all types is indicated by the fact that they are a feature not only of the Sarcopterygii but also of the most primitive of living actinopterygians. Lungs may perhaps be still more ancient in vertebrate history. There is no trace of them in the modern Chondrichthyes, but fossil evidence indicates the presence of a pair of ventral pharyngeal pouches that may have been primitive lungs in a Devonian member of the Placodermi; if these were lungs, they may have been common to all primitive jawed fishes.

The mode of origin of lungs, phylogenetically, is none too certain. However, as we have noted, air-breathing can be carried on through any moist and permeable membrane, and it may be that in some ancient fishes, much as in amphibians today, the lining of the mouth and pharynx functioned in this manner. Any pocket-like outgrowth of the pharynx increasing the respiratory surface would, of course, be advantageous. In amphibians, the lungs sometimes appear in early embryonic stages as pockets that have the appearance of an extra pair of branchial pouches. Quite probably, the oldest lungs were a modified pair of posterior pharyngeal pouchs. They might, on the other hand, be a new pair of sacs in that area, much like those found in a few modern teleosts (cf. Fig. 288 *A*).

Lung Versus Swim Bladder. If we assume, as have most students of the subject, that lungs and swim bladder are homologous structures, which is the ancestral type? Early writers took it for granted that, because the swim bladder is a fish structure and the lung is a characteristic tetrapod feature, the swim bladder was the progenitor of the lung. But consideration of the evidence now available strongly indicates that, despite the seeming reasonableness of the older theory, the reverse is the case: the lung is the more primitive of the two. The swim bladder is found only in one subdivision of the bony fishes, the actinopterygians, whereas the most primitive member of that group, and members of the Sarcopterygii as well, have a lung. The swim bladder, thus, would appear to have evolved from a more ancient and widespread lung-like structure during the course of actinopterygian history.

The shift from lung to swim bladder in actinopterygians can be correlated with their history. The lung was an adaptation for survival of the individual among the early Devonian bony fishes during seasons of drought. Lungs are still of advantage to the few fishes that have them, living as they do in tropical areas in which such conditions are present today. For actinopterygians generally, the lung became of little advantage. As a ventral structure, a lung makes a fish top-heavy; if a dorsal shift of the air sacs to the swim bladder position could be accomplished, greater stability would be gained; this has happend in lungfishes as well as in actinopterygians. More important, later ray-finned fishes migrated in great numbers to the seas, and teleosts probably arose there. In the oceans, there is no drought and no need, and generally no opportunity, for air-breathing. On the other hand, a swim bladder, if present, would prove highly useful to dwellers in the deep as a hydrostatic organ.

We have no satisfactory structural intermediates in living actinopterygians between the lung of *Polypterus* and the swim bladder of more progressive forms. However, the lungfishes, although of course not belonging to this series, suggest the method of transition. The dipnoan lungs open by a ventral duct but are actually in the dorsal position of a swim bladder, and, in *Neoceratodus*, only a single symmetric dorsal lung is present. The further changes needed to parallel the presumed development of a swim bladder are shifts in the opening of the duct and in the blood supply; one existing teleost shows a transition in the first regard (Fig. 259 *C*).

Lungs in Lower Tetrapods.   In amphibians, the lungs show little advance over the fish condition and are relatively inefficient structures. In some urodeles, the inner walls remain smooth; generally, however, there is, in amphibians, a development of septa—ridges containing connective tissue that somewhat increase the respiratory surface and give the internal surface of the lung a honey-combed appearance (Fig. 262 *A*).

366
Chapter 11
Mouth and
Pharynx;
Teeth,
Respiratory
Organs

Among reptiles, the lungs of *Sphenodon,* snakes, and many lizards are not much further advanced (Fig. 262 *B*). They are still saclike structures, with a modest internal subdivision of the walls into alveolar pockets. Parenthetically, we may note that in a variety of limbless lizards, snakes, and in apodous amphibians as well, only a single lung, usually the right, attains any degree of development in the slender body; amphisbaenians, uniquely, retain the left as the largest or only lung. In some lizards and in turtles and crocodiles, the lungs become more complicated. The septa, originally confined to the outer walls of the sac, have grown inward and subdivided in complicated manner to give the lung as a whole a rather spongelike texture. A main central tube runs the length of the lung as a continuation of the bronchus leading to it from the trachea. From this, branches lead outward between the major septa to subdivide and finally reach the alveoli.

In amphibians, owing to reduction of ribs, the lungs are filled by swallowing air, a force pump mechanism; in reptiles (and amniotes generally), the action is that of a suction pump. In most reptiles, the elongate lungs lie in the general body cavity. The action of axial muscles raises the ribs and expands the capacity of the trunk as a whole; air is sucked into the lungs, which expand to fill the additional space. In some reptiles, there is a modest development of air sacs of the type highly developed in birds and described below. In crocodilians, there develops a structure somewhat comparable to the mammalian diaphragm. Turtles, of course, cannot move their ribs and mostly rely on movements of the pectoral limbs and girdles, which act to expand and contract the body cavity and the contained lungs.

The Lung in Birds.   In birds, with their high metabolic rate and consequent demand for great amounts of oxygen, the pulmonary apparatus has developed into a complex

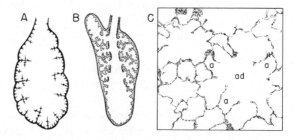

Figure 262.   Diagrammatic sections through the lung of *A,* a frog, and *B,* a lizard. *C,* Section of a small area of a human lung, × about 50, showing its complex construction. Abbreviations: *ad,* an alveolar duct, smallest component of the duct system; *a,* the alveoli to which it leads. (*A* after Vialleton; *B* after Goodrich.)

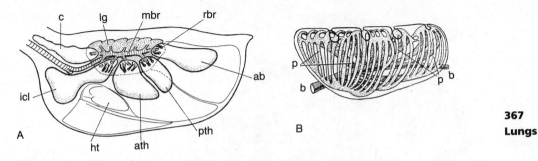

Figure 263. *A,* Diagram of the trunk of a bird, left side view, to show the position of the respiratory sacs. Abbreviations: *ab,* abdominal air sac; *ath,* anterior thoracic sac; *c,* cervical air sac; *ht,* heart; *icl,* clavicular air sac; *lg,* left lung; *mbr,* bronchus running through the lung and leading from trachea to various air sacs; *pth,* posterior thoracic air sac; *rbr,* recurrent bronchi that return air from sacs to respiratory areas of lung. *B,* Lateral view of the left lung of a bird. The primary bronchus, *b,* traverses the length of the lung to the abdominal sac. Branches are given off, leading to other air sacs and to parabronchi *(p).* These connect at both ends with the bronchi. Notches in the dorsal outline of the lung are rib impressions. (*A* from Goodrich; *B* after Locy and Larsell.)

series of structures. The lungs themselves are relatively small and compact. Out from them, however, grow numerous extensions as paired **air sacs** that invade every major part of the body (Fig. 263 *A*). A pair of cervical sacs grows upward along the neck; clavicular sacs (usually fused) are found beneath the wishbone; two pairs of lateral sacs are present in the thoracic region; a pair of abdominal sacs extends far back among the viscera. With the development of these sacs is associated a complicated subdivision of the body cavity (cf. Fig. 225). From the sacs, further air spaces invade certain of the bones in most birds and partially replace the bone marrow. The lining of the air sacs is, in general, smooth; they are thus of little value as respiratory surfaces (they may aid, as cooling devices, in the regulation of body temperature). They nevertheless play a major role in respiration. In the birds, as in other amniotes, the breathing mechanism is that of a suction pump, but the lung itself does little in the way of expansion or contraction. When the bird is at rest, air is drawn in by a forward and upward movement of the ribs and a forward and downward movement of the sternum. These movements together expand the trunk. During flight, the action of the wing muscles causes movements of skeletal elements that similarly result in expansion and contraction of the chest. Most of the inspired air passes through the lungs, by channels continuous with the bronchi, directly to the more posterior air sacs without much respiratory exchange taking place on the way. The sacs, however, are connected with the lung proper by smaller tubes in addition to the continuations of the bronchi through which the air enters them; when, on expiration, the air is forced out of the sacs, it passes through these openings to tubes in the lung and the respiratory surfaces and then to the more anterior air sacs. Maintenance of this one-way flow is complex; structural valves are lacking.

The structure of the avian lung itself is unique among vertebrates. In other amniotes, the respiratory membranes are in "dead-end" alveoli. Nothing of the sort is found in bird lungs; every major passage is open at both ends so that there is a true circulation of air (Fig. 263 *B*). Within the lung, a series of tubes, the **parabronchi,**

368
Chapter 11
Mouth and
Pharynx;
Teeth,
Respiratory
Organs

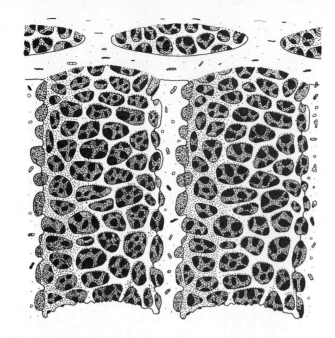

Figure 264. Diagram of the fine structure of a bird's lung, based on scanning electron micrographs. Two parabronchi are cut longitudinally and others, at the top, are shown in cross section. The respiratory exchange takes place in the many small pockets and sponge-like areas along the wall of the parabronchi; airflow through the parabronchi occurs in one direction only. (After Schmidt-Nielsen.)

loop from the bronchi through the lung tissue and back to the bronchi; the respiratory exchange takes place in the spongy tissue surrounding the parabronchi (Fig. 264).

The Mammalian Lung. In mammals, the lungs are large but with a rather simpler pattern than that found in birds. They are relatively short anteroposteriorly but expanded, frequently with a varied subdivision into lobes; together with the heart, lying between them, they occupy most of the thorax. The lungs are finely subdivided into tiny but exceedingly numerous alveoli (Fig. 262 *C*), to which the air passes by means of a branching series of larger and smaller bronchi and finally **bronchioles**; the alveoli cluster above the terminal alveolar ducts like grapes on a stem. Phylogenetically, as indicated by conditions in amphibians and more primitive reptilian types, increasing complexity of lung structure was attained by increasing development of septa, large and small, which brought about finer subdivisions of the cavities of the sac. By the time the mammalian condition is reached, however, it is, so to speak, the alveolar "hole" rather than the septal "doughnut" that is prominent. The tubes and terminal alveoli form the obvious pattern; the mesodermal tissues about them appear to play a relatively minor part in pulmonary ontogeny. In the embryo, the primary lung bud invades a compact mass of mesenchyme (Fig. 265) within which, by repeated budding and subdivision of cavities, the adult pulmonary structure finally emerges.

Breathing is accomplished by a highly efficient suction mechanism. In contrast with the condition in primitive amniotes, the lungs are not in the general coelomic cavity; instead, a tendinous and muscular partition, the diaphragm (cf. p. 321), separates the thorax from the abdominal region, and each lung lies anterior to the diaphragm within an individual **pleural cavity**, a subdivision of the embryonic coelom (Fig. 266). Expansion of the thorax is accomplished in part by upward and outward movement of the ribs, as in reptiles and birds. More important, however, is a poste-

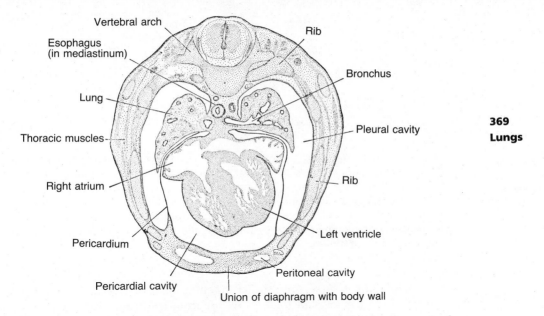

Figure 265. Cross section of a human embryo (at 8 weeks) to show the development of the lungs and of pleural and pericardial cavities and heart. (From Arey.)

rior or downward movement of the diaphragm. Each lung lies free within its pleural cavity, attached only at the anteromedial surface where the bronchus and blood vessels enter. Movements of the ribs and diaphragm expand the chest and consequently the pleural cavities; air is pulled into the lungs as they expand to fill the vacuum that would otherwise exist.

Larynx (Figs. 231, 267). With the increased importance of the lungs in terrestrial life, the entrance to the lungs differentiates to form the larynx, an enlarged vestibule antecedent to the trachea. The laryngeal cavity is surrounded by a complex of cartilages or bones (cf. Fig. 164). Associated muscles adjust the laryngeal cartilages and

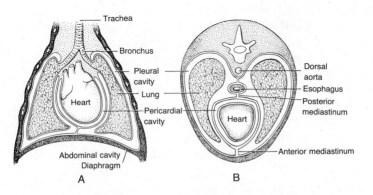

Figure 266. Diagrammatic longitudinal, *A,* and cross sections, *B,* of the thorax in mammals to show the position of heart and lungs and of pleural and pericardial cavities.

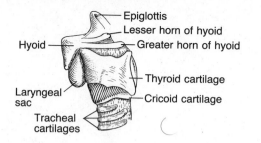

Epiglottis
Lesser horn of hyoid
Hyoid
Greater horn of hyoid
Thyroid cartilage
Laryngeal sac
Cricoid cartilage
Tracheal cartilages

370

Chapter 11
Mouth and
Pharynx;
Teeth,
Respiratory
Organs

Figure 267. The larynx of the rhesus monkey, in side view. The hyoid apparatus, with its greater and lesser horns, is seated above and anterior to the larynx. The rhesus monkey has a resonating chamber in the laryngeal sac. Shown are the thyroid and cricoid cartilages, tracheal cartilages, and a membrane connecting the hyoid and thyroid cartilages. A muscle runs from thyroid to cricoid and covers most of the latter; there are several other small muscles proper to the larynx more deeply situated and not shown. (After Geist in Hartman and Straus: Anatomy of the Rhesus Monkey. Williams & Wilkins.)

open and close the glottis. This slitlike opening from the pharynx is protected in many forms by a flap of skin folding over it from the front, and, in mammals, this develops as the **epiglottis**, containing a special cartilage.

Many tetrapods are voiceless in any true sense. This is true of the salamanders and Gymnophiona among the amphibians and of the great majority of reptiles, although certain of these forms can make hissing or roaring noises by a violent expulsion of air through the glottis. In frogs, a few lizards and, notably, most mammals, the larynx is a vocal organ. Voice production is accomplished through the presence of a pair of **vocal cords**, ridges containing an elastic tissue, which are stretched across the larynx. The two cords can be closely appressed to one another, and set in vibration by the passage of a current of expired air between them.*

In birds, the larynx is present, but vocal cords are absent; voice production takes place in a special organ, the **syrinx** (Fig. 268). This is a structure somewhat comparable to the larynx but situated farther down the air passage, typically at the point at which the trachea divides into the two major bronchi.

Trachea and Bronchi. In *Polypterus,* the lungs expand directly from the pharyngeal opening, but in other lung-bearers, there is usually, for a greater or lesser distance, a median air passage, the trachea; this usually has a ciliated epithelium, and mucous cells and glands. In tetrapods, it leads from the larynx to the point where the two primary bronchi diverge (Fig. 260). In anurans and typical salamanders, in the absence of a developed neck, the trachea is of negligible length; a formed tracheal tube is, however, present in some long-bodied salamanders and in the Gymnophiona. In amniotes, in which head and trunk are separated by a distinct neck, a trachea is always present, its length, naturally, varying with the degree of elongation of the neck. Some birds (the swan, for example) have an extra length of tracheal tube coiled beneath the sternum; this may serve as a resonating device but more probably is

*In frogs, the sound may be enhanced by resonating sacs, commonly paired, which open into the floor of the mouth. Resonating chambers may also develop from the larynx itself, as in the case of the gibbon or the South American howler monkeys.

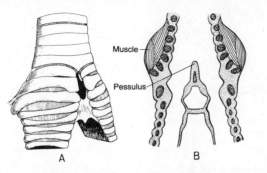

Muscle

Pessulus

A

B

Figure 268. The syrinx of a songbird (magpie). *A,* External view; *B,* in section. Vibratory membranes on the inner aspect of the two bronchi meet at the base of the trachea to form the median pessulus; further membranes may develop between the expanded rings at the tracheal forks. The syrinx is controlled by musculature of the hypobranchial group, which can be very complex and varies greatly among different groups of birds. (After Haecker.)

related to the slow wing-beat of a large flying bird being coupled with the frequency of breathing, even though a relatively immense volume of air must be passed through the lungs.

Snakes may have the back part of the trachea vascularized to act, in effect, as part of the lung. In almost every case, the trachea is stiffened by cartilaginous structures within its walls. In amphibians, these are rather irregular nodules; in amniotes, they are rings, generally incomplete except in some archosaurs and birds, in which they may be ossified as well as complete (cf. Figs. 164, 231, 266, 268).

In amphibians and some reptiles, the lungs spring almost directly from the end of the trachea; in most amniotes, however, the air duct itself bifurcates before reaching the lungs, giving rise to a pair of primary bronchi that may also be stiffened by cartilaginous rings. When the internal structure of the lung attains the complicated pattern found in some reptiles and in all birds and mammals, the primary bronchi run

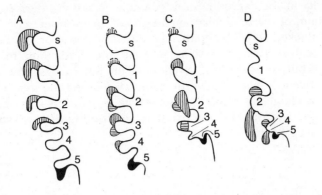

Figure 269. Diagrams of the gill pouches of the left side of the pharynx in *A,* a shark; *B,* a urodele; *C,* a lizard; *D,* a typical mammal; to show the derivation of thymus, parathyroid, and ultimobranchial bodies. The dorsal part of each gill pouch is, for purposes of the diagram, at the upper side. Broken outline, variable thymus derivatives; vertical hatching, thymus; horizontal hatching, parathyroid; solid black, ultimobranchial body. Abbreviation: *s,* spiracular pouch; the numbers indicate post-spiracular branchial pouches. (Mainly after Maurer.)

372

**Chapter 11
Mouth and
Pharynx;
Teeth,
Respiratory
Organs**

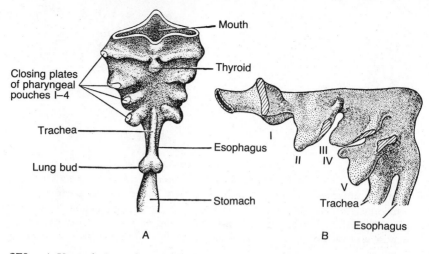

Figure 270. *A,* Ventral view of a model of the anterior part of the gut and its outgrowths in an embryo mammal *(Homo). B,* Lateral view of the pharynx in a slightly later embryo. Closing plates separating tips of branchial pouches from the surface are hatched; pouches, including the spiracular pouch (middle ear), indicated by Roman numerals. (From Arey.)

on into the lung substance before dividing into smaller bronchi, which in turn divide into bronchioles and subsidiary ducts.

Pharyngeal Derivatives

Despite its reduction in the higher vertebrate classes, the pharynx is important in every group as the embryonic source of glandular structures. In fishes, glands arise from the endodermal pharyngeal epithelium, and such structures are just as important in tetrapods. We have noted that the branchial epithelium in teleosts may take on glandular activity in the excretion of salts. Other glands are derived from the endo-dermal epithelium of the branchial pouches in all gnathostome classes. These glands are formed by thickening of the epithelium at either dorsal or ventral borders of the pouches—thickenings that later separate from the epithelium to become varied masses of tissue buried in the throat region of the adult (Figs. 269, 270). Certain of these pouch-derived glands, notably the parathyroids, and the thyroid, formed in the pharyngeal floor, are glands of internal secretion; the thymus is concerned with lym-phocyte formation and the immune reaction. Description of these structures will be deferred to later chapters.

Chapter 12

Digestive System

The mouth and pharynx, described in Chapter 11, are part of the digestive tube. They constitute, however, only its forward outposts, with gathering of food materials as their original major duty. The business of digestion is the function of the remainder of the digestive tract, to which the ancient and simple Anglo-Saxon term **gut** may properly be applied. The gut, so limited, together with its outgrowths—liver and pancreas—will be considered in this chapter.

Functions of the Gut. The functions of the gut may be considered under four heads: (1) **Transportation**. Once food materials are gathered, they must be carried along the "dis-assembly line" of the successive sectors of the gut, and wastes must be ultimately disposed of posteriorly as feces. (2) **Physical treatment**. Although food may be taken in as small particles or brought down to small size in the mouth by chewing, food materials frequently pass back into the gut in large masses, the size of which must be reduced before an efficient chemical attack on them can be made. Further, fluid materials must be added to bring the food to a pasty condition and thus facilitate digestive activity. (3) **Chemical treatment**. This is digestion in the technical sense, the breakdown of potentially useful raw materials in the food to relatively simple substances that can be used by the cells of the body. (4) **Absorption**. When this chemical breakdown has been accomplished, the useful products are absorbed by the intestinal wall for circulation to cells and storage areas.

Although cilia may in certain instances be present in anterior sections of the digestive tract, transportation is mainly the function of the visceral musculature that surrounds the digestive tube for its entire length. At the anterior end of the gut proper (the region of the esophagus), striated musculature, an extension of that of the pharynx, may be present to a variable degree. Along the remainder of the tube, the musculature consists of sheets of smooth muscle cells, typically including layers of both longitudinally and circularly arranged fibers that pass in a somewhat spiral manner about or along the gut (cf. Fig. 273). These muscles are innervated by nerves of the autonomic (involuntary) nervous system, which may either stimulate or inhibit their movement, and they may also be affected by hormones. To a considerable extent, however, they operate quite independently of any central control, being stimulated by the contraction of neighboring fibers or through a local nerve net. The major muscular activity moving food is **peristalsis**—successive waves of muscular contraction causing constrictions of the gut, which travel backward and push the food before them.

Food taken into the digestive tract must be reduced to a soft pulp, the **chyme**, before effective chemical attack can be made. This reduction is mainly accomplished by muscular action. Most effective here are (in contrast to peristalsis) rhythmic contractions of the gut musculature in food-containing areas, especially the stomach. The epithelium of the gut includes mucous cells or mucus-producing glands throughout its length; the slime they produce facilitates the passage of food and, further, gives an appropriate consistency.

The cells of the body need oxygen, water, a variety of inorganic salts, glucose, fats, amino acids, and small amounts of other organic compounds—the vitamins—which the cells themselves cannot manufacture. Oxygen is mainly obtained via the gills or lungs; for all other items, the animal must rely on the food materials brought to the digestive tract and, after absorption, carried on to the cells (or to the liver, for storage or reworking) by the circulatory system. Certain of the requirements—water, simple salts—can be obtained from nonliving sources: sea water, for example. The organic materials, however, can be obtained only from other organisms—from the plants by which alone most needed substances are manufactured or from animals that have already obtained these substances from plants.

Water, necessary salts, and vitamins, as well, are readily absorbed into the circulation by the lining of the intestine. But the situation is different in regard to the carbohydrates, fat materials, and proteins needed in the body. The cells of the body use these in large part as building stones to form more complex materials, and, in the animal food eaten, they are present, naturally, in similar complex form. As such, they would not, of course, be immediately useful to the body if they were taken into the circulation; but more than that, they are incapable of being absorbed into the lining of the gut while in this state. A chemical breakdown is requisite; this is accomplished by the action of the digestive enzymes, together with accessory substances that activate these catalytic agents or facilitate their work.

In every cell of the body, varied enzymes bring about the internal chemical changes associated with vital processes. In the gut, we have a special situation, because the enzymes formed in the digestive tube and glands act outside the confines of the individual cell. Such enzymes are discharged into the gut and attack the organic compounds in the food materials, reducing the complex carbohydrates to simple sugars, most fats to glycerol and fatty acids, and the proteins to amino acids. When the work of the enzymes has been accomplished, the organic materials are reduced to molecules that may pass through the cells lining the intestine and enter the circulation through the blood vessels and lymphatics surrounding that organ. Sugars and amino acids pass via veins to the liver and thence to all other parts of the body; most of the fats, reconstituted once the intestinal lining is passed, reach the general circulation by way of lymphatic vessels.

Development. In Chapter 5, we described the earliest stage in the development of the digestive tract: the formation, in one manner or another, of an archenteron, or primitive gut.

In amphioxus (cf. Fig. 78), the gut begins as a simple cylindric pouch, lined by a single layer of endodermal cells that contain a modest amount of yolk. In mesolecithal eggs, such as those of amphibians, the early gut is similarly constructed, except that the floor of the gut is not a simple epithelium, but a thick mass of endodermal cells rich in yolk. The embryo's belly is in consequence a paunchy, distended structure (cf. Fig. 83) until, at a later stage, this yolk has been absorbed into the animal's circulation and has contributed to its growth.

In macrolecithal eggs, the great amount of yolk has caused, we have seen, a radical change in the nature of the embryonic gut. Instead of forming a mass of yolky endodermal cells, the yolk fails to cleave at all, and the lining of the gut is at first simply a flat plate lying above the inert yolk. As the embryo develops, the endoderm grows down around the yolk, to enclose it eventually in a yolk sac that is an extension of the gut (cf. Figs. 76, 84 D, E, 87). The yolk is gradually digested, the sac being progressively reduced in size as absorption proceeds. Above the yolk sac, the embryo proper has meanwhile begun to take form; the endoderm lying below and within its other structures gradually separates from its yolk sac extension, although the two regions remain connected. The gut proper eventually becomes tubular, comparable at last to that of mesolecithal types. Mammals, despite the secondary absence of yolk, follow the macrolecithal pattern of gut development and produce a yolk sac, although an empty one (Figs. 76 D, 87, 271). In amniotes, we noted (cf. Fig. 76 B, D) the presence of a second diverticulum from the embryonic gut, the allantois.

For part of its development, the embryonic gut, whether in mesolecithal or macrolecithal types, usually has the essential form of a tube closed at both ends. Anteriorly, a stomodeal depression of the ectoderm develops; it forms the mouth and eventually breaks through to connect with the pharyngeal region of the gut (Fig. 228). Posteriorly, mesolecithal types have in the gastrula stage a posterior opening, the blastopore, in approximately the region of the anal or cloacal aperture of the adult. But even in most mesolecithal types, the blastopore closes over in an early postgastrular stage; in macrolecithal eggs, the blastopore is at best transitory and atyp-

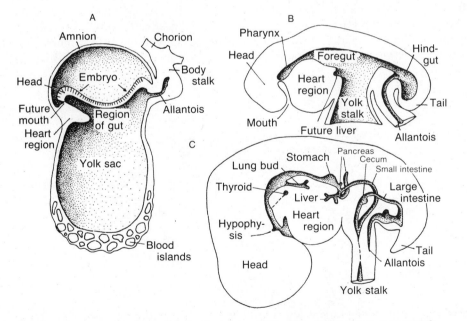

Figure 271. Diagrams to show the development of the digestive tract in a mammal (*Homo*). *A,* Stage somewhat later than that of Figure 76 *C; B,* stage slightly later than that of Figure 76 *D; C,* an embryo of about the age of that in Figure 87. (After Arey.)

ical in its appearance, and, posteriorly as anteriorly, the gut tube ends blindly for much of embryonic life. Analogous to the development of the stomodeum anteriorly is the appearance posteriorly of the **proctodeum** (cf. Fig. 228 A), an indipping pit of ectoderm beneath the tail, which comes to be closely apposed to the distal terminus of the embryonic gut, separated from it by a membrane that eventually disappears.

The postgastrular development of the gut consists primarily of a lengthening of the tube, a serial differentiation of successive regions into adult organs, and the building up of the complex, histologic structure of the wall (Figs. 271, 272). The most anterior compartment of the gut, the pharynx, has been described previously. The most posterior subdivision is the cloaca; this becomes associated with urinary and genital as well as digestive systems, and its consideration is best postponed until these other systems have been described. Subtracting these specialized terminal areas, remaining here for consideration are the major segments of the endodermal tube to which the term gut is for present purposes restricted.

Gut Regions. A study of the higher vertebrates gives one the impression that the succession of structures along the gut is consistent and uniform: esophagus, stomach, small intestine, large intestine, rectum, and anus. But when we extend our view to the lower groups, the picture becomes confused (cf. Fig. 274). As inspection of a shark or cyclostome gut makes evident, the distinction between large and small intestine breaks down completely in primitive vertebrates; the esophagus may be absent as a distinct region, and, in some fishes, the stomach is not differentiated. In amphioxus and cyclostomes, the postpharyngeal gut is essentially a single tubular unit; most familiar landmarks are absent.

It is, however, possible to establish in most vertebrates one constant and definite point of division along the length of the gut. Between stomach and intestine, there is generally a distinct **pylorus**, a constriction in the digestive tube that guards the gateway into the intestine and that is usually furnished with an efficient sphincter muscle. In some fish, the pylorus may be poorly marked or absent, but the fact that the bile

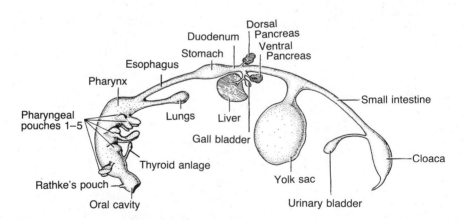

Figure 272. A diagram of the digestive tract and its outgrowths in an amniote embryo similar to that of Figure 271 *C,* but with the structures concerned shown as solid objects rather than in section. The first pharyngeal pouch here is the spiracle; the successive pouches are those numbered 1 through 4 elsewhere. (After Turner.)

duct from the liver always enters the intestine a short distance beyond the pylorus enables us, even so, to establish a line of demarcation.

The region anterior to the pylorus may be termed the **foregut**; the intestinal area beyond, the **hindgut**. In most vertebrates, there is some overlapping of function between the two areas, but primitively, it would seem, the two served different purposes: the foregut was merely a short and unimportant connecting link with the pharynx; the intestine, forming the hindgut, was alone responsible for chemical treatment of food as well as its absorption. In the course of vertebrate evolution, however, the foregut has assumed a greater importance; stomach and, in terrestrial forms, esophagus have grown to be regions of prominence, and the former has assumed a true digestive role.

Structure of the Gut Tube. Although highly variable from region to region and from form to form, the gut is generally formed, from the inside outward, of four successive layers, or "tunics" (Fig. 273). (1) The **tunica mucosa** includes the inter-

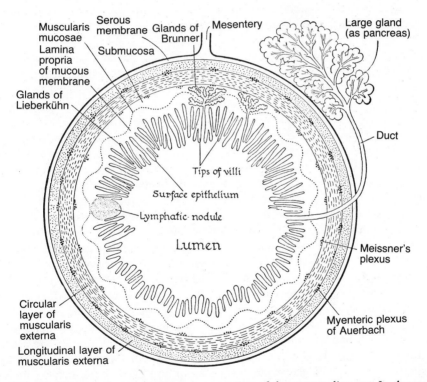

Figure 273. A diagram of a generalized cross section of the mammalian gut. In the upper half of the drawing, the mucous membrane is provided with glands and villi; in the lower half, it contains only villi. Meissner's plexus (submucous plexus) and the myenteric plexus are part of the autonomic nervous system; the former consists of sympathetic fibers, the latter, cells and fibers of the parasympathetic system. The tunica mucosa extends outward as far as the muscularis mucosae; it is followed outward in the section by the submucous tissue, the muscular tunic, in two layers, and finally, the serous tunic. The glands of Brunner and of Lieberkühn are characteristic of the mammalian small intestine. (From Maximow and Bloom.)

nal epithelium and a layer of connective tissue beneath it; this latter often contains smooth muscle fibers (the muscularis mucosae). The epithelium is the sole part of the tube formed from the endoderm. Anterior and posterior boundary zones (esophagus, rectum) may develop stratified epithelium; over the greater length of the gut, however, the lining is a mucous membrane of a simple columnar type. The cells of this epithelium have varied functions, as secretory cells or (posteriorly) absorptive areas. The epithelium may produce a variety of glandular structures, some the seat of formation of enzymes or other chemically active materials, others producing (as do some, at least, of the superficial cells in every region of the gut) a mucous material useful in softening and transporting food. (2) The **submucous tunic** is usually a thick layer, mainly connective tissue, but containing numerous small blood vessels and nerve cells and fibers. (3) The **muscular tunic** generally includes two prominent layers of smooth muscle: internally a circular layer capable of constricting the gut, externally a longitudinal layer capable of locally shortening its length. These muscles, in most cases, are especially well developed in the stomach. Between and adjacent to the two layers, there may be diffuse plexuses of cells and fibers of the autonomic nervous system (cf. p. 549). (4) For most of its length, the gut, lying in the peritoneal cavity, is surrounded externally by a **serous tunic** consisting of coelomic epithelium underlain by connective tissue.

Esophagus.   Not only in amphioxus and cyclostomes, but in a fair number of jawed fishes as well—chimaeras, lungfishes, certain teleosts—the entire foregut consists merely of a simply constructed length of tube interjected between pharynx and intestine (Fig. 274 *A, C, D*). Its epithelium is abundantly supplied with mucous cells and is often ciliated. No stomach is present in the forms listed; the term esophagus may perhaps be applied to this simple part of the gut. In all other living fishes—elasmobranchs and most ray-finned forms—a stomach is present in the foregut, leaving a short and ill-defined area anterior to it to be considered an esophagus. In many fishes, a sphincter may be found between the esophagus and stomach.

It is not until the tetrapods that the esophagus assumes prominence. Here, with the loss of gill-breathing, the pharynx becomes restricted in length to a short area within the confines of the head, and a distinct neck is developed in amniotes. With pharyngeal reduction, the esophagus elongates proportionately, as the neck develops, to form a tubular connection between the pharynx and the remaining portions of the gut, concentrated posteriorly in the abdominal region of the trunk back of the lungs.

Only exceptionally has the esophagus any function beyond that of transportation of food from mouth and pharynx to the more posterior parts of the gut in the abdomen. The epithelium is in general of a tough stratified squamous type, able to withstand rough food particles; longitudinal internal folds generally render the esophagus distensible. Glands are in general little developed except for mucous cells, but, in a few cases, there is greater glandular development and even some enzyme formation. In a number of elasmobranchs, amphibians, and reptiles (except for turtles), the epithelium may be ciliated; but, in general, food transport is due to peristaltic movements, and there is a stout muscular coat. This is, of course, visceral musculature (cf. p. 301), but the esophagus lies at the boundary zone between the regions of the gut with striated and smooth muscle, and the nature of the muscle fibers varies. In a majority of vertebrate groups, smooth muscle fibers are present. In many fishes, however, and again in mammals (particularly ruminants), there is a notable development

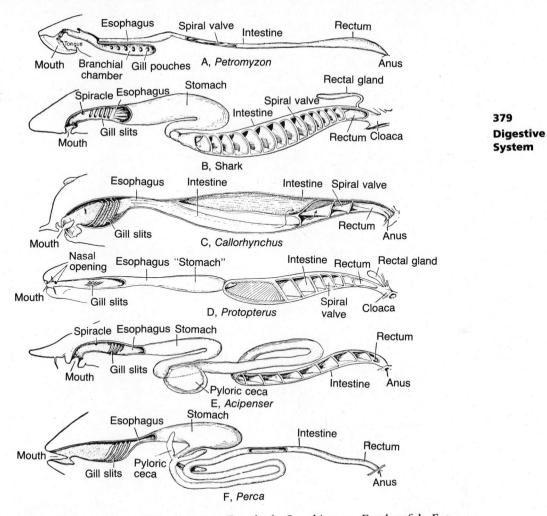

Figure 274.   Digestive tracts of *A,* a lamprey; *B,* a shark; *C,* a chimaera; *D,* a lungfish; *E,* a sturgeon; *F,* a teleost (perch). The "stomach" of the lungfish is nonglandular and is simply a somewhat enlarged section of the esophagus. (From Dean.)

of striated musculature; in some fishes, this extends to the stomach. Most of the digestive tract lies more or less free within the abdominal cavity; the esophagus, however, lies, for most of its length, within the compact structures of the neck and chest.

Two unusual developments may be noted. In the adult lamprey (cf. Figs. 256 *B,* 274 *A*), the pharynx is split into two parts. As we have seen, the lower segment is a blindly ending pouch, the upper part a tube leading directly from mouth to gut and hence giving the appearance of an elongate esophagus, although most of it is actually pharyngeal in origin. In birds, there develops part way down the esophagus a distensible sac, the **crop** (Fig. 275 *C*), which serves as a place for the temporary storage of grain or other food; in doves, the crop lining exudes, in both sexes, a milky material with which, regurgitated, the young are fed.

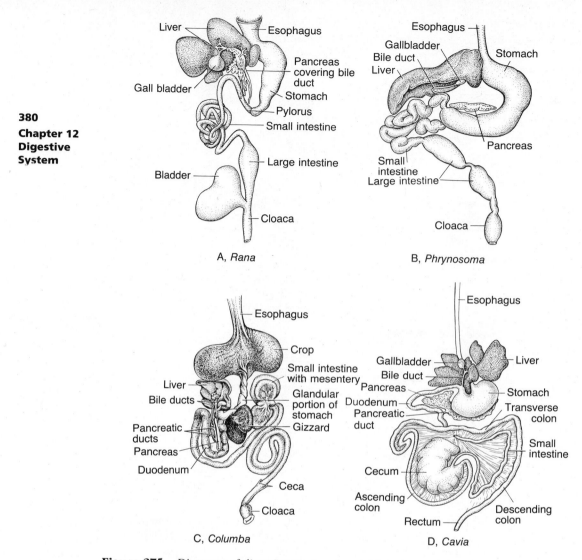

Figure 275. Diagrams of digestive tract and appendages, seen in ventral view, in *A,* a frog; *B,* a lizard; *C,* a bird (pigeon); *D,* a mammal (guinea pig). (*A* after Gaupp; *B* after Potter; *C* after Schimkewitsch.)

Stomach. The stomach, found in most (but not all) vertebrates, is a muscular, pouchlike expansion of the foregut, lying in the anterior part of the peritoneal or abdominal cavity. Its epithelial walls are generally highly folded and studded with mucous cells; in specific areas, deep-seated glands produce digestive secretions as well. Accustomed as we are to thinking of the stomach as a normal and useful part of the gut, its absence in a variety of types seems at first an anomalous situation. It is absent in amphioxus; in true vertebrates, it is absent not only in the cyclostomes but also in chimaeras, lungfishes, and a number of teleosts. All other vertebrates have a stomach, which, despite great variations in form, always has much the same function.

The stomach serves primarily as a place for storage of food awaiting reception into the intestine and also for the physical treatment of this food and for initial chemical treatment of proteins. These functions give some suggestion of the history of the stomach. In filter-feeders such as amphioxus (and, we believe, the ancient jawless ostracoderms), food particles were small and collected more or less continuously; they could be passed on directly to the intestine without the necessity of storage or preliminary treatment. Quite in contrast are the food habits of such a predaceous jawed fish as a shark. The food intake is irregular; a large quantity may be taken in at one time and must be stored until the intestine can take care of it. Further, food may be bolted in large chunks upon which the intestinal juices cannot act effectively without proper preparation. Presumably, the primary functions of the stomach, when developed early in the history of jawed fish, were, first, food storage and, second, physical treatment of the stored materials. The introduction of digestive enzymes was presumably a phylogenetic afterthought. Quite possibly, the absence of a stomach in cyclostomes is primitive. Chimaeras and lungfishes do not bolt their prey but cut it into bits before swallowing and, hence, have less need of a stomach as a storage area. Its absence in these groups may well be secondary as it is in some teleosts.

In regard to external shape (Figs. 275, 276), the central type of gastric structure is that seen in forms as far apart phylogenetically as sharks and man. At the end of

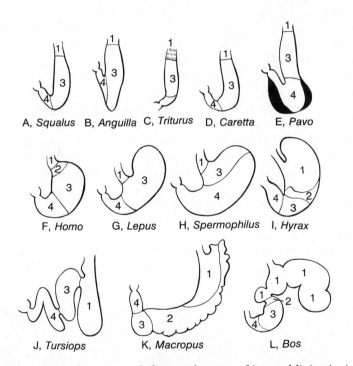

Figure 276. Diagrams to show stomach form and nature of internal lining in *A*, a shark; *B*, a teleost (eel); *C*, a salamander; *D*, a turtle; *E*, a bird (peacock—the thickened wall of the gizzard is indicated); *F*, man; *G*, a hare; *H*, a ground squirrel; *I*, the coney of Africa; *J*, a whale; *K*, a kangaroo; *L*, a cow. *1*, Epithelium of esophageal type (ciliated in *C*) which may penetrate into stomach, particularly in mammals; *2*, cardiac epithelium (found only in some mammals); *3*, fundic epithelium; *4*, pyloric epithelium. (After Pernkopf.)

the esophagus, the digestive tube curves to the left, expands into a major sac, and then ascends at the right to end at the pylorus, thus forming a J. Topographically, the proximal end of the stomach, closest to esophagus and heart, is termed the **cardiac region**; the expanded middle part, the **fundus**; the distal limb, the **pyloric region**. Gastric shapes may vary greatly from group to group, and the shape of the stomach in a dead, dissected animal may be quite different from that in life, since violent motions with marked changes of shape may occur from moment to moment in an active stomach (Fig. 277).

Highly important is the nature of the epithelium lining the various regions of the stomach. Beyond the presence throughout of mucus-secreting cells, four epithelial types may be distinguished. (1) **Esophageal**. The proximal end of the stomach in many mammals has a nonglandular, stratified epithelium similar to that of the esophagus; this region of the stomach is essentially an expanded part of the esophagus. (2) **Cardiac**. In mammals alone, there is usually distinguished a transitional region at the proximal end of the stomach in which the epithelium contains compound tubular glands with a mucous product. In this and the following regions, the endodermal lining of the gut is a simple columnar epithelium, although often folded in a complicated way. (3) **Fundic** (Fig. 278). Although the intestine is entirely competent to deal with chemical digestion by itself, the fundic region of the stomach is characterized by the presence of numerous tubular glands. In them (in addition to mucous secretion at the glandular neck) there are produced digestive enzymes, prominently **pepsin**, which aids in the breakdown of proteins preliminary to intestinal digestion, and a fat-splitting lipase. Hydrochloric acid, which gives the gastric juice an acid condition favorable for the action of pepsin, is also produced here. In most vertebrates, both enzymes and acid appear to be produced by cells of a single type. But in mammals, two sorts of cells are distinguishable, both often contained in the same tubule—**chief cells** producing enzymes, and **parietal cells** yielding hydrochloric acid. (4) **Pyloric**. This epithelial type, occurring in the distal part of the stomach, contains branched tubular glands, comparable to those of the cardiac region.

Although three of the four epithelial types just described are given names used in the topography of the stomach, their distribution is not necessarily correlated with the regions so named. External appearance may be misleading in regard to histologic conditions. Thus, for example, the stomach of a mouse is shaped much like that of a man, but, in contrast to human conditions, the greater part of the stomach lining is of the esophageal type.

There are numerous variants among vertebrates in gastric form and epithelial linings (cf. Fig. 276). The shape may vary greatly even in forms with a simple type of

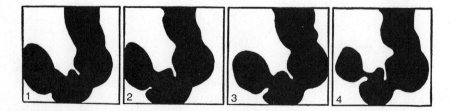

Figure 277. Skiagrams of a human stomach taken at intervals after eating, showing its variable shape when active. (After Cole, from Fulton.)

Figure 278. *Left,* a section of a portion of the fundic region of a mammalian stomach, showing the gastric pits *(p)* and the glands deep to the pits and opening into them. Mucous cells *(n)* are present in the neck region of the glands; below, two cell types can be distinguished: parietal cells *(a),* secreting hydrochloric acid, and smaller and more lightly staining chief cells, producing pepsin; *m,* smooth muscle cells. *Right,* a section through the pyloric epithelium, showing pits and glands of an essentially mucous type. (From Windle, Textbook of Histology, The McGraw-Hill Company.)

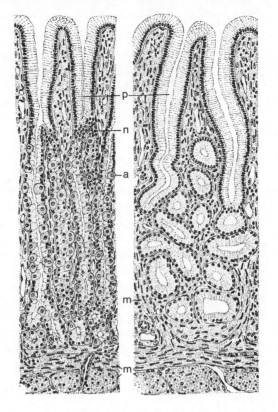

stomach. Teleosts, for example, often develop a deep V-shaped fundus, sometimes with a blind sac extending on from the apex of the V; on the other hand, a variety of fishes, long-bodied amphibians, lizards, and snakes have perfectly straight, cigar-shaped stomachs. The stomach has stout muscular walls, which, through repeated regional contraction, may squeeze and soften the food and aid in preparing it for intestinal treatment. Usually this muscular tissue is more or less evenly distributed; in crocodilians and birds, however, it is concentrated in a distal muscular compartment, the **gizzard**, in contrast to a proximal region, the **proventriculus** (Figs. 275 *C*, 276 *E*). The avian proventriculus is notable for the arrangement of the fundic glands in a series of pockets, giving the appearance of compound glands. In grain-eating birds, small stones are swallowed to lodge in the gizzard and aid this rough-walled chamber in its function as a grinding mill substituting for the lost teeth (and thereby keeping the weight of the grinding organ nearer the center of gravity—an important factor in a flying organism).

Among many specialized mammalian stomach types, that seen in the cud-chewing artiodactyl ungulates—the ruminants, such as cow, sheep, and deer—is notable (Fig. 279). In the stomachs of these forms, four chambers are present. Food when eaten descends to engage the attention of the first two of these chambers, the **rumen** and the **reticulum**. The former is a large pouch, the latter a smaller accessory chamber whose interior surface consists of a crisscross series of ridges with deep intervening pits (honeycomb tripe!). In these chambers, the food is reduced to a more workable pulp by the addition of liquid, is kneaded through the action of the muscular

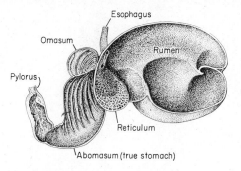

Esophagus

Omasum

Rumen

Pylorus

Reticulum

Abomasum (true stomach)

Figure 279. The stomach of a sheep, sectioned to show the four compartments characteristic of higher ruminants. (After Pernkopf.)

walls, and is subjected to fermentation through the action of bacteria and protozoans for the breakdown of complex plant materials, particularly cellulose. These microorganisms further aid the ruminants' metabolism by manufacturing useful organic compounds, including amino acids, proteins, and even vitamins; still further, they manufacture, from ingested carbohydrates, fatty acids that can be directly absorbed in the rumen.

The rumen and reticulum having done their best, the animal regurgitates the "cud" for further chewing, or rumination, if you will. On its second descent, the food bypasses both the rumen and reticulum, by way of a deep fold in the anterior wall of the latter chamber, and enters the **omasum** (or psalterium) whose parallel internal ridges reminded early investigators of the leaves of a hymn book. After further physical reworking here, the food at long last enters a final compartment, the **abomasum**. Here, all three types of epithelium distinctive of the mammalian stomach—cardiac, fundic, and pyloric (Fig. 276 *L*)—are present. Obviously, this compartment alone is the "proper" stomach; the other three in advance of it are essentially elaborations of its proximal part, in which, as we have noted, an esophagus-like epithelium is present in various other mammals. Kangaroos and some related marsupials show basically similar, although of course quite independent, adaptations.

The Intestine (Figs. 274, 275, 280, 281, 282).   The major stages in the true digestive process normally occur in the hindgut, the variably built intestine. More anterior segments of the digestive tract receive, transport, store, and prepare food materials. In the intestine, however, occur most—and primitively, all—the chemical processes that constitute digestion in a narrow sense. And here alone occurs, in most cases, the final crucial step, the absorption of the digested food for body use.

Figure 280. Lining of a portion of the small intestine of a turtle. The many, wavy ridges that run more or less lengthwise along the intestine are shown here enlarged about eight times.

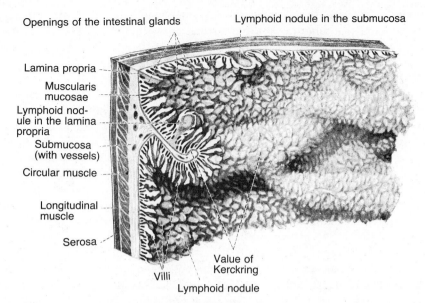

Openings of the intestinal glands

Lymphoid nodule in the submucosa

Lamina propria

Muscularis
mucosae

Lymphoid nod-
ule in the lamina
propria

Submucosa
(with vessels)

Circular muscle

Longitudinal
muscle

Serosa

Villi

Value of
Kerckring

Lymphoid nodule

Figure 281. Part of the wall of the mammalian small intestine. (After Maximow and Bloom.)

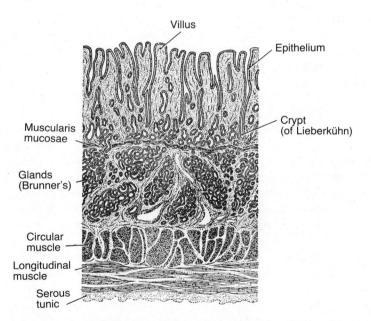

Villus

Epithelium

Muscularis
mucosae

Crypt
(of Lieberkühn)

Glands
(Brunner's)

Circular
muscle

Longitudinal
muscle

Serous
tunic

Figure 282. A section through the lining of the mammalian duodenum, showing intestinal villi, the crypts between them, a thin muscle layer (the muscularis mucosae) at the base of the mucosa, Brunner's glands (mucus-secreting), muscular tunic, and the serous tunic, facing the coelomic cavity. (From Young, The Life of Mammals, Oxford University Press.)

Although the production of digestive enzymes may have been, in the vertebrate ancestors, an exclusive function of the intestinal epithelium itself, that is not the case in living vertebrates. A special glandular outgrowth of the intestine, the pancreas, has evolved to become the place of origin of a considerable part of these enzymes, poured from it into the anterior end of the intestine; pepsin and other enzymes have come to be formed in the stomach; an enzyme may even be produced by the salivary glands of the oral region. Despite all this, the intestine itself is still important as a seat of enzyme production. It secretes a series of these substances, forming an intestinal juice that is instrumental in the final breakdown of all types of organic foods. Several enzymes aid in preparing sugars for absorption; others split starches and fats, activate trypsin, break down peptides into amino acids, and split phosphoric acid from nucleotides.

Most production of intestinal enzymes takes place in a series of small glandular outgrowths in the intestinal walls. The intestinal epithelium proper (Fig. 282) is of a simple columnar type. Occasional goblet cells supply mucus to lubricate the system, and, despite the development of glands, a certain amount of enzyme formation may occur in the superficial cells; these cells, further, produce hormones that, via the blood stream, stimulate to activity the pancreas, gallbladder, and intestinal glands (cf. p. 617). In general, however, the intestinal epithelium has as its dominant function the absorption of digested food. Into its cells pass the water, salts, simple sugars, fats as glycerol and fatty acids (or in part as neutral fats), amino acids, and vitamins needed by the cells of the body; these are discharged from the inner surface of the epithelium into the capillaries and lymphatics with which the intestine is richly supplied.

To allow sufficient absorption, a large area of intestinal epithelium is necessary; a short, straight, smoothly lined tube in general will not suffice. Methods of increasing this internal surface are demanded by every vertebrate. This is particularly so in large animals; for here (as in so many cases), surface-volume relations enter the picture, and the area capable of absorption must be roughly proportional to the volume of the tissues to be nourished. Increases in area may be brought about on three levels of magnitude: (1) Countless small foldings of the gut lining, essentially of microscopic size, which involve only the innermost, mucous tunic of the gut, are found in all vertebrates. Primitively these folds of the mucous layer appear to have been in the form of a network of tiny ridges with pockets between them; a more specialized condition, prominent in mammals, is the presence of countless **villi**, small finger-like projections from the gut wall (Fig. 281). These are not the same as microvilli, which are *much* smaller projections of individual cells present on most epithelia in the gut and elsewhere. (2) In numerous instances, somewhat larger folds, in which the submucous layer is involved, may be present, often as a series of ridges (Fig. 280) or rings (as in the mammalian folds of Kerckring; Fig. 281). (3) Major structural developments occur that greatly increase the area of the intestinal surface. Two are most notable: the intestine with a spiral valve, short, but of complicated internal structure, characteristic of primitive vertebrates; and the slender tubular intestine, developed by teleosts and by tetrapods, in which length, and hence surface area, is increased by coiling.

The Spiral Intestine (cf. Figs. 23, 26, 31, 274 *A–E*).   In some members of every major group of fishes, there is a type of intestine seemingly primitive for vertebrates: the spiral intestine. This, as characteristically developed in elasmobranchs, is typically a large cigar-shaped body, extending almost straight anteroposteriorly and occupying most of the length of the peritoneal cavity. Internally, it is seen to have a complex structure. In addition to minor epithelial folds, the surface area is greatly increased by the presence of a **spiral valve**. This is a fold of epithelium and connective tissue extending from one end of the intestine to the other, the attachment of which in typical forms twists numerous times in a spiral manner along the walls of the gut. A somewhat similar structure would be formed if we took a carpenter's auger, or bit, and enclosed it in a tube. In some sharks, the spiral valve may become exceedingly complex, with dozens of turns along its course. The development of the spiral fold greatly increases the internal area of the gut, and also its functional length, because the food must follow a long course down around the twists of the spiral staircase to reach the end of the seemingly short intestine. In a minority of sharks, the spiral valve twists only a little and hence is relatively short; the valvular fold, however, is highly developed and rolled up into a great scroll running lengthwise along the intestine (Fig. 274 C).

The primitive nature of the spiral structure is demonstrated by its widespread presence in other fish groups, although the number of turns is never as high as in typical sharks. The spiral intestine is present in chimaeras, and fossils demonstrate its presence even in the ancient placoderms. Further, it was present in the early bony fishes, because living lungfishes and all lower actinopterygians retain a spiral valve, and even a few teleosts have a vestige of this structure. Cyclostomes are more problematic; the lamprey has a slightly spiraled ridge along its intestinal wall, which may be a degenerate spiral valve, a partly formed spiral valve, or a structure paralleling a true spiral valve. Hagfish lack even this ridge.

The intestinal region, here termed as a whole the hindgut, is often divided into two parts, which in mammals form the small and large intestines. Conditions in fishes show that this division, based on human conditions, is not a general one. In sharks, the spiral intestine alone includes almost the entire length of the digestive tube between the stomach and cloaca. Anteriorly, there is merely a short connecting piece, at the most, between stomach and spiral valve region. Posteriorly, there is only a short terminal segment, somewhat comparable to the rectum of higher types, leading to the cloaca. Sharks have in this rectal region, a finger-shaped (or digitiform) **rectal gland**, which has much the appearance of the cecum of amniotes. It is, however, not homologous with that structure but secretes a thick solution of sodium chloride and is thus important in ridding the shark blood of excess salts (a problem discussed in the following chapter). Lungfishes, too, have a rectal gland, but it is not homologous with that of elasmobranchs.

The Intestine in Higher Vertebrates. The teleosts, and terrestrial vertebrates as well, have abandoned the old-fashioned spiral intestine for a new type of hindgut. In these forms (Figs. 274, 275), the intestine is a slender tube, without major internal folds. In compensation, however, it is generally much elongated; in teleosts, its length is, on the average, half again that of the whole body. A long, coiled part of the teleost hindgut is highly active in digestion, functionally comparable to the spiral intestine of their ancestors, on the one hand, and on the other roughly comparable to the small intestine of their tetrapod cousins. Distal to this, we find a relatively short and straight terminal segment of the gut, somewhat similar to the colon of tetrapods, leading to the anus. Apart from coiling of the gut, most teleosts and their lower actinopterygian relatives add to their intestinal surface by the development of distinctive **pyloric ceca**—pouches, sometimes very numerous (up to 900 have been reported!), from the proximal end of the intestine, into which food materials may enter and be absorbed.

In the vertebrate embryo, the intestine is supported dorsally by a straight and simple dorsal mesentery. When, in teleost or tetrapod, intestinal coiling occurs, this mesentery becomes a complex, tangled structure, fanning out here, crumpled there; frequently, the sheet supporting one intestinal coil may fuse with an adjacent one, or an intestinal loop may fuse with the dorsal coelomic wall (cf. Figs. 223, 226).

In both teleosts and tetrapods, there are enormous differences from form to form in the degree of convolution of the intestine and consequently in the amount of absorptive surface as well as in the length of the food channel. Two factors may be correlated with intestinal length: absolute size, which we have already discussed, and food habits. Plant food, often in the form of compact masses of complex carbohydrates, particularly cellulose, is difficult to digest and its products are difficult to absorb, and, in general, the intestine is longer in herbivores than in flesh-eaters. From teleosts to mammals (although with many variants), small carnivores tend to have the shortest intestines, large herbivores the longest.

Although a spiral intestine was present in the sarcopterygian ancestors of the terrestrial vertebrates, no living tetrapod has the slightest trace of this organ. Instead, tetrapods have paralleled the teleosts in the development of a rather simple tubular intestine, coiled to a variable degree. A seemingly primitive and generalized tetrapod type is that found in a variety of amphibians and reptiles (Fig. 275 A, B). The hindgut is usually divided into two distinct segments. The proximal, and longer, part is a **small intestine**, of narrow calibre, usually considerably coiled and often much longer than the entire body of the animal. At the end of the small intestine, there is, even in amphibians, an **ileocecal valve**, a constriction beyond which may lie a well-defined region, broadly homologous with the **colon**, or large intestine, of mammals; this is a relatively short, straight segment, expanded proximally and well furnished with mucous lubrication. "Large" and "small" refer to the usual diameters of the parts of the intestine; lengths, and even volumes, are the reverse, with the small intestine bigger. In amniotes, there is usually a small outpocketing, a **cecum**, at the proximal end of the colon; that of reptiles is median and usually dorsal in origin.

Materials that were not absorbed in the small intestine pass into the colon to be collected into the feces. Digestive juices carried down from the small intestine may continue their work while food materials are in the colon. This region is, further, populated with bacteria that putrefy the food residues, break down proteins into a variety of substances that are of dubious value to the organism or are actually toxic, and may effect the difficult task of digesting the complex plant carbohydrate, cellulose. Through the walls of the colon is absorbed much of the water remaining in the prospective feces as well as some further food and part of the products resulting from putrefaction.

In birds (Fig. 275 C), the small intestine is similar to that of lower tetrapods in build but is more highly convoluted; on the average, the length of its coils is more than eight times that of the whole body. This great elongation is presumably correlated not merely with the herbivorous habits of a large percentage of birds, but even more with their great activity and high metabolic rate, which results in an increased demand for food and the consequent requirement of additional absorptive surface in the intestine. The large intestine is relatively short and not expanded proximally. Distally, it is expanded and not sharply marked off from the cloaca. Because birds have no functional urinary bladder, urine and feces tend to be mixed, to be discharged as a slushy material. Paired ceca are generally present.

In mammals (Fig. 275 *D*), the small intestine has undergone a development comparable to that of birds. In relation, again, to metabolic needs, the intestine is complexly coiled, with a length that averages seven to eight times that of the body, and is several times greater than this figure in some large herbivores; the intestine of an ox may be more than 50 m long! In man, it is customary to distinguish three successive regions of the small intestine (duodenum, jejunum, ileum), but the differences between them are not marked, and, in many mammals, these terms can be applied only in a rather arbitrary manner. In mammals, the embryonic yolk stalk (cf. Fig. 271) lies in the region that becomes the ileum in the adult; as an individual variant, a short length of the stalk may persist as **Meckel's diverticulum**, which sometimes requires surgical attention in man.

In some small mammals, the large intestine, or colon, usually (but not always) separated by an ileocecal valve from the small intestine, may be a short, straight structure; in most mammals, however, this segment is highly developed, much more so than is usual in other tetrapod classes. It is a tube of considerable diameter, but of irregular shape, usually with a series of outpocketings and longitudinal muscle bands (absent, however, in carnivores and ruminants). Its characteristic shape, as seen from the ventral surface, is that of an inverted U, or of a question mark. The morphologically proximal segment is the ascending colon, running anteriorly along the right side of the body; the small intestine enters this segment at a sharp angle. After a flexure, the descending colon runs posteriorly along the left wall of the body cavity. In primates, a transverse segment is present between ascending and descending portions, and, in most herbivores, further coiling occurs.

Almost every mammal has a cecum, furnishing an additional area for colonic functions and a reservoir of intestinal bacteria. Ceca frequently contain numerous lymph nodes in their walls. As its development shows, the cecum is usually a ventral pocket of the gut; but, in most mammals, the adult condition gives it the appearance of being a part of the ascending limb of the colon, with the small intestine opening into its side. Hyracoids, oddly enough, resemble birds in having paired ceca. Such variation casts serious doubt on the homology of the cecum among amniotes. Frequently its diameter is as great as that of the colon itself, or greater. Its length is variable; it may be long and even much coiled in herbivorous animals. In man, it terminates in the narrow **vermiform appendix**. This is frequently cited as a vestigial organ; however, a terminal appendix is a fairly common feature in the cecum of mammals, occurring in a host of primates and a number of rodents. Its major importance would appear to be financial support of the surgical profession.

The mammalian **rectum**, endowed with muscular sphincters of both smooth and striated types, is a terminal piece of the digestive tract, leading to the anus. As will be seen (cf. p. 437), it is a part of the embryonic cloaca and hence not strictly comparable to the terminal portion of the gut to which this name is sometimes given in groups in which a cloaca is retained in the adult.

The Liver. In concluding this chapter, we will discuss two organs, the liver and pancreas, derived embryologically from the gut endoderm, which are important both for secretions that they furnish to the intestine and for their functions in the metabolism of food already digested.

Amphioxus has a large saclike structure, often called an **hepatic diverticulum**, branching off the gut in a position comparable to that of the liver (cf. Fig. 4) and having a venous portal system corresponding to the hepatic portal. But this sac, with a ciliated glandular epithelium, seems to be a seat of enzyme production and food absorption and lacks the characteristic metabolic functions of the liver. If this sac is truly antecedent to the liver, there was certainly a major evolutionary change above the amphioxus stage. All vertebrates possess a definitive liver which, as the most massive of the viscera, always occupies a considerable volume in the more ventral and anterior part of the abdomen. As shown by its embryonic history, the liver develops as an outpocketing of the gut in the ventral mesentery below the anterior part of the intestine (cf. Fig. 2 *D*) and is, hence, covered by coelomic epithelium. It remains connected to the stomach by a part of the ventral mesentery, termed the **lesser omentum**. Below the liver, a persistent part of the ventral mesentery, the **falciform ligament**, ties it to the ventral body wall. Its major attachment, however, is an anterior one. In lower vertebrates, as we have seen, a transverse septum develops, separating the pericardial cavity from the remainder of the coelom (cf. p. 317); in mammals, this septum is incorporated in the diaphragm. The liver is attached to septum or diaphragm—broadly, or by a narrowed **coronary ligament**—and from this point extends backward to expand into the abdominal cavity.

From the point of view of gross morphology, the liver is of little interest. It has no constant form and needs none; sufficient bulk and an appropriate internal arrangement, on a microscopic level, of its cells and blood vessels are all that are necessary for its proper functioning. In most types, more or less distinct right and left lobes are developed, but snakes and similarly shaped animals often have undivided, cigarshaped ones. Details of arrangement and lobation vary greatly from form to form and even among individuals. Ancient priests divined the future from the liver patterns of animal sacrifices; there was plenty of scope for interpretive imagination in such haruspication. Essentially the liver accommodates its form to that of the other viscera and expands into any part of the abdominal area not needed by other structures.

In general, as we have noted, the endoderm forms only the thin lining of the gut and its glands. Only in the liver (and to a lesser extent in the pancreas) does endodermal material bulk large in body composition. Except for a framework of connective tissue and the varied vessels that weave through its substance, the liver is composed entirely of the countless **liver cells** of endodermal derivation. These are polyhedral structures, morphologically similar in all parts of the liver. Despite a deceptive simplicity in microscopic appearance, these cells play a complex series of varied roles in the life of the body.

A minor function of the liver is that of a gland tributary to the intestine. However, historically the liver presumably arose as a glandular structure (whether or not homologous with the pouch in amphioxus) and embryologically arises as a glandlike outgrowth from the gut tube (Figs. 228 *A*, 271 *C*, 272). The liver appears at an early stage in the development of the primitive gut as thickened masses of tissue that push out ventrally and anteriorly from the intestinal wall, surrounding a median tubular outgrowth. The main mass of liver tissue extends forward in the developing ventral mesentery to the region of the transverse septum, leaving behind a side pocket of the tube that forms the gallbladder. From its anterior base in the septum or diaphragm the liver tissue rapidly expands posteriorly into the coelom.

In the adult, the primary glandular structure of the liver is for the most part lost. Throughout the liver there remains, however, a network of tiny **bile tubules** with which every cell is in contact. These tubules gather hepatic secretion, the **bile**, into one or more **hepatic ducts** draining the liver (Fig. 283). Part way toward the intestine, this main duct is joined by the **cystic duct**, connected with the **gallbladder**, in which bile may be stored. Beyond this junction, a **common bile duct** (ductus choledochus) typically empties into the small intestine, sometimes in association with a duct from the pancreas. This normal arrangement is subject to great variation. Generally, there is a valve at the outlet of the bile duct regulating its discharge into the intestine. The gallbladder may be absent (as, for example, in lampreys and many birds and mammals), or several separate bile ducts may open into the intestine. Although adult lampreys are often said to lack bile ducts completely, a small one has recently been reported.

Of the bile contents, most are purely excretory, including products of decomposition of protein materials and bile pigments derived from the hemoglobin of worn-out red blood cells. The only actively useful materials are the bile salts, which, discharged into the intestine, aid a pancreatic enzyme in the splitting and absorbing of fats. Cholesterol, present in the bile, may solidify to form gall stones.

The major duties of the liver, however, are not those directly connected with digestion, but with the treatment, within the confines of the liver, of food materials after their digestion and absorption into the body. In part, this activity is that of a storage depot. All classes of food materials may be stored in the liver, but particularly notable from a quantitative point of view is storage of carbohydrates. Large quantities of the simple sugars received into the body are stored in the liver as glycogen and released again into the blood stream when needed. Some fish, such as the dogfish and the cod, store much fat (or, since it is rather fluid, oil) in the liver. Every cell of the body is capable of producing, for its own needs, various changes in the food materials it receives; the hepatic cells do more, acting as chemical factories for the body as a whole. Proteins may be synthesized, fats may be altered in composition, proteins and fats may be made into carbohydrates, blood cells may be done away with, and nitrogenous cell wastes such as ammonia may be transformed into uric acid or urea to be eventually excreted by the kidneys.

For its major functions as a storage depot and a manufacturing plant, the liver is strategically situated along a "main line" of the transportation system formed by the blood vessels. The liver receives the oxygen necessary for its activities by way of an arterial blood supply. Bulking much larger is the venous flow through the liver. As

Figure 283. Diagram of an early stage of liver development to show the relationship of the ducts.

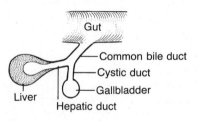

will be described later (cf. pp. 471–472), all the venous vessels from the intestine carrying absorbed food collect as the hepatic portal system of veins that carries blood to the liver; after passing through a series of sinusoids in that organ, this venous blood finally reaches a hepatic vein that runs on toward the heart (Fig. 284). From the vessels that they border, the cells of the liver thus have the first opportunity to select, for storage or chemical modification, food materials newly arrived in the body.

In relation to the overwhelming importance of its metabolic rather than its secretory functions, the liver has tended to abandon its original glandular structure in favor of one dominated by the proper arrangement of cells in relation to the circulatory vessels. The pattern varies from group to group, but in some mammals, much of the liver appears to consist of **lobules** of microscopic size (Figs. 284, 285, 286). These are, crudely, polyhedral blocks of cells arranged in what one may term a three-dimensional honeycomb. Surrounding each lobule are connective tissue sheets in which are located a triad of vessels: branches of the bile duct and of both the hepatic portal vein and the hepatic artery bringing blood inward. Bile tubules and capillaries from the artery penetrate into the lobule from outside, as do sinusoids from the portal system. Inside each lobule, a central cavity contains a **central vein**, which carries the blood from the sinusoids onward to the hepatic vein. The liver tissue between the sinusoids has the appearance of strands or cords of cells. Actually, however, these apparent cords are cut sections of plates of cells (Fig. 286) lying in a tangled network about the sinusoids. These plates are generally not more than two cells thick, so that every cell faces a sinusoid on at least one side; in birds and mammals, they are only one cell thick.

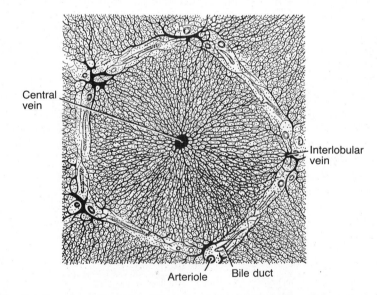

Figure 284. Section of a mammalian liver, showing a lobule and portions of others. The portal system of veins has been injected (black), showing the course of the blood from interlobular veins via a multitude of sinusoids through the sheets of liver cells to central veins of the lobules; branches of the bile duct and of the hepatic artery are seen in the interlobular septa. (After Young, The Life of Mammals, Oxford University Press.)

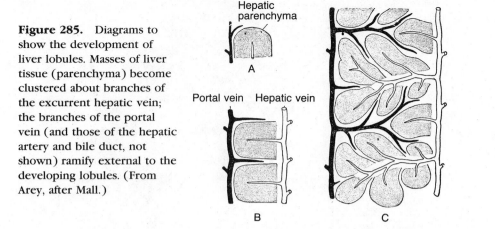

Figure 285. Diagrams to show the development of liver lobules. Masses of liver tissue (parenchyma) become clustered about branches of the excurrent hepatic vein; the branches of the portal vein (and those of the hepatic artery and bile duct, not shown) ramify external to the developing lobules. (From Arey, after Mall.)

The Pancreas.   Common to all vertebrates, this organ is generally found as a mass of soft tissue lying in the mesentery not far distal to the stomach and morphologically dorsal to the intestine; it opens by one or more ducts into the proximal part of the intestine near the point of entrance of the bile duct. It is a major item among the glandular materials that appear on the table as sweetbreads. In most vertebrates, it is a moderately compact if somewhat amorphous body, but in various cases, notably teleost fishes (but also in many other things—rabbits, for example), it may be diffuse, spreading out thinly along the mesenteries and even invading the substance of the adjacent liver or spleen. In many instances, there is only a single pancreatic duct. The opening of this duct is variable in position; it may enter either morphologically dorsal or ventral surfaces of the intestine, and, if ventral, may be combined with that of the common bile duct. Multiple ducts may occur; three, for example, are usual in birds.

Histologically, the greater part of the pancreas shows a typically exocrine glandular structure, with fine branching ducts collecting secretions produced by the cells of terminal acini and passing them on to the duct system (Fig. 287). In almost all forms, however, areas of tissue of another type can be seen distributed through the gland as isolated pancreatic islands; these form an endocrine organ, discussed in Chapter 17.

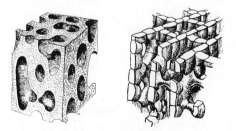

Figure 286.   Diagrammatic enlargements of a small portion of a liver lobule (such as is shown in Fig. 284) to show the structure of the plates of liver cells. The plates are separated and perforated by channels, which in life are occupied by sinusoids. *Left,* typical structure in a lower vertebrate; the plates are in general two cells thick. *Right,* mammalian structure; the plates are in general one cell in thickness. (After Elias and Bengelsdorf, and Elias.)

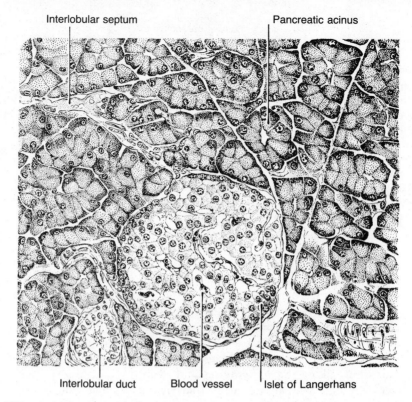

Interlobular septum   Pancreatic acinus

Interlobular duct   Blood vessel   Islet of Langerhans

Figure 287. A section through the pancreas of a rat, to show both exocrine tissue (pancreatic acini) and a pancreatic island. A duct of the exocrine part of the gland and a connective tissue septum between lobes are also shown. (From Turner.)

The exocrine functions of the pancreas are well known. The pancreas secretes a number of enzymes (or rather proenzymes) that are discharged into the proximal end of the intestine and are responsible for a great part of the digestive activities of the gut. The pancreatic enzymes are important for the digestion of all three major types of organic food. Only one is known for fat digestion, but there are two or more for both carbohydrates and proteins, and perhaps as many as ten or so are present.

The functional "reason" for the evolution of the pancreas is readily deduced. Originally, the intestine itself was presumably the place of production of all the enzymes that it utilizes. Some, as we have seen, are still produced in the gut, but the segregation of much of this type of activity in a distinct organ would appear to have been an advantageous development. Conditions in amphioxus and the cyclostomes render such a story reasonable. In the former, there is no separate pancreas, but groups of cells lying in the wall of the anterior part of the intestine show the characteristics of pancreatic cells. The lamprey shows a transitional condition. There is still no formed pancreas, but these cells have formed a series of tiny separate glands around the gut near the opening of the bile duct. Lungfishes show a similar condition, with the pancreatic tissue embedded in the wall of the intestine.

The embryologic picture throws light on the varied nature of the pancreatic ducts (Fig. 288). The pancreas is not a single structure, but a compound one. In early

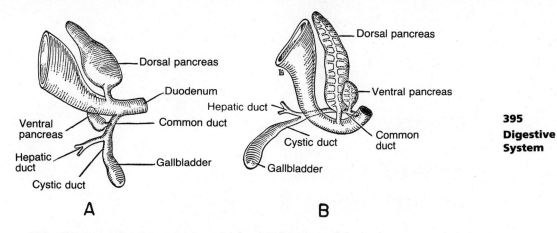

A   B

Figure 288.   Diagrams showing two stages in the development of the human pancreas. *A,* An early stage, with both dorsal and (smaller) ventral pancreas developing. *B,* Later stage in which dorsal and ventral portions are beginning to fuse. (After Arey.)

developmental stages, clusters of pancreatic cells are often found in the wall of the gut near the area in which the liver diverticulum is developing. A major outpocketing of pancreatic tissue takes place at the dorsal edge of the gut. Part, however, of the pancreatic tissue may become involved in the mouth of the future bile duct, and one or a pair of pancreatic outgrowths forms in the walls of that tube. These tend to grow upward and fuse with the dorsal pancreas. Not always do all three structures persist in the adult, but usually the dorsal outgrowth and one, at least, of the ventral pair contribute to the adult pancreas. Each part may retain a separate duct; on the other hand, the duct systems usually fuse, so a single outlet, dorsal or ventral, may serve the entire gland. As still another variant, the dorsal pancreas may fail to separate as distinctly as is usual from the gut, and its various parts may all drain separately into the intestine, somewhat as in the presumed primitive condition.

Chapter 13

Excretory and Reproductive Systems

From a functional point of view, a joint consideration of urinary and genital systems seems absurd, because excretion and reproduction have nothing in common. Morphologically, however, the two systems are closely associated in their mode of development and their use of common ducts, and it is impossible to describe one without numerous cross references to the other. This association appears to be due mainly to embryonic propinquity; in the embryo, the major organs of the two systems arise in areas of the mesoderm, which lie close to one another in the walls of the trunk near the upper rim of the coelomic cavity (Fig. 302).

Urinary Organs

Kidney Tubule Structure and Function. Paired kidneys (Greek *nephros*, Latin *ren*), developed in varied form in all vertebrates, are the major organs of the urinary system. The basic unit within the kidney is the minute **kidney tubule**, **renal tubule**, or **nephron**; the numerous tubules connect with a system of ducts that eventually leads, posteriorly, to the body surface.

Among vertebrates as a whole, the most generalized type of tubule is that shown diagrammatically in Figure 289 *A*; tubules of this type are present in such diverse forms as sharks, freshwater and many marine teleosts, and amphibians. The proximal part of the unit is the spherical **renal corpuscle** (Malpighian corpuscle). If sectioned, this is seen to consist of two parts, a **glomerulus** and a **capsule** (Bowman's capsule). The glomerulus is formed by the circulatory system; it is a compact little cluster of small blood vessels the size of capillaries, which coil through the mass in intricate, sometimes branching and anastomosing, loops (Fig. 290). The capsule is a proximal part of the tubule; its shape is that of a double-layered hemisphere. The inner layer, closely adherent to the lining of the vessels of the glomerulus, consists of complexly stellate cells, **podocytes**, with narrow spaces between their branches, so that in places only the basement membrane separates the blood vessels from the cavity within the capsule. This cavity is continuous with that of the **convoluted tubule**.* The tubule may attain considerable (relative) length; usually distinct proximal and

---

*Note that the term tubule is generally used in two ways: (1) as a synonym of nephron, i.e., a name for the entire nephric unit of renal corpuscle and convoluted tubule; and (2), more properly, for the latter structure only.

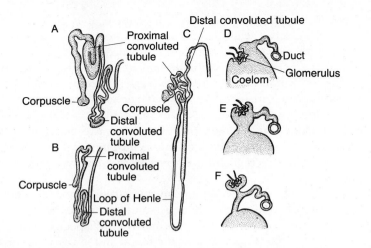

Figure 289. Renal tubules. *A* to *C,* The three major types common in adult vertebrates. *A,* Presumably the most primitive, with corpuscle of good size, found in elasmobranchs, freshwater bony fishes, amphibians. *B,* Corpuscle reduced or absent, characteristic of marine teleosts, reptiles. *C,* Corpuscle large; a loop of Henle inserted; found in mammals, birds. *D* to *F,* Primitive tubule types found in lower vertebrates, principally in the embryo, and perhaps illustrating the early evolution of renal tubules. *D,* Tubule runs from coelom to renal duct; glomerulus, if present in coelom, not associated with tubule. *E,* Special small coelomic chamber formed for glomerulus. *F,* This chamber has become the capsule of a renal corpuscle; tubule still connects with coelom; closing the coelomic opening leads to more progressive tubule type.

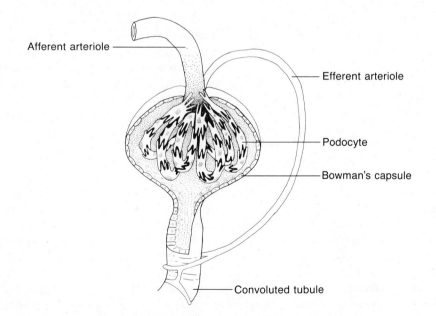

Figure 290. A mammalian renal corpuscle. Here the vessels around the tubule come from the efferent arteriole; in animals with a renal portal circulation, capillaries from this system surround the tube; they are frequently associated with ones from the efferent arteriole, but this is not always the case. The glomerulus would normally contain more capillary loops than are diagrammed here.

distal segments can be distinguished. Along its course, it is closely bordered by a network of capillary vessels. Distally, each tubule connects, in a manner that varies in different groups, to ducts leading to the exterior of the body.

The result of the activity of the tubule and its corpuscle is the production of urine, destined for excretion and derived, obviously, from the associated circulatory vessels. Urine is composed mainly of water but contains variable amounts of other substances in solution. These include salt ions, particularly sodium, potassium, chloride, and sulfate; bulking much larger, however, are waste materials, mainly simple nitrogenous compounds, generally urea or uric acid. The major functions of the kidney tubules are twofold: (1) regulation of the internal environment, and (2) elimination of waste.

398
Chapter 13
Excretory
and Repro-
ductive
Systems

The cells of the body must live in a proper environment, and a most important feature of this is the continued presence, in solution in the interstitial fluids that surround them, of appropriate amounts and proportions of specific simple salts. These ions are present in the same proportions in the blood and in the interstitial fluid with which it readily interchanges through capillary walls. The maintenance of proper salt content in the body liquids as a whole demands a proper balance between body intake, mainly via the intestine, and output. Such structures as the salt-excreting glands of the lamprey and teleost gills, the shark rectal gland, the nasal glands of marine birds, or the sweat glands of the mammalian skin may aid in the removal of salt. The kidney tubules, however, furnish the main means for the elimination of excess salts on the one hand or, on the other, for the removal of excess amounts of liquid that tend to dilute too greatly the body fluid.

Apart from regulation of salt and liquid content, however, there is a further important function of the kidney tubule: the elimination of wastes and harmful materials from the body. Digestion may bring into the body fluid compounds that are, for cellular utilization, inert or may be actually destructive or poisonous; the kidneys are frequently effective in removing them. Most important of materials needing removal are the broken-down products of metabolism, particularly of proteins. Metabolism of any food produces carbon dioxide and excess water, both of which are readily eliminated. Protein metabolism, however, produces other substances, primarily simple compounds containing nitrogen. The bulk of the nitrogenous waste as it leaves the cell consists of ammonia, which is highly toxic. In many vertebrates inhabiting fresh water, such as freshwater teleosts, lungfishes, larval amphibians, and crocodiles, where there is plenty of water to flush out the ammonia, much or all this waste remains in this form. In saltwater fishes and most land dwellers, however, almost all the ammonia is transformed in the liver, almost as fast as it appears in the blood, into less toxic compounds: urea and uric acid. Urea includes the greater part of the excreted nitrogenous waste in sharks, marine teleosts, the surviving crossopterygian *Latimeria*, adult amphibians, and mammals; uric acid (an almost insoluble material) is the bulk of the material in lizards, some turtles, and birds, excreted as chalky sludge (with useful conservation of water). Some nitrogenous materials may be eliminated by the gills in teleosts, and a minor amount by the sweat glands in mammals. For the most part, however, the kidneys bear the responsibility for the excretion of waste nitrogen.

Two distinct operations take place in the kidney, one having to do with the renal corpuscle, the other with the tubule proper. The structure of the corpuscle suggests a relatively simple filtering device: through the membranes separating the glomerulus from the cavity of the capsule there passes, under pressure, a filtrate of the blood plasma. This is indeed the case; liquid drawn from the capsule of amphibians by micropipette proves on analysis to have in it many of the materials found in the blood, and in the same proportions. Blood corpuscles, of course, do not pass through this filter, nor do large molecules, such as proteins. All other materials, however, including not only wastes but also valuable food materials, particularly glucose, pass through readily. Still further, the amount of liquid filtered is excessive. It has been calculated that if all the liquid that passed through the glomeruli of a frog were to be eliminated from the body, almost half a liter of urine would be produced daily by this small animal and that a man would produce almost 200 liters!

Obviously, nothing of the sort happens. If the only kidney action were simple filtration from the glomerulus, the body would soon be dehydrated. Further, glomerular filtration would do little to alter the proportions of materials found in the body fluid; wastes and harmful or excess materials would still be present in the remaining fluids in the same relative amounts, and valuable food elements would be sacrificed.

The corrective to this situation lies in the work of the convoluted tubule. From the nephric corpuscles, the filtrate traverses the length of these tubules to various ducts and, in many groups, to a storage reservoir, the bladder. Analysis of the resulting urine after this passage shows that we are dealing with a different substance. First, the volume is greatly reduced. In a frog, not more than 5 per cent of the liquid filtered from the glomeruli reaches the bladder, and, in man, hardly 1 per cent. Further, the composition is much changed. Useful food materials such as glucose, although readily filtered into the tubule, do not ordinarily reach the bladder, and of other substances, the relative proportions are radically altered.

Thus, the kidney tubules obviously play a major role in the process of excretion. They are lined with epithelia that change from one part of the tubule to another in patterns that are quite variable in different vertebrate types, and, in many groups, the specific function of different segments of the tubules is uncertain. As a whole, the tubule cells have two activities. The major one in higher vertebrates is that of reabsorption of most of the filtrate. Much of the water is obviously resorbed, and food materials are withdrawn from the fluid as it passes by—principally glucose, it appears, in the proximal part of the tubule, salt distally. On the other hand, there is, even in higher vertebrates, further active excretion of waste substances through cells of the tubule into the urine as a supplement to the filtrate, and in lower forms, which possess a renal portal system, the major role in kidney functioning appears to be played by the tubule capillaries; the glomerular function in such forms is a minor one.

In a nephron, the crude work is, so to speak, done by filtration from the glomerulus; the tubule proper adds the necessary refinement to the process.

Tubule Types and Vertebrate History. Whether fresh or salt water formed the original home of vertebrates has long interested students of kidney function as well as students of classification and phylogeny. The paleontologic record is ambiguous and open to controversy, but all the earliest known vertebrates (from the Cambrian and Ordovician periods) are found in marine deposits and probably, though less certainly, lived there. A study of renal tubule structure and function can, however, lead to a very different conclusion.

Three main types of nephric units may be found in the functional adult kidney of one group or another (Figs. 289, 291). All have a well convoluted tubular region.

400
Chapter 13
Excretory
and Repro-
ductive
Systems

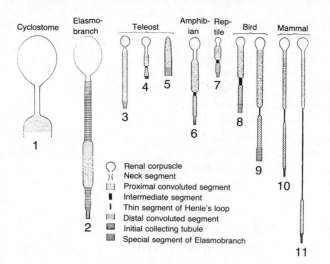

Figure 291. Diagram of kidney tubules of various vertebrates, all reduced on the same scale, to show the relative size of the components in the different groups. The glomeruli are at the upper end in each case, and the tubules are represented as if straightened out. The glomeruli are well developed in most groups and of enormous size in cyclostomes and elasmobranchs but of reduced dimensions in reptiles and are done away with in some marine teleosts (5). All have a proximal convoluted segment of the tubule; an intermediate segment, followed by a distal convoluted tubule, appears in some fishes and is present in all tetrapods. The intermediate segment becomes the loop of Henle in birds (in part) and mammals; this loop may be much elongated in the latter group. 1, hagfish; 2, skate; 3, sculpin; 4, catfish; 5, toadfish; 6, frog; 7, painted turtle; 8 and 9, chicken; 10 and 11, rabbit. (After Marshall, Kempton, from Prosser.)

(*a*) In one type (Fig. 289 *A*), found characteristically developed in such varied forms as amphibians, most bony fishes, and elasmobranchs, there is a renal corpuscle of good size, and the amount of filtrate (and consequently of water output) is high. (*b*) A second type (Fig. 289 *B*) is that found, with variations, in various marine teleosts and in reptiles. Here the corpuscle is small or absent (and, in marine teleosts, the distal convoluted tubule may be absent as well); obviously water output is low. (*c*) A third type (Fig. 289 *C*) is that seen in mammals and, in less extreme form, in birds. Here we see the interjection, into the middle of the convoluted tubule, of a long, slim extra loop, the **loop of Henle**. Although, with a large corpuscle, a large amount of fluid is produced, relatively little reaches the bladder through this type of tubule.*

From the distribution of these types of tubule among the vertebrates, a consistent story can be constructed regarding the environmental history of vertebrates. It is assumed that (*a*) is the primitive type of tubule, possessed by early freshwater vertebrates and retained by forms that still inhabit such waters. Such an animal lives in a medium that is more dilute than its own body fluids; in consequence, water tends to invade the body by osmosis at any membranous surface. To prevent overdilution of the body content and death, large amounts of water must be eliminated, and this

---

*Water resorption can be so effective that some desert mammals, such as kangaroo rats, can survive without drinking and rely solely on water obtained by metabolic breakdown of organic materials.

is afforded by the large corpuscle and its large output of water in the dilute urine; additionally, it appears that some salt may be absorbed through the gills.

In a marine fish, on the other hand, the sea water has a higher salt concentration than the body fluids, and the result tends to be loss of water into the surrounding brine, dehydration, and an overconcentration of salts in the body. In marine teleosts, lampreys, and the surviving crossopterygian *Latimeria*, salt excretion from the gills (cf. p. 354) partially remedies the situation; in addition, however, reduction or loss of the renal corpuscle from the tubule system conserves a vast amount of water in many marine teleosts, and considerable nitrogenous waste is excreted from the gills to make up for decreased kidney function.

Thus, it may be argued that the presence of a large glomerulus is a primitive character; that this is a result of its efficiency as a pump for the discharge of quantities of water; and that the desirability of such discharge is due to the freshwater environment of early vertebrates. However, possibly a marine environment was primitive and the development of a large glomerulus came later, when vertebrates moved into fresh water. A problem with this idea is the sharks, which have a large glomerulus and a tubule of type (*a*) and yet are almost exclusively marine!

The typical shark lives in a medium in which the salt concentration is greater than that in its own body fluids; yet it does not lose water to its environment and can pump water out of its system through large glomeruli just as does a freshwater fish. For although its salt concentration is that of any other higher vertebrate, and low compared to sea water, its total osmotic pressure (and that is what counts) is comparable to that of the sea, or even slightly higher. To some degree, the shark has solved the problem of salt concentration by the development of a rectal gland, mentioned in the last chapter, which secretes a considerable amount of sodium chloride. But more especially, the shark can retain, without apparent harm, much of its nitrogenous waste, in the form of urea, in the blood stream and thus can raise its total concentration of materials in solution without increasing its salts. A special section of each shark renal tubule has the function of resorbing urea from the urine flowing through it. Unexpectedly, the living coelacanth appears to have independently adopted the same strategy.

We have thus in sharks and in saltwater teleosts and lampreys two radically different renal adaptations to salt water. If the sea had been the original home of fishes, the argument goes, we would hardly expect such differences to exist. It is more reasonable to assume that the ancestors of both sharks and marine teleosts lived in inland waters; that they invaded the seas independently; and that in the two cases, quite different methods were evolved for combating the dangers of too great salinity.

There is still a further quirk to the story. Among vertebrates, the hagfish fail to regulate their salt content, and simply adopt the composition of the sea water surrounding them as that of their own internal environment. From this, it has been argued that the primitive vertebrates, like the living hagfish, lived in the sea, with no regulation of their salt content, and that such regulation—and life in fresh waters— came later. Proponents of a freshwater origin, however, objected that this implies that salt regulation, with a variety of different methods of attaining this, was "invented" independently by a number of different fish groups: lampreys, sharks, bony fishes. Thus, they suggested that probably early vertebrates generally regulated their salt content but that the hagfishes, invading the oceans, failed to find an answer to the problem of keeping salt concentration down and, so to speak, gave up in despair.

Despite these arguments, almost all investigators now agree that vertebrates originated in the oceans. After all, all known lower chordates are marine, and almost all major groups of animals appear to have arisen in the seas. Certainly, the structure and physiology of the kidneys suggest that both the Osteichthyes (ignoring the acanthodians) and the Chondrichthyes were originally inhabitants of fresh water (that this is true of the bony fish is also indicated by their possession of lungs, apparently as a primitive character). However, between the time of the earliest known vertebrates and the appearance of these groups in the fossil record, there is a gap of well over 100 million years—clearly much could happen in that time. We know virtually nothing of the history of cyclostomes and can only speculate on their significance in this story. For that matter, there is no real reason to assume that the earliest known fossil vertebrates were actually the *first* vertebrates, so whether the former were marine or not is probably irrelevant. A marine origin seems inherently more probable so is generally favored, although the evidence we have is incomplete and contradictory.

Terrestrial vertebrates have the same basic problem as a marine fish. They live in a dry environment, where water is continually lost through the body surface, with danger of dehydration and consequent increase in salt concentration within the body. Conservation of water is necessary, particularly because there can be no excretion of salt or waste through gills. Ancestral tetrapods presumably had kidney units of type (*a*), with a large corpuscle and normal tubule, and a free flow of watery urine. Some saving of water is effected by urine retention in a cloacal bladder (cf. p. 436) with absorptive walls. In living reptiles, further reduction of water output has been attained by reduction of the size of the corpuscles and a consequent reduction of the amount of water filtered through them. Oceanic reptiles and birds, which have only salt water to drink, have a difficult problem. Marine turtles have a salt gland in the orbit; the marine birds have a large salt gland, frequently above the eye, which empties into the nasal cavity. Mammals have developed a different method of conserving water. The normal corpuscle is present, and a large quantity of urine is poured into the tubule. The tubule, however, is an "Indian giver." There is a considerable resorption of water in any type of tubule; in these forms, there is inserted between proximal and distal convoluted tubules the long, thin loop of Henle, which descends into the medulla of the kidney (cf. Fig. 300). Here processes involving sodium exchanges and osmotic relations result in further water resorption in the distal convoluted tubule and collecting tubule and, consequently, in conservation of body water. Birds have some development of a loop of Henle. In addition, nitrogenous waste is excreted, as, in many reptiles, as uric acid; this is almost insoluble and needs little water for its carriage.

Primitive Tubule Structures.   The types of nephric units described in the foregoing discussion are those most characteristic of adult vertebrate kidneys. Still other types, however, may be noted; they are more often found in embryos, particularly in the first units developed, than in adults, and are more common in lower than in higher vertebrate types; there is hence considerable reason to consider them primitive in nature.

These types (Fig. 289 *D–F*) have in common the fact that they have openings into the coelomic cavity that are commonly ciliated funnels. In some cases, there is, in addition to the coelomic opening, a typical renal corpuscle connected with the tubule. In a few instances, we find a glomerulus that protrudes into the coelomic

402
**Chapter 13
Excretory
and Repro-
ductive
Systems**

cavity near the funnel instead of having a position within the tubule. And, in still other cases, no corpuscle or glomerulus of any sort is present.

One theory of the origin of the vertebrate nephric structures suggests that the last-described condition was the truly ancestral one, that the tubules first functioned in draining waste products and excess liquid from the coelomic cavities. A subsequent development of blood vessels, as glomeruli, in the walls of the coelom afforded a greater supply of liquid for the tubules to work upon. A third stage, the theory suggests, was the incorporation of the glomerulus in a corpuscle forming an integral part of the nephric unit; the final stage, the loss of the primitive coelomic connection, and dependence upon the glomerulus alone for a supply of filtrate.

Most major invertebrate groups have excretory structures of various types, termed **nephridia**. These, however, are not homologous with the vertebrate nephron, which appears to have evolved independently, possibly as a water pump for life in fresh water, with elimination of wastes as a secondary function. Amphioxus strikes a seemingly discordant note, for this creature has a segmental series of little nephridial structures that appear similar to those of many flatworms. However, recent studies, using the electron microscope, have shown that the supposed flame cells of these nephridia are really related to the vertebrate podocytes and that the nephridia do connect to the coelomic cavity as do renal tubules.

Organization of the Kidney System.   So far we have discussed merely the nature of the vital microscopic elements within the kidney. Now to be considered is the organization of these secretory units and the ducts leading from them to form the gross structure of the urinary system.

As seen in the higher, amniote classes of vertebrates, the pattern of the renal system is fairly uniform and apparently simple, with paired, compact kidneys projecting into the abdominal cavity from its dorsal walls, and paired ureters carrying urine from the kidneys and often emptying into a median bladder. Rather comparable structures are seen in various lower types. Study, however, shows that there are great differences from group to group in the construction of the urinary system. The kidneys themselves are varied in structure; the ducts vary; the bladder is variable.

These variations have two major causes: (1) Unlike many organs, the kidney must begin its operations at an early stage of development, to take care of the metabolic wastes of the rapidly growing embryo. In consequence, there is no possibility of the peaceful and undisturbed development of a kidney suitable for postnatal life alone; there must be rapidly constructed a functioning embryonic kidney that, however, is subject to modification or replacement in later developmental stages and adult existence. (2) The gonads, testis and ovary, lie adjacent to the kidneys; these organs, particularly the testis, tend to take over part of the tubes and tubules of the urinary structures as a conducting system for their products. The urinary organs have been markedly modified in most vertebrate groups as a result of this invasion.

Holonephros and Opisthonephros.   We may begin our discussion of types of urinary systems by the description of the structure and embryonic development of an idealized primitive kidney, which may be termed a **holonephros**.

In our embryologic story, we noted that in every trunk segment, the mesoderm includes, on either side, an intermediate region, often segmentally distinct as a **mesomere** (or **nephrotome**), a small discrete block of tissue interposed between somite

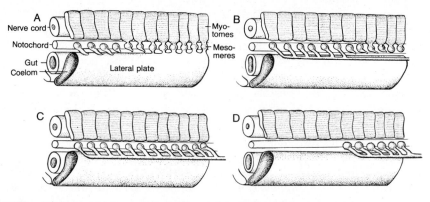

404

Chapter 13
Excretory
and Repro-
ductive
Systems

Figure 292. Diagrams of the anterior part of the trunk of an embryo (skin removed) to show the development of the archinephric duct. *A,* Most anterior—pronephric—mesomeres are budding out tubules that tend to fuse posteriorly. *B,* The pronephric tubules have formed the duct; some of the mesomeres farther posteriorly are forming tubules that are to enter the duct. *C,* The more posterior tubules have joined the duct. *D,* The pronephros is lost, but the archinephric duct formed by it persists to drain the more posterior part of the kidney.

and lateral plate (Figs. 75 *D,* 80 *C,* 292). In the ancestral vertebrate, we believe, each mesomere gave rise to a single nephric unit; the small coelomic cavity within it became the cavity of a renal corpuscle, and out from this grew a kidney tubule. There would thus be established a row of nephric units, segmentally arranged, down either side of the body. As with the somites, the differentiation of mesomeres takes place in embryos from front to back; in consequence, the oldest members of the series of tubules in our ideal ancestral type would be those at the anterior end of the series, the last-formed those at the posterior end of the trunk. The nephrotome lay above the coelom, and the developing kidney region would in later stages lie dorsal to the coelomic region, well to either side of the midline (Fig. 302). As noted earlier, conditions in embryos of lower vertebrates suggest that the nephric units in an ancestral vertebrate retained the coelomic connection now seen in early embryonic stages and occasionally in adults in various lower groups.

In many invertebrates, the excretory elements open directly to the exterior. But in vertebrates, the region in which the kidney tubules develop is separated from the surface by downgrowth from the myotomes of the lateral trunk musculature. The solution has been the development on either side of a longitudinal duct, gathering the urine from the series of segmentally arranged units; the two ducts often unite before emptying to the exterior in the region of the cloaca. The primitive kidney duct has been variously named; it is most commonly termed the **archinephric duct**.* This duct is, like the tubules, of mesodermal origin. Most typically, it forms by

---

*It is often called the **wolffian duct**, after its discoverer. **Pronephric duct** and **mesonephric duct** are terms applied to it on embryologic grounds. When serving as a kidney duct in the adult of lower vertebrates, it is sometimes incorrectly called a ureter or, more properly, a holonephric duct or **opistho-nephric duct**. If taken over by the genital system, it becomes the duct of the epididymis and the ductus deferens (pp. 431–435). Although the use of such eponyms as wolffian duct is now discouraged, that is often the most convenient term as it can be used for the duct in any vertebrate without implying, usually quite incorrectly, a particular function.

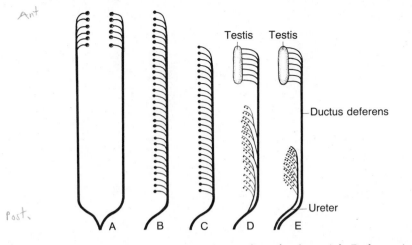

Ant

Post.

Testis Testis

—Ductus deferens

—Ureter

A B C D E

Figure 293. Diagrams of kidney types. *A,* Pronephros (embryonic); *B,* theoretical holonephros (each trunk segment with a single tubule), much as in a young hagfish or gymnophione amphibian; *C,* primitive opisthonephros: pronephros reduced or specialized, tubules segmentally arranged, as in hagfish. *D,* Typical opisthonephros: multiplication of tubules in posterior segments, testis usually taking over anterior part of system, trend for development of additional kidney ducts (most anamniotes). *E,* Metanephros of amniotes: an opisthonephros with a single additional duct, the ureter, draining all tubules. In *A,* both sides of the body are included; in *B* to *E,* one side only (cf. Fig. 309).

a fusion of the tips of the tubules of the most anterior and first formed nephric units (Figs. 292 *A, B,* 293 *A*). It grows backward along the lateral surface of the mesomeres, frequently using further mesomeric material in its formation. As tubules develop farther back, they grow outward to connect with this previously formed structure. The end result (Figs. 292 *C,* 293 *B*), ideally, is a **holonephros**: a kidney with a single nephric tubule in each trunk segment on either side of the body, the series draining through a pair of archinephric ducts. Such an ideal is approached only in the larvae of hagfishes and of the Gymnophiona among the amphibians.

The adult hagfish has a kidney that does not differ greatly from this type, but, even here, the most anterior and first-formed kidney tubules, those that form the duct, are specialized or degenerate; and the same is true of all higher vertebrates. These anterior tubules are termed the **pronephros** (Figs. 292 *B,* 293 *A,* 294 *A*). The remaining, major part of the kidney system, from which is formed in one manner or another the kidney of adult living vertebrates, may be termed as a whole the **opisthonephros**, the "back kidney" (Fig. 293 *C, D*). This opisthonephros generally differs from the theoretic holonephros in three main particulars: (1) The anterior tubules, the pronephros, do not form part of this structure. (2) The simple segmental arrangement is no longer present (above the hagfish level). For most, at least, of the length of the adult vertebrate kidney numerous tubules may develop in each segment; in fact, separate mesomeres may fail to develop, and, particularly toward the back end of the trunk, there may be a development of a mass of tubules from a longitudinal band of unsegmented mesomeric tissue. (3) In most vertebrate groups, the archinephric duct is used by the male for the transport of sperm, and part or all the trans-

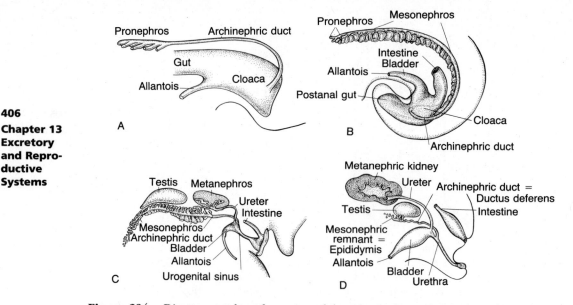

Figure 294. Diagrams to show formation of the metanephros of an amniote (male) embryo as seen from the left side. *A,* Pronephros and duct formed; *B,* mesonephros partly formed; *C,* pronephros reduced, posterior part of mesonephros functional, ureter formed, and metanephros beginning to differentiate; *D,* definitive stage; mesonephros reduced, and tubules and duct used only for sperm transport; metanephros, the functional kidney.

mission of urine may be performed by a newer type of duct, a ureter, or some equivalent.

Amniote Kidney Development. To show an extreme contrast with the ideal holonephros, we may describe the development of the kidney in mammals (Fig. 294); the story is similar in reptiles and birds. Our departure here from a logical sequence is due to the fact that study of the mammalian kidney has been influential in the development of concepts of renal types and their nomenclature.

As the differentiation of mesodermal somites progresses in the mammalian embryo, simple, segmental, rounded nephric vessels, or mesomeres, appear below them in a limited number—not over a dozen—of the anterior body segments, at the back of the head and along the future neck. These vesicles form the pronephros. In the nephrogenic tissue adjacent to these rudimentary tubules, the anterior end of the archinephric duct forms; this grows rapidly backward toward the cloacal region. In a mammal, the tubules constituting the pronephros disappear at an early stage; indeed, the first ones begin to degenerate before the last members of even this short series are formed.

Segmental growth of the kidney continues backward without interruption beyond the end of the pronephric series of tubules. The tubules now formed are considered to belong to a second embryonic nephric structure, the **mesonephros**, which is the functional kidney for a considerable period of embryonic life in mammals (and may persist until after birth in reptiles). Still segmental, to begin with, mesonephric tubules continue to form along the trunk as far as the lumbar region.

Pronephric units in mammals are at best simple structures, without glomeruli and often rudimentary; the mesonephric elements, on the other hand, have characteristic glomeruli and well-coiled tubules. At first, there is only one tubule to a segment in the mesonephric region; later, however, additional tubules develop, and the segmental condition is obscured.

As the back portions of the mesonephros are developed, the more anterior tubules are degenerating. Long before birth, in many mammals, the entire mesonephric series has lost its original urinary function, as has the archinephric duct, although, as will be seen, mesonephric tubules and duct persist in modified form as part of the reproductive system of the adult male (cf. pp. 431–435, Figs. 294, 301, 302).

The most posterior part of the nephrogenic tissue is not used, however, for the mesonephros. Instead it forms an unsegmented spherical mass in the roof of the lumbar region of the body cavity. In this mass of mesoderm, there presently appear large numbers of renal tubules. These tubules are to form the **metanephros**, the functional kidney of the late embryo and the mature mammal. Unlike the mesonephric units, these tubules do not empty into the archinephric duct; a new tube, the **ureter**, develops to drain them. Although phylogenetically new, the ureter precedes and even induces the tubules embryologically. From near the point at which each archinephric tube enters the cloaca, this new duct buds off and grows forward and upward to enter the metanephric tissues. Here it forks repeatedly into smaller tubules. At the tips of the smallest branches of the ureter, the **collecting tubules**, clusters of nephric tubules come to open. Further growth and differentiation of this metanephric mass produce, on either side of the body, the adult kidney, which is thus in form, position, and mode of drainage quite distinct from the pronephros and mesonephros.

Pronephros, Mesonephros, and Metanephros. We see the development in the amniote embryo of three successive nephric structures: pronephros, mesonephros, and metanephros. It is often stated or implied that these three are distinct kidneys that have succeeded one another phylogenetically as they do embryologically. However, there is little reason to believe this. The differences are readily explainable on functional grounds; the three appear to be regionally specialized parts of the original holonephros, which serve different functions.

In many vertebrates, the pronephric tubules differ in structural features from those of the mesonephros, but these differences are not constant; sometimes part or all the pronephric units are identical with those behind them. The one really distinctive feature of the pronephros is that the archinephric duct is formed by it. But there is nothing really significant or mysterious about this; it is a practical matter. An actively growing embryo has wastes to excrete. When kidney tubules are formed anteriorly, they begin the process of urine formation. The formation of a urinary tube for drainage cannot be delayed until the entire kidney is formed; the anterior tubules just cannot wait that long.

In amniotes, the metanephros is readily distinguished from the mesonephros. It is a discrete formation; it attains a vastly greater size; its drainage by the ureter rather than by the archinephric duct is definitive. But although the metanephros assumes a distinctive nature, it is, to begin with, simply a greatly enlarged posterior portion of the band of mesomeric tissue from which mesonephros and pronephros were also formed. Its large size is, of course, due to the presence of large numbers of tubules; the need for large numbers of tubules in the amniotes is correlated in part with the

408
**Chapter 13
Excretory
and Repro-
ductive
Systems**

relatively large size of the late embryo and adult, in part with the high metabolic activity of birds and mammals. The efficiency of a kidney depends upon the amount of surface present in its glomerular filters and tubular walls, and the same surface-volume relationship confronts us here as in the case of many other organs. The presence of complex collecting ducts is required, owing to the impossibility of attaching all the countless tubules directly to a single unbranched archinephric duct; analogous collecting ducts may occur in the kidneys of anamniotes for the same reason. Further, the utilization of the archinephric duct for the transport of sperm in the mature male makes a separation of the urinary transporting system desirable.

The adult kidney of fishes and amphibians is frequently termed a mesonephros and assumed to be homologous with only the embryonic kidney in amniotes. This position is untenable, because it implies that there is nothing in the kidney of a shark or frog comparable to the metanephros of amniotes. But the embryologic story shows that the entire length of the band of nephrogenic tissue is used in the development of the kidney of lower vertebrates, and hence the back portion of their kidney is formed from the same materials as the metanephros. Never, it is true, is the adult kidney of a lower vertebrate so completely concentrated in one short region as is the case with the metanephric type, but very similar conditions may be observed in some forms. Although the anamniote kidney may be drained by the archinephric duct, as in the case of the embryonic mesonephros, various cartilaginous fishes and amphibians have structures analogous (though not, presumably, homologous) to ureters (Fig. 301). All in all, it is perhaps best to confine the term mesonephros solely to the kidney of embryonic amniotes. The adult anamniote kidney may be termed the opisthonephros, with the understanding that the amniote metanephros is a special variety of the posterior part of this structure.

Pronephros and Head Kidney.   We have described the pronephros as the most anterior and first developed part of the embryonic kidney—essentially that set of tubules associated with formation of the archinephric duct. In amniotes and Chondrichthyes, the pronephros is an exceedingly short-lived structure; it never survives in the adult, and, except for its association with formation of the duct, it hardly deserves a separate name. In contrast are conditions in other fishes and amphibians, with small-yolked eggs, in which the embryo generally becomes an active food-seeking larva at an early stage. The pronephros persists in these larvae to satisfy excretory needs; it is, however, highly specialized and is frequently termed a **head kidney**, in reference to its anterior position. In most cases, a fair number of tubules from the more anterior body segments are involved; however, the number is usually reduced during embryonic development, so that the larval head kidney has but from one to three large convoluted tubules, and in frogs and salamanders only two or three units are involved at all.

In the head kidney, the tubules frequently have ciliated funnels opening from a coelomic space containing a single large glomerulus. This nephric part of the coelom may be closed off from the remainder of the coelomic cavity as a closed pocket—a sort of large renal corpuscle. In the cyclostomes, the portion of the coelom drained by the head kidney is the pericardial cavity; as a further peculiarity in this group, we find that the liquid filtered from the pericardial cavity by the head kidney does not pass into the archinephric duct, but, instead, into an adjacent vein. The larval head kidney disappears as a functional organ in the adult amphibian and in most fishes. It persists, however, throughout life in a few teleosts and in hagfishes as a mass of lymphoid tissue that functions in blood formation. Unfortunately, more posterior parts of the kidney may also serve this function in teleosts and may be given the same name.

The Opisthonephros of Anamniotes

(Figs. 295–297). We shall here survey the varied forms that the opisthonephros assumes as the functional adult kidney in lower vertebrates. In larval hagfishes, as we have seen, the kidney is essentially a holonephros; in adult cyclostomes, this is not the case, because the pronephric part has disappeared from the renal system. The structure is nevertheless simple. In an adult hagfish, the opisthonephros consists simply of a long series of tubules arranged in a segmental manner most of the length of the trunk, each tubule draining directly into the archinephric duct. In a lamprey, the tubules are more numerous, and the histologic structure is unique, with all the capsules fused into a large trough.

In gnathostomes, the kidney is universally more complex. Two major developments are seen in the various jawed fishes and amphibians: (1) There is a notable increase in the number of tubules, with loss of segmental arrangement and growth of renal bulk, often associated with a concentration of the functional kidney into a small portion of its original anteroposterior extent. (2) Part of the opisthonephric kidney system in the male is used for the passage of sperm. These two features are correlated; in many groups, the testis tends to take over some one specific region of the kidney, usually at its anterior end, for its special use; this region may lose all or part

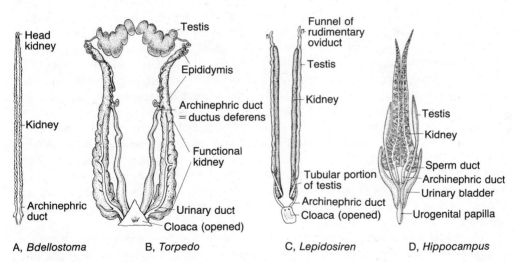

A, *Bdellostoma*　　　B, *Torpedo*　　　C, *Lepidosiren*　　　D, *Hippocampus*

Figure 295. Urogenital systems in ventral view of males of *A,* the slime hag *Bdellostoma; B,* the elasmobranch *Torpedo; C,* the lungfish *Lepidosiren; D,* a teleost, the seahorse *Hippocampus.* In *A,* the testis, not shown, is pendent from a mesentery lying between the two kidneys and has no connection with them. In *B,* the testis has appropriated the anterior part of the kidney as an epididymis, much as in most tetrapods, and uses the entire length of the archinephric duct as a sperm duct. In *C,* the sperm ducts drain, on the contrary, only into the posterior part of the kidney and thence to the archinephric duct. In *D,* the sperm duct is entirely independent of the kidney system. (*A* after Conel; *B* after Borcea; *C* after Kerr, Parker; *D* after Edwards.)

410
Chapter 13
Excretory
and Repro-
ductive
Systems

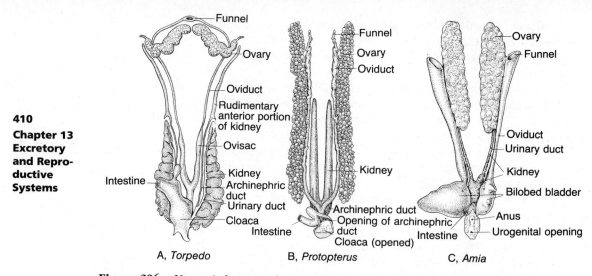

Figure 296. Urogenital systems in ventral view of females of *A,* the elasmobranch *Torpedo; B,* the lungfish *Protopterus;* and *C,* the primitive actinopterygian *Amia.* In *Torpedo,* the shell gland is not developed. (*A* after Borcea; *B* after Parker, Kerr; *C* after Hyrtl, Goodrich.)

of its original excretory function, and the remaining portion becomes in compensation greatly hypertrophied.

The anamniote kidney is generally elongate; in its length are included most of the trunk segments behind the pronephric region. Seldom, however, is the entire length of the kidney well developed and functional as a urinary organ. In sharks and chimaeras, the anterior part is slender, and the great bulk of the renal substance is concentrated posteriorly in a dozen body segments or less, a condition obviously comparable, morphologically, to the development of the metanephric kidney. The reason for this concentration appears to be the fact that the testis has become associated with the anterior part of the kidney and has come to use its tubules for sperm transport to the archinephric duct (cf. Fig. 301 *C, D*). This portion of the kidney loses part or all of its urinary function; in the male, it may become essentially an accessory to the testis (comparable to the mammalian epididymis) and may degenerate in the female. In correlation, the functional posterior part of the kidney tends to grow and lose its original metameric structure, with a considerable number of tubules developing in each segment. Clusters of tubules may join distally to form collecting tubules emptying into the archinephric duct.

The general trend in kidney evolution in anamniotes has been toward posterior concentration of urinary function in the opisthonephros and an approach to the metanephric pattern of amniotes. Such transitional types are found in the cartilaginous fishes, salamanders, and gymnophionans; the anurans, in which only a few segments are present in the entire trunk, have, as one would expect, a short, compact kidney. The actinopterygians have struck out along a line of their own in nephric evolution as in other regards. In teleosts (and *Polypterus* as well), the testis has acquired a duct quite independent of any connections with the urinary system, and an elongate opisthonephros often persists (Figs. 295 *D*, 301 *G*, 309 *d*). The lungfishes (Figs. 295 *C,* 301 *F,* 309 *c*) show a morphologic stage antecedent to this condition.

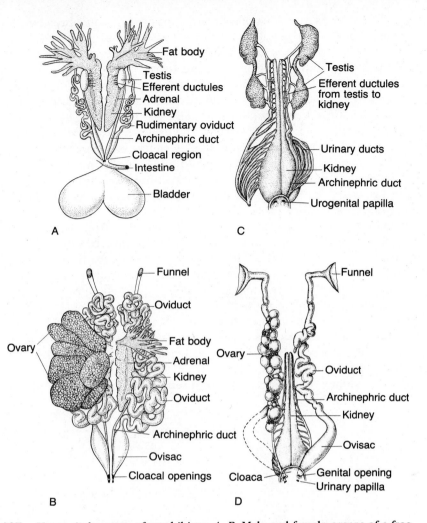

Figure 297. Urogenital system of amphibians. *A, B,* Male and female organs of a frog *(Rana);* in *B,* the ovary (shown only on the right side of the body) is in a condition close to breeding maturity. The bladder and intestine are not shown in *B. C, D,* Male and female organs of the urodele *Salamandra.* In *C,* the urinary ducts of the right side are detached and spread out to show their connections with the kidney. In *D,* the ovary is shown only on the right side; the oviduct of the same side is partly removed to show the more posterior urinary ducts. Ventral views. (*A* and *B* after McEwen.)

The Amniote Kidney. As indicated by our story of mammalian development, the amniote kidney is a specialized end type in which the trends seen in lower classes toward compact structure and separation from the genital system have attained a peak development. There remains almost no connection between the formed kidney and the male genital organs, because the old archinephric duct has been, except for its extreme posterior end, completely abandoned to testicular use in the male, and the kidney is drained by the ureter and its branches (cf. Fig. 301 *E*). The anterior part of the opisthonephros is functional in the embryo as the mesonephros but disappears in the adult except for vestiges attached to the testis. The mature kidneys, the metanephroi, are formed by a great expansion of nephric material in the region of the most posterior segments of the trunk and are relatively short and stout structures, bulging downward into the abdominal cavity in the lumbar region.

The reptilian kidney (Fig. 298) frequently has a crenulated appearance due to the fact that it is built of numerous small lobules. The lobulation, in turn, reflects the arrangement of the urine-collecting system. The ureter has, typically, numerous long branches into which series of collecting tubules empty; each branch of the ureter is the center of formation of a lobule. Compared with other amniotes, the number of tubules is low; estimates of 3,000 to 30,000 are recorded for various lizards. This may perhaps be associated with the generally low rate of metabolic activity in reptiles compared with birds and mammals.

Adult avian kidneys (Fig. 299) arise embryologically from nephrogenic tissue belonging to only a single segment at the end of the trunk, but these organs are of relatively enormous size, typically filling large paired cavities in the roof of the abdomen, shielded by the complex synsacrum. Each of the rather elongate kidneys is usually divided into three or more irregular lobes, and these into a large number of small lobules, each containing a branch of the ureter. The glomeruli are small, and tubules are exceedingly numerous; a fowl, for example, has been calculated to have some 200,000 of them, or twice as many as a mammal of comparable size.

The mammalian kidney (Fig. 300) generally has a smooth external surface. Internally, it shows a markedly different build from that of other amniotes. Within the substance of the kidney, the ureter, instead of sending off branches as in reptiles and

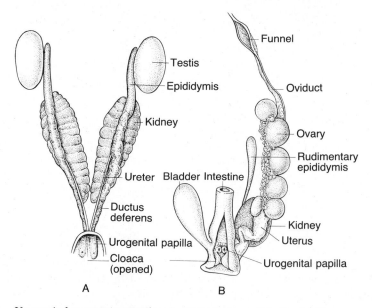

Figure 298. Urogenital organs in reptiles. *A,* Male organs of the lizard *Varanus. B,* Female organs of *Sphenodon.* In *A,* the bladder is omitted; in *B,* it is shown turned to one side. In *B,* the organs of the left side only are figured. Ventral views. (*A* after van den Broek; *B* after Osawa.)

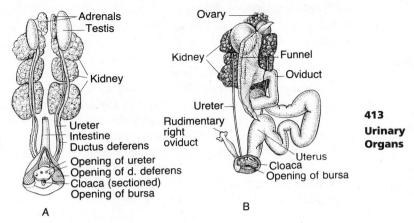

Figure 299. Urogenital organs of pigeon; *A,* male; *B,* female. The bursa (of Fabricius) is a pouch of lymphoid tissue opening dorsally into the cloaca of birds and discussed in the following chapter. (*A* after Röseler and Lamprecht; *B* after Parker.)

Labels, A:
Adrenals
Testis
Kidney
Ureter
Intestine
Ductus deferens
Opening of ureter
Opening of d. deferens
Cloaca (sectioned)
Opening of bursa
A

Labels, B:
Ovary
Kidney
Funnel
Oviduct
Ureter
Rudimentary right oviduct
Uterus
Cloaca
Opening of bursa
B

birds, expands into a cavity, the **renal pelvis**, into which the collecting tubules drain. Frequently, the pelvis is partially subdivided to form small cup-shaped cavities, **calyces**. The kidney is most commonly a bean-shaped structure, and these cavities lie along its inner curve (the **hilus**). Into each calyx converge collecting tubules in a radiating structure termed a **renal pyramid**. In some cases, however, the kidney is externally divisible into lobes; this is true of cattle and a variety of other mammals. Even more spectacularly, the kidney may, in some carnivores (mainly, but not entirely, aquatic ones), elephants, and whales, be divided into a number of distinct **reniculi**, each attached to a separate branch of the ureter; there may be several thousand of these in some whales. The advantages of such subdivision are obscure.

We have earlier (Fig. 289 *C*) noted the structure of the mammalian nephric unit; characteristic are large glomeruli and the long straight loop of Henle between proximal and distal convoluted regions. In section, the mammalian kidney usually shows distinct **cortical** and **medullary** portions. The former contains principally the renal corpuscles and the convoluted portions of the tubules. The medulla is mainly composed of long, straight collecting tubules and loops of Henle; it is these that give a striated appearance to the pyramids. Although the number of nephrons is not so great

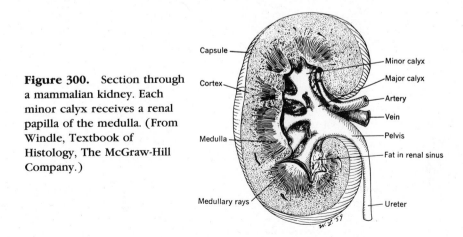

Figure 300. Section through a mammalian kidney. Each minor calyx receives a renal papilla of the medulla. (From Windle, Textbook of Histology, The McGraw-Hill Company.)

Labels:
Capsule
Cortex
Medulla
Medullary rays
Minor calyx
Major calyx
Artery
Vein
Pelvis
Fat in renal sinus
Ureter

as in birds of similar size, it is nevertheless high; even a mouse appears to have about 20,000, and in such large mammals as a man or a cow, the number may run into the millions.

414
**Chapter 13
Excretory
and Repro-
ductive
Systems**

Blood Supply to the Kidney. The kidney demands a profuse blood supply for the filtering activities of the glomeruli and for the activity of the tubules. The nature of the renal circulation will be treated in Chapter 14 and will be briefly noted here.

In cyclostomes and in mammals, the entire blood supply to the kidney is arterial. In all intermediate groups, however, from jawed fishes to reptiles and birds, we find an accessory blood supply in the renal portal system (cf. pp. 473–475). Venous blood from the tail or hind legs, or both, in its course to the heart is forced to pass through a venous capillary system within the kidney. In some forms, it has a "choice": it may pass through the sinusoids of the liver instead.

In vertebrates with only an arterial blood supply, the arteries furnish blood both to the glomeruli and to capillaries that cluster about the tubules. In forms with a renal portal system, the functions of the two blood supplies are distinct. The glomeruli are always supplied by arterial blood, and there may also be arterial capillaries around the tubules—notably in birds. Venous blood, on the other hand, is never concerned with the glomeruli but forms a network around the convoluted tubules. No fully satisfactory reasons have been advanced for the development of the renal portal system and its later abandonment in mammals, nor for the curious division of labor between arterial and venous systems when both are present.

Evolution of Urinary Ducts (Figs. 301, 309). In the primitive vertebrate kidney, as seen in cyclostomes, a pair of simple archinephric ducts sufficed for the drainage of urine. In gnathostomes, however, a complicating factor enters the situation. Sex rears its head; the archinephric duct is invaded in the male by the genital system.

In cyclostomes, the sperm are shed into the coelomic cavity and reach the exterior through a posteriorly placed genital pore. But in primitive gnathostomes, it appears, there arose connections between the sperm tubules and kidney tubules; the path of sperm discharge came to be via the archinephric duct. As a result, this duct came to serve a dual purpose. Among modern vertebrates, it functions fully for both sperm and urine conduction in a few scattered forms: the Australian lungfish, lower actinopterygians such as sturgeons and gar pikes, the common frog, and some gymnophionans and urodeles such as *Necturus* (the mud puppy).

A condition of this sort is, however, apparently unsatisfactory. In the course of gnathostome history, there has been a struggle between urinary and genital systems for possession of the archinephric duct.

Among bony fishes, the urinary system has been the winner. In African and South American lungfishes and *Polypterus* (Figs. 301 *F*, 309 *c*) sperm enter the archinephric duct but only near its posterior end; for most of its course, it is a purely urinary tube. In teleosts, development of a discrete sperm tube has been completed; there is a separate route for sperm conduction back from the testis, and the archinephric duct is triumphantly restored to its original function of conduction of urine alone (Figs. 301 *G*, 309 *g, h*).

In all other gnathostomes—sharks on the one hand, most tetrapods on the other—the fight has gone the other way; in most, the archinephric duct has been

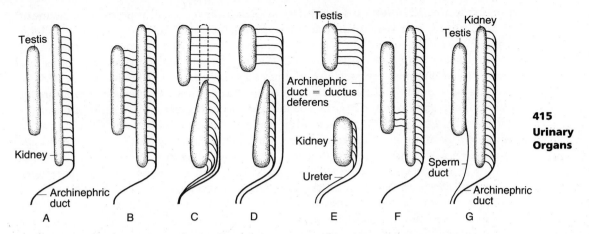

Figure 301. Diagrams to show the differentiation of urinary and genital ducts in the males of various vertebrates. Ventral views, left side only shown. *A,* Condition seen in cyclostomes; archinephric duct solely for urinary system; gonad not involved. *B,* Condition presumably primitive for gnathostomes, preserved in the sturgeon and gar pike. Testis connects at various points with kidney and thence with archinephric duct. *C,* Stage beyond *B* as seen in many sharks and urodeles. The testis has taken over the anterior part of the original kidney; the functional posterior part of the opisthonephros tends to drain by a series of ureter-like, accessory urinary ducts. *D,* Stage more advanced toward the amniote condition, found in some sharks and urodeles; the kidney drainage is by a single ureter-like duct. In females (not shown) of the types represented by males in *C* and *D,* the tendency toward development of new ducts for kidney drainage is generally not so marked as in the males; the condition shown in *D* is reached by few female forms below the amniote level. *E,* Amniote condition; a single definitive ureter formed in both sexes. *F* and *G,* Lungfish and a teleost, showing a type of specialization peculiar to modern bony fishes. The testis tends to concentrate its connections toward the back end of the kidney (as in *F*), and, in teleosts evolves a separate sperm duct, releasing the archinephric duct for its original urinary functions. Note that the two ducts for sperm and urine in *G* are *not* respectively homologous with ductus deferens and ureter, as might be thought at first sight.

completely taken over in the male sex as a sperm duct. In both sexes, new tubes have appeared, to drain urine from the kidney.

Amphibians show a number of stages in the development of functional replacements for the archinephric duct (Fig. 301 *C, D*). In some members of all three orders, the archinephric duct still carries both products; a series of short collecting ducts empties into it along the length of the kidney. In other members of both the Anura and Urodela, however, the most posterior collecting ducts make their own way to the cloaca independently of the archinephric duct, and in males of such anurans as *Alytes,* the "obstetrical toad," and of the common newt *Triturus*, these tubules have joined to form a single duct draining the entire kidney quite independently of the archinephric duct. Such a duct appears to be essentially comparable (but not, we assume, homologous) to the ureter of amniotes. In female amphibians, a similar but rather more retarded development of an accessory urinary duct can be seen, despite the lack of other use of the archinephric duct.

416
Chapter 13
Excretory
and Repro-
ductive
Systems

In the Chondrichthyes, a parallel development has occurred. In no male shark or skate is the archinephric duct concerned with urine transport. A fan of tubes collects urine from the opisthonephros and unites posteriorly into a common stem—an accessory urinary duct, often, but we think improperly, called a ureter. In sharks and skates, as in amphibians, such development is not so marked in the female, where there is no conflict with sex organs, as in the male. In some female elasmobranchs, the archinephric duct still drains the kidney; in others, part of the drainage is by a fan of ureter-like tubes emptying posteriorly into the end of the original duct. And in advanced elasmobranchs of both sexes, the most anterior part of the kidney is reduced or absent, and the entire organ is drained by accessory ducts.

In all amniotes, as we have seen (cf. p. 407), the end result of a similar evolutionary process has been the same—a definitive ureter has been developed; the archinephric duct has been abandoned completely to the use of the genital system as a ductus deferens for sperm (Fig. 301 E).

Urinary Bladder in Fishes.   In a majority of vertebrates, there is developed a bladder of some sort, a distensible sac in which urine may be stored before being voided. In elasmobranchs and primitive bony fishes, such structures, usually small and inconspicuous, are developed from the urinary ducts themselves. In female elasmobranchs, the posterior end of each archinephric duct may become enlarged for urine storage; because the two ducts fuse near their external opening, there may be a median portion to this simple type of bladder. In the male elasmobranch, these primary ducts are used for the passage of sperm, and no urinary bladder develops along their course, but, as we have just seen, the sharks and rays tend to develop on either side new tubes distinct from the primary duct, for urine passage. Much like the primary ducts in the female, these urinary ducts may expand in a bladder-like way.

In primitive bony fishes generally—lungfishes and lower actinopterygians—a small urinary bladder develops out of the conjoined ends of the archinephric ducts and hence seems essentially similar to that of female sharks. In teleosts, there may be similarly a bladder just inside the urinary opening (Fig. 295 D); here, however, a pocket of the cloaca—present in the embryo, but not in the adult—takes part in the formation of the bladder walls. In the lamprey, there is a comparable small bladder or sinus formed from a pinched-off portion of the cloaca, into which the urinary duct opened in the embryo (cf. Fig. 316 A).

The Bladder in Tetrapods.   In tetrapods, a bladder is useful as a rudimentary sanitary measure; it is, further, useful in water conservation, and in some lower tetrapods—anurans and possibly turtles—resorption of water by the bladder is thought to be important in preventing desiccation under terrestrial conditions. The tetrapod type of bladder appears among the amphibians. It is a new structure, not seen in fishes, which develops as a pocket in the floor of the cloaca and often lacks a direct connection with the kidney ducts (cf. Fig. 317 A). Voiding of urine from the bladder takes place through the common cloacal outlet.

The reptilian bladder is of similar construction and position. It is found in *Sphenodoñ*, in turtles, and in most lizards. It is lost, however, in a few lizards, in snakes, and in crocodilians; the bladder has also disappeared in birds except in the ostrich. In the absence of a bladder, the urine is poured directly into the cloaca and may be mixed in the feces.

We have noted earlier (cf. p. 132) that the tetrapod type of bladder plays an important part in amniote embryology. In all forms with this developmental pattern, the allantois, a pouchlike structure arises from the floor of the most posterior part of the primitive gut. This outgrowth is obviously elaborated from the urinary bladder. At birth, the allantois itself disappears, but the adult bladder develops as an expansion of the allantoic stalk.

In mammals, the urinary bladder is better incorporated into the urinary system. The ureters come to enter directly into the wall of the bladder, and no passage of urine through a cloacal cavity (lost in most mammals) is necessary to reach it.

The tetrapod bladder is a highly distensible organ with stout walls. It is endowed, particularly in mammals, with thick coats of smooth musculature. The inner lining is of a peculiar type known (inappropriately) as **transitional epithelium**. When the bladder is empty, it appears to be of a thick, stratified nature. When the bladder is filled and the epithelium is expanded in area, it becomes thinned down to a layer or two of flat squamous cells. Epithelium of such a readily distensible type is especially necessary in bladder construction but is often found in other parts of the urinary system as well.

Genital Organs

Sexual reproduction is almost universal in vertebrates; only a very few lizards normally reproduce parthenogenetically—without needing (or having) mates. Among certain groups of invertebrates, the hermaphrodite condition, in which the same individual may function as both male and female, is commonly found; in vertebrates, the sexes are functionally separate in the overwhelming majority of cases.

The basic reproductive structures are the **gonads**—ovary or testis. In these organs are produced the **gametes**—eggs (ova) or sperm—by the union of which, in the process of fertilization, the new generation is begun. In all gnathostomes, there are associated with the gonads tubes or ducts (with accessory structures) for the transport of gametes and in certain cases for the protection and nourishment of the growing young within the female body. In addition, copulatory organs, which aid in internal fertilization of the eggs, may develop in the region of the external opening of the genital tract. And, finally, secondary sexual characters are frequently found in other bodily structures; sex may affect general body size or proportions, and the development of such features as the plumage in birds, mammary glands in mammals, antlers and horns in ruminants, and beards in our own species.

Sex Development.   Although in gnathostomes the genital systems differ greatly in the two sexes, one sex frequently possesses rudiments of the structures characteristic of the other. This is due to the fact that for some time during embryonic life there is an **indifferent stage**, during which gonads and ducts proceed far in their development without any indication of which sex the individual is to become (Figs. 302 *C*, 304 *A*, *B*). The gonads may attain considerable size without showing specific features of either ovary or testis, and both male and female duct systems may differentiate to a considerable degree in potential members of both sexes. Eventually, however, there appears a definite sexual stage; the gonads become specifically testes or ovaries, and only the ducts and other accessory structures appropriate to one sex or the other continue their development. The nonpertinent structures of the opposite sex gener-

418

**Chapter 13
Excretory
and Repro-
ductive
Systems**

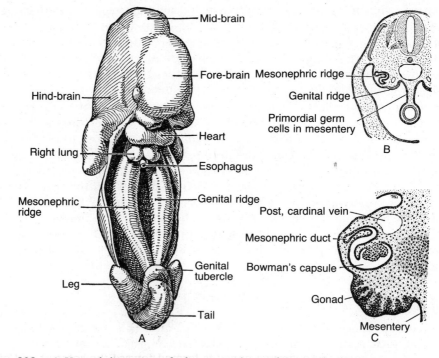

Figure 302. *A,* Ventral dissection of a human embryo of 9 mm, digestive tract removed, to show genital ridge and kidney (mesonephric ridge) projecting downward into the (opened) coelomic cavity. *B,* Cross section of an embryo at an earlier stage (7 mm); and *C,* a slightly later one (10 mm). In the latter, the primary sex cords are forming in the still "indifferent" gonad, and capsule and glomerulus are forming in the kidney tubules. (From Arey.)

ally cease to develop and may be resorbed, but they are sometimes merely arrested in their growth, to persist as rudiments in the adult.

The presence of the indifferent stage in sexual development is correlated with the nature of vertebrate sex determination. As discussed in Chapter 5, the early development of the embryo is mainly due to the organization already present in the unfertilized egg; the influence of the sperm and the hereditary characters that it introduces are not appreciable until a relatively late stage. Whether an individual is to become a male or female depends for the most part upon the chromosome complement of the fertilized egg. The egg has, so to speak, no knowledge of which sex it will become, and the early embryo is prepared for both possibilities. The influence of the sex chromosome mechanism is eventually manifested in the gonad. Once this has become definitely either testis or ovary, male or female patterns in other sexual organs appear in the later stages of embryonic life. These events are apparently affected by hormones produced by the developing gonad, although these hormones do not appear identical with those produced by ovary or testis in later life. In a variety of vertebrates, the mechanisms of sex determination are so delicately balanced that the gonads hesitate (so to speak) between male and female pathways, and both eggs and sperm may develop in a single individual. In hagfishes, both eggs and sperm begin development in different regions of the same gonad. For the most part, one sex or the other dominates in maturity, but a few hagfishes remain sterile intersexes throughout life. Among the teleosts, members of four marine families exhibit in the young a development of both eggs and sperm in the same gonad, and, in these families alone among vertebrates, functional hermaphrodites are reported. Amphibians show a delicate balance between sexes, and variants of one sort or another are numerous. For example, in the common toad (*Bufo*), the males have an ovary-like structure—Bidder's organ—in addition to a testis. In the common frog (*Rana*), young males show female tendencies, and genetic females may in old age shift to the male side of the balance and produce sperm.

There is in vertebrates a great difference in the rapidity with which different organ systems develop. The nervous system, for example, grows exceedingly rapidly in early stages; the genital organs, primary or secondary, are, on the other hand, the slowest of any to develop. After all, whereas most of the bodily structures must be put to use at birth or even during embryonic existence, the sex organs do not function until the maturity of the individual has been attained.

Gonads.   The gonads first appear at a stage in embryonic history when most of the main features of other organ systems have been blocked out, and when the coelomic cavities are well developed. At this time, paired longitudinal swellings, the **genital ridges**, are formed along the roof of the coelom, lying lateral to the root of the mesentery and medial to the embryonic kidney (Fig. 302). Elongate to begin with, the gonads developed from such ridges often become relatively short and compact in later stages. Usually, the definitive gonad is formed toward the anterior end of the abdominal cavity; fat bodies or other structures may arise from abandoned portions of the genital ridge. The **germinal epithelium** of the ridge, continuous with the mesodermal lining of the rest of the coelom, forms the more important structural elements of the gonad (Figs. 302 *B, C,* 303 *A*); mesenchyme lying beneath the epithe-

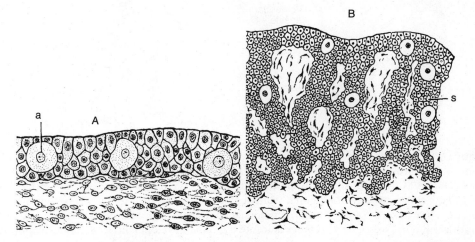

Figure 303.   Early stages in mammalian gonad development. *A,* A section of the epithelium of the gonad at an early indifferent stage; primordial germ cells are present *(a),* but sex cords are not developed. *B,* A somewhat later stage, with primary sex cords *(s)* growing inward from the epithelial layer. (From Maximow and Bloom.)

lium forms connective tissues and in higher vertebrates, at least, gives rise to special **interstitial tissues**.

Before the end of the indifferent stage, the gonad develops in many cases into a compact, swollen structure extending out into the coelomic cavity from its dorsal wall in the neighborhood of the developing kidney and is often supported by a special mesentery. From the germinal epithelium covering its surface, finger-like structures, the **primary sex cords** (Figs. 302 *C*, 303 *B*, 304 *A*, *B*), grow inward into the substance of the gonad. These cords contain germ cells as well as less specialized supporting elements.

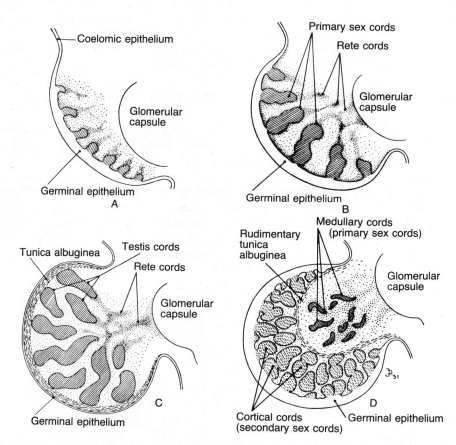

Figure 304. Development of testis and ovary of a mammal. *A*, Genital ridge, with incipient primary sex cords growing inward from the germinal epithelium; the stage is slightly earlier than Figure 303 *B*. *B*, Gonad still in the indifferent state; the primary sex cords are well developed, and cords are forming that will become the rete testis if the gonad becomes a testis. *C*, Development of the testis; the germinal epithelium degenerates and is replaced by a sheath around the testis; testis tubules and rete continue development. *D*, Early development of an ovary, with reduction of primary sex cords and rete rudiments, but, on the other hand, great development externally of secondary cords in which eggs develop. (From Burns, in Willier, Weiss and Hamburger, Analysis of Development.)

Origin of Germ Cells. Primordial germ cells from which eggs or sperm develop—large cells with a characteristic clear cytoplasm—may be identified at a relatively early stage in the germinal epithelium of the gonad, and it seems logical to assume that the germ cells are of local, mesodermal origin.

Possibly the functional eggs and sperm of the adult do develop in this manner. However, the first germ cells to appear in ovary or testis seem, actually, to be derived in most cases from the endoderm. In embryos of many vertebrates, from cyclostomes to mammals, there have been observed in the lining of the gut or yolk sac at an early stage, cells that histologically are quite distinct from the ordinary cells of the digestive tract. These distinctive cells appear to leave the gut walls and migrate, either in amoeboid fashion through the intervening tissues (Fig. 302 *B*) or by way of the blood, into the coelomic epithelium from which the genital ridges form. These cells are primary germ cells. In some forms, such as urodeles, the primary germ cells appear to be mesodermal, not endodermal, in origin.

Biologists have often emphasized the fact that the germ cells form an exceedingly independent tissue; the rest of the body, the "soma," is, from this point of view, merely a temporary structure shielding and conserving the potentially immortal germ plasm. But there are puzzling features in this story. Most investigators believe that the primordial germ cells, once arrived at the gonads, give rise to eggs or sperm. Some, however, believe that this is not the case and that these cells degenerate after a fruitless migration. Many hold an intermediate viewpoint that these first germ cells give rise to a first-developed series of gametes but that the functional eggs and sperm of the adult are derived from other germ cells, which arise *in situ* from the mesodermal epithelium of the gonad. Under this view, the function of the immigrant primary germ cells may be merely the initiation and stimulation of the process of gamete formation.

Ovary: Egg Production (Figs. 305, 306). In the development of an ovary beyond the indifferent stage, there is generally found as a diagnostic feature the proliferation inward from the germinal epithelium of **secondary sex cords** (Fig. 304 *D*). The primary cords, already formed, degenerate in the female, and it is in the succeeding series that the ova of the adult arise.* In the developing ovary, the germ cells contained in these cords divide repeatedly and develop eventually, by a complicated maturation process—**oogenesis**—into definitive egg cells. Each egg may become surrounded by a cluster of other cells derived from the sex cords to form a **follicle**; connective tissue cells may form a further external sheath. The follicular cells aid in the sustenance of the growing egg; in addition, the follicle is the major seat of estrogen formation. As maturity is approached, egg and follicle increase in size. In forms with large-yolked eggs, the follicle becomes relatively enormous and forms a major expansion on the ovarian surface. In viviparous mammals, the egg is small, but the follicle surrounding it nevertheless enlarges, causing the development of a liquid-filled space in its interior.

At seasons of reproductive activity, follicles ripen at the surface of the ovary and

---

*It is from these primary cords, as we shall see, that sperm develop in the male. The gonad, one might say, always assumes it is destined to become a testis; when it discovers that it is fated to become an ovary, it has to "back up and try again."

422
Chapter 13
Excretory
and Repro-
ductive
Systems

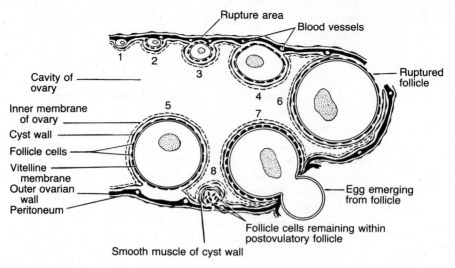

Figure 305. Diagrammatic section through a lobe of the frog's ovary. *1* to *5* illustrate stages in the growth of the follicle; *6* and *7*, rupture of a follicle and emergence of egg; *8*, postovulatory follicle. (From Turner.)

burst, releasing the egg into the surrounding coelomic space—the process of **ovulation**. In most lower vertebrates, the follicular material is quickly resorbed. In mammals, however, and in some elasmobranchs, the follicle persists for some time after the egg has been released, its cavity filled with a mass of yellow material from which its name of **corpus luteum** is derived. This "yellow body" secretes the hormone **progesterone** (cf. p. 616).

The number of eggs in a mature state in the ovary at any one time is, in most groups of vertebrates, small—from two to a dozen in many cases. In amphibians, however, hundreds or even thousands of ripe eggs may be present at breeding time, and, in actinopterygians, there may be hundreds of thousands or even millions (the codfish is estimated to lay 4,000,000 eggs in one season). At an early stage of ovarian development, a large number of tiny eggs are produced as a stock from which those matured at successive seasons may be drawn. It was once believed that there is no further production of egg cells beyond this early stage; but the large number of eggs produced by many fishes and amphibians indicates that there must be later proliferation of germ cells, presumably from the germinal epithelium on the surface of the ovary. On the other hand, forms in which the number of eggs that mature is small obviously do not use up the abundant supply of egg cells; in such types, a large proportion of the eggs degenerate, even if far advanced in follicle construction, without reaching maturity.

The ovary is usually a paired structure, often with an oval shape when in a resting phase, but frequently distended and irregular in outline at the breeding period. In cyclostomes, the pair of gonads fuse into a single median structure, and the two ovaries fuse in many teleosts as well. In some elasmobranchs, the left ovary remains undeveloped; in almost all birds and in the primitive mammal *Ornithorhynchus* (the duckbill) the left alone matures. In amphibians and reptiles, the ovary is thin-walled

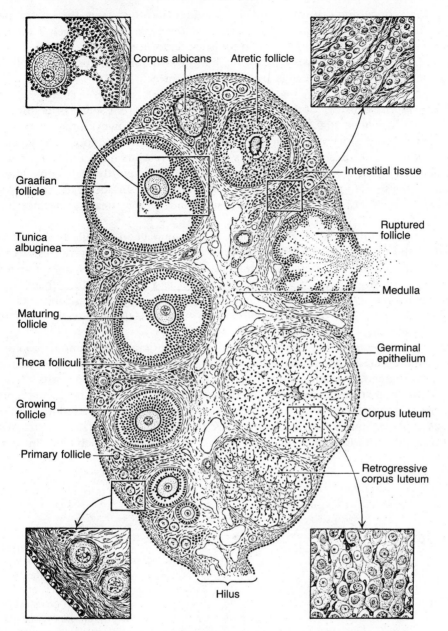

Corpus albicans

Atretic follicle

Interstitial tissue

Ruptured follicle

Graafian follicle

Tunica albuginea

Medulla

Maturing follicle

Germinal epithelium

Theca folliculi

Corpus luteum

Growing follicle

Retrogressive corpus luteum

Primary follicle

Hilus

Figure 306. Diagram of a composite mammalian ovary. Progressive stages in the differentiation of a follicle are indicated on the left. The mature follicle may ovulate and form a corpus luteum *(right)* or degenerate without ovulation (atretic follicle.) (From Turner.)

and hollow, containing central lymph-filled cavities. In other cases, the central part—the medulla—is mainly a connective tissue structure; the eggs and their follicles, together with the germinal epithelium on the surface, form the cortex. As explained on page 427, the ovary in teleosts may form a hollow structure connecting directly with the oviduct; in this case, the eggs are shed to the inside.

424
Chapter 13
Excretory
and Repro-
ductive
Systems

Testis. The early history of the testis, through the indifferent stage, is identical with that of the ovary (Fig. 304 *A–C*). From this point onward, however, the history of the two organs diverges. In contrast to the ovary, no secondary sex cords are developed. On the other hand, there is, in amniotes, further development, deeper in the gonad, of strands of tissue already present in the indifferent stage; these become the rete testis (cf. p. 433). No germinal epithelium remains at the surface of the testis, which, beneath a normal coelomic epithelium, is surrounded by a sheath of connective tissue.

The testis tends to be, on the whole, a more compact and regularly shaped structure than the ovary; it is subject to seasonal change, but generally does not exhibit such marked differences between breeding and resting stages as does the female gonad. In cyclostomes, the testis, like the ovary, is a median rather than a paired structure. In sharks, the two testes may fuse posteriorly, and the right member of the pair may be the larger. In various birds and mammals, the left testis tends to be slightly larger than the right.

The testis generally has in its structure a considerable amount of connective tissue, which includes interstitial cells producing male hormone. Its reproductive tissues are produced from the primary sex cords. The material of the cords breaks up in the substance of the testis into numerous hollow structures. In fishes and amphibians, these are usually more or less spherical **sperm ampullae**; in amniotes and some teleosts, elongate **seminiferous tubules** are formed (Fig. 307). These structures are lined by epithelium in which countless sperm are produced at each breeding season.

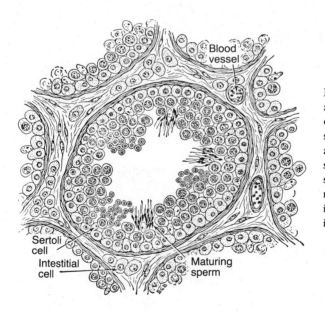

Figure 307. A small area of a mammalian testis, showing one tubule and parts of several others in cross section and the interstitial tissue. All stages in the development of sperm shown here would not normally appear at one time in any tubule. (From Hooker in Fulton-Howell.)

The epithelium of tube or ampulla is stratified and contains both reproductive cells and relatively rare supporting cells—**Sertoli cells**. At the base of the epithelium are germ cells that are little differentiated, the **spermatogonia**. From these primary cells, after repeated divisions, there are formed mature cells that are released at the surface of the epithelium as **spermatozoa**. They include in their structure little except a head, containing the nuclear material, and a long, motile tail. Spermatozoa are exceedingly small but exceedingly numerous, and, even in small animals, the total production may be measured in hundreds or thousands of millions.

In most vertebrates, sperm production—**spermatogenesis**—like that of eggs, is a cyclic phenomenon. The sperm ampullae of lower vertebrates are rather comparable to ovarian follicles. All the germinal materials of a given number of ampullae ripen at a specific breeding season, the sperm are discharged, and the ampullae resorbed, to be replaced by others that have meanwhile developed more slowly. The seminiferous tubules of higher vertebrates, on the other hand, are characteristically permanent structures, producing successive crops of sperm and persisting little changed throughout the breeding life of the individual.

Descent of the Testis (Fig. 308).   In most vertebrates, the gonads of the adult retain their primitive position in the upper part of the coelomic cavity. In a majority of mammals, however, a radical change in the position of the testes occurs during embryonic development.* Paired pouches form in the floor of the abdominal cavity and cause external swellings on the ventral surface of the body just anterior to the pelvis. These are the **scrotal sacs**, lined internally with coelomic epithelium, and constituting separate portions of the coelomic cavities. Each testis moves posteriorly and ventrally from its original position and descends into a sac, accompanied, of course, by the ductus deferens for sperm transport, its proper blood supply, and a fold of its proper mesentery—the **gubernaculum**. In some cases, the scrotal sacs remain in open connection with the abdominal cavity through an **inguinal canal**, and the testes may be withdrawn from the sacs into the body between breeding seasons through the action of a muscle (cremaster) in the scrotal wall. In other mammals, the

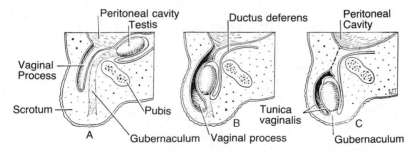

Figure 308.   Descent of the testis in mammals; ventral surface of body at left. A vaginal process develops from the body cavity; its peritoneum forms the vaginal tunic of the scrotal sac. Broken line in *C*, position of inguinal canal in mammals in which sac is not completely closed. (From Turner.)

*In mammals the ovaries typically descend in a similar manner into the lower part of the coelom, but only for a short distance.

sacs may be closed off from the abdominal cavity by a fold of tissue. Rupture of this fold—a weak spot in the abdominal wall—may lead to the condition known in man as inguinal hernia. Most commonly, the two scrotal sacs push downward and backward to fuse externally below and behind the penis; more rarely, as in many marsupials, they may be anterior to that organ.

Experiments and the study of abnormalities have revealed the functional basis for this unusual phenomenon of testicular descent. The delicate process of sperm maturation cannot go on at high temperatures. The temperatures maintained internally in the bodies of mammals are in general too great for sperm formation; temperatures in the exposed scrotal sacs are several degrees lower.*

The Oviduct and Its Derivatives in Lower Vertebrates (Figs. 296–298).   In amphioxus, the sex cells (formed in segmentally arranged gonads) are shed directly to the surface (via the cavity of the "hood" surrounding the gills). In cyclostomes, both eggs and sperm are shed into the coelom and must find their own way to the posterior end of this cavity, whence **genital pores**, which open at the breeding season, afford them passage to the exterior. In all gnathostomes, the sperm are conducted to the outside by way of closed tubes, but the eggs are still shed in most cases into the coelom. They are not, however, really freed into this cavity, because they are (except by accident) received, through a funnel-like structure close beside the ovary, into a tube that carries them on the remainder of their journey. This is the primitive **oviduct** (or müllerian duct), present, with various regional modifications, in most jawed vertebrates. Along the course of the oviduct, there may be formed specialized enlargements for various purposes: storage of eggs before laying, the deposition of a shell, or the retention of the egg during embryonic development and a subsequent "live" birth—the **viviparous** condition.†

The oviduct and specialized structures derived from it arise in the dorsolateral wall of the coelomic cavity and may come to be suspended in the coelom by special mesenteries. The embryonic course of the oviduct parallels that of the archinephric duct. In elasmobranchs and urodeles, it is formed by a longitudinal splitting in two of this duct. In some tetrapod groups, there are indications of a similar origin, and in some lower actinopterygians—sturgeons and paddlefishes—the oviduct continues, even in the adult, as a side branch of the archinephric duct. On the other hand, in a majority of tetrapods, the oviduct arises as a distinct structure, formed by an infolding of the coelomic epithelium or the development of a cord of tissue closely associated with the archinephric duct. The evidence on the whole thus indicates that the female genital duct (like that of the male, described below) is derived from the urinary system but has become so specialized that embryonic evidence of its origin may be lost. As would be expected, the oviduct, present in all embryos in the "indifferent"

---

*Despite the fact that temperatures are generally somewhat higher than in mammals, there is no descent of the testes in birds. However, the avian testes lie close to air sacs (cf. p. 367), which may give a cooling effect. Some mammals (whales, for example) have "learned" to produce viable sperm at body temperature.

†This contrasts with the primitive **oviparous** method, in which the egg is laid and development takes place externally. Some distinguish from viviparity an intermediate **ovoviviparous** type, in which development is internal, but it takes place without the formation of a placenta or other means of obtaining nourishment from the mother.

stage, usually disappears in males; exceptionally, however, it persists in the adult male, as in some amphibians and lungfishes.

The oviduct is seen in its simplest form in lungfishes and amphibians—groups with reproductive habits that apparently are primitive, the female laying at each breeding season a moderate number of shell-less eggs of modest size. It is here, for the most part, a simple ciliated tube, with glands or gland cells that secrete mucus or a gelatinous covering for the eggs. In the resting phase, the tube is usually small in diameter and relatively straight; at the breeding season, it may elongate to become highly convoluted, and its diameter may increase greatly.

There is no direct connection between ovary and oviduct in primitive forms (nor, for that matter, in most advanced types except teleost fishes). The anterior end of each oviduct, however, lies close beside the ovary; it consists of a ciliated funnel, the **infundibulum**, with a frilled margin. The funnels are usually separate; in the Chondrichthyes, exceptionally, the two are fused. Eggs released from the ovary are normally caught up by the funnel to begin their journey down the tube. We have noted that anterior kidney tubules in embryos and in lower vertebrates frequently connect with the coelomic cavity by ciliated, funnel-shaped openings (Fig. 289 F); if the oviduct is derived from kidney structures historically, its funnel may have originated from a greatly enlarged opening of this sort.

In such vertebrate groups as the more primitive bony fishes and amphibians, there is little specialization of the oviduct posteriorly. There is generally an expanded distal part of each tube, an **ovisac**, that serves for the storage of eggs. A few salamanders and toads are viviparous; fertilized eggs are retained within this sac until embryonic development has been completed and the young are "hatched." The living coelacanth, *Latimeria*, is also ovoviviparous, producing few but extremely large "pups." The oviducts open posteriorly into the cloaca, when present, or to the surface of the body. In variable fashion, the two tubes may remain separate over their entire length, as in salamanders and some anurans, or fuse into a common tube, as in other anurans and in lungfishes.

Among actinopterygians, the chondrosteans and *Amia* have a normal oviduct system (Fig. 309 *f*). But an exceptional situation is that which must be dealt with by the teleosts, in which countless thousands or even millions of eggs may be released during a short breeding season.* The normal system of egg reception through an open funnel would be inadequate; there would be the danger that the entire body cavity would be hopelessly choked with eggs. In most teleosts (and in the related garpikes as well), the difficulty has been solved by a complete enclosure of the ovary and an associated duct so that no escape of eggs into the general body cavity is possible (Fig. 309 *g*). The ovary itself becomes folded so as to form within itself a closed cavity. Into this, the eggs are shed, and directly from it a funnel-like duct, formed by peritoneal folds and probably not homologous with the usual vertebrate oviduct, leads backward to the exterior; no eggs are released into the peritoneal cavity.†

*In a few viviparous teleosts, the eggs develop within the ovarian cavity. In such forms, the number of eggs is secondarily reduced, as it is in forms that have parental care of young and other specialized reproductive habits.

†In salmon and a few other forms, the ovary is sheathed but is not directly connected with the duct (Fig. 309 *h*).

428
Chapter 13
Excretory
and Repro-
ductive
Systems

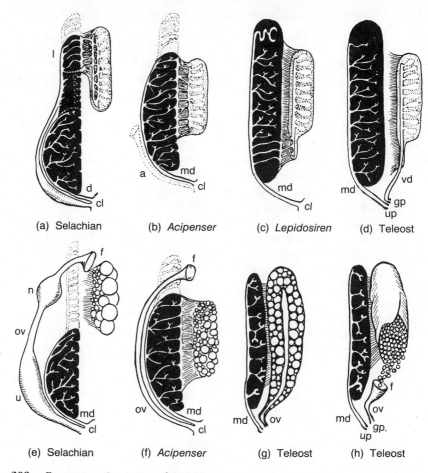

(a) Selachian   (b) *Acipenser*   (c) *Lepidosiren*   (d) Teleost

(e) Selachian   (f) *Acipenser*   (g) Teleost   (h) Teleost

Figure 309. Representative types of urogenital systems in fishes. Upper series, males, lower series, females. *Acipenser* is a primitive actinopterygian, *Lepidosiren,* the South American lungfish. Black structures, opisthonephros; testis stippled; organs with circles (representing eggs) are ovaries; mesenteries hatched; vestigial pronephros, and vestigial male oviduct (in *b*) in stippled lines. Abbreviations: *a,* vestigial oviduct; *cl,* cloaca; *d,* special duct draining kidney in advanced selachians; *f,* open funnel of oviduct; *gp,* genital papilla and pore of teleosts; *l,* gland in anterior region of selachian kidney analogous to epididymis; *md,* archinephric duct as kidney duct; *n,* shell gland; *ov,* oviduct; *u,* selachian ovisac, or "uterus"; *up,* urinary pore of teleosts; *vd,* sperm duct of teleosts, analogous to ductus deferens of amniotes. In male selachians, as in amniotes, the testis tends to take over the original duct for sperm transport (cf. Fig. 301 *D*), but in other fishes, this duct continues to drain the kidneys, and in teleosts a separate sperm duct *(vd)* develops. Specialized methods of egg transport are present in teleosts. (After Portman and Goodrich, from Hoar, in The Physiology of Fishes, M. E. Brown, editor, Academic Press.)

In other groups of vertebrates—cartilaginous fishes, reptiles, and birds—a large, shelled egg has been evolved, and special regions of the tube are developed that produce the egg white and shell. In elasmobranchs and chimaeras, there is, part way down the tube, an enlarged and specialized region, the **shell gland** (nidamental gland). Although this usually appears to be a uniform structure externally, there are internally two types of epithelia present when the gland is well developed (Fig. 309 e). In the upper part of the nidamental region of the tube, albuminous material—"egg white"—is secreted about the egg; in the lower part a hard, horny egg case is formed. Fertilization must, of course, occur before the shell is added to the egg,* and in the Chondrichthyes (and at least some placoderms), internal fertilization takes place. With the aid of claspers on the pelvic fins of the male, the sperm are injected into the cloaca. Thence they travel "upstream" in the oviduct and fertilize the egg before it descends to the shell gland.

We have noted that the young are born alive in a few amphibians, a fraction of the teleosts, and the coelacanth. Viviparity requires internal fertilization of the egg and is hence the exception in vertebrates with primitive reproductive habits; external fertilization appears to have been the original method of insemination and, even though in many salamanders and the gymnophionans the sperm are brought into the cloaca for fertilization, there is little trend toward viviparity. With the development of shelled eggs and consequent internal fertilization in cartilaginous fishes, opportunities for the development of viviparous conditions are increased. Distally, there is present in the elasmobranch oviduct an ovisac, which in many forms is simply a repository for eggs before discharge. In a variety of sharks and skates, however, viviparity has arisen; in extreme cases, the shell fails to develop around the egg, and the ovisac serves as an organ in which the embryo is sheltered and nourished until its birth as a "pup." The ovisac in such forms supplies food materials in one manner or another to the young. Most common is the development from the inner wall of the ovisac of highly vascular leaflike or filamentous outgrowths that gain contact with the embryo's yolk sac; a connection is formed somewhat comparable to the placenta of a mammal.

The Oviduct in Amniotes—Uterus, Vagina (Figs. 298 *B*, 299 *B*, 310, 313). In amniotes, as in sharks, a shelled egg has brought about specializations of the tube for the production of albumen and shell. In reptiles and birds, the successive divisions of the oviduct are frequently given names comparable to those used in mammals, although the functions differ. The greater part of its length is the oviduct proper, or **uterine tube**. This is a broad, muscular tube, capable of further distention at the breeding season. The **uterus**† lies close to the distal, cloacal end of the tube; it is in the resting phase somewhat greater than the tube in internal diameter and may be greatly expanded during the breeding season. Its walls contain a heavy coat of smooth muscle. Glands are present in the epithelia of both tube and uterus; those in the tube secrete albumen (egg white), those of the uterus form the shell. Beyond the uterus, a short

*Hagfish eggs have a thin horny shell and yet have external fertilization; a soft spot (micropyle) in one end of the shell allows the sperm to enter. In the absence of any oviduct system, the capsule is formed about the egg before it leaves the ovary.

†The distal expansion of the oviduct in elasmobranchs and amphibians, here termed the ovisac, is frequently called a uterus. Because its homology with the mammalian uterus is none too certain, it is probably better not to extend the use of this term to these lower vertebrates.

430
**Chapter 13
Excretory
and Repro-
ductive
Systems**

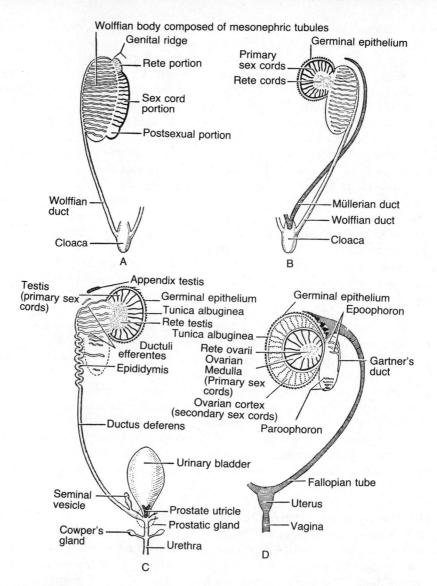

Figure 310. Embryonic development of the genital system of amniotes. *A,* Early indifferent stage, showing sex cords, developing rete testis, mesonephric kidney, and archinephric duct (wolffian body, wolffian duct). *B,* Somewhat later stage, in which the embryonic oviduct (müllerian duct) has appeared. *C,* The adult male (cf. Fig. 314). *D,* Adult female (cf. Fig. 313). (From Turner.)

terminal segment leads to the cloaca; this is often termed a vagina, although it lacks the special function of that region in mammals. The two tubes of reptiles open separately into the cloaca. Birds, turtles, and crocodilians are universally oviparous, but various lizards and snakes (and various extinct reptiles such as ichthyosaurs) bear their young alive and, in some cases, have paralleled elasmobranchs and mammals in the development of structures through which nutriment may pass from mother to young.

In birds, the right oviduct, like the right ovary, is absent, but the remaining structure has retained the reptilian pattern. Among mammals, the monotremes, which produce a shelled egg, although one of small size, have a uterine structure essentially similar to that of reptiles.*

Markedly different from those of more primitive amniotes are conditions in marsupials and placental mammals, for here, with viviparity, shell formation has ceased; instead, the uterus becomes the site for the development of the embryo, and the ends of the two oviducts, usually conjoined, develop as the **vagina** for reception of the penis (Fig. 313).

Because the mammalian ova are small, the upper part of the original oviduct, the uterine or Fallopian tube, is a slender structure. It lacks the albumen-secreting glands found in other amniotes. Fertilization of the egg usually takes place at the upper end of the tube, and the trophoblast (cf. p. 413) has already developed before the uterus is reached.

The uterus, or womb, is a thick-walled muscular structure, within which the embryo is destined to pass its embryonic existence. It has a richly vascular epithelium, the **endometrium**, which may be relatively thin during resting phases, but is greatly thickened at seasons of reproductive activity; its growth rhythms are controlled by hormones of ovarian origin. The placenta, affording maternal nourishment of the developing embryo, is formed by a union of the uterine epithelium with the external membranes of the embryo. In the most primitive mammalian situation, the two uteri are quite separate, a **duplex** condition; this is found, for example, in marsupials and in many rodents and bats (Fig. 311). In most mammals, however, the distal parts of the two uteri are fused together to give a **bipartite** or **bicornuate uterus**, and, in higher primates, there is a complete fusion to a **uterus simplex**.

Although the uteri remain distinct, the terminal portions of the two oviducts are invariably fused in placental mammals to form a vagina. The cloaca, into which the genital tubes opened in ancestral types, is absent in viviparous mammals but is represented by a **urogenital sinus** (cf. Fig. 318 *E*). This sinus is frequently shallow; when of some depth, as in a considerable number of mammalian types, it has the appearance of a further continuation of the vagina itself.

In marsupials, vaginal construction is of a specialized sort (Fig. 312). The vaginal tubes are paired but are fused at proximal and distal ends, and a vaginal sinus between them sometimes extends outward to form a third, median exit from the uterus to the exterior.

Sperm Transport: Epididymis, Ductus Deferens.

In the male cyclostome, as in the female, there is no duct system for sex products. Ripe sperm ampullae open to the coelomic surface of the testis; the spermatozoa are discharged into the coelom and, like the eggs, must make their own way to the outer world through genital pores.

In all higher vertebrates this seemingly inefficient mode of transport has been abandoned; the male has developed a system of ejaculatory ducts. This, however, contrasts strongly with the female duct system in being always closed throughout its course. Further, though the female duct may be of somewhat questionable origin, the male transport system is demonstrably a modified part of the urinary structures. As

---

*Whether or not the various extinct Mesozoic mammals were also egg-layers is, of course, uncertain; however, in some, the pelvic canal would appear too small to allow passage of a normal amniote egg, suggesting the early attainment of viviparity in at least some groups.

432
Chapter 13
Excretory
and Repro-
ductive
Systems

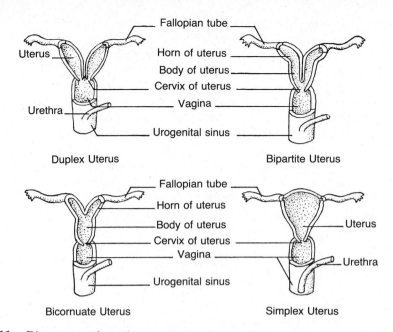

Fallopian tube

Uterus

Horn of uterus
Body of uterus
Cervix of uterus
Vagina

Urethra

Urogenital sinus

Duplex Uterus

Bipartite Uterus

Fallopian tube

Horn of uterus
Body of uterus
Cervix of uterus
Vagina

Uterus

Urethra

Urogenital sinus

Bicornuate Uterus

Simplex Uterus

Figure 311. Diagrams to show the progressive fusion of the posterior ends of the oviducts (Fallopian tubes) in placental mammals. The uterus and part of the vagina have been cut open. (From Walker, after Wiedersheim.)

the testis develops in any vertebrate, it lies close beside (and medial to) the embryonic kidney (cf. Fig. 302). A short distance away from the ampullae or seminiferous tubules are the tubules of the kidney, which empty into the archinephric duct leading to the cloaca. If the short gap between testis and kidney is bridged, there is available a closed system of ducts leading to the exterior, which could be followed by the sperm in safety, avoiding the vicissitudes of travel through the coelomic wilderness. This gap was bridged by ancestral gnathostomes, and reproductive functions were superimposed on the urinary system.

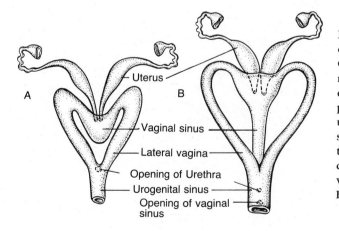

A

B

Uterus

Vaginal sinus

Lateral vagina

Opening of Urethra

Urogenital sinus

Opening of vaginal sinus

Figure 312. Female genital organs of marsupials. *A,* Opossum; *B,* kangaroo. From a median vaginal sinus there develop, in the opossum, a pair of lateral vaginae that unite distally in the urogenital sinus. In the kangaroo figured, the vaginal sinus has developed into a median vaginal tube. (After van den Broek.)

Although there are variations of one sort or another, the connections between the seminiferous structures of the testis and the archinephric duct follow a basically similar pattern in many vertebrates (Fig. 315). Ripe ampullae or the tips of the seminiferous tubules may be connected with one another by a longitudinal **central canal** of the testis or, in birds and mammals, by a network of small canals, the **rete testis**. From this, a number of parallel tubules develop to run across the short distance to the edge of the kidney. In many vertebrates, these ductules, when the kidney is reached, enter a **lateral kidney canal**, which connects with adjacent kidney tubules. This canal does not develop in mammals; in any event, however, the sperm pass through the erstwhile kidney tubules, now termed **ductuli efferentes**, and finally reach the archinephric duct.

Presumably the most primitive condition in gnathostomes is one in which the archinephric duct continues in the male to serve impartially both urinary and genital functions as, for example, in frogs and sturgeons (Fig. 309 *b*). As we have noted, however, in our discussion of the urinary system, most vertebrates have developed new ducts so that sperm and urine travel separate courses. The events in this history have been noted in discussing the urinary organs; they may be reviewed here from (shall we say) the point of view of the testis (Fig. 301).

In most bony fishes, the kidney has retrieved its original duct; the spermatozoa pass back to an external opening behind the anus by way of a discrete **sperm duct**, probably developed from the central canal of the testis, which by-passes the kidney (Fig. 309 *d*). In all other groups of vertebrates, sperm transportation has triumphed in deciding the fate of the archinephric duct. Parallel developments are witnessed in cartilaginous fishes and in tetrapods. Urine conduction in elasmobranchs and many amphibians is found, at the most, only in the back end of the duct, and in all amniotes,

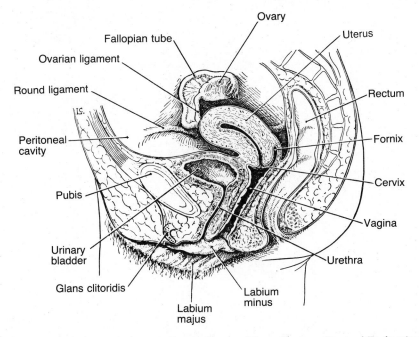

Figure 313. Female reproductive organs in *Homo*. (From Turner, General Endocrinology.)

434
Chapter 13
Excretory
and Repro-
ductive
Systems

it has been completely excluded. With such partial or total exclusion from urinary function, the archinephric duct has become a **ductus deferens**, a tube specific for sperm transport; the testicular connections are universally with the proximal, or anterior, end of the duct.

For much of its length, the ductus deferens remains a simple structure, but it may undergo major modifications in various regions and in various regards. Most notable are the modifications in the anterior portion of its length, where in both Chondrichthyes and amniotes, it takes part in the formation of an appendage to the testis termed the **epididymis**. In sharks, skates, and chimaeras a limited number of efferent ductules from the testis enter the anterior end of the embryonic opisthonephros and reach the former archinephric duct (Figs. 294 *D*, 301 *C, D*). The anterior end of this duct is much convoluted and may be termed the **ductus epididymidis**. As it passes backward, before reaching the functional kidney, it is joined in sharks by a series of nephric tubules that have been modified into glandular structures; these secrete a fluid medium that is believed to stimulate the sperm.

A parallel development has occurred in amniotes; in them, the epididymis forms a compact body resting, as its name implies, close beside or upon the testis. The ductules that cross from the testis to this former nephric structure enter, in reptiles, a longitudinal canal whence the efferent ductules continue to the former archinephric duct; in birds and mammals, the longitudinal canal is absent, and the sperm continue directly via the much convoluted ductules to the archinephric duct, which is here, as in elasmobranchs, highly convoluted (Figs. 310, 314, 315).

Distally, the ductus deferens may expand into an ampulla for sperm storage be-

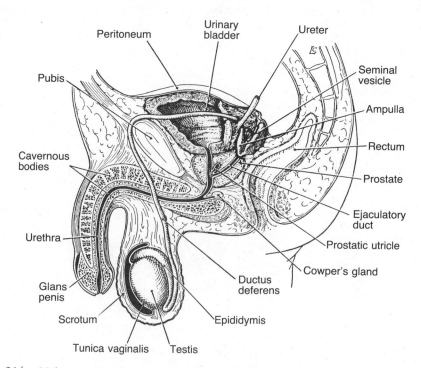

Figure 314. Male reproductive organs in *Homo*. (From Turner, General Endocrinology.)

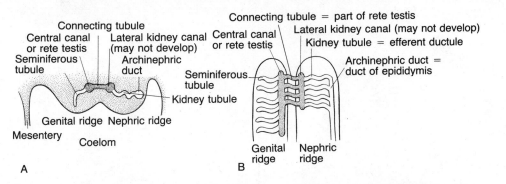

Connecting tubule
Central canal or rete testis
Lateral kidney canal (may not develop)
Seminiferous tubule
Archinephric duct
Kidney tubule
Genital ridge
Nephric ridge
Mesentery
Coelom

A

Connecting tubule = part of rete testis
Central canal or rete testis
Lateral kidney canal (may not develop)
Kidney tubule = efferent ductule
Seminiferous tubule
Archinephric duct = duct of epididymis
Genital ridge
Nephric ridge

B

Figure 315. *A,* Cross section of an amniote embryo, to illustrate the fact that testis and kidney are adjacent (cf. Fig. 302) and may thus readily come into connection by the development of tubules bridging the gap between seminiferous and kidney tubules. *B,* Diagrammatic ventral view of a section of nephric and genital ridges to show mode of connection, usually with a central testis canal or rete testis between the sperm tubules, with connecting tubules, and frequently with a lateral kidney canal.

fore ejaculation; such a structure is common, but not universal, in Chondrichthyes, amphibians, and amniotes. In many mammals (Figs. 310, 314), there are further present **vesicular glands** (seminal vesicles), which do not store semen (as the alternative name implies) but secrete a thick liquid as part of the seminal fluid. In most mammals, the **prostate gland**, surrounding the urethra and secreting a thinner fluid, and the more distally placed **bulbo-urethral glands** (Cowper's glands) are further secretory structures along the path of the sperm duct.

The Cloaca and Its Derivatives

In a great variety of fishes and tetrapods, a ventral pocket is present at the back end of the trunk region, communicating with the exterior, into which open the orifices of the digestive, genital, and urinary systems. This structure, appropriately termed the **cloaca** (the Roman name for a sewer), appears to have been a primitive vertebrate feature. We shall in this section follow, through the vertebrate groups, the history of the cloacal region and the varied disposition of the outlets of the systems concerned. Inseparable from this general story is that of the urinary bladder of the tetrapods and that of the external genitalia of the amniotes.

The Cloaca In Fishes.   The cloaca is typically developed in elasmobranchs (Fig. 316 *B*), where it forms a depression, which may be provided with a sphincter muscle, on the ventral surface of the body behind the pelvic girdle. The major opening into it is that of the posterior end of the intestine. Into it, further, open the urinary ducts and the vasa deferentia of the male or the paired oviducts of the female.

Embryologically, the cloaca has a twofold origin (cf. Fig. 318 *A*). Its major part in the embryo of any vertebrate consists of an expansion of the posterior end of the endodermal gut tube. This cavity is, until a relatively late stage of development, closed off from the exterior by a membrane, external to which lies a depression of the ectoderm, the **proctodeum**. When the membrane disappears, this ectodermal

436
Chapter 13
Excretory
and Repro-
ductive
Systems

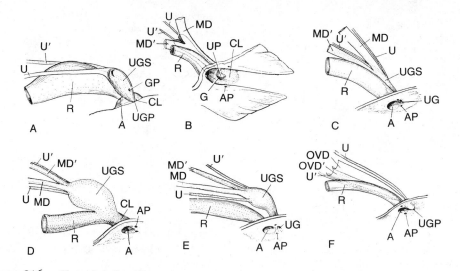

Figure 316. Cloacal and anal region in fishes. *A*, The lamprey, *Petromyzon*; *B*, a female shark; *C*, a young female chimaera; *D*, a female Australian lungfish, *Neoceratodus*; *E*, a female sturgeon; *F*, a female salmon. Abbreviations: *A*, anus; *AP*, abdominal pore; *CL*, cloaca; *G*, genital opening; *GP*, genital pore; *MD, MD'*, left and right oviducts (müllerian ducts); *OVD, OVD'*, left and right oviducts of teleost; *R*, rectal region of intestine; *U, U'*, left and right urinary ducts; *UG*, urogenital opening; *UGP*, urogenital papilla; *UGS*, urogenital sinus; *UP*, urinary papilla. (From Dean.)

area is incorporated into the cloaca. It is generally assumed that the ectoderm plays little part in cloacal formation. However, once the membrane breaks, it is difficult to tell which part of the cloacal lining comes from ectoderm, which from endoderm; in at least some cases, the ectodermal contribution to the cloaca and its derivatives is of considerable magnitude.

Among other fishes, we find a well-developed cloaca only in the lungfishes (Fig. 316 *D*) and in the male of the sole surviving crossopterygian, *Latimeria*. In the chimaeras (Fig. 316 *C*), the cloaca has been entirely eliminated; anal, urinary, and genital openings all reach the surface separately. Among cyclostomes, hagfishes have a cloacal pocket into which anus, urinary tube, and genital pore all open, but, in lampreys, the anus has an opening separate from that of the conjoined urinary and genital tubes (Fig. 316 *A*).

In actinopterygians (Fig. 316 *E, F*), the cloaca has likewise disappeared. In the female *Latimeria*, lower ray-finned fish types, and some teleosts, urinary and reproductive tubes empty by a common sinus representing part of the cloaca, but the anus has a separate opening. In other teleosts, we find that three quite distinct openings are present: anal, genital, and urinary.

The Cloaca In Lower Tetrapods (Fig. 317 *A*). In the amphibians, a cloaca is present, presumably a direct inheritance from piscine ancestors. The archinephric ducts and, in the female, the oviducts open as paired structures into the inner end of the cloacal pouch. Ventrally, there develops from the cloaca a urinary bladder of varied shape, often distensible to large size (cf. p. 416). In reptiles, the cloacal structure is similar;

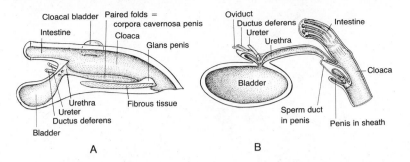

Figure 317. Section of the cloacal region of *A,* male tortoise, *B,* a monotreme mammal *(Echidna).* In *A,* a penis-like structure is contained in the floor of the cloaca; paired folds may meet to form a tube at the time of sperm emission. In the monotreme, a formed penis is present within the cloaca; it contains a tube, divided into several branches, for sperm transport, but urine passes out via the cloaca. In most reptiles, the ureter opens dorsally into the cloaca at a point far from the bladder; in its chelonian position, there is a closer approximation to the mammalian condition in the ventral shift of this opening. The tube labeled "urethra" in the two figures is equivalent to that structure in the female of higher mammals but is here essentially a ventral proximal subdivision of the cloaca. A turtle specialization is the presence of a pair of small "supernumerary" bladder-like structures in the side walls of the cloaca. (*A* partly after Moens; *B* after Keibel.)

the paired openings of the ureters are typically on the dorsal wall of the cloaca, above the anal entrance; the urinary bladder, usually present and (as in all amniotes) formed from the base of the allantois, is a distinct ventral pocket not connected with the ureters, except in turtles. In birds, entrances into the cloaca are similarly placed, but no urinary bladder is present except in the ostrich.*

Fate of the Cloaca In Mammals. In mammals, the lowly monotremes still have a cloaca. Higher types have done away with this structure and have a separate anal outlet for the rectum. Monotremes show the initiation of this subdivision (Fig. 317 *B*). The cloaca as such includes only the distal part, roughly comparable to the proctodeum. The more proximal part is divided into (1) a large dorsal passage into which the intestine opens, the **coprodeum**, and (2) a ventral portion, the **urodeum**, with which the bladder connects. Ureters open into the proximal end of the urodeum, which is partially equivalent to the placental urethra (Fig. 318), and the urine may thus enter the bladder without being associated with fecal material.

In marsupials, the cloaca is persistently represented by a slight proctodeal depression, with a sphincter muscle, but, in most higher placental mammals, the cloaca has disappeared completely in the adult; coprodeum and urodeum have distinct orifices. The coprodeum becomes the rectal region of the gut, opening at the anus. The urodeum, however, has a more complex history and differs much in the two sexes. Conditions here are best understood by considering the developmental story; the development of the placental mammal recapitulates in many respects the phylogenetic story (Fig. 318).

---

*Birds have, uniquely, a pocket, termed the **bursa of Fabricius**, opening into the cloaca from above. It is a lymphoid organ discussed in Chapter 14.

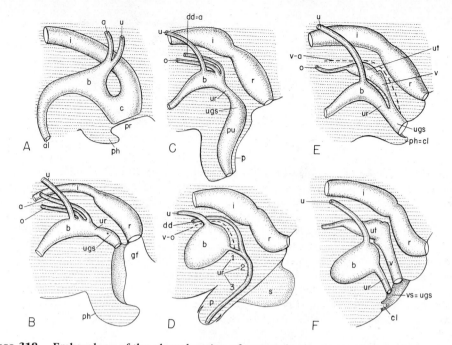

Figure 318. Embryology of the cloacal region of mammals; diagrammatic lateral views. *A,* Sexually indifferent stage, with intestine and allantois opening into undivided cloaca, archinephric duct and ureter opening together into base of allantois. *B,* Later indifferent stage; embryonic oviducts developed, ureter and archinephric duct separated, cloaca divided into rectum and urogenital sinus; phallic organ has begun development. *C,* Early stage of development of male, and *D,* adult male structure. In contrast to female, the oviduct disappears (broken line in·*D*), the archinephric duct becomes the ductus deferens, and an extensive urethra includes, in addition, (1) the base of the allantois, (2) the urogenital sinus, and (3) a duct traversing the penis. *E,* Early stage of development of female, and *F,* adult female structure; disappearance of archinephric duct (broken line in *E*), development of bladder from base of allantois with short urethra distal to it, differentiation of uterus and vagina from embryonic oviduct, development of phallic organ as clitoris. Abbreviations: *a,* archinephric duct; *al,* stalk of allantois; *b,* bladder; *c,* cloaca; *cl,* clitoris; *dd,* ductus deferens; *gf,* genital fold; *i,* intestine; *o,* oviduct; *p,* penis; *ph,* phallic organ; *pr,* proctodeum; *pu,* penile urethra; *r,* rectum; *s,* scrotum; *u,* ureter; *ugs,* urogenital sinus; *ur,* urethra; *ut,* uterus; *v,* vagina; *v-a,* vestige of archinephric duct; *v-o,* vestige of oviduct; *vs,* vestibulum.

In the sexually indifferent stage of a placental mammal, there is a cloaca, formed as a simple, distal endodermal expansion of the gut. A membrane is present to separate the endodermal proctodeum. Into the walls of the endodermal cavity open paired archinephric ducts (paralleled later by the oviducts); out from its floor extends the stalk of the allantois.

While the indifferent stage still persists, a horizontal septum develops, and extends out to the closing membrane. This divides the cloaca into two chambers: a coprodeum continuous with the gut above, and a urodeum or **urogenital sinus** below. The archinephric ducts—the ducti deferentes of the male—enter the lower division, as do the female genital ducts. As development proceeds, part of the allantoic stalk enlarges as the beginning of the typical amniote bladder, and a narrower tubular region develops between the urogenital sinus proper and the bladder, which forms part or all of the **urethra** of the adult. The genital ducts remain connected with the sinus. The ureters, however, as they develop, separate from the parent archinephric ducts and gain distinct distal openings. These openings migrate to the region of the bladder, their definitive position in placental mammals; urine in mammals (in contrast with most lower tetrapods) enters the bladder directly, without having to traverse any part of the cloaca or its subdivisions.

From this stage onward, the course of development of the two sexes diverges in regard to the structures associated with the urogenital sinus. In the female, when the cloacal membrane disappears, this ventral part of the cloaca becomes the **vestibulum** of the urinary and genital organs. This may attain a considerable depth (as in typical carnivores) or be a relatively shallow depression (as in primates). Into this vestibulum open the conjoined distal ends of the oviducts as the vagina and, separately, a relatively short urethral tube from the bladder; the archinephric ducts degenerate.

In the male, the sinus has a different history. Instead of becoming shallow and broad, it becomes an elongate tube that continues into the penis. Into its proximal end open the ducti deferentes (the oviducts degenerate) and the tube from the bladder. In the female, this short proximal tube constitutes the entire urethra, but, in the male, the entire extent from bladder to end of penis is termed **urethra**. Female and male urethrae are, thus, not comparable structures; that of the male includes the homologues of both female urethra and vestibulum (sinus).

External Genitalia. In primitive aquatic vertebrates with shell-less eggs, external fertilization is the rule; sperm and eggs are shed into the water, and fertilization takes place in that medium. But in forms with shelled eggs or with viviparous habits—such types include the Chondrichthyes and all amniotes—internal fertilization takes place, because the sperm must reach the egg before it has descended the oviduct to the shell gland or uterus. Special structures are usually formed to facilitate the entrance of the sperm into the female genital tubes.

In sharks, rays, chimaeras, and at least one placoderm—fishes with heavily shelled eggs—**claspers** are developed in the males to aid in fertilization. These are posterior extensions of the pelvic fins, which are stiffened by rods from the fin skeleton (Fig. 145 *C*). One clasper is inserted into the cloaca of the female; rolls of skin folded into tubes along the clasper form a channel for the sperm. In a limited number of teleosts, internal fertilization (and viviparity) is present; one specialization to facilitate this is that the anal fin may develop a clasper-like intromittent organ, a **gonopodium**, with a groove or tube to carry the sperm (Fig. 319 *A*).

In the amniotes, internal fertilization is rendered necessary, apart from the presence of the shelled egg or of a viviparous type of development, by terrestrial life and the consequent absence of a watery environment for sperm and eggs at the time of breeding. In the ancestral amniotes, direct contact of male and female cloacae would seem to have been sufficient for the transfer of sperm, for *Sphenodon* lacks copulatory organs, which are also absent in most birds (here their absence is almost certainly secondary).

In many reptiles, however, the male has some type of accessory organ, a **penis**, to aid in sperm transfer. In snakes and lizards, there are unique structures, the **hem-**

440

**Chapter 13
Excretory
and Repro-
ductive
Systems**

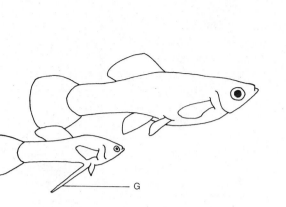

A

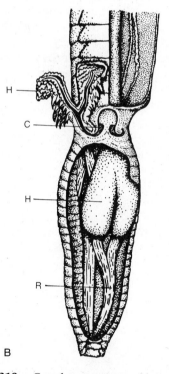

B

Figure 319. Copulatory organs of lower vertebrates. *A,* Poeciliid teleosts; the male has a gonopodium, formed from certain rays of the anal fin, which can be inserted into the female's reproductive tract. *B,* Rattlesnake; one of the hemipenes is everted from the cloaca and the other is in the resting position within the base of the tail. Abbreviations: *C,* cloaca; *G,* gonopodium; *H,* hemipenis; *R,* retractor muscles of hemipenis. (After Rosen and Gordon, and Hoffmann.)

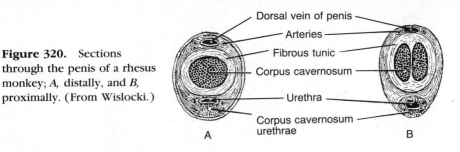

Figure 320. Sections through the penis of a rhesus monkey; *A,* distally, and *B,* proximally. (From Wislocki.)

Dorsal vein of penis
Arteries
Fibrous tunic
Corpus cavernosum
Urethra
Corpus cavernosum urethrae

A

B

ipenes. These are a pair of pockets, often containing thornlike spines, which lie in the skin adjacent to the cloacal opening; at the time of copulation, the hemipenes may be turned inside out, extruded, and one of them inserted in the female cloaca to guide the sperm (Fig. 319 *B*). In turtles and crocodilians, there are structures that may be morphologic forerunners of the penis of mammals (Fig. 317 *A*). In the ventral wall of the cloaca are a pair of longitudinal ridges, the **corpora cavernosa penis**, with a groove between them; they are composed (as the name implies) of spongy fibrous tissue. A further spongy structure, the **glans penis**, lies at the outer end of the groove. On excitation, these structures are filled with blood and thus distended; the glans is protruded and inserted into the female cloaca; the cavernous bodies expand to meet one another and turn the groove between them into a tube for the passage of the sperm. These structures appear in the embryo in the sexually indifferent stage and persist in the adult female in a relatively undeveloped and functionless state as the **clitoris**. The ratites, geese, and a few other birds have modest development of penis and clitoris, but these structures have been lost in most birds.

The phallic organs of typical mammals are more highly developed (Figs. 314, 320); the monotreme penis, however, is transitional in some regards, and the mammalian clitoris is simpler than the penis in structure. Glans and corpora cavernosa are present and form the main erectile structures; they are, however, permanently fused in the adult penis, together with an additional erectile body continuous with the glans penis, the **corpus cavernosum urethrae**. These structures enclose a continuation of the urethra as a tube carrying both sperm and urine. The corpora cavernosa are in reptiles embedded in the walls of the cloaca; in typical mammals, however, the penis as a whole extends outside the body, and the glans is surrounded with a skin fold, the **prepuce**. As mentioned earlier, a bony structure, the os penis, may develop—most notably in carnivores—in the connective tissues of the penis.

Chapter 14

Circulatory System

In many small invertebrates, a circulatory system is not necessary. Distances are short, structures are simple, and transportation of materials within the body may be effected by diffusion and by such flow of internal fluids as may result from body movement. With greater size and complexity, a definite circulatory system is required. A comparison with human communities is a fair analogy. In a village, a transportation system is unnecessary; stores, school, and church are all close to the homes they serve. With growth of the community, this is no longer the case, and an organized transportation system is a public necessity.

Among invertebrates, we see the development of two types of circulation: open and closed systems. In the former, a heart forms a pump whence blood is forced out through a series of vessels, the arteries, to various parts of the body. From the point at which these vessels terminate, however, the blood oozes back to the heart through tissue spaces; return vessels are absent. In the more advanced closed system—the type present in the vertebrates—the blood almost never leaves the vessels and thus lacks direct contact with the body cells that lie in the separate interstitial fluid. The blood serves the tissues by diffusion of materials through small, thin-walled terminal vessels, the capillaries, and returns to the heart again through a second series of closed channels, the veins. Amphioxus shows an intermediate stage in development, for although both outgoing and returning vessels and equivalents of capillaries are present, these all lack the continuous endothelial lining found in true vertebrates (cf. p. 453).

Functions.   The foremost functions served by the vertebrate circulatory system are those of transportation to the cells (via the interstitial fluid) of the materials needed for their sustenance and activity, and of the removal of wastes. The tissues must be constantly supplied with oxygen from gills, skin, or lungs, and fed with a small but steady stream of food materials from intestine or from storage and manufacturing centers, notably the liver. Wastes must be removed: carbon dioxide, destined for gills or lungs; nitrogenous wastes, to be excreted in the kidneys; and excess metabolic water.

A second important function of the circulation lies in the influence that it exerts over the maintenance of the stable and narrowly defined internal environment necessary for cellular welfare. The constant circulation of the blood makes for uniformity of composition in the interstitial fluids of every region and aids in the maintenance of relatively uniform temperatures.

The vascular system has still further functions. It aids in the struggle against disease and in the repair of injuries; through the circulation of hormones, the blood acts as an accessory nervous system.

Components. The basic structural features of the vertebrate circulatory system are familiar and readily demonstrated. A muscular pump, the heart, forces the blood stream by way of arteries to gills or lungs and to the varied tissues of the body. The tiniest circulatory units are networks of capillaries, through whose walls exchange of materials with the interstitial fluid surrounding body cells takes place. From the capillaries, blood returns to the heart via veins; lymphatic vessels, originating in the tissues and unconnected with the arteries, form a route by which fluid can reenter the circulatory system. In mammals, this system also includes nodes, described below, which are important in immune reactions. In certain instances, blood from capillaries does not return directly to the heart but is forced through an intervening second capillary bed in another organ. The veins between two capillary systems are termed a **portal system** and are named for the organ supplied (e.g., the hepatic portal system consists of the veins transporting blood from the gut to the liver).

Blood

The blood, filling the vessels of the circulatory system, may be regarded as a tissue— a tissue with very remarkable properties, but one that is, in certain aspects, comparable to the connective and skeletal tissues, to which, ontogenetically, the blood system is closely related. Each of these tissues consists of cellular elements derived from mesenchyme and lying in a "matrix." In bone, the matrix is a hard substance; in connective tissue, it is gelatinous in consistency; in blood, there is a liquid matrix in which the cellular components float freely.

Blood Plasma. The liquid part of the blood, plasma,* is a watery fluid of complex composition, of which—in mammals, for example—some 7 to 10 per cent of the total volume consists of materials in solution. The blood is to be regarded as essentially a part of the complex interstitial fluid enclosed within the walls of the blood vessels, and both have a practically identical salt content.

In addition, however, the blood contains materials peculiar to itself in the form of special **blood proteins**—albumin, globulin, and fibrinogen—which make up the greater part of the nonaqueous fraction of the plasma. These molecules are so large that they are normally unable to pass through capillary walls and thus leave the blood stream. Because the salt concentration of the blood is comparable to that of the interstitial fluids lying outside the blood vessels, the presence of these blood proteins (notably the **albumins**) raises the osmotic pressure of the blood above that of the interstitial fluids—a point of importance in capillary function. **Fibrinogen** makes up only a small percentage of the total, but is important in clot formation when a blood vessel is cut. One type of **globulin** also takes part in the formation of the blood clot. Other globulins are sensitive to protein materials of "foreign" origin and will, for example, cause the clumping (agglutination) of red blood cells of another species or

*The term **serum** refers to the liquid remaining after the protein clot material (fibrin) has been removed from clotted blood plasma.

even (as in man) of individuals of another blood type if the bloods are admixed. Certain globulins are active agents in disease protection as the antibodies effective against invading organisms or chemicals.

The salts and blood proteins are stable, permanent, plasma constituents. In addition, however, analysis of the blood reveals various materials in transit. A quantity of glucose is always present, as well as smaller amounts of fats and of amino acids, on their way to the cells of the body. Waste materials, on the way to kidney, gills, or lungs, are also present, particularly carbon dioxide (mainly as a bicarbonate), nitrogenous waste (usually in the form of urea or uric acid), and lactic acid. In more minute amounts, enzymes may be identified in the plasma, and endocrine secretions—the hormones.

Blood Cells (Fig. 321). Cellular blood components are absent in amphioxus but are present in all vertebrates. They normally include (1) red blood corpuscles or erythrocytes, (2) white blood corpuscles or leukocytes, and (3) thrombocytes.

Constant transportation of oxygen is one of the most important functions of the circulatory system. Various metallic compounds, particularly of iron or copper, are found in the blood of almost every form with a developed circulatory system. These compounds have the property of attaching to themselves large amounts of oxygen and, thus, vastly increase the oxygen load that the blood can carry. No such material is found in amphioxus; almost all vertebrates, however, possess **hemoglobin**, a protein molecule that usually includes four iron atoms, each capable of carrying an oxygen ion. Hemoglobin is also a factor in the return of carbon dioxide from the tissues.

In most invertebrates, the oxygen-carrying substances are loose in the blood stream; in vertebrates, the hemoglobin is carried by the red blood corpuscles, or **erythrocytes**. In most vertebrates, these corpuscles are properly nucleated cellular structures, flattened ovals in form, with their cytoplasm tightly packed with hemoglobin. In bulk, the erythrocytes, as the name suggests, appear red; singly, however, an erythrocyte is actually greenish yellow. The mammalian erythrocyte is unusual in

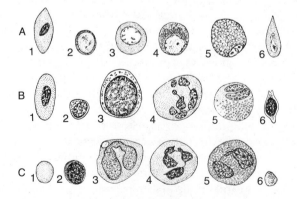

Figure 321. Blood cells of *A*, a teleost *(Labrax); B*, a frog *(Rana); C*, a mammal (man). *A1, B1, C1,* Erythrocytes; *A2, B2, C2,* lymphocytes; *A3, B4, C4,* neutrophilic granulocytes; *A4,* fine-grained acidophilic granulocyte; *A5,* coarse-grained acidophilic granulocyte; *A6, B6,* thrombocytes; *B3, C3,* monocytes; *C6,* blood platelet. *A,* × about 1800; *B, C,* × 1200. (*A* after Duthie; *B* after Jordan; *C* after Maximow and Bloom.)

structure. Except for camels and llamas, the shape is round rather than oval, and furthermore, the mature mammalian erythrocyte extrudes its nucleus.

There is great variation in the size of red blood corpuscles. In man, for example, the diameter of the erythrocyte is only 7 or 8 micrometers, and, in a goat, only half this figure; the erythrocytes are larger in all lower classes than in mammals and birds, and some amphibians have corpuscles with a volume 100 times or more as great as that of human red blood cells. Nevertheless, the amount of hemoglobin present in the blood is in general fairly constant in animals of any given volume, because the larger the corpuscles, the fewer their numbers. Exceptions are to be found among teleosts, however; some small forms dwelling in very cold, well-oxygenated Antarctic waters have no hemoglobin or red blood cells.

The white corpuscles characteristic of the blood and, to a lesser extent, the lymph streams are the **leukocytes**. They are much less numerous than erythrocytes. In amniotes, they almost never compose more than 1 per cent of the blood corpuscles; in fishes, however, the figure may rise to 10 per cent or over. Leukocytes vary greatly, in appearance and in types present, from one group of vertebrates to another, and it is difficult to establish any classification of them that will hold throughout. In general, we may distinguish two major groups of white cells—lymphoid types, with a simple, single nucleus and a nongranular cytoplasm; and granulocytes, in which the nuclear materials are irregularly arranged and often subdivided, and the cytoplasm is granular in appearance.

Among **lymphoid leukocytes**, the common forms are the **lymphocytes**, medium-sized to small, rounded cells with a large nucleus and a relatively small amount of clear cytoplasm. These are present in all classes of vertebrates and are especially abundant in the lower groups. They derive their name from the fact that they abound in the lymph. Lymphocytes are commonly present in connective tissue and, like granulocytes, are abundant in injured or infected tissues. They play an important role in immunologic protection of the organism. Two kinds of lymphocytes have different immunologic properties and origins but look much the same. **Monocytes**, once regarded as merely large lymphocytes, have a larger amount of cytoplasm and a nucleus that may become somewhat irregular or kidney-shaped. They are potentially phagocytic cells migrating, via the blood, to various sites in the tissues.

The second characteristic group of leukocytes is that of the **granulocytes** or **polymorphonuclear leukocytes**. These, like the monocytes, are large cells. The nucleus, as the alternative name suggests, is irregular in structure and more or less subdivided into lobes. Most characteristic is the fact that the abundant cytoplasm is highly granular. In some cases, the granules are stainable with acid dyes; in others, basic dyes are effective; in still others, both types of dyes take hold to some degree. These three types of "polys" are hence reasonably termed, respectively, **acidophilic** (eosinophilic), **basophilic**, and **heterophilic** (neutrophilic) **granulocytes**. Heterophilic elements are the most abundant granulocytes in all vertebrate classes except, apparently, reptiles; acidophils are uncommon in humans but are widespread among vertebrate groups; basophils are generally still fewer in numbers and are seldom identified in fishes.

Granulocytes appear to be essentially spherical while in the blood stream but are capable of ameboid motion and seem to be able to wriggle in and out of capillaries between endothelial cells. They tend to accumulate rapidly in injured, inflamed, or infected tissues. The heterophilic granulocytes are capable phagocytes, "eating-cells," and a major means of reducing bacterial infection. Eosinophils play a role in immune and allergic reactions.

Thrombocytes are elements associated with the process of blood clotting. In mammals, these structures take the form of **blood platelets**, tiny discs, lacking a nucleus, that are formed by the fragmentation of giant cells—**megakaryocytes** (Fig. 322)—present in the blood-forming tissues. In most nonmammalian vertebrates, the thrombocytes take the form of **spindle cells**—small, oval, pointed structures with a central nucleus, which appear to be related in origin to other white blood cells. When thrombocytes leave the environment of the blood stream (as in a cut vessel), they tend to disintegrate, and release materials chemically concerned with blood clotting; in mammals, the platelets themselves form much of the clot.

Blood-Forming Tissues

In body tissues generally, most cellular differentiation takes place in embryonic stages once and for all; proliferation and differentiation of blood cells continue throughout life. The life of an individual blood cell is short; mammalian erythrocytes, for example, appear to live, at the most, only a few months, and the life of some leukocytes appears to be measured in days or even hours. In consequence, an animal must retain throughout its existence **hemopoietic tissues** in which blood cells are constantly formed and must, in addition, provide for the removal of worn-out cells.

Students of the blood have devoted much study to blood cell formation, and some features of the process are still in doubt (Fig. 323). Most other differentiated cells are parts of tissues that are fixed in position; their development takes place on

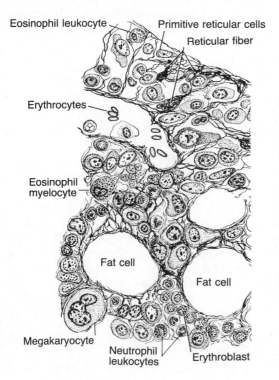

Eosinophil leukocyte — Primitive reticular cells
Reticular fiber

Erythrocytes

Eosinophil
myelocyte

Fat cell

Fat cell

Megakaryocyte

Neutrophil
leukocytes

Erythroblast

Figure 322. A blood-forming tissue—bone marrow from a mammalian femur. Much of the reticular framework and two reticular cells are seen, as well as two types of granular leukocytes in process of differentiation, and erythroblasts from which erythrocytes are formed. The megakaryocyte is a giant cell type from which blood platelets are formed. (After Maximow and Bloom.)

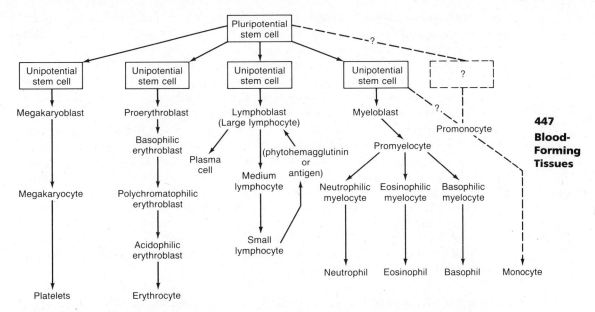

Figure 323. Development of the various types of blood cells from pluripotential stem cells. This, representing the stages found in mammalian bone marrow, is still not completely worked out (note the question marks along the possible paths to monocytes). We shall not attempt to describe or define the many types of cells that are involved. (From W. Bloom and D. W. Fawcett, A Textbook of Histology, 10th ed., W. B. Saunders Co., Philadelphia, 1975.)

the spot. But blood cells are free agents, and it is immaterial where they are formed, as long as the area is in communication with the blood vessels. Thus, blood-forming tissues are found in the most varied places in various vertebrates. Unlike most organ systems, little or none of the circulatory system is derived from any of the epithelial sheets laid down in the early embryo. Instead it is derived from mesenchyme cells—vessels and blood corpuscles alike—and thus from the same elements that give rise to the connective tissues.

Except for certain sites used in early embryonic stages, areas in which blood cells arise or are stored have, in general, common structural features (Fig. 322). All are spaces that are enlargements of circulatory vessels or lie adjacent to such vessels. A network of fibers forms the "skeleton" of the tissue. Along the course of the fibers lie **reticular cells**, which are responsible for the formation of the fiber network. They are part of the **macrophage** system of the body—cells of phagocytic nature, having the power of engulfing waste or foreign materials, such as bacteria or damaged blood cells. Macrophages are not restricted to blood-forming tissues but may be found along the course of blood vessels, notably in the liver and spleen, or occur as free, ameboid, scavenging elements in the connective tissues.

Enmeshed within the reticular framework of a hemopoietic tissue are found masses of blood cells in process of multiplication and differentiation. The **pluripotential stem cell** or **hemocytoblast**—derived from the embryonic mesenchyme—is the primary unspecialized cell type, from which blood corpuscles of all sorts (erythrocytes, leukocytes, megakaryocytes) are derived (Fig. 323). It is a cell of modest size, with a relatively small amount of clear cytoplasm.

Embryonic Blood-Forming Sites. The first blood vessels that form in the embryo are those engaged in bringing food materials into circulation, and the first blood cells are erythrocytes formed in connection with them. In forms with a mesolecithal egg type, the food lies in the yolk-filled gut, and the first vessels and blood cells form in the mesenchyme of the belly floor. In large-yolked eggs, vessels and cells form in the yolk sac, typically as clusters of cells termed **blood islands** (Fig. 271 *A*), and in mammals, vessels and cells are early formed in the chorionic portion of the placenta as well. At a somewhat later stage, as vessels extend through the body, blood cells may form locally in the mesenchyme of various regions and may long continue to arise from the walls of blood vessels. In later phases of embryonic development, blood cell formation becomes more localized, and the embryonic blood-forming tissues are often situated in areas quite different from those found in adult life. The embryonic kidney is an important seat of this activity in many forms from sharks to reptiles and birds. The tissues of the pharyngeal region and the thymus in particular are an important source of blood cells in many embryos and young animals. In mammals, the liver and spleen are no longer important as hemopoietic organs in the adult but play prominent roles in the embryo.

Thymus Gland. Almost universally present and highly important in early life is the thymus gland. This consists of clusters of soft glandlike material found in variable arrangement in the branchial or throat region. When well developed, the thymus tissue is arranged in lobules, each containing a cortex and medulla (Fig. 324). Characteristic are **reticular cells** that form a network throughout the gland and constitute most of the bulk of the medulla. The cortex mainly consists of lymphocytes. In addition, there may be other histologic features—elongate muscle-like cells, spherical corpuscles that perhaps are masses of degenerating cells in the medulla, and so on. A thymus has not been definitely identified in hagfishes, although cells in some of the pharyngeal muscles show certain thymic characteristics. A thymus is, however, found in all higher fishes (including lampreys), usually as a paired series of soft, irregular masses of tissue situated deep to the surface above most or all of the gill slits. In amphibians, the thymus is found deep within the lateral surface of the neck behind the pharynx. In frogs, it consists of a single pair of compact oval bodies; in addition, a pair of **jugular bodies** are thymus-like lymphoid structures that are derivatives of branchial pouches. In the other two amphibian orders, the thymus is generally a rather diffuse, lobulated mass of tissue. In young reptiles, it is frequently in the form

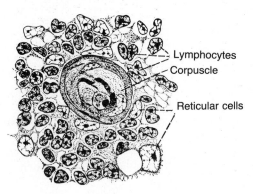

Lymphocytes
Corpuscle
Reticular cells

Figure 324. A small part of a mammalian thymus, showing the numerous lymphocytes, the cells forming the reticulum, and one of the peculiar corpuscles characteristic of the organ. (After Dahlgren and Kepner.)

of a strand of tissue running the length of the neck on either side. It retains this form in adult crocodilians but usually in reptiles forms more compact bodies—two pairs of them in *Sphenodon,* lizards, and snakes; a single pair in most turtles. In birds, as in their crocodilian relatives, the thymus is generally a long strand of tissue extending the length of the neck; it is somewhat irregularly and variably lobate. Among mammals, the most common condition of the thymus is that of a pair of structures that have migrated from the throat back to a position in the anteroventral part of the thorax, deep to the sternum and near the point at which the major vessels leave the heart. However, the thymus is sometimes cervical in position, and, in some mammals, both cervical and thoracic bodies may be present.

The reticular portion of the thymus is derived from the endodermal lining of the embryonic gill pouches, generally from epithelial thickenings formed at the dorsal side of such pouches. There is, however, much variation in the specific pouches concerned (Fig. 269). In fishes, every typical gill pouch may produce a thymic bud, and even the spiracle is implicated in many sharks. Among the amphibians, the Gymnophiona produce thymic material from every embryonic pouch; but, in salamanders, the number concerned is usually reduced, and, in frogs and toads, only the first posthyoid cleft contributes. Reptiles also show a variable condition, but usually only two pairs of pouches contribute to the adult thymus. In birds, the thymus usually comes from the second and third posthyoid pouches. In mammals, the second postspiracular cleft is the usual seat of thymic origin. The mammalian thymic buds are, in contrast to the usual situation, derived from the ventral rather than the dorsal parts of the branchial pouches. Another odd feature in mammals is that the cervical part of the thymus, which we have mentioned as present in various forms, does not come from the endoderm at all, but from an inpouching of skin in the neck of the embryo; it is, seemingly, a new development, comparable in structure to a true thymus, but not homologous with it. The lymphoid cells of the thymus develop (as do other blood cells) from mesenchyme, which migrates into the thymus from other blood-forming organs.

The thymus grows rapidly during embryonic life and is frequently a massive structure of the throat region of young or larval vertebrates (the calf thymus forms a good proportion of commercial sweetbreads). By the time the adult stage is reached, however, the thymus has ceased to grow and often shrinks; it thus becomes relatively, and often absolutely, smaller than in earlier stages. Its tissues tend to degenerate, and, in many adult mammals, the thymus has disappeared entirely. Its function includes the rapid production of lymphocytes during development and early life; in older animals, lymphocytes may proliferate in the spleen, bone marrow, and (in mammals) lymph nodes. The thymus is always essential in the establishment of immunity reactions in the developing vertebrate; these reactions are, as we have noted, associated with the lymphocytes.

Adult Blood-Forming Tissues. A great variety of organs may give rise to blood cells in adult vertebrates of one group or another. In fishes, much of the maturation of erythrocytes takes place in the blood vessels. Except in birds and mammals, the spleen is an important area of blood-cell formation. In lampreys, elasmobranchs, many teleosts, and amphibians, the kidney continues throughout life as a great source of blood cells. In cartilaginous fishes, the gonads are areas in which white blood corpuscles are formed. In teleosts, amphibians, and turtles, hemopoietic tissue is associated with the liver, and, in urodeles, this is a large source of leukocytes. In the sturgeon and paddlefish, blood-forming tissue surrounds the heart. As a persistence of embryonic conditions, masses of lymphoid tissue are present in the intestinal walls, and many vertebrates show in adult life blood-cell formation in the throat region. Masses of lymphocytic tissue are present in the walls of the mouth and pharynx in amphibians and in many mammals as **tonsils**, palatine, lingual, and pharyngeal. Peyer's patches in the wall of the intestine and, in some forms (like ourselves), in much of the wall of the appendix are further knots of lymphoid tissue. The esophagus of sharks shows a large mass of lymphocyte-forming tissue, and similar esophageal "tonsils," as well as the thymus, may be found in a number of amniotes.

The **bone marrow** is a great center of blood cell formation in higher vertebrates. The hollowing out of large bones appears to be primarily an adaptation for greater efficiency in the use of bony tissue and weight reduction. The resulting "waste" space is, however, available for other purposes; fat storage and blood cell formation are the uses to which the marrow cavities are put. In frogs, some hemopoiesis takes place here, especially in the males at the breeding season, and, among reptiles, the lizards are notable for considerable utilization of bone marrow. In birds and mammals, this site is universally used and highly important as a source of blood cells. In reptiles and birds, the bone marrow produces blood cells of all sorts; in mammals, however, although some lymphoid cell formation occurs in bone marrow, the lymph nodes are the center for their proliferation, and production in marrow cavities is largely of erythrocytes, granular leukocytes, and monocytes.

Bursa of Fabricius.   Another lymphoid organ quite similar to the thymus in many ways, the **bursa of Fabricius**, is found in birds (Fig. 299). This develops as a dorsal pouch off the cloaca. Its history—an early development from near the end of the gut, rapid growth, production of lymphocytes in the embryo, and reduction or loss with maturity—parallels that of the thymus closely. Histologically, the lymphoid tissue is like that of the thymus, but the bursa also retains fairly normal mucosal epithelium; this epithelium may be the source of the lymphocytes, but (as usual) their origin is open to dispute. In birds, the lymphocytes produced here differ in their function in the immune system from the thymic lymphocytes. Interestingly, mammals, which lack the bursa, still produce, from an unknown source, lymphocytes of the bursal type.

Spleen (Fig. 325).   It is only in the spleen and in the lymph nodes characteristic of mammals that adult hemopoietic tissues assume the condition of discrete organs. The spleen is the largest mass of reticular tissue in the vertebrate body and is an important organ in the formation, storage, and destruction of blood corpuscles; it is, further, of importance in defense against disease. In most vertebrates, it is a reddish body situated in the dorsal mesentery close to the stomach; it is surrounded by a connective tissue capsule. Inwardly extending connective tissue partitions form an extensive reticular network, in the spaces of which are contained tissues of two types, a **white pulp** and a **red pulp**. As may be inferred, the former contains clusters of white corpuscles; the red pulp consists of thick masses of all the elements of the blood, including, of course, a large percentage of erythrocytes. In fishes and some amphibians, the pulp is dominantly red; in higher tetrapods, white pulp is also quite common. The spleen is fed by an artery and is drained by a vein; lymphatic vessels are rarely well developed. The internal circulation of the spleen is obviously complicated, with

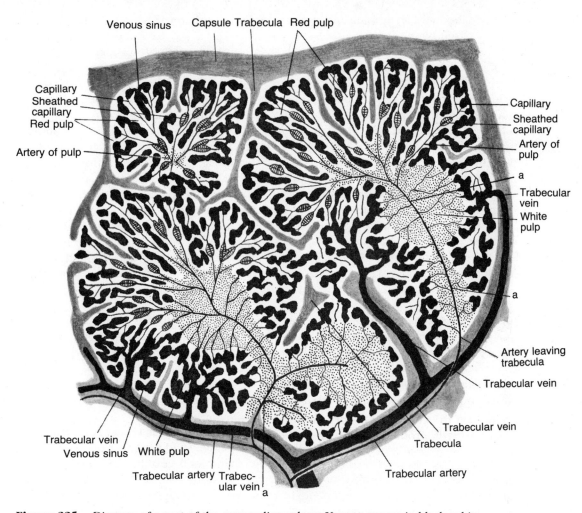

Figure 325. Diagram of a part of the mammalian spleen. Venous spaces in black; white pulp, heavy stipple; connective tissue capsule and trabeculae, light stipple; red pulp, unstippled. "Sheathed arteries" are surrounded by white pulp, here shown as hatched areas. (From W. Bloom and D. W. Fawcett, A Textbook of Histology, 10th ed., W. B. Saunders Co., Philadelphia, 1975.)

much of the blood passing through an "open" circulatory system. The white pulp clusters about the terminal arterioles; the red pulp lies near the venous outlets.

The spleen is always an important center of blood cell production. In the embryos of all vertebrates, erythrocytes as well as granulocytes are formed there, and this function persists in the adult in every group except mammals. In this last class, in which bone marrow has become the important seat of red blood cell formation, hemopoiesis in the adult spleen appears to be confined to lymphocytes. Red corpuscles are stored in great quantities in the spleen, and destruction of such cells occurs in this organ through the agency of numerous macrophages found there. The macrophages of the spleen are further of great importance in destroying infectious materials

451

in the blood stream passing through this organ; in this regard, the spleen resembles an oversized lymph node.

The spleen was developed early in the course of vertebrate history; it is present in most fishes and universal among tetrapods. Historically, it appears to be connected with the primary seat of blood cell formation in the embryo, because it arises in connection with the gut tube. Cyclostomes and lungfishes lack a formed spleen. In the former, it is represented by a layer of reticular blood-forming tissue surrounding much of the gut; in the latter, by a mass of tissue of a somewhat more compact nature, but still ensheathed within the outer lining of the gut. In the remaining fishes—Chondrichthyes, actinopterygians—this tissue has separated to form a definitive spleen. Its form is variable, and it appears to modify its outlines readily in conformity with those of adjacent organs. An elongate shape, paralleling much of the gut length, is presumably primitive and is retained in a majority of fishes, urodeles, and reptiles. A more compact structure is, however, present in some fish, some urodeles, anurans, some reptiles, many birds, and all mammals above the monotreme level.

In many mammals, there are occasional small red structures termed **hemal nodes** interposed between arterioles and venules. Their structure is that of a spleen in miniature, and they may be accessory organs of a similar nature.

Lymph Nodes (Fig. 326). For the most part, lymph nodes are a specialized mammalian type of structure. Although lymphatic vessels are present in all vertebrates, we rarely find any notable aggregation of hemopoietic tissues connected with them until we reach mammals. In a few anurans, crocodilians, and some water birds, a few nodular lymphatic developments may be present; in mammals, there are numerous and variable lymph nodes, both in the subcutaneous tissues and in the interior of the body.

These are small, rounded encapsulated masses of reticular tissue, comparable to a small spleen in general structure, but sharply distinguished from that organ by being situated along the course of lymphatic rather than blood vessels. Their cellular contents consist almost entirely of lymphocytes of various sizes. These nodes are, in mammals, a special seat of lymphocyte proliferation; in adult lower vertebrates, the lymphocytes develop in the same organs as the other types of corpuscles. Macro-

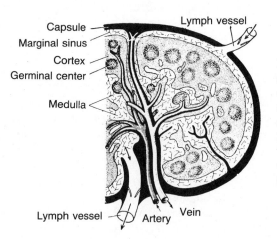

Figure 326. Diagram of the structure of a lymph node. In addition to the lymph vessels, a node is supplied by a small artery and vein. (After Portmann, Einführung in die vergleichende Morphologie der Wirbeltiere, Benno Schwabe & Co.)

phages in the reticulum deal with bacteria and other foreign or waste materials that have entered the body and reached the lymphatics; under such conditions, enlargement of the node may occur. Some forms, including the cat, have the lymph nodes in the mesentery fused into a large mass often called the **pancreas of Aselli**.

Circulatory Vessels

The vessels of the circulatory system, like the blood cells, are derived from the embryonic mesenchyme. As food-containing liquid begins to flow through the body of the early embryo, adjacent mesenchyme cells gather about such channels and surround them with a thin but continuous wall. All early formation of major blood vessels takes place in this manner. Later in development, minor vessels arise by outgrowths from the cells lining the blood channels already established, and throughout the life of an animal, new vessels arise frequently in correlation with changes in body tissues; after injury or damage, breaks may be repaired and circulation reestablished, and, with growth of a tissue or organ, there is a corresponding growth of additional circulatory vessels.

The inner lining of circulatory vessels is a special type of epithelium that arises *de novo* and is not derived directly from any epithelial formation originally present in the embryo; it is termed the **endothelium**. It consists of a single layer of thin, leaf-shaped cells, continuous with one another at their margins. There are no openings or pores except at the ultrastructural level, and even these usually have an exceedingly fine membrane over them. In vertebrates, quite in contrast with various invertebrates, the circulating fluid is nowhere, under normal conditions, in continuity with the interstitial liquid and cells outside its vessels, although most of the plasma contents can freely interchange through the thin endothelial membrane.

The vessels of the circulatory system include (1) the heart; (2) arteries, by definition vessels carrying blood from heart to body tissues; (3) capillaries and comparable structures, typically small vessels, connecting arteries and veins; (4) veins, returning blood to the heart; and (5) lymphatics, auxiliary vessels, which aid in the return of fluid from the tissues.

Capillaries (Figs. 327, 330). These smallest of vessels typically have only the bore necessary to allow an erythrocyte to pass; in relatively inactive tissue, many capillaries may be constricted and let little or no blood through. Capillaries are simple in structure. They deploy from the ends of the arterioles or the veins of portal systems and twine in among the cells of most of the body tissues so that no cell is far from a capillary. The degree of "vascularization" of a tissue by capillaries is generally proportionate to the metabolic activity of the tissue concerned. At their distal ends, the capillaries are re-collected into larger vessels, usually veins, in which the blood continues to flow onward. The capillary walls in most cases consist simply of a thin, essentially continuous layer of endothelial cells, although connective tissue cells or other mesenchyme cells may adhere to them and, in some instances, a smooth muscle cell is occasionally present.

Capillaries are too small to be dissected by ordinary means and hence are often neglected in gross anatomy. But it must never be forgotten that functionally the capillaries are the most important part of the circulatory system. Elsewhere, blood is merely in transit; here it is at work. The endothelium is in general a barrier to cells

ven.

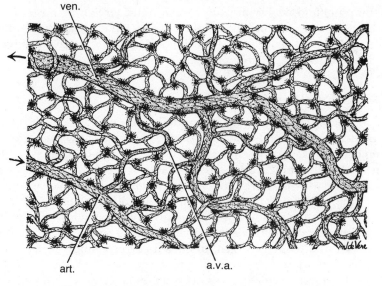

art.   a.v.a.

Figure 327.   Portion of a capillary bed from the web of a frog's foot, showing small arteriole *(art.)* and venule *(ven.),* a capillary network, and a direct arteriovenous anastomosis *(a.v.a.).* (From Young, The Life of Mammals, Oxford University Press.)

and large molecules, although white blood corpuscles can push out between the endothelial cells. Small molecules and ions pass readily through the endothelium, and, in the walls of the capillaries, an active exchange of materials takes place. Typically the capillary in the proximal part of its course gives out oxygen and food materials; in its distal part, it takes on waste and carbon dioxide from the tissues. This shift of direction of transit, vitally necessary, is made possible by a change of balance between physical and osmotic pressures in the blood as it passes through the capillaries. The presence of the special blood proteins creates a tendency for materials to flow from the interstitial fluid into the blood stream. At the proximal end of a capillary, however, the hydrostatic pressure put on the blood by the heart pump more than balances this tendency for inflow; distally in the capillary network, the hydrostatic pressure has been decreased by the friction of the capillary walls, and osmotic forces dominate over physical pressure.

Capillary networks in most areas of the body tissues are interposed between arterial and venous vessels. But the breaking up of a blood stream into capillaries and its reconstitution into larger vessels may take place also along the path of either the arteries or the veins in special regions. In gill-bearing vertebrates, the course of the arterial flow is interrupted by a capillary system for aeration of the blood in the gills.* The return of the venous blood from the tissues may also be interrupted by its forced passage through the capillary networks of a portal system. In all vertebrates, blood

*We are accustomed to think, in visual images, of arterial blood as oxygenated and hence "red"; but arterial blood between heart and gills in a fish is, of course, of the "blue" venous type; the arteries and veins leading to and from the tetrapod lung likewise have a reversal of "blue" and "red" blood types.

from the gut passes by a venous portal system to capillaries in the liver; a majority of vertebrates (the highest and the lowest are exceptions) route part of the venous blood returning from the posterior part of the body through a portal system to capillary networks in the kidneys. Smaller, but functionally important, portal systems occur in other organs, such as the pituitary.

Although capillaries are the major type of arteriovenous connection, there are other modes. Microscopic study sometimes reveals direct connections of larger caliber, **anastomoses**, between small arteries and veins that can vary the amount of blood passing through the capillaries of a given organ. Sometimes the exchange of materials between tissues and blood may take place in small, irregularly shaped, thin-walled "ponds" of blood termed **sinusoids**, which are essentially larger, often flattened capillaries; they occur in the liver, spleen, and various other organs. In the spleen, these sinusoids may have incomplete walls, forming an "open" circulatory system. In some instances, usually connected with special organ functions, a blood vessel breaks up into a complicated, coiled mass of tiny blood vessels reasonably termed "a marvellous network"—**rete mirabile**; a kidney glomerulus is such a structure, and others are found, for example, in the red body of the teleost air bladder and the heat-sensing organ of pit vipers. A special type of network may be present in the distal part of the limbs, particularly in wading birds, and the flippers of aquatic mammals. A maze of tiny channels from the arteries leading to the foot or paddle may surround the returning venous vessels. This appears to be a heat-conserving device, warmth being short-circuited from the arteries to the veins and back into the body, at the expense of the chilly limb. These, and many other retia, may consist of many parallel vessels and work to concentrate substances or conserve heat by a "countercurrent multiplier" system (Fig. 328). The tails of many mammals (such as dog, cat, rat) show small tissue masses, termed **caudal glomeruli**, which contain retia mirabilia; man retains one such structure as a **coccygeal glomus**. These and other retia may simply be tangled masses of vessels and not, apparently, countercurrent multipliers; this type is seen in many aquatic mammals that may have coelomic retia formed of large vessels.*

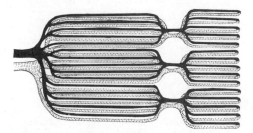

Figure 328. Diagram of a rete mirabile, from the red body of the swim bladder in an eel. The parallel arterial (lines) and venous (solid black) vessels provide a countercurrent multiplier, in this case one enabling the fish to secrete gas into the swim bladder at pressures well above those in the blood. (After Giersberg and Rietschel.)

*Such retia do not occur in all aquatic mammals but do occur in sloths and some primates, which makes their function hard to imagine.

Arteries and Veins (Fig. 329). Sheathing materials, in addition to the ubiquitous endothelium, surround not only the larger vessels of the body, **arteries** and **veins**, but also their smaller branches, the **arterioles** and **venules**, and likewise the major lymphatic vessels. Reinforcing elements include connective tissue fibers, elastic fibers, and smooth muscle cells in variable amounts and arrangements; nerve fibers for the smooth muscles are also present, and small nutrient blood vessels are present in the walls of large arteries or veins.

The walls of blood vessels are customarily described as consisting of three layers or "tunics": the **tunica intima**, **tunica media**, and **tunica externa** or **adventitia**. As seen in simplest form in small arteries or veins, the intima may consist only of the endothelium, the media a few muscle fibers (absent in the smallest veins), the externa a bit of connective tissue. In larger vessels, however, thick sheaths of complicated and variable arrangement are present.

In a typical large artery, the intima includes, in addition to the endothelium, a thin sheet of fibrous connective tissue just beneath it and a sheath of elastic tissue that completely surrounds the vessel as an **internal elastic membrane**. The tunica media contains some connective tissue but is dominantly the muscular layer, although smooth muscle fibers are sometimes found in the other two layers. When well developed, the muscle may be arranged in two layers, of circular and longitudinal fibers. The tunica media is often bounded externally by a second elastic tissue structure, an **external elastic membrane**. Beyond this is the adventitia—connective tissue, often rather loose, that binds the vessel to adjacent structures.

Veins have a somewhat similar but simpler build; in general, there is much less muscle and elastic tissue in proportion to connective tissue. Much of the contrast between arteries and veins is due fundamentally to the fact that the blood flows under higher pressure in the arteries. That pressure may be better withstood, arterial walls are thicker and more complex than those of veins. Again, the lumen of an artery is much smaller than that of a vein of comparable importance; this is correlated with

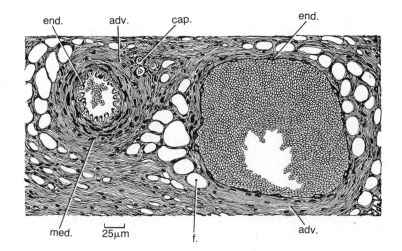

Figure 329. Section through a small artery and its accompanying vein, showing contrast in size and thickness of walls. Abbreviations: *adv.,* adventitia; *cap.,* capillary; *end.,* endothelium; *f,* fat cell; *med.,* media (muscle layer) of artery. (From Young, after Maximow and Bloom.)

the faster flow of arterial blood (just as, for example, a high-speed through highway needs fewer lanes than a street where, with the same amount of traffic, the speed is slower). Frequently, in histologic sections, twin vessels can be seen beside one another (cf. Fig. 329), and artery and vein can be clearly distinguished by these criteria. The bore of an artery tends to remain constant, diminishing only as branches are given off; a vein, particularly in sharks and cyclostomes, may expand along its course to form a large sac, or **sinus**. Again associated with speed and pressure differences is the fact that the arterial system presents relatively few individual anomalies, but veins are highly variable. A fast-flowing mountain stream tends to take a direct and undeviating course, whereas a sluggish stream meanders, branches and re-forms, and produces islands; so in the embryo, a vein often appears as a network of variable channels (cf. Fig. 357). Which channels of the network are to form the definitive vein is not fixed, hence the frequent anomalies. In limbs, arteries are deeply placed; veins may in part be superficial in position.

Lymphatic Vessels.　The lymphatic system, well developed in most vertebrates as a route for liquids (lymph may contain leukocytes, but lacks erythrocytes) from tissues to heart, paralleling the veins, consists of vessels that are comparable to capillaries and veins of the blood system. Lymphatic capillaries, however, tend to be larger in diameter and more irregular than blood capillaries (Fig. 330). They have, of course, no connection with arteries and so arise blindly in the tissues. Larger lymphatic vessels, in which the lymph flows sluggishly toward the heart, have walls still thinner than those of veins and may expand into thin-walled pockets, **lymphatic cisterns**.

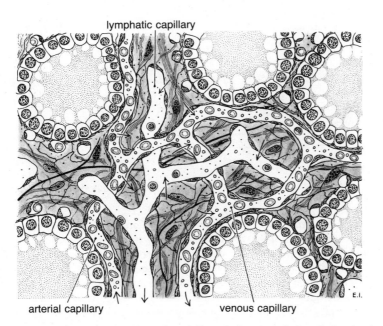

Figure 330.　Diagram of blood and lymph capillaries, showing their relation in the mammalian thyroid. Arrows indicate the direction of flow of materials into and out of the capillaries. (From Kampmeier.)

Valves and Sphincters. Expansion and contraction of blood vessels is controlled by the action of the muscular and elastic tissues present along their walls; more positive control of blood and lymphatic flow is frequently exercised by the development of sphincters and valves. In small arteries and veins, circular muscle fibers may be greatly developed in a sharply localized region to form a sphincter capable of closing off the blood flow in the vessel concerned. A backwash of blood or lymph through a vessel may be prevented by the presence of valves. These are folds of endothelium and connective tissue, behind which lie pocket-like depressions. A backward flow of liquid distends the pocket and pushes the valve flap out into the lumen of the vessel; two such flaps generally suffice to close a vessel (three or four, however, are occasionally found). Valves are common in the veins in tetrapods (but not in fishes) and in the lymphatics of birds and mammals (and lymph hearts of lower forms). They are, of course, highly developed in the heart. A valve is almost never found in an artery, because arterial blood cannot ordinarily work back against the strong pressure under which it flows from the heart.

Arterial System

Aortic Arch System in Fishes (Figs. 332, 335). In primitive, gill-breathing vertebrates (and in amphioxus as well), all the blood from the heart courses forward in a **ventral aorta** (Fig. 331). This median trunk, bifurcated anteriorly and lying in the floor of the throat, gives off a series of paired vessels that arch upward on either side of the pharynx between successive gill clefts. Each typical **aortic arch*** breaks up into a series of capillary vessels in the branchial membranes for aeration of the blood, which is thence re-collected dorsally into further arterial vessels that pass to the

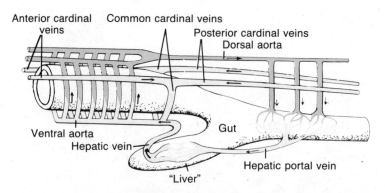

Figure 331. Diagram of the major circulatory vessels of amphioxus. Except for the absence of a heart at the posterior end of the ventral aorta, the system is closely comparable to that of vertebrates. Only a few of the numerous aortic arches are shown, and paired segmental vessels to the body wall are omitted completely. The vessels are colored as though oxygen intake were entirely through the gills; in fact much of it may be through the skin, although most does occur at the gills; however, obviously not via the large vessels as here indicated—the color scheme is largely to facilitate comparison with the following two figures.

*The triple use of the word **arch** in the branchial region was commented on earlier; in this chapter, *aortic* arch is generally implied.

various tissues of the body and head. In tetrapods, the gills are lost, but the aortic arches are found in every embryo. The history of these aortic arches is one of the most interesting chapters in the structural evolution of the vertebrates.

In a diagrammatic primitive vertebrate (Fig. 335 A), we may picture the aortic circulation as including, in addition to a ventral aorta running forward from the heart, paired aortic vessels passing upward laterally in front of each branchial slit or pouch. The major vessel reached by the blood after passage through the gills is the **dorsal aorta**. Posteriorly, the aorta is a single median trunk; anteriorly, it consists of a pair of vessels, one along either side of the head. The number of aortic arches was presumably high and variable in the ancestral vertebrate; it is very high in amphioxus, and there may be as many as 15 in hagfishes. In most living jawed forms, however, there are only five normal gill slits plus a spiracle, and hence potentially six pairs of aortic arches, usually designated by Roman numerals.* In the vertebrate embryo, these arches are, to begin with, continuous vessels passing without interruption from ventral to dorsal aorta (Fig. 336 A); only as the branchial slits open and become functional does the branchial capillary system develop, with consequent interruption of the loop of the arch. In all vertebrates, the aortic arches develop in order, in the embryo, from front to back. The first, or mandibular, arch is, in the early embryo, the only passage from ventral to dorsal aorta (Fig. 339 A).

In the cyclostomes, the gills are subspherical, pouchlike structures. Afferent branchial vessels ascend between successive pouches and divide to supply the adjacent halves of the neighboring pouches; further afferent vessels supply the front half of the first pouch and the back half of the last. In all other living fishes, the branchial openings are slit-shaped, and the afferent vessel of an aortic arch typically runs up between each successive pair of slits (Fig. 336). In selachians and lungfishes, the afferent vessel continues as a single structure far up the gill, giving off capillaries as it goes. The efferent vessels of each gill are a pair, anteriorly and posteriorly placed; at the top of the gill slit, the two members of the pair fuse to a single efferent structure. Actinopterygians have a different pattern of vessels: in them both efferent and afferent vessels are single throughout.

Vertebrates invariably refuse to adapt themselves to a man-made structural diagram, and, in the aortic arches, as elsewhere, departures from the "idealized" arrangement must be noted. Afferent and efferent branchial arteries are usually represented, diagrammatically, as entering and leaving the branchial region at a point opposite the middle of each gill bar. This is usually the case; but, in typical adult selachians (Fig. 335 B), each efferent branchial artery lies opposite a slit, not a branchial bar, and, in cyclostomes, both afferents and efferents have this position. Again, the efferent vessels in successive arches are in theory discrete; but actually in many forms, including all cartilaginous fish and many of the actinopterygians, there may be connections on either side between the efferents of successive arches; these connections are more common dorsally but may loop below the gill slits as well. Enlargement of certain of these connections allows the apparent resegmentation of the elasmobranch efferent arteries. In all gnathostome fishes, there is commonly a ventral **hypobranchial artery** (not illustrated), which is developed from the efferent system and runs posteriorly along the throat to supply that region and the heart with aerated blood. Characteristically, this takes its root from the fourth arch. Still further, in fishes, we find an

*Three rare sharks have one or two "extra" gill slits and, hence, have "supernumerary" aortic arches.

artery running downward and forward from the efferent branchial arteries to the mandibular region. This, the **external carotid artery**, is embryologically the anterior end of the ventral aorta; during development, it becomes associated with the efferent part of the system, a necessary change if it is to supply oxygen to the lower jaw.

In cyclostomes, all the aortic arches are similarly developed; in jawed fishes, in which six arches are usually developed in the embryo, the posterior four usually attain a typical adult structure, but the first two may be highly modified. The first arch, the **mandibular**, runs dorsally in the embryo between mouth and spiracle; the second, the **hyoid arch**, lies between the spiracle and the first normal branchial slit.

Presumably ancestral vertebrates had a well-developed arch I running up behind the mouth. But although a first aortic arch is found in the embryo, it is never found in typical fashion in the adult of any living jawed vertebrate (the numbering in cyclostomes is debatable so we omit them here). Its afferent portion almost universally disappears. In primitive fishes, there is a gill-like structure on the anterior margin of the spiracle. In theory, such a gill should be supplied by arch I. But its blood supply comes, in the adult, from the efferent system of the second arch or from a branch of the dorsal aorta. The blood that reaches it has, thus, already been aerated. The spiracular gill is therefore not a true gill, but a "false gill" or pseudobranch (cf. p. 351). The efferent portion of arch I persists in cartilaginous and a few bony fishes and runs up from the pseudobranch into the orbital region, enters the braincase to connect with the arteries beneath the brain, and supplies the eye with blood (Fig. 332, spiracular artery; cf. Fig. 335 B).

The second aortic arch is somewhat variable but more persistent. It is generally present in typical fashion in Chondrichthyes and is also found in some bony fishes. But it is lost in most actinopterygians, and the same is true of the lungfish *Neoceratodus*.

Arches III to VI are in general, as we have seen, both normally and fully developed in jawed fishes. In the lungfish *Protopterus*, an interesting exception is that

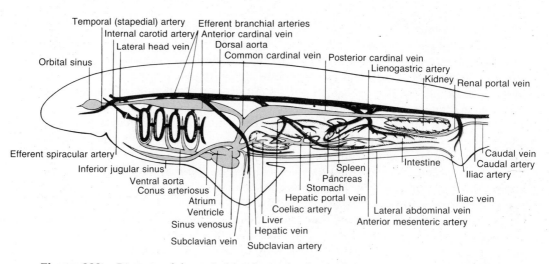

Figure 332. Diagram of the main blood vessels of a shark as seen in lateral view. The renal portal vein here (and in Fig. 333) should appear dorsal to the kidney; it is drawn ventrally to show the pattern more clearly.

arches III and IV run without break through the branchial region. This condition is associated with the reduced gills and developed lung of that genus and shows a partial parallel to developments in the amphibian relatives of the dipnoans. A few teleosts also have reduced the number of functional gills.

The introduction of a lung in certain bony fishes makes, for most forms still in the fish stage, no great change in the circulatory system of the gills. The lung in *Polypterus* and lungfishes (Fig. 335 *C*) is supplied by an artery from the efferent portion of the sixth aortic arch or the aorta beyond this point; hence the blood reaching it has already passed through the normal aerating device of the branchial capillaries.

Aortic Arches in Amphibians (Figs. 333, 335 *E*, 337 *A*).   In the amphibian stage, marked changes are present, even in the larval forms. Arches I and II disappear early in development, although a lingual (or external carotid) artery may continue forward ventrally to the lower jaw. The remaining four arches may persist in the adult as continuous tubes in urodeles, since internal gills fail to develop (the external gills of the larva are supplied by accessory capillary loops). In frogs, the extensive development of the peculiar gills leads to a larval interruption of continuity of the arches, which, however, is restored for the most part at metamorphosis. In the adult amphibian, we generally find, in consequence, a system of complete tubular aortic arches, which always include the third, fourth, and sixth pairs*; arch V often persists in urodeles but is absent in frogs and all adult amniotes. In fishes, the dorsal ends of all these arches connect with the uninterrupted dorsal aorta. Variations in these connections, however, are already present in amphibians. Even in fishes, the blood flowing up through the third arch tended to pass anteriorly toward the head rather than backward with the main aortic flow, and cranial blood supply becomes the sole func-

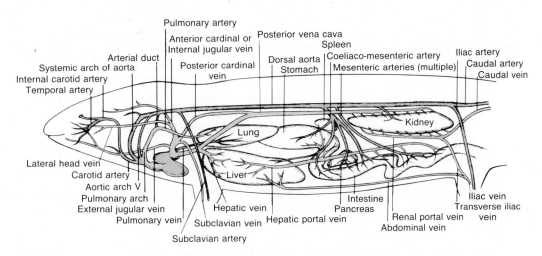

Figure 333.   Diagram of the main blood vessels of a urodele amphibian as seen in lateral view.

Necturus, the mud puppy, is exceptional, indeed unique, in losing the ventral part of the sixth arch and supplying the lungs from its dorsal section.

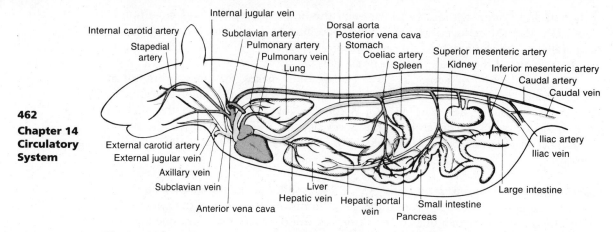

Figure 334. Diagram of the main blood vessels of a mammal (rat) as seen in lateral view.

tion of this arch in tetrapods. The part of the dorsal aorta between arches III and IV tends, in correlation, to disappear in adult tetrapods, thus completely separating cranial and body segments of the original aorta. However, this connection, termed the **carotid duct**, still persists in the adult in the Gymnophiona, some urodeles, and even in various reptiles. The third arch and the associated anterior part of the dorsal aorta is the **internal carotid artery**; the ventral aortic stem leading forward to it becomes the **common carotid** (cf. Fig. 340).

The fourth arch is always a large, bilaterally developed vessel in lower tetrapods; it is termed the **systemic arch**, because it is the main channel for blood flowing from heart to body. The fifth arch, on the other hand, tends to disappear in tetrapods; as was stated above, it may persist in urodeles but is found in no other tetrapods beyond the embryonic stage; even in the embryo, it is usually small and transient in appearance and may fail to develop at all. Why the fourth arch is selected as the blood channel to the body rather than the seemingly shorter and more direct route via arch V is an unsolved puzzle.

As we have noted, the lung is supplied with blood via arch VI. The lung is functionless in the amphibian larva (as in the embryos of amniotes). During the larval period, the major part of the blood in this arch travels straight on upward into the dorsal aorta as part of the main blood stream to the body. But when in the amphibian (or in the amniote at birth or hatching), the lung begins to function, the dorsal part of this arch should, for the sake of efficiency, be eliminated to prevent admixture of blood streams. This connection, the **ductus arteriosus** (or duct of Botalli), persists in reduced form in the adult among urodeles and Gymnophiona, in *Sphenodon,* and in some turtles; in frogs and most amniotes, it closes completely after metamorphosis or birth.

The original ventral aorta, as a result of the processes just described, is reduced to a channel of supply for arches III, IV, and VI. The last, as a bearer of venous blood to the lung, merits a separate channel from the heart. We find that in living amphibians, the ventral aorta as such is done away with; it has been split to its base to separate the **pulmonary trunk** from the remainder of the original aorta.

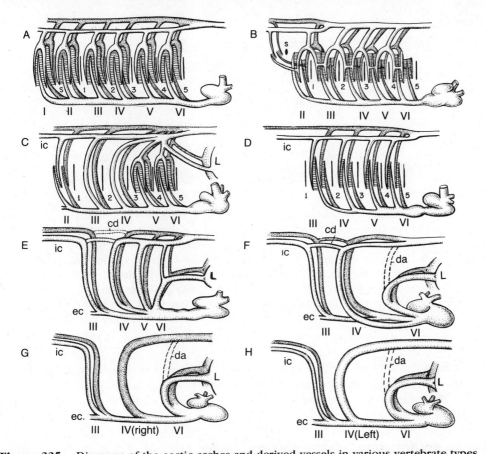

Figure 335. Diagrams of the aortic arches and derived vessels in various vertebrate types. *A,* Theoretical ancestor of the jawed vertebrates with six unspecialized aortic arches; *B,* typical fish condition as seen in a shark; *C,* the lungfish *Protopterus; D,* a teleost; *E,* a terrestrial salamander; *F,* a lizard; *G,* a bird; *H,* a mammal. Various accessory vessels are omitted. The vessels of the right side are heavily shaded. In terrestrial forms, the position of the vessels is made (for purposes of the diagram) to correspond more or less to that of the arches from which they are derived. Aortic arches in Roman numerals. Abbreviations: *s,* spiracular slit; following branchial slits in Arabic numerals. *cd,* Carotid duct; *da,* embryonic ductus arteriosus; *ec,* external carotid artery; *ic,* internal carotid artery; *L,* lung. The carotid duct shown in the lizard is absent in other reptiles; in turtles, the carotids arise by a separate stem directly from the heart. In *H,* the embryonic arterial duct by-passing the lungs is shown in broken lines.

In urodeles, there persists a relatively complete and "fishy" system of arches, but in anurans, as a result of the changes discussed, we find a pattern of a sort characteristic of a majority of reptiles and, indeed, more advanced than in many of that group. In this pre-amniote stage, we have remaining (1) a carotid system, including the anterior end of the ventral aorta, the third arch, and the anterior end of the paired dorsal aorta; (2) a paired systemic arch, including the major basal stem of the ventral aorta and the fourth pair of arches; (3) a pulmonary system, including a basal stem

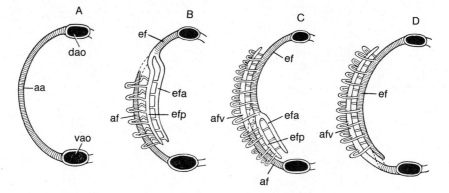

Figure 336. Diagram of circulation in a fish gill, left side, from behind. *A*, Embryonic condition, with continuous aortic arch, *aa*, from ventral aorta, *vao*, to dorsal aorta, *dao;* amphioxus is basically similar. *B*, Shark condition; the afferent gill vessel, *af*, is formed from the aortic arch; the paired efferent vessels, *efa, efp*, are new formations (later embryonic changes in sharks make the efferent arteries appear to alternate with the afferents—as in Figure 335 *B*—but this is a secondary modification). *C*, Condition transitional to *D*, seen in sturgeons. *D*, Teleost condition. The embryonic arch gives rise to the efferent vessel, *ef*; the afferent vessel, *afv*, is a new formation. (After Sewertzoff, Goodrich.)

split off from the ventral aorta, part of the sixth arch, and a **pulmonary artery** leading from arch to lung. A peculiarity of modern amphibians is the utilization of the skin for "breathing," and the frogs have a special branch of the pulmonary arch to bring blood to the skin.

Amniote Aortic Arches. In the amniotes, further developments in the aortic arch system have to do mainly with variations in the fourth arch. In earlier evolutionary stages, the arches were in general strictly bilaterally symmetric; in the amniotes, asymmetry appears.

Probably in the ancestral reptiles (as in the frog), there were two ventral trunks from the heart: one to the lungs, the other a common trunk to both systemic arches and both carotids. In living reptiles, however, we find that three vessels, rather than two, open forward from the heart (Fig. 337 *C, D*). These include (1) the pulmonary arch; (2) a separate tube leading only to the left systemic arch; and (3) one for the right systemic arch, with which both carotids and arteries to both front legs are associated. The three are so situated that in the ventricular cavity of the heart, incompletely divided in most reptiles, the vessel for the left fourth arch would appear to receive largely venous blood; however, recent physiologic work has shown that, in fact, it may contain either arterial or venous blood. The problem of separation of oxygenated and nonoxygenated streams is best considered when we discuss the heart (cf. p. 484). The birds, descended from reptiles related to the crocodiles, have eliminated the left systemic arch completely (Fig. 335 *G*). Apart from the pulmonary stem from the right ventricle, there thus remains in birds only a single trunk leaving the heart—that from the left ventricle. This carries aerated blood to both the carotids and thence to the head, to both front legs, and to the body by a single systemic arch—the right member of the original pair. In correlation with the length and slen-

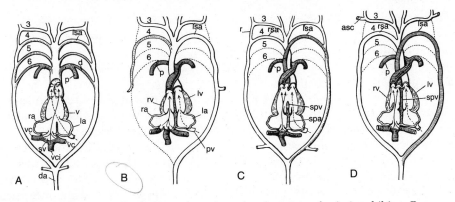

Figure 337. Diagram of the heart and aortic arches in tetrapods. *A,* Amphibian; *B, a* mammal; *C,* typical modern reptiles; *D,* crocodilian. Ventral views; the heart (sectioned) is represented as though the chambers were arranged in the same plane; the dorsal ends of the arches are arbitrarily placed at either side. Solid arrows represent the main stream of venous blood; arrows with broken line, the blood coming from the lung. Vessels apparently carrying aerated blood are unshaded; those that appear to contain venous blood, hatched (physiologic work, as noted in the text, shows that the situation is not this simple). The two vessels at the top of each figure are the internal carotid (laterally) and external carotid (medially). In amphibians without a ventricular septum, the two blood streams are somewhat mixed; subdivision of the arterial cone tends to bring about partial separation, but some venous blood is returned to the dorsal aorta. In mammals (and birds), ventricular separation is complete, the arterial cone is subdivided into two vessels, and the arches are reduced to the left systemic and pulmonary. The mammalian condition has apparently arisen directly from the primitive type preserved in the Amphibia, because in modern reptiles, the conus arteriosus shows a division into three vessels, rather than two; the heart, as noted later, is more complex than this diagram indicates. In crocodilians, the ventricular septum is almost complete, and the elimination of the left fourth arch would give the avian condition. Abbreviations: *asc,* anterior subclavian; *d,* ductus Botalli; *da,* dorsal aorta; *la,* left atrium; *lsa,* left systemic arch; *lv,* left ventricle; *p,* pulmonary artery; *pv,* pulmonary vein; *r,* portion of lateral aorta remaining open in some reptiles; *ra,* right atrium; *rsa,* right systemic arch; *rv,* right ventricle; *spa,* interatrial septum; *spv,* interventricular septum; *sv,* sinus venosus; *v,* ventricle; *vc,* anterior vena cava; *vci,* posterior vena cava. (From Goodrich.)

derness of the bird neck, the two carotids run close beside one another below the cervical vertebrate, and one of the pair is often reduced and may be absent.

The history of the mammalian arches seems to have been simpler (Figs. 334, 335 *H,* 337 *B*). Mammalian ancestry diverged from that of other reptiles at an early date, and there is no reason to believe that the system of three arterial trunks seen in modern reptiles was ever present in the mammalian line. Presumably, as in anurans, the early reptilian ancestor of mammals had, in addition to a pulmonary arch, a single trunk leading from the left side of the ventricle to the two carotids and the two systemic arches. But a double systemic arch is unnecessary and, somewhere along the line of development to mammals, the right fourth arch disappeared (except that its base remained as a connection with the subclavian artery to the arm); thus the entire blood supply to the trunk follows the curve of the left member of the embryonic pair. The mammal, like the bird, has simplified the systemic blood supply; but the two groups contrast as to which member of the pair of fourth arches is retained.

In mammals, there is great variation in the way in which the pairs of carotids and vessels to the front legs (**subclavians**) branch off from the arch of the aorta (Fig. 338). The carotids may leave the aorta separately, by a common stem, or jointly with the neighboring subclavian; and, as a final variant, all four vessels may depart as a single, large trunk.

The embryonic development of the aortic arches of a mammal recapitulates to a considerable degree the phylogenetic story outlined above (Fig. 339). The first vascular channels established from heart to body in the embryo are a pair of anteriorly placed arches, which, subsequent history shows, are the first pair—the mandibular arches—of the fish series. Branchial pouches develop in sequence behind this point, and successive aortic loops are formed between them as arches II, III, and IV. As the more posterior arches develop, they tend, each in turn, to become the most prominent pair; synchronously, the more anterior arches tend to be reduced in importance, so that the first two cease to be functional in a short time. Behind arch IV, arch VI presently develops, but the fifth arch is sometimes absent and at the best is small and transitory. In the embryo, arch VI runs directly upward to the dorsal aorta, and little blood passes through the pulmonary artery branching from it. At birth, however, the upper part of this arch, the ductus arteriosus, becomes almost immediately occluded, and the full flow of blood passes to the lung. Meanwhile, the pulmonary trunk has separated from the aorta, the right fourth arch has disappeared, and the carotid duct has closed, to produce the adult mammalian pattern.

Blood Supply to the Head (Fig. 340). Though the main dorsal flow of arterial blood is in a posterior direction, to the trunk, tail, and limbs, an important element in the circulation, if a minor one in quantity, is a forward flow to the head.

The anterior part of the dorsal aortic system is a paired vessel that, in the cranial region, primitively runs along either side of the braincase, receiving the efferent branchial arteries from more anterior gills. Posteriorly, this vessel is considered a paired part of the dorsal aorta; anteriorly, it is an obvious homologue of the **internal carotid artery** of higher vertebrates. If we follow this artery forward in fishes, we find that after giving off a major branch—the **orbital** (temporal or stapedial) **artery**—to

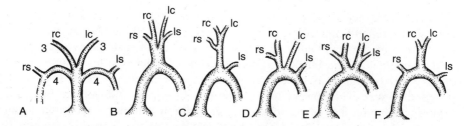

Figure 338. Diagrams, in ventral view, to show variations in the branching of the main blood vessels from the mammalian aortic arch. *A*, Embryonic condition, with ventral trunk of aorta and third (carotid) and fourth pairs of arches, of which the right fourth arch is later lost beyond the point of departure of the subclavian. By differential growth of the vessels, the various arrangements shown in *B* to *F* are brought about (*D* is the human type).
Abbreviations: *lc,* left carotid; *ls,* left subclavian; *rc,* right carotid; *rs,* right subclavian arteries. (After Hafferl, in part.)

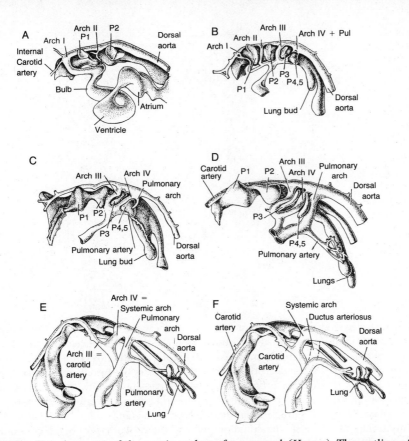

Figure 339. Development of the aortic arches of a mammal *(Homo)*. The outline of the gut cavity, branchial pouches, and lung buds are shown in addition to blood vessels; in *A,* the cavities within the heart are included. Arch I is developed. *B,* Arch I already reduced, II and III formed. *C,* Arch II reduced, IV (systemic) formed, VI (pulmonary) arch and pulmonary artery forming (Arch V does not develop in man). *D,* Pulmonary arch well developed. *E.,* Carotid arch (III) separated dorsally from aorta, pulmonary arch becoming distinct at root from ventral aorta. *F,* Diagram to show reduction at birth of upper end of arch VI (ductus arteriosus). P^1 to P^5 = pharyngeal pouches. (After Congdon.)

supply much of the general facial and jaw region, its main stem passes upward into the braincase in front of the pituitary. Here it furnishes the main blood supply for the brain, gives off an artery to the eyeball, and connects with the efferent pseudobranchial artery mentioned earlier. Ventrally, a small **external carotid** (or lingual) **artery** extends from the base of the most anterior efferent branchial artery to the lower jaw.

The condition just described is a primitive one, which holds, with little variation, among the various groups of fishes. In amphibians, the system tends to be modified in relation to the disappearance of the internal gills; all that remains of the aortic arch system anterior to the fourth arch is the proximal part of the carotid system, and the pseudobranchial artery of fishes is absent.

In typical reptiles and amphibians, the carotid gives off near its base, while ventral in position, a small external carotid artery, which supplies tissues of the tongue,

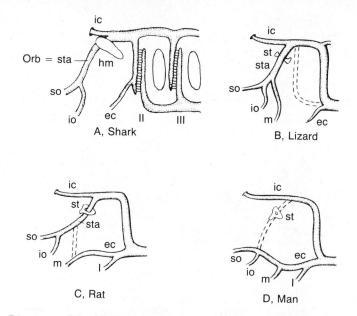

Figure 340.   Diagrams of the left side of the head to show the evolution of the carotid system. To make homologies clear, all are shown as if no neck ever developed and the fish pattern were retained. In a relatively primitive fish stage, *(A)*, the direct forward continuation of the dorsal aorta is the internal carotid artery, which enters the braincase near the pituitary. This gives off a major branch, the orbital artery, which, passing close to the hyomandibular, supplies most of the more superficial features of the skull and upper jaw. An accessory blood supply to the head from the spiracle is omitted from the diagram.

In many tetrapods *(B)*, a similar situation persists, the orbital artery being commonly called the stapedial, because it passes close to or through the stapes (= hyomandibular). However, the small external carotid present near the root of the carotid extends forward and, in mammals, may take over part *(C)* or all *(D)* of the functions of the stapedial.

Abbreviations: *ec,* external carotid; *hm,* hyomandibular; *ic,* internal carotid; *io,* infraorbital artery; *l,* lingual artery; *m,* mandibular artery; *orb,* orbital artery; *so,* supraorbital artery; *st,* stapes; *sta,* stapedial artery; II, III, second and third aortic arches. In *B* to *D,* modified aortic root = common carotid.

throat, and occasionally other structures of the lower jaw. The main carotid stem, much as in fishes, passes upward and forward as the internal carotid, to the middle ear region of the skull. Here the internal carotid artery continues anteriorly toward its opening into the braincase; but it gives off, close by the stapes, a major branch, the large **stapedial artery**, which supplies all the structures of the outer parts of the head and most of the area of the jaws. This last artery is identical with the orbital artery of the fish ancestors.

In primitive mammals and all mammalian embryos, the arterial system of the head persists in much the same manner. In most mammals, however, major changes occur. The originally small external carotid grows anteriorly and dorsally and taps the branches of the stapedial artery. As a result, the external carotid is large in most mammals, the stapedial artery reduced or absent; a new and shorter route has developed for supplying blood to the face and jaws. The process is analogous to "stream piracy," whereby one river taps the headwaters of another. This reaches its extreme in cats; the base of the internal carotid closes, and blood is carried to the brain largely by branches of the external carotid.

Blood Supply to the Body and Limbs (Figs. 332–334, 341).   In every vertebrate, the major blood supply to the trunk, tail, and limbs is furnished by the dorsal aorta and its branches. Its most anterior, cranial part may be paired, but, in the trunk, the aorta becomes a median vessel, lying beneath notochord or vertebrae and above the root of the mesentery; posteriorly, it continues as a median **caudal artery**. The branches of the aorta tend to be of three types: (1) median ventral branches running downward in the mesentery to the gut and associated structures, (2) paired ventrolateral branches to the urogenital system, and (3) paired lateral branches to the outer areas of the body and to the limbs.

The median ventral, splanchnic, vessels may be numerous in certain lower vertebrates and in embryos; in general, however, there is a somewhat variable tendency for the concentration of these vessels into a few main trunks, usually including a **coeliac artery** to the region of the stomach and liver, and **mesenteric arteries**, generally two, to the intestines.

Of paired vessels, short lateral branches are present for the gonads and the kidneys—the visceral branches. Separate from these is a series of paired segmental—intersegmental—blood vessels that run laterally from the aorta to supply the axial musculature, the skin, and the spinal cord. At several points, there may be longitudinal connections between the paired vessels of successive segments. These may form lengthwise arterial channels, supplementary to that afforded by the aorta, especially in the dorsal region of the body and the skin of the back and flanks. These channels may, further, afford the opportunity for originally segmental arteries to coalesce into

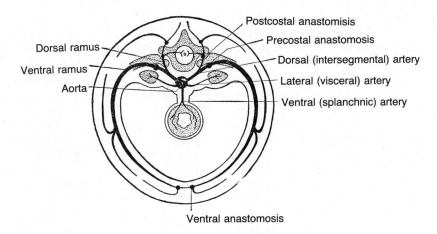

Figure 341.   Diagrammatic cross section of the body of a higher vertebrate to show various types of branches that may be given off by the aorta. Most prominent are median ventral branches descending in the mesentery to the gut and associated structures and paired intersegmental arteries, the main ventral ramus of which descends the flanks between the myomeres or adjacent to successive ribs. Longitudinal anastomoses may occur between successive segments in various regions. (From Arey.)

a smaller number of large vessels stemming from the dorsal aorta; thus, in higher vertebrates, the segmental arrangement is increasingly obscured.

In the embryonic development of paired fins or legs, there is seen, at an early stage, a network of small arteries continuous with the segmental arteries of the neighboring region of the trunk (Fig. 357). During ontogeny one or another of these tends to become dominant and to form a main channel from aorta to limb. In the pectoral limb, this main trunk is usually given its mammalian name of **subclavian artery**, although it is improbable that it is exactly the same vessel in every case. Giving off branches to the shoulder and chest region, this same main trunk is termed the **axillary artery** as it enters the limb, and the **brachial artery** as it proceeds down the arm. The point of departure of the subclavian from the primitive aortic trunk seems to have been rather variable; in consequence, its relation to carotids and main aorta varies in tetrapods (Figs. 333, 334, 338). Subclavians may arise from the unpaired stem of the dorsal aortic trunk (salamanders) or from the paired systemic arch (frogs) or from the right systemic (reptiles, birds). In mammals, the right systemic arch is done away with, but the base of the arch is retained as a proximal part of the right subclavian. In crocodiles, the origin of the subclavians is still farther forward, from a point near the top of aortic arch III, and, hence, in the adult, the subclavians and carotids have a common trunk (Fig. 337 D); the same situation persists in their avian relatives.

In the pelvic appendage, as in the pectoral, there is a tendency for the concentration of the blood supply to the limb into a single vessel. Primitively this was an **ischiadic artery** running outward behind the ilium; in mammals, the major stem becomes the **iliac artery**, emerging in front of the girdle. Further down the limb, the main trunk is termed the **femoral** in the thigh, the **popliteal** at the knee region, and the **peroneal** in the shin.

Venous System

The veins, vessels that bring blood from capillary systems to the heart, have a complicated and variable arrangement. If, however, their embryonic history is studied, it is seen that they can be logically sorted out into a small number of systems. The discussion of the venous system here will be based largely on the developmental story. So treated, we may consider the following principal components of the venous system (Fig. 342).

1. An embryonic **subintestinal system** of veins flowing forward below the gut, forming in the adult the **hepatic portal system** running from the gut to the liver and the **hepatic veins** from liver toward the heart.
2. Veins situated dorsal to the coelom or gut and carrying blood toward the heart from the dorsal parts of the body and the head (and from the paired limbs of higher vertebrates as well); they include the **cardinals**, or the **venae cavae**, which replace them, and their affluents.
3. A relatively minor group, the **abdominal vein** or veins, draining the ventral part of the body wall and, primitively, the paired limbs.
4. In typical lung-bearing vertebrates, the **pulmonary veins** from lungs to heart.

The first and fourth of these four components form the drainage of the gut tube and its outgrowths; they form an essentially visceral venous system. The second and

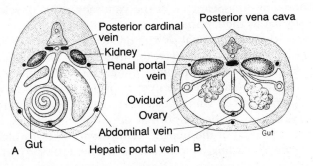

Figure 342. *A,* Cross section of the abdominal region of a shark, to show the position of the main veins. The renal portal vein is usually dorsal rather than lateral to the kidney. *B,* The same in a urodele amphibian.

third components are, on the contrary, mainly somatic venous systems, draining the outer wall of the body, although the third becomes partly visceral in tetrapods.

Hepatic Portal System and Hepatic Veins (Figs. 332–334). The hepatic portal system, common to all vertebrates (and even amphioxus—see Figure 331), is composed of veins that collect blood from the intestine and transport it to the sinusoids of the liver. Functionally, the system is of great importance; the intestinal capillaries take in all the food materials absorbed from the gut (with the exception of part of the fats), and the presence of a hepatic portal system guarantees that the liver has "first chance" at such materials, for their storage or transformation, before they are turned out into the general circulation.

Beyond the liver, the blood from the intestine is re-collected into a **hepatic vein** or veins. In most fishes, this is a large median vessel that empties directly into the sinus venosus of the heart. In the lungfishes and in all tetrapods, however, we find that, as discussed later (cf. p. 475), part of this hepatic vein has been used in the formation of the posterior vena cava, which carries the main blood stream forward from the dorsal part of the trunk to the heart. In consequence, the term "hepatic vein" is restricted in these forms to the vessel or vessels that empty from the liver into the terminal part of the posterior vena cava.

In the embryo, the hepatic portal and hepatic veins arise as a unit system. Generally, the first blood vessels to appear in the embryo in forms with mesolecithal eggs are a pair of veins that form in the floor of the gut and coalesce into a single channel running forward ventrally as a **subintestinal vein** (Fig. 358). From the far anterior end of this trunk develop the heart and ventral aorta, structures with which we are not concerned at this point. The remainder of this vessel gives rise to the hepatic and hepatic portal veins.

In large-yolked types, such as elasmobranchs, reptiles, and birds, in which the early embryo is spread out over the yolk surface and there is no formed midventral line, the blood channels that correspond to the subintestinal vein of other forms are paired structures that collect blood from the surface of the yolk. As the body of the embryo takes shape, the parts of these vessels that lie within it fuse to form a typical subintestinal vessel; those parts remaining external to the body persist as large and important embryonic structures, the **vitelline** or omphalomesenteric **veins** (Fig. 359). Similar veins develop in mammals in early stages, although the yolk is absent.

With the reduction and disappearance of the yolk sac during development, the parts of this system external to the embryo disappear.

In early stages, the subintestinal or vitelline system runs forward below the gut directly to the heart. Presently, however, the liver grows out ventrally from the gut. With its growth, liver and vein become intermingled; the venous trunk breaks up into small vessels and finally into a system of liver sinusoids (cf. p. 392), with the resulting formation of a separate portal trunk posteriorly and a hepatic vein anteriorly (Fig. 360). In typical fishes, no further important development occurs; in the embryo of higher vertebrates, a branch of the hepatic vein reaches dorsally along the mesenteries to tap the posterior cardinal system and form the anterior part of the posterior vena cava (cf. p. 475). The hepatic portal vein remains a large and important vessel, collecting blood not only from the intestine but also from the stomach, pancreas, spleen, and (in teleosts) the swim bladder, for conduction forward to the liver.

Dorsal Veins—Cardinals and Venae Cavae. The principal blood drainage from the "outer tube" of the body is cared for by important longitudinal vessels situated dorsally above the gut and mesenteries. In lower vertebrates, these veins are the cardinals; in higher forms, major modifications of these vessels produce the venae cavae.

In the embryo of every vertebrate (and of amphioxus as well), paired veins appear at an early stage in the tissues above the coelomic cavity, one on either side of the midline (Fig. 359). These are the primitive **cardinal veins**. The **posterior cardinals** run forward along the trunk on either side of the aorta to a point in the body wall dorsal to the heart. Paired **anterior cardinals** begin as head veins on either side of the developing braincase and run back dorsally above the gills or along the neck to meet their posterior mates. From this point of junction, on either side a major vessel descends to enter the sinus venosus of the heart; this is the **common cardinal** (or duct of Cuvier).* This characteristic embryonic cardinal system is retained in almost diagrammatic form in the adult of elasmobranchs (Fig. 332) and is also retained in ray-finned fishes. We may discuss separately the history of the anterior and posterior parts of this system.

In all vertebrate classes except the mammals, the main stem of each anterior cardinal (or anterior vena cava) is the **lateral head vein** (Fig. 343). This characteristically arises deep within the orbit, where it is often present as an expanded sinus, and receives vessels from various areas of the anterior part of the head. Traveling posteriorly along the side of the braincase and otic region, the lateral head veins receive tributaries from the brain and, continuing backward, enter the common cardinals. In lungfishes and tetrapods (where the posterior vena cava has taken the place of the posterior cardinals), the common cardinals are incorporated in the trunk of the anterior cardinals, including in their course, therefore, the openings of the subclavian veins from the pectoral limbs. With this modification, the anterior cardinals come to resemble the vessels termed anterior venae cavae in mammals and are frequently called by this name.

In mammals (and, to a certain extent, crocodilians and birds), there is an important change in the cranial circulation. A system of intercommunicating venous sinuses develops within the expanded cranial cavity. The lateral head veins disappear, and the blood from most of the vessels of the front part of the head enters the braincase,

*The adult lamprey retains only a single common cardinal (the right).

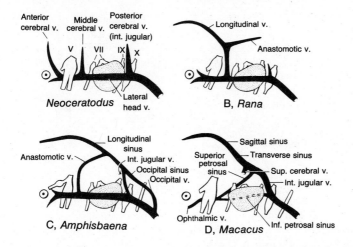

Figure 343. Diagrams of the left side of the head to show stages in the evolution of the venous drainage. Roots of some cranial nerves are indicated (Roman numerals); the position of the eye is shown, and the otic capsule is stippled. In lower vertebrates, the main drainage is by a lateral head vein, which forms in the orbital region and runs backward to become the anterior cardinal. This primitively receives several successive veins from the interior of the skull. A series of sinuses develops within the braincase; the lateral head vein is abandoned in mammals, and blood from the orbital region enters the sinus system, all finally draining from the skull as the internal jugular vein. *A,* Lungfish; *B,* frog; *C,* amphisbaenian; *D,* mammal (macaque monkey). (After van Gelderen.)

to leave posteriorly on either side, after collection of the venous blood from the brain, as the **internal jugular veins**. As these travel posteriorly, they are joined by vessels, the **external jugulars**, which gather blood from the more superficial parts of the head, to form **common jugulars**; joining the subclavian vein, the further course of each of these vessels to the heart is a trunk termed an **anterior vena cava**. Such veins are easily recognizable as the old anterior cardinals and common cardinals, except that intracranial channels supplant the lateral head veins that originally formed the most anterior part of the cardinal stems.

In many mammals (including man), the terminal part of the system is further modified (Fig. 344). In the embryo, a connection develops between the two anterior cardinals a short distance anterior to the heart, and the left common cardinal disappears. All the blood from the left cardinal then flows over to enter the heart through the right cardinal, so that in the adult only a single anterior (or superior) vena cava is present. A comparable development is seen in birds.

The story of the **posterior cardinals** is more complex (Figs. 345, 346). It begins with a pair of rather simple dorsal vessels draining forward into the heart via the common cardinals; it ends in mammals with the draining of the same region by a single but complex vessel, the posterior vena cava. In between lies a considerable history.

In cyclostomes, the posterior cardinals are simple paired vessels, receiving blood from a median caudal vein, kidneys, gonads, and dorsal parts of the body musculature and running forward uninterruptedly to the common cardinals. In the jawed fishes, however, a new feature appears in the system that was destined to persist upward

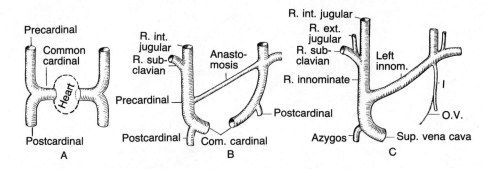

Figure 344. Ventral views of the veins anterior to the heart in successive developmental stages to show the formation, in man and certain other mammals, of a single anterior (or superior) vena cava from the two anterior cardinals (precardinals). An intercostal vein *(I)* and a small vein from the wall of the left atrium (*O. V.,* oblique vein) are persisting vestiges of the original left anterior cardinal. (From Arey.)

into the avian stage—the development of a **renal portal system**. The blood from the posterior part of the trunk and from the tail no longer flows directly to the heart. The cardinal channels are interrupted beside the kidneys, and this blood is diverted into a network of capillaries around the kidney tubules (*not* through the glomeruli). After this passage, the blood is received by new vessels that carry it forward to the truncated stumps of the two posterior cardinals and thus to the heart. It is customary to term the vessels from kidney to heart in fishes the posterior cardinals over their entire length; but it is, of course, only the more anterior part that is truly homologous with the "original" cardinals of the cyclostomes or of the fish embryo before the renal portal system was established.

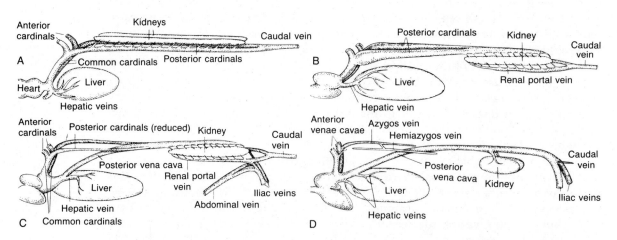

Figure 345. Diagrams in lateral view to show the evolution of the posterior cardinals and the development of a posterior vena cava. *A,* Lamprey (larva); *B,* typical fish condition; interjection of renal portal system. *C,* Lungfish or primitive tetrapod; a shortened route to the heart is established by using part of the hepatic vein system in the initiation of a posterior vena cava. *D,* Mammal; the renal portal eliminated and the posterior cardinals separated from the posterior vena cava. Vessels of the right side are shown in deeper shading.

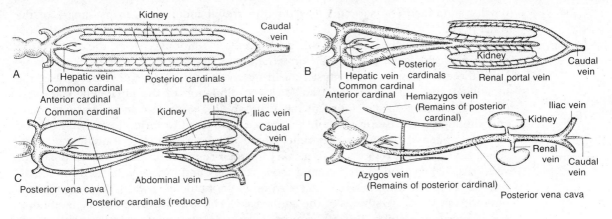

Figure 346. Diagrams in ventral view to show the evolution of the posterior cardinals and the development of a posterior vena cava. Stages as in Figure 345.

In the Sarcopterygii, as shown by the living lungfishes, a second progressive change initiates the development of a **posterior vena cava**. A branch of the hepatic veins draining the liver grows upward past that organ along a pulmonary fold to the dorsal side of the body cavity and taps the right posterior cardinal vein some distance back of its entrance into the common cardinal. Once this is done, blood from the right cardinal may follow this new short cut to the heart; further, because there tends to be fusion in the kidney region between the vessels draining the two kidneys, blood from the left vessel may be shunted over and follow the same course. In lungfishes and urodeles, some of the blood still uses the old posterior cardinal connections to the heart, but in frogs and all higher tetrapods, both posterior cardinals lose connection with their more posterior trunks; one or both remain merely as outlets for the small and variable series of **azygos veins**, draining the blood from part of the flanks. The new major trunk, from kidneys past the liver to the heart, may be properly termed the posterior vena cava from the lungfish stage upward.

Among reptiles, and even in some amphibians, the renal portal system begins to degenerate, and much of the portal blood may be shunted through or past the kidney without entering the capillary network around the tubules. In birds, the renal portal system is very complex; much of the blood reaching the kidney plunges straight through that organ on its way to the heart without entering a capillary system, but muscular valves can close the shunts and force the blood through the capillaries around the tubules. In mammals, the renal portal system has been completely abandoned. Blood from the hind legs and tail, if any, passes directly forward, into the posterior vena cava; the posterior part of this vessel may be paired, but for most, if not all, of its length, the posterior vena cava is a single major vessel extending without interruption forward from the caudal vein in the sacral region to the heart.

The posterior vena cava of the mammal is a vessel of seeming simplicity. As its history shows, however, it is a composite patchwork structure, including part of the original posterior cardinals, much of the system of veins draining the kidneys in the renal portal stage, and anteriorly, a branch of the original hepatic veins. In its evolutionary development from the primitive posterior cardinals, there have been three main steps: (1) introduction of a renal portal; (2) tapping of the right cardinal by the hepatic vein, and consequent degeneration of much of the anterior ends of the car- **475** dinal system; and (3) subsequent reduction and abandonment, in mammals, of the renal portal.

As might be expected from this complicated evolutionary story, the embryology of the posterior vena cava in higher tetrapods is likewise complicated. Since, however, the circulatory system must remain functional throughout development of the individual, the story is far from an ideal recapitulation of the phylogenetic history (Fig. 347). Postcaval development has been studied with care in several mammals, and although there is considerable difference from form to form, the general outline of developmental history is the same. Three successive pairs of longitudinal veins appear in the dorsal region of the trunk: (1) a pair representing the original posterior cardinals; (2) a second pair, situated more ventrally and medially, termed **subcardinals**, and representing the veins seen, from elasmobranchs upward, draining the kidneys forward into the stumps of the cardinals; (3) finally, **supracardinals**, mediodorsal in position, which do not appear clearly in the phylogenetic story. As in cyclostomes, full length posterior cardinals arise; then, through the intervention of the subcardinals a renal portal system is established and, almost concomitantly, the connection with the hepatic veins is made as in lungfish. The role of the supracardinals is variable. In some forms, they contribute little except drainage of the flanks through the system of azygos veins; in other cases, as in man, they contribute a fair length of the posterior part of the vena cava trunk.

Abdominal Veins; Limb Veins.   In the Chondrichthyes, longitudinal **abdominal veins** are present low down in the body wall on either side (Figs. 332, 342 *A*), draining blood from the lower part of the axial musculature and running forward to empty into the common cardinals. Anteriorly, each receives a small vein draining the pectoral fin; posteriorly, a still smaller one from the pelvic fin. Abdominal veins are absent in ray-finned fishes, but a single median ventral abdominal vein is found in lungfishes and persists in amphibians (Figs. 333, 342 *B*) and in reptiles. In these tetrapods, however, it does not reach the heart directly, but runs upward anteriorly into the liver, to become essentially part of the hepatic portal system.

The abdominal veins persist in the embryos of all amniotes, even when absent in the adult, for the significant reason that the veins from the allantois empty into paired abdominals. In reptiles and birds, these veins are important for bringing oxygenated blood from the allantoic "lung" into the body; in placental mammals, they are the carriers of the entire food supply from the placenta (Fig. 359). They are in the mammalian embryo termed the **umbilical veins**, since, fused into one vessel for part of their course, they form the one venous afferent in the umbilical cord of the late embryo. As in the abdominal veins of adult amphibians and reptiles, the embryonic umbilical veins drain into the liver (Fig. 360).

We may mention here in passing a pair of anterior veins in fishes that perform for the anterior end of the body the same function as the abdominal veins in the trunk; they are found in sharks and other lower vertebrates, as paired ventrolateral structures draining the walls of the pharyngeal region and running back into the common cardinals (Fig. 332). They are frequently termed external jugulars but appear not to be homologous with the vessels of that name in tetrapods, which are part of the anterior cardinal system.

Of the limb veins, those from the pectoral fin lose their slight affiliation with the abdominal vein in actinopterygians to enter the common cardinals directly, and in

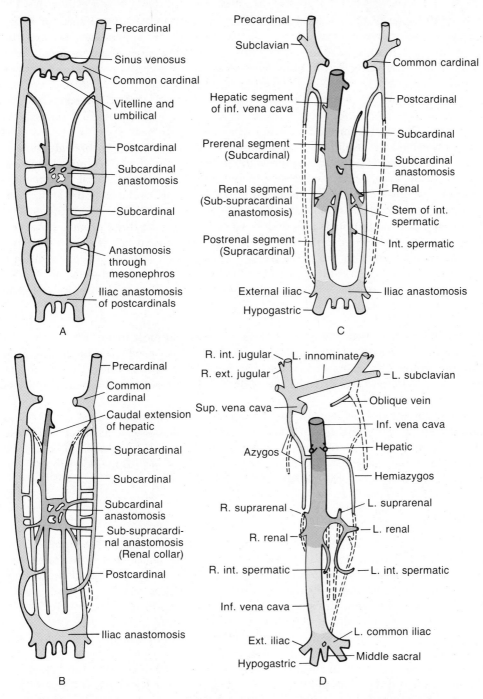

Figure 347. Development of the posterior vena cava of man, showing the embryologic processes roughly paralleling the evolutionary history. The original cardinal system is shown in blue. Presently there develop subcardinal vessels (red), corresponding to the veins draining the kidney when a portal system is established in sharks. In *B*, this venous system is tapped (as in lungfishes) by a branch of the hepatic veins (purple). A third system of embryonic veins, the supracardinals (yellow), is not exactly paralleled in phylogeny. In green are vessels that develop in amniotes to by-pass the kidney and eliminate the renal portal system. As seen in *D*, the definitive posterior vena cava includes fractions of all these structures. (From Arey, after McClure and Butler.)

tetrapods drainage of the pectoral limb becomes associated with the anterior venae cavae as the **subclavian veins**. The veins from the pelvic appendages, termed the **iliac veins** in tetrapods, have a complex history. In sharks, they drain, as mentioned, into the abdominal veins. But in bony fishes, they also develop a connection dorsally with the renal portal system—a connection retained in higher forms that develop a posterior vena cava. In actinopterygians, this is the only channel present, but the ventral as well as the dorsal route persists in lungfishes, amphibians, and reptiles. As a consequence, blood from the limb can travel toward the heart by either one of two routes in these forms. In either case, it must, in lower tetrapods, go through another capillary (or sinusoid) network, either in the kidney or the liver. But more than this, the connection may work in reverse—blood from the tail and sacral region may, instead of entering the renal portals, be shunted downward and follow the abdominal vein forward to the heart in lungfishes or to the liver in amphibians and reptiles. Ray-finned fishes and (adult) birds lack abdominal veins. In both these groups, however, a vein, perhaps derived from the old abdominal, runs downward and forward from the renal portal system in the pelvic region to enter the system of hepatic sinuses, and thus gives the same "choice" of routes for blood from hind limbs and the hinder part of the body. In mammals, neither abdominal vein nor any substitute for it exists; the posterior vena cava is the sole exit.

Pulmonary Veins. These veins are, of course, absent in most living fishes, in which lungs are nonexistent. In the actinopterygian *Polypterus,* in which a functional lung exists, the pulmonary veins empty into the hepatic vein; hence their blood mixes with the general blood stream from the body. Similar arrangements occur in teleosts with swim bladders.

In the lungfishes, however, the pulmonary trunk by-passes the sinus venosus and enters the atrial cavity of the heart directly and at the left side. This separate course of the aerated blood from the lungs to the heart persists in all tetrapods. The entrance of the pulmonary veins into this region of the heart allows the subdivision of the atrium and the final subdivision of the entire heart.

Lymphatics

In most, if not all, vertebrates, we find, supplementary to the venous system, a second series of vessels returning fluids from the tissues to the heart—the **lymphatic system**. Although paralleling the veins in many functions (and often paralleling them topographically), the lymphatics differ from them in major respects. A fundamental difference is that the lymphatics are not connected in any way with the arteries but arise from their own capillaries, which are closed blindly at their tips (Fig. 330). There is thus no arterial pressure behind the fluid contained in the lymphatics; the flow of materials in them is generally sluggish.

The contained liquid, the **lymph**, gains entrance into the lymphatic vessels by diffusion through their walls and hence is essentially similar to the general tissue fluid and (except for the absence of blood proteins) to the plasma of blood vessels into which the lymphatics empty. There are, in most cases, few blood corpuscles in the lymph; ameboid white blood cells, however, can enter from adjacent tissue spaces. The lymphocytes owe their name to the fact that in mammals the main site of their formation is in the lymph nodes (cf. p. 452) placed along the course of the lymphatic vessels so they are found abundantly in the lymph. This, however, is almost exclusively a mammalian peculiarity, and, in other vertebrate groups, there is little or no special association of the lymphocytes with the lymphatic system.

Related to the low pressure under which lymph travels is the fact that the walls of lymphatics are thin; they are consequently difficult to find and dissect unless specially injected. Even the largest lymph vessels have only a thin coat of musculature and connective tissue comparable to that of the smaller veins. In general, movement of the lymph is brought about by movement of the body and its various organs; in mammals and birds, the lymphatics are supplied with numerous valves, preventing a backward flow. In amphibians, reptiles, and some birds, the lymphatic system is supplied with **lymph hearts**—small, two-chambered, muscular structures that lie at points at which lymph vessels empty into veins and actively pump lymph onward into the general circulation (Fig. 348 A). These hearts are usually few in number, but, in the Gymnophiona, there may be 100 or so in a paired series. A pair in the pelvic region is prominent in both anurans and urodeles and persists in reptiles and some birds (notably the ostrich), and lymph hearts are found in the tail of teleosts. No

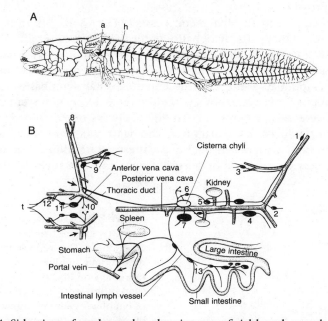

Figure 348. *A*, Side view of a salamander, showing superficial lymph vessels. Dorsal, lateral, and ventral longitudinal vessels are present; a series of lymph hearts *(h)* is present along the lateral vessel. Lymph from this vessel enters the venous circulation through an axillary sac *(a)*; lymph from the ventral vessel enters through an inguinal sac. *B*, Diagram of the deep vessels of the lymphatic system of the rat, anterior end at left. Lymphatics in solid black; the neighboring veins are also shown. Nodes are numbered according to the region in which they lie; *1*, knee; *2*, tail; *3*, inguinal; *4*, lumbar; *5*, kidney; *6*, nodes about the cisterna chyli; *7*, intestinal node; *8*, elbow; *9*, axilla; *10*, thoracic; *11*, cervical; *12*, submaxillary; *13*, mesenteric nodes; *t*, plexus of lymphatics around tongue and lips. Arrows indicate the point of entrance of lymph into the veins near the junction of jugular and subclavian and into the portal vein. (*A* after Hoyer and Udziela; *B* after Job.)

lymph hearts are, however, found in most birds or in mammals. Along the course of lymphatics, enlargements of the vessels may occur as **lymph sinuses**.

The Agnatha and Chondrichthyes possess a series of thin-walled sinusoids that drain into veins but have few or no arterial connections. Although some have denied that these are true lymphatics, they seem to represent that system of vessels, albeit in a rather simple and presumably primitive way. Typical lymphatics are present, however, in bony fishes and all tetrapods. They are highly developed in amphibians, where they are remarkably abundant in the subcutaneous tissues as a protection against potential desiccation; owing to the development of the lymph hearts, amphibians have a relatively active lymphatic circulation. Reptiles have very complex lymphatic systems, with many plexuses surrounding the veins (Fig. 349).

Lymphatic vessels in higher vertebrates penetrate most of the body tissues; the central nervous system, liver, cartilage, teeth, bone, and bone marrow are exceptions. Lymphatics to the gut are highly developed. The intestinal ones are of great importance, because, while carbohydrate and protein food materials pass into the hepatic portal vein, much of the fats enters the circulation by way of the lymphatics; this is perhaps associated with the relatively large size of the molecules involved and the difficulties encountered by them in entering the intestinal capillaries of the blood vessels against pressure.

In their arrangement, the lymphatic vessels (Figs. 348, 349) vary greatly from group to group. Terminally, the lymphatics enter the veins, generally the dorsal veins, cardinals, or venae cavae. Primitively, the points of entrance may have been numerous and even more or less metameric; conditions of this sort prevail in urodeles and gymnophionan amphibians, and, even in amniotes, small lymphatics enter the veins at a variety of points. In general, however, there tends to be a concentration of lymph discharge in three areas: (1) Anterior: There are always one or more main entrances into the anterior cardinals or jugulars near the heart, where pressure in the veins is, of course, lowest and unopposed entrance easiest. (2) Middle: Entrances along the course of the posterior cardinals or posterior vena cava are variable; they are rare in

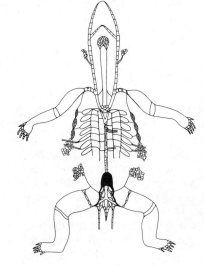

Figure 349. Diagram of the major lymphatic trunks in a crocodilian. This represents, despite its complexity, an extreme simplification. In this (and other reptiles), networks of lymphatics surround many blood vessels (the net shown in the midline is around the dorsal aorta), surround or lie in various organs (that in the throat is around the larynx), and underlie the skin (shown in the posterolateral part of the trunk). The lacteals and many other important parts of the system are omitted. (After Ottaviani and Tazzi.)

mammals, numerous in urodeles. (3) Posterior: In fishes and birds, but not mammals, there may be a major flow of lymph into the veins in the pelvic region.

The fatty lymph, or **chyle**, leaving the intestine runs up the mesenteries in vessels termed **lacteals** from their milky appearance when distended. There tend to develop a pair of longitudinal lymphatics, **thoracic ducts**, along the back of the body cavity, paralleling the posterior cardinals or vena cava, with anterior and sometimes posterior openings. These ducts serve as collectors for the lacteals. Paired ducts are common in fishes, reptiles, and birds, but, in some fishes and reptiles, and in mammals generally, the pair tend to fuse into a single asymmetric duct; a large cistern, **cisterna chyli**, may develop along its course or at its back end in the lumbar region.

The lymphatic vessels frequently run parallel to the veins. Some, in the embryo, apparently arise as outgrowths of veins and have been thought to represent a part of the embryonic venous system. This, however, is debated, and, in postembryonic life, growth of lymphatics is independent of the veins.

A functional explanation of the development of lymphatics is suggested by the fact that they are highly developed in tetrapods, in which the gills have been eliminated from the arterial circuit, and the blood, hence, courses at a higher pressure than in fishes. In consequence, entrance of liquids from the tissues into the capillaries, against pressure, is more difficult than in the original fish condition. The lymphatics offer a relatively low pressure system of drainage into the veins, where the blood pressure is at its lowest.

The Heart

Some type of muscular pumping device is necessary for the efficient circulation of the blood. A variety of hearts are found in invertebrates, and, in amphioxus, there is a whole series of paired "heartlets" along the bases of the aortic arches, and the ventral aorta itself is contractile. In hagfishes, there is an accessory caudal "heart," but otherwise in vertebrates the heart is a single structure, situated ventrally and well anteriorly in the trunk, drawing in posteriorly venous blood and pumping it anteriorly, in lower vertebrates, to the aortic arch system and the gills. Primitively, it consisted of four successive chambers, termed, from back to front, sinus venosus, atrium, ventricle, and conus arteriosus; in advanced groups, the first and last lose their identity, but atrium and ventricle tend to subdivide.

The heart is situated ventral to the gut in a special anterior region of the coelomic cavity; it is in the adult attached to the walls of this cavity only at the points of entrance and departure of the blood vessels and, thus, is able to change its shape readily during its powerful pumping movements. This pericardial cavity (Fig. 352, also p. 316) is usually completely separated from the general body cavity in the adult, although very small connections persist in the cyclostomes and Chondrichthyes.

The heart is essentially a series of expansions developed along the course of a main vascular trunk; its histologic structure, although much modified, is essentially comparable to that seen in other blood vessels. There is a thin internal lining, the **endocardium**; an outer covering is a thin, mesodermal epithelium, the **epicardium**, similar to that lining the rest of the coelom. The main bulk of the heart consists of connective tissues and muscle—the **myocardium**. The connective tissue may be compact (particularly about the ventricles) and serve as a type of skeleton; in some instances (various ungulates, for example), there may even be a development of cartilage or bone. There are frequently cross strands of connective tissue or muscle across the cavity of the ventricle (and sometimes of the atrium as well), preventing undue expansion under pressure. The muscular tissue has been discussed earlier. Between the chambers of the heart and at the points of entrance or exit of blood vessels lie a series of heart **valves**, similar basically to those in veins or lymphatics, but usually powerful and of complex structure. These yield freely to the forward propulsion of blood but prevent a backflow when a chamber contracts. Frequently there are tendinous strands of tissue that limit the extent to which the valves may be pushed backward into the orifice to be closed. In some higher vertebrates, the valves between the atria and ventricles may even be furnished with small muscles that contract with the rest of the heart muscles; these ensure that the strands remain the correct length even when the ventricle changes shape during contraction.

The heart is, of course, itself supplied by variable **coronary vessels** that ramify in its muscular walls. Though small, these vessels are of crucial importance; occlusion of the coronary arteries that supply the heart may result in sudden death through heart stoppage.

Heart Beat. Except in hagfishes, fibers from the autonomic nervous system reach the heart (at the sinus venosus or atrium) and may affect the rhythm of its beat; the heart, however, is essentially on its own, as is shown by the fact that its muscles will continue a rhythmic contraction even when cultured apart from the body. The rate is highly variable, partly in relation to the needs for oxygen supply, tending to be low in large animals, high in warm-blooded animals. Thus, the heart beats only 20 or so times a minute in a codfish, 30 in an elephant, about 70 in a man or ostrich, 350 in a rat, and 600 or above in a hummingbird. In lower vertebrate classes, the four successive cardiac chambers contract in sequence from posterior to anterior: the beat starts with a contraction of the sinus venosus; the wave of contraction of the sinus musculature acts as a pacemaker, stimulating the atrium to activity, and so on down the remaining chambers. An accessory center of stimulation in fishes lies in the muscles of the atrioventricular valve, and in teleosts, in which stimulation of the sinus venosus is generally unimportant, an accessory pacemaker is present in the atrium. In amniotes, continuity between the muscles of atria and ventricles is broken and there develops a unique conducting system for this stimulus, the **sinoventricular system**—essentially a local simulation of a nervous system (Fig. 350). It consists of strands or aggregations of fibers that are specialized muscle fibers—specialized not for contraction but for carriage of an impulse and thus simulating nerves. A **sinus node** of these tissues, situated in the right atrium, initiates the beat. The stimulus is mainly transferred to a second node in the septum between the atria, whence a fiber bundle—the **atrioventricular bundle**—descends to disperse among the muscles of the ventricles and set up a contraction of these chambers. The asymmetric point of origin of the beat is correlated phylogenetically with the fact that the sinus venosus, in which the stimulus began in the primitive heart, is incorporated in amniotes into the right atrium.

The Primitive Heart. The heart of typical fishes is a single tube consisting of four consecutive chambers. The heart of a bird or mammal is also four-chambered, but the four chambers do not correspond to the original four; it is, rather, a double pump, with two chambers in each of its two parts. Great changes have occurred in the

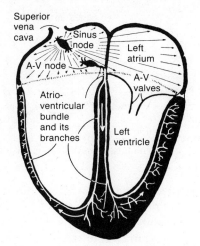

Figure 350. Diagram of a mammalian heart opened from the ventral surface to show the conducting system. (From Carlson and Johnson, The Machinery of the Body, University of Chicago Press.)

history of the heart, changes correlated with the shift from gill-breathing to lung-breathing and the necessary associated changes in the course of the circulation.

In the primitive vertebrate heart (Figs. 351, 352 A) the four chambers are:

1. **Sinus venosus**, a thin-walled sac, with little muscular tissue, essentially a place for the collection of venous blood, lying posteriorly in the region of the septum between pericardial cavity and general coelom. It receives the hepatic vein or veins posteriorly and the paired common cardinal veins laterally.
2. **Atrium** (or **auricle***), the next anterior chamber, still relatively thin-walled and distensible.
3. **Ventricle**, thick-walled; the main contractile portion of the heart.
4. **Conus arteriosus**,† thick-walled, but small in diameter and frequently containing several sets of valves.

These four chambers are arranged in posteroanterior series in the embryo of lower vertebrates. But during development, the anterior part of the heart tube tends to fold back ventrally (and somewhat to the right) in an S-shaped curve, thus combining length with compact structure. As a result, the more posterior heart chambers tend, even in a fish heart, to be situated dorsal to the anterior ones, and the atrium may be anterior as well as dorsal to the ventricle. The ventricle tends to be so placed that it forms a pocket with a "free" posterior end and with both atrial entrance and arterial exit via the conus placed near one another anteriorly. The folded position makes it difficult to visualize heart construction, and frequently (as in Figs. 337, 351) diagrams of heart structure arbitrarily represent the organ as though "pulled out" into its embryonic longitudinal arrangement.

---

*Auricle, although often used as a synonym for atrium, refers properly to the supposedly earlike flaps on the atria of the mammalian heart.

†Often confused with the **bulbus arteriosus**, developed as a proximal part of the ventral aorta just in front of the heart. The term **bulbus cordis** is frequently used for the embryonic equivalent of the conus in mammals.

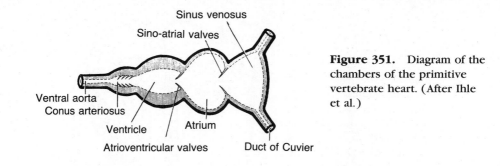

Figure 351. Diagram of the chambers of the primitive vertebrate heart. (After Ihle et al.)

Sinus venosus
Sino-atrial valves
Ventral aorta
Conus arteriosus
Ventricle
Atrium
Atrioventricular valves
Duct of Cuvier

The primitive type of heart here described is typically developed in the cartilaginous fishes, and most other fishes have a heart similarly constructed. However, there may be minor modifications. In cyclostomes, for example, the sinus venosus tends to be reduced in size; in teleosts, the conus is practically absent and is more or less replaced by an arterial bulb.

Evolution of the Double Heart Circuit (cf. Fig. 337). In the lungfishes and amphibians, a major problem arises connected with the substitution of lungs for gills as breathing organs. The heart now receives blood of two different types: "spent" venous blood from the body, and "fresh" oxygenated blood from the lungs. These two streams should be kept separate, as far as possible, and sent to two different destinations—the venous blood to the lungs, the "fresh" blood to the body. Beyond the heart, a splitting of the arterial channels is made for these two destinations. But how to keep the two streams distinct in a single-barreled pump?

The perfect solution of this difficulty was not attained until the avian and mammalian stages were reached, but the lungfishes and amphibians have made some prog-

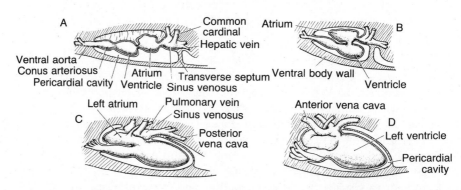

Figure 352. Diagrammatic views of the left side of heart in various vertebrates, to show its position in the pericardial cavity and phylogenetic modification of the heart chambers. *A,* Hypothetic ancestral condition, found essentially repeated in embryos (cf. Fig. 351). The four primitive chambers are in line anteroposteriorly, and a dorsal mesentery is still present. *B,* Selachian stage; the mesentery is gone; the atrium has pushed forward above the ventricle, but the sinus venosus is still posteriorly placed. *C,* Amphibian stage; the sinus and accompanying blood vessels have moved anteriorly. *D,* Amniote stage; sinus and conus arteriosus have lost their identity; the heart attaches to the walls of the pericardium only anteriorly. (After Goodrich, 1930.)

ress toward attainment of a separation of the two blood streams. In the lungfishes, we find that the blood coming from the lungs by way of the pulmonary veins does not enter the sinus venosus, as does the typical venous blood (and, indeed, the pulmonary veins of *Polypterus*), but enters the left side of the atrium directly and separately. The sinus venosus, thus, is restricted to the typical venous blood circuit; the sinus remains large in lungfishes and urodeles but is reduced in size in anurans, represented by a vestige or lost in reptiles and birds, and completely absorbed into the atrial structure in mammals.

A second change lies in the gradual subdivision of the orginally single atrium. The pulmonary venous trunk enters the left side of the atrial chamber. A partition develops that cleaves the atrium in two in such a way that the left chamber contains only pulmonary blood; the right (with which the sinus becomes confluent), typical venous blood. In lungfishes, there is a partial septum between the two sides of the atrium; in typical amphibians, the division is complete. In these forms, however, there is still a single opening from the atrial chamber into the ventricle. Primitively, this was guarded by a pair of valves. In lungfishes, instead of valves, a special cushion of tissue **atrio-ventricular plug**, from the posterior (originally ventral) wall of the heart helps close this opening (now functionally double); in amphibians, four valves are present, although the opening is still single.

But atrial subdivision is in vain if the two blood streams meet and mix in the ventricle. In lungfishes, there has developed a partial septum, essentially a continuation of that in the atrium, along the posterior wall of the ventricle. The heart in modern amphibians is a simpler structure (Figs. 337 A, 353 A), because there is no apparent subdivision of the ventricle. One would tend, at first sight, to think of this condition as primitive, and the partial separation in lungfishes a parallelism to conditions in higher vertebrates. It is, however, possible that the simpler structure of the modern amphibian heart is due to secondary reduction; the available evidence does not permit either choice to be made with confidence. In ancestral fishes and the ancient amphibians, all air-breathing was done by the lungs. But in most living amphibians, much of the "breathing" is done through the skin, from which the oxygen-

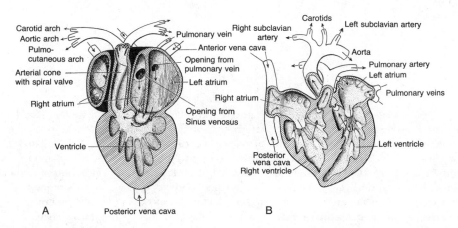

Figure 353. Diagrammatic section through the heart of *A,* a frog; *B,* a typical mammal. (Partly after Jammes.)

ated blood enters the right atrium; there is hence little point in separation of the blood streams. There is, however, a variable degree of functional separation; the amphibian ventricle is a spongy structure that may prevent a too free mixing of the two blood streams. In many cases, the carotid and systemic arches receive oxygenated blood primarily from the left atrium; the vessels to lung and skin are fed from the right atrium. In other cases, considerable mixing appears to occur.

A septum or, better, two partial septa is developed within the ventricle of reptiles. In turtles, lizards, and snakes, however, a major gap persists near the points of entrance and exit of the blood streams, so that admixture can still take place. The structure of the reptilian ventricle is extremely complex—and much oversimplified in Figure 337 C. It is divided into dorsal and ventral parts, with the former subdivided into right and left halves (Fig. 354); the latter correspond, in some ways, to the separate ventricles of higher forms, although blood from both atria enters the dorsal chamber. Anatomic inspection suggests that the right systemic arch should receive oxygenated blood from the left side of the heart, whereas the left systemic arch, like the pulmonary, should receive nonoxygenated blood from the right. However, the situation is not that simple. Although the left systemic arch may, under certain conditions in various forms, receive "mixed" or even venous blood, this is rare—usually, it receives *only* oxygenated blood. As an aside, we may note that the idea that mixing is necessary because the pulmonary circulation lacks the capacity to handle half of the circulating blood is not substantiated by experimental work.

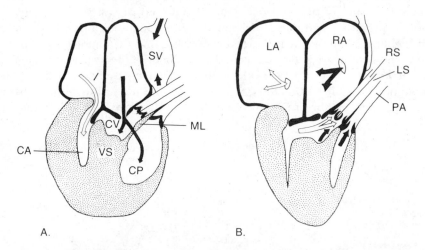

A.   B.

Figure 354. Diagrams of the circulation of blood through the heart of a varanid lizard; not all modern reptiles are similar and the pattern can vary considerably. *A,* The heart during atrial contraction (systole) and ventricular relaxation (diastole); *B,* during atrial diastole and ventricular systole. The open arrows represent oxygenated blood and the black arrows venous blood. Here, the ventricle is partly divided by a vertical septum *(VS)* into a cavum arteriosum *(CA)* and a second part, which is further subdivided by another incomplete septum, often called the Muskelleiste *(ML),* into a cavum venosum *(CV)* and cavum pulmonale *(CP).* The black arrow in *A* goes around the free edge of the Muskelleiste. Abbreviations: *LA,* left atrium; *LS,* left systemic arch; *PA,* pulmonary arch; *RA,* right atrium; *RS,* right systemic (and carotid) arch; *SV,* sinus venosus. (After Webb, Heatwole, and deBavay.)

The crocodilians have, technically, a complete ventricular septum, with venous blood entering the pulmonary and left systemic arches from the right ventricle and aerated blood supplied to the right systemic from the left ventricle. But even here, there is a gap, the **foramen of Panizza**, at the base of the arterial trunks (Fig. 337 *D*). Physiologic work indicates that, under most conditions, almost all the blood entering the left systemic arch does come from the left ventricle via the foramen.

The two most progressive vertebrate classes, the birds and mammals (Figs. 337 *B*, 353 *B*), have completed the ventricular septum and at long last have completely separated the two blood streams along the length of the major heart chambers. This development has obviously been brought about independently in the two cases, since mammals and birds have evolved independently from primitive reptiles. Indeed it is uncertain, even improbable, that the interventricular septa are completely homologous in birds and mammals; as expected, the interventricular septum of birds appears much the same as that of crocodilians, but whether that of mammals is similar is currently much debated.

There is no point in establishing separate heart circuits unless the blood, when it leaves the heart, is directed into appropriate arterial channels—aerated blood to head and body, venous blood to the lung. We have earlier (cf. p. 462) described subdivisions that occur in the ventral arterial trunk and separate these two streams. We must now, finally, consider the history of the most anterior region of the heart, that of the **conus arteriosus**, which is the primitive connecting piece between ventricle and aorta.

Primitively, as seen in a shark, the well-developed conus was a valved, tubular structure with a single terminal opening into the ventral aorta, and the aorta itself was a single stem (at least posteriorly). The aorta, as noted previously, becomes subdivided in tetrapods so that at least the pulmonary arteries of the sixth aortic arch are separated from the more anterior arches that circulate blood to head and body. There remains as a problem the course of the blood through the conus arteriosus, most anterior of heart chambers. In lungfishes and amphibians, a change occurs in the nature of the conus. Originally it contained simple, typical heart valves; in these primitive lung bearers, there develops, instead, a twisted "spiral valve" that runs the length of the conus. This valve partially subdivides the conus for its entire length into two channels (Figs. 337 *A*, 353 *A*), and in the South American lungfish and amphibians, the conus is completely subdivided for part of this distance. The structure and functioning of the heart are such that the venous blood is directed mainly into one of these two channels, and the oxygenated blood into the other; distally, the two channels tend to direct their streams mainly into, respectively, the pulmonary arch and the arches leading to trunk and head. Thus, in lungfishes and amphibians, despite the incomplete division of the heart, there is a moderate degree of separation of the two blood streams.

In amniotes, the conus arteriosus is reduced and not identifiable in the adult. In part, it may be incorporated into the ventricles, in part subdivided into the roots of the discrete aortic vessels: three in reptiles, two in birds and mammals (Fig. 337 *B–D*). With this splitting of the aortic vessels and their direct attachment to the completely divided ventricles in birds and mammals, complete separation of the two blood streams is attained, that through the right side of the heart flowing from body to lungs, that through the left from lungs to body.

To sum up, the introduction of the lung into the blood circuits in advanced fishes threw out of order the simple circulatory system of primitive vertebrates and caused a "problem" that advanced vertebrates found difficult to solve. Lungfishes, amphibians, and reptiles even today have not solved it completely, although their partial solutions obviously allow them to survive and even thrive. For that matter, there is no "perfect" solution for lungfishes (which use gills as well as lungs) or amphibians (which use the skin as well as lungs); under almost any system, mixing will occur when two or more separate respiratory organs are involved. Under certain conditions, the possibility of shunting blood from one side to the other is probably advantageous, because different mechanisms to allow it are retained or developed. In only birds and mammals is complete separation of circuits attained. The result is surprising in its efficiency. The single pump of the original heart has become a double one; each half of the heart performs effectively its own distinct task.

Heart Development (Fig. 355).   We have noted that the first blood vessels to develop in typical vertebrate embryos are paired vitelline vessels, which fuse to form a subintestinal vein running forward beneath the gut. The heart develops along the course of this longitudinal tube, strategically situated in the embryo and in the adult of lower vertebrates, near its anterior end, just behind the gills. The necessity for a circulatory pump develops at an early stage of embryonic history, and in large-yolked types, heart function may begin even before the two vitelline veins have fused; the heart may thus appear in transitory fashion as a paired structure. Coelomic cavities develop within the lateral plate mesoderm lying on either side of the embryonic heart. The peritoneal linings of the two cavities grow to surround the heart tube; they supply it with its epicardial outer coat, furnish mesenchyme from which the heart muscle differentiates, and form mesenteries—**mesocardia**. A ventral mesocardium, if developed at all, is short-lived, and the dorsal mesocardium disappears at an early stage, so that a single pericardial cavity is created.

Posterior to the heart, a transverse septum is presently formed (cf. p. 317), which in most vertebrates shuts off the pericardial cavity from the general coelom. With this septum is associated the entrance into the subintestinal vessel of a pair of veins, the common cardinals, descending from the dorsal regions of the body. Their entrance causes an enlargement at the back of the heart, which becomes the sinus venosus. Meanwhile, the more anterior part of the heart tube shows a series of en-

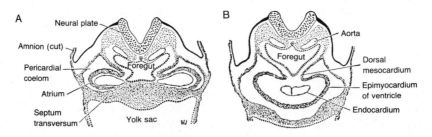

Figure 355.   Cross sections of a mammalian embryo to show an early stage in heart formation, before the fusion of the two subintestinal vessels from which the heart forms. In the atrial region, *A,* the two are still widely separated; farther forward, in the ventricular region, *B,* the two tubes are apposed inside a single pericardial sac, but are not yet fused. (From Arey.)

largements and intervening constrictions that delimit, in linear order from back to front, atrium, ventricle, and conus. Simultaneously there begins the elongation and twisting of the tube noted in the description of the adult chambers, so that the ventricle comes to lie posteroventral to the atrium, and somewhat to the right, rather than in its morphologically primitive anterior position.

With these developments, the embryonic heart has attained the general structural features found in typical fishes and needs little except valvular development to reach the adult piscine condition. In higher vertebrates, however, further differentiation is needed. We have noted phylogenetically a series of progressive stages in the subdivision and rearrangement of heart chambers in adjustment to lung breathing and the consequent necessity of separating two blood streams. In the various tetrapod classes, ontogenetic events essentially recapitulate the phylogenetic stages. In a mammalian heart, for example, principal developments include (1) subdivision of both atrium and ventricle and of the original single opening between them; (2) absorption of the sinus venosus into the right atrium; (3) the segregation of the pulmonary veins so that they open into the left atrium; and (4) subdivision of the conus arteriosus into a pulmonary trunk leading from the right ventricle and a major aorta leading from the left. As noted later, however, the different breathing mechanisms before and after birth in amniotes cause certain embryologic developments in the heart that have no phylogenetic significance.

Blood Circuits

In earlier sections of this chapter, the components of the blood circuit have been described piecemeal. We shall here briefly review the general evolutionary history of the circulation as a whole, with particular reference to blood pressures and capillary nets (Fig. 356).

As in any passage of liquid through tubes, the friction of the liquid on the vessel walls tends to lessen the pressure given by the pump; the capillaries are, of course, the parts of the system in which the loss of pressure is the greatest. In fishes, in general, every drop of blood that leaves the heart must pass through at least two capillary systems before its return to that organ, because blood first passes through a capillary network in the gills, where there is a sharp drop in pressure, before it can reach the aorta and pass thence to the capillary system of the ordinary body tissues. But before its return to the heart much of the blood passes, in addition, through a portal system—the hepatic portal in all forms, and a renal portal system in fishes above the cyclostome level. In this situation, the blood passes through three, rather than two, capillary systems, losing much pressure in each.

With the introduction of the pulmonary circuit into the circulatory system and the abolition in adult amphibians of gill-breathing, circulatory efficiency is greatly promoted. The branchial capillary system is eliminated, and hence all body tissues may be reached directly with little loss of pressure. Blood passing from heart to lungs and much of the body tissues completes the circuit back to the heart without encountering more than the one capillary system en route. Blood to the posterior end of the body or to the intestine, however, still has to pass through a portal system, renal or hepatic, on its return; but, even here, there are only two capillary drops in pressure, rather than three.

With reduction of the renal portal system in reptiles and birds and its abolition

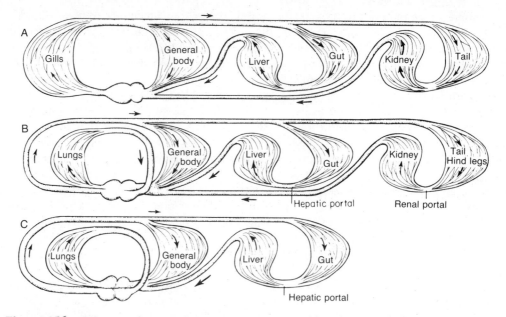

Figure 356. Diagrams showing the general nature of the blood circuits and capillary networks encountered in *A,* a typical fish; *B,* a terrestrial amphibian or a reptile, with elimination of gill circulation and introduction of pulmonary circuit; *C,* a mammal, with elimination of renal portal system and complete separation of the systemic and pulmonary circuits.

in mammals, the efficiency of the circulation is further increased. In mammals, only the intestinal circulation passes two capillary networks; blood to every other region* goes directly from and to the heart, the only capillary system encountered being that in the tissues supplied.

The substitution of lungs for gills has, seemingly by a happy accident, brought about, in the long run, the development of a more efficient circulatory system. On the other hand, lymphatic development in higher vertebrates may perhaps be a response to the increased difficulties in drainage of the tissue, with increased pressure in the vessels of the blood-vascular system.

Embryonic Circulation

In earlier sections of this chapter, we mentioned the formation of various blood vessels in the embryo. Here we may attempt to gain a general picture of the development of the circulatory system (although this necessarily involves some repetition), with attention to certain of the vessels not retained in the adult but necessary in the sacs and membranes of embryos of large-yolked types. It must be emphasized that the circulatory system cannot develop in the embryo merely with the aim of producing the adult structure; it must be functionally effective at every moment of every

---

*Minor portal systems, like that in the pituitary, do exist, but the amount of blood concerned is minimal.

embryonic or larval period. In general, too, we may note that in many areas (as in the limbs, Fig. 357), the circulation first develops in the form of a diffuse network, in which major vessels are sorted out only at a later stage.

A generalized pattern of vascular development (Fig. 358) is shown by such lower vertebrates as amphibians and lungfishes, in which the egg is not heavily yolked, normal body form is soon attained, and the picture is not complicated by the presence of intricate extra-embryonic membranes. The basic plan of the vertebrate circulation is one in which (in contrast with the annelid scheme) the blood flows forward ventrally beneath the gut, turns dorsally, and then runs posteriorly above the gut. The earliest blood vessels to appear in an amphibian or lungfish are portions of the ventral part of this system. Beneath the yolk-laden gut, small vessels form and run forward, assembling into a pair of vitelline veins that fuse, anteriorly at least, to form a subintestinal vessel. The most anterior part of this vessel forms the ventral aorta; the succeeding section soon develops into a heart. Posterior to this point, the ventral vessel is later invaded by liver tissue and disrupted here into a system of sinusoids. Between this liver "portal" and the heart, the ventral trunk becomes the hepatic vein; posterior to it, there develops the hepatic portal system, which continues to drain the gut in the adult.

But before the ventral vessels have proceeded far in their differentiation, there appears a dorsal arterial trunk that, coursing backward, completes the return circuit of the blood to the walls of the gut. From the front end of the subintestinal vessel, in front of the developing heart, blood vessels arch upward around either side of the foregut to the dorsal part of the body. Here, below notochord and nerve cord, they turn backward above the gut and send off branches by which the blood descends around the digestive tube to enter once more the vitelline system. The anterior part of the ventral vessel becomes the ventral aorta. The pair of loops ascending around

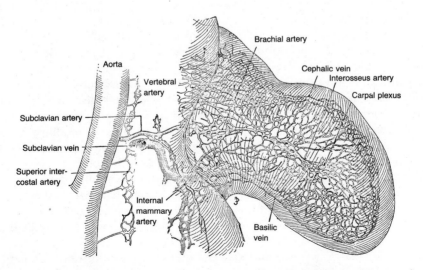

Figure 357. An early stage in the development of the forelimb of a pig embryo, to show the manner of formation of patterns in limb circulation. There is a network of interweaving small vessels from which the main adult vessels develop. Choice of one definitive channel or another allows for the occurrence of variants as anomalies. (From Woollard.)

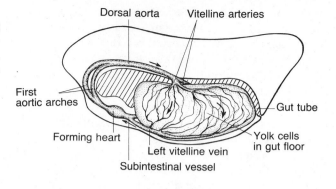

Figure 358. Diagram of the general circulation in a young frog tadpole. The food supply still lies in the yolky gut cells, and the vitelline circulation is of great importance.

the foregut are the first of the aortic arches of the pharynx (Fig. 339). As development proceeds, additional aortic loops develop more posteriorly. These may break up along their course into capillary systems as the gills develop; the anterior arches usually atrophy or undergo modification even in early stages of development. The dorsal vessel is the dorsal aorta, and its anterior continuations become the internal carotids. At its first appearance, the aorta may be a paired structure, and its derivatives remain paired in the most anterior part of the cephalic region in the adult; for most of the body length, the two vessels fuse, however, at an early stage to form a single median longitudinal aortic vessel. Its first branches, which descend around the gut, are paired to begin with; as the gut diminishes in size relative to the body as a whole, these vessels fuse to descend the mesentery as median arteries, which become the coeliac, mesenteric, and so on, of the adult.

With the development of the subintestinal and aortic vessels and their connections, a primary circulatory system is completed. This, it will be noted, is essentially a visceral circulation, associated primarily with the gut. With growth of mesodermal structures and of the nervous system, there develops the necessity for circulatory vessels in the outer body tube—a somatic circulation. The afferent arterial supply in this new system is still the dorsal aorta, which sends out lateral branches into the axial musculature, kidneys, and gonads, and upward to the neural tube. From the capillary networks developed at either side of the aorta arise return venous vessels as the paired cardinal veins, anterior and posterior; these gain a discharge into the heart by forming a descending trunk, a common cardinal, down either side of the thoracic region. In higher vertebrates, the anterior cardinals are slightly modified to form the anterior vena cava—double or single. The posterior cardinals, we have noted, undergo important changes in lungfishes and tetrapods to become the posterior vena cava; these changes include, in every embryo, the interpolation of a renal portal system, which is eliminated again in the adult stage of mammals. Other elements in the somatic circulation include an arterial supply to, and venous return from, the fins or limbs, and, in most lower vertebrates, the development of abdominal veins that drain the ventral part of the body wall.

In large-yolked types—elasmobranchs, amniotes—we find an extra-embryonic circulation composed of vessels that run from and to the yolk sac and, in amniotes, connect the embryo with allantois or placenta. The presence of these two structures complicates the embryonic story (Fig. 359).

In elasmobranchs and amniotes alike, there is at first no "floor" of yolk-filled cells

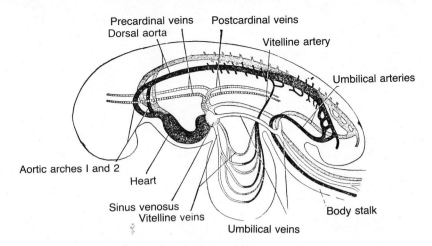

Precardinal veins   Postcardinal veins
Dorsal aorta
                              Vitelline artery

                                          Umbilical arteries

Aortic arches I and 2

Heart

Sinus venosus
Vitelline veins   Body stalk

                    Umbilical veins

Figure 359.   The circulatory vessels in a mammalian embryo. (After Arey)

to the gut cavity but, instead, a mass of nutritive yolk; in consequence, the paired **vitelline veins**, carrying nutriment inward to the body, arise on either side in the roof of the yolk sac. There can be no formation of a single ventral subintestinal trunk until a relatively late stage in embryonic development; ventral vessels (and often the heart as well) persist for a time as paired elements. The return vessels from the dorsal aorta likewise long remain paired structures as **vitelline arteries**. A highly developed system of vitelline veins and returning vitelline arteries is present throughout embryonic life in elasmobranchs, reptiles, and birds. In mammals, despite the absence of yolk in the yolk sac, similar vessels develop early and persist for some time.

In amniotes, a second series of embryonic vessels of note extend along the stalk of the allantois, which in reptiles and birds serves as an embryonic lung. Paired **allantoic** or **umbilical veins** enter the body of the embryo and course forward in the ventral part of the lateral body walls; they appear to represent, within the body, the lateral abdominal veins of lower vertebrates. In the early embryo, these veins course directly to the heart, as do the abdominal veins of a shark. In later development, however, they turn inward anteriorly to enter the liver and become (together with the vitellines) part of its portal system (Fig. 360), as does the abdominal vein of the adult amphibian or reptile. Apparently, however, the flow of blood of the combined vitelline and umbilical veins is too great for proper handling by the liver sinusoids, and, in embryonic amniotes, much of the blood passes through the liver by a direct channel—the **ductus venosus**. Corresponding to the umbilical veins, paired **allantoic** or **umbilical arteries** arise from the dorsal aorta posteriorly and, in strong contrast to the vitelline vessels, descend to the allantoic stalk in the body walls rather than via the gut. In mammals, the allantoic stalk becomes the connection between placenta and embryo, and the allantoic vessels are of the utmost importance, bringing food as well as oxygen to the embryo; in this class of vertebrates, the yolk sac circulation diminishes with the gradual absorption of the yolk, and the vitelline vessels disappear.

The allantoic or umbilical vessels of amniotes persist until birth but abruptly cease their function at that time. They are shut off at the points where they leave the

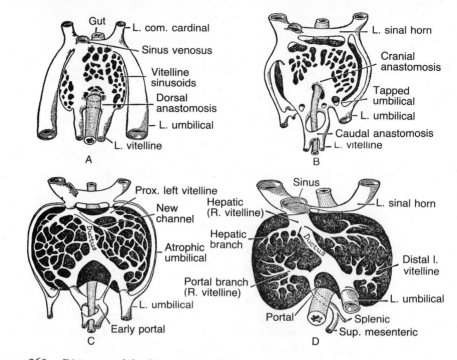

Figure 360. Diagrams of the liver region of human embryos at successive stages (4.5, 5, 6, and 9 mm in length), seen from the ventral surface, to show developmental changes in the vitelline and umbilical veins. The gut tube is stippled. In *A,* the vitelline vessels from the yolk sac are well developed and pass through the liver tissue; in *B–D,* their transformation into the portal system is seen. The umbilical veins (from the placenta) are already well developed in *A* but run directly to the sinus venosus. In later stages, this blood flow is diverted to the hepatic circulation, much of it flowing through this organ via a ductus venosus. The right umbilical vein is reduced; the left persists until birth, when this vessel and the ductus venosus undergo reduction. The posterior cardinal system would be dorsal to the liver and is not shown. (From Arey.)

body, and their internal courses are soon obliterated, as is the channel of the ductus venosus through the liver.

In amniotes, oxygen is supplied to the embryo by the umbilical veins. The abrupt entrance of the pulmonary circuit into the picture at the time of hatching or birth calls for special modifications in the embryonic circulation. In the embryo, the pulmonary arteries and veins develop at a fairly early stage, but the blood for the most part by-passes these vessels. We have noted that the upper part of the pulmonary arch, from which the pulmonary arteries proper take origin, remains open in amniote embryos so that blood that takes this course from the heart need not enter the lungs but, instead, may proceed to the dorsal aorta and enter the general body circulation. This ductus arteriosus is paired in reptiles and bird embryos and present only on the left side in mammals (cf. Fig. 339 *E, F*). At the other end of the circuit, we find that, in the amniote heart, the left atrium has no proper blood supply in the embryo, because the only vessels entering it are the pulmonary veins, which carry almost no blood. This situation is remedied by the fact that, in all amniotes, the septum between the two atria is imperfect in the embryo: blood from the right atrium may cross over and fill the left atrium sufficiently to enable it to function properly.

At birth there is a sharp change in the pulmonary circuit. The ductus arteriosus closes rapidly in birds and mammals and most reptiles, so that all blood entering the pulmonary arch is forced into the lungs. The return of this blood to the heart furnishes the left atrium with its proper share of the blood stream, and the pressure within it helps quickly to close the opening between the two atria.

Chapter 15

Sense Organs

All cells, one may believe, are capable of receiving and responding to stimuli that measure, so to speak, some condition in the environment or a change in such condition. If, however, the proper response to a sensory stimulus is one that should be performed by a distant region of a vertebrate, or by the animal as a whole, reception of that stimulus is in vain unless there is some channel of communication between the sensory receptor and the organs—muscles or glands—that should make the appropriate response. Such communication may be made by hormonal action, but, in general, the mechanism used is that of the nervous system. The tips of nerve fibers are themselves capable of direct excitation, but, more often, the reception of sensation in vertebrates is the function of specialized **sensory cells**, mainly of ectodermal origin and generally grouped in more or less complex organs. These are attuned to physical or chemical stimuli of specific types and are associated with nerves that relay these stimuli to specific centers in the brain or to the nerve cord.

Anatomists divide such sensory nerves into two groups: the **somatic sensory nerves**, carrying impulses of a sort that in ourselves usually reach the level of consciousness, from the "outer tube" of the body—the skin and body surfaces and the muscles; and **visceral sensory nerves**, whose impulses, seldom reaching our consciousness (with the obvious exceptions of stomach-aches and the like), arise from the viscera. The physiologist customarily classifies sensory receptors in a manner that for the most part fits readily into the neurologic scheme. **Exteroceptors** are those sensory structures of the skin and special senses that receive sensations from the outer world; **proprioceptors** include those situated in the striated voluntary muscles and tendons; **interoceptors** are those located in the other internal organs. The first two of these correlate fairly well with the somatic system of sensory nerves, the third with the visceral sensory system of the anatomist.

Simple Sense Organs

One ubiquitous sensation, that of **pain**, seems not to need any special organ for its reception. Pain, as it appears in our own perception, is not specific but may be due to a variety of physical and chemical causes. This sensation is produced by direct stimulation of the end fibers of sensory nerves. Such nerve endings, in the skin at least, appear to be also sensitive to touch in vertebrates generally. In fishes and amphibians, most cutaneous sensations, apart from those from the special sensory organs, appear to be picked up by free nerve endings.

Sense Corpuscles. Many "simple" sensations are received by small sensory structures that may be present in almost any part of the body—in skin, muscles, viscera. These are most highly developed in mammals and birds; in contrast, amphibians have only a few sensory corpuscles, and none are known in fishes. Most corpuscles are of such tiny size that their study lies in the realm of the histologist with his microscope rather than that of the dissector with his scalpel. Their appearance and structure vary greatly (Fig. 361). In general, we see a series of specialized cells with which are associated tangled skeins of terminal nerve fibrils. Such corpuscles are commonly present in the connective tissues, but similar sensory structures may be present in epithelia (special tactile cells in the mammalian epidermis, for example).

With small and varied bodies of this sort, accurate determination of function is difficult. Some knowledge can, of course, be gained from our own sensory experience and can be extrapolated to give us some idea of the sensory situation in other mammals and, with lesser chance of correctness, in other vertebrate classes. Among mammals, at least four types of sensations, in addition to pain, may be distinguished in the dermis: warmth, cold, touch, and pressure. It has been thought that these types of reception are associated with corpuscles of different sorts, but this conclusion appears to be uncertain in many cases. Quite surely, however, the widespread pacinian corpuscles (Fig. 361 *C*) are associated with pressure. They are relatively large, up to as much as a millimeter in length, and are built in onion-like fashion of concentric layers of connective tissue. In mammals, complexly coiled nerve endings surrounding hair bases, notably vibrissae, are important in touch reception. A number of types of

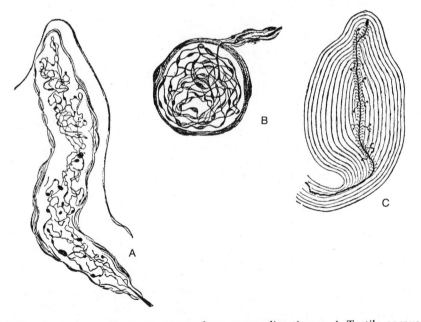

Figure 361.   Some types of sensory organs from mammalian tissues. *A,* Tactile corpuscle (Meissner's corpuscle) from the connective tissue of the skin; *B,* an end bulb of Krause, sensitive to cold; *C,* a Pacinian corpuscle, which registers pressure and tension. (From Ranson, after Dogiel, Sala, Böhm-Davidoff, Huber.)

interoceptive corpuscles are present in and among the viscera, but it is difficult to tell the nature of the sensations to which they respond. Structures along vessels of the aortic arch system (particularly the carotids in mammals) monitor the blood, probably responding to both pressure and chemical (carbon dioxide) stimuli.

Proprioceptive sensations, of the physiologist's terminology, include those obtained from sensory structures located in the striated muscles and tendons as **muscle spindles** and **tendon organs**. Proprioceptive organs are unknown in fishes but are present in tetrapods generally. In lower tetrapods, sensory nerve fibrils simply twist about individual muscle fibers or tendon fibers. In mammals, however, muscle spindles consist of small and somewhat specialized muscle fibers surrounded by a tangle of sensory nerve endings and enclosed in an elongate connective tissue sheath (Fig. 362). These muscle and tendon structures are the seat of "muscle sense." They furnish to the central nervous system a report on the state of the contraction of the muscles—data necessary before any further action of the muscle can be properly ordered by way of motor nerves. In addition, these spindles are the source of the animal's information concerning the position in space of the various parts of the body—information, as we are ourselves well aware, that can be furnished without the aid of other sensory structures but confined (in the absence of contact with other objects) to parts of the body containing striated muscles or their tendons.

Taste. Most of the simpler senses are responses to physical stimuli; taste is a response to chemical stimulation. The skin of all primitive water-dwelling vertebrates is sensitive to some degree to chemical stimuli, the receptors apparently being simply the tips of sensory nerve fibers. In the more highly developed sense of taste, there are developed small organs, the **taste buds**, which usually consist of barrel-like collections of elongated cells sunk within the epithelium, usually ectodermal (Fig. 363 *A*). Two cell types are present, supporting and sensory. Each taste cell has a sensitive, hairlike projecting process. Taste buds are for the most part confined to the mouth and pharynx. In mammals, they are in great measure concentrated on the tongue, where they are commonly present in folds surrounding small lingual elevations, the **lingual papillae**. Taste buds are rare on the tongue in reptiles and birds; instead, most are found in the pharynx. In fishes and amphibians, they may also develop in the skin. In a sturgeon, for example, they are abundant on the under side of the projecting rostrum, so that as the fish glides over the bottom, it can obtain a foretaste of potential food before the mouth reaches it. In some catfishes, notably the "bull-head" *Amiurus,* there is a tremendous spread of taste buds over the surface of the body, giving a phenomenal possibility of pleasant (or unpleasant) gustatory sensations. It must be noted that much of what we casually think of as taste is actually a

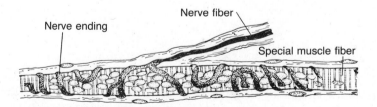

Figure 362. A muscle spindle. (After Windle, Textbook of Histology, McGraw-Hill.)

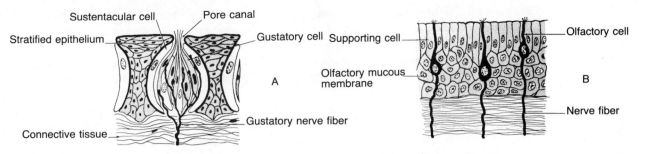

Figure 363. Microscopic sections through chemical sense organs. *A*, A taste bud; *B*, nasal mucosa. (From De Coursey, The Human Organism.)

smelling of oral contents (foods do not "taste" as good when a head cold clogs the nose) and, in addition, the "feel" of the food in the mouth. All taste buds look alike, but in such animals (including ourselves) as have been experimentally tested, a limited number of types, distinct in regard to reception, are present. Their differences lie in their responses to specific sorts of molecules or ions in solution in the liquids of the mouth (or body surface). In ourselves, for example, sensations of sour, salty, bitter, and sweet can be distinguished and some fishes, at least, have similar discriminatory powers; but the frog, it seems, distinguishes sour and salty only. Acids and acid salts are responsible for sour taste; inorganic salts, particularly the chlorides, produce salty tastes. Bitter taste is caused particularly by alkaloids; sugars and alcohols give the sweet sensation.

Complex Sense Organs. In the following sections, the more prominent of the complicated sensory structures of vertebrates, including nose, eye, ear, and lateral line organs, are described. The first three of these are familiar as part of our own sensory apparatus; the last type of organ we do not ourselves possess, but it is so important in fishes that it cannot be overlooked.

Presumably there is a variety of still other sensory structures in other vertebrates, particularly the lower classes of them, which give responses of types unfamiliar to us and hence difficult for us to understand. One such sense, about which we do have data, lies in the **pit organ** of the so-called pit vipers among snakes, of which the rattlesnake is a familiar (although not popular) example. Between eye and nose on either side of the face is a pit filled with a peculiar, highly vascular tissue with numerous nerve endings. Experiments show that this is a heat-sensitive organ; a pit viper can perceive the movement of a moderately warm body past its head at a distance of a meter or two—a sensory power extremely useful to this animal, which makes its living mainly by capturing warm-blooded rodents and is not too well endowed with the other sense organs. A series of tiny pits along the jaw margins of pythons and some boas have similar sensory properties.

The Nose

In tetrapods, the nose, with an internal opening, has become associated with breathing, but its primary function, of course, is that of olfactory reception. In certain vertebrate groups, smell is relatively unimportant; although there are some conspicuous

499

exceptions, it is not in general highly developed in teleost fishes, and is feeble in most birds, whales, most bats, and higher primates, including man; most are groups in which the eyes have become the dominant sense organs. Among vertebrates in general, however, smell is in many ways the most important of all senses. From cyclostomes and sharks up through the more primitive bony fishes, amphibians, reptiles, and especially among most mammals, smell is a main source of information concerning the outside world—the location of enemies, friends, and food—and produces responses based on previous olfactory experiences. That we ourselves have this sense feebly developed should not blind us to its general importance; the fact that, as we shall later see, the most highly developed mammalian brain centers arise in a part of the brain primarily connected with smell is testimony to the part olfaction has played among the vertebrates.

In a majority of fishes, the nasal structures consist of a pair of pockets placed well anteriorly on the head and without any internal opening to the mouth. Typically, they are lateral in position; in sharks they are placed on the under surface of the snout. The external opening of each sac is paired, as in typical teleosts, where water may flow inward through an anterior and out through a posterior opening, or partially subdivided into two openings by a fold of skin, as in sharks. The floor of the nasal sac commonly bears a series of ridges, the **olfactory lamellae**, arranged in some type of rosette pattern (Fig. 364). In many teleosts, there may be accessory sinuses off the

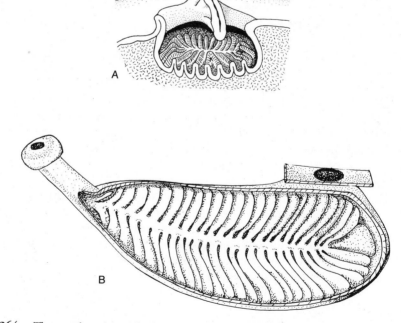

Figure 364. The nasal cavities of teleosts, opened to show the olfactory lamellae in the floor of the sac. *A,* A minnow (*Phoxinus*); *B,* an eel (*Anguilla*); in both, the anterior external nostril is to the left, and the posterior to the right. The vertical flap in the minnow helps deflect a current of water through the nasal cavity. (After Giersberg and Rietschel, and Liermann.)

main nasal sac; these serve as aspirators to help pull water over the sensory area. A few fish show very different structures—both *Polypterus* and the living coelacanth (*Latimeria*) have an odd system of parallel tubes. The olfactory lamellae are typically covered by a simple columnar epithelium. This epithelium contains, in all vertebrates, **olfactory cells**, interposed with supporting elements (Fig. 363 *B*) and, except in fishes, generally includes mucous cells as well. Each olfactory cell has a rodlike extension to the surface, which, at its tip, bears a radiating brush of short, hairlike processes. In one remarkable feature these cells differ from typical receptor organs in vertebrates. Other receptor organs depend upon nerve fibers to relay inward the sensations received. The olfactory cells, on the contrary, do their own work; a long fiber extends from each cell into the olfactory bulb of the brain. This type of organization, with a sensory cell itself serving for nerve conduction as well, is found in various instances among invertebrates and even in amphioxus and is suggestive of the antiquity of the sense of smell in vertebrate history.

The olfactory cells are sensitive to minute amounts of chemical materials, derived from distant objects, present in solution in the liquid covering the nasal epithelium. Despite much work on the part of physiologists, we have few data as to the precise nature of olfactory sensations of different sorts. Possibly olfactory discrimination is due to the presence of submicroscopic pockets of different shapes at the tips of the sensory cells; stimulation of a cell may be caused by the presence of a molecule whose shape enables it to fit into one or another type of pocket.

The jawless vertebrates present a puzzling structural situation. In contrast to that of all other living vertebrates, the cyclostome olfactory organ is a single median pouch, opening either at the tip of the snout, as in hagfishes, or far back on the upper surface of the head, as in lampreys (Figs. 17, 256). A further peculiarity, already noted, is the fact that the hypophyseal sac is combined in a common opening with the nostril. In the hagfish, the nasal pouch breaks through internally into the pharynx, presumably an aid to breathing when, in this parasitic type, the mouth is otherwise engaged.

Is the dorsal single-pouched structure of the cyclostome primitive or specialized? The answer is none too clear. In the ancient ostracoderms, two of three main groups show the lamprey condition; but in a third group (the Heterostraci), the nostrils were probably ventral and paired, so that both situations were present at an early stage in vertebrate history. Some embryologic evidence can be adduced for a belief that the cyclostome condition is not primitive. In the embryo lamprey (Fig. 229), the nasal pocket is at first ventral in position and only later migrates dorsally; in the larval lamprey, the nasal sac is bilobed, although not distinctly paired. Even in the adult, it is partly divided into right and left halves, and its innervation is paired.

In earlier chapters, we commented upon the evolution in the typical crossopterygians of a type of nostril with an opening into the roof of the mouth as well as to the exterior,* and the utilization, in tetrapods, of this passage as an adjunct to air breathing. In typical fishes, the nasal sac is filled with water; in tetrapods, air replaces water, but the airborne chemical particles are absorbed into a film of liquid that covers the olfactory mucous membranes. Specific glands develop to keep the membranes moist with water and fatty material. Liquid brought from the eye by the lacri-

---

*Such a pattern has also arisen quite independently in a very few teleosts; whether the connection between the mouth and nose in lungfish is comparable is currently under debate.

mal ducts (probably also nasal in origin—the old posterior opening of the teleost nose) aids as well.

In amphibians (Fig. 365 A, B), the main air channel is usually simply constructed. Each small **external naris** opens into a somewhat elongated sac; posteroventrally, a large opening, the **choana** or **internal naris**, opens directly into the front part of the roof of the mouth. In fishes, practically the entire ventral surface of the olfactory pouch is covered with sensory epithelium, which is thrown into numerous small folds. These folds are absent in amphibians; the surface of the main nasal sac, used in great measure as an air passage, is almost smooth, and part of the surface is covered by a nonsensory epithelium. Such use of the nasal passage in respiration occurs in no living jawed fish—lungfish and some teleosts do have connections between the nasal and oral cavities, but they appear never to be used in respiration. We may note, in passing, peculiar olfactory tentacles developed from the nostrils of the wormlike gymnophionan amphibians.

In typical reptiles, the nasal region begins to assume a more complicated structure (Fig. 365 C). The air passage is longer, the choana generally more posteriorly placed, and there is usually a distinct, small anterior sac, or **vestibule**, through which inspired air passes before it reaches the main olfactory chamber. Of this main chamber, the ventral part, the direct air passage toward the choana, is usually lined by a nonsensory epithelium. The expanded dorsal part contains the sensory area; here we find typically developed one or more swollen, curved ingrowths, **conchae** or **turbinals** from the lateral (never the medial) wall of the chamber; this increases the surface over which sensory epithelium may form. Such conchae are not present in turtles. Beyond the main chamber a constricted area, the **nasopharyngeal duct**, leads to the choana. This duct is short in most lizards, snakes, and turtles. In some of the last group, however, there may be some development of a secondary palate and, in relation to this, some elongation of the duct. In the crocodilians, with a long snout and

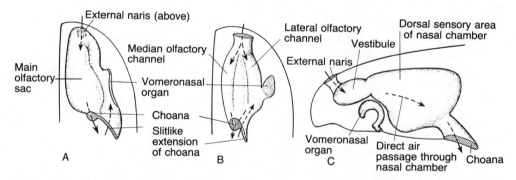

Figure 365. *A,* Ventral view of anterior part of the left side of the palatal region of the salamander, *Triturus,* with the nasal chambers shown as solid objects, the remainder as transparent structures (cf. Fig. 230). *B,* Similar view of the toad *Pipa. C,* Longitudinal section of the nasal region of a lizard, cut somewhat to the right of the midline, to show the cavities of the nasal apparatus. In the embryo lizard, the vomeronasal organ was a medial pocket of the main nasal channels; in the adult (as in many mammals), this organ has separated to open independently into the roof of the mouth by a nasopalatine duct. Arrows show the main air flow inward in all figures and the outward flow toward the vomeronasal organ in the amphibians. (*A* after Matthes; *B* after Bancroft; *C* after Leydig.)

an elongate secondary palate, the nasal chamber, containing three conchae, is itself considerably elongated, and the duct is a long tube.

In birds generally, smell appears to play an exceedingly small role in sensory reception. As would be expected, vision is dominant, smell being of little possible value to a flier. The nasal structures, built on the general reptilian plan, are of modest size, and the olfactory epithelium is generally restricted to a small area in the posterodorsal part of the nasal chamber. Birds, however, retain three nasal conchae similar (but not completely homologous) to those seen in their crocodilian relatives, and these may develop into complicated scrolls. The external nares, generally situated at the base of the beak, are of varied form and, as in fowls and doves, are often protected by a horny structure something like a small eyelid. We have mentioned previously (cf.p. 402) the presence of salt-secreting glands in the avian nose; similar glands occur in almost all tetrapods, but they secrete salt only in birds and a few reptiles.

Among mammals, there is reduction of olfaction in whales, microchiropteran bats, and higher primates, but nasal structures reach a peak, as far as size is concerned, in typical forms (Figs. 230, 366). In lower vertebrates, they are generally confined to a short region at the anterior end of the head; in typical mammals (higher primates are exceptional), they push far backward to the orbital region and may occupy more than half of the cranial length. Their size is augmented through the fact that there has developed in mammals a secondary palatal structure. As a result, the original vault of the mouth is enclosed in the nasal area; this former region of the mouth forms the ventral part of the main nasal cavity and posteriorly forms the long nasopharyngeal duct, carrying air backward to the choana.

These new additions to the nasal area function primarily as aids to respiration. There is, however, a considerable development of olfaction. In reptiles, there tends to be a concentration of sensory epithelium in the upper part of the nasal chamber. This is true in mammals as well. The lower part of the system, forming a straight passage from external naris to nasopharyngeal duct and choana, is an air passage; the sensory areas are concentrated in the posterodorsal parts of the chamber.

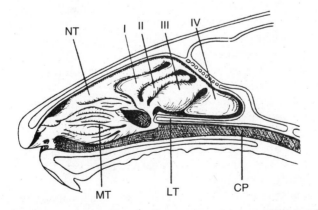

Figure 366. A section through the nasal region of a guinea pig, slightly to the right of the midline, to show the development of the turbinals, including the maxilloturbinal, *MT,* nasal turbinal, *NT,* and ethmoturbinals (I-IV). *CP,* choanal passage; *LT,* lamina transversalis, a remnant of the original primary palate, separating the sensory area of the nasal fossa from the choanal passage. (After Cave.)

The nasal chamber is, however, by no means a simple sac. We have noted the development in reptiles and birds of folded ingrowths from the lateral wall of the nasal chambers, the conchae or turbinals. In mammals, these are developed to a high degree as coiled scrolls of bone, covered with epithelium, which subdivide the air passage into a series of partially separated channels. Typically there are two to four turbinals developed from the ethmoid bone, and one each from the maxilla and nasal. Ventrally, these structures act as "air conditioners," filtering and warming incoming air; dorsally and posteriorly, they afford, as in reptiles, additional sensory surface.

Most therian mammals (and a few reptiles, such as crocodilians) possess extensions of the nasal air cavities into the adjacent bones as **sinus pneumatici** (plural), frequently present in the adjacent maxillary, frontal, ethmoid, and sphenoid regions, and reaching to the occiput in the high-domed head of the elephant and in the heads of a variety of other mammals. These sinuses effect a lightening of the skull; other distressing human attributes are perhaps all too familiar to some readers.

In numerous tetrapods, there is a specialized part of the olfactory system termed the **vomeronasal organ** (organ of Jacobson, Fig. 365), which in most cases appears to have as its main function the picking up of olfactory sensations from the food in the oral cavity. Its predecessor is seen in simple form in urodele amphibians. There a partially separated grooved channel runs along one side of each nasal sac and opens out into the roof of the mouth in a slit continuous with the choana. This groove bears an area of olfactory epithelium quite separate from that of the main part of the nose; its sensory fibers form a distinct branch of the olfactory nerve, which leads to a distinct, accessory olfactory lobe of the brain. In the other two amphibian orders, this pocket is almost completely separated from the rest of the nasal system as a blind pouch opening into the main nasal cavity.

The vomeronasal region is not well separated from the rest of the nose in turtles, and it is lost in adult crocodilians and birds. In *Sphenodon*, it occupies a club-shaped blind pouch that opens into the choana. In lizards and snakes, the vomeronasal organs are well developed and during ontogeny come to be completely separated from the normal nasal structures; the two organs are separate pouches that open independently into the roof of the mouth anterior to the choanae. Use in snakes is peculiar. The tongue, cleft into two prongs and darting in and out of the mouth, serves as an accessory olfactory organ. When the tongue is withdrawn, the tips are placed near (but *not* into) the entrance to the vomeronasal pockets; chemical particles that adhered to them in the air are transferred to the moisture on the sensory epithelium of these pouches.

Among mammals, the vomeronasal organ is absent in man and other higher primates, in many bats, and in various aquatic forms. In almost all other groups of mammals, however, it persists as a functional organ. It is a small, cigar-shaped structure, buried in the floor of the nasal region toward the midline. In many rodents, it opens into the main nasal cavity. In most other cases, however, the vomeronasal organs connect independently with the mouth, much as in lizards and snakes, by paired **nasopalatine ducts** that pierce through the secondary palate and appear to retain their original mouth-smelling functions.

The Eye

The vertebrate body is subjected constantly to radiations that may vary from the extremely short but rapid waves of cosmic rays and those from atomic disintegration to the long, slow undulations used for radio transmission. Many of these radiations affect protoplasm, but specific sensitivity to them is usually limited to a narrow band part way between the two extremes; for knowledge of other wavelengths we must resort to mechanisms that transform their effects into terms receivable by our own limited senses. It is reasonable that the animal band of sensitivity corresponds in great measure to the range of radiations reaching the earth from the sun, because that body is the source of the vast bulk of the radiations that normally reach us. Of this band, slower waves are received as heat; faster, shorter waves are perceived as light—the process of **photoreception**.

Specific sensitivity to light is widespread through the animal and plant kingdoms. In many invertebrates, such sensitivity is concentrated in special cells, or cell clusters, frequently with associated pigment, which form "eye spots." Often there is evolutionary progress to an organized photoreceptive structure, with a lens for concentrating light upon sensitive cells contained in a closed chamber. Many such organs are merely receptors of light; with better organization and the arrangement of the sensory cells in a definite pattern so that they receive light from specific external areas, true vision results. Well-developed eyes with many common features, but surely independently evolved, are found in forms as far apart as molluscs, arthropods of various sorts, and vertebrates.

The general composition of the vertebrate eye (Figs. 367, 368) may first be noted before consideration of details. The essential structure is the roughly spherical **eyeball**, situated in a recess, the **orbit**, on either side of the braincase, and connected with the brain by an **optic nerve** contained in a stalk emerging from the internal surface. The eyeball has an essentially radial symmetry, with a main axis running from inner to outer aspects. Internally, there is a set of chambers filled with watery or gelatinous liquids. In the interior, well toward the front, lies a **lens**. The walls of the hollow sphere are formed basically of three layers, in order, from the outside inward, the **scleroid** and **choroid** coats, and the **retina**. The sclera is a complete sphere; choroid and retinal coats are incomplete externally. The first two are of mesenchymal origin and are essentially supporting and nutritive in function; the retina (actually a double layer) includes the sensory part of the eye where visual stimuli are received and transferred to the brain via the optic nerve. At the outer end of the eyeball, the scleroid coat is modified to form, with the overlying skin, the transparent **cornea**. Externally, choroid and retinal layers are fused and modified. Opposite the margins of the lens, the conjoined layers usually expand to form a **ciliary body**, from which the lens may be suspended. Forward beyond this point, the two fused layers curve inward parallel to the lens to form the **iris** but leave a centrally situated opening, the **pupil**.

The general nature of the operation of the eye is commonly (and reasonably) compared to that of a simple box camera. The chamber of the eyeball corresponds to the dark interior of the camera box. In the eye, the lens (and in terrestrial vertebrates, the cornea as well) operates, like the camera lens, in focusing the light properly upon the sheet of sensitive materials in the back of the chamber; the retina functions in reception of the image as does the camera film. The iris of the eye is comparable to the similarly named diaphragm of the cameras, regulating the size of the pupil.

Development (Fig. 369). Embryologically, the most important functional parts of the eyeball arise from the ectoderm (including neurectoderm), but mesenchyme also enters prominently into the picture.

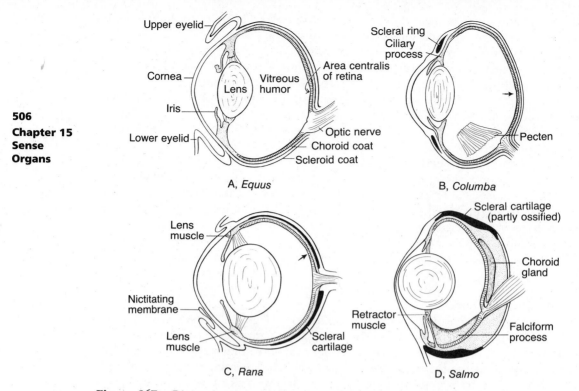

Figure 367. Diagrammatic vertical sections through the eye of *A,* a horse; *B,* a dove; *C,* a common frog; *D,* a teleost (salmon). Connective tissue of sclera and cornea unshaded; scleral ring or cartilage black; choroid, ciliary body, and iris stippled; retina hatched. Arrows point to the fovea. In *B* is shown the pecten, lying to one side of the midline. In *D,* the section is slightly to one side of the choroid fissure through which the falciform process enters the eyeball. (After Rochon-Duvigneaud, Walls.)

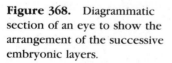

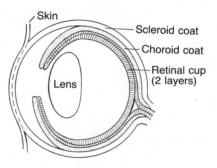

Figure 368. Diagrammatic section of an eye to show the arrangement of the successive embryonic layers.

The first indication of an eye comes at about the time of completion of the brain tube. Outward from the forebrain on either side grow spherical vesicles that remain connected by a stalk with the brain. As each **optic vesicle** develops, its outer hemisphere folds into the inner to form a double-layered **optic cup**. Ventrally, however, there is a slit in the cup, the **choroid fissure**, through which blood vessels enter

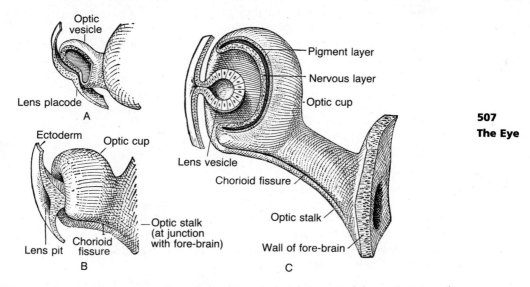

Figure 369. A series of diagrams to show the embryonic development of the optic cup and lens. (From Arey.)

during development; normally, except in teleosts, this fissure is later obliterated. The optic cup becomes the retina, which is thus primarily a two-layered structure; from the retina comes part of the ciliary body and iris.

The lens is the second main optic structure to appear. As the optic vesicle approaches the surface, the overlying ectoderm of the skin thickens, and a spherical mass or pocket of ectoderm detaches itself to sink into the orifice of the cup and forms the lens. It has been proved in a number of amphibians that the stimulus for lens formation is provided by the approach of the optic vesicle to the epidermis. If the optic vesicle in these forms is removed experimentally, no lens forms; if the vesicle is transplanted to (say) the flank, formation of a lens from the ectoderm of that region occurs. In certain other forms, however, induction by the optic vesicle is not necessary for lens formation.

The mesoderm contributes further structures of the eyeball. From the mesenchyme arises an inner sheathing layer, primarily vascular, to enclose the retinal cup as the choroid and to form much of the ciliary body and iris. Subsequently, an external coat of connective tissue forms a complete sphere about the organ as the sclerotic coat (or sclera) including the cornea. These sheaths correspond in great measure to the wrappings (meninges) that surround the brain—reasonably, since the developmental picture shows that the retina is in reality a peculiarly developed part of the brain itself.

Sclera and Cornea. The outermost sheath of the eye is the sclera, a stiff external structure that preserves the shape of the eyeball and resists pressure, internal or external, which might modify this shape. In cyclostomes and mammals, at the two extremes of the scale of living vertebrates, it consists entirely of dense connective tissue. In most groups, however, this is reinforced by the development of cartilage or bone. Often there is a cartilaginous cup enclosing much of the eyeball posteriorly; in birds, this sometimes ossifies. Further protection against pressures is afforded by the frequent presence of a **scleral ring** (Fig. 370), a series of bony plates, fixed in position, lying in the sclera in front of the "equator." Fossil evidence shows that such a ring was present in many primitive vertebrates and continued onward along all the main lines of vertebrate evolution. It has, however, disappeared in many groups, and is found today only in actinopterygian fishes, numerous reptiles, and birds. Primitively, it appears, the ring consisted in fishes of four plates, but it is reduced in living ray-finned forms to two elements (or to one or none). In the crossopterygian ancestors of land vertebrates and the older fossil amphibians, the number of plates was greatly increased, often to a score or more, and a high count is still present in reptiles and birds.

The superficial, external part of the sphere of sclerotic tissue is the translucent cornea, through which light enters the eyeball. Primitively, the cornea appears to have lain beneath the skin as a distinct structure, and this is still the case in lampreys. In all gnathostomes, however, the scleral coat and skin fuse inseparably in the adult; the cutaneous component of the cornea and the area of sensitive skin beneath the lids constitute the **conjunctiva**.

The refractive index of the cornea—its ability, that is, to deflect the course of light waves—is practically the same as that of water. Hence, in primitive water-dwelling vertebrates, the cornea has no power to act as a lens. In air, however, the curved cornea does much of the work of focusing and relieves the lens of much of this task. Its importance is shown by the fact that in ourselves many defects calling for optical correction, such as astigmatism, are due to departures of the cornea from its proper form as a segment of a sphere.

Choroid. The inner of the two mesodermal layers of the eyeball is the choroid coat. This contrasts strongly with the sclera, because it is a soft material containing numerous blood vessels that are particularly important for the sustenance of the retina. The choroid is pigmented and absorbs most of the light reaching it after penetrating the retina. But in addition, the choroid of many vertebrates, from elasmobranchs to a variety of mammals, includes a light-reflecting device, most familiar to us as seen in the ghostly eyes of a night-prowling cat illuminated by automobile headlights. This phenomenon is due to the **tapetum lucidum**, generally developed in nocturnal terrestrial animals and in fishes that live deep in the water. With plenty of light coming to the eye, internal reflection is not needed and is in fact harmful; it may confuse the details of the visual image. Where light is scarce, however, conservation of light rays, turned back to the retina by this mirror, more than makes up for the disadvantages.

Tapeta are variously constructed. In one type, common in hoofed mammals, the inner part of the choroid develops as a sheet of glistening connective tissue fibers that act as a mirror of sorts. In a second type, exemplified in carnivorous mammals,

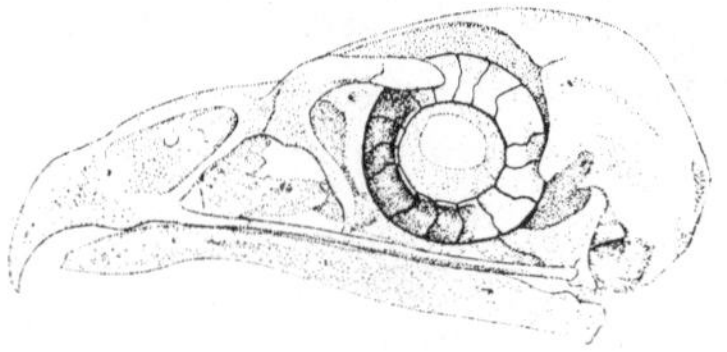

Figure 370. A skull of an eagle (*Aquila*), showing the scleral ring in place. (From Edinger.)

elasmobranchs, and some marine teleosts, the cells of this part of the choroid form an epithelium filled with fiber-like crystals of guanine (a material already mentioned in connection with pigmentation of the skin). Although a tapetum is normally formed in the choroid, some teleosts develop a guanine mirror in the pigment layer of the retina.

Iris. This structure, universally present, is formed by a combination of modified segments of both choroid and retinal layers of the eyeball. Both layers have, in this region, lost their most characteristic functions—vascular supply and photoreception, respectively. In attenuated form, they join to furnish an external covering for the lens and to outline the pupil—the restricted opening through which light penetrates into the inner recesses of the eyeball. The iris is always pigmented. Its inner (retinal) layer, for example, gives in man a blue effect, whereas brown pigments that may mask the blue are added by the outer (choroid) component. In a number of fish groups, the iris is fixed in dimensions, except where forward or backward movement of the lens may affect its distention and the consequent size of the pupil. However, in sharks, some teleosts, and tetrapods generally, muscle cells are present—striated fibers in most reptiles and birds, smooth in amphibians and mammals, and mixed in crocodilians. Arranged in circular and radial patterns, as sphincters and dilators, these fibers may contract or expand the pupillary opening. This gives the iris the function of a camera diaphragm; the pupil may be expanded in dim light for maximum illumination, contracted in bright light for protection of the retina and for more precise definition. Nocturnal forms frequently have a slit-shaped pupil that closes more readily to exclude bright light.

The nature of the muscle fibers in the iris is of theoretic interest. They are formed from the retinal component of the iris. This is a derivative of the neural ectoderm of the embryo. But muscle, smooth or striated, is supposedly formed by the mesoderm alone, and an ectodermal origin of muscle fibers is highly unusual. However, these cells in the iris have all the attibutes of muscle fibers, structurally and functionally. As is often the case, structures form where needed from whatever is available.

The Lens and Accommodation. In terrestrial vertebrates, light-rays are strongly refracted as they enter the cornea, which consequently does much of the work of focusing. In fishes, this is not the case, and the entire task must be performed by the lens. In consequence, we find that the fish lens is spherical, which gives it the highest possible power, and is situated far outward in the eyeball to afford the maximum distance for the convergence of rays onto the retina. In tetrapods, in which the cornea relieves it of its optical task except for "fine adjustment," the lens is less rounded (the primate lens is exceptionally flat) and is situated farther back in the eyeball. The lens is formed of elongated lens fibers (actually simplified cells, sometimes even enucleate), wound about in a complicated pattern of concentric layers; it nevertheless has excellent optical properties and is completely transparent. It is firm and, especially in lower vertebrates, resistant to distortion.

In cyclostomes, the lens is not attached to the walls of the eyeball; pressure from the vitreous humor behind it and the cornea in front fixes it in place. In all other groups, however, the lens is attached peripherally by a belt of tissues of some sort. This may be (as is elasmobranchs) a circular membrane or zonule, shaped like a plumber's washer; in most cases, however, the suspensory structure consists of a circular belt of zonule fibers. The area of attachment of these fibers is a ring-shaped region of conjoined choroid and retinal layers, the **ciliary body**, lying opposite the equator of the lens. The inner surface of this body is an epithelium formed by retinal tissue; its substance is derived from the choroid.

As every user of a camera is aware, it is impossible to obtain exact definition of objects at varied distances without adjustment of the lens focus. Such adjustment in the eye is termed **accommodation**. The eyes of most vertebrates are capable of accommodation; but curiously, it is attained in a different way in almost every major group. The methods used may be broadly classified as follows:

A. Lens moved to achieve accommodation:
1. "Resting" position for near vision; actively moved backward to accommodate for distant objects (lampreys, teleosts)
2. "Resting" position for far vision; actively moved forward to accommodate for near objects (elasmobranchs, amphibians)
B. Shape of lens modified; "resting" form for far vision, expanded shape for near objects (amniotes)

For distant vision, the lens is, in lampreys, pushed back by a flattening of the cornea, which it directly underlies; the flattening is accomplished by an external somatic muscle peculiar to these animals. In teleosts, there is an unusual situation in that the embryonic choroid fissure in the floor of the eye remains open. Through it projects an elongated vascular structure, the **falciform process** (Fig. 367 *D*), serving a nutritive function for the interior of the eyeball. From the front edge of this process, there generally originates a small **retractor lentis muscle** (of mesodermal origin) that attaches to the lower edge of the lens and pulls it backward.

In elasmobranchs, the lens, most strongly suspended by a dorsal ligament, is normally fixed for distant vision; for close sight, it is swung forward by the pull of a small protractor muscle attached to the ventral rim of the lens. This muscle is, like those of the iris, of ectodermal (retinal) origin.

Movement of the lens in amphibians is the same as that in sharks—a forward pull of the lens for close-up focus. All amphibians (except, of course, those with greatly reduced eyes) have a small ventral muscle that pulls forward on the lens, and, in anurans, there is a second, dorsal muscle as well. These muscles form in the substance of the ciliary body and hence are of mesodermal derivation.

In all amniotes, the second principal mode of accommodation is present. The lens is somewhat flexible and capable of changing its shape from a flattened condition adjusted to far vision to more rounded contours for near sight. These changes in shape are accomplished by smooth muscles of mesodermal origin situated in the ciliary body. But despite this basic agreement among amniotes, we find that (perversely) reptiles and birds, on the one hand, and mammals, on the other, do the trick in two different ways. In typical reptiles and in birds, the greatly developed ciliary body has a ring of padlike processes that extend inward to gain a contact with the periphery of the lens. When circular muscle fibers of the ciliary body contract, these processes push in on the lens and force it to bulge into the more rounded shape suitable for near vision.

In reptiles and birds, the zonule fibers are unimportant. In mammals, however, the ciliary body does not touch the lens; this is suspended by the fibers, which attach at the back edge of the ciliary region (Fig. 371). The pull of the fibers holds the lens, when the ciliary muscles are relaxed, in a flattened condition suitable for distant vision. Contraction of the ciliary muscles brings the region to which the fibers attach outward, closer to the lens; the fibers are no longer under tension and the elastic lens, released from tension, takes a more rounded shape. The amniote type of mechanism is effective only if the lens retains its elasticity. In man, as older readers know, it often stiffens with age, accommodation diminishes, and, without artificial aids, a book can be read only at arm's length, if at all. Further, the ciliary musculature is poorly developed in many mammals, and a large variety of forms, ranging from rats to cows, lack powers of accommodation and are permanently far-sighted.

Why this great variety of methods to attain a single result? One may more than suspect that it is due to the absence of any device of this sort in the ancestral vertebrates (some fish still have none today). Over the course of hundreds of millions of years, various vertebrate groups have solved the problem of accommodation but have solved it independently of one another, each in its own way. Loss of the ability to accommodate is always possible with reduced use of the eyes; later forms could then redevelop means of accommodation, but in a very different way.

Cavities of the Eyeball. Much of the eyeball, as far as function goes, is merely a blank space that need only be filled by some substance that will not block or distort light rays passing through. This filling is in the form of liquids, here known under the old-fashioned name of humors. The principal cavity of the eyeball, between lens and retina, is filled by the **vitreous humor**, a thick, jelly-like material. In front of the lens is the **aqueous humor**, a thinner watery liquid (as it needs to be particularly in lower vertebrates where the lens moves to and fro in this area). The cavity filled by aqueous humor between the cornea and iris is termed the **anterior chamber** of the eye; the **posterior chamber** is not, as one might think, that filled by the vitreous humor, but the area, never of any great dimensions, that the aqueous humor may occupy between iris and lens (Fig. 371).

Intrusive structures are sometimes found in the cavity occupied by the vitreous

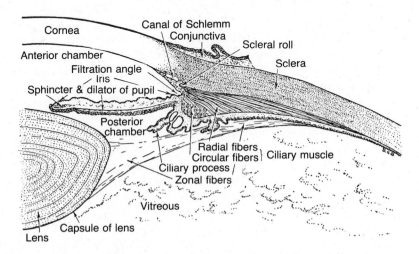

Figure 371. Details of the outer segment of the human eye. (From Fulton.)

humor. We have already noted the intrusion of the falciform process of teleosts. In reptiles, a **papillary cone** projects into this cavity from the region of attachment of the optic nerve. This is a highly vascular structure that presumably aids in the supply of nutritive substances to the retina by diffusion through the vitreous humor. In birds, this projection has developed into the **pecten** (Fig. 367 *B*), a characteristic structure of this class. The pecten is, like its reptilian homologue, a source of nutritive or oxygen supply, but its usual shape, with pronounced parallel ridges, suggests that it is, in addition, a visual aid. Possibly the shadows of these ridges, falling on the retina, act as a grille ruling, and small or distant moving objects are more readily discerned as their images pass from one component to another on this grille. The presence of a pecten may be responsible, in part at least, for the high visual powers attributed to many birds.

Retina. All other parts of the eye are subordinate in importance to the retina; their duty is to see that the light rays are brought in proper arrangement and focus to this sensory structure for its stimulation and the transfer of these stimuli inward to the brain. Embryologically, the retina is a double-layered structure; in the adult, however, the two layers are fused. The outer one is thin and unimportant; it contributes nothing but a set of pigment cells to reinforce the pigment present external to it in the choroid. The complex sensory and nervous mechanisms of the retina all develop from its inner layer.

The nature of the retina varies greatly from form to form and from one part to another of a single retina. Frequently, however, a sectioned retina has the general appearance seen at the left in Figure 372. Externally, next to the choroid, there is the thin pigmented layer; just inside this, a zone showing perpendicular striations; inside this, again, three distinct zones containing circular objects recognizable as cell nuclei. Special methods of staining reveal the nature of this stratification. The striated zone contains the elongated tips of the light-receiving cells, the rods and cones; the outer nuclear zone contains the cell bodies and nuclei of these structures. The nuclear zone next in order is that of the bipolar cells, which transmit the impulses inward from the rods and cones, and of accessory types of retinal nerve cells. The innermost nuclei are those of ganglion cells, which pick up the stimuli from the bipolar elements and send fibers along the optic nerve to the brain.

The **rods** and **cones** are the actual photoreceptors of the system. They receive their names from their usual shape in mammals. Each cell includes a sensory tip directed outward toward the choroid, a thickened section, and (beneath a membrane lying at this level of the retina) a basal piece containing the nucleus. In most mammalian eyes, the two types are readily distinguishable; the rod cell is slenderly built throughout; the cone cell has generally a short broad tip and broad "body." In other groups, however, rods and cones vary greatly in shape and are sometimes difficult to differentiate.

One is immediately impressed by the fact that rod and cone cells in the vertebrate retina are *pointing the wrong way*! In a "logically" constructed retina, the tips of the photoreceptors should point toward the light source, and the parts of the retina that carry the impulses inward toward the brain should be farther removed. Some invertebrate eyes, such as those of squids and octopi, are so built, but not those of vertebrates, in which the light has to plunge through the full depth of the retina before reaching the point of reception. Various theories have been made to account

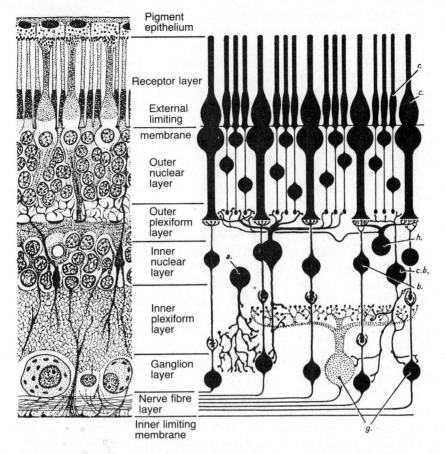

Figure 372. *Left,* Vertical section through the retina of a mammal; *Right,* retinal connections are revealed by silver impregnation. The center of the eyeball (vitreous humor) is toward the bottom of the figure. *a* and *h,* Nervous elements of more or less uncertain function. Other abbreviations: *b,* bipolar cells associated with single cones; *c,* cones; *cb,* bipolar cells, connecting with a series of rods; *g,* ganglion cells; *r,* rods. (From Walls, partly after Polyak.)

for this anomalous situation. Possibly visual cells first appeared in the floor (or side wall) of the brain cavity and were able to function with upwardly (and inwardly) pointing tips, owing to a translucent nature of the head in ancestral chordates; such cells appear to be present in the spinal cord of amphioxus. When such cell groups became outfolded optic vesicles, the retinal elements simply retained their original orientation.

Rods and cones differ markedly in their functions, as one can determine from one's own eyes, where the cones are concentrated in the center of the field of vision, and the rods are situated mainly peripherally. (1) Good illumination is necessary before the cones come into play, but rods are effective in faint light; indeed, a rod may be stimulated by a single quantum, the smallest theoretic unit of energy. At night one can often catch a glimpse of a faint star in the margin of the field of vision but fail to see it if looking directly at it. (2) Cones as a group give good visual details;

rods give a more blurred picture. For accurate detail, we focus the cone-bearing center of the eye on the object; things seen peripherally with the rods are blurred and indistinct. (3) Cones give color vision; rods give black-and-white effects only. In one's visual field, peripheral objects are gray and colorless. The difference in acuity of vision between cones and rods is explained by the nerve fiber connections of these two types, noted below. We do not know what makes the difference between the high and low thresholds of cone and rod cells.

Cell morphology offers little aid in the interpretation of color vision. In turtles and birds, colored oil globules in the cones may filter light for reception of one or the other of the assumed basic colors, but, in other groups, all cones in a given animal appear to be anatomically of a single type.

Color vision is widespread in vertebrates but far from universal. Obviously there is little if any color vision in forms in which cones are rare or absent. But color vision may not be present or similar to our own, even where cones or conelike structures are found in the retina. Experiments on living animals are necessary to determine whether color sensitivity is present. Many teleosts have color vision, as do many reptiles and most birds, but there is no proof of color vision in elasmobranchs or amphibians. In mammals as a whole, relatively few of those tested show much response to color; the ancestral mammals, it is believed, were mainly nocturnal forms, with rods predominating in the eyes. The high color sensitivity of higher primates, including man, is an exceptional situation in mammals; to a dog or cat, the world is probably gray or has at the most faint pastel tints.

The distribution of rods and cones in different animals and in different retinal regions is highly variable. As would be expected, rods dominate in nocturnal animals and in fishes dwelling in deeper waters; cones are more abundant in forms active by day. Because, however, members of a given order or class may differ widely in their habits, the distribution of rod and cone types of retinas does not sort out well taxonomically. In an "average" vertebrate, probably not more than 5 per cent or so of the total photoreceptive cells are cones. Retinas containing rods alone are found in many sharks, some deep sea teleosts, and a relatively few nocturnal forms in the amniote classes. On the other hand, eyes that have cones only are also known in a few amniotes of each class. Most reptiles and birds have eyes relatively rich in cones, and hence are mainly suited for daytime life. Teleosts may have "twin" cones, with partially fused cell bodies.

Rods and cones may both be present in any retinal region. Often, however, in forms in which both types of cell are common, there develops a distribution, as in our own eyes, with the cones only sparingly present over most of the retinal area and concentrated, to the practical exclusion of rods, in a central region in which vision is most acute and perception of detail best developed. This is the **area centralis** (sometimes termed **macula lutea** because, exceptionally, the area has a yellow tinge in man). Frequently, there is present here a **fovea** as well, a depression in the surface of the area centralis from which blood vessels and the inner layers of cells have been cleared away.

In many birds (and in some lizards as well), there is, curiously, a development of two central areas in each eye, one centrally situated, the other well toward the posterior or outer region. The bird's eye is normally pointed well to the side, and the primary central area is aimed at this lateral visual field. In bird flight, vision straight ahead is of great importance; the secondary center gives each eye a perception of detail in the anterior part of each visual field.

Inward from the rods and cones is a layer that consists for the most part of **bipolar cells**; and, well to the inner surface of the retina, a third cell layer is that of **ganglion cells**. The bipolar cells have short processes serving to connect with rods and cones on the one hand and with ganglion cells on the other; the latter cells are neurons from which long fibers relay to the brain the impulses initiated by light reception. Both bipolar and ganglion layers include in addition peculiar types of cells that appear to make cross connections between the various direct pathways. Often only one cone cell connects with a given bipolar cell, and only one such bipolar cell connects with an associated ganglion cell; thus each cone may have an individual pathway to optic nerve and brain. On the other hand, a considerable number of rods always converge into a single bipolar cell. In consequence, the brain obtains no information as to which of the group of rods has been stimulated, a condition that accounts for the lack of precision in vision using rods as compared with that of cone areas.

Optic Nerve.   Although the brain and nerves are the subject of the next chapter, we may perhaps permit ourselves to violate proper order and discuss the central connections of the eye in relation to vision. The ganglion cells of the retina produce long nerve fibers that travel inwardly along the course of the original optic stalk to the forebrain. The fibers originate on the inner surface of the retina, converge to a point opposite the attachment of the stalk, and there plunge through the substance of the retina. At this point, there can, of course, be no rods or cones, and there is, hence, a **blind spot** in the field of vision. In ourselves, at least, this blank area is fudged by the brain, which fills it in with the same materials seen in the surrounding region.

Although the connection of retina with brain proper is customarily called a nerve, it differs from any normal nerve in two regards. Here, somewhat as in the olfactory nerve, we are dealing with fibers running inward toward the brain rather than outward from cells in or near that organ. Further, because the retina is, from an embryologic point of view, properly a part of the brain itself, the optic "nerve" is actually not a true external nerve, but rather a fiber tract connecting two regions of the brain.

Reaching the floor of the forebrain, the optic nerve fibers enter via the X-shaped **optic chiasma** (Fig. 373). We find that frequently fiber bundles cross from one side to the other within the substance of the brain, without any immediately obvious reason, and this is the case also with the optic nerves. In the crossroads of the chiasma, in primitive vertebrates, almost all the fibers of the right optic nerve cross to the left side of the brain, and vice versa; such a crossing of fibers is termed a

Figure 373.   Diagram of the optic chiasma in a mammal with good stereoscopic vision. All fibers from the corresponding half of each eye pass to the same side of the brain. (From Arey.)

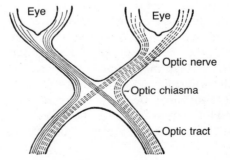

decussation. The two sets of fibers in lower vertebrates run to paired structures in the roof of the midbrain (tectum); in mammals, most are relayed to special areas in the gray matter of the cerebral hemispheres. Details are little known in lower vertebrates, but it appears that, as is known to be the case in mammals, the points in the brain receiving sensations from the different parts of the retina are arranged topographically in many if not all cases in such a way as to give a brain pattern—and a resulting mental "picture"—reproducing the arrangement of objects as "seen" by the retina.

In a majority of vertebrates, the eyes are directed almost straight laterally, and the two fields of vision are partially or totally different; the brain builds up two separate pictures of two quite separate views. In a number of higher types, however, such as birds of prey and many mammals, the eyes are turned forward, the two fields of vision overlap to a greater or lesser degree, and the two sets of impressions transmitted to the brain are more or less the same. Most notable is the case in primates, where, from *Tarsius* to man, the two fields of vision are practically identical.

Strangely, two duplicate mental pictures seem to be formed, as far as can be discovered, in birds and other nonmammalian forms with overlapping visual fields. In mammals, however, we find a new development. **Stereoscopic vision** appears. The field of vision is unified; sensations from objects received in common by the two eyes are superposed. As a result, such a form as man is aided, by the slight differences in point of view of the two eyes, in gaining effects of depth and three-dimensional shape of objects—effects otherwise impossible of attainment.

An important anatomic phenomenon concerned with this development is **incomplete decussation** at the optic chiasma. In many mammals—and in mammals alone—we find that, for overlapping parts of the field of the two eyes, fibers from the areas that both retinas view go to the same side of the brain. In consequence, certain groups of fibers do not cross (i.e., decussate), but turn a right angle at the chiasma, to accompany their mates from the opposite eye. In man, for example, where the overlap of visual fields is almost complete, practically all the fibers from the left halves of both retinas enter the left side of the brain, and all fibers from the right halves of the retinas enter the right side of the brain (Fig. 373). As a result, the visual area of each side of the brain builds up a half-picture of the total visual field as a "double exposure"; by further complicated interconnections between the hemispheres, the two halves of the picture are welded together to emerge into consciousness as a single stereoscopic view.

Accessory Structures.   External to the eyeball are various visual aids. We have previously described (cf p. 290) the six striated muscles that move the eyeball. This muscle group may occasionally form additional muscles, notably in many tetrapods a **retractor bulbi muscle** that pulls the eyeball inward. There may also be a **levator bulbi** to restore the eyeball to its proper position. In some vertebrates, there even develops in the orbit a derivative of the jaw musculature that, when contracted, bulges with the effect of popping the eye outward.

In fishes, there may be present folds of skin around the margins of the orbit, but there is little further development of **eyelids** in most fishes. In some teleosts, however, the cornea may be partly covered by fixed folds of skin, and some sharks have upper and lower lids, the latter movable.

Most tetrapods have movable lids. A dry cornea would become opaque; the lids, by closing at intervals, moisten and clean the corneal surface. Upper and lower lids are always present; they are usually thick and opaque. In most groups, the lower lid is more prominent than the upper; in mammals, however, and crocodilians, the upper lid is the major structure. In birds and in many reptiles and mammals, there is a third eyelid, the **nictitating membrane**, a transparent fold of skin lying deep to the other eyelids and drawn over the cornea from anterior (or medial) to posterior margins.

No muscles were originally present in the region of the eyelids, and various methods have been evolved to effect their movements. In many tetrapods, as noted above, the eyeball can be withdrawn and again protracted outward; such motions may produce a passive closing and opening of the lids. The operation of the nictitating membrane is effected by the development of a tendon that passes inward from it around the eyeball and becomes associated in variable fashion with the eyeball muscles; a similar tendon may arise in connection with the lower lid. Slips from these same muscles in some cases grow forward to attach to the lids. In mammals, use is made of a new development in that class. We have noted that facial muscles have here grown forward over the head; a ring of these muscle fibers operates as a sphincter to close the eye (Fig. 215).

In snakes and some lizards, there lies external to the cornea a second superficial eye-covering of transparent tissue. This protective **spectacle** is formed from a fusion of the eyelids, giving the snake its unwinking stare.

In terrestrial vertebrates are developed tear glands, **lacrimal glands**, whose function is to furnish liquid for moistening and cleaning the cornea. As seen in some urodeles, there is primitively a row of small glands running the length of the eyeball within the lower lid. In most groups, however, there is a concentration of glandular development at either the front (medial) or the back (lateral) margin of the orbit. Characteristic of the former position is the **harderian gland**, which tends to secrete a rather oily liquid, in contrast to the true lacrimal gland. The latter is typically developed at the back or outer corner of the eye. In amphibians, reptiles, and birds the harderian glands are usually prominent, the lacrimal glands often poorly developed or absent; in mammals, on the other hand, the lacrimal gland proper is characteristic, and a harderian gland relatively rare. In marine turtles, the lacrimal gland secretes a hypertonic salt solution, thus allowing the animals to drink sea water and survive.

A minor if useful adjunct to the tetrapod eye is the **tear duct** (lacrimal duct), which generally carries surplus fluid from the anterior or medial corner of the eye to the nasal cavity. The lacrimal fluid has thus an accessory function in moistening the nasal mucosa and, in lizards and snakes, of reaching the mouth via the nose and forming an addition to the salivary fluid. The duct is lacking in turtles, and in lizards and most snakes, it empties into the vomeronasal organ (cf. p. 504). Although fish appear to lack this duct, it actually is there—the tetrapod duct develops from the more posterior of the two external nostrils on each side in bony fish.

Eye Variations. Our previous comments on optic structures have given some faint idea of the great amount of adaptive variation seen in vertebrate eyes as a whole. Throughout the vertebrate series, the basic pattern persists with remarkable consistency; nevertheless, entire volumes could be (and have been) devoted to an account of the wealth of changes rung on the basic structures in different groups. A few additional notes on variations may be given here.

Eyeballs are generally almost spherical, but there are striking departures from this. Tubular eyes—deep and narrow—are seen in deep-sea teleosts on the one hand, and in owls on the other. Both these forms operate in dim light; in this type of eye, all available illumination is concentrated on a narrow patch of sensory retina to bring the light intensity up to the minimum necessary for vision. Typical diurnal birds, in contrast to owls, have a broadly expanded retinal surface (as in Fig. 367 *B*), making perception of small details possible. A curiosity is the little "four-eyed fish" *Anableps* of the American tropics, which has two separate corneas and retinal areas in each eye, so that, floating at the surface, it can simultaneously see both above and below water.

Eyes are relatively small in large animals, relatively large in little forms. This one might reasonably expect, because if an eye of a certain size furnishes satisfactory vision for a given animal, there is no selective advantage in a larger eye. The function of the eye in no way depends upon the volume of animal behind it.

Eyes have degenerated and lost much of their structural niceties in a variety of forms from hagfishes to mammals. Such degeneration is, of course, usually correlated with a mode of life in which light is dim or absent. Among marine teleosts, those living at fairly considerable depths almost universally have large and specialized eyes that make the most of the faint illumination present; but, in the great deeps, some fishes have, so to speak, abandoned the struggle, and the eyes have degenerated. Cave life, too, has led to optic degeneration in various fishes and amphibians, some closely related to forms with normal eye structure. Among tetrapods, burrowing forms—gymnophionans, amphisbaenians, moles, and others—usually exhibit a similar tendency. Snake eyes are in general well developed but are peculiarly built and lack various structures, such as the scleral ring and certain of the usual eye muscles. It is suggested that the ancestral snakes were burrowers with reduced eyes and that the typical ophidian eye of today has been secondarily built up from the remains of the degenerate ancestral organ.

Median Eyes. Our attention thus far has been exclusively directed toward the paired, laterally placed eyes common to all vertebrates. In their early history, however, the ancestral vertebrates had a third eye, medially situated on the forehead and directed upward. In the oldest ostracoderms (cf. Fig. 19), a socket that obviously contained such an eye is usually present and often conspicuous, although always much smaller than the paired eyes. This socket was generally present in placoderms but not in the sharklike fishes. It is found in the Devonian in primitive representatives of all three major groups of bony fishes and was especially common in crossopterygians (Fig. 170). In early tetrapods, this third eye persisted; it was present in almost all the older amphibians and early reptiles, including most of the mammal-like types (Figs. 177 *A, B,* 186 *A,* 187 *A*). But by Triassic times, median eyes had apparently gone out of fashion. Even in fishes, they disappeared at an early stage in most groups, and in modern amphibians, most reptiles, and all birds and mammals they are likewise absent. Today we find well-developed median eyes present only in lampreys, on the one hand, and *Sphenodon* and some lizards on the other (Fig. 424); buried beneath the skin, they can do little more than detect the presence or absence of light, although a miniature cornea, lens, and retina may be developed (Fig. 374). In frogs, there may be a vestigial external structure of dubious nature; in all other groups, the remains of such organs are no more than glandular materials within the braincase.

A functional interpretation of this story can be made in terms of the habits of the

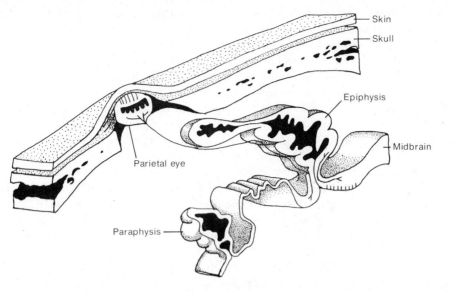

Figure 374. Median eye and associated structures in a lizard. The epiphysis (or pineal) and parietal organ (parapineal) are both dorsal outgrowths of the diencephalon, which may, in different animals, form "eyes"; the paraphysis is another outgrowth, but never eyelike. In some lizards (and *Sphenodon*), the parietal eye lies within the parietal foramen in the skull and is a well-formed structure with a lenslike upper epithelium and retina-like lower part. It is attached to the diencephalic roof by a thin nerve. (After Wurtman, Axelrod, and Kelly.)

vertebrate ancestors. Presumably they were bottom-dwelling mud strainers. In this mode of life, knowledge furnished by a median eye of what threatened from above was vital. With the development of a more active mode of life, vertical vision declined in importance.

Such dorsal eyes spring from the roof of the diencephalon and hence, like lateral eyes, are essentially parts of the brain. There is a curious quirk in the story; the median eye, it appears, is not the same structure throughout, but may develop from either of two outgrowths, the **parietal organ** (or parapineal organ) and **pineal organ**. Both may be present in the same animal and, although adjacent, are readily distinguishable through differences in the position of the attachments of their stalks. In lampreys, both form eyelike structures, but the pineal is the dominant element. In *Sphenodon* and lizards, however, the parapineal is the functional eye. The situation suggests that possibly the remote ancestor of the vertebrates may have had paired dorsal eyes as well as paired lateral ones, and conditions in some fossil placoderm skulls tend to strengthen this belief.

Despite loss of visual function, the pineal organ persists in higher vertebrates as a glandular structure. As noted later, it appears to have endocrine functions related to its original light-perceiving ability.

Lateral Line Organs

A highly developed sensory system of a type quite unknown in land vertebrates is that of the lateral line organs found in fishes and in aquatic and larval amphibians.

The receptors are clusters of sensory cells termed neuromasts; these may be found distributed in isolated fashion in the skin but are generally located along a series of canals or grooves on the head and body.

A main element of the system is the **lateral line** in the narrower sense of the term (Fig. 375 *A*). In typical fishes, this is a long canal running the length of either flank (parallel secondary lines sometimes develop). The canal continues forward onto the head, where similar canals form a complex pattern. Generally, a major canal runs forward over the temporal region, then curves downward and forward beneath the orbit to the snout as an infraorbital canal. A second cranial canal is supraorbital in position in its forward course. The canals of the two sides may have a transverse connection across the occiput; a canal generally crosses the cheek and then runs downward and forward along the lower jaw. Other accessory canals may be present on the head and gill region, and, in some cases, there are accessory body lines (Fig. 376). The patterns vary considerably among the different groups of aquatic vertebrates (Fig. 377). Isolated neuromasts, **pit organs**, may also be found on the head, often in linear arrangements.

In most fishes, the lateral line organs are contained in closed canals, opening at intervals to the surface. In many bony fishes, this canal system may be sunk within the substance of the bony plates of the skull and within or below the scales; in cyclostomes, the neuromasts, on the contrary, are in isolated pits, although the pits are arranged in roughly linear order. In a few cartilaginous and bony fishes, open grooves take the place of canals, and the same appears to have been true among all but the earliest of the fossil amphibians. In modern amphibians, the neuromasts lie in the skin in a more or less isolated fashion, although preserving in part a linear arrangement.

The sense organs of the lateral line system are the **neuromasts**, consisting of bundles of cells, often with much the appearance of taste buds. The neuromast cells are elongated, and each bears a projecting hairlike structure. Usually there is present above the "hairs," and enclosing their tips, a mass of gelatinous material that may be

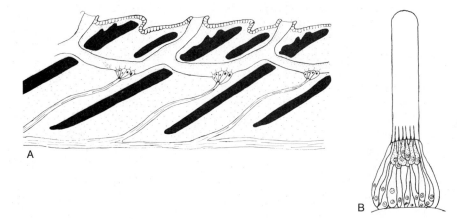

Figure 375. Lateral lines of fish. *A,* Part of the main lateral line canal of the perch (*Perca*) showing three pores to the surface and three neuromast organs. *B,* Diagram of a single neuromast organ showing the sensory processes and the gelatinous cupula (cf. Fig. 381). (After Goodrich and Yapp.)

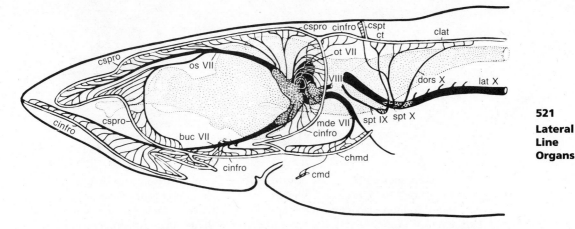

Figure 376. Left side of the head of a shark to show the lateral line canals (parallel lines) and the nerves (black) supplying them. Pit organs are not shown. Abbreviations: *buc VII,* buccal ramus of nerve VII; *chmd,* hyomandibular canal; *cinfro,* infraorbital canal; *clat,* lateral line canal proper; *cmd,* mandibular canal; *cspro,* supraorbital canal; *cspt,* supratemporal (or occipital) canals; *ct,* temporal canal; *dors X,* dorsal ramus of nerve X; *lat X,* lateral line ramus of nerve X; *mde VII,* external mandibular ramus of nerve VII; *os VII,* superficial ophthalmic ramus of nerve VII; *ot VII,* otic ramus of nerve VII; *spt IX,* supratemporal ramus of nerve IX; *spt X,* supratemporal ramus of nerve X. The brain is stippled in and the ganglia of the cranial nerves shown by a pattern of small circles. (From Norris and Hughes.)

termed a **cupula**, secreted by the neuromast cells, which waves freely in the surrounding water (Fig. 375 *B*). Lacking structures at all comparable ourselves (except, as we shall see, in the internal ear—Fig. 381), the nature of the sensations registered has been difficult for us to determine. It appears that most neuromasts respond to waves or disturbances in the water through movement of the cupula and consequent bending of the hairlike processes, and supplement vision by making the fish aware of nearby moving objects—prey, enemies—or of fixed "obstructions to navigation." A fish seldom has visible landmarks against which to measure its progress through the water; sensory structures of this sort will at least tell it of water movements and the pressure of currents against its body and thus be of great aid in locomotion.

The neuromasts and their nerve cells are derived embryologically (as are several other nervous and sensory structures) from thickenings of the ectoderm on either side of the neural tissues of the head, termed **placodes**. The lateral line placodes become associated with three of the cranial nerves. Much of the system of cephalic canals is formed from an anterior placode associated with cranial nerve VII. A short segment in the temporal region is formed from a small placode related to nerve IX. A posterior placode forms the posterior part of the cephalic system and grows backward the length of the trunk; it is associated with cranial nerve X (the vagus), which sends a branch along the flank to accompany it. In their development, the lateral line canals are closely associated, in those forms with dermal bones in the skull, with the centers of ossification of many of the bones, and provide important evidence on the homologies of many of them (Fig. 378).

Although the lateral line system is present in larval and aquatic amphibians (Fig.

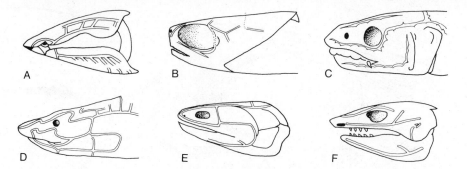

Figure 377. Lateral lines on the heads of various forms for comparison with that of the shark shown in Figure 376. *A, Poraspis,* a heterostracan ostracoderm (after Denison); *B, Rhinosteus,* an arthrodiran placoderm (after Stensiö); *C, Hybopsis,* a teleost (after Reno); *D, Protopterus,* a lungfish (after Orvis); *E, Eusthenopteron,* a rhipidistian crossopterygian (after Jarrik); *F, Palaeoherpeton* (with lower jaw after Eogyrinus), an anthracosaurian labyrinthodont (after Panchen).

379), it was lost, it seems, in the earliest reptiles. Many reptiles and mammals have returned to a water-dwelling life, but, useful as it would be to them, this sensory system, once lost, never reappears.

Another type of sense organ, at least functionally, seems to be a part of the lateral line system. Many fossil fish, most notably some of the ostracoderms and crossopterygians, have a complex system of channels, the **pore canals**, within the armor or scales; they are especially characteristic of well-developed layers of cosmine (the triangular spaces in the cosmine layer of Fig. 111 *C* are a good example). Such a system was probably present in all early vertebrates and is now believed to have contained organs sensitive to electrical charges.

This presumably primitive electrosensory capacity is retained by modern lampreys and elasmobranchs. In the latter, the structures concerned are the **ampullae of Lorenzini**, clusters of small, jelly-filled tubules, scattered over the head, especially the snout, and containing what appears to be modified neuromasts, without the usual cupulae, in their swollen bases. These ampullae can be made to react to stimuli other than electric currents but appear to be specialized primarily for electroreception.

Although such electroreceptors were apparently lost in early actinopterygians and are not now known in most of them, several groups of teleosts have, presumably independently, redeveloped somewhat similar receptors, also from parts of the lateral

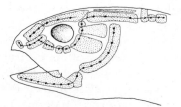

Figure 378. The head of *Amia,* showing the relationship between the cephalic lateral line canals and some dermal bones of the skull. There are, of course, other bones that lack such relationships and are not shown here. (After Pehrson.)

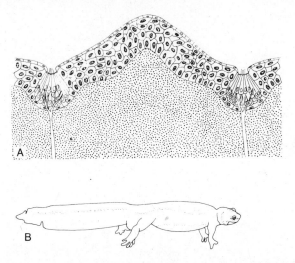

Figure 379. Lateral line organs of amphibians. *A*, Two neuromast organs in the skin of an aquatic frog *(Xenopus); B,* general pattern of neuromast organs in the skin of a urodele (*Pleurodeles*). The neuromasts are not enclosed within canals but are separate organs on the skin; however, they do tend to form lines along the body and head. (After Murray and Escher.)

line system. These are especially prominent in a variety of relatively primitive freshwater teleosts, such as the African mormyrids, which produce electric impulses as a type of "radar" for use in muddy water, in which vision is of limited use (cf. p. 314).

The Ear

First thoughts as to the primary anatomic or functional aspects of the vertebrate ear are liable to be misleading when based on familiar human features. One tends to think, when the word is mentioned, of the ornamental pinna of the mammalian external ear or, perhaps, of the middle ear cavity behind the drum, with its contained ear ossicles. These items, however, are entirely lacking in fishes; the basic structures of the ear in all vertebrates are those of the internal ear, the sensory organs buried deep within the otic capsule. We naturally think of hearing as the proper ear function; but, in the ancestral vertebrates, audition was apparently unimportant and perhaps absent; equilibrium was the basic sensory attribute of the "auditory" organ.

The Ear as an Organ of Equilibrium. Before considering the hearing function, which becomes increasingly important as we ascend the scale of vertebrates, we may discuss the ear as an organ of equilibrium, a basic and relatively unchanged function from fish to man. Equilibrium is a type of sensation produced by the internal ear alone; all accessory ear structures are related to the hearing sense and need not concern us for the moment.

In a variety of fishes, amphibians, and reptiles, the paired internal ears are built upon a relatively uniform pattern in which most of the structures present are related to equilibrium (Fig. 380 *A–D*). The **membranous labyrinth** consists of a series of sacs and canals contained within the otic region on either side of the braincase. These

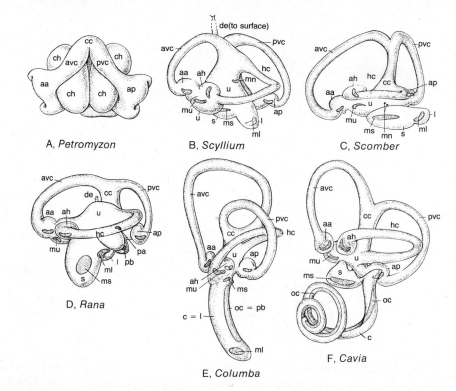

A, *Petromyzon*   **B,** *Scyllium*   **C,** *Scomber*

D, *Rana*   **E,** *Columba*   **F,** *Cavia*

Figure 380. Membranous labyrinth of *A,* lamprey; *B,* shark; *C,* teleost; *D,* frog; *E,* bird; *F,* mammal; all external views of the left ear. Sensory areas are shown (except in *A*) as if the membrane were transparent. Abbreviations: *aa,* ampulla of anterior canal; *ah,* ampulla of horizontal canal; *ap,* ampulla of posterior canal; *avc,* anterior vertical canal; *c,* cochlear duct; *cc,* crus commune with which both vertical canals connect; *ch,* chambers in the lamprey ear lined with a ciliated epithelium; *de,* endolymphatic duct; *hc,* horizontal canal; *l,* lagena; *m,* macula of lagena; *mn,* macula neglecta; *ms,* macula of sacculus; *mu,* macula of utriculus; *oc,* organ of Corti; *pa,* papilla amphibiorum; *pb,* papilla basilaris; *pvc,* posterior vertical canal; *s,* sacculus; *u,* utriculus. (After Retzius.)

form a closed system of cavities, lined by an epithelium and containing a liquid, the **endolymph,** similar to the interstitial fluid.

Two distinct, major, saclike structures are generally present in jawed vertebrates, the **utriculus** and the **sacculus.** A slender tube, the **endolymphatic duct,** usually extends dorsomedially from them to terminate within the braincase in an **endolymphatic sac.** In general, the sacculus lies below the utriculus. In both there is found a large oval "spot" consisting of a sensory epithelium associated with branches of the auditory nerve; these areas are the **utricular macula** and **saccular macula.** The utricular macula lies on the floor of that sac, in a horizontal plane; the saccular element is typically in a vertical plane on the inner wall of the sacculus. A pocket-like depression is found in the floor of the sacculus near its posterior end; this is the **lagena,** which contains a smaller **lagenar macula.**

The sensory cells of these maculae (and, indeed, of the entire internal ear) are comparable to the neuromasts of the lateral line system (Fig. 381). During develop-

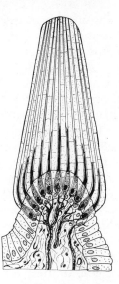

Figure 381. Crista (neuromast organ) from a human semicircular canal. Each sensory cell, here heavily stippled, has a hairlike process extending into the flexible, gelatinous cupula; bending of the cupula stimulates the cells. (After Bargmann.)

ment, there forms over the combined tips of the sensory hairs of these cells a gelatinous membrane or cupula. In the utricular and saccular maculae, and often that of the lagena as well, this material becomes a thickened structure, weighed down by the deposition in it of a mass of crystals of calcium carbonate termed an **otolith**. The nature of this "ear stone" varies. It may remain a rather amorphous mass of crystals. In ray-finned fishes, however, the otoliths develop into compact structures. The saccular otolith is generally the best developed; in teleosts, it is so large as practically to fill the sacculus and reflect in its outlines the shape of that cavity. This shape varies from genus to genus and from species to species; a single tiny otolith will often furnish positive identification of the fish that bore it.

The utricular macula and, to a much lesser degree, those of the sacculus and lagena, register, by the tilt of the otolith and its enclosed sensory "hairs," the tilt of the head and linear acceleration. They do not, however, furnish data as to turning movements. This is the function of another series of structures, the **semicircular canals**.

These tiny tubes are dorsal elements of the endolymphatic system, springing out from the utriculus and connected at either end with this sac. In every jawed vertebrate, without exception, three such canals are present; each of the three lies in a plane at right angles to the other two, so that one is present for each of the three planes of space. Two canals lie in vertical planes, an **anterior vertical** (or superior) **canal**, angled anterolaterally from the upper surface of the utriculus, and a **posterior vertical canal**, running posterolaterally; a **horizontal canal** extends laterally from that body. Often the two vertical canals arise at their proximal ends by a common stalk, a **crus commune**, from the upper surface of the utriculus. Each semicircular canal has at one end a spherical expansion, an **ampulla**. The vertical canals bear these ampullae at their distal ends, anteriorly and posteriorly; the horizontal canal (for no known reason in particular) has its ampulla anteriorly placed. Within each ampulla is a sensory area, usually elevated, termed a **crista** (Fig. 381). Here we find again the familiar hair cells, or neuromasts; their tips are embedded in a common membrane or mass of gelatinous substance, a tall cupula, which nearly blocks the ampulla and can move to and fro like a swinging door.

The arrangement of the canals and their sensory structures indicate their function: to register turning movements of the animal in the several planes of space. Displacement of liquid in one or more of the canals displaces the cupulae, with a consequent bending of their sensory hairs.

The organization of the parts of these otic sacs and canals is in general constant through the jawed vertebrates, high or low. There are, however, certain variables in the system. As in many other features, the cyclostomes exhibit an unusual and puzzling condition. The lamprey has only two semicircular canals, equivalent, apparently, to the vertical canals of other vertebrates, and, in addition, a peculiar system of ciliated sacs beneath them (Fig. 380 A); the hagfish has only a single canal, presumably representing a combination of these two. As in the case of the nose, we cannot be sure whether this is a primitive or a degenerate condition. The lamprey type of ear (as well as nostril arrangement) is known to have been present in some of the most ancient fossil vertebrates (the Osteostraci). In cartilaginous fishes, sacculus and utriculus are merely two lobes of a common cavity from which canals and endolymphatic duct arise, and the posterior vertical canal may be associated, at its ampullar end, with the saccular region. In these forms, the crus commune is little developed, and, in skates, the canals are connected only by narrow ducts with the remainder of the system.

The endolymphatic duct usually terminates in a sac of modest proportions lying within the braincase. In Chondrichthyes—presumed to be primitive but possibly neotenic in this regard—the endolymphatic ducts may extend upward to open on the top of the head. The endolymphatic sac is absent in some teleosts. In some bony fishes, the reptiles, and particularly the amphibians, the endolymphatic sacs may be of great size and contain calcareous matter; those of the two sides may connect either above or below the brain in some lower tetrapods, and, in the typical frogs, they may extend the length of the spinal canal as well.

Origin of the Vertebrate Ear.   Embryologically, the internal ears first appear, like lateral line organs, as ectodermal thickenings—placodes—on either side of the head (Fig. 382). These sink inward to form a pair of sacs, which for some time may retain a tubular connection with the exterior (adult elasmobranchs retain these connections as endolymphatic ducts). Typically, each sac then subdivides into utricular and saccular portions; from the former, the canals are gradually separated off, and from the latter arise further structures, discussed later.

This embryologic story, together with the nature of the sensory endings of the ear, suggests that the internal ear originated phylogenetically as a specialized, deeply sunk part of the lateral line system.* Its sensory maculae and ampullae are typical neuromast organs, with hair cells embedded in a gelatinous cover and responding to pressure or to movements in the liquid-filled spaces in which they lie. The nerves from the two sets of organs are closely associated (as will be noted in the next chapter) to form an acoustico-lateralis system. Functionally also, the two sets of sensory structures are closely knit; together they offer a primitive fish the major part of the data by which its locomotion is regulated; still further, there is even some slight evidence of hearing function in the fish lateral line organs themselves.

*Some investigators prefer to say that both the lateral line and ear are parts of an acoustico-lateralis system, but this is merely quibbling over the definition; there is no difference in the ideas.

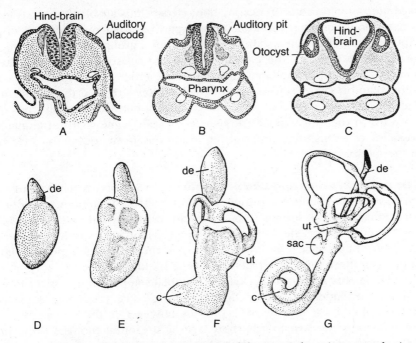

Figure 382. Diagrams to show the development of the internal ear in mammals. *A* to *C,* Cross sections of the head of early embryos; an ectodermal placode sinks inward on either side to form an otic vesicle. *D* to *G,* Successive stages in the development of the various parts of the membranous labyrinth from the otic vesicle; Abbreviations: *c,* cochlear duct; *de,* endolymphatic duct; *sac,* sacculus; *ut,* utriculus. (*A* to *C* from Arey; *D* to *F* after His and Bremer.)

Hearing in Fishes. Whether or not fish hear is a topic that was long disputed. However, hearing is certainly present in many teleosts (many of which *make* a variety of noises), but in elasmobranchs there appears to be merely response to very low vibrations. Localization of the sensory uptake of vibrations is difficult. In most cases, the saccular macula appears to be the major receptor, but, in some forms, the utriculus is apparently concerned as well. A suspect is the **macula neglecta** (Fig. 380 *B,* *C*), a small sensory spot found in the utriculus in a variety of fishes and lower tetrapods. The hearing organs of terrestrial vertebrates develop in connection with the lagenar pocket, but, in fishes, the macula of the lagena appears to be concerned only with gravity.

Fishes in general lack the various accessory devices by which in higher vertebrates sound waves reach the internal ear. However, the conduction of vibrations through the cranial skeleton to the internal ear can produce some degree of hearing (and lateral line sensations are closely allied to, and may overlap with, hearing, of course). In most fishes, water vibrations, to be heard, must set up head vibrations, and these in turn produce endolymphatic vibrations that can be picked up by the hair cells of the internal ear.

Some bony fishes, however, have accessory structures that parallel in a way the "hearing aids" found in terrestrial vertebrates, although evolved quite independently and along other lines. In these fishes, it appears, the swim bladder is used as a device for the reception of vibrations. In herring-like teleosts, this air bladder sends forward a tubular extension that comes to lie alongside part of either membranous labyrinth and can thus induce vibrations in the endolymph. In a group of teleosts termed the Ostariophysi, which includes the catfishes, carp, and relatives, in which hearing is unquestionably developed, another method is used. Processes of the most anterior vertebrae develop on either side as small detached bones termed the **Weberian ossicles** (Fig. 383). These articulate in series to form a chain extending from the air bladder forward to the otic region. The ossicles operate somewhat in the manner of the ear ossicles of a mammal, transmitting vibrations from the air bladder to the liquids of the internal ear.

The Middle Ear in Reptiles and Birds.

Hearing is an important sense in tetrapods. A different problem in sound reception exists in terrestrial vertebrates as contrasted with the fish situation. The sounds to be heard are relatively faint air waves that can have ordinarily little direct effect in setting up endolymphatic vibrations in the internal ear. Devices for amplification of these waves and their transmission to the internal ear are important; these form the structures of the middle ear.

The developments found in the living amphibians appear to be specialized and quite possibly independently developed; they will be considered later. A more generalized type of middle ear structure was present, however, in the ancient labyrinthodonts, and numerous modern lizards show a structure often thought to be not far removed from this primitive tetrapod condition (Fig. 384). The spiracular gill cleft and the hyomandibular bone are the sole elements concerned. The spiracular pouch of the embryo never breaks through to the surface; the corresponding depression in the surface, where developed, becomes an **external auditory meatus**. A thin membrane between external depression and pouch becomes the eardrum or **tympanic**

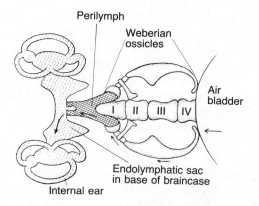

Figure 383. Diagrammatic horizontal section of the posterior part of the head and anterior part of the body of a teleost with Weberian ossicles. Vibrations in an anterior subdivision of the air bladder set up corresponding vibrations in a series of small ossicles, which in turn set up waves in a perilymphatic sac. This, again, sets up vibrations in an endolymphatic sac at the base of the braincase. Arrows show the course of transmission of the vibrations. Roman numerals indicate the vertebrae from which the Weberian ossicles are derived. (After Chranilov.)

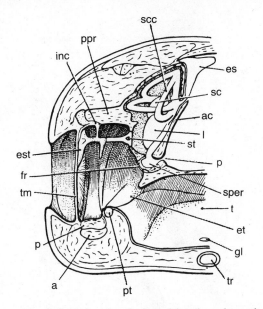

Figure 384. Posterior view of the left side of a lizard head to show the auditory apparatus. A shallow external depression leads to the ear drum (*tm*). Internal to the drum, the stapes is seen, divided into two parts, the "extracolumella" (*est*) and columella or stapes proper (*st*); processes from the former articulate above with the skull (*inc*) and below with the quadrate anterior to the middle ear cavity. This cavity opens by a broad eustachian tube (*et*) to the throat. The internal ear is shown in diagrammatic fashion. Other abbreviations: *a,* articular bone of lower jaw (= malleus); *ac,* inner wall of auditory capsule; *es,* endolymphatic sac; *fr,* fenestra rotunda; *gl,* glottis; *l,* lagena; *p,* perilymphatic duct connecting inner ear with cranial cavity; *ppr,* paroccipital process of otic region; *pt,* pterygoid; *q,* quadrate (= incus); *sc,* sacculus; *scc,* semicircular canals; *sper,* position of perilymphatic sac; *t,* tongue; *tr,* trachea. (From Goodrich, after Versluys.)

membrane, which picks up air vibrations. This membrane may incorporate the operculum of the fish. The spiracular pouch is the **middle ear cavity**; its connection with the pharynx is the **eustachian** or **auditory tube**. The pouch must enlarge, of course, to surround the hyomandibular. The hyomandibular bone of fishes changes its function to become a rodlike **stapes** (or columella), which crosses the middle ear cavity. Externally, it attaches to the eardrum and picks up the vibrations received by it. Internally, an expanded footplate of the stapes fits into an opening in the otic capsule, the **fenestra ovalis**; through the agency of the stapes, air vibrations are carried inward to set up vibrations in the liquids of the internal ear and eventually reach its sensory structures.

The fish spiracular opening lies high up on the side of the head (Fig. 385 *A*), the position in which the otic notch and presumably eardrum lay in the older fossil amphibians (Fig. 385 *B, E*). In amniotes, however, the drum usually lies farther posteriorly and ventrally (Fig. 385 *F*); although in reptiles the drum may be supported by the adjacent skull bones, it seems certain that these supports are new developments, not directly inherited from the primitive otic notch of early amphibians. Further, it is not at all certain that the drum is strictly homologous in all cases or even if one was present in many fossil forms. Primitively, the drum was almost flush with the surface

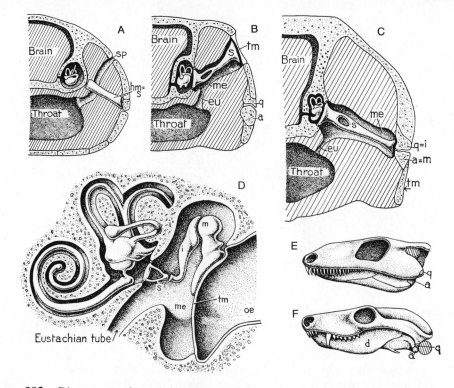

Figure 385. Diagrams to show the evolution of the middle ear and auditory ossicles. Diagrammatic sections through the otic region of the head of *A,* a fish; *B,* a primitive amphibian; *C,* a primitive reptile; *D,* a mammal (showing the otic region only); *E,* side view of the skull of a primitive land vertebrate; *F,* of a mammal-like reptile to show the shift of the eardrum from the otic notch of the skull to the region of the jaw articulation. Abbreviations: *a,* articular; *d,* dentary; *eu,* eustachian tube; *hm,* hyomandibular; *i,* incus; *m,* malleus; *me,* middle ear cavity; *oe,* outer ear cavity; *q,* quadrate; *s,* stapes; *sp,* spiracle; *tm,* tympanic membrane. (From Romer, Man and the Vertebrates, University of Chicago Press.)

of the head. It is still in this position in anurans and turtles; in most reptiles and in birds, however, it attains some protection by being placed at the inner end of a short external meatus.

The eardrum is in general a thin membrane capable of rapid vibration in response to air waves. It contains a sheet of fibrous tissue and is bounded outside and inside by epithelia derived from the skin and from the lining of the first gill pouch, respectively. Its presence is very probably due to the failure of the spiracular pouch to break through to the surface. The pouch itself expands lateral to the braincase to form the air-filled middle ear cavity—a cavity that, in crocodiles, birds, and mammals, may extend its air spaces into the neighboring bones of the skull (as in the mammalian mastoid). Primitively, as in anurans and many lizards, the middle ear cavity opens broadly into the pharynx; in most amniotes, however, the connection is narrowed to form the eustachian (or auditory) tube. In many amphibians (as noted later), the drum and middle ear cavity are lacking, probably secondarily. Such a loss is also found in snakes and many lizards. It has been suggested that in snakes this is due to the probability that ancestral forms were burrowers, but the specialized jaw mechanisms developed in ophidians may be responsible. A specialization of the turtles and snakes is the development of a secondary outer wall of the otic capsule around the shaft of the stapes; in the space beneath this is an extension of the fluid-filled cavities of the inner ear.

In the fish ancestors of the tetrapods, the hyomandibular, once part of the gill apparatus, functioned as a jaw prop, bracing the quadrate on the otic region of the skull, and lay just behind the spiracular pouch. In most tetrapods, jaw suspension is adequately taken care of by the skull itself and the muscles of the temporal region. The hyomandibular, as the stapes, functions in transmitting vibrations from eardrum to internal ear and comes to lie within the bounds of the expanded middle ear cavity. Exactly when this change in function occurs is not certain; in numerous early tetrapods, the element is still massive and appears more suited for support than sound transmission. The stapes, essentially a rodlike structure, has its main outer attachment to the eardrum; medially, there has developed, underneath an original attachment to the braincase, an opening into the internal ear capsule, the fenestra ovalis. In reptiles and birds, the stapes often has a complicated build. The structure as a whole is often called the **columella**, and the stapes proper is considered to be the proximal part of the shaft. The latter is always ossified, sometimes pierced by an arterial foramen, and ends medially with an expanded footplate covering the oval window. The outer part, often termed the **extracolumella**, may remain cartilaginous and have a complicated structure. The direct outer termination lies, of course, on the eardrum, but there may be extra processes connecting, as in fishes, with the quadrate region of the jaw articulation, with the skull dorsally, and ventrally with the hyoid arch (of which the stapes was originally a part).

External and Middle Ear in Mammals (Fig. 385 *D*).   In mammals, the external ear region is well developed. There is a deep, tubular external meatus. In some reptiles, the meatus may be surrounded by cartilage; this is expanded in mammals, and, in addition, there is almost always a projecting ear **pinna** of elastic cartilage; this structure, often elaborately folded, may be of value as a collector of sound waves.

The mammalian middle ear is, as noted in an earlier chapter (cf. p. 259), generally enclosed in a bony bulla. Instead of a single auditory bone, the stapes, it contains an articulated chain of three **auditory ossicles** leading from eardrum to oval window—**malleus**, **incus**, and **stapes** (hammer, anvil, and stirrup)—named in relation to their fancied shapes.

The origin of this series of auditory ossicles was long debated. It was thought at one time that they might be due to a subdivision into three parts of the reptilian columella or stapes. Embryology, comparative anatomy, and paleontology combined have, however, revealed the true story. The inner element, the stapes, although much shortened, is equivalent to the whole columellar apparatus of a lower tetrapod. The other two elements are homologous with the articular and quadrate bones, which in lower vertebrates form the jaw joint. Mammals evolved a new joint system for the jaw, and the older skeletal elements of this region, now superfluous, have been put to new use. The reptilian eardrum lay close to the old jaw joint. The articular formed an attachment to the membrane and became the malleus. The quadrate, connecting in reptiles with articular on the one hand, and stapes on the other, has become the incus, and the homologue of the jaw articulation of a reptile lies in mammals at the joint between two ear ossicles (Fig. 194). These bones, originally branchial elements and then parts of the jaws, afford a good example of the changes of function that homologous structures can undergo. Breathing aids have become feeding aids and finally hearing aids.

Small muscles attach to two of the ossicles and help regulate the system. Interestingly, and properly, the muscle associated with the malleus is of mandibular origin, innervated by the trigeminal nerve; that associated with the stapes is hyoid and innervated by the facial nerve.

The Internal Ear in Reptiles (Fig. 386).   In tetrapods, as already indicated, the parts of the internal ear devoted to equilibrium remain little changed from the fish condition. The auditory apparatus, however, gradually develops into structures that attain such size and importance that the older regions of the sacs and canals are often termed (rather slightingly) the **vestibule** of the internal ear.

It is the lagenar region in which this expansion takes place. The macula of the lagena, mentioned earlier as present in fishes, persists in all tetrapods except mammals above the monotreme level but never develops to any degree; it has little if any auditory function in terrestrial vertebrates. The lagena itself, however, is important as a basal structure in the development of the cochlea.

The sensory area vital in the development of hearing in tetrapods is that termed the **basilar papilla** (Fig. 380 *D, E*). This area is unknown in fishes but is characteristically developed in many amphibians and in typical reptiles. In these forms, it is an area of hair cells, covered by a common gelatinous membrane, situated on the posterior wall of the sacculus at the base of the lagena. It is here that vibrations, typically brought in from without by the stapes, are received.

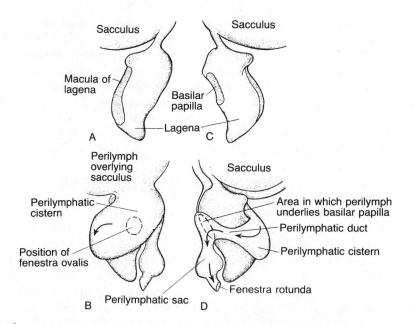

Figure 386.   The ear of a late embryo of a lizard (*Lacerta*). *A,* Left ear, lateral view of the membranous labyrinth in the floor of the sacculus and the lagena. *B,* The same with the perilymphatic system shown in addition. *C, D,* Medial views comparable to *A* and *B,* respectively. Arrows indicate course of conduction of vibrations from stapes to basilar papilla and on to the "round window" at the distal end, beyond the perilymphatic duct.

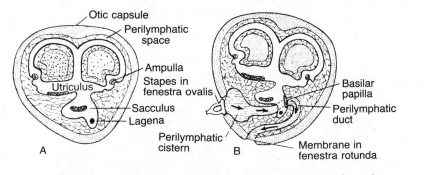

Figure 387. *A,* Schematic section through the ear capsule of a fish, to show the perilymphatic space surrounding the membranous labyrinth containing endolymph. *B,* Similar scheme of a tetrapod, in which a part of the perilymph area (arrows) is specialized to conduct sound from the fenestra ovalis to and past the auditory sensory area. The maculae are darkly shaded. (After de Burlet.)

In general, the sacs and canals of the inner ear are not adherent to the skeletal walls of the auditory capsule that surround them. They are separated from them by spaces filled with liquid and crossed by connective tissue strands. This liquid is the **perilymph** (Fig. 387). It surrounds the membranous labyrinth and thus is quite distinct from the endolymph within. From the perilymphatic spaces, there develops in tetrapods a conduction system that leads from the fenestra ovalis to the basilar papilla and forms the last link in the chain of structures by which potentially audible vibrations reach the auditory sensory areas.

Internal to the oval window in lower tetrapods, there develops a large **perilymphatic cistern**, against which the stapes plays (Figs. 386, 387 *B,* 388 *A*). Vibrations received here are, in typical reptiles, carried around the lagena to its posterior border in a perilymph-filled canal. This duct passes just outside the area of the basilar papilla and is separated from the base of its sensory cells only by a thin **basilar membrane**. Vibrations in this membrane agitate the hair cells and at long last, in this roundabout manner, the sensory organ is reached. This situation—an auditory sensory structure agitated by vibrations of the membrane at its base—is a fundamental feature in the construction of the hearing apparatus in all amniotes.

As a final point here, it must be noted that for vibrations set up by the stapes in the perilymph and carried along by the perilymphatic duct, some release mechanism, allowing a corresponding vibration, must be set up at the far end of the perilymphatic system. In many amphibians, this consists of a **perilymphatic sac**, which projects from the auditory capsule into the braincase. In some amphibians and in reptiles (except turtles), a further development takes place. There is a large, rather tubular foramen (primarily for the vagus nerve) in the wall of the braincase that runs out from a point near this sac. We find that the perilymphatic duct may, instead of terminating in a sac, run outward through this foramen—or a separate one formed close by, the **fenestra rotunda*** (Fig. 387)—and terminate in a membrane that vibrates in phase with the impulses received through the fenestra ovalis at the other end of the perilymph system. Primitively, this membrane is buried in the tissues at the margin

---

*There is, unfortunately, also a foramen rotundum in the skull—they are *not* synonymous. The same unhappy situation occurs with the oval "hole" and "window" (foramen ovalis and fenestra ovalis).

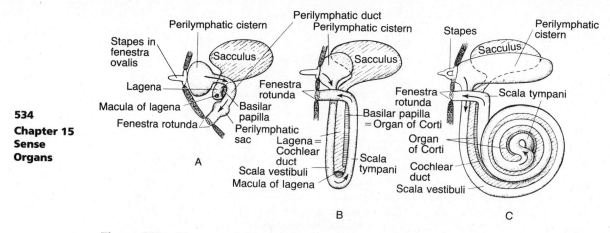

Figure 388. Diagrams of the saccular region to show the evolution of the cochlea. *A,* Primitive reptile with a small basilar papilla adjacent to the perilymphatic duct. *B,* The crocodile or bird type; the lagena has elongated to form a cochlear duct, the basilar sense organ with it, and a loop of the perilymphatic duct follows the cochlear duct in its elongation. *C,* The mammalian type; the cochlea is further elongated and coiled in a fashion economical of space.

of the skull; in more advanced forms, this round window opens back into the middle ear chamber.

Development of the Cochlea. Both birds and mammals have greatly extended their hearing ability by the evolution of a **cochlea**, a highly developed structure for auditory reception. The crocodilians, related to the birds, demonstrate the manner in which the cochlea was developed. Three structural features are involved: the lagena, the perilymphatic duct, and the basilar papilla.

In the crocodiles and birds (Fig. 388 *B*), the finger-like lagena has become expanded into a long but essentially straight tube, filled of course with endolymph; this is the **cochlear duct** or **scala media**. At the tip in these forms (and in monotremes as well) persists the original macula of the lagena, of doubtful function. The important sensory structure is the basilar papilla, which is here greatly expanded into an elongate area running the length of the cochlear duct and generally termed the **organ of Corti**. It has a complicated series of sensory hair cells and supporting cells; over them folds a membranous flap, the **tectorial membrane**, corresponding to the cupula in other neuromast structures; beneath is an elongate fiber-supported and vibratile basilar membrane. With the lengthening of this sensory organ, the perilymphatic duct, which remains closely applied to its base, expands in a double loop. The part of this tube leading inward from the oval window is termed the **scala vestibuli** (so-called because the fenestra ovalis is considered to lie in the vestibular part of the inner ear). The distal limb of the tube, leading to the round window with its membrane, or "tympanum," is the **scala tympani**. It is in this distal segment of the loop that the perilymphatic duct underlies the basilar membrane and the sensory organ.

These modifications of the primitive reptilian lagena and associated structures result in the development, in birds and their crocodilian relatives, of a true, although simply constructed, cochlea. Crudely one may visualize the formation of this type of cochlea by imagining that he has grasped the region of the lagena and the perilymphatic duct crossing its base and pulled this area outward. This pull has stretched the lagena out into a long blind tube as the cochlear duct; the perilymphatic duct, attached at both ends, has pulled out as a double-looped structure, the two scalae.

The mammalian cochlea (Figs. 388 *C*, 389), we believe, has developed independently of that of birds, but in closely parallel fashion. In monotremes, it is still a simple, uncoiled structure, readily comparable to that of birds. The typical mammalian cochlea is, however, much more elongated than that of birds or crocodiles. If it were to remain straight, it could not be comfortably accommodated within the otic capsule, and, in relation to this, we find that (as the name, which means snail, implies) the mammalian cochlea of three tubes has been coiled into a tidy spiral structure.

The mammalian organ of Corti is, like that of a bird, readily derivable, by elongation and increased complexity of structure, from the primitive basilar papilla. Beneath it, adjacent to the scala tympani, is the basilar membrane, crossed by numerous fibers. Above, as in birds and crocodiles, is a tectorial membrane in which the tips of the hair cells are engaged. Centrally there is a curious little triangular tunnel, braced by pillar cells on each side; on both sides of this structure are rows of sensory hair cells, running the length of the cochlea—typically three or four rows toward the center of the cochlea, a single row externally. Rows of supporting cells are present

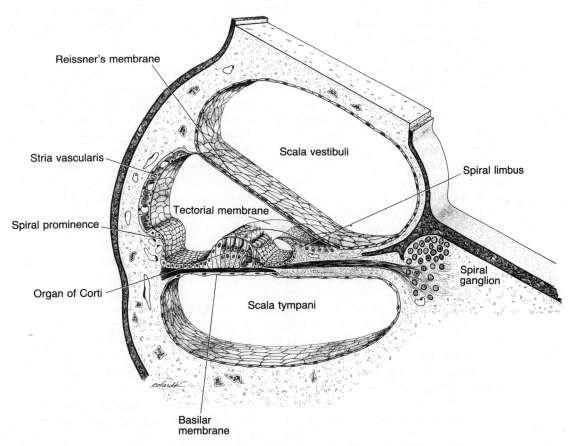

Figure 389. A much enlarged and schematic section through one turn of a mammalian cochlea to show the organ of Corti (cf. Fig. 388 *C*). (From Bloom and Fawcett.)

farther toward either edge of the membrane. As in lower tetrapods, vibrations in the basilar membrane caused by waves in the perilymphatic system cause hearing sensations through displacement of the hairs of the sensory cells above the membrane. Although in a primitive tetrapod hearing system the vibrations reach the basilar membrane and basilar papilla (organ of Corti) only from below (cf. Fig. 388 A), here vibrations entering this triple system via the scala vestibuli can "short circuit" the double loop and send impulses through the vestibular membrane and down to the basilar structures through the liquid of the cochlear duct. The functional "reason" for elongation of the basilar papilla into the formed organ of Corti appears to be the discrimination of sounds of different pitch, i.e., of different wavelengths. Differences in the transverse span of the basilar membrane, which grades in width from one end of the cochlea to the other, appear to result in making the membrane sensitive to different wavelengths at different parts of its extent and thus rendering tonal distinctions possible. The apex of the cochlea, where the basilar membrane is broadest, is most sensitive to the lowest tones, the narrow region of the membrane at the base of the cochlea to high notes. Even in the relatively short avian cochlea, there is excellent discrimination of tone, and, in mammals, the range of vibrations heard may be extremely wide. The human ear, for example, can respond to frequencies ranging from about 15 to 16,000 or so cycles per second; typical bats (Microchiroptera) can emit and receive sounds in pitch far above those audible to a human ear and have evolved an echo location system that aids them in flying in the dark and in capturing insect prey on the wing.

The Middle Ear in Amphibians. We have omitted almost all reference to ear construction in living amphibians, for the reason that conditions in these forms are not, in general, primitive but are highly specialized and extremely varied. Three main points may be noted: (1) the drum and middle ear cavity are often missing; (2) the stapes is frequently reduced or absent; and (3) a second ossicle, the operculum (not the operculum of the fish), is frequently present in the oval window.

A majority of frogs and toads have a well-developed middle ear, with a large, superficially placed eardrum and a good stapes,* carrying vibrations inward from drum to oval window. But in urodeles and the Gymnophiona, and even in a number of anurans, there is no drum or middle ear cavity (possibly, but debatably, a primitive feature; possibly a result of secondary loss), and typical sound transmission is impossible. The stapes may persist in the absence of drum and cavity and is well developed in the Gymnophiona and some salamanders. Its outer end articulates with the quadrate region of the skull, and it may in some way pick up for transmission vibrations in this region of the head (Fig. 390 A). In a great number of both urodeles and anurans, there is a second ear ossicle, the **operculum**, peculiar to amphibians. This is a flat plate fitting into the oval window. It may be present in the fenestra in company with a well-developed stapes, as in typical frogs and toads; in many forms, however, the stapes is reduced, and the operculum becomes the major sound transmitter. The operculum appears to have no precise homologue in the skeleton of other groups of vertebrates; it is thought to be a separate part of the wall of the otic capsule itself. In all terrestrial and even some aquatic urodeles and anurans, the operculum is connected with the shoulder blade by a small **opercular muscle**, a specialized axial

---

*The stapes, or columella, is sometimes termed the plectrum in anurans.

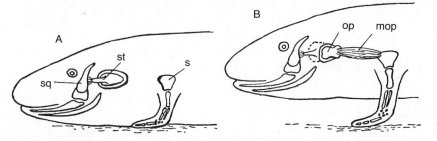

Figure 390. Diagram to show the operculum (*op*) and its muscle (*mop*) in urodeles. *A,* An aquatic form in which the stapes, or columella, picks up vibrations by a ligamentous attachment to the squamosal. *B,* A terrestrial form in which the operculum and opercular muscle are well developed; in this case, the stapes is reduced, but both it and the operculum may be well developed in other forms. Abbreviations: *s,* scapula; *sq,* squamosal; *st,* stapes. (After Kingsbury and Reed.)

muscle (Fig. 390 *B*). This muscle seems responsible for maintaining proper tension in the operculum; it probably does *not* actually transmit vibrations from the front leg to the inner ear.

The Inner Ear in Amphibians. In inner ear as in middle ear, the amphibians were not discussed earlier because they tend to exhibit specializations that obscure the general story. There is typically developed in modern amphibians, as in reptiles, a perilymphatic cistern, perilymphatic ducts, and a perilymphatic sac or release mechanism at the base of the skull. But though amphibians may have a basilar papilla sensitive to high-frequency sounds, this is often absent. There is usually a special amphibian sensory structure, the **papilla amphibiorum**, which appears to play the main role in hearing low-frequency sounds (Fig. 391). This is situated near the upper margin of the sacculus. It is structurally similar to the simpler types of basilar papillae. However, the perilymphatic duct does not lie beneath the base of this structure; instead, it approaches it, so to speak, laterally, being separated from the endolymph in a neighboring region by a thin membrane through which vibrations are transmitted to the latter fluid. The same condition holds for the relations of the perilymph to the basilar papilla in amphibians.

Evidently, auditory structures were in an experimental stage in the ancestral amphibians. Here, as in many other features, the ancestors of the modern amphibians chose a series of different paths than did their ancient relatives who gave rise to the amniotes.

Figure 391. Schematic section of the internal ear in a salamander. The basilar papilla is absent here but is found in addition to the papilla amphibiorum in many amphibians. (After de Burlet.)

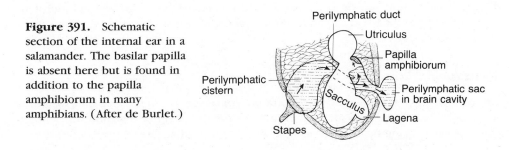

Chapter 16

The Nervous System

In a protozoan, the single cell in itself receives sensations and responds to them. In higher, metazoan organisms, there tends to be, to an increasing degree, a differentiation between cells specialized for the reception of sensations—**receptors**—and those that make the appropriate response—the **effectors.** In lowly multicellular organisms, the relations between these two types of cells may remain relatively simple, and cells receiving sensations may, by their physical and chemical activities, arouse their neighbors to respond. In the vertebrates, the circulation of hormones in the blood stream is a retention of such an essentially primitive method of stimulation. But such a method of arousing responses tends to be slow, is often nonspecific, and is relatively ineffective for swift and precise reaction. With the development of a nervous system, special tissues designed to transmit stimuli rapidly from sensory structures to specific end organs (muscles or glands) are formed, which furnish the response.

In primitive metazoans, such as coelenterates, the nervous system may be a diffuse network of cells and fibers spread through the tissues. But in most animals of any degree of complexity, the system is an integrated, well-organized one. Instead of scattered cells and fibers, we find collections of fibers formed into nerve trunks and centers where transfer of nerve impulses between fiber systems takes place, much as a telephone system brings its numerous lines into major cables and central exchanges. In most groups, a dominant center, a brain in some manner or other, makes its appearance, usually in a situation strategically located with regard to major sense organs. In the vertebrates, there is a well-organized tubular brain anteriorly and a single dorsal hollow nerve cord, the spinal cord, running backward from this along the body. These form the **central nervous system.** Running outward from brain and spinal cord are numerous paired nerves that constitute the **peripheral nervous system.** Clusters of nerve cells, **ganglia,** may lie along these nerves and are also part of the **peripheral nervous system.** We have noted that embryologically the tissues of the nervous system are of ectodermal origin—mainly from the neurectoderm that forms the neural tube and the adjacent neural crests, plus some addition from nearby ectodermal placodes.

Structural Elements

The Neuron. Nervous tissues, in the brain, spinal cord, and ganglia, contain numerous cell bodies. Prominent in the nervous system, however, are slender but elongate nerve fibers, making up the nerves and much of the substance of the central nervous system as well. Most nerve fibers are exceedingly long, and connections between them and cell bodies are hard to see clearly; however, these fibers are universally processes of cells rather than independent structures. The basic units of the nervous system are **neurons**; each consists of a cell body and its processes, long or short.

Most of the cell bodies of the neurons (Fig. 392) are situated within the walls of the central nervous system, a relatively small number in peripheral ganglia. The shape of the neural body is variable; in many cases, however, it is stellate, owing to the fact that a number of processes extend out from it. With appropriate stains, microscopic preparations show in the protoplasm of the cell body various characteristic structures not present in other tissues. Notable are clusters of darkly staining materials, **Nissl bodies**; these contain large amounts of RNA and numerous ribosomes. Clearly the chemical manufacturing work for the entire neuron is performed in the cell body, the products formed here flowing outward to supply the various processes. In the adult, there is no evidence of mitosis in the neurons, indicating, as a notable peculiarity of the nervous system, that a full complement of nerve cells has been attained, for the most part at least, by about the time of birth or hatching. In consequence, destruction of nerve cells through injury or disease is a permanent loss; cell processes may regenerate if the cell body is intact, but a neuron, once lost, cannot be replaced.

Nerve Fibers. Extending outward from the cell body of the neuron are slender processes along which the nerve impulses pass; the cell body is in great measure a center for their maintenance and nutriment. The distribution and length of such pro-

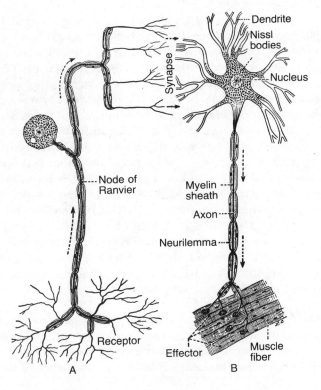

Figure 392. Two types of nerve cells. *A,* Afferent (sensory) neuron of spinal and cranial nerves; long axon-like processes run from sensory receptor to cell body in spinal ganglion and thence into cord, where ramification occurs. *B,* Efferent (motor) neuron, with cell body in cord and a long axon extending out to the effector (muscle fiber). In both parts, the peripheral processes (axon of the motor cell) are greatly abbreviated; at this scale they should be several meters long. (From Millard, King and Showers, Human Anatomy and Physiology.)

cesses are quite variable. Most often thought of as typical are such neurons as those that innervate striated body muscles (Fig. 392 *B*). In these motor neurons, short processes—slender, numerous, and branching—that carry impulses inward toward the cell body, are termed **dendrites**, from their treelike appearance. There is a single **axon**, a relatively stout and elongate process, that may be in large animals as much as several meters in length, carrying impulses away from the region of the cell body. A second common type (Fig. 392 *A*) is that of the afferent neurons that carry sensory stimuli inward to the central nervous system; here both the long process, leading in from the point of reception to the cell body adjacent to the cord, and a second long process that enters the cord, are comparable in structure to the axon of a motor neuron. Because of the presence of several dendrites as well as a single axon, a motor neuron may be termed **multipolar** (such neurons are also present in the central nervous system). Typical sensory neurons are **bipolar**, with two long processes.

Functionally, the most important part of a major nerve fiber is its central structure, the **axis cylinder**, a thin strand of protoplasm continuous with that of the cell body. Its appearance is homogeneous in unstained materials; however, appropriate stains show the presence within it of numerous threadlike longitudinal neurofibrils; these are continuous with fibrils found within the cell body of the neuron. The axis cylinder is bounded by a very thin plasma membrane; beyond this, however, there are almost always further wrappings. The cylinder is sheathed by a series of cells that produce a tough, inelastic membrane, the **neurilemma**. In the central nervous system, the sheathing cells are part of the neuroglia (cf. p. 565); outside the cord the covering is formed by a series of sheath cells, **Schwann cells**, derived from the neural crest. These cells, further, produce a **myelin sheath** of fatty material surrounding the fiber.

When the sheath, an effective insulating material, is well formed, it gives a shiny, glistening appearance to the fibers. Although most peripheral axons have myelin sheaths, this is not always the case; there are some unmyelinated fibers. Generally, the sheath cells wrap themselves around segments of the axis cylinder in a manner crudely comparable to a jelly roll except that the turns are very numerous and very thin. In the intervals between the territory of two successive sheath cells, the myelin sheath of peripheral fibers is interrupted, and there is a **node of Ranvier**.

If nerve fibers are cut, the part distal to the cut degenerates, and the proximal region and cell body may show evidence of damage. But frequently peripheral fibers regenerate, growing out again from the stump still connected with the cell body. It appears that they are aided in retracing their former paths by the persistence of the sheath cells that surrounded the former axis cylinder. Fiber degeneration is a feature of great value in neurologic work. Particularly in the central nervous system, it is difficult to trace groups of long fibers through from origins to termini. However, a group of problematic fibers may be cut experimentally and allowed to degenerate. If the material is stained in an appropriate way and sectioned, microscopic study enables one to distinguish the degenerating fibers from their normal neighbors and to trace them to their destination.

The Nerve Impulse. Investigation of the nature and properties of the impulses transmitted by nerve fibers has been the object of a vast amount of physiologic work. By analogy, one tends to compare nerve transmission with an electrical impulse, and, as an impulse travels along a fiber, there is a change of the electric potential at the fiber surface associated with a shift in sodium and potassium ion concentrations within and without the membrane. The energy needed is obtained (as in the case of muscle contraction) from the breakdown of ATP in the axis cylinder. Hence, electric and metabolic phenomena go hand in hand in the promulgation of the impulse.

A nerve impulse, though rapid, cannot be compared for speed with an electric impulse. On the whole, mammalian fibers give speedier transmission than do those of lower vertebrates—about twice the speed of fibers in a frog, for example. But even in mammals, the highest recorded speeds are only about 130 meters per second, and some mammalian fibers have a speed of only about half a meter a second. The faster speeds are found in the fibers with greater diameter; these diameters range in general from 1 μm to 20 μm. Although the speed of transmission along some fibers is considerable, there is, nevertheless, a distinct if small time lag between stimulation and response in such a large animal as an elephant, for example, because of the time necessarily consumed by the stimulus in even the simplest reflex in travelling many meters along a fiber.

A nerve impulse is anonymous and nonspecific. The nature of a sensation "felt" in the brain depends upon the centers that received it, not on any difference in the type of impulse received.

Nerve impulses normally travel in only one direction along the fibers, although the fiber is equally capable of transmitting impulses in either direction, as can be shown experimentally. The unidirectional transmission actually found in the working nervous system is due to the pattern of connections among fibers; neurons are morphologically "polarized."

A nerve impulse is an "all or none" phenomenon. There are no strong or weak impulses; either the fiber reacts fully, or it does not react at all.* True, there may be variations in the strength of the impulses sent along a nerve, but these variations are due to factors of other sorts. There may be differences in the number of fibers in a nerve that are stimulated at one time or another. Most important is the fact that impulses do not come singly, but follow one another in variable rapid series, so that their effect may be cumulative.

The Synapse. Never, among vertebrates, does a single neuron span the entire distance between the sensory organ (receptor) and the motor or glandular structure stimulated (effector). The action takes place through a chain of neurons, at least two and usually more. The point of transfer between successive neurons is termed a **synapse**. The end of the axon of the first of the two neurons breaks up into fine branches, typically with knoblike ends, or boutons that intertwine with the dendrites of the second neuron or twine about the cell body of that element. Although processes of the two cells concerned may be closely appressed (to as close as 20 nanometers) in vertebrates, there is no protoplasmic continuity between them. In many cases among invertebrates, the transmission of the impulse across the synapse is electrical. This, however, appears to be relatively rare among vertebrates. The timing of nerve transmission shows that generally a distinct, although very small, time interval is taken up by the bridging of a synapse. This is related to the fact that most vertebrate synapses are bridged chemically. Minute amounts of chemical material, a **neurohumor**, are produced at the terminal branches of the "incoming" fiber and diffuse across the gap

*In the higher brain centers, special types of neurons may act, exceptionally, in different fashion.

to stimulate the second neuron. Acetylcholine is the most common neurohumoral material.

The Reflex Arc. Before considering more complex anatomic structures, we may note the general nature of a simple type of nerve action of the sort known as a **reflex** (Figs. 393, 394 *A*). Reflex actions are those of such type as the "automatic" withdrawal of a bare foot that has trod on a tack or of a finger that has touched a hot stove. In general, a sensory stimulus is picked up from receptor cells and carried in toward the central nervous system by a long **afferent** nerve fiber. The cell body of the **sensory neuron** to which this fiber belongs lies, along with many others, in a ganglion close to cord or brain; the fiber, however, continues directly onward into the central nervous system. Here, normally, it does not connect simply with one successive neuron but branches so that a stimulus may be produced in a whole series of neurons with which the sensory neuron synapses. Conversely, each neuron with which it connects may receive impulses from numerous afferent fibers, so that a considerable amount of interplay between receptors and effectors may take place. Presumably one single incoming impulse will rarely "set off" the second neuron; excitation only occurs when impulses arrive from a number of sources.

In the simplest of reflexes, the neurons stimulated by incoming impulses may be **efferent** or **motor neurons**. Their cell bodies lie within the central nervous system, and long axons run out from them to the effector end organs (usually muscle fibers). But almost all reflexes other than those involved in maintaining postural muscle tone are usually one stage more complicated than this, and are three-neuron rather than two-neuron chains. In them, the afferent fibers entering the central nervous system do not synapse directly with motor cells, but with neurons contained entirely within the central nervous system. These **association neurons** (or internuncial neurons) send out, like the afferents, branched processes that synapse with numerous motor cells. This further multiplies the number of possible responses that a sensory impulse

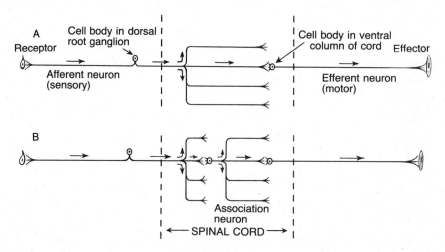

Figure 393. Diagrams to show simple reflexes. Area between broken lines is part of arc lying within the cord (cf. Fig. 394). *A,* Two-neuron reflex; *B,* association neuron interpolated, increasing the number of possible paths. This diagram is greatly oversimplified in that most efferent (and association) neurons would actually be receiving input from many different neurons, thereby producing convergence and summation.

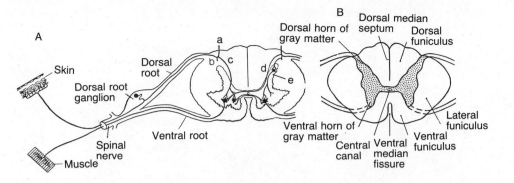

Figure 394. *A,* Diagram of mammalian spinal cord and nerve to show path of reflex arc. A sensory fiber entering by the dorsal root may send branches, *a, b,* up and down the cord. At various levels, the sensory fiber may connect with motor neurons of the same side, *c,* or opposite side of the cord, *d,* or with association neurons, *e. B,* Diagrammatic section of a mammalian spinal cord, to show distribution of white and gray matter and funiculi. (After Gardner.)

may cause and, conversely, also increases the number of sensory impulses that might produce any given motor effect. Summations of stimuli may occur, and choices, so to speak, may be made as to possible reactions to stimuli. We have, previously, assumed that the result at a synapse is stimulation of the next cell. But frequently the result is inhibitory, tending to counter stimuli that this cell may have received from other sources. Still further, even at this relatively simple level of nervous activity, there are quite surely feedback mechanisms that increase complexity. The interposition of a single association neuron in the system is apparently as far as any normal spinal reflex usually goes in building up end-to-end neuron chains but gives some clue to the way in which more complicated brain mechanisms may have evolved. The intercalation of still further association neurons and more complex feedbacks, modulations, and reverberating circuits, could well produce higher association centers, into which a wide variety of sensory impressions may drain and from which may come a wide variety of responses after the various messages received have been sorted, compared, and integrated.

Spinal Nerves

In contrast to the central nervous system, the peripheral nervous system, which we shall first consider, is simply constructed. It consists essentially of the nerves that penetrate to almost every region of the body: groups of fibers along which the sensory impulses are brought into the spinal cord and brain by afferent pathways, and out through which resulting efferent impulses pass to affect muscular or glandular structures. Included, too, in the peripheral system are the ganglia found along the course of the nerves and containing the cell bodies of peripheral neurons. We shall discuss spinal nerves, relatively simple and uniform in structure, before considering the more specialized nerves of the cranial region.

The typical spinal nerves of most vertebrates (Fig. 394 *A*) are paired structures present in every body segment. Each nerve arises from the spinal cord by two roots, dorsal and ventral (posterior and anterior in specialized human terminology). The

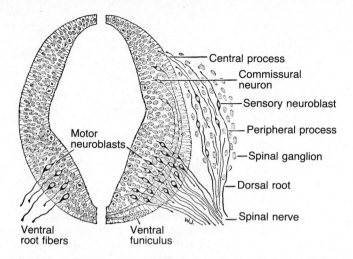

Figure 395. Sections of the spinal cord of early mammalian embryos. *Left,* axons are growing outward from motor nerve cells. *Right,* association or commissural neurons are developing within the cord and sensory neurons are developing externally from neural crest cells (cf. Fig. 77). These sensory neuroblasts are at this stage bipolar, i.e., with separate proximal and distal processes; later the two processes fuse proximally in higher vertebrates to give a unipolar condition to the mature ganglion cell (but one with a T-shaped process). (From Arey.)

ventral root runs straight outward from the ventral margin of the lateral wall of the cord; the **dorsal root**, which bears a prominent ganglion, enters the cord on its lateral wall. In most vertebrates, the ventral and dorsal roots unite to form the main trunk of the nerve before leaving the canal of the vertebral column. Emerging from the vertebral column, the nerve divides into a variety of branches, or rami. Neglecting for the time a ramus running toward the viscera, we may note the frequent presence of a major **dorsal ramus** supplying fibers to the dorsal part of the axial musculature and the skin of the back, and a **ventral ramus** to the more lateral and ventral parts of the skin and hypaxial musculature of the body wall.

The main trunk of the spinal nerve and its principal branches carry both afferent and efferent fibers. The two roots, however, show a sharp division of functions in higher vertebrates. The ventral roots carry only outward bound impulses through efferent fibers, whose cell bodies are contained within the cord (Fig. 395). In all typical spinal nerves of higher vertebrate groups, the dorsal root, in contrast, carries mainly afferent fibers with incoming impulses. The cell bodies of these neurons generally lie in the dorsal root ganglion; they send out a stem that, almost immediately, branches in a T-fashion; one process brings impulses inward from sensory structures, and a second carries them centrally into the cord.

Spinal Nerve Plexuses. * In many typical segments of the trunk, each spinal nerve is a discrete structure, without connections with its neighbors fore or aft; it inner-

---

*In Latin, singular and plural forms of the word plexus are identical in spelling; we use an anglicized plural here for the sake of clarity.

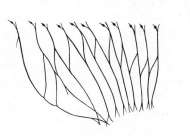

Figure 396. *Left,* nerve plexus supplying the left pelvic fin of a chimaera, showing interchange of fiber bundles between members of the series of spinal nerves concerned. The two roots and the dorsal ganglia of the nerves concerned are shown at the top of the figure; the base of the fin is at the bottom. *Right,* on a larger scale, the brachial plexus supplying the pectoral limb of a mammal. The nerve roots are not shown, and the branches of the plexus are cut short of their terminations; the largest trunk is that of the radial nerve supplying most of the forearm and front foot. The number of nerves involved is less than in the case of the fish fin, but the pattern of the plexus is more complex. (Right figure, partly after Walker.)

vates the trunk muscles that have arisen from the myotome pertaining to its segment and on the sensory side supplies a corresponding strip of skin, although with some overlap with the nerves of the adjacent segments. In certain regions, however, there is an interweaving of branches of a number of successive spinal nerves to form a **plexus** (Fig. 396). In forms with well-developed paired limbs, there is associated with each a well-developed nerve plexus, the **brachial plexus** and **lumbosacral plexus** for front and hind legs, respectively; they are less complex in fishes. As a result, we find that the muscles of any limb may be supplied by fibers from several spinal nerves. This condition is perhaps to be associated functionally with the development of complex limb reflexes, and structurally with the fact that the appendicular musculature in tetrapods has lost all trace of its original segmental nature.

Constancy of Innervation; Nerve Growth. Even in complicated limb plexuses, we find that there is a high degree of constancy in the innervation of a given muscle in various animals by twigs from seemingly comparable nerve stems—a feature of practical value in the study of muscle evolution and homology (cf. p. 279). This and other facts tended to create the doctrine that innervation is an absolutely constant feature—that a given muscle is always, in every generation and over long evolutionary lines, innervated by the same nervous elements and by similar pathways. Actually, however, there are notable (although relatively rare) cases where such a doctrine cannot be maintained. Any discussion of this problem leads to a consideration of nerve growth.

The peripheral nerves are derived embryologically from two sources (Fig. 395). The fibers of efferent neurons emerge from cell bodies within the cord and grow out from the cord toward the muscles that they innervate. In the vertebrate ancestors, the afferent fibers may have had cell bodies in the cord; neurons of this sort are found in the young of many anamniotes. Even in the adults of cyclostomes and some teleosts, we find some cell bodies of afferent neurons in the dorsal part of the cord rather than in a dorsal root ganglion. Generally, however, these cells do not develop in the cord, but in the neural crest, a versatile structure that we noted in Chapter 5 to have been produced along the line of closure of neural tube from skin ectoderm. Cells from the neural crest migrate ventrally to form the spinal ganglia and grow fibers both inward to the cord and outward to the periphery. In addition, the crest supplies part, at least, of the cells sheathing peripheral nerve fibers and the cells in the ganglia. The neural crest represents, in a sense, a "spilling over" of neural materials from the restricted area that forms the cord. In the head there is a still further expansion of the area of origin of neural tissues, because the sensory cells of many of the cranial nerve ganglia may arise from thickenings, placodes in the ectoderm lateral to the forming neural tube and crest.

As can be observed experimentally, peripheral nerve fibers grow out as long processes that finally reach their end organs: sensory structures, muscle fibers, or gland cells. How do they do this? Some investigators have assumed that there is a sort of specific, mystic affinity between each nerve fiber and the special organ to which it attaches, so that the nerve fiber "finds its mate." That there is a degree of specific association is certain, because efferent fibers do not attach to sensory structures, nor afferent fibers to muscle cells. Experimental work demonstrates that the presence of materials to be innervated does attract nerve fibers in some broad way; but such work also furnishes good evidence against precise specific affinities. For example, a salamander limb bud, moved in the embryo to an unusual position, may become connected with nerves; but these may be nerves arising from body segments quite different from those that normally supply it. Though the manner of growth of nerve fibers is not completely understood, clearly many fibers tend to push out from the nerve cord along paths of least resistance through the surrounding materials. Since in successive generations the topography of a region will tend to be the same, given nerve fibers will tend to find and follow similar courses and hence reach and innervate similar structures. Essential constancy of innervation may thus be explained without the necessity of postulating precise specificity.

Nerve Components and Spinal Nerve Composition (Fig. 397).   Highly useful in the study of both peripheral and central nervous systems is the doctrine of nerve components. Both afferent and efferent types of nerve connections can be divided into two types: **somatic** and **visceral**. We have, in discussing sensory structures, noted this type of subdivision. **Somatic afferent (sensory) fibers** are those carrying impulses in from the skin and the sense organs of muscles and tendons—the exteroceptive and proprioceptive groups of the physiologist. **Visceral afferents (sensory)**, on the other hand, carry sensations from interoceptive sensory endings in the gut and other internal structures. On the motor side, **somatic efferents (motor)** are fibers to the somatic musculature—striated voluntary muscles of the body and limbs derived (directly or, at any rate phylogenetically) from myotomes (cf. Chap. 9). **Visceral efferents (motor)**, on the other hand, carry stimuli primarily to the visceral musculature of the gut, blood vessels, etc., and glands in various regions. There are thus four components present in most spinal nerves: the two afferents in the dorsal root, and two efferents in the ventral. All four components are present in the trunk of a typical spinal nerve as it emerges from the vertebral column. Beyond this point, however, the larger and more characteristic trunks are composed mainly of the two

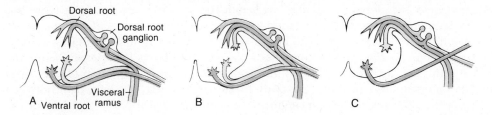

Figure 397. Diagrams showing the distribution of nerve components in dorsal and ventral spinal roots. Somatic sensory, blue; visceral sensory, green; visceral motor, yellow; somatic motor, red. *A,* Mammalian condition; the dorsal root is almost purely sensory, and almost all motor fibers are in the ventral root. *B,* More primitive type, common in lower vertebrates; some visceral motor fibers emerge through the dorsal root. *C,* Probable primitive condition; dorsal and ventral roots are separate nerves, visceral motor fibers are part of the dorsal nerve, and the ventral nerve is purely somatic motor (cf. Fig. 398).

types of somatic fibers; most visceral fibers turn ventrally in a visceral ramus, and their further distribution will be discussed in the next section.

The arrangement of the spinal nerve roots just described is usually assumed to be that present in vertebrates as a whole. As a matter of fact, however, as we descend the scale, we find increasing evidences of a different structure, which give us a clue to some of the peculiar features of the cranial nerves studied later.

In higher vertebrate classes, both afferent components are present dorsally in every known case, and both efferent components are in general confined to the ventral root. But, in at least some cases among amniotes, a limited number of visceral efferent fibers are carried by the dorsal root as well as the ventral, and in amphibians and jawed fishes, where investigated, dorsal visceral efferent fibers appear to be common (Fig. 397 *B*). In the higher vertebrates, the two roots emerge from the column at about the same level of the cord, one above the other; but in some fishes, there tends to be an alternation of dorsal and ventral roots, and, in sharks and hagfishes, the union of the two roots is less intimate than in higher forms. The lowest evolutionary stage appears to be that seen in the lampreys and amphioxus (Fig. 397 *C*). Here dorsal and ventral roots do not connect but are quite separate nerves. Further, visceral efferent fibers in amphioxus are confined to the dorsal roots; these roots thus carry three components, and the ventral root is purely somatic motor in nature. In lampreys, some visceral efferents emerge with the ventral roots, but a majority lie within the dorsal root. Still further, dorsal and ventral roots alternate in position in both amphioxus and lampreys.

Probably this condition of separate roots was that present in vertebrate ancestors; the union of the two roots may not have occurred until some stage in early fish history. That the visceral efferent fibers were primitively carried by the dorsal root and only later tended to shift to emerge by the ventral root is suggested by the stages cited; further, we will see later a retention of the seemingly primitive three-to-one split of components in various cranial nerves in all vertebrates.

In the alternating position of the separate dorsal and ventral spinal nerves of the lamprey or amphioxus, the ventral roots emerge opposite the middle of each body segment, and the dorsal roots are intersegmental in position (Fig. 398). This is a logical situation for the somatic components, at least. The sole function of the primitive ventral root is the innervation of musculature formed from the somites, and the most effective position for the emergence of a group of fibers is opposite the middle of a somite. On the other hand, the most important primitive function of the dorsal root is innervation of the skin; the body surface is best reached by a path between successive somites, with an intersegmental position a logical point of origin of the nerve from the cord. Spinal nerves are segmentally arranged; but it would seem that this segmental arrangement is not inherent in the nervous system but has, historically, been imposed upon it by the mesodermal somites. We shall see, in the case of the brain, a partial imposition of a second type of segmental nerve arrangement in the brain stem because of the presence of the branchial structures, although this may also reflect the position of the somites.

Visceral Nervous System

In the remote ancestors of the vertebrates, the nervous system presumably consisted (as in acorn worms) of two parts that were only loosely connected with one another: one, a slightly organized set of superficial structures that responded to external stimuli; the other, a net of cells and fibers around the gut and other internal organs that enabled them to adjust themselves directly to internal stimuli. In the vertebrates, the more superficial portion, within which brain and nerve cord are developed, has become highly organized and dominant. The old nerve nets of the gut still persist, however, and, to some degree in all vertebrates, the gut may still make its own responses to internal stimuli. But with the rise of the central nervous system of cord and brain, these new structures have tended, so to speak, to conquer the nervous system of the viscera and in great measure to do away with its independence; the somatic animal takes over the visceral (cf. Fig. 14). Efficient connections, both afferent and efferent, have been established, and much visceral activity is mediated by the cord or even hypothalamic centers in the brain. But, as we are well aware from personal experience, visceral sensations and motor responses to them are not generally associated with the higher centers of the brain; we "know" little of what our viscera feel and have little control over them. The hypothalamic region of the brain (cf. p. 586) is the highest center directly related to visceral sensations and motor responses to them.

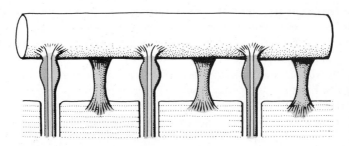

Figure 398. Diagram of the spinal cord and nerves of the left side of a lamprey seen in dorsal view (anterior end to the left), to show the alternating arrangement of separate dorsal and ventral spinal nerves, related to intermyotomic spaces and myotomes. Nerve components colored as in Figure 397.

The afferent pathways of the visceral system call for little remark. Visceral sensory endings are present in internal organs, blood vessels, and so forth. Fibers from these ascend to the spinal cord and brain through special visceral nerves of the trunk, discussed in the following section, or through the vagus nerve of the cranial system, which extends over much of the length of the coelom.

The Autonomic System (Figs. 399–402). Certain specialized efferent visceral pathways from the brain to the striated muscles of the branchial arches by cranial nerves will be discussed later. The remainder of the visceral efferent fibers innervate the smooth muscles and glands of the body. This system of efferent fibers was formerly called the sympathetic system; the term sympathetic, however, is now commonly restricted to one subdivision of the visceral efferents,* and the efferents as a whole are usually referred to as the **autonomic system**, with reference to the generally self-governing nature of the reflexes affecting smooth muscles and glands. A classification is as follows:

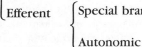

Visceral nervous system — Afferent; Efferent — Special branchial; Autonomic — Sympathetic; Parasympathetic

The fiber pattern followed by autonomic impulses from the spinal cord differs in one notable regard from that described earlier as typical for motor impulses to striated muscles (including the striated muscles of the special visceral system in the head and branchial region). Somatic motor impulses are carried from cord to effector by the long axon of a single neuron, but the typical visceral efferent is, in contrast, a two-neuron chain. The first neuron, one whose cell body lies in the cord, is termed a **preganglionic neuron**. Its axon is typically myelinated. This axon, however, does not extend the full length of the pathway to the effector organ. At some point along its course, it enters a ganglion of the autonomic system. Here motor impulses are relayed to a second series, **postganglionic neurons**, whose axons (with little or no myelin sheath) complete the passage to the end organ. Although this second series of neurons is often far removed ventrally from the region of the cord, it appears to be derived (like dorsal root ganglion cells) from cells of the neural crest; these descend in the embryo along the developing nerves. The only exception to a two-neuron chain in autonomic innervation is the very special case of the adrenal medulla (cf. p. 612).

A combination of physiologic and anatomic investigations indicates that in mammals the autonomic system may be sorted out into two subdivisions: (1) the **sympathetic** (in a narrow sense) or **thoracolumbar system,** and (2) the **parasympathetic** or **craniosacral system** (Figs. 400, 401). The two differ both functionally and topographically.

---

*But there is still wide variation in usage of the term sympathetic. By various authors it may be defined (*a*) narrowly, as here; (*b*) as equivalent to autonomic; (*c*) as equivalent to the entire visceral ("vegetative") system, both afferent and efferent. As still another confusing variant in terminology, "autonomic" is sometimes used for the entire visceral system, afferent as well as efferent.

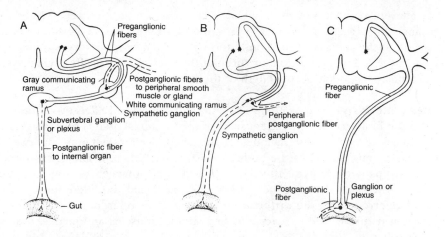

Figure 399. Diagrammatic cross sections to show the path of autonomic fibers. *A,* Sympathetic (thoracolumbar) distribution in a mammal, with autonomic ganglia both in a lateral chain and in a subvertebral position. Preganglionic fibers may be relayed in either position and run either to superficial structures via the major nerve trunks or to the viscera, in both cases with a long postganglionic neuron. *B,* Sympathetic distribution as found in many lower vertebrates; there is little development of a sympathetic chain and no distinction of ganglia into two groups; fibers to peripheral structures course independently or with blood vessels, rather than with the major (somatic) nerve trunks. *C,* Course of parasympathetic fibers; the preganglionic fiber makes the entire run from cord or brain to a point in or near the organ concerned, where there is a relay to a short postganglionic neuron.

Stimulation of true sympathetic nerves tends to increase the activity of the animal, speed up heart and circulation, slow down digestive processes, and, in general, to make it fit for fight or frolic. On the other hand, the action of the parasympathetic tends to slow down activity in general but promotes digestion and a "vegetative" phase of existence. In both systems, neurohumors are produced by the tips of the nerve fibers. In the sympathetics proper, the materials produced are in almost all cases two amines of similar composition, **noradrenaline** (norepinephrine) and, to a much lesser degree, a related substance, **adrenaline** (epinephrine). These neurohumors are almost identical with the materials put into the blood stream by the medulla of the adrenal gland (cf. pp. 612–613). In the case of the parasympathetic, the substance given off (as appears to be the case in most other parts of the nervous system) is **acetylcholine**. Warning must be given, however, that this pretty picture of clearcut differentiation, chemical and functional, of sympathetic and parasympathetic systems does not fully hold even in a mammal.

Anatomically, the two autonomic types can be distinguished in mammals as to *(a)* the point along the course of the nerve at which the relay to the postganglionic fiber occurs, and *(b)* the point along the length of the neural tube at which the nerves leave the central nervous system. In true sympathetic fibers, the relay typically takes place close to the spinal cord, and the preganglionic fibers are in consequence short, the postganglionic long (Fig. 399 *A, B*). In the parasympathetic system, in contrast, the preganglionic axons are long, the postganglionics short, and the relay occurs in or close to the organ concerned.

Topographically, in a mammal (Figs. 400, 401) the sympathetic system proper (as its alternative name, thoracolumbar, suggests) is confined at its origins to the

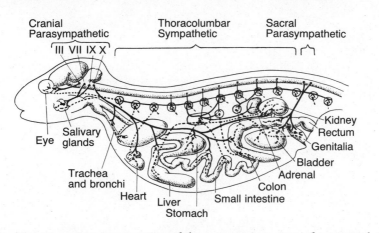

Cranial
Parasympathetic

Thoracolumbar
Sympathetic

Sacral
Parasympathetic

III VII IX X

Eye   Salivary
glands

Kidney
Rectum
Genitalia
Bladder
Adrenal
Colon
Small intestine

Trachea
and bronchi
Heart   Liver   Stomach

Figure 400. Diagrammatic representation of the autonomic system of a mammal. Only a fraction of the true number of trunk segments is represented. A sympathetic chain is developed, allowing exchange of fibers between segments. Sympathetic ganglia are represented by circles; short nerves projecting from them represent gray rami rejoining the main segmental nerve trunk and running to peripheral structures. There is here a regional sorting out of autonomic nerves into parasympathetic elements associated with cranial and sacral nerves, and sympathetic nerves arising from trunk segments. The two systems almost completely overlap, both reaching nearly every organ. In the sympathetic system, the relay to postganglionic fibers takes place in the sympathetic ganglia for peripheral structures and those in the head and chest; the relay to abdominal viscera occurs in a series of ganglia—coeliac, superior mesenteric, inferior mesenteric—which are more ventrally situated. Preganglionic nerves or fibers in full line; postganglionic in broken line.

thoracic and lumbar regions of the cord. The parasympathetic fibers emerge from the central nervous system both anteriorly and posteriorly. Anteriorly, they are associated with certain cranial nerves, most particularly the important vagus nerve; posteriorly, they are present in the spinal nerves of a few segments in the sacral region. From their origins, both sets of nerves spread out widely, so that glands and smooth muscle* receive, in every major organ, a double innervation from the two antagonistic systems.

The true sympathetic fibers in mammals leave the vertebral canal in the main trunk of the nerve of each thoracic and lumbar segment. Just beyond the vertebra, however, they leave this trunk and turn ventrally in a short **visceral ramus (white ramus,†** ramus communicans) to enter a small **sympathetic ganglion**. In most mammals, as in all other tetrapods and in teleosts, the ganglia on either side of the vertebral column are connected by lengthwise strands of fibers to form a **sympathetic chain** along which preganglionic fibers may course some distance before reaching the level at which they terminate. In the sympathetic ganglia occurs the relay to postganglionic neurons in the case of impulses destined for the smooth muscles of peripheral blood vessels and for smooth muscles and glands in the skin. In mammals, such an outward-bound axon rejoins the main trunk of the nerve from the ganglion by way of an accessory **communicating** (or **gray**) **ramus**, gray because of

---

*Including cardiac muscle as a specialization from the smooth type.

†Termed this because the included preganglionic fibers have thick myelin sheaths, giving a glistening white appearance to the ramus.

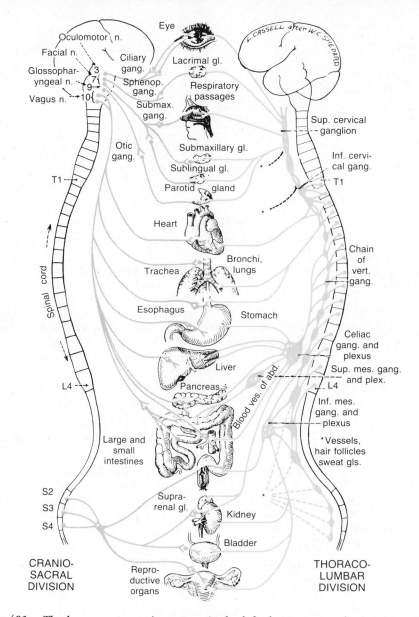

Eye

Oculomotor n.

Facial n.
Ciliary
gang.

Glossophar-
yngeal n.
3
Lacrimal gl.

7
Sphenop.
gang.

9
Respiratory
passages

Vagus n.
10
Submax.
gang.

L. CASSELL after W.C. SHEPARD

Sup. cervical
ganglion

Otic
gang.
Submaxillary gl.

Inf. cervi-
cal gang.

T1
Sublingual gl.
T1

Parotid gland

Heart
Chain
of
vert.
gang.

Spinal cord
Trachea
Bronchi,
lungs

Esophagus
Stomach

Celiac
gang. and
plexus

Liver
Sup. mes. gang.
and plex.

L4
Pancreas
L4

Blood ves. of abd.
Inf. mes.
gang. and
plexus

Large and
small
intestines
*Vessels,
hair follicles
sweat gls.

S2
Supra-
renal gl.

S3
Kidney

S4
Bladder

CRANIO-
SACRAL
DIVISION
Repro-
ductive
organs
THORACO-
LUMBAR
DIVISION

Figure 401. The human autonomic system. At the left, the parasympathetic system; right, the sympathetic system. (From Millard, King and Showers, Human Anatomy and Physiology.)

the paucity of myelin around its fibers. In many lower vertebrates, however, the peripheral sympathetic nerves pursue an independent course, running outward along the blood vessels.

But the sympathetic fibers relayed in the ganglia of the chain are only a part of the sympathetic system in mammals. Most of those bound for the deep internal organs pass uninterruptedly through these ganglia and turn ventrally and medially. Beneath the vertebral column, sympathetic fibers from the two sides meet in an interlacing network of fibers and ganglion cells, the **subvertebral ganglia** or subvertebral

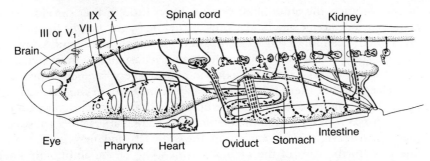

Figure 402. Diagrammatic representation of the autonomic system of a shark. As in Figure 400, only a fraction of the true number of trunk segments is represented, and "samples" only of types of innervation of abdominal viscera and blood vessels are given. Sympathetic ganglia, indicated by white circles, are developed in the trunk and are usually associated with bodies of chromaffin tissue (stippled). There is no development of a sympathetic chain, no development of gray rami to peripheral structures, and no regional sorting out of sympathetic and parasympathetic systems (cf. Fig. 400). Preganglionic autonomic nerves in full line, postganglionic nerves or fibers in broken line. (After J. Z. Young, modified.)

plexuses (such as the **coeliac plexus**, lying beneath the anterior lumbar vertebrae of mammals, and the **mesenteric plexuses**, situated farther posteriorly). Here each preganglionic fiber may relay its impulses to much more numerous postganglionic cells whose long axons extend downward, mainly via the mesenteries, to reach the various parts of the gut.

The greater part of the parasympathetic distribution is by way of the vagus nerve, a major branch of which follows the gut for much of its length (hence the name vagus—wandering); apart from minor elements associated with other cranial nerves, the remaining parasympathetic nerves come from a few segments in the sacral region. In this system, there are no proximal ganglia; long, myelinated preganglionic fibers extend the entire distance to, or nearly to, the various viscera; the relay takes place in plexuses generally embedded in the organ concerned.

Our knowledge of the autonomic system has been gained chiefly from a study of mammals. Autonomic structure in lower vertebrates is relatively poorly known, because both anatomic and physiologic studies are needed to elucidate conditions in any animal, and both types of work are difficult in this system. Although the story of autonomic evolution is still imperfectly known, some major points in its development seem clear. We shall cite conditions seen in three types that appear to represent some of the principal evolutionary stages.

1. In amphioxus, connections between central nervous system and gut are already established, but the system is a diagrammatically simple one. For the entire length of the "head" and trunk, every dorsal segmental nerve sends fibers to the viscera, in which there exists a plexus of nerve cells. There are, however, no relays in the system until the organs concerned are reached and thus the autonomic nerves resemble anatomically those of the parasympathetic type. There are no known functional distinctions between sympathetic and parasympathetic systems. Further, there is no known autonomic innervation of the skin or blood vessels, and hence there is no need for ganglia corresponding to those in the mammalian chain.

2. In cyclostomes, the autonomic system shows little further development except that autonomic fibers are present in the vagus nerve. In sharks (Fig. 402), there is a more complicated pattern, although there is no known functional distinction between sympathetic and parasympathetic systems. Autonomic fibers emerge from the cord in almost every segment from head to pelvic region, but there is a short break in continuity in the cervical region. In the head, the autonomic fibers emerge (with one exception) with nerves of the dorsal root type described later. Most cranial fibers are associated with the vagus nerve, which reaches to the stomach and may even reach the intestine. In the case of the cranial and in some of the autonomic fibers of the trunk, the relay still takes place at the organ concerned in the fashion of mammalian parasympathetic fibers; but in other autonomics of the trunk, the relay takes place close to the backbone, as in mammalian sympathetics. There is still no motor innervation of the skin, but there are autonomic fibers to the blood vessels. We consequently find a series of small ganglia, essentially segmental in disposition, along the trunk where the relays to these fibers take place; but there is little development of a sympathetic chain connecting them, and there are no gray rami, the fibers running to the blood vessels proceeding independently of the somatic nerve trunks.

3. Although teleosts are a side branch, these forms show advances comparable to those found in typical lower tetrapods. Here autonomic fibers to the skin are present, and gray rami developed; further, a sympathetic chain is developed connecting the ganglia in the trunk. In a shark, there was little overlapping of areas innervated by nerves of the head and trunk, but, in teleosts, postganglionic fibers from the trunk chain "invade" the stomach, pharynx, and even the head, formerly served exclusively by the vagus and other cranial nerves. In the teleost trunk, most nerves to the viscera are relayed proximally, in the manner of mammalian sympathetics. Posteriorly, however, the urinary bladder is innervated by long preganglionic fibers whose relays are in a plexus in its walls; this is suggestive of the mammalian sacral parasympathetic. Viewed as a whole, the teleost type of organization, possibly characteristic of Osteichthyes in general, because it is repeated essentially in amphibians, is one that with further anatomic and functional specialization could lead to that seen in mammals.

Cranial Nerves

In the head of vertebrates, there is a special series of varied nerves that are, particularly at first sight, difficult to compare with those of the body (Fig. 403). They deserve close consideration and analysis. The cranial nerves were first studied in man and were early given names and numbers based on their mammalian functions and positions. Although, as we shall see, the human arrangement does not hold throughout, we shall introduce the study of cranial nerves by listing them:*

I. **Olfactory**: sensory, from the olfactory epithelium.
II. **Optic**: sensory, from the eye.

*The initial letters of the names of the cranial nerves are the initial letters of the words of the following choice bit of poetry: "On old Olympus' towering top, a Finn and German viewed a hop."

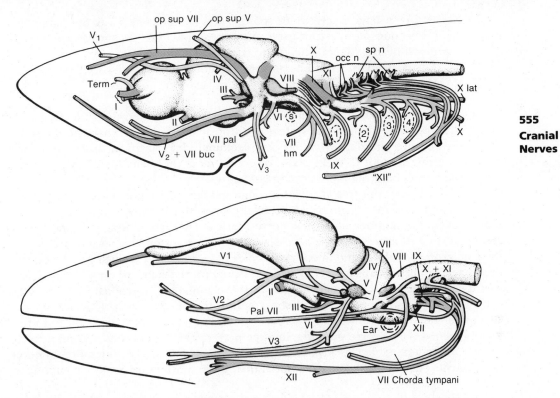

Figure 403. *Above,* diagram of the distribution and components of the cranial nerves of a shark (*Squalus*). *Below,* diagram of the distribution and components of the cranial nerves of a lizard (*Anolis*). Special somatic sensory nerves (I, II, lateral line, VIII), purple; somatic sensory components, blue; visceral sensory, green; visceral motor, yellow; somatic motor, red. Roman numerals refer to the cranial nerves. Abbreviations: *buc,* buccal; *hm,* hyomandibular; *lat,* lateral line trunk; *occ n,* occipital nerves; *op sup,* superficial ophthalmic; *pal,* palatine; *S,* position of spiracle; *sp n,* anterior spinal nerves; *Term,* terminal nerve; *1* to *4,* position of gill slits; V_1, ophthalmic (profundus) ramus of trigeminal; V_2, maxillary division; V_3, mandibular division of trigeminal; *Pal VII,* palatine ramus of facial; *"XII,"* trunk of conjoined occipital and anterior spinal nerves corresponding to amniote hypoglossal XII. In *Squalus,* special somatic sensory components of VII and X = lateral line nerves. (*Squalus* after Norris and Hughes; *Anolis* data from Willard, Watkinson.)

III. **Oculomotor**: innervates four of the six extrinsic muscles of the eye.

IV. **Trochlear**: to the superior oblique muscle of the eye (sometimes termed the trochlear muscle).

V. **Trigeminal**: a large nerve with three main branches, mainly bringing in somatic sensations from the head, with motor fibers to the jaw muscles.

VI. **Abducens**: to the posterior rectus muscle (which abducts the eye).

VII. **Facial**: partly sensory, but mainly important, in mammals, as supplying the muscles of the face.

VIII. **Acoustic** (= **Auditory** or **Stato-acoustic**): sensory, from the internal ear.

IX. **Glossopharyngeal**: a small nerve in mammals, mainly sensory, and innervating (as the name implies) much of the tongue and pharynx.

X. **Vagus**: a large nerve, both sensory and motor, which (as the name suggests) does not restrict itself to the head but runs backward to innervate much of the viscera—heart, stomach, and so forth.

XI. **Accessory** (= **Spinal accessory**): a motor nerve accessory to the vagus.

XII. **Hypoglossal**: a motor nerve to the muscles of the tongue.

One can brutally memorize such a list of cranial nerves and their functions, but no one interested in the nervous system can stop at this point. We have here a series of nerves that are amazing in variety and seemingly haphazard in distribution, and one cannot but attempt to make sense out of them. Is there any logic in their distribution? Can they be grouped in any sort of natural categories?

Stimulus to attempt this is added through the fact that comparative study shows that the fixed, orthodox mammalian scheme just presented is not found in all vertebrates. Even the number of cranial nerves varies. The hypoglossal, for example, is absent as such in fishes, and the accessory is in lower vertebrates an integral part of the vagus; on the other hand, most vertebrates have anteriorly a small terminal nerve that is very small and was not recognized in man until the standard numbering system was well established. One main branch of the mammalian trigeminal appears originally as a quite distinct element. In fishes, there is frequently a variable series of extra "occipital" nerves that emerge at the back of the braincase; these may fuse to form a hypobranchial nerve corresponding, in large part, to the hypoglossal of amniotes.

A clue to a possible classification of cranial nerves lies in a consideration of nerve components (cf. Table 3). We have noted that, in the postcranial region, components

TABLE 3. Table of Nerve Components of Cranial Nerves

Nerve Types	Special Sensory	Branchial (Dorsal)					Ventral
Components	Special Somatic Sensory	General Somatic Sensory	General Visceral Sensory	Special Visceral Sensory	Special Visceral Motor	Visceral Motor (Autonomic)	Somatic Motor
O. Terminalis		X					
I. Olfactory	X						
II. Optic	X						
III. Oculomotor						(X)	X
IV. Trochlear							X
V_1. Profundus		X				(X)	
$V_{2,3}$. Trigeminal proper		X			X		
VI. Abducens							X
VII. Facial	L	(X)	X	X	X	X	
VIII. Acoustic	A						
IX. Glossopharyngeal	L	(X)	X	X	X	X	
X and XI. Vagus (and accessory)	L	X	X	X	X	X	
XII. Hypoglossal							X

Proprioceptive fibers (muscle sense) are not included but are present in all somatic motor nerves. *L*, Lateralis sensory components of lower vertebrates (in X, the vagus, alone in amphibians); *A*, acoustic components of lateralis-acoustic system. Components in parentheses: variable or negligible. The three areas between vertical double-ruled lines indicate the components proper to each of the three types. Except for the usual presence of autonomic fibers accompanying the oculomotor nerve, the distinctions are clear-cut.

of four types are present. In the head region, these can be made out together with three further subcategories. In addition to the ordinary smooth visceral muscles found in the trunk, the head and throat have special striated visceral muscles connected with the visceral skeleton, and a **special visceral motor** category has been established for nerve components supplying them. A **special visceral sensory** component is distinguished for fibers from the taste organs. Further, there are found in the head special somatic sensory structures: nose, eye, and acoustico-lateralis organs; their nerves merit a **special somatic sensory** category. We can group cranial nerves into three types that (as indicated by the double vertical rulings in the table) show essentially clean-cut distinctions as to the components present. These three types are *(a)* **special sensory nerves** of somatic type; *(b)* complex **dorsal root** or **branchial nerves** that may contain sensory components, both somatic and visceral, and visceral motor elements, including both autonomic and special visceral motor components associated with the branchial region; and *(c)* **ventral root nerves**, containing almost exclusively somatic motor fibers. The first category is peculiar to the cranial region; the other two are comparable to the dorsal and ventral roots of spinal nerves of lower vertebrates, especially to the separate dorsal and ventral spinal nerves of lampreys and amphioxus.

Special Sensory Nerves. In all vertebrates, the three main sense organs—nose, eye, and ear—are innervated by special nerves; in primitive aquatic vertebrates, we find in addition nerve trunks, intimately associated with that for the ear, that innervate the lateral line organs.

Olfactory (I). As mentioned previously (cf. p. 501), the olfactory is not a typical nerve. A normal sensory nerve usually has the cell bodies of its neurons located in a ganglion close to the spinal cord or brain. In this case, however, the nerve fibers are not formed by normal elements of the nervous system at all; they are, instead, formed by the cells of the sensory epithelium, and run inward to the olfactory bulb at the front of the brain.*

In most vertebrates, the olfactory is not a compact nerve; rather, it consists of a number of short fiber bundles that pass back from the olfactory organ, in mammals through the cribriform plate of the ethmoid bone. In some forms, the nose and the olfactory bulb of the brain are farther apart, and a distinctive pair of olfactory nerves is formed; usually, however, a part of the brain is elongated and the nerve remains short. In typical amphibians and reptiles and in many mammals with a well-developed vomeronasal organ (cf. p. 504), a discrete branch of the olfactory nerve develops for the innervation of this structure. Cyclostomes have only a single olfactory pocket, but the nerve is paired.

Optic (II). The optic nerve, entering the brain at the chiasma, was discussed in connection with the eye (cf. p. 515). It is, properly, not a nerve at all, but a brain tract, since the retina is formed as an outgrowth of the brain; the fiber-forming cells are situated peripherally—in the ganglion cell layer of the retina. We noted earlier that the two optic nerves cross each other (at least to some degree) at the optic chiasma before entering the brain.

---

*Prior discussion of cranial nerves is necessary for a proper consideration of brain structure, but, conversely, some mention of prominent brain landmarks is necessary in the description of cranial nerves.

Acoustic (VIII).   More normal in construction is the acoustic nerve, connected with the internal ear. Here the nerve fibers are formed by true ganglion cells that (as in the cochlear branch of the nerve in mammals) may be situated peripherally, in the otic capsule, close to the sensory structures. The acoustic nerve is not, of course, visible in an external dissection of the braincase, because its fibers run directly from the capsule into the endocranial cavity, where it enters the medulla oblongata at a dorsal position. Two main branches are present in higher vertebrates, each with a separate ganglion. The **vestibular nerve** serves the anterior parts of the system of canals and sacs; the **cochlear nerve** is so named because, in mammals, it serves the auditory organ of the cochlea as well as the posterior part of the organs of equilibrium.

Lateral Line Nerves (cf. Fig. 376).   The lateral line organs of primitive water-dwellers are, we have noted, intimately related to the auditory sense; hence it is reasonable to find that the nerves innervating these structures are closely associated with the acoustic nerve. In fishes, a pair of stout lateral line nerves flanks the root of the acoustic nerve anteriorly and posteriorly on either side of the medulla. These nerves enter the braincase with certain of the branchial nerves and, therefore, frequently are reckoned as part of these nerves. This association, however, seems to be merely one of convenience, and the essential independence of the lateral line nerves should be kept clearly in mind. The special nature of these nerves is emphasized by the fact that their nerve cells do not arise, like typical afferent elements, from the neural crest, but from epithelial thickenings, placodes, on the lateral surface of the embryonic head.

The **anterior lateral line nerve** enters the skull in company with the facial nerve after supplying the lateral line organs of much of the head. Its main branches—superficial ophthalmic, buccal, hyomandibular—are often quite independent of the branches of the facial nerve proper. Some fibers of the **posterior lateral line nerve** may fuse with the glossopharyngeal (IX), but most or all of the posterior nerve accompanies the vagus. It innervates the neuromast organs of the occipital region but primarily forms a stout trunk accompanying the lateral line the length of the body.

Branchial Nerves.   We have noted that primitively (as in amphioxus and the lamprey today) dorsal and ventral roots were probably quite distinct nerves; the dorsal roots contain both somatic and visceral sensory components and, in addition, appear to have carried originally the visceral motor fibers as well. Table 3 shows that a large series of cranial nerves appear to belong to this category of dorsal root nerves, all containing sensory fibers and in most (but not all) cases visceral motor components as well. This series includes the terminal nerve, profundus nerve, trigeminus proper, facial, glossopharyngeal, and vagus (Fig. 404).

These dorsal cranial nerves all, however, differ in various ways from dorsal root nerves of the trunk. Most of the differences are obviously due to their close association with the branchial system, prominently developed in this region of the body in lower vertebrates. The presence of the gill slits tends to bring about a serial, possibly segmental, pattern of distribution of these nerves. This is enhanced by the serial pattern of the striated branchial muscles that are innervated by special visceral motor components.

Most typical of these dorsal root or branchial nerves is perhaps the glossopharyngeal in fishes (Fig. 405), which is associated with the first typical branchial slit. Like

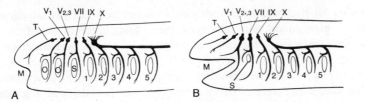

Figure 404. Diagrams showing the distribution of branchial (dorsal root) cranial nerves. *A,* Hypothetic primitive condition, with typical nerves to each of two anterior gill slits lost in jawed vertebrates, and a terminal nerve to anterior end of head. *B,* Condition in jaw-bearing fishes. Abbreviations: *M,* mouth; *O, O',* anterior gill slits lost in gnathostomes; *S,* spiracular slit; *T,* terminal nerve; *1* to *5,* typical gill slits of gnathostomes.

all members of this series, it arises from the dorsal part of the medulla oblongata and bears a prominent ganglion on its sensory root. A major trunk, containing both sensory and motor fibers, runs down behind this slit as a **post-trematic ramus**. A smaller **pretrematic ramus** runs down in front of the branchial slit, and a **pharyngeal ramus** turns inward and forward beneath the lining of the pharynx; these two branches contain visceral sensory fibers only. A small **dorsal ramus**, often absent, may bring in somatic sensations from the skin.

Posterior to the first, each of the branchial slits has a similar pattern of associated nerve branches; all, however, unite to connect with the brain via a single large compound nerve, the vagus. Anteriorly, the development of the jaw apparatus in gnathostomes has greatly modified the pattern of the branchial nerves, which may have been present in the ancestral jawless vertebrates. The facial nerve, associated with the spiracular slit, is not too atypical but has reduced or lost its somatic sensory component. The development of jaws, it is thought, may have eliminated two anterior gill slits. The trigeminal nerve complex appears to have been originally associated with these lost gills, and the little terminal nerve may be a vestigial anterior member of this series.

These dorsal root nerves may be described in anteroposterior sequence.

Terminal Nerve. In many vertebrates, representing every class except cyclostomes and birds, there is a tiny nerve that parallels the olfactory in its course inward from the nasal epithelium; its fibers run back to the diencephalon. Little is known of its nature; it is apparently sensory but has nothing to do with olfactory reception. It is present only in the embryo of man. Quite possibly it is the last remnant of a most

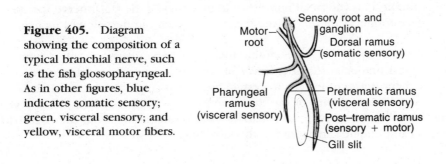

Figure 405. Diagram showing the composition of a typical branchial nerve, such as the fish glossopharyngeal. As in other figures, blue indicates somatic sensory; green, visceral sensory; and yellow, visceral motor fibers.

anterior member of the branchial series that innervated the primitive small mouth area—no matter how many gills may have been lost, there must, of course, always have been an area in front of the most anterior pair of branchial pouches.

Profundus Nerve (V_1).   The **ophthalmicus profundus** nerve* is a stout trunk that runs forward through the deeper part of the orbit and functions as a somatic afferent, receiving sensations from the skin of the snout. In most vertebrates, this nerve is intimately associated with the trigeminus and counted as one of its three main trunks. In lower vertebrates, however, it has frequently a separate ganglion and may even emerge from the brain independent of the trigeminus. Presumably, therefore, it was originally a separate nerve of the branchial series, associated with an anterior branchial slit that has been lost with the development of the jaws and expansion of the mouth.

Trigeminal Nerve (V_2, V_3).   The trigeminal nerve proper is believed to have been originally associated, like its profundus component, with an anterior branchial slit lost with the expansion of the mouth. This nerve has not, however, been reduced; it has become the main nerve of the mouth and jaws in all jawed vertebrates. Two main branches, somewhat comparable to pretrematic and post-trematic rami of a typical branchial nerve, are always present in the trigeminus proper: a **maxillary ramus** (V_2) to the upper jaw region, and a **mandibular ramus** (V_3) to the lower jaw. Both branches include somatic sensory elements from much of the surface of the head and oral cavity. The fibers innervating the jaw muscles are typically associated with V_3. Because of its anterior position the trigeminal has no visceral sensory component.

Facial Nerve (VII).   This is the nerve proper to the spiracular gill slit. Its main, post-trematic trunk in fishes is the **hyomandibular nerve**, which descends alongside the hyomandibular cartilage or bone behind the spiracle. This supplies the branchial muscles proper to the hyoid arch and contains visceral sensory fibers as well. Anterior branches are purely visceral sensory in nature and include a **palatine ramus** that corresponds to the pharyngeal rami of more posterior nerves. The visceral sensory fibers innervate much of the mouth, including some of the taste buds. As we have noted, the anterior lateral line nerve is associated with the facial.

In tetrapods, the nerve preserves its various branches in modified form. The main ventral trunk, after looping through the region of the middle ear, turns forward as the **chorda tympani** to give sensory innervation to taste buds and lower jaw. In mammals, as we have noted, the musculature of the hyoid arch has expanded over the surface of the head as the facial muscles, and branches of the facial nerve spread over skull and face to innervate them—the feature to which the nerve owes its name.

Glossopharyngeal Nerve (IX).   As noted, this small nerve is associated in fishes with the first typical branchial slit; it is never of any great size or importance. A somatic sensory element is absent except in certain fishes and amphibians. A small amount of striated visceral musculature of the throat and larynx and part of the salivary glands are innervated by this nerve, and it carries visceral sensations from the back part of the mouth (and tongue) and some of the taste buds.

---

*In mammals, it is termed simply the ophthalmic; in most vertebrates, however, it is termed the deep ophthalmic to distinguish it from a more superficial nerve in the same general region.

Vagus Nerve (X, XI).   The vagus is the largest and most versatile of all cranial nerves. The **accessory nerve** of amniotes is essentially a motor portion of the vagus that continues its area of origin back from the medulla into the cervical region of the spinal cord; in lower groups, it appears to be an integral part of the vagus. Usually there is a minor cutaneous sensory component from the neck; for the most part, however, the vagus is a visceral nerve. The vagus nerve supplies, in fishes, a typical series of branches, one to each of the gills behind the first, innervating the pharynx and the striated branchial muscles. This branchial part of the vagus survives, in modified form, in the tetrapod pharynx and its musculature. In fishes, the vagus carries (like VII and IX) sensations of taste, but this function is reduced in terrestrial forms. In every vertebrate, the main trunk of the vagus runs posteriorly into the body as an **abdominal ramus**, which extends in all forms to the stomach and heart, to the lungs when present, and, in some cases, may run the entire length of the gut. This nerve carries visceral sensations to the brain and (as noted previously) is the main route of the fibers of the parasympathetic system. In fishes and many amphibians, the posterior lateral line nerve emerges from the braincase with the vagus.

Somatic Motor Nerves.   The nerves in this category—III, IV, VI, and XII of the human series—are highly comparable in most regards to ventral roots of spinal nerves. They are almost entirely composed of somatic motor fibers that emerge ventrally from the brain stem. Their arrangement is closely correlated with the distribution of the myotomes in the head. As we have seen earlier (cf. p. 290; Fig. 196), the myotomal succession found in the trunk is interrupted anteriorly in the otic region; farther forward, three further somites form the muscles of the eyeball. Nerves III, IV, and VI innervate the eye muscles formed by these somites (Fig. 205). Nerve XII, when present, supplies musculature arising from somites formed in the occipital region; in anamniotes, the hypobranchial nerve is basically similar, although not always cranial in origin.

Oculomotor (III).   The greater part of this nerve, which arises ventrally from the midbrain, consists of somatic motor fibers innervating the four ocular muscles developed from the first cranial somite: the superior, anterior, and inferior rectus, and the inferior oblique. A seeming exception is the innervation in cyclostomes of the inferior rectus muscle by fibers associated with the abducens nerve. However, the fibers concerned are probably a discrete branch of the oculomotor that becomes associated with the abducens. Autonomic fibers, partly derived from nerve V_1 and hardly to be considered an integral part of the oculomotor, generally accompany the nerve to the eyeball where they innervate the smooth muscle fibers of the ciliary body and iris; they are concerned with accommodation and pupillary reflexes. A few similar fibers may accompany nerves IV and VI.

Trochlear (IV).   The trochlear is a small nerve with a most unusual course. It arises ventrally within the midbrain on either side; but instead of emerging directly outward, its fibers turn upward within the substance of the brain and cross to emerge dorsally on the opposite side. It innervates solely the superior oblique muscle of the eyeball, derived from the second pre-otic somite. The name trochlear refers to the pulley through which the muscle passes in mammals.

Abducens (VI).   This small nerve emerges ventrally from the anterior end of the medulla and supplies the posterior rectus muscle, which abducts the eye and is derived from the third pre-otic somite. A muscle retracting the eye is developed in many forms from the same somite and is also innervated by the abducens.

Hypoglossal (XII).   In fishes (particularly among the sharks), the posterior end of the skull, and hence of the cranial nerve series, is not a fixed point. The number of potential vertebrae that are incorporated in the occipital region of the skull is variable; hence, we find a variable number of would-be spinal nerves emerging from the occiput. Such **spino-occipital nerves** tend to lose their dorsal roots and are mainly motor nerves supplying muscles formed from the somites of the occipital region. In amniotes, the condition is stabilized; posterior (and ventral) to the vagus-accessory region there is a final cranial nerve, the hypoglossal, usually formed from three ventral roots that represent three segments fused into the occiput. There is no hypoglossal nerve in modern amphibians. However, fossil amphibians and crossopterygians show a cranial foramen for it; it would seem that a definitive hypoglossal was formed in the ancestors of the tetrapods and that its absence in frogs and salamanders today is a derived characteristic of these forms.

The myotomes in the region close to the back of the skull migrate in the embryo backward and downward around the gill region to form the hypobranchial muscles in fishes and the tongue musculature of higher vertebrates (Fig. 196). Trunks of the occipital nerves of fishes follow this migration route to innervate the hypobranchial muscles (and are hence often termed the hypobranchial nerve), and the hypoglossal nerve of amniotes follows a similar path, back, down, and then forward around the pharyngeal region to innervate the musculature of the tongue.

Segmentation of the Head

It is convenient at this point to interrupt our discussion of the nervous system to consider a problem to which study of the cranial nerves provides some of the evidence. Obviously, the trunk of vertebrates is segmentally arranged, but such clear division into repeated units is not seen anteriorly in the head. Nevertheless, many workers have tried to demonstrate that the segmentation, however masked, is actually there. Most of the earliest attempts were based on supposed resemblances between the skull and a series of vertebrae; but these, although fascinating exercises in ingenuity, are, for a variety of reasons, not seriously considered today. However, we have just been referring to certain cranial nerves as comparable to dorsal or ventral roots of the spinal nerves in primitive chordates—and spinal nerves are clearly segmentally arranged. Can a similar pattern be demonstrated in the head? Unfortunately, no unequivocal answer is possible. Completely convincing demonstration appears impossible, but we can present a more or less plausible story (Fig. 406 and Table 4).

Ventral root nerves are those containing only somatic motor fibers and innervating somatic muscles; the latter develop from the myotomes of the somites, and somites are the primary segmental structures in vertebrates. Therefore, ventral roots should be segmental. In the head, as elsewhere, these nerves do in fact line up with the myotomes. Three small, rather aberrant pre-otic somites (often called head cavities) give rise to the extrinsic muscles of the eye; each has its own nerve, numbers III, IV, and VI. The more posterior somites in the head give rise to the hypobranchial muscles and are innervated by the hypobranchial or hypoglossal nerve; this nerve goes to muscles formed from several somites (three in the dogfish) and arises from several separate roots. The segmental pattern is maintained. The problem is that, as the ear develops, it occupies the space that would normally contain the somites, and the sequence of myotomes is interrupted. In some forms, the missing ones never appear; in others, they start to form but soon degenerate. At least two pairs of somites are eliminated here. There *could* be more, but we will assume that only two were lost, which, as we shall soon see, makes things balance fairly well.

Dorsal root, or branchial, nerves also form a series. However, this series, as we have already noted, lines up not with myotomes or other somatic structures, but with the visceral elements, both skeletal and muscular, associated with the gills. Thus, the trigeminal is the nerve of the mandibular arch, the facial is the nerve of the hyoid, and so forth. The pattern is not completely simple: the vagus has a series, usually four, of separate branchial branches to successive branchial arches; it also arises as a series of separate roots along the side of the brain. Anteriorly, we may assume (as in Table 4) only a single premandibular visceral arch and consider the terminalis as the nerve of the region anterior to the gills. An alternate theory is that there were originally two premandibular arches and the terminalis was the nerve of the more anterior one. In any event, the relation between these nerves and the visceral arches is clear.

But is there any reason to believe that dorsal and ventral roots must line up neatly here as they do in the trunk? Not really. It seems plausible that they should; after all, if the nerves are similar, the simplest assumption is that the whole pattern is similar. Also, although it is a weak argument, it just seems tidier and simpler to most

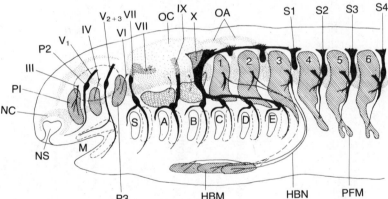
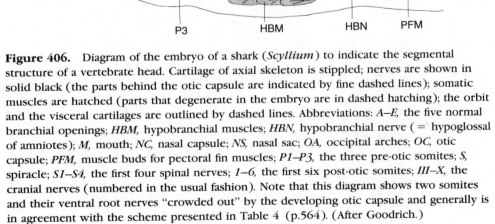

Figure 406. Diagram of the embryo of a shark (*Scyllium*) to indicate the segmental structure of a vertebrate head. Cartilage of axial skeleton is stippled; nerves are shown in solid black (the parts behind the otic capsule are indicated by fine dashed lines); somatic muscles are hatched (parts that degenerate in the embryo are in dashed hatching); the orbit and the visceral cartilages are outlined by dashed lines. Abbreviations: *A–E*, the five normal branchial openings; *HBM*, hypobranchial muscles; *HBN*, hypobranchial nerve (= hypoglossal of amniotes); *M*, mouth; *NC*, nasal capsule; *NS*, nasal sac; *OA*, occipital arches; *OC*, otic capsule; *PFM*, muscle buds for pectoral fin muscles; *P1–P3*, the three pre-otic somites; *S*, spiracle; *S1–S4*, the first four spinal nerves; *1–6*, the first six post-otic somites; *III–X*, the cranial nerves (numbered in the usual fashion). Note that this diagram shows two somites and their ventral root nerves "crowded out" by the developing otic capsule and generally is in agreement with the scheme presented in Table 4 (p.564). (After Goodrich.)

Table 4. Segmentation of the Head

Segment	0 = Snout	1	2	3	4	5	6	7	8	9
Myotome	(none)	First pre-otic	Second pre-otic	Third pre-otic	(crowded out by developing ear)		First post-otic	Second postotic	Third postotic	Fourth postotic
Ventral Root Nerve	(none)	Oculomotor	Trochlear	Abducens ←	(lost) →		Hypobranchial = Hypoglossal →			→Ventral root of second spinal nerve
Dorsal Root Nerve	Terminalis	Profundus	Trigeminal	Facial	Glosso-pharyngeal	Vagus (including Accessory) →				→ Dorsal root of second spinal nerve
Visceral Arch	(none)	Premandibular	Mandibular	Hyoid	First branchial	Second branchial	Third branchial	Fourth branchial	Fifth branchial	(none)
Branchial Opening	(none)	(lost with development of large, jawed mouth)		Spiracle	First branchial	Second branchial	Third branchial	Fourth branchial	Fifth branchial	(none)

Compare with Figure 406 and description in the text. This table is based largely on a dogfish, in which the first spinal nerve is largely vestigial (although fibers that "should" be in it contribute to the hypobranchial and vagus nerves).

people to have both repetitive series, that of myotomes and that of visceral arches, possess the same interval between successive units; arranging the parts otherwise tends to be rather awkward. Another rather weak argument is that it appears to work; we can point to pictures like Figure 406 and draw up charts like Table 4. This may seem convincing, but remember, we only *assume* two somites are lost as the ear develops. Also, at the back of the head, the numbering may not be clear; different vertebrates clearly include a variable number of somatic segments within the head (that is, the number of occipital arches varies); again we can juggle things so that they come out even. Despite all these ifs and maybes, we do suspect that the head was originally segmented in much the same way as the trunk and that this pattern was obscured by the various specializations, such as the formation of major sense organs, that inevitably occur at the front end of the animal.

565
Cranial
Nervous
System —
Accessory
Elements

Central Nervous System—Accessory Elements

In addition to the functionally important neurons, other types of cells are developed as part of the nervous system. We have already noted the cells sheathing nerve fibers and ganglion cells; within the central nervous system are various other cell types termed, in a broad sense, the **neuroglia**. In most organs, supporting functions are performed by connective tissues. These are present only to a very minor degree in the central nervous system, and the neuroglial cells take their place as protective and supporting structures—the nervous system, in a sense, makes its own brand of connective tissue.

Most neuroglial elements arise embryologically as a part of the true nervous tissues, from which they become differentiated during development. They are found throughout the brain and spinal cord as small but exceedingly numerous elements scattered among the neurons. Typically, there is a rather small cell body from which radiate fine processes, giving the neuroglia a starlike appearance. Relatively large cells with numerous processes are termed **astrocytes**; smaller, with fewer processes, but much more abundant are **oligodendroglia**. **Ependymal cells** retain the epithelial character true of all the tissue of the embryonic nerve tube and persist in a layer of epithelium (the **ependyma**), frequently ciliated, lining the cavities of the brain and spinal cord.

The brain and spinal cord, enclosed within the braincase and the neural arches of the vertebrae, are further protected and sustained by one or more wrappings, or **meninges** (Fig. 407). Most fishes have only a single meninx of compact tissue, external to which a loose mucous or fatty connective tissue lies between cord or brain and neural arches or braincase. In all tetrapods, at least two meninges are present. The outer, the **dura mater**, is a stout sheath, connected by numerous slender ligaments with an inner membrane, which is closely applied to the brain or cord. In mammals, there is a separation of the softer, inner wrapping into two delicate structures, an outer **arachnoid** and an inner, vascular **pia mater**; the two are separated by a fluid-filled **subarachnoid space**, which is crossed by a spidery cobweb of delicate tissue threads. All the meninges, and also microglial cells within the central nervous system, appear to develop from normal connective tissue.

The vertebrate nervous system develops as a hollow tube, and this tubular arrangement persists throughout life; within the brain are the ventricles and in the spinal cord there is a central canal. Present in these cavities and also found between

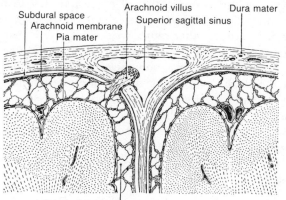

Subdural space
Arachnoid membrane
Pia mater
Arachnoid villus
Superior sagittal sinus
Dura mater

Subarachnoid space

Figure 407. Transverse section of a portion of the brain of a mammal to show the meninges. The section is taken through the partition (falx cerebri) between the two cerebral hemispheres. The superior sagittal sinus is a prominent venous channel following a longitudinal course backward between the hemispheres; sections of smaller blood vessels are seen in meninges and brain. The subarachnoid space is occupied by cerebrospinal fluid; the arachnoidal villi offer a minor means of transfusion of material between this fluid and the blood. (After Weed.)

arachnoid and pia mater in the brain wrappings is a clear liquid, the **cerebrospinal fluid**, similar in composition to the interstitial fluid or the perilymphatic liquid of the ear. Nutrient or other materials reach the cerebrospinal fluid from the blood through the medium of special vascular structures, the choroid plexuses of the brain (cf. p. 571). The membranes of these plexuses, separating blood vessels from the ventricles of the brain, are more impervious to large molecules or other structures of any magnitude than are ordinary capillary walls, and thus serve as a barrier against the introduction of foreign materials into the brain. The cerebrospinal fluid further serves as a water-bed suspending brain and cord and protecting them from injury.

Spinal Cord

The spinal cord (Figs. 394, 397, 408 A), extending most of the length of the body, is a little-modified adult representative of the nerve tube formed in the early embryo. It still contains, as did that of the embryo, a centrally situated, fluid-filled **central canal**; this, however, has become relatively tiny in diameter, owing to the great growth of the nervous tissues surrounding it.

The spinal cord is subcircular or oval in section in lower vertebrates, but, in higher groups, it tends to expand in bilateral fashion, and pronounced grooves may be present in the midline dorsally and ventrally. The cord tends to taper distally; in many forms (notably in mammals), it may be shorter than the vertebral column, and, in the distal part of the vertebral canal, there may be merely a series of nerves running back from the termination of the cord to serve the most posterior segments. Two layers of material can be readily distinguished in the cord, a central area of **gray matter** surrounding the canal of the cord, and a peripheral region of **white matter**. The former consists mainly of cell bodies, the latter of countless myelinated fibers coursing up and down the cord.

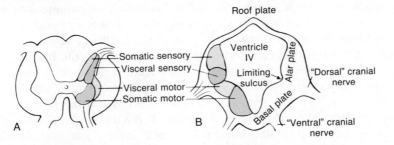

Figure 408. Diagrams showing the distribution of sensory and motor columns. Somatic sensory association column, blue; visceral sensory, green; visceral motor column, yellow; somatic motor, red. *A,* The spinal cord of the adult of certain lower vertebrates. *B,* The embryonic medulla oblongata; the embryonic spinal cord shows a similar arrangement of the columns. The plate of tissue lying below the limiting sulcus is termed the basal plate; from this, motor centers arise. The sensory region above is the "wing" or alar plate. (Partly after Herrick.)

Primitively, the gray substance was, it seems, arranged in a fairly even manner about the central cavity; in most vertebrates, however, there is a bilaterally symmetric arrangement, as seen in section, into an H-shape, or that of a butterfly's wings. Such a pattern gives the appearance of a pair of "horns" on either side, dorsally and ventrally. Actually, of course, each "horn" is merely a section of a longitudinal structure, and we should speak rather of a dorsal column and a ventral column.

The **ventral column** is the seat of the cell bodies of the efferent or motor neurons of the spinal nerves, their axons typically passing out through successive ventral nerve roots. The number of neurons in any given part of the cord will, naturally, vary with the volume of musculature at that level of the body, and in tetrapods, the ventral column (indeed, the entire cord) may be much expanded in the regions supplying the limbs. In some dinosaurs, the sacral enlargement could be much larger than the brain. Most of the motor supply is to somatic muscles, but visceral efferents may be present over much of the length of the trunk. The cell bodies of these latter neurons are situated above and lateral to those of the somatic motor type, and sometimes are distinguishable as a **ventrolateral column**.

The **dorsal column** of the gray matter is associated with the dorsal, sensory roots of the spinal nerves; it is the seat of the cell bodies of association neurons through which impulses brought in from sense organs may be relayed and distributed. These neurons have processes that may ascend and descend the cord to connect with motor neurons of the same side, that may cross to connect with motor neurons of the opposite side of the cord, or, still further, that may ascend the cord to the brain. The arrangement of various clusters of these association cells in the dorsal column is complex and variable, but in some cases (particularly in certain embryos), it appears that we can distinguish a larger series associated with somatic sensory reception, situated dorsally and medially, and a smaller, visceral sensory group situated more ventrally and laterally. There thus appear to be in the gray matter four areas on either side related to the four nerve components, the four being in sequence, from dorsal to ventral: somatic sensory, visceral sensory, visceral motor, and somatic motor. It is of interest that the same arrangement is found in the gray matter of the brain stem (Fig. 408 *B*).

The **white matter** is composed, as has been said, of innumerable myelinated fibers. These include ascending and descending fibers of sensory nerve cells that enter the cord through the dorsal nerve roots and of similar fibers from association cells. Also present are fibers that carry sensory stimuli forward to the brain and fibers returning from brain centers to act (usually via association neurons) on motor neurons. Fibers of these latter categories are especially abundant in higher vertebrate groups, in which the trunk loses the semi-autonomous nature that it has in fishes and comes more directly under the influence of the brain. Topographically, the white matter is more or less subdivided by the dorsal and ventral "horns" into major areas: **dorsal**, **lateral**, and **ventral funiculi**. Anatomic and physiologic investigation enables one to distinguish in these areas **fiber tracts** with the varied types of connections noted above. These tracts, however, vary too greatly in nature and position from group to group to be described here in detail, although in general the dorsal funiculi mainly carry ascending, sensory fibers, the lateral and ventral ones descending fibers to motor neurons as well as other ascending fibers.

The Brain

In all vertebrates, as well as in the more highly organized invertebrates, we find a concentration of nervous tissues at the anterior end of the body in the form of a brain of some sort. Such a concentration is to be expected. In an actively moving, bilaterally symmetric animal, this region is that which first makes contact with environmental situations to which response must be made. It is, in consequence, the region in which the major sensory structures come to be situated, and in which, hence, it is most advantageous to locate correlation and integration of sensory impulses.

Primitively, we may believe, the vertebrate brain was merely a modestly developed anterior region of the neural tube in which, in addition to facilities for local reflexes to the head and throat, special sensory stimuli were assembled and "referred for action" to the semi-autonomous posterior parts of the body via the spinal cord. Such is the situation in amphioxus. Within the vertebrates, however, there has occurred a strong trend for the concentration in the brain of command over bodily functions (except for the simplest reflexes), with the development of many complex, intercommunicating brain centers. As already noted, the intercalation of association neurons into simple reflex arcs greatly broadens the field of possible responses to a sensory stimulus and, conversely, greatly increases the variety of stimuli that may excite a specific motor response. The pattern within the brain is essentially an elaboration of this principle—the interposition of further series of neurons between primary areas of sensory reception and final motor paths. These intermediate neurons are clustered in functional centers. In such centers, afferent impulses may be correlated and integrated for appropriate responses, or motor mechanisms may be coordinated; on still higher levels, there may develop association centers of whose activity memory, learning, and consciousness may be the products.

Brain Development (Fig. 409). The general topography of the brain and its parts is best understood through a consideration of its development. The brain develops rapidly in the embryo—much more rapidly than almost any other organ—and there is early established a generalized structural pattern upon which the numerous variations seen in the adult brains of different groups are superposed. In early stages, the future

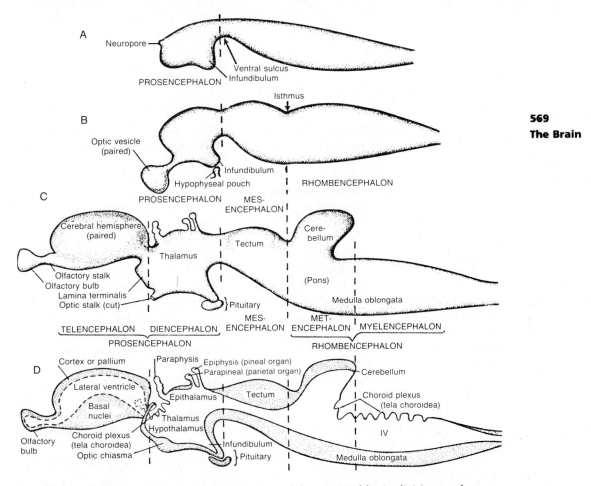

Figure 409. Diagrams to show the development of the principal brain divisions and structures. *A*, Only prosencephalon (primitive forebrain) distinct from remainder of neural tube. *B*, Three main divisions established. *C*, More mature stage in lateral view. *D*, The same in median section. (Partly after Bütschli.)

brain is merely an expanded front part of the neural tube. With continued increase in size, its anterior end tends to fold downward at the **cephalic flexure**. This distinguishes a median, terminal saclike structure, the primitive forebrain, or **prosencephalon**, from the remainder of the tube. Somewhat later a second flexure, in a reverse direction, is found more posteriorly at the **isthmus**. This separates the midbrain, or **mesencephalon**, from the primitive hindbrain, the **rhombencephalon**, in which develops the **medulla oblongata** of the adult. Posteriorly, the rhombencephalon tapers gradually, without any abrupt change, into the spinal cord.

Prosencephalon, mesencephalon, and rhombencephalon are the three primary subdivisions of the brain. The three successive regions of the neural tube that constitute them in early stages are still recognizable in the adult, where they all contribute to the **brain stem**—that part of the brain that is presumably phylogenetically the oldest and in which are persistently located centers for many simple but basically

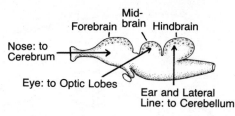

Figure 410. Diagram to show the relation in lower vertebrates of the three major sense organs to the three dorsal areas of gray matter in the three major subdivisions of the brain.

important reactions within the nervous system.* Subdivision of the brain stem into three parts may be related to the appearance, at an early stage in phylogeny, of three major sense organs: nose, eye, and ear + lateral line. In primitive vertebrates, each of the three becomes associated with one of the three subdivisions of the brain, and for each of the three, there tends to develop a dorsal outgrowth of laminated gray matter from its proper section of the stem (Fig. 410). These three outgrowths are, respectively, cerebrum, midbrain roof (tectum), and cerebellum.†

At the three-vesicle stage, midbrain and hindbrain are simple in construction, but, in the primitive forebrain, special structures appear early. Most notably, the optic vesicles (whose history was discussed in Chap. 15) push out at either side; the optic stalks, which form the optic nerves of the adult, remain attached anteriorly to the base of the primitive forebrain at the optic chiasma. More posteriorly, there is a down-growing median projection from the diencephalon, the **infundibulum**. Concomitantly, a pocket of epithelium, the **hypophyseal pouch** (Rathke's pouch) grows upward from the roof of the embryonic mouth. In later stages, modified infundibular tissues and those derived from the pouch combine to form the pituitary gland, discussed in Chapter 17. Dorsally, there grows from the roof of the primitive forebrain a series of median processes, the paraphysis and a median eye (sometimes two), which are described elsewhere.

Further developments occur to transform the tripartite brain stem into a brain of five regions. The midbrain shows little important change except for paired dorsal swellings that form the **tectum** (prominent in lower vertebrates), but both the hindbrain and forebrain become subdivided. In the hindbrain, a dorsal outgrowth from the roof of the front of the rhombencephalon becomes the **cerebellum**. The region of the medulla oblongata beneath the cerebellum is little changed in most vertebrates but, in mammals, is expanded into the structure termed the **pons**. Pons and overlying cerebellum are distinguished as the **metencephalon** from the more posterior part of the medulla, the **myelencephalon**.

Still more striking is the development anteriorly of paired outgrowths from the primitive forebrain. These are hollow pockets (at first somewhat analogous to the

*The cerebellum and the cerebral hemispheres are "new additions" not developed at this stage and not included in the brain stem. Also excluded by neurologists in defining the brain stem are certain structures (such as the pons, cf. p. 578), lodged in the brain stem but intimately connected with the functioning of cerebellum or hemispheres.

†In mammals, the eye, as we shall see, has "deserted" the midbrain in favor of the hemispheres as its major brain connection.

optic cups) that grow forward toward the nasal region; from them develop the **cerebral hemispheres** and, still farther anteriorly, the **olfactory bulbs**. These structures constitute the **telencephalon**, the anterior terminal segment of the brain; the unpaired part of the forebrain is the **diencephalon**, although, by definition, the most anterior tip of the median forebrain pocket, between the foramina leading to the paired vesicles of the hemispheres, is considered to be part of the telencephalon.

The principal structures of the adult brain may be tabulated according to the divisions established in the embryo as shown in the diagram.

Prosencephalon
- Telencephalon.......Cerebral hemispheres, including olfactory lobes, basal nuclei (corpus striatum) and cerebral cortex (pallium); olfactory bulbs
- Diencephalon........Epithalamus; thalamus; hypothalamus; various appendages

Mesencephalon...........................Tectum, including optic lobes (corpora quadrigemina in mammals); tegmentum; crura cerebri (cerebral peduncles) in mammals

Rhombencephalon
- Metencephalon......Part of medulla oblongata; cerebellum; pons of mammals
- Myelencephalon.....Most of medulla oblongata

Ventricles (Figs. 411, 415, 429).   The original cavity of the embryonic neural tube persists in the adult brain in the form of a series of cavities and passages filled with cerebrospinal fluid. A cavity, or **lateral ventricle**, is present in each of the cerebral hemispheres. Each of these cavities connects posteriorly through an **interventricular foramen** (foramen of Monro) with a median **third ventricle**, situated in the diencephalon. Within the midbrain, there is in lower vertebrates a well-developed ventricle, but, in amniotes, this becomes a narrow channel termed the **cerebral aqueduct** (aqueduct of Sylvius). Within the medulla oblongata is a **fourth ventricle**; this tapers posteriorly into the central canal of the spinal cord. Some lower vertebrates also have a separate ventricle within the cerebellum. Over most of the extent of the brain, the ventricles are surrounded by thick walls of nervous tissue. The walls are commonly thin, however, in two dorsal regions, one at the junction of the hemispheres with the diencephalon, the other forming the roof of the fourth ventricle. In each of these areas, there develops a **choroid plexus** (or tela choroidea)—a highly folded area of richly vascular tissue. Through these plexuses, exchange of materials takes place between the blood and cerebrospinal fluid.

Figure 411.   Diagram showing position of brain ventricles (From Gardner.)

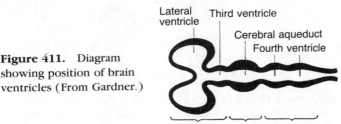

Brain Architecture. In the present elementary account of the vertebrate brain our attention will be mainly centered on external features and gross structures (Figs. 412–420). Although such superficial aspects of brain anatomy are significant, an adequate understanding of the working of the brain can no more be gained from them than a knowledge of the working of a telephone system can be gained from an acquaintance with the external appearance and room plan of the telephone exchange. What is of importance in a telephone system is the arrangement of switchboards and wiring; in a brain, the centers act as switchboards, in a sense, and tracts of fibers form the wiring between them.

Possibly the brain wiring was primitively much like that of the spinal cord with fibers crisscrossing to interconnect all areas. Generally, however, there is a strong tendency for the clustering together of nerve cells of specific functions in centers and the assembling of fibers with like connections into definite (though often very hard to see) bundles.

Although certain special centers have special names, most are termed **ganglia** or **nuclei** (making an unfortunate, duplicate, biologic use of this latter word). Nuclei, or centers, in a broad sense, range in size from tiny clusters of cells embedded in the gray matter of the brain stem and discernible only on microscopic study, to such massive structures as the cerebellum or cerebral cortex. Bundles of fibers connecting nuclei with one another are in general termed **tracts**; the fibers constituting such a

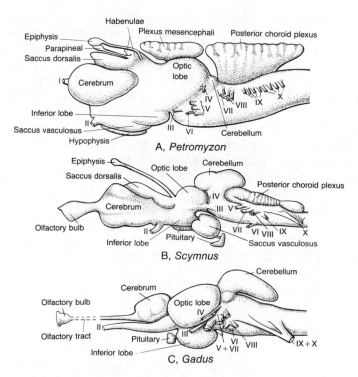

Figure 412. Lateral views of brain of *A*, a lamprey; *B*, a shark; *C*, a codfish. In the lamprey, an exceptional condition is the development of a vascular choroid area, the plexus mesencephali, on the roof of the midbrain. Roman numerals indicate cranial nerves. (After Bütschli, Ahlborn.)

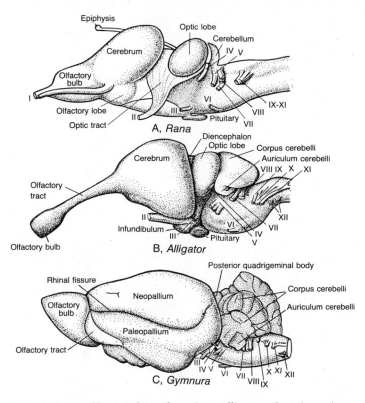

Figure 413. Lateral views of brain of *A,* a frog; *B,* an alligator; *C,* an insectivore representing a primitive mammalian type. In normal head posture, the front end of the alligator brain is tilted upward. Roman numerals indicate cranial nerves. (After Bütschli, Clark, Crosby, Gaupp, Wettstein.)

tract are, of course, axons of neurons whose cell bodies lie in the nucleus of origin. A tract is generally given a compound name, the two parts of which designate its origin and termination; thus, the corticospinal tract carries impulses from the cortex of the cerebral hemispheres to or toward the motor cells of the spinal cord.

Despite the general tendency of brain cells and fibers to organize into clean-cut centers and tracts, a primitive condition persists in the **reticular formation**. This is a system of interlacing cells and fibers associated with the motor columns in the brain stem and anterior part of the spinal cord; it is particularly well developed in the anterior part of the brain stem. It receives a variety of sensory input. This system is important in motor coordination. Still further, the reticular network appears to function in carrying stimuli downward from the anterior regions of the brain to the motor centers of the medulla and cord. This function is an essential one in lower vertebrates, in which motor tracts giving the brain control over activities of the trunk are poorly developed (cf. Figs. 432, 433); the reticular formation is still important in mammals, not only as a persistent low level path to motor centers but also (working in the opposite direction) as an agent that tends to promote activation of the higher centers of the cerebral cortex. It also controls complex reflexes such as sneezing and coughing.

The brain is constructed on a bilaterally symmetric pattern; in consequence,

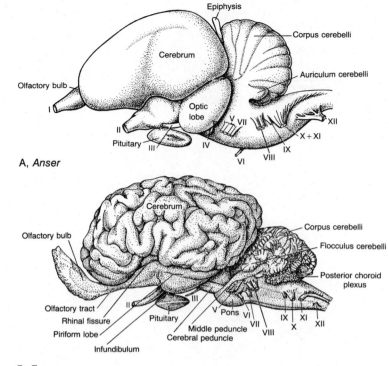

A, *Anser*

B, *Equus*

Figure 414. Lateral views of brain of *A,* a goose; *B,* a horse. The goose brain, like that of the alligator, is tilted upward anteriorly in life. Roman numerals indicate cranial nerves. (After Bütschli, Kuenzi, Sisson.)

Figure 415. Right half of the brain of a shark (*Scyllium*) in median aspect. Unshaded areas are those sectioned. (After Haller, Burckhardt.)

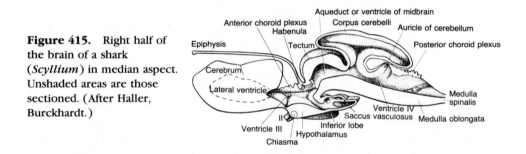

cross connections must be established in order that an animal may not have a literally dual personality. Even in the spinal cord, there are numbers of association neurons whose fibers cross to the opposite side, and, in the brain, such connections are numerous. There are a number of **commissures**, fiber tracts that connect corresponding regions of the two sides. In addition, there are cases where tracts in their course along the brain cross over from one side to the other (i.e., decussate), sometimes without apparent reason. We have described such a decussation in the case of the optic nerves (which are really brain tracts). Another example is recalled by the well-

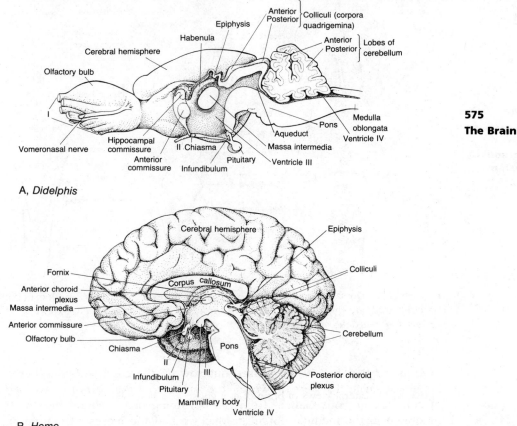

A, *Didelphis*

B, *Homo*

Figure 416. Right half of the brain, in median aspect, of *A,* an opossum; *B,* man. Unshaded areas are those sectioned. The internally bulging side walls of the diencephalon may meet and fuse in the midline, forming a "massa intermedia," which, however, has no functional importance. (*A* after Loo.)

known fact that movements of one side of the body are controlled by the gray matter of the brain hemisphere of the opposite side—a situation caused by a decussation of the corticospinal tracts of the two sides as they pass backward along the brain stem.

The discussion above of the "wiring" of the brain was stated in terms comparable to those of a telephone system, a pattern of end-to-end connections of series of neurons straight through the system from receptor to effector. But actually, as has been increasingly realized in recent years, the general patterns of activity in the brain are far more intricate in nature and include feedbacks, resonating circuits associated with memory and learning, and other complexities, into which we shall not attempt to enter here, which are being increasingly used (and well-known) in the development of computers.

Medulla Oblongata. Approach to the study of brain architecture is best made by first considering those parts that are simplest in construction and most closely resemble the spinal cord. The brain stem, including the three primary vesicles of the em-

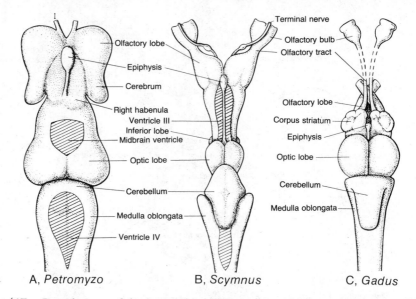

Figure 417. Dorsal views of the brain of *A,* a lamprey; *B,* a shark; *C,* a teleost (codfish). Hatched areas are those in which a choroid plexus has been removed, exposing the underlying ventricle. (After Bütschli, Ahlborn.)

bryo, is simpler than its specialized outgrowths, the cerebral hemispheres and cerebellum. Even in the stem, however, the anterior part appears to have been the seat of complex nervous centers from the beginning of vertebrate history. It is in the hindbrain in the medulla oblongata that we find the closest structural approach to the cord. Further, with the medulla and the adjacent parts of the midbrain are connected all the cranial nerves except the atypical ones from nose and eye and the terminalis.

The medulla oblongata is, particularly in lower vertebrates, closely comparable to an anterior section of the spinal cord—enlarged, however, through the expansion of the central canal to form the fourth ventricle, and with the columns of gray matter of either side widely separated dorsally as a consequence. For most of its length, the roof of the medulla is thin, membranous, and infolded to form the **posterior choroid plexus**; anteriorly, the medulla is covered by the cerebellum.

In the gray matter of the medulla and the posterior part of the midbrain, there is a series of columns or nuclei of gray matter basically similar to those present in the cord (Figs. 408 *B,* 423). In that region, we have noted the presence of a dorsal column, containing association centers for the reception and distribution of sensory impulses, and a ventral column containing motor neurons; we have further noted evidence that both columns may be subdivided into somatic and visceral components. A similar situation is found in the medulla and the posterior part of the midbrain, although the widely expanded ventricle makes for a somewhat different appearance. A horizontal groove, the **sulcus limitans**, running along the inner surface of the brain stem on either side, separates a dorsal sensory region from a ventral motor region. Further, as in the cord, each of these two can be subdivided into somatic and visceral components.

In the embryo, each subdivision appears to be formed as an essentially continu-

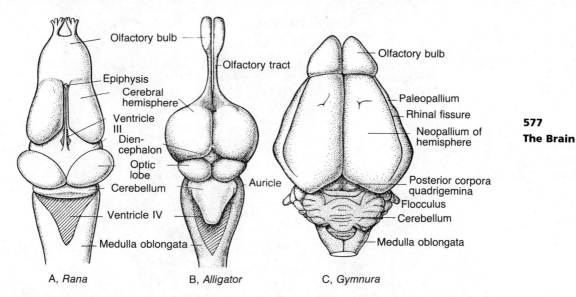

Figure 418. Dorsal views of the brain of *A*, a frog; *B*, an alligator; *C*, a tree shrew. Hatched areas are those in which a choroid plexus has been removed, exposing the underlying ventricle. (After Gaupp, Crosby, Wettstein, Clark.)

ous fore and aft column. Most ventral is a somatic motor column. Dorsal to this, but still below the limiting sulcus, are visceral motor columns for both special branchial and autonomic efferents. Above the sulcus are, in order, visceral sensory and somatic sensory columns, the primary areas of reception for sensations from gut and skin, respectively. The ventralmost column is that from which the ventral cranial nerves—III, IV, VI, and XII—arise; the others are the areas of central connections of the dorsal root or branchial nerves, including V, VII, IX, and X.

In the adults of lower vertebrate groups, much of the embryonic longitudinal continuity of the column is preserved; in higher types, however, there is a strong trend for a breaking down of the columns into discrete nuclei for the cranial nerves concerned and as centers for various body activities that are controlled or influenced by the brain stem. Thus, in mammals, the somatic motor column is fragmented into (1) several small anterior nuclei in the midbrain and in the anterior end of the medulla for the eye muscle nerves, and (2) a more posterior nucleus for the hypoglossal. The special visceral motor column for branchial musculature is broken up into separate special visceral motor nuclei for nerves V, VII, and IX and X (**nucleus ambiguus**), and, in this column, there develops a center concerned with respiratory rhythm. Small autonomic nuclei are present in the midbrain for optic reflexes, farther back in the medulla for salivary glands, and still farther back for the autonomic component of the vagus. In vertebrates generally, the nuclei associated with the normal visceral sensory fibers remain concentrated in a **nucleus solitarius** of the medulla, but a parallel **gustatory nucleus** is present for the special visceral sense of taste; the latter is so closely associated with the nucleus solitarius that it is sometimes considered simply a part of that center. The somatic sensory column remains as a single elongate nucleus, which, primarily associated with the trigeminal nerve, extends much of the length of the brain stem and even back into the cord.

In mammals, a great mass of fibers connecting the cerebellum and cerebrum

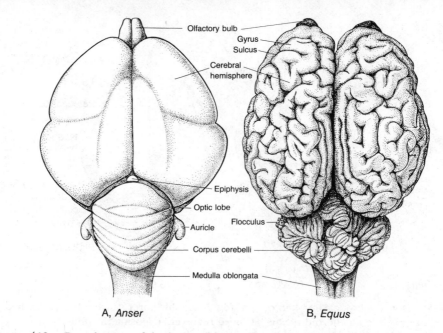

A, *Anser*   B, *Equus*

Figure 419.   Dorsal views of the brain of *A,* a goose; *B,* a horse. (After Bütschli, Kuenzi, Sisson.)

bridge over the base of the medulla anteriorly, causing so prominent a swelling that this region is termed the **pons.** Even apart from this, the gray matter of the medulla is thickly sheathed, in amniotes, by white tracts of fibers passing between the body and more anterior brain centers. Like the gray matter, the white matter of the medulla is organized much like that of the spinal cord. An added feature, especially in higher forms, is that large numbers of ascending and descending fibers decussate within the medulla. In lower classes, the sheathing is relatively thin, because trunk and tail are semi-autonomous and perform most movements reflexly, without referring them to the brain.

In most fishes, only a limited number of sensory fibers from the body reach the brain stem, cerebellum, and midbrain; in amniotes such fibers reach centers in the forebrain. Conversely, there is little positive control by the higher brain centers over body movement below the stage of birds and mammals except through the rather indirect means of the reticular system mentioned above. A striking exception is the presence in fishes and tailed amphibians of a pair of spectacular **giant cells of Mauthner** in the medulla. Their cell bodies, closely associated with the acoustico-lateralis centers, lie in the floor of the medulla; their large axons extend the entire length of the cord. As discussed before (cf. p. 195), locomotion in fishes and primitive tetrapods is mainly accomplished by rhythmic undulations of the body. To some extent this could be (and presumably is) regulated by local spinal reflexes. This pair of cells, however, appears to exercise general control over these movements. In adult frogs, typical reptiles, and higher forms, this type of locomotion disappears, and so do these giant cells.

The series of motor and sensory columns and nuclei in the medulla give us all the elements required for reflex circuits in the brain between sensory reception and

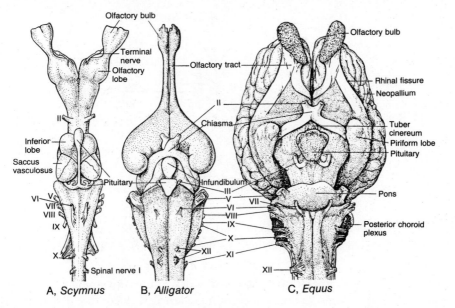

Figure 420. Ventral views of the brain of *A,* a shark; *B,* an alligator; *C,* a horse. Roman numerals indicate cranial nerves. (After Bütschli, Wettstein, Sisson.)

the responding effector organs of the head and branchial region. But brain mechanisms are not built solely on such a simple plan. In addition to ordinary sensory receptors in the skin, muscles, and gut, such as are found throughout the body, we have in the head special sensory organs. Centers must be present for the primary reception of sensations from these organs, and higher centers must be built up for the association and correlation of these sensations before final "directions" can be issued to the motor columns of brain stem and cord. Much of this apparatus is situated elsewhere in the brain. But even in the relatively simple region of the brain stem here considered, we find the primary area of reception of one of the main sensory systems, the acoustico-lateralis system.

The lateral line organs and the ear, parts of a single primitive sensory system, are somatic sensory structures; as such, sensations from them were primitively received, we may reasonably assume, in the somatic sensory column of the medulla. So special are they, however, that, in fishes, a specific **acoustico-lateralis area,** often of considerable size, develops above the normal somatic sensory column in the anterior part of the medulla. In terrestrial vertebrates, the lateralis system disappears but acoustic nuclei persist; in mammals, there are distinct centers for the vestibular and cochlear parts of the ear.

Cerebellum (Figs. 421, 422). Rising above the brain stem at the anterior end of the medulla oblongata is the cerebellum, a brain center, often of large size, which is of extreme importance in the coordination and regulation of motor activities and the maintenance of posture. The cerebellum acts in a passive, essentially reflex manner in the maintenance of equilibrium and body orientation. In addition, it plays a major positive role, particularly in mammals, in locomotion. Its function in regulating mus-

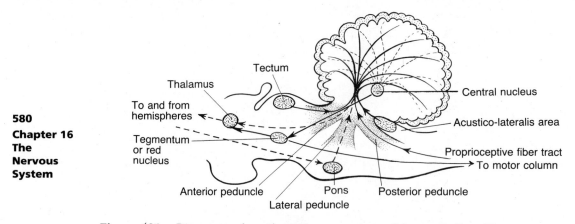

Tectum

Thalamus

To and from hemispheres

Tegmentum or red nucleus

Central nucleus

Acustico-lateralis area

Proprioceptive fiber tract

To motor column

Anterior peduncle

Pons

Posterior peduncle

Lateral peduncle

Figure 421. Diagram to show the main connections of the cerebellum. The connections with the cerebral cortex, peculiar to mammals, are shown in broken lines.

cular activity may be compared to that of staff work in the movement of an army. To carry out the general orders of an army commander, it is necessary that there be in hand information as to the position, current movements, condition, and equipment of the bodies of troops concerned. A directive from the higher brain centers (hemispheres or tectum) for a muscular action—say the movement of a limb—cannot be carried out efficiently unless there are available data as to the current position and movement of the limb, the state of relaxation or contraction of the muscles involved, the general position of the body, and its relation to the outside world. Such data are assembled in the cerebellum and synthesized there, and resulting "orders" issued by efferent pathways render the movement effective.

The data upon which the cerebellum acts are derived from two main sources: One is the acoustic area of the medulla, acoustico-lateralis area of lower vertebrates, in which are registered sensations concerning equilibrium from the ear and lateral line. The base of the cerebellum lies immediately above this area, and the cerebellum presumably originated as a specialized part of this sensory center. The second main source of cerebellar data is the system of muscle and tendon spindles. Into the cerebellum are directed fibers of this proprioceptive system carrying data regarding the position of parts of the body and the state of muscle tension.

In addition to these two primary sources, the sensory picture assembled in the cerebellum is rounded out by additional fiber relays from cutaneous sensory areas, from the optic centers, and, in lower vertebrates, even from the nose. Information, further, is supplied regarding muscular movements that are directed by higher brain centers but upon which the cerebellum has influence. In lower vertebrates, these data are furnished by fibers from the midbrain, where such impulses in great measure originate; in mammals, where the cerebral cortex dominates motor functions, strong fiber tracts connect cerebrum with cerebellum via the pons.

After integration of data by the cerebellum, outgoing fiber tracts carry impulses forward and downward on either side of the brain stem to the lateral walls of the midbrain (the tegmental region), whence they continue (in part after a relay there) to the appropriate motor nuclei of the head or body and, in mammals, to the thalamus and thence also to the cerebral cortex.

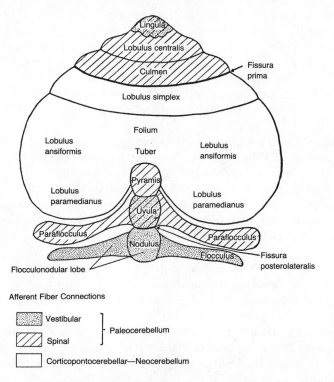

Afferent Fiber Connections

Vestibular
Spinal } Paleocerebellum

Corticopontocerebellar—Neocerebellum

Figure 422. Diagram of a surface view of a mammalian cerebellum (showing details not discussed in the text). The stippled and hatched portions, associated with equilibrium (vestibular) and with muscle sensations (spinal), are the phylogenetically oldest parts of the cerebellum; the white area is a mammalian addition associated with the cortex of the cerebral hemispheres. (From Fulton, after Larsell.)

In forms in which the cerebellum is well developed, the fiber tracts leading to and from it are prominent features in the architecture of the brain stem. In mammals, for example, these fibers form three pairs of pillar-like structures, the **cerebellar peduncles** (or **brachia,** Fig. 421). Anterior peduncles, mainly composed of efferent fiber tracts, connect cerebellum and midbrain. A middle, or lateral, pair of peduncles rise straight upward along the sides of the metencephalon from the pons; these carry fibers, running from cerebral cortex to cerebellum, which cross and are relayed ventrally in the pons. Posterior peduncles rise up through the medulla and carry proprioceptive fiber bundles from the cord.

The cerebellum varies greatly in size and shape from group to group. Its degree of development is roughly correlated with the intricacy of bodily movements; it is large and elaborately constructed in many fishes and in birds and mammals, small in reptiles, and little developed in cyclostomes and amphibians, in which it is little more than a pair of small centers lying just above the acoustico-lateralis nuclei and hardly distinct from them. This most ancient part of the cerebellum persists in all groups as the **auricles** (or flocculi), especially concerned with equilibrium and closely connected with the inner ear.

Even in cyclostomes and amphibians, however, there is some development of a median cerebellar region between the two auricles above the front end of the fourth ventricle. To this come fibers from the muscle spindles and from sensory centers in the brain for integration and correlation with stimuli from the organs of equilibrium in the ear. This primary structure, the **corpus cerebelli,** constitutes the main mass of the cerebellum in typical fishes, reptiles, and birds. In mammals, the rise of the cerebral cortex to command of motor functions and the development of large tracts from motor cortex to cerebellum and return are correlated with the appearance of large paired and convoluted structures, cerebellar hemispheres, which make up much of the bulk of the mammalian organ.

In contrast to every other area of the central nervous system except the cerebral hemispheres and roof of the midbrain, the cerebellum is a region in which the gray cellular material is superficially placed as a laminated cortex, with the white matter internal. In a well-developed cerebellum, the gray matter is spread out as a superficial sheet that is often highly convoluted and thus gains greater area. Beneath this is the white matter, principally a fan-shaped radiation of incoming fibers spreading out in all directions to the surface and of fibers returning from the gray matter. The cerebellar cortex contains cells of several quite distinctive types, arranged in layers, between which sensory data are interchanged by a complex fiber network. Most remarkable of cellular elements are the **Purkinje cells,** whose dendrites form highly branched "trees," collecting data from a large area of the cortex and sending the outgoing impulses via their axons to a relay in a **central nucleus.** The histologic pattern is quite constant among all vertebrates and even more so among all regions of the cerebellum. There is considerable localization of function in the cerebellum, but, especially in mammals, in many respects the corpus cerebelli appears to act as an integrated unit.

Midbrain and Diencephalon.   In contrast with the posterior part of the brain stem, the mesencephalon and diencephalon show specialized functional features in vertebrates of all classes. Sensory and motor nuclei connected with cranial nerves extend some distance into the midbrain from the medulla (Fig. 423). Farther forward, however, such structures are absent. The anterior centers in the brain stem have, in general, no direct connection with afferent or efferent impulses apart from those of the optic nerve. This region serves two main functions. In higher vertebrates, more particularly, it is a principal way station between more posterior areas of the brain and the cerebral hemispheres. In all groups, it is important to at least some degree as a locus of centers of nervous correlation and coordination. In mammals, the latter functions are overshadowed by those of the cerebral cortex, but, in many lower groups, the most highly developed association mechanisms lie in the midbrain and thalamus. In a mammal, destruction of the hemispheres results in functional disability; a frog, on the other hand, can go about its business in almost normal fashion without cerebral hemispheres as long as the brain stem and the structures surmounting it are intact.

The topography of the midbrain and diencephalon and their annexes will be described before dealing with the functions and connections of the various nuclei of their gray matter (Fig. 424).

The midbrain is that part of the brain tube traversed by the cerebral aqueduct; the diencephalon is a region lying about the third ventricle. Diencephalon and midbrain lie on the lines of communication between anterior and posterior parts of the brain. In mammals, the diencephalon is buried below and between the expanded

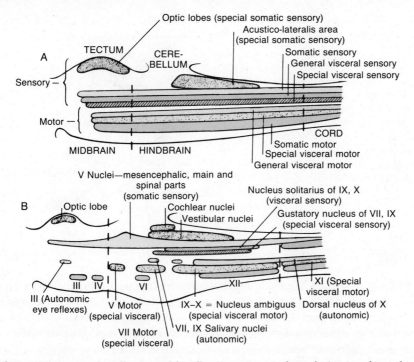

Optic lobes (special somatic sensory)

Acustico-lateralis area
(special somatic sensory)

Somatic sensory

General visceral sensory

Special visceral sensory

A

TECTUM

CERE-
BELLUM

Sensory —

Motor —

CORD

Somatic motor

Special visceral motor

General visceral motor

MIDBRAIN   HINDBRAIN

V Nuclei—mesencephalic, main and
spinal parts
(somatic sensory)

Nucleus solitarius of IX, X
(visceral sensory)

B   Optic lobe

Cochlear nuclei

Vestibular nuclei

Gustatory nucleus of VII, IX
(special visceral sensory)

III   IV   VI   XII   XI (Special
visceral motor)

III (Autonomic
eye reflexes)   V Motor
(special visceral)   IX–X = Nucleus ambiguus
(special visceral motor)   Dorsal nucleus of X
(autonomic)

VII Motor
(special visceral)   VII, IX Salivary nuclei
(autonomic)

Figure 423. Diagrams of midbrain and hindbrain regions in lateral view to show the arrangement of sensory and motor nuclei. Somatic sensory, blue; special somatic sensory, stippled; visceral sensory, green; special visceral sensory, hatched; visceral motor, yellow; special visceral motor, stippled; somatic motor, red. *A,* Hypothetic primitive stage, in which brain stem centers were continuous with one another and with the columns of the cord. Even at such a stage, however, it would be assumed that special somatic centers would have developed for eye and ear. The brain includes a special visceral motor column for the branchial muscles. *B,* Comparable diagram of the mammalian situation. The somatic sensory column is still essentially continuous (almost entirely associated with nerve *V*), but the other columns are broken into discrete nuclei. The visceral sensory column includes both a general visceral nucleus (mainly for afferent fibers from the viscera via the vagus) and a special nucleus for the important sense of taste. Of efferent visceral nuclei, there are small anterior ones for autonomic eye reflexes and the salivary glands and a large nucleus for parasympathetic fibers to the viscera via the vagus. There are important branchial motor nuclei for *V, VII,* and *IX, X* (*ambiguus*). The somatic motor column includes small nuclei for eye muscles anteriorly and a hypoglossal nucleus posteriorly.

hemispheres, and fiber tracts of white matter form the greater part of the midbrain walls and floor, these including great bundles of motor fibers (the pyramidal tracts descending from the hemispheres), which form the **cerebral peduncles** (crura cerebri).

In the midbrain, the gray matter above the aqueduct is greatly thickened to form the mesencephalic **tectum;** thinner areas of gray matter in the side walls are termed the **tegmentum.**

Of the areas bounding the third ventricle, the anterior wall, the **lamina terminalis** in which the brain stem ends, is technically considered to form part of the telencephalon; the remainder forms the diencephalon. Accessory structures, described be-

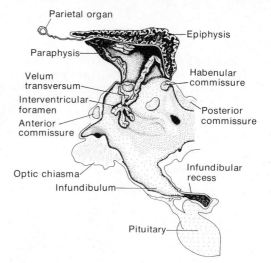

Figure 424. Sagittal section through the diencephalon of *Sphenodon,* showing the various structures projecting into and out of the third ventricle. Anterior is to the left. (After Gisi.)

low, are present in the roof and floor of the diencephalon; the main substance of this region consists of the lateral walls of the ventricle, which contain in their gray matter a host of nuclei. These lateral walls are termed collectively the **thalamus,** because of the fancy that the diencephalon forms a "couch" upon which rest in higher vertebrates the great cerebral hemispheres. Dorsal and ventral parts—roof and floor—of the diencephalon are called the **epithalamus** and **hypothalamus.** Thickened parts of the hypothalamus may cause external swellings as **inferior lobes** in fishes (Fig. 412) or **lateral lobes** in amphibians and reptiles. The thalamus proper is further divided, on the basis of the centers contained within it, into a **dorsal thalamus** and a **ventral thalamus**.

The roof of the third ventricle is for the most part a thin, non-nervous structure. In it develop a varied series of outgrowths. Most anteriorly, in that part of the roof that pertains to the telencephalic region, there grows upward in embryos of most groups a thin-walled sac, the **paraphysis**; this structure disappears in the adult in most cases (but not all; cf. Fig. 424). Almost nothing is known of its function except that it may produce glycogen to be passed into the cerebrospinal fluid. Adjacent to it, in the anterior part of the diencephalic or telencephalic roof, is the region of the **anterior choroid plexus**. In higher vertebrate classes, this is typically invaginated into the cavity of the ventricles; in lower groups, however, it is often an extroverted dorsal sac. More posteriorly in the diencephalic roof, there may develop one or both of a pair of median, stalked, eyelike structures, the parietal and pineal organs, discussed elsewhere (cf. pp. 518–519). Between these and the choroid plexus, a velum transversum may project downward into the ventricle.

In the floor of the diencephalon, the **optic chiasma** is a prominent feature anteriorly. Posteriorly, we find in most fishes the **saccus vasculosus** (or infundibular organ), a thin vesicle that may reach considerable size. It contains neurosensory cells and is believed to register fluid pressure within the ventricles of the brain; the sac is generally best developed, it may be noted, in deep-sea fishes (Fig. 412). Below the diencephalon lies the **pituitary gland**, or **hypophysis cerebri**, most important of all the endocrine structures of the body; it is described in Chapter 17.

The eyes appear to have been responsible in great measure for the development, in lower vertebrates, of important association centers in the anterior region of the brain stem (just as the acoustico-lateralis system is related to cerebellar development in the hindbrain and the olfactory sense to cerebral development). The optic nerves enter anteroventrally at the chiasma. In all vertebrates except mammals, however, most of their fibers do not tarry in the diencephalon, but proceed, almost without exception, upward and backward to the roof of the midbrain, where the primary visual center is located. The mesencephalic roof, the tectum, became, early in vertebrate history, the seat of an important association center; it is the dominant brain region, it would appear, in fishes and amphibians but became of lessened importance in amniotes with increased development of the cerebral hemispheres (cf. Figs. 432–434).

This center developed in connection with visual reception. A greater part of it develops as a pair of **optic lobes**, varying in size with the importance of the eyes, and becoming particularly large in birds and many teleosts. The histologic pattern of the optic lobes of nonmammalian vertebrates is complex, with successive layers of cells and fibers, giving them a structure broadly comparable to the laminated gray matter of the cerebellar or cerebral cortex. Presumably, a visual pattern is here laid out in a manner similar to that developed in our own case in the cerebral gray matter, although there are fibers running from the tectum to the cerebrum that could be involved.

The presence of this visual center in lower vertebrates was responsible, it would seem, for the attraction to the tectal region of stimuli from other sensory areas. Fiber paths lead hither from the acoustico-lateralis area, from the somatic sensory column, and from the olfactory region via the diencephalon, and connections with the cerebellum are developed. Sensory stimuli from all somatic sources are here associated and synthesized, and motor responses are originated. Primitively, these motor stimuli were relayed, it would seem, by way of the reticular formation lying more ventrally in the tegmental region of the midbrain. In amphibians and reptiles, however, direct motor paths are developed, with the formation of definite tracts from the tectum to the motor columns of the brain stem and to the cord.

In fishes and amphibians, the tectum appears to be the true "heart" of the nervous system—the center that wields the greatest influence on body activity. In reptiles and birds, the tectum is still an area of great importance but is rivaled, particularly in birds, by the development of higher centers in the hemispheres.

In mammals, the tectum has undergone a great reduction in relative importance; most of its functions have been transferred to the gray matter of the cerebral hemispheres, and most of the sensory stimuli that are integrated in the midbrain in lower vertebrates are, instead, projected to the hemispheres in mammals. Most auditory and other somatic sensations that reach the midbrain are, in mammals, relayed onward by way of the thalamus to the hemispheres. It is especially notable that few of the optic fibers in mammals follow the original course to the midbrain; most are interrupted in the thalamus, and visual stimuli are shunted forward to the hemispheres. The tectum still serves in a limited way as a center for visual and auditory reflexes. In mammals it takes the form of four small swellings in the roof of the midbrain, the **corpora quadrigemina**. Of these, the anterior pair (**superior colliculi**) deal with visual reflexes and represent the optic lobes of lower vertebrates. In amphibians, with the development of hearing functions in the ear, secondary auditory centers develop on either side of the midbrain adjacent to the optic tectum. These develop into the posterior pair of elements of the corpora quadrigemina (**inferior colliculi**), which attain considerable size in some groups of mammals; they function as a relay station for auditory stimuli on their way to the thalamus and thence to the cerebral hemispheres.

The region on either wall of the midbrain termed the tegmentum is essentially an anterior continuation of the motor areas of the medulla. As such, it functions as a region in which varied stimuli from diencephalon, tectum, and cerebellum are coordinated and transmitted downward to the motor nuclei of the brain stem and spinal cord. In lower vertebrate groups, it consists in general of the rather diffuse series of nerve cells and fibers of the reticular system (cf. p. 573). In some instances, however, well-defined nuclei may be formed; we may note, for example, the presence in this region in mammals of the **red nucleus** through which are relayed efferent impulses from the cerebellum.

The epithalamus is of little importance as a brain center. We may note the constant presence here in all vertebrates of the **habenular body**, a group of small nuclei through which olfactory stimuli pass on their way back from hemispheres to brain stem. We have mentioned elsewhere that from this area develop the paraphysis, the anterior choroid plexus, and the pineal and parietal (parapineal) bodies.

The hypothalamus contains olfactory centers, notably the **tuber cinereum** and, in mammals, the **mammillary bodies** adjacent to the pituitary region. Its main importance, however, is that of a visceral center; other major centers are almost exclusively somatic in their activities. Many of the visceral nervous functions are carried out by reflexes in the cord or medulla, but the hypothalamus is in all vertebrate classes a region that serves as a major integrative center for the body's visceral activities, and there are fiber connections with the autonomic centers of the brain stem and cord. Stimuli from olfactory and taste organs as well as sensations from various visceral structures of the body pass to a number of nuclei in this region. Particularly important are connections with the more strictly olfactory areas of the cerebral hemispheres; the sense of smell plays a large role in visceral nervous activities. These nuclei have efferent connections, posteriorly, with autonomic centers. The hypothalamus is involved in the control of the activity of the pituitary; as noted in the next chapter, some hypothalamic nuclei actually secrete hormones that pass to the neurohypophysis. The range of hypothalamic regulatory functions is incompletely known. It is, however, of interest that (for example) temperature regulation in reptiles, birds, and mammals is accomplished by a sensory "thermostat" in the hypothalamus, and that in mammals, heart beat, respiratory rate, blood pressure, sleep, and activities of the gut are controlled or influenced by the hypothalamus.

The thalamus proper is in lower vertebrate classes an area of modest importance. Its ventral, motor part may be considered an anterior outpost of the motor column of the brain stem; it is in every class a motor coordinating center and is further a relay center on the motor path from the basal nuclei of the hemispheres back to the brain stem.

In lower vertebrates, the dorsal, sensory region of the thalamus appears to be merely an anterior extension of the sensory correlation areas connected with the tectal region of the midbrain. Its importance increases proportionately with the increased development of association centers in the cerebral hemispheres. Even in primitive vertebrates, considerable sensory data may be passed forward from the brain stem, particularly the reticular system, through thalamic relays to the cerebral hemispheres for synthesis there with olfactory stimuli. With a high degree of development of association centers in the hemispheres of amniotes, there is a great development of nuclei in the dorsal thalamus for relay purposes. In the mammalian stage (with the practical abandonment of the tectal association center), the thalamus reaches the height of its development. In mammals, all somatic sensations are assembled in the gray matter of the cerebral cortex; and all this sensory material (except for, of course, olfaction) is relayed to it via the dorsal thalamus. Conspicuous elements in its structure are the **lateral geniculate body** and **pulvinar**, whence optic stimuli are projected upward to the hemispheres, the **medial geniculate body**, which relays auditory stimuli, and the **ventral nucleus**, which transmits somatic sensory stimuli.

Cerebral Hemispheres. The evolution of the cerebral hemispheres is one of the most spectacular stories in comparative anatomy. These paired outgrowths of the forebrain began their history, it would seem, largely, possibly entirely, as loci of olfactory reception. Early in tetrapod history, they became large and important centers of sensory correlation; by the time the mammalian stage is reached, the greatly expanded surfaces of the hemispheres have become the dominant association center, seat of the highest mental faculties. The development of centers of such importance in this anterior, originally olfactory, segment of the brain emphasizes the importance of the sense of smell in vertebrates generally. The acoustico-lateralis system and vision are, as we have seen, senses upon which important correlative mechanisms were erected early in vertebrate history, but, in the long run, smell has proved dominant. Smell is of little account in higher primates, such as ourselves. But in following down our ancestral line to early mammals and on down to their early vertebrate ancestors, it appears that this sense has been throughout a main channel through which information concerning the outside world has been received. It is thus only natural that its brain centers should form a base upon which higher correlative and associative mechanisms have been built.

Presumably the brain tube of the earliest vertebrates was, like that of amphioxus, a single, unpaired structure all the way forward to its anterior end; the cavity of the telencephalon was a median, unpaired terminal ventricle, as it is today in the early embryo of every type. In all vertebrates, the most anterior part of the wall of the third ventricle is considered as belonging to the telencephalon, but in tetrapods, most of that brain segment consists of distinct paired structures, the **lateral ventricles** and the tissues surrounding them. Most fishes—cyclostomes, Chondrichthyes, actinopterygians—show a transitional condition, because there is, for part of the length of the telencephalon, a single ventricular cavity that bifurcates only anteriorly.

In even the lowest of living vertebrates, the cyclostomes, each half of the telencephalon is subdivided into two parts, the **olfactory bulb** and the **cerebral hemisphere**. The bulb is a terminal swelling in which the fibers of the olfactory nerve end; its size varies with the acuteness of the sense of smell; it is, for example, small to minute in birds. Here is located the primary olfactory nucleus. From its cells, fibers pass back (sometimes with a relay en route) to be distributed to various regions of the hemisphere. Depending upon the configuration of the head and braincase, bulb and hemisphere may be in close contact or may be well separated, with a distinct **olfactory tract** between them.

Primitively, as seen in cyclostomes, the hemisphere is mainly an **olfactory lobe**—an area in which olfactory sensations are assembled and learned olfactory reactions are relayed to more posterior centers for correlation with other sensory impulses (Figs. 425 *A*, 426 *A*). At this level of development, relatively few fibers ascend from the stem to the hemispheres for correlation there, although recent work indicates that the hemispheres are not as completely olfactory as was once thought. In general, fiber paths from olfactory areas follow two routes: ventrally to end in visceral centers in the hypothalamus, or dorsally to the habenulae in the epithalamus and thence to the tectum or motor areas of the brain stem.

A higher stage, although still primitive, is that seen in the amphibians (Figs. 425 *B*, 426 *B*); sharks and lungfish are at approximately the same stage of cerebral development. Here most of the tissues of the hemispheres can be divided regionally into three areas, of interest because of their history in more progressive types. All three areas receive olfactory stimuli and exchange fibers with one another; all three discharge to the brain stem. Ventrally lies the region of the **basal nuclei**, essentially equivalent to the corpus striatum of mammals and destined in higher stages to move into the central parts of the hemisphere. The basal nuclei form a correlation center at an early evolutionary stage and become increasingly more important in this regard in more advanced groups. Sensory impulses are projected upward into the basal nuclei from the thalamus for correlation with olfactory sensations; descending fibers

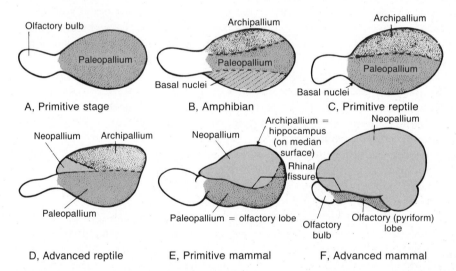

Figure 425. Diagrams to show progressive differentiation of the cerebral hemisphere (cf. Fig. 426). Lateral views of left hemisphere and olfactory bulb. In *A*, the hemisphere is merely an olfactory lobe. *B*, Dorsal and ventral areas, archipallium (= hippocampus) and basal nuclei (corpus striatum) are differentiated. *C*, The basal nuclei have moved to the inner part of the hemisphere. *D*, The neopallium appears as a small area (in many reptiles). *E*, The archipallium is forced to the median surface, but the neopallium is still of modest dimensions, and the olfactory areas are still prominent below the rhinal fissure (as in primitive mammals). *F*, The primitive olfactory area is restricted to the ventral aspect, and the neopallial areas are greatly enlarged (as in advanced mammals). The various cellular components of the hemispheres are distinguished by color.

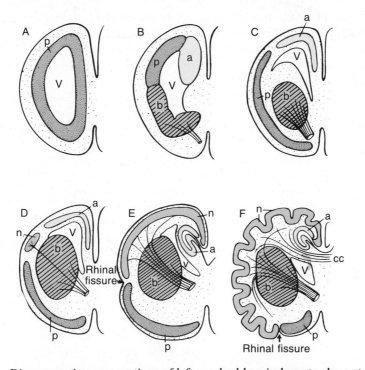

Figure 426. Diagrammatic cross sections of left cerebral hemisphere to show stages in the evolution of the corpus striatum and cerebral cortex. *A,* Primitive stage, essentially an olfactory lobe; gray matter internal and little differentiated. *B,* Stage seen in modern amphibians. Gray matter still deep to surface, but differentiated into paleopallium (= olfactory lobe), archipallium (= hippocampus), and basal nuclei (= corpus striatum), the last becoming an association center, with connections from and to the thalamus (indicated by lines representing cut fiber bundles). *C,* More progressive stage, in which basal nuclei have moved to interior, and pallial areas are moving toward surface. *D,* Advanced reptilian stage; beginnings of neopallium. *E,* Primitive mammalian stage; neopallium expanded, with strong connections with brain stem; archipallium rolled medially as hippocampus; paleopallial area still prominent. *F,* Progressive mammal; neopallium greatly expanded and convoluted; paleopallium confined to restricted ventral area as pyriform lobe. The corpus callosum developed as a great commissure connecting the two neopallial areas. Abbreviations: *a,* archipallium; *b,* basal nuclei; *cc,* corpus callosum; *n,* neopallium; *p,* paleopallium; *V,* ventricle. The different types of "gray matter" are colored as in Figure 425.

carry impulses from the basal nuclei to centers in the thalamus and mesencephalic tegmentum.

The gray matter of all parts of the hemisphere except the basal nuclei tends progressively to move outward toward the surface, and thus becomes the **cerebral cortex**, or **pallium** ("cloak"). In fish and amphibians, the gray matter is still largely internal, but these terms may nevertheless be used in the light of later history. A band of tissue along the lateral surface of the hemisphere is termed the **paleopallium**. This area remains largely olfactory in character, and the paleopallial region is that of the olfactory lobes in higher stages. Dorsally and medially lies the **archipallium**, which is antecedent to the hippocampus of mammals. This area is to a minor extent a correlation center in all tetrapods, with ascending fibers from the diencephalon as well as fibers from olfactory bulb and lobe, and appears to be related to "emotional" behavior. The tract from this region to the hypothalamus is the main component of the fiber bundle termed the **fornix** in mammals.

An aberrant, "everted" type of forebrain is seen in most bony fishes (Fig. 427). Here the gray matter of the hemispheres has been crowded downward and inward by the lateral and then ventral growth of its originally dorsal part, to form massive structures bulging up into the ventricles from below. These masses include both the basal nuclei or striatum and tissues above it representing the pallial areas, which in some teleosts include a cortex of complex structure. The roof of the hemispheres is only a thin, non-nervous membrane. This development is typical of teleosts and also occurs in the more primitive actinopterygians (including *Polypterus*). As one might expect, since tetrapods do not show this structure, lungfish have normal not everted, hemispheres; however, the only living coelacanth, the rather aberrant *Latimeria,* displays an odd, rather intermediate condition.

The reptile hemispheres are advanced over the amphibian type in both complexity of organization and relative size compared with other brain parts. The dorsal and

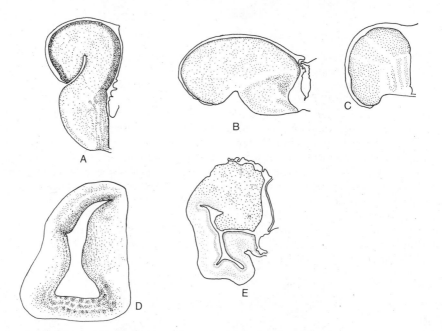

Figure 427. Transverse sections through one cerebral hemisphere of various Osteichthyes; the midline is on the right of each drawing. *A, Polypterus,* a chondrostean; *B, Amia,* a holostean; *C, Carassius,* a teleost; *D, Protopterus,* a lungfish; *E, Latimeria,* a crossopterygian. The hemisphere of the lungfish resembles that of amphibians (cf. Fig. 426 *B*); in all the actinopterygians and, to a lesser extent, *Latimeria,* the dorsal wall of the hemisphere is membranous, the ventricle is dorsally displaced, and the dorsal and lateral parts of the hemisphere are rolled laterally and then ventrally, a pattern termed "everted." (After Nieuwenhuys and Kuhlenbeck.)

lateral walls of the hemispheres show in the main an essentially primitive arrangement of the pallial areas, but some of the gray matter has spread outward toward the surface. The basal nuclei are large and have moved inward to occupy a considerable area in the floor. Strong fiber bundles project to the basal nuclei from the thalamus and back from them to the brain stem; the basal nuclei are obviously correlation centers of importance.

In birds (Fig. 428 A), the hemispheres are further enlarged. Their evolutionary development, however, has taken place in a manner radically different from that seen in mammals. Presumably correlated, as in teleosts, with the great reduction of the sense of smell in birds generally, there is relatively little development of the cerebral cortex; there is a very small area of paleopallium, and a modest development, at the medial posterior areas of the hemispheres, of a cortical area of archipallial nature; areas corresponding to the mammalian neopallium, described below, are filled in to form, with the enormously expanded basal ganglia or **corpus striatum**, a large projection into the ventrolateral side of the ventricle. The corpus striatum proper is a large, solid, and complex mass of cells and fibers occupying much of the inside of the hemispheres. Birds are notable as having born in them a complex series of stereotyped action patterns that may be called forth to meet a great variety of situations. Presumably, these are lodged in the highly developed corpus striatum. However, many birds, notably the crows and ravens, have been shown to possess, in addition to these innate patterns, a considerable ability to learn by experience. In mammals, memory and learning are associated with the cerebral cortex; in birds (as proved experimentally), these powers reside in the.**hyperstriatum,** an internal development of the hemispheres situated above the striatum proper. Extirpation of this region (particularly its uppermost part) destroys the bird's memory and learning ability, while leaving its normal reflex activities undisturbed. Recent work indicates that the hyperstriatum is, indeed, the homologue of the mammalian neocortex and that similar structures occur in many reptiles.

The first faint traces of mammalian cortical development are to be seen in certain reptiles (Figs. 425 D, 426 D). In the hemispheres of these forms, we find, between paleopallium and archipallium, a small area of superficial gray matter of a new type, that of the **neopallium**. Even at its inception, it is an association center, receiving,

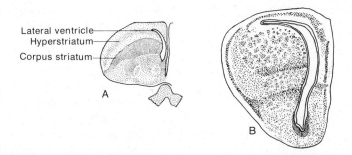

Lateral ventricle
Hyperstriatum
Corpus striatum

A

B

Figure 428. Transverse sections through the cerebral hemisphere of a sparrow, A, and an alligator, B. In birds, the basal nuclei form a greatly expanded corpus striatum and, above it, hyperstriatum; this last structure is the equivalent of the mammalian neocortex. The alligator shows some similarities but appears more "normal." (After Kappers, Huber, and Crosby.)

like the basal nuclei, fibers that relay to it sensory stimuli from the brain stem and, in turn, sending "commands" directly to the motor columns.

The evolutionary history of the mammal brain is essentially a story of neopallial expansion and elaboration. The cerebral hemispheres have attained a bulk exceeding that of all other parts of the brain, particularly through the growth of the neopallium, and dominate functionally as well. This dominance is apparent in mammals of all types but is particularly marked in a variety of progressive forms, especially in man. In even the more primitive mammals (Figs. 425 E, 426 E), the neopallium has expanded over the roof and lateral walls of the hemisphere. It has crowded the archipallium on to the median surface above; the paleopallium is restricted to the ventrolateral part of the hemispheres, below the **rhinal fissure**—a furrow that marks the boundary between olfactory and nonolfactory areas of the cortex. With still further neopallial growth (Figs. 425 F, 426 F), the archipallium is folded into a restricted area on the median part of the hemispheres, where it remains as the **hippocampus**,* and the olfactory lobes, which include the paleopallium, come to constitute only a small ventral hemisphere region, the **pyriform lobe**. The corpus striatum persists as a relay center for the more automatic reactions.

As it develops in mammalian evolution, the neopallium assumes newer and higher types of neural activity in correlation and association and also takes over many of the functions previously exercised by centers in the brain stem and many of those of the basal nuclei. The mesencephalic tectum loses its former importance and is reduced to a reflex and relay center. Auditory and other somatic sensations are relayed anteriorly to the thalamus, most of the optic fibers are intercepted there, and all are projected from thalamus to hemispheres by great fiber tracts. We have seen that thalamic connections of this sort for the basal nuclei had evolved in lower vertebrate groups, and, in birds, powerful projection tracts evolved in connection with that dominant brain region. In mammals, however, most of these fibers plunge on through the basal nuclei, here termed the **corpus striatum**,† to radiate out to the neopallial surface. With all the sensory data thus made available, the appropriate motor "decision" is made by the cortex. As mentioned earlier (cf. p. 580) one set of stimuli is sent from cortex to cerebellum by way of the pons for appropriate regulatory effects; there are cortical connections to the corpus striatum and even some discharge to the hypothalamus and thence to the autonomic system. The main motor discharge, however, is by way of the **pyramidal tract**, a fiber bundle that extends directly, without intervening relay, from the neopallial cortex to somatic motor regions of brain stem and cord—a feature emphasizing the dominating position of the cerebral cortex in mammals.

With expansion, the cerebral hemispheres tend to cover and envelop the other brain structures. In a primitive mammalian brain (Fig. 416 A), the cerebrum leaves much of the midbrain exposed; but in a majority of living mammals (as in Fig. 414 B), the midbrain and part of the cerebellum are overlapped. The paired ventricles and the more primitive areas associated more purely with olfaction—olfactory bulb, pyriform lobe, hippocampus, and related tracts and nuclei—have been shifted and distorted, with the growth of the mammalian hemispheres, into patterns that are difficult to compare with those of lower and more simply built brains (Fig. 429).

---

*So-called because of its fancied resemblance (in section) to a sea horse, with its coiled tail.

†The name is due to the striated appearance caused by the passage of the fiber bundles through the gray matter of the basal nuclei.

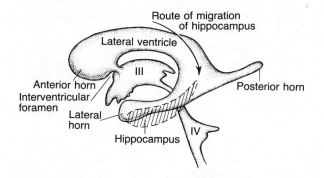

Figure 429. The brain ventricles of an advanced mammalian type (*Homo*) in lateral view from the left. The ventricles are represented as solid objects, the brain tissue being removed. With expansion of the cerebral hemisphere, the lateral ventricle has expanded backward to a posterior horn in the occipital lobe, and downward and forward laterally to a lateral horn in the temporal lobe. With this backward and downward expansion, various shifts in position of brain parts occurred. The hippocampus, which developed dorsally on the median surface of the hemisphere (cf. Fig. 426 *F*) has been rotated, in advanced mammals, backward and downward into a ventral position near the midline.

Because the neopallium is essentially a thin sheet of laminated cellular material, underlain by the white fibrous mass of the cerebrum, simple increase in hemispheric bulk fails to keep cortical expansion in step with increase in the volume of fibers; folding of the surface is necessary. In small or primitive mammals, the neopallial surface is often smooth; in large or more progressive types, the surface is generally highly convoluted—thrown into folds that greatly increase the surface area. That this folded pattern arose independently many times in the evolution of mammals is clearly shown by the fossil record, since mammalian cranial cavities usually reflect closely the shape of the contained brain (Fig. 430). The folds are termed **gyri**, the furrows between, **sulci**. These are prominent landmarks on the surface of the brain, and it was once believed that, in some cases, they are structural boundary markers for spe-

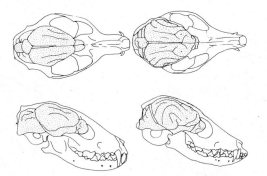

Figure 430. Brains (stippled) in skulls of a fossil and a modern dog to show the increase in size and complexity of the brain, especially the cerebral hemisphere. *Hesperocyon,* on the left, is an Oligocene form (roughly 30 million years old); *Fennecus,* on the right, is a modern foxlike form of about the same size. The actual brain of the fossil is, of course, not preserved, but the form of the cranial cavity reflects its structure in considerable detail. (After Radinsky.)

cific cortical areas. Further study, however, shows that (apart from the rhinal fissure and, to some degree, a central sulcus in primates) there is no fixed relationship between the pattern of convolutions and the structural subdivisions of the cortex.

In man, most particularly, the hemisphere is often described as being composed of a series of lobes—a **frontal lobe** anteriorly, a **parietal lobe** at the summit, an **occipital lobe** at the posterior end, and a lateral **temporal lobe**. These terms are, however, purely topographic and have no precise meaning in regard to the architecture or functioning of cortical areas.

The gray matter of the neopallium has a complex histologic structure with, in eutherians, six superposed layers of cells and masses of intervening fibrils—this in contrast to paleopallial and archipallial regions, in which only two to four cell layers are present. It is estimated that in some of the larger mammalian brains, the number of neopallial cells may run into the billions. The white matter internal to the gray includes, in addition to a fan of cortical connections to and from lower brain regions, a great interweaving meshwork of fibers that connects every part of the cortex with every other. An **anterior commissure** (Fig. 424), connecting olfactory portions of the two hemispheres, is present in all vertebrates, and other commissural fibers are present in the fornix; to connect the neopallial structures a massive new commissure, the **corpus callosum** (Fig. 416 *B*), develops in placental mammals, functioning to allow both hemispheres to share memory and learning.

The complex "wiring" system connecting all parts of the cortex with one another would suggest that the gray matter is essentially a unit, equipotent in all its parts for any cerebral activity. This is to a limited extent true; experiments show that in laboratory animals a good part of the neopallium may be destroyed without permanently interfering with normal activity, and the results of injury and disease conditions show that the same holds for certain areas of the human brain. On the other hand, it is clear that certain cortical areas are normally associated with specific functions (Fig. 431). We have previously discussed the primitive cortical regions devoted mainly to olfactory sensations—the paleopallium and archipallium, represented in mammals in the pyriform lobe and hippocampus, respectively. Regional differentiation is present in the neopallium as well. The front part of the hemispheres includes a motor area. The posterior part is associated with sensory perception. Special regions

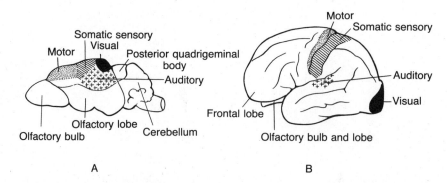

Figure 431. Lateral view of *A,* the brain of a shrew; *B,* the cerebrum of man; to show cortical areas.

are associated with eye and ear, in occipital and temporal lobes, whereas the areas for sensations received from the skin and proprioceptive organs are situated farther forward, close to the motor area. In primates, a **central sulcus**, which crosses the top of the hemisphere from medial to lateral surfaces, divides (although not exactly) motor and sensory areas. Along its anterior margin are specific motor areas for each subdivision of the body and limbs, arranged in linear order; along the posterior margin is an exactly parallel arrangement of loci for sensory reception from the various members. In many mammals, almost the entire surface of the neopallium is occupied by areas associated more or less closely with specific sensory or motor functions. Although the central sulcus may not be present, placental mammals generally have a similar linear arrangement of sensory and motor areas opposite one another. Marsupials (and also the placental edentates) have a similar layout of body regions, but sensory and motor areas are combined rather than separate. In man, particularly, however, we find that these specific functional areas occupy only a relatively small part of the neopallium. Between them have developed large areas of gray matter, most conspicuously one occupying much of the frontal lobe, which are not associated with specific sensory or motor functions. They are, in consequence, sometimes termed "blank areas," although, as shown by injury to these areas, they are the seat of our highest mental properties, including learning ability, initiative, foresight, and judgment. However, some areas may be removed without greatly apparent changes in functioning.

Size of cerebral hemispheres might be thought of as giving a clue to the mental abilities of a mammal. This is true in a sense but is subject to strong qualification. If the amount of cortical surface present is related in any way to intelligence, it is obvious that of two brains of the same size, one with convoluted hemispheres is better than one with a smooth surface. The bulk of the animal concerned affects brain size, presumably because of the need of greater terminal areas for the increased sensory and motor connections, but the increase in brain size as a whole is not absolutely proportionate to the mass of the body, and large animals tend to have relatively small brains without, it seems, any loss of mental powers. That absolute brain size is not a criterion of intellect is indicated by the fact that the brain of a whale may have five times the volume of that of a man. Nor is the proportion of brain size to body size a perfect criterion, because small South American monkeys may have a brain one fifteenth or one twentieth of body weight, whereas the brain weight of an average man is only one fortieth that of the body. Smaller animals will almost always have *relatively* large brains. A final and most important factor that must be taken into account is the relative complexity of the cortical sheet.

Brain Patterns—Summary. In earlier sections of this chapter, we described the principal structural and functional elements of the brain and certain of their interconnections. The more important features may be summarized here.

In the posterior part of the brain stem, the medulla oblongata, there is present a series of sensory and motor columns closely comparable to those of the spinal cord and connected with them by the reticular system. As in the cord, direct reflexes may take place between sensory and motor columns. The history of brain evolution, however, has been mainly one of the development, above and in front of the medulla, of higher centers of coordination and association, interposed between sensory and motor areas. In such centers, sensory data are assembled and synthesized, and the resultant motor stimuli sent out to the motor columns of the brain stem and cord. These centers have been mainly built up, as dorsal outgrowths of laminated gray matter, about areas primarily concerned with the special senses of the head: acoustico-lateralis, visual, and olfactory.

1. The primary area of reception for equilibrium and stimuli from the lateral line lies in the sensory columns of the medulla oblongata; above this developed the cerebellum, important in all vertebrates. This organ initiates no bodily movement, apart from adjustments in posture, but ensures that motor directives initiated in other centers be carried out in proper fashion. It may receive fibers from all the somatic senses but is principally informed by the adjacent acoustico-lateralis centers and by fibers from the proprioceptive system of the muscles and tendons. Its outgoing, regulatory stimuli are for the most part sent to the midbrain and thence distributed to the motor areas. In mammals, in which the cerebral cortex assumes direct control over most motor responses, powerful circuits are established connecting cerebellum and cortex in both directions (Fig. 421).

2. In lower vertebrates, the main centers dominating nervous activity are situated in the anterior regions of the brain stem. *(a)* A great center of coordination in which motor activity is initiated is established in the tectum of the midbrain, in which the primary visual center is situated (Fig. 432). To this region are relayed also stimuli from the nose, ear, and other somatic senses; from it are sent out stimuli to the motor columns. As we ascend the vertebrate scale, the tectal area becomes rivaled and then exceeded by the association centers of the cerebral hemispheres; in birds, despite the great de-

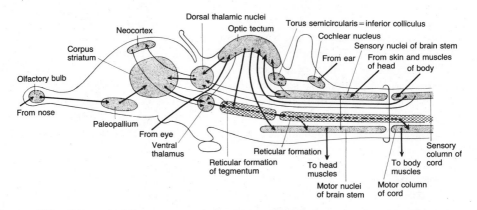

Figure 432. Diagram of the main centers and "wiring" arrangement of a primitive reptile, in which the tectal region of the midbrain plays a dominant role; the corpus striatum (basal ganglia) is of some importance as a correlation center, but the neocortex (neopallium) is unimportant. The reticular formation of the brain stem (cross hatched) is important in carrying motor impulses to nuclei of the stem and cord. In this oversimplified diagram, only a limited number of paths between somatic receptors and effectors are included; visceral centers and paths are omitted, as are cerebellar connections (shown in Fig. 421).

velopment of the corpus striatum and hyperstriatum, the tectum is still prominent, but, in mammals, most of the sensory data formerly assembled in the tectum are, instead, relayed to the neopallium via the thalamus, and the tectum is reduced to a reflex center. *(b)* The tectal centers are somatic in nature; corresponding centers for visceral sensations and visceral motor responses were early established farther anteriorly and ventrally, in the hypothalamus. This situation remains little changed throughout the vertebrate series.

3. As the vertebrate scale is ascended, the cerebral hemispheres, originally largely a center for olfactory sensation, have become more and more important as association centers. *(a)* The first of cerebral areas to gain importance is that of the basal nuclei, the corpus striatum. Fiber tracts from the thalamus relay somatic sensations to this body, and return fibers carry motor stimuli back to the midbrain and thence to the motor columns. In reptiles, the corpus striatum is a prominent structure that rivals the older tectal center in importance, and, in birds, the corpus striatum, with the new addition of the hyperstriatum, becomes a large, complex, and dominant center (Fig. 433). *(b)* In mammals, descended along a different evolutionary line from that leading to birds, a different development has occurred. The new master organ is the neopallium, a greatly expanded gray cortical area for correlation, association, and learning, partly, at least, equivalent to the avian hyperstriatum. This assumes the greater part of the higher functions once concentrated in the tectum or corpus striatum, gains a complete array of somatic sensory data through projection fibers from the thalamus, and develops direct motor paths to the motor columns of the brain stem and spinal cord (Fig. 434).

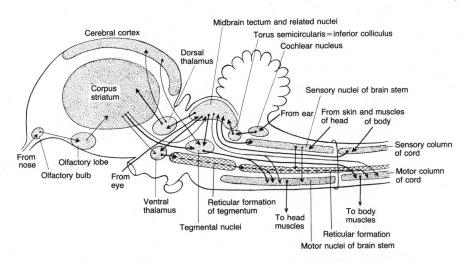

Figure 433. A "wiring diagram" of a bird brain, comparable to that of Figure 432. The mesencephalic tectum is still of importance, but the corpus striatum (which, as the hyperstriatum, includes the equivalent of the mammalian neocortex) is the dominant center in many regards.

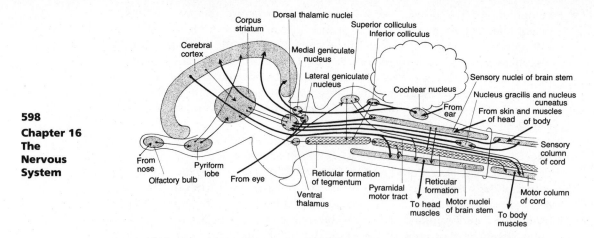

Figure 434. A "wiring diagram" of a mammalian brain comparable to Figures 432 and 433. The mesencephalic tectum is reduced to a minor reflex center, and the corpus striatum is relatively unimportant; most sensory impulses are projected "upward" to the cerebral cortex, whence a direct motor path (pyramidal tract) extends to the motor centers of brain stem and cord.

Obviously wiring diagrams like those in Figures 432–434 are greatly oversimplified. The actual situations, as studied in a variety of forms, are extraordinarily complex and quite variable among vertebrates. To do them anything approaching justice would require a volume as long as this one, or even longer, as in many of those cited in the list of references.

Chapter 17

Endocrine Organs

In the last chapter, we described, in the nervous system, an exceedingly complicated but highly efficient method of coordinating bodily activities by "messages" received from and sent to specific areas of the body with speed and precision. We shall here consider a second integrative system, under which information and directives are carried through the blood stream by chemical "messengers," the **hormones** produced by **endocrine glands**—glands, that is, that have no ducts but instead pass their products into circulatory vessels. This method of transmission is, of course, slower than transmission by neural impulse, and hormonal effects are often broadly distributed over the body to a variety of organs and tissues, in contrast to the pinpointing possible in the case of the nervous system. But despite these apparent drawbacks, many of the hormones are absolutely essential for the maintenance of the life of the organism.

For convenience, in this chapter, we have gathered data on all the known hormone-producing structures of the body, although they do not actually form an organ system but are scattered here and there throughout the body—literally from stem to stern—and may derive from a variety of organs or may develop as discrete units. The situation is somewhat similar to that of blood-forming organs, just as it makes no difference in which part of the body blood corpuscles are produced, so the area of production of hormones is inconsequential, as long as they can be passed into some element of the circulatory system and thence be distributed throughout the body.

Neural and hormonal systems of communication, although distinct, are far from independent of one another. Directly or indirectly, the nervous system may be powerfully affected by hormones. On the other hand, the "master gland" of the endocrine system, the hypophysis, is strongly influenced by the adjacent hypothalamus, and some of its hormones are actually produced by ganglia in that region of the brain. Again, the adrenal medulla, although an endocrine organ, is actually composed of modified nerve cells.

Which is the older regulatory system, nervous or endocrine? There is no clear answer to this; probably both evolved in parallel fashion. Elementary nervous systems are present in some of the most primitive of metazoan animals; numerous hormonal systems are known in invertebrates, and undoubtedly many more await discovery. Also, many organs with well-known functions are now being found to be endocrine glands on the side, as it were—the gonads are the most obvious example, but the stomach and intestine are equally good examples.

The Hypophysis

Below the diencephalic region of the brain lies a small but essential structure, the major endocrine organ of the body, the **pituitary gland** or **hypophysis cerebri** (Figs. 231, 412–416, 435–438). In most vertebrates, the pituitary tissues form a single compact mass, contained in a pocket (the **sella turcica**) in the floor of the braincase. Actually, however, the gland is a dual structure, its two portions having very different embryologic origins and functioning in different ways. The position of the gland, below the hypothalamus, is significant, because the production of pituitary hormones is strongly influenced by that area of the brain and, as will be seen, certain of these hormones are actually created by hypothalamic cells.

Downward from the embryonic diencephalon extends a hollow, finger-like process, the infundibulum. Upward from the embryonic mouth there grows an ectodermal pocket, the hypophyseal pouch (Rathke's pouch). This latter usually closes over in adult life but remains open in a few fishes (notably the cyclostomes). From both these embryonic structures, there proliferate masses of tissue that unite to form the adult pituitary.

The terminology applied to the subdivisions of the pituitary has been (and is) variable and confusing. It is best considered as consisting of two parts or lobes in accordance with embryologic origins. The greater part of its substance, and that most important in hormone formation, is included in the **adenohypophysis**, derived from the hypophyseal pouch. This consists of masses of secretory cells, of which the major portion is (a) the **pars distalis**; there is often, particularly in mammals, (b) a distinct **pars tuberalis**, an upgrowth around the stalk of the infundibulum, and (c) a **pars intermedia**, which tends to fuse with the neural part of the gland.* The other "half" of the gland is the **neurohypophysis**. The glandular part of the neurohypophysis is the **pars nervosa**, or neural lobe, lying close beside the adenohypophysis in the sella turcica; but the **infundibulum**, the stalk connecting the neural lobe and the hypothalamus, is also to be included in the neurohypophysis, and, as we shall see, a fraction of the hypothalamus forms an integral part of the neurohypophyseal mechanism and might properly be included as well.

The adenohypophysis, histologically, consists of masses and cords of secretory epithelial cells separated by sinusoids and supported by a loose framework of connective tissue. The greater part of its mass is included in the pars distalis. Two major types of gland cells are present; various histochemical tests allow these to be subdivided into several groups, each presumably responsible for the production of a different hormone. The intermediate lobe consists of polygonal secretory cells. The pars tuberalis, when present, consists of cells that appear to have little if any secretory function.

Very different in nature is the neural lobe. This contains, as a storage depot, amounts of hormonal material that can be passed on into the small blood vessels that ramify here. Present, too, are branched cells that were once thought to produce the secretions of the neurohypophysis; now they are believed to be supporting elements, comparable to the neuroglia in other parts of the central nervous system. Actually, the secretions found in this "gland" are not produced there but are neurosecretions

*The term "anterior lobe" is frequently used to designate the pars distalis, with or without the pars tuberalis; "posterior lobe" may designate the neurohypophysis, or neural lobe only, but may include the pars intermedia as well.

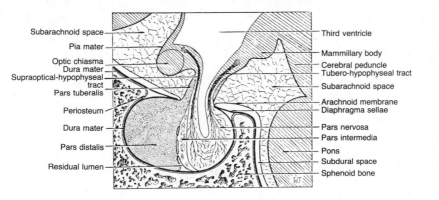

Figure 435. Section through a human pituitary and the adjacent structures at the base of the brain. (From Turner, General Endocrinology, W. B. Saunders Company.)

produced by cells within the substance of the hypothalamus and given off at the tips of their axons in the neural lobe of the pituitary.

Among the various small nuclei of the hypothalamus, two are of concern here. Anterior to the infundibulum, just above the optic chiasma, are the pair of appropriately named **supraoptic nuclei**. In an analogous position in the tuber behind the infundibulum are the **paraventricular nuclei**. From both these pairs of nuclei, bundles of stout axons pass down into the neural lobe, and a few similar fibers appear also to originate from other hypothalamic centers. The cell bodies in these nuclei can be seen histologically to have secretory powers. Their axons may have some degree of ability to transmit ordinary nerve impulses, but it is clear that their main function is to enable hormonal materials produced in the cell bodies to flow down their lengths into the neural lobe, where they are stored pending release into the blood stream.

The hypophysis as described above is of the type found in typical mammals; very different anatomic features are found in some lower vertebrates (Fig. 438). In the lamprey, there is no distinct infundibulum and a rudimentary neural lobe is represented by a thickened plate of tissue lying in the floor of the diencephalon. Again, the hypophyseal pouch is not closed, but, as discussed in an earlier chapter (p. 325, Fig. 256), is a tube lying below the diencephalon and opening to the surface in conjunction with the nasal pouch. The lamprey adenohypophysis consists of clusters of epithelial tissue detached dorsally from this tube and underlying the neural "lobe"; two subdivisions of the pars distalis and a pars intermedia are distinguishable. The parts of the pituitary are separated by tissues corresponding to the arachnoid and pia mater that cover other parts of the brain and that contain blood vessels as well as connective tissue.

Cartilaginous fishes and actinopterygians possess a more or less distinct infundibulum (from which a vascular sac projects at the posterior margin), and there is a well-formed mass of tissue representing the neural lobe; below, the hypophyseal pouch has closed in almost every case,* and its epithelium forms a massive

---

*Among living fishes it is still open in *Latimeria* and *Polypterus*, and fossil specimens suggest that it was still open in a number of Paleozoic fishes.

Figure 436. Diagrams showing stages in the embryonic development of the mammalian pituitary from brain tissue (hatched) and from the hypophyseal (Rathke's) pouch. As will be seen, only the neurohypophysis is formed from neural material; the three other parts, constituting the adenohypophysis, are derived from the pouch epithelium. (From Turner.)

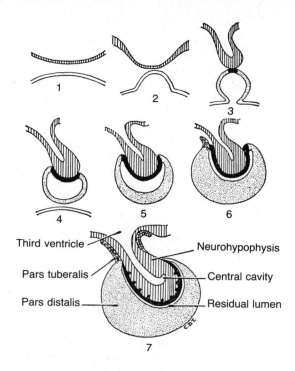

adenohypophysis. Peculiar to thse two groups is the fact that the lower surface of the neural lobe is irregular, and finger-like processes may extend down from it into the adenohypophyseal tissue; the two parts, however, are still distinctly separated throughout by a vascular cleft. Unlike other vertebrates, teleosts appear to have reduced the hypophyseal portal system but to possess hypothalamic nerve fibers innervating the adenohypophysis.

We would expect that in a lungfish we would find conditions approximating those expected in the ancestors of land vertebrates. This is the case. There is no vascular sac like that in most fish, and the neural lobe is a rather flat sheet of tissue, little thickened except posteriorly. Part of the adenohypophyseal tissue is closely applied posteriorly to the neural lobe in a manner comparable to the mammalian pars intermedia and is separated by a cleft from the major portion of the adenohypophysis, the pars distalis. However, there is still some interdigitation between neural and adenohypophyseal tissues. Amphibian hypophyses are similar in nature, but the neural lobe is somewhat better defined, and the hypophyseal portal system, described below, is well developed. In reptiles, there is some growth of the neural lobe, but it is only in mammals that it tends to assume a compact and more or less spherical structure. The development of a distinct pars intermedia is somewhat variable in amphibians and reptiles. It is absent in birds, and, even in mammals, this area may be poorly defined in such groups as insectivores, edentates, and carnivores.

The pituitary, particularly the adenohypophysis, is subject to strong influence by other endocrine organs; this is readily accomplishable, since both major parts of the gland are supplied with blood vessels from the internal carotid system, and the hypothalamic ganglia forming the neural lobe secretions are reached by intracranial vessels. The adenohypophysis also is greatly influenced by the brain, and the manner

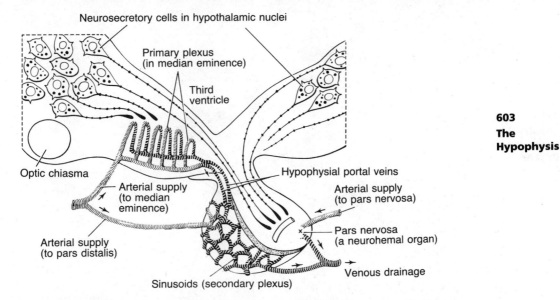

Figure 437. Diagram of anatomic connections between the hypothalamus and the pituitary gland. Cells in certain hypothalamic nuclei secrete the hormones of the pars nervosa; the secretion passes down the axons of the neurosecretory cells to the pars nervosa. Other secretory cells release materials in the median eminence at the base of the diencephalon; they appear to be picked up here by the primary plexus of the hypophyseal portal system and, passing to the sinusoids in the pars distalis of the adenohypophysis, stimulate hormone production there. (From Turner, General Endocrinology, W. B. Saunders Company.)

in which this influence is effected is of interest. Some nerve fibers enter the adenohypophysis, and one would assume that glandular activity there is affected by nervous control. But although fibers enter the pars intermedia, there is little evidence in any vertebrate other than teleosts and lungfishes that (apart from some that appear to be autonomic fibers for blood vessels) they penetrate to the pars distalis, where almost all the hormones are formed. In this condition, how does the brain influence the gland?

Apparently indirectly, through a curious local portal system of small blood vessels (Fig. 437). Some vessels supplying arterial blood to the adenohypophysis pass close to or through the floor of the diencephalon anterior to the infundibulum. The tract carrying neural secretions to the neural lobe passes through this same area, which, particularly in mammals, tends to form a bit of swelling on the lower surface of the brain, termed the **median eminence**. In lungfishes, in many other fishes, and in all tetrapods, these blood vessels to the adenohypophysis do not run directly to that gland but enter the region of the median eminence, break up into a capillary system, and then re-collect as a set of small portal veins to pass on to the pars distalis. In its passage through the median eminence, the blood presumably picks up neuro-humoral materials, which are the agents transmitting "information" from the brain to the gland. This seems like a curiously roundabout way of accomplishing an important function, but, if nerves are indeed lacking, no other effective method is known.

Attempts have been made to identify predecessors of the hypophysis, or of its

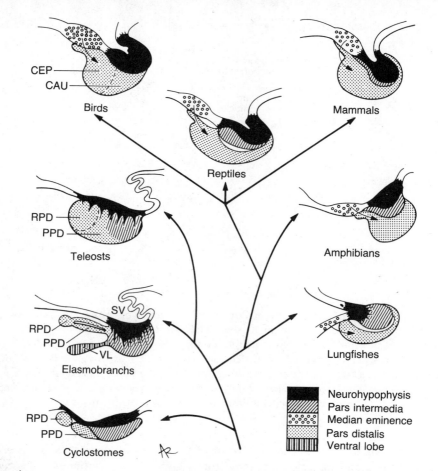

Figure 438. Schema to show probable evolutionary changes in the vertebrate pituitary gland. Arrows extending from the median eminence to the pars distalis represent the hypophyseal portal system. Abbreviations: *CAU,* caudal division of the avian anterior lobe; *CEP,* cephalic division of the same; *PPD,* proximal pars distalis; *RPD,* rostral pars distalis; *VL,* ventral lobe of the elasmobranch pituitary; *SV,* saccus vasculosus. (From Turner, General Endocrinology, W. B. Saunders Company.)

components, in lower chordates. In amphioxus, there is a group of cells in the anterior part of the neural tube (where a brain should be, if amphioxus had a brain) that has been compared with an infundibulum, but this is uncertain. Below, in the roof of the mouth, there is a small pit suggesting the pouch of embryonic vertebrates from which the adenohypophysis is formed. But this pit appears to be merely a mucus-secreting, rather than a hormone-secreting, structure. In adult tunicates (Fig. 6), there is a **neural gland**, which is in certain regards suggestive of a pituitary. This gland opens into what is morphologically the dorsal side of the entrance to the pharynx and lies close to the ganglion, which is the closest approach to a brain present in the simple nervous system of the adult tunicate. Morphologically, the neural gland could be reasonably considered as antecedent to the hypophysis. But at present there is no positive evidence that it is an endocrine gland.

Some nine hormones are known to be produced by the pituitary and still others have been, or are, suspected to be present. The greater part of the work concerned with them has been done on the mammalian pituitary, but most of the whole series are present throughout the vertebrate classes (although, for example, the hormones of the neurohypophysis are numerous but poorly understood in elasmobranchs, and the cyclostome situation is incompletely known). Most of the pituitary hormones are produced by the adenohypophysis, principally by the pars distalis. All are glycoproteins, proteins, or polypeptides. They are listed and briefly described as follows:

Growth Hormone or Somatotropin (GH or STH).   This has a very broad influence on growth and metabolism in general, with marked influence on growth of skeleton and muscle, metabolism of fats and carbohydrates, and protein synthesis; it further acts to enhance the effect of other hormones on the activity of thyroid, adrenal cortex, and reproductive organs.

Adrenocorticotrophic Hormone (ACTH).   This is vital for adrenal cortical activity in secretion of hormones.

Thyrotropin (TSH).   This is essential for the stimulation of the thyroid to form and release thyroid hormones.

Prolactin.   Prolactin stimulates the secretion of milk and, in a few mammals, prolongs the functional life of the corpus luteum (and consequent secretion of progesterone; cf. p. 616). The functions of prolactin in lower vertebrates appear varied and are far from completely known. In pigeons, it stimulates the gland in the crop that produces "pigeon milk"; in some urodeles, it causes the animal to return to the water for reproduction (essentially a secondary metamorphosis); in fish, it may be involved in osmoregulation, may serve as a growth hormone,* or may be involved in the regulation of pigmentation. There is now good evidence that prolactin-producing cells predominate in the anterior pars distalis of many teleosts.

Luteinizing or Interstitial Cell–Stimulating Hormone (LH or ICSH).   This and the following hormone are mainly associated with sexual structures and activities and hence are termed **gonadotropic hormones**. They are best known (and named) from their observed effects in mammals, although they are obviously important in lower vertebrate classes (cyclostomes are a partial exception). Luteinizing hormone influences maturation of the gonads and production of sex hormones (cf. pp. 614–616); acts in formation of corpora lutea and secretion of progesterone in the ovary; and stimulates the interstitial cells of the testis, thus promoting the production of androgens and the maturation of sperm.

Follicle–Stimulating Hormone (FSH).   This stimulates the growth of ovarian follicles, promotes spermatogenesis and, in conjunction with the last, promotes estrogen secretion and ovulation.

---

*Prolactin shows certain structural similarities to proper growth hormone; each may produce certain effects of the other.

Intermedin or Melanocyte–Stimulating Hormone (MSH). This polypeptide acts to disperse pigment granules in melanophores, to aggregate reflecting organelles in iridophores, and to darken the skin. In contrast to the hormones above, this (as the name indicates) is produced in several varieties by the intermediate portion of the gland when this is distinctly developed. It has no clearly known function in birds and mammals; however, it has been implicated in several effects on the brain.

In contrast to the wealth of hormones produced by the adenohypophysis, the secretions given off by the mammalian neurohypophysis consist of a limited series of polypeptides similar in chemical composition, prominently **vasopressin** or antidiuretic hormone (ADH) and **oxytocin**; there are minor chemical differences in various lower vertebrates. They contain a series of eight amino acids, six being common to the two hormones. The effects of the two overlap to some degree, as might be expected. Vasopressin is principally associated with controlling water output or intake in a variety of ways in different forms. Oxytocin is best known from its effects on the female mammal in promoting contraction of uterine muscle and mammary development and ejection of milk after the birth of young. It appears, however, that it has sexual effects on at least certain other vertebrates—spawning in minnows, for example. In contrast to the importance of the numerous hormones of the adenohypophysis, those of the neural lobe play a relatively modest part in the body economy; their effects are mainly on contractile elements, principally smooth muscle, and on semipermeable membranes, such as renal tubules.

Parathyroid Glands

We have noted earlier the trend for the development of outgrowths of tissue from the branchial pouches of the embryo in every class of vertebrates (Figs. 269, 270, 272). Notable in tetrapods is the presence in the adult of small glandular structures, termed the **parathyroid glands**. They are variable in position and number; there are usually two pairs in mammals, derived from the ventral portions of the third and fourth pairs of branchial pouches; in other tetrapods, there are often three pairs, derived from pouches two, three, and four (Fig. 439). Under the microscope, the mature parathyroid exhibits densely packed masses of cells usually arranged in cords (Fig. 440). The greater part of them are **principal cells** (with a pale, clear cytoplasm), which appear to be the essential elements of the gland; in addition, there is, in some mammals at least, a small percentage of **oxyphil cells**, filled with granules that stain with acid dyes. The name "parathyroid" is a misleading one, because, functionally, these little glands differ widely from the thyroid, although found in the adult in the neighborhood of that gland and sometimes (as in man) embedded in it. Attention was called to them when it was discovered that extirpation of parts of the human thyroid in which these little bodies were enclosed resulted in the death of the patient from disturbance of calcium balance. Only recently has the pure parathyroid hormone, a protein that has been named **parathormone**, been extracted. The hormone is principally involved in the regulation of calcium and, to a lesser degree, of phosphorus concentrations in various tissues. Although the amount of calcium in the body fluids is small, the presence of a proper amount is essential for neuromuscular functioning; if the level is too low, tetanus and death may occur. Highly important is the relation of this hormone to maintenance of the bony skeleton. The greater part of the substance of bone is calcium phosphate; bone materials are being constantly re-

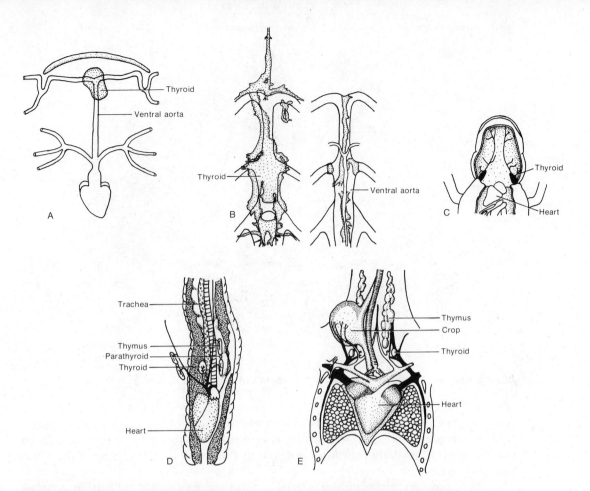

Figure 439. Thyroid and adjacent organs in various vertebrates. All are in ventral view except *B*, in which the ventral view is left and a dorsal view on the right. *A, Raja,* a ray; *B, Salmo,* a salmon; *C, Ambystoma,* a urodele; *D, Natrix,* a snake; *E, Gallus,* a chicken. (After Ferguson, Hoar, Weichert, Clark, and Ede.)

newed, and unless adequate supplies of "new" calcium and phosphorus are available, skeletal degeneration occurs. Calcium and, hence, this gland are also involved in the excitability of nerves, clotting of the blood, permeability of cells, and functioning of various other glands.

No structures exactly comparable to parathyroids are known in fishes; in actinopterygians, the corpuscles of Stannius (cf. p. 614) have been suggested as comparable structures; however, these do not seem to be homologous. The need for calcium regulation would seem to be as great in fishes with bony tissues as in tetrapods, and the seeming absence of glands secreting an appropriate hormone has been a puzzling situation. A partial equivalent in fishes (and also in lower tetrapods) may be the **ultimobranchial gland**. In many vertebrates, tissues may be budded off the last branchial pouch of the embryo in much the same way that parathyroid and thymus tissues are derived from more anterior branchial pouches. These tissues come to form a pair

607

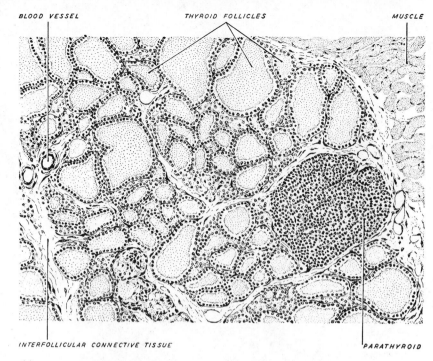

BLOOD VESSEL   THYROID FOLLICLES   MUSCLE

INTERFOLLICULAR CONNECTIVE TISSUE   PARATHYROID

Figure 440.   Thyroid and parathyroid tissues of the rat. (From Turner.)

of small glandular bodies, found beneath the esophagus of most fish groups (cyclostomes excepted), and, in one instance, abnormalities in this gland were found to be paralleled by skeletal abnormalities, suggesting a function similar to that of the tetrapod parathyroids.

There has recently been discovered a hormone, **calcitonin**, which, in contrast to parathormone, tends to lower the level of calcium. It appears to be produced by the ultimobranchial bodies in submammalian forms and by parafollicular cells of the thyroid in mammals. In all cases, apparently, the actual calcitonin-secreting cells (C-cells) are derived from the neural crest of the embryo.

Thyroid Gland

Of glandular derivatives from the pharynx, such structures as the thymus and parathyroids are paired structures, formed by outgrowths from the branchial pouches. In contrast is the **thyroid gland**, likewise of pharyngeal origin, but originating as a median ventral pouchlike outgrowth from a point well forward in the pharyngeal floor (Figs. 270–272). Found in every group of vertebrates, the thyroid is generally situated in fishes below the gill chamber close to the ventral aorta; in tetrapods, it is generally ventral to the windpipe at some point along the length of the neck (Fig. 439). The gland remains a single structure in many fishes, some reptiles, and mammals, although distinctly bilobed in many cases; its name comes from its shield-shaped appearance in man. In most amphibians, lizards, and birds, the thyroid becomes paired. In teleosts, it is often separated into right and left halves, and these in turn may be broken up into smaller units; in some teleosts, bits of thyroid tissue come to lie in such unlikely places as the spleen, kidneys, heart, and even the eyes. Small accessory masses of thyroid tissue are also found in a variety of amniotes; they are sometimes termed parathyroids, making for regrettable confusion with the radically different glands described above.

Rich in blood vessels, the gland is composed of numerous small, spherical **thyroid follicles**, bound together by connective tissue (Fig. 440). Each follicle consists of a cuboidal epithelium of secretory cells, surrounding and discharging into a central cavity filled with a gelatinous **colloid substance**. In this colloid are found, in storage, quantities of an iodine-containing protein, **thyroglobulin**, from which, through hydrolysis, are formed iodine-bearing hormones, **thyroxine** and **triiodothyronine**— the former the more common. Hormonal output from the thyroid is in great measure controlled by the pituitary, but a variety of environmental stimuli may also be influential.

The thyroid hormones are highly important in maintaining general tissue metabolism. In birds and mammals, the degree of metabolic activity necessary to maintain body temperatures is in great measure maintained through the influence of the thyroid hormones. In a hibernating mammal, activity of the thyroid decreases; in a hibernating amphibian, the reverse is true. Thyroid hormones are influential in the development of the nervous system and in behavior; they are concerned with reproductive functions and with growth phenomena; the most spectacular of these thyroid functions is its control of metamorphosis in amphibians.

The thyroid has a pedigree that stretches far back in chordate history. In both amphioxus and tunicates, we have noted, there are present ciliated and partly glandular channels along the pharynx that are concerned in the feeding mechanism. Such a channel in the floor of the pharynx of amphioxus is the **endostyle**, and a like structure is found in tunicates (Figs. 5–6). In the endostyles of both amphioxus and tunicates, iodine compounds, as well as mucus, are produced, although the compounds differ from those of the thyroid hormone; they are obviously carried on into the digestive tract with the food materials.

In the ammocoete larva of lampreys, alone among living vertebrates, we find a similar feeding habit and a series of ciliated grooves much like those of amphioxus and tunicates. The ventral groove terminates posteriorly, however, in a deep pouch in the pharyngeal floor (Fig. 256 A). This pouch is subdivided in a complicated manner, and contains a number of types of epithelial cells, some of which are glandular, and there is produced an iodine-bearing material that is carried on down the digestive tract. The term endostyle has been applied to this pouch (although, of course, the homology of this gland with the groove from which it arises is not an exact one).

The endostyle, including the ammocoete gland, is a median ventral pharyngeal structure, comparable in position to the thyroid gland. Is it truly homologous? The lamprey gives us a positive, conclusive answer. At metamorphosis, the larval endostylar pouch closes off from the gut and breaks up into a series of follicles indisputably thyroid in structure.

It would appear that the history of the thyroid gland, in phylogeny as in lamprey ontogeny, has been one of a feeding structure being transformed into a gland of internal secretion. It is significant that clinically the thyroid hormone can be taken by mouth, whereas other hormones, when introduced into the digestive tract in an unaltered state, are destroyed by gastric and intestinal enzymes. Early in chordate evolution, secretion of iodine compounds began in the endostyle while this was still a part of the feeding mechanism, the "hormone" being carried into the gut with food materials. With the abandonment of filter-feeding, the endostyle lost its original function as a constituent of the pharynx; the thyroid gland has, as an endocrine structure, continued its glandular activity.

Pancreatic Islands

Although the greater part of the glandular tissue of the pancreas is devoted to the production of enzymes, which pass through ducts to the intestine, areas of tissue of another type can be seen, in almost all forms studied, distributed through the gland as isolated islands (Fig. 287). These **islets of Langerhans** consist of cells of at least two types, which are secretory in nature, but are not furnished with ducts; they thus obviously form an endocrine organ, sending secretions into the blood stream. The insular material is usually diffused among the ordinary tissues of the formed pancreas. In cyclostomes pancreatic endocrine follicles are present in the walls of the anterior part of the gut and in teleosts small clusters of islet cells are spread here and there in the general region of the gut; in a few fishes, these tissues form a small special organ of their own.

It has long been known that disease or injury to the pancreas interferes with carbohydrate metabolism and produces the disease known as diabetes mellitus. We are now familiar with the fact that the islands produce a specific protein hormone, **insulin**. This material has an important regulatory action on metabolism, particularly of carbohydrates, by increasing their utilization by the tissues, moderating glucose formation in the liver, facilitating the entry of glucose into certain types of cells (such as muscle), and maintaining the level of sugar in the blood. There is little nervous influence on insulin secretion, which appears to be mainly controlled by the level of sugar in the blood passing through the islands. A second substance, the polypeptide **glucagon**, formed in a cell type differing from those secreting insulin, is also produced in the pancreatic islands; abundant in birds and reptiles, this tends to increase blood sugar by the breaking down of glycogen stored in the liver.

The Adrenal Cortex and Homologous Tissues

In most tetrapods there is found, adjacent to the kidneys and often capping them, a pair of endocrine structures termed the **adrenal glands** (epinephric glands, suprarenal glands, Figs. 297, 299). Microscopic examination shows the presence in them of two different types of tissues, intermingled or juxtaposed in lower tetrapods, but, in mammals they form distinct cortical and medullary layers of a single compact organ (Fig. 441). Both parts are endocrine glands, but of very different sorts. The medullary tissue is a modified part of the nervous system, as discussed on page 612. The cortical substance is of quite another nature. In fishes, there are almost never formed adrenals. In sharks, the two components are quite distinct; in other fishes diffuse cell masses—sparsely present in cyclostomes and lungfishes—representing both components can be found between and around the kidneys and along the course of the major blood vessels dorsal to the coelom. The cortical materials are sometimes termed **interrenal tissues**, but are probably better called **adrenocortical tissues**. Bodies within the kidneys of some teleosts, the corpuscles of Stannius (cf. p. 614),

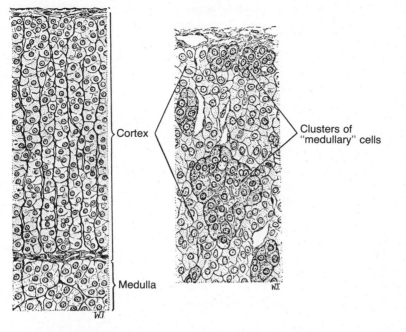

Cortex

Medulla

Clusters of "medullary" cells

611
The
Adrenal
Cortex and
Homologous
Tissues

Figure 441. A section through part of the adrenal gland (outer surface above) in a mammal (rat), with division into cortical and medullary layers, and a reptile *(Heloderma)*, in which the two tissues are intermingled. (From Turner.)

were sometimes said to be adrenocortical tissue; however, they develop from urinary ducts and hence are not homologous with such tissues from which they also differ histologically and in their secretions.

That the cortical material is vital for the maintenance of life was recognized more than a century ago, when human deaths from an ailment known as Addison's disease were invariably associated with deterioration of the adrenal cortex. The cortical cells have since been discovered to secrete a series of steroid hormones that have widespread influence over bodily functions in ways that as yet are not fully understood. Some 50 or so steroids have been found to be present in this organ in mammals. Certain of them appear to be inactive; others, present in small amounts, are chemically similar to sex hormones produced in the gonads; still others are active hormones peculiar to the adrenal cortex, of which half a dozen or so appear to be most important. In general, the total effect of cortical hormones is in aiding the body to meet long-continued environmental stresses, in contrast to the function of the adrenal medulla (as described below) in coping with brief emergencies.

The steroids secreted by the adrenal cortex, the **adrenocortical steroids**, fall into two main types. **Glucocorticoids** include cortisol, cortisone, corticosterone, and others. They have a variety of functions, important ones including increases in the breakdown of proteins, the storage of glycogen in the liver, and the concentration of glucose in the blood; they thus act in opposition to insulin. The synthesis and secretion of glucocorticoids, which are found in all groups of vertebrates, are regulated by the adrenocorticotrophic hormone. The other type, the **mineralocorticoids**, include 11-deoxycorticosterone and aldosterone; they increase the absorption of sodium and excretion of potassium by the kidneys. Aldosterone is known in lungfishes and tetrapods. Its production is controlled, not by adrenocorticotrophic hormone, but by the renin-angiotensin system (cf. p. 617) and by the concentration of potassium in the blood. Although the functions, as briefly noted here, of the two sorts of adrenocortical steroids are quite different, their metabolic effects do overlap, and each shows some of the activities of the other.

The importance of several cortical hormones in water and salt regulation suggests a relationship of some sort between the cortical materials and the kidneys. That this physiologic association and the topographic propinquity of kidney and adrenal are not accidental, but have historic significance, is indicated by the embryologic origin of the cortical tissues. They appear as strands of cells that bud off from the epithelium of the roof of the coelom, medial to the developing kidney tubules (and lateral to the gonads). Kidney and adrenal cortex are thus derived from adjacent regions of the embryonic mesoderm. Of further significance, noted later, is the fact that gonadal tissues, producing hormones similar chemically to those of the adrenal cortex, are also derived from this same region. Gonads and cortex are the two tissues that produce steroid hormones.

The Adrenal Medulla and Chromaffin Tissues

Very different in origin and function from the cortical portion of the adrenal gland are its medullary portion and structures antecedent to it in lower vertebrates. Here we are dealing with a portion of the nervous system that has been modified to perform an endocrine function. We have seen that the visceral motor nerve supply to the internal organs of the body is of a peculiar type, in which the impulses do not directly reach the smooth muscles or glands concerned but are relayed through a series of postganglionic neurons. The cells of the adrenal medulla and homologous structures in lower vertebrates are much modified postganglionic neurons.

There are various clusters of cells here and there in the vertebrate body, particularly along the region near the dorsal aorta or adjacent to sympathetic ganglia, which are termed **chromaffin cells** because of the readiness with which they stain with certain biochromate salts. Embryologically, they arise from cells migrating downward along the path of the visceral nerve rami from the neural crest and are thus identical in origin with the postganglionic neurons of the sympathetic system. In fishes (Fig. 442), small masses of such cells, often associated with interrenal tissues, are present between the kidneys and along the dorsal wall of the body cavity. They are often and appropriately termed **paraganglia** because they are embryologically identical with the sympathetic ganglia, which they may adjoin, particularly in sharks (Fig. 402). In tetrapods, occasional small cell clusters of this type persist,* but, except in urodeles, the chromaffin material is concentrated into a compact mass of tissue that forms part of an adrenal gland; in amniotes, this is generally found capping the kidney. In mammals, the chromaffin cells are concentrated in the center of the adrenal body forming its medulla; in lower tetrapods, they are more diffuse and interspersed with the cortical component (Figs. 441–442).

These masses of cells are innervated by preganglionic autonomic nerve fibers; on

---

*Bodies along the mammalian carotids and other arterial "arches" above the heart are sometimes referred to as paraganglia but are actually sensory structures.

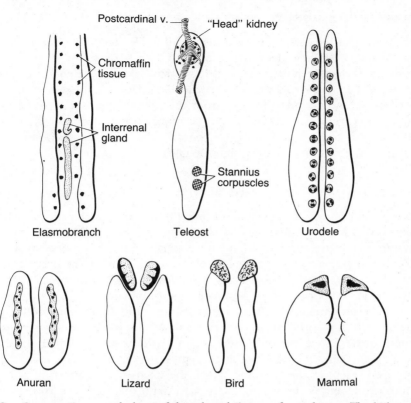

Figure 442. Comparative morphology of the adrenal tissues of vertebrates. The kidneys are shown in outline; solid black indicates chromaffin (medullary) tissues; stippling indicates cortical tissues. The two tissues are typically separated spatially in elasmobranchs. In bony fishes, the two tissues tend to be concentrated in the head kidneys, around branches of the postcardinal veins; they are frequently intermingled. (Based on Gorbman, A., and Bern, H. A., Textbook of Comparative Endocrinology, New York, John Wiley & Sons, 1962.)

stimulation, they give off into the blood stream two related chemicals that are identical with those given off by the postganglionic fibers of the sympathetic system—**adrenaline** and **noradrenaline** (or **epinephrine** and **norepinephrine**), with the difference, however, that here the former is the more abundant product. The cells of the adrenal medulla do not look like nerve cells, because they lack fibers. But since they are homologous with postganglionic sympathetic neurons, it is not surprising that they produce comparable neurohumors. The difference is that, whereas the true postganglionic sympathetic cell produces only a tiny amount of adrenalin-like material that affects only structures immediately adjacent, the mass of adrenal cells is capable of rapidly releasing large quantities of these materials, which may have a strong, immediate "shotgun" effect on all parts of the body when carried about by the circulatory system and prepare the organism to meet emergencies.*

*Although these massive emergency outflows, mainly of adrenalin, from the adrenal medulla have been of major interest in the study of this organ, it appears that there is a normal, constant but much smaller, output of medullary hormones.

Corpuscles of Stannius

All holosteans and teleosts appear to have small bodies embedded in the ventral surface of the kidneys; they are not present in other bony fishes. Typically they are numerous and small in holosteans and fewer and larger in teleosts. Although once considered to be parathyroid or adrenocortical homologues, they seem quite distinct. These **corpuscles of Stannius** arise from the tubules and ducts of the embryonic kidney. Their secretions, presumably one or more hormones (but *not* adrenocortical steroids), appear to be involved in the control of fluid balance, blood pressure, and the absorption of calcium ions through the gills.

Urophysis

Just as in all vertebrates a special anterior secretory area of the central nervous system develops as the neurohypophysis, so in most, if not all, fishes a posterior secretory system may arise, to which (not unreasonably) the term **urophysis** has been given. This is highly developed in many teleosts (Fig. 443). In the nerve cord, toward the end of the tail, can be found large cells, obviously secretory in nature. Fibers—axons—extend posteriorly from these cells and terminate in bulbous tips, which are filled with secreted material. These bulbs are most commonly clustered together on the underside of the cord, causing here a slight swelling or a wartlike structure, readily visible to the naked eye on dissection. In elasmobranchs, similar secretory cells are found in the tail; the end bulbs are absent, however, although masses of secreted material are present in the area into which the axons of the secreting cells run. The hormones produced by the urophysis, termed **urotensins**, appear to influence regulation of salt content of the blood.

Sex Hormones

Far more than in the case of any other activity, reproduction in vertebrates is powerfully influenced by hormones, with regard to both anatomic structures and behavior

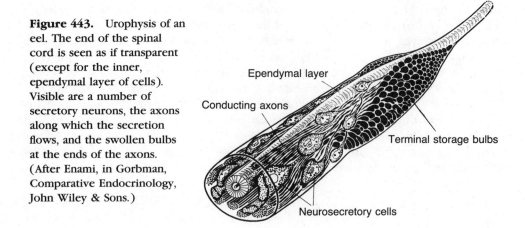

Figure 443. Urophysis of an eel. The end of the spinal cord is seen as if transparent (except for the inner, ependymal layer of cells). Visible are a number of secretory neurons, the axons along which the secretion flows, and the swollen bulbs at the ends of the axons. (After Enami, in Gorbman, Comparative Endocrinology, John Wiley & Sons.)

Ependymal layer

Conducting axons

Terminal storage bulbs

Neurosecretory cells

patterns. The subject merits a volume in itself (and there are several such); here we will only briefly note the aspects of this complex picture that are of morphologic significance. A "chain of command" is involved in the action of sex hormones. Gonadal hormones stimulate to development and sexual differentiation secondary and accessory sexual characters, including not only reproductive ducts and copulatory organs, but such sexually distinctive features as the plumage of birds, hair types, and mammary glands in mammals, and, in many instances, even major sexual differences in body size and proportions. But the gonadal hormones do not operate purely on their own initiative, for, as noted earlier, their production is in considerable measure regulated by pituitary hormones. A step further, the pituitary, we have seen, is strongly influenced by the adjacent hypothalamic region of the brain and is further influenced, reflexly, by hormones from the gonads and other organs. A final step is that, in regard to the conduct of sexual matters, the brain is subject to strong influences, both from within the body and from a variety of environmental factors.

Our present concern, however, is not with these interesting but far-flung actions and interactions but is to deal simply and briefly with the nature of the sex hormones and the gonadal structures that produce them.

With one exception,* the sex hormones produced by the gonads are steroids, termed **androgens** in the case of those primarily produced by the testis, and **estrogens** in the case of the ovary. These steroids, as mentioned earlier, are very similar in chemical composition to the hormones of the adrenal cortex. So close, in fact, are the relationships that, so to speak, the gonads and cortex are unable to completely sort out their respective roles as hormone producers. A limited amount of sex hormones may be detected among the cortical secretions, and a small fraction of the gonadal production is of substances proper to the adrenal cortex. Still further, male and female gonads are not completely differentiated in regard to their products, for the male gonads, where tested, are found to secrete a certain amount of female hormone and vice versa.

This similarity of hormonal products between adrenal cortex and gonads correlates with the similarity in origin of the cells that produce them. Apart from the actual sex cells (which are not implicated in hormone production), all the materials making up the gonads are derived, we have seen (cf. p. 419), from the mesoderm lining the dorsal rim of the coelomic cavity on either side of the midline, and the cortical adrenal cells, as noted earlier in this chapter, are derived from the adjacent region of the mesoderm, between gonad and kidney.

In the case of the male gonad, the major androgens, where analysis has been made, are the steroids **testosterone** and **androstenedione**; other compounds are recovered from testes, but these may be metabolic products derived from these potent and probably primary substances which, when experimentally injected, are generally highly effective in producing male structural growth and behavior patterns. In the case of the female gonadal products, the estrogens **estradiol** and **estrone** are the major steroids concerned, the former much the more potent. What cellular elements in the gonads produce these sex hormones? In the testis, the source appears to be

*A female hormone, the protein **relaxin**, which in mammals relaxes the pelvic symphysis and facilitates birth of offspring, also has functions aiding in reproduction in lower vertebrates. Actually, because it is produced by the uterus and placenta as well as the ovary, it is only a partial exception.

distinctive **interstitial cells** (Fig. 307). Cells of this type, derived from the mesodermal epithelium surrounding the gonad, are unknown in lower chordates but are present in every major vertebrate group, although varying in abundance from group to group and also, in some cases at least, varying in abundance at different phases of the breeding season. These cells, as their name indicates, are quite distinct from the formed seminiferous tubules or ampullae and lie (together with connective tissue) in the interstices between them. In the lining of the tubules or ampullae, there are found, beside the sperm-producing elements, supporting cells (**Sertoli cells**, Fig. 307) that have the same embryonic origin as the interstitial cells. Although these Sertoli cells do not themselves produce hormones, their secretions may be involved in binding androgens and thus maintaining a high concentration of them within the testis.

In the ovary, there may be a certain amount of interstitial tissue (cf. Fig. 306), but most of the ovarian cellular materials—of common embryonic origin with the interstitial cells—are engaged, as **follicle cells**, in furnishing sustenance to the developing eggs. In addition to this function, the follicle cells are the major producers of estrogens, although the minor amount of interstitial tissue present may also be concerned in this process.

Apart from the production of the primary sex hormones, we find that in the case of the ovary, a second type of formation of steroid hormones may occur. When an egg bursts from the ovary, one would expect the follicle in which it had been enclosed to degenerate. This appears to be the case in many groups. But in mammals, most notably, there is no immediate degeneration. Instead, the follicular cells long persist and hypertrophy to form a yellow-colored tissue, the **corpus luteum**, filling the empty follicle. The corpus luteum in mammals secretes an important steroid hormone, **progesterone**; this prepares the uterine epithelium for implantation of the ovum and, if fertilization and implantation are accomplished, stimulates the development of a placenta. Although major interest in the corpus luteum and its hormonal function has been centered on mammals, with their placental type of development, a corpus luteum that is apparently similarly productive of progesterone develops after the bursting of the follicle in elasmobranchs. Even further, although no typical corpus luteum is formed in other vertebrate groups, progesterone has nevertheless been discovered in the ovary of certain other vertebrates (notably birds). A small percentage of elasmobranchs and a few lower tetrapods bear their young alive, but, in general, no major specific function is known for progesterone in lower vertebrates. This suggests that we have in progesterone an example of the process of hormonal evolution proposed at the beginning of this chapter—that of a chemical product given off by a tissue that originally may have had little positive function, but later found a useful role in the economy of the body and, in mammals, has become an important hormone.

The nature and functions of the mammalian placenta are part of the story of vertebrate embryology rather than vertebrate anatomy. We may, however, note here that, during the course of pregnancy, the mammalian placenta itself becomes an endocrine-producing organ, secreting a number of steroids, including not only estrogen and progesterone but also a special hormone of its own, **gonadotropin**, which is important in maintaining pregnancy. Common tests for pregnancy are based on the presence of this hormone in the urine.

Gastrointestinal Hormones

The stomach and intestine produce hormonal substances that play a part in the regulation of digestion. However, we are here dealing with matters rather different from those discussed earlier in this chapter. Whereas most hormones form an interlocking system, often with checks and balances on one another, the **gastrointestinal hormones** do not appear to be affected by other hormones or even by other members of their own group. Further, they do not arise from specific endocrine glands but from tissues that serve other functions as well. Again, our knowledge of these hormones is almost entirely confined to mammals. Although there is some evidence that they are present in other vertebrate classes, little is known about their physiology in these forms.

The mucosa of the pyloric region of the stomach produces a hormone termed **gastrin** in response to the local presence of food; this material is carried through the blood stream to the fundic region, causing secretion of hydrochloric acid. The passage of food into the upper end of the mammalian small intestine causes the mucosa of the region to release into the blood stream a polypeptide termed **secretin**; this causes the flow of pancreatic juice into the intestine and also appears to have some effect on the flow of bile and production of intestinal juices. There is evidence of the presence of a second intestinal hormone, **pancreozymin-cholecystokinin**, which tends to stimulate the pancreas to increased enzyme formation, with consequent enrichment of the pancreatic juice. Production of this hormone is also stimulated by the entrance of fats and other food materials; its effect is to evacuate the gallbladder. These three gastrointestinal principles are well established, and their chemical and physiologic characteristics are at least partly known. Similar hormones have been postulated on the basis of physiologic activity, but they have not yet been defined chemically.

Fringes of the Hormone System

Some borderline matters may be noted briefly. For example, the peptide **angiotensin**, which may be found in the blood stream, is highly effective in elevating the blood pressure, and its presence is primarily the result of kidney activity. The "hormone" is not produced in the kidney, however; a kidney enzyme, **renin**, is responsible for initiating chemical changes in proteins in the blood stream that cause eventual production of the "hormone." Again, evidence indicates that red blood cell formation is influenced by a hormonal agent, and there are observations suggesting that here, again, the kidney is involved in some way. The mast cells, mentioned in our discussion of connective tissues, secrete such important chemicals as heparin, histamine, and serotonin, and, by stretching a definition a bit, we might regard the mast cells as unicellular endocrine "organs."

There are a number of body structures whose tissues have a glandular appearance and hence have been suspected, despite the absence of convincing evidence, of having endocrine properties. Most often cited in this category are the pineal and parapineal structures, which, as in the case of the mammalian pineal organ, tend to persist even when their original visual function has been lost (cf. p. 519). In such forms as sharks and frogs, the pineal, although not open to the surface, contains sensory cells and still acts as a photoreceptor. In mammals, this is no longer the case. The pineal, however, appears to receive data on the presence or absence of light by way of nerves from the autonomic system and contains secretory cells that produce a hormone, **melatonin**. This receives its name from the fact that when injected into tadpoles, it causes lightening of the skin due to concentration of the pigment within the melanophores; more than that, however, the pineal in mammals and some other vertebrates appears, through its secretion, to influence the gonads. This effect is generally antigonadal; some evidence suggests that melatonin is the active agent; other evidence implicates polypeptides. We probably have much more to learn in the future about the functions of this supposedly vestigial structure.

We have noted earlier that one major function of the thymus is the establishment of immune reactions in the body; it is suspected that the thymus is also the source of a factor, carried by the blood, that induces the differentiation of lymphoid stem cells. Again, ultimobranchial bodies, budded off from the last gill pouch, have been thought to have some possible hormonal nature. In mammals, they become embedded in the thyroid and parathyroid. C-cells are present in the ultimobranchial bodies of lower forms, some of which also have parathyroids, and in the parathyroids of tetrapods. The secretory cells of ultimobranchial bodies are not formed from branchial epithelium, like the rest of the organ; they appear to come from neural crest cells, as do the cells of the parathyroids that produce calcitonin.

Appendix 1

A Synoptic Classification of Chordates

The classification given here is presented primarily for the purpose of allowing the student to place in their proper position the forms discussed in the text. In consequence, no attempt is made to list the genera or even families of vertebrates, and, in many cases, suborders and orders are neglected when such subdivisions of groups lack interest for present purposes. Most of the major fossil forms are mentioned, although our anatomic knowledge of them is practically confined to the skeletal system.

In addition to the major terms in the classification given below, (1) the urochordates and cephalochordates, with or without the hemichordates, are often termed **Protochordata**, lower chordates, in contrast to the **Vertebrata**; (2) among the vertebrates, the term **Gnathostomata**, jawed vertebrates, may be used to contrast all other higher vertebrate groups with the class **Agnatha**; (3) **Tetrapoda** is frequently used for amphibian and higher four-footed types in contrast with **Pisces**, fishes in a broad sense; (4) **Amniota**, referring to features of embryonic development found in reptiles, birds, and mammals, may be used for these three classes, all fishes and the amphibians being grouped as **Anamniota**.

PHYLUM HEMICHORDATA

Little or no development of notochord or dorsal nerve cord; probably a separate phylum, distinct from but closely related to the Chordata.

Class Pterobranchia
Simple, sessile, plantlike animals that gather food with ciliated lophophores (Fig. 8 *A*, *B*).

Class Enteropneusta
Acorn worms; worm-shaped burrowers with well-developed pharyngeal gills (Fig. 8 *F*).

PHYLUM CHORDATA

SUBPHYLUM UROCHORDATA

Tunicates and related forms, with notochord and nerve cord well developed in the larvae in many cases, but adults sessile or floating organisms, consisting mainly of an elaborate branchial apparatus (Figs. 6, 8 *C, D, E*).

SUBPHYLUM CEPHALOCHORDATA

Amphioxus *(Branchiostoma)* with notochord, nerve cord, and pharyngeal gills all well developed in the adult stage (Fig. 4).

SUBPHYLUM VERTEBRATA

Generally with developed backbone, skull, brain, and kidneys.

Class Agnatha
Jawless vertebrates.

> *†Order Osteostraci.*
> †Order Anaspida.
> †Order Heterostraci.
> †Order Coelolepida.

"Ostracoderms"; usually armored forms from the Cambrian through Devonian Periods; gills well developed in the well-known osteostracans, like *Cephalaspis,* which were probably bottom-dwelling filter-feeders (Figs. 18, 19, 21).

Order Petromyzontia. Lampreys; freshwater or marine forms "parasitic" or predaceous on other fishes, these are often lumped with the following order as the Cyclostomata (Fig. 16 *C*).

Order Myxinoidea. Hagfishes; marine scavengers (Fig. 16 *A, B*).

Class Elasmobranchiomorphi
Cartilaginous fishes and certain more primitive and bony relatives.

†Subclass Placodermi.
Jawed fishes, usually with heavy bony armor and often with peculiarly constructed fins; almost all Devonian in age.

Order Arthrodira. Predaceous forms, often with a well-developed joint between the skull and thoracic armor (Fig. 22 *A, B*).

Order Ptyctodontida. Mollusc-crushing forms resembling the cartilaginous Holocephali (Fig. 22 *F*).

> *Order Phyllolepida.*
> *Order Petalichthyida.*
> *Order Rhenanida.*

Odd, usually flattened forms that tend to have a reduced amount of bone and superficially resemble the modern Batoidea (Fig. 22 *C, D, E*).

Order Antiarchi. Small bottom dwellers with bony armor covering the pectoral fins (Fig. 22 *G*).

Subclass Chondrichthyes.
Cartilaginous fishes.

*†Indicates extinct.

Infraclass Elasmobranchii. Sharks and related forms; upper jaw not fused to braincase, gills with separate external openings, and relatively rapidly replaced teeth.

    †*Order Cladoselachii.* Primitive Paleozoic sharks (Fig. 24 *A*).

    †*Order Pleuracanthodii.* Odd, freshwater, Paleozoic sharks (Fig. 24 *B*).

    Order Selachii. Typical sharks, Paleozoic to Recent, with claspers and narrow-based fins (Fig. 24 *C*).

    Order Batoidea. Skates and rays; flattened forms basically similar to sharks (Fig. 24 *D*).

Infraclass Bradyodonti. Chimaeras and related forms; teeth are replaced more slowly than in elasmobranchs, and gills are covered by an operculum.

    †*Superorder Paraselachii.* Poorly known extinct forms with numerous teeth; jaw attachment varies (Fig. 25 *A*).

    Superorder Holocephali. Upper jaw fused to braincase; teeth represented by crushing tooth-plates.

        Various extinct orders are poorly known (Fig. 25 *B*).

    Order Chimaeriformes. Chimaeras or ratfish (Fig. 25 *C*).

Class Osteichthyes

Generally called bony fishes; however, many other fishes also contain bone (Fig. 28).

†Subclass Acanthodii.

"Spiny sharks"; Paleozoic forms of doubtful relationships, once placed with Placodermi but probably related to the ancestral Osteichthyes (Fig. 27).

Subclass Actinopterygii.

Ray-finned fishes.

    Superorder Chondrostei. Primitive ray-finned fishes, with heterocercal tails; represented by the varied and mainly Paleozoic †palaeoniscoids and the living *Polypterus,* sturgeons, and paddlefishes (Figs. 32, 33).

    Superorder Holostei. Dominant ray-finned forms of the Mesozoic, with abbreviated heterocercal tails; the only living forms are *Amia* and *Lepisosteus* (Fig. 34).

    Superorder Teleostei. Dominant fishes of the Cenozoic and Recent, with homocercal tails; they include many thousands of forms classified in about 6 to almost 50 orders (Figs. 36, 38).

Subclass Sarcopterygii.

Fleshy-finned forms (sometimes called Choanichthyes).

    Order Crossopterygii. Predaceous forms with the upper jaw not fused to the braincase.

        †*Suborder Rhipidistia.* Paleozoic forms probably ancestral to tetrapods (Fig. 29 *A*).

        Suborder Coelacanthiformes. Aberrant marine forms, including the living *Latimeria* (Fig. 29 *B*).

    Order Dipnoi. Lungfish; Paleozoic to Recent, mollusc-crushing forms with the upper jaw fused to the braincase and, in most cases, large tooth-plates (Fig. 30).

Class Amphibia

Tetrapods, but without the development of an amniote egg (Fig. 39).

†Subclass Labyrinthodontia.

Primitive forms in which the vertebral centra are formed by pleurocentra and intercentra.

Order Ichthyostegalia. The earliest amphibians, from the late Devonian and Mississippian, which retained a fishlike tail (Fig. 41 *A*).

Order Temnospondyli. The common larger amphibians of the Carboniferous, Permian, and Triassic; the intercentrum is larger than the pleurocentrum (Fig. 41 *B*, *C*).

Order Anthracosauria. Relatively rare Paleozoic amphibians, in which the pleurocentrum is larger than the intercentrum; they include the ancestors of reptiles (Fig. 41 *D*).

†Subclass Lepospondyli.

Paleozoic forms in which the vertebral centra may form as single, often spool-shaped structures.

Order Aistopoda. Limbless, snakelike forms (Fig. 40 *A*).

Order Nectridea. Salamander-like forms, some with "horns" (Fig. 40 *C*).

Order Microsauria. More salamander-like forms (Fig. 40 *B*).

Order Lysorophia. Forms with reduced limbs and highly modified skulls.

Subclass Lissamphibia.

The modern, smooth-skinned amphibians.

Order Anura. Frogs and toads; specialized for hopping.

Order Urodela. Salamanders and newts; a primitive body form, but many degenerate characters.

Order Gymnophiona. Wormlike burrowing forms from the tropics; also called **Apoda**.

Class Reptilia

Amniotes, but without advanced avian or mammalian characters (i.e., no feathers or hair) (Fig. 42).

Subclass Anapsida.

Forms with no temporal fenestrae.

†Order Cotylosauria. Primitive "stem reptiles" of the late Paleozoic and Triassic (Fig. 43 *A*).

Order Testudines. Turtles and tortoises, with a bony shell; also called **Chelonia** or **Testudinata** (Fig. 43 *B*, *C*).

Subclass Lepidosauria.

Primitively diapsid reptiles without archosaur specializations.

†Order Eosuchia. Permian and Triassic ancestral diapsids.

Order Rhynchocephalia. The living *Sphenodon* of New Zealand and fossil relatives (Fig. 46 *A*).

Order Squamata. Forms with reduced temporal arches.

Suborder Lacertilia.   Lizards; primitive and very diverse forms, including numerous limbless ones (Fig. 46 *B*).

Suborder Amphisbaenia.   Specialized, usually limbless, burrowing forms (Fig. 46 *C*).

Suborder Serpentes.   Snakes; limbless forms (Fig. 46 *D*).

Subclass Archosauria.

"Ruling reptiles"; diapsid forms, usually with extra cranial fenestrae and adaptations for bipedal locomotion (Fig. 47).

†*Order Thecodontia.*   Triassic ancestors of dinosaurs, birds, and others (Fig. 48 *A*).

Order Crocodilia.   Crocodiles and alligators; amphibious survivors of the primitive archosaurs (Fig. 48 *B*).

†*Order Pterosauria.*   Flying reptiles of the Mesozoic with membranous wings; often called pterodactyls (Fig. 48 *E*).

†*Order Saurischia.*   "Reptile-like" dinosaurs with a triradiate pelvis (Fig. 48 *C*).

Suborder Theropoda.   The bipedal, carnivorous dinosaurs.

Suborder Sauropodomorpha.   The largest of the quadrupedal herbivores.

†*Order Ornithischia.*   "Birdlike" dinosaurs with a tetraradiate pelvis; all were herbivores (Fig. 48 *D*).

Suborder Ornithopoda.   Bipedal, unarmored forms.

Suborder Stegosauria.   Quadrupeds with odd plates down the back.

Suborder Ankylosauria.   Another group of armored quadrupeds.

Suborder Ceratopsia.   The horned dinosaurs; rhinoceros-like forms.

†Subclass Euryapsida.

Diverse, probably unrelated reptiles with a single dorsal temporal fenestra on each side; their classification is much debated at present; also called **Parapsida** and **Synaptosauria**.

Order Araeoscelidia.   Various obscure terrestrial forms, mostly rather lizard-like; also called **Protorosauria** (Fig. 45 *A*).

Order Sauropterygia.   Plesiosaurs and relatives; marine forms of the Mesozoic with the limbs transformed into powerful paddles (Fig. 45 *C*).

Order Placodontia.   Armored marine mollusc-eaters of the Triassic (Fig. 45 *B*).

Order Ichthyosauria.   Ichthyosaurs, forms paralleling the mammalian porpoises in their marine adaptations (Fig. 45 *D*).

†Subclass Synapsida.

Mammal-like reptiles with a single lateral temporal fenestra on each side.

Order Pelycosauria.   Primitive Carboniferous and Permian forms, often very similar to cotylosaurs, but with a temporal fenestra (Fig. 52 *A*).

Order Therapsida.   Advanced, often very mammal-like forms from the later Permian and Triassic (Fig. 52 *B*).

Class Aves

Birds; winged descendants of the archosaurs, with feathers and temperature control.

†Subclass Archaeornithes.

Archaeopteryx, the Jurassic bird with a basically reptilian skeleton, plus feathers (Fig. 49 *A*).

Subclass Neornithes.

All other birds; skeleton is "modernized" (Fig. 51).

†Superorder Odontognathae. Toothed birds of the Cretaceous (Fig. 49 *B*).

Superorder Palaeognathae. The ostrich-like birds or ratites, usually with reduced wings and relatively primitive skeletons; probably an artifical group (Fig. 50).

Superorder Neognathae. All other birds; placed in a large number of separate orders, but all essentially similar in most anatomic features.

Class Mammalia

Animals with hair; the habit of nursing their young; a single bone on each side of the lower jaw (Fig. 53).

Subclass Prototheria.

Primitive mammals defined by certain technical characters, such as a small alisphenoid bone and no tritubercular teeth.

Infraclass Allotheria. Forms with widened braincases and no jugals.

Order Monotremata. The duckbill and spiny anteaters of Australia and New Guinea.

†Order Multituberculata. Jurassic to Eocene forms, perhaps comparable in habits to the later rodents.

†Infraclass Eotheria.

Order Triconodonta. ⎫   Various poorly known, small, prim-
Order Docodonta.   ⎭   itive mammals.

Subclass Theria.

Normal mammals with well-developed alisphenoids.

†Infraclass Patriotheria. Small, primitive, ancestral forms.

Order Symmetrodonta. Forms with primitive teeth.

Order Pantotheria. Forms with tritubercular teeth; ancestral to higher mammals.

Infraclass Metatheria. The pouched mammals or marsupials; young are born alive, but at an immature stage (Fig. 54).

Order Polyprotodonta. Primitive, mainly carnivorous forms, including the American opossum and many Australian forms; also called the **Marsupicarnivora**.

Order Peramelida. Bandicoots; Australian omnivorous forms.

Order Caenolestoidea. Carnivorous South American forms, mainly extinct.

Order Diprotodonta. Herbivorous Australian forms such as wombats and kangaroos.

Infraclass Eutheria. The higher mammals, with an efficient placenta (Fig. 55).

Order Insectivora. Various primitive eutherians; usually small and insectivorous.

Suborder *Proteutheria.* Very primitive extinct forms and a few living groups, including the tree shrews (Fig. 56 *A*).

Suborder *Macroscelidea.* The African elephant shrews (Fig. 56 *B*).

Suborder *Dermoptera.* The "flying lemur."

Suborder *Lipotyphla.* Moles, shrews, hedgehogs, and relatives (Fig. 56 *C, D, F*).

Suborder *Zalambdodonta.* A small group of insectivores with odd teeth (Fig. 56 *E*).

†*Order Tillodontia.* Large, early, and aberrant herbivores.

†*Order Taeniodonta.* Very large, archaic, somewhat rodent-like forms.

Order *Chiroptera.* Bats; the only true flying mammals.

Suborder *Megachiroptera.* Large, fruit-eating forms.

Suborder *Microchiroptera.* Ordinary small bats.

Order *Primates* (Fig. 57).

†*Suborder Plesiadapoidea.* Early Tertiary forms with rodent-like adaptations; poorly understood.

Suborder *Lemuroidea.* Lemurs and relatives.

Suborder *Tarsioidea.* *Tarsius* and extinct relatives; transitional between lemurs and monkeys.

Suborder *Platyrrhini.* South American monkeys and marmosets; the nostrils open to the sides.

Suborder *Catarrhini.* Old World monkeys, apes, and men; the nostrils open downward.

†*Order Creodonta.* Archaic carnivores.

Suborder *Deltatheridia.* Small, insectivore-like forms.

Suborder *Hyaenodontia.* Large carnivorous forms.

Order *Carnivora.* The true carnivores (Fig. 58).

Suborder *Fissipedia.* Terrestrial carnivores.

†*Infraorder Miacoidea.* Extinct ancestors of the modern forms.

Infraorder *Aeluroidea.* Cats and their relatives, including civets, mongooses, hyenas, and the like; also called **Feloidea.**

Infraorder *Arctoidea.* Dogs and their relatives, including weasels, skunks, otters, raccoons, bears, and the like; also called **Canoidea.**

Suborder *Pinnipedia.* Marine carnivores; seals, sea lions, and walruses.

†*Order Condylarthra.* Primitive forms, probably ancestral to the ungulates and many related forms (Fig. 59 *A*).

†*Order Pantodonta.*

†*Order Dinocerata.*

†*Order Embrithopoda.*

†*Order Notoungulata.*

Suborder *Notioprogonia.*

Suborder *Toxodontia.*

Suborder *Typotheria.*

†*Order Xenungulata.*

†*Order Pyrotheria.*

†*Order Astrapotheria.*

†*Order Litopterna.*

All of these are generally archaic ungulates in a very broad sense; the first two orders are basically characteristic of the Northern Hemisphere, the third is African, and the remainder are South American (Fig. 59 *B–E*).

Order *Hyracoidea.* The conies of Africa and the Near East; rabbit-like in general appearance, but actually ungulates; this and the next two orders appear related and of African origin.

Order Proboscidea.   The elephants and their relatives.

†Suborder Moeritherioidea.   An early African relative of elephants.

Suborder Euelephantoidea.   Elephants, mammoths, and mastodonts.

†Suborder Deinotherioidea.   Elephant-like forms with recurved lower tusks.

Order Sirenia.   The sea cows, manatees, and dugongs; an aquatic offshoot of some ungulate stock.

†Order Desmostylia.   Forms vaguely like sirenians.

Order Perissodactyla.   Odd-toed ungulates (Fig. 60).

Suborder Hippomorpha.   Horses plus various fossils, including titanotheres—large, ungainly, horned ungulates.

†Suborder Ancylopoda.   Chalicotheres; odd ungulates, generally horselike but with claws.

Suborder Ceratomorpha.   Tapirs, rhinoceroses, and the like.

Order Artiodactyla.   Even-toed ungulates (Fig. 61).

†Suborder Palaeodonta.   Ancestral forms.

Suborder Suina.   Relatively primitive forms with simple stomachs and bunodont teeth; pigs, and relatives such as hippos.

Suborder Ruminantia.   Advanced forms with complex stomachs and selenodont teeth.

Infraorder Tylopoda.   Camels and various extinct groups; they are intermediate between the Suina and Pecora in some ways.

Infraorder Pecora.   Advanced ruminants, often with horns or antlers; deer, giraffes, pronghorn, bison, sheep, goats, antelopes, and the like.

Order Edentata.   So-called "toothless" mammals; a South American group also called the Xenarthra.

Suborder Pilosa.   The hairy edentates.

†Infraorder Gravigrada.   Ground sloths.

Infraorder Tardigrada.   Tree sloths.

Infraorder Vermilingua.   South American anteaters.

Suborder Loricata.   The armored edentates; armadillos and the extinct glyptodonts.

Order Pholidota.   The Old World pangolin, another "anteater."

Order Tubulidentata.   The aardvark of Africa, still another type of anteater.

Order Cetacea.   The whales and porpoises; they may represent two independent groups.

†Suborder Archaeoceti.   Probably ancestral forms.

Suborder Odontoceti.   The toothed whales, porpoises, and dolphins.

Suborder Mysticeti.   Whalebone whales; the big filter-feeders.

Order Rodentia.   Gnawing animals (except the rabbits); their classification is extremely poorly understood and that given here is conservative and oversimplified.

Suborder Protrogomorpha.   Primitive forms, including the living mountain beaver (which is not a beaver).

Suborder Sciuromorpha.   Squirrels and their allies.

Suborder Myomorpha.   Rats, mice, and many more.

Suborder Caviomorpha. Guinea pig and many other South American forms such as porcupines; certain African and other Old World forms (more porcupines) may be related.

Order Lagomorpha. Rabbits and a few relatives; gnawing forms, but not closely related to the rodents.

Appendix 2

Scientific Terminology

In anatomic terminology, common Latin or Greek words are used, more or less, as such for any part of the body for which the ancients had a name. For numerous other structures, scientific names have been invented (1) by using, in a new sense, some classical word that seemed to be descriptive of the part concerned, or (2) commonly, by combining Greek or Latin roots to form a new compound term. The student frequently attempts to memorize such terms without understanding their meaning and with consequent mental indigestion. We give here the roots from which many of these descriptive terms and compounds are derived, as an aid to comprehension. As will be seen, some names formed by the anatomists are rather fanciful or farfetched; some are none too appropriate. This list is not intended, of course, as a glossary or dictionary of scientific words. We have not, for instance, included common names of bones and muscles. Most of the terms used in this book are defined or discussed in the text.

For a wider vocabulary, use of a standard biologic or medical dictionary* is recommended, but the larger editions of Webster, Oxford, and others are satisfactory in most regards. Other books, although not dictionaries, consider many of the roots used in scientific terms; excellent examples include L. Hogben, The Vocabulary of Science, New York, Stein and Day, 1970, and E.C. Jaeger, A Source Book of Biological Names and Terms, 3rd ed., Springfield, Illinois, Charles C Thomas, 1978.

In the list below, we give the root or word, the language from which it is derived, the root or word in the original (unless it is identical to the root used in which case it is omitted), and the original meaning; we do not give the anatomic meaning, which may differ quite wildly from the original. For example, *coraco-* is defined as "crowlike"; anatomically it refers to the ventral part of the shoulder girdle and neighboring structures—because that portion of the girdle in us, the coracoid process, was thought to resemble a crow's beak. Abbreviations: adj. adjectival form; dim., diminutive; F., French; G., Greek; gen., genitive; L., Latin; M.E., Middle English; n., noun; NL., "New" Latin; pl., plural; p.p., past participial form; Sp., Spanish.

---

*Such as Dorland's American Illustrated Medical Dictionary, 26th ed., Philadelphia, W.B. Saunders Company, 1982; or Henderson's Dictionary of Biological Terms, 9th ed., New York, Van Nostrand, 1979.

A-, ab-. L., implying separation.

Abdomen. L. (from *abdere* (?), to hide), abdomen.

Abducens. L., *ab,* away, + *ducens,* leading.

Abductor. L., *ab,* away, + *ducere,* to lead.

Acanth-. G., *akantha,* spine, thorn.

Acelous. G., *a,* not, + *koilos,* hollow.

Acetabulum. L., vinegar cup.

Acido-. L., *acidus,* sour.

Acinus. L., berry.

Acrania. G., *a,* without, + *krania,* heads.

Acro-. G., *akros,* height or extremity.

Acromion. G., *akros,* height or extremity, + *omos,* shoulder.

Actino-. G., *aktis* (gen. *aktinos*), a ray.

Ad-, af-. L., to, toward, at, or near.

Adeno-. G., *aden,* gland.

Adipose. L., *adeps* (gen. *adipis*), fat.

Adrenal. L., *ad,* near, + *renes,* kidneys.

Adventitia. L., *adventitius,* extraordinary.

Albus, -a, -um. L., white.

Alisphenoid. L., *ala,* wing, + G., *sphen,* wedge, + G., *eidos,* form.

Allantois. G., *allas* (gen. *allantos*), sausage, + *eidos,* form.

Alveolus. L., little cavity.

Amello-. M.E., enamel.

Amnion. G., fetal membrane.

Amphi-. G., on both sides; hence, around or double.

Amphioxus. G., *amphi,* both, + *oxys,* sharp.

Ampulla. L., a flask or vessel swelling in the middle.

An-, A-. G., without or not.

Ana-. G., upon, up, throughout, back again, similar to, frequently, or reinforcing a meaning.

Analogy. G., *ana,* similar to, + *logos,* discourse, ratio.

Anastomosis. G., *ana,* again, + *stoma,* mouth.

Andro-. G., *aner* (gen. *andros*), a male.

Ankylosis. G., stiffening of the joint.

Annulus. L., *anulus, annulus,* a ring.

Anura. G., *an,* without, + *oura,* tail.

Anus. L., fundament.

Apo-. G., from.

Apoda. G., *a,* without, + *pous* (gen. *podos*), foot.

-apophysis. G., (*apo,* away, + *physis,* growth), process on a bone.

Appendicular. L., *appendere,* to hang upon.

-apsid. G., *apsis,* an arch.

Arachnoid. G., *arachnes,* spider, + *eidos,* form.

Arch-, archi-. G., first or chief; hence primitive or ancestral.

Archenteron. G., *arch,* first, + *enteron,* gut, intestine.

Archipterygium. G., *archi,* first, + *pterygion,* a little wing.

Arcualia (pl.). L., *arcualis,* bow shaped.

Arrector. L., *arrigere,* to raise.

Arthrosis. G., *arthron,* joint.

Arytenoid. G., *arytana,* jug, + *eidos,* form.

Asper, -a, -um. L., rough, thorny.

Astragalus. G., *astragalos,* ankle bone, used as a die (commonly pl.).

Atrium. L., court, entrance hall.

Auditory. L., *auditum,* hearing.

Auricle. L., *auricula,* external ear.

Auto-. G., *autos,* self.

Autonomic. G., *autos,* self, + *nomos,* law.

Axial. L., axis of a wheel, the line about which any body turns.

Axillary. L., *axilla,* armpit.

Axon. G., axle.

Azygos. G., *a,* not, + *zygon,* yoke.

Baculum. L., rod.

Basal. L., *basis,* footing or base.

Basi-. L., pertaining to the base.

Bi-. L., two, twice, or double.

Biceps. L., *bi,* two, + *caput,* head.

Bio-. G., life.

Blasto-. G., bud, germ, or sprout.

Brachial. L., *brachialis,* belonging to the arm.

Brachy-. G., *brachys,* short.

Branchial. L., *branchiae,* or G., *branchia,* gills.

Branchiostegal. G., *branchia,* gills + *stege,* a covering.

Bronchus. G., *bronchos,* trachea.

Buccal. L., *bucca,* cheek.

Bulbus. L., bulb, swollen root.

Bulla. L., a bubble; hence spherical in shape.

Bunodont. G., *bounos,* mound, + *odous* (gen. *odontos*), tooth.

Bursa. L., a purse.

Caecum, cecum. L., *caecus, -a, -um,* hollow.

Calice. L., cup.

Calc-. L., *calx* (gen. *calcis*), lime, heelbone.

Callosum. L., *callosus, -a, -um,* thick-skinned.

Calyx (pl. **calices**). L., husk, cup-shaped protective covering.

Canine. L., *canis,* dog.

Capillo-. L., *capillus,* hair.

Capitulum. L., *caput* (dim. *capitulum*), head.

Caput (pl. **capita**). L., head.

Carapace. NL., *carapax,* bony or chitinous covering.

Cardiac. G., *kardia,* heart.

Cardinal. L., *cardinalis,* pertaining to a hinge, chief, principal.

Carina. L., keel.

Carno-. L., *caro* (gen. *carnis*), flesh.

Carotid. G., *karos,* heavy sleep.

Carpus. G., *karpos,* wrist.

Cartilago (pl. **-agines**). L., gristle, cartilage.

Caudal. L., *cauda,* tail.

Cava. L., *cavus, -a, -um,* hollow.

Cavernosus. L., *caverna,* a hollow or cave.

Cephalo-, -cephalon. G., *kephale,* head.

Cerato-. G., *keras* (gen. *keratos*), horn.

Cercal. G., *kerkos,* a tail.

Cerebellum. L., *cerebrum* (dim. *cerebellum*), brain.

Cerebrum. L., brain.

Cervical. L., *cervix* (gen. *cervicis*), neck.

Chiasma. G., figure of X.

Chiro-. G., *cheir,* a hand.

Choana. G., *choane,* funnel.

Choledochal. G., *chole,* bile, + *doche,* a container.

Chondro-. G., *chondros,* cartilage, granule.

Chondroclast. G., *chondros,* cartilage, + *klastos,* broken.

Chorda. G., *chorde,* string of gut, cord.

Chorion. G., embryonic membrane.

Chromaffin. G., *chroma,* color, + L., *affinitas,* alliance, relationship.

Chyme, chylo. G., *chylas,* juice.

Ciliary. L., *cilium,* eyelash.

Cinereum. L., *cinus* (gen. *cineris*), ashes.

Cingulum. L., girdle.

Circum-. L., around.

Clavicle. L., *clavis* (dim. *clavicula*), key.

Cleithrum. G., *kleithron,* a door latch or bar.

Clitoris. G., *klitos,* slope, hill.

Cloaca. L., sewer.

Cnemial. G., *kneme,* lower leg.

Coccyx. G., *kokkyx,* cuckoo.

Cochlea. L., *cochlea* (from G., *kochlias*), spiral, snail shell.

Coel-, Cel-. G., *koilia,* the belly.

Coeliac. G., *koilos,* hollow.

Coll-. L., *collis,* hill.

Coll-. L., *collum,* neck.

Collagen. G., *kolla,* glue, + *genos,* descent, begetter of.

Colon. L., and G., *kolon,* large gut.

Columella. L., *columna* (dim. *columella*), pillar.

Comma. G., *komma,* a fragment, that which is cut off.

Commissure. L., *commissura,* connection, seam.

Concha. L., (G., *konche,* mussel), bivalve, oyster shell.

Condyle. L., *kondylos,* knuckle.

Conjunctiva. L., *conjunctivus,* connecting.

Copro-. G., *kopros,* dung.

Coraco-, corono-. G., *korax* or *korone* (gen. *korakos*), crow.

Corium. L., leather.

Cornus. L., *cornu,* horn.

Corono-. L., *corona,* wreath, crown.

Corpus (pl. **corpora**). L., body.

Cortex. L., bark, rind.

Cosmine. G., *kosmos,* orderly arrangement.

Costa. L., rib.

Cotyle. G., *kotyle,* cup.

Coxa. L., hipbone.

Cranial. G., *kranion,* skull.

Cre-, crea. G., *kreas,* flesh.

Cribro-. L., *cribrum,* sieve.

Cricoid. G., *krikos,* ring, + *eidos,* form.

-crine. G., *krinein,* to separate.

Crista. L., crest.

Crosso-. G., *krossoi,* fringe.

Crus (pl. **crura**). L., leg.

Ctenoid. G., *kteis* (gen. *ktenos*), a comb, + *eidos,* form.

Cucull-. L., *cucullus,* cap, hood.

Cuneiform. L., *cuneus,* wedge, + *forma,* shape.

Cupula. L., *cupa* (dim. *cupula*), a cask or tub.

Cutis. L., skin.

Cycloid. G., *kyklos,* circle, + *eidos,* form.

Cyno-. G., *kyon* (gen. *kynos*), dog.

Cystic. G., *kystis,* a bag or pouch.

Cytos. G., *kytos,* vessel or bowl, cell.

-dactyl. G., *daktylos,* finger.

De-. L., signifying down, away from, deprived of.

Deciduous. L., *deciduus* (*de,* away from, + *cado,* drop), falling off.

Decussatio. L., crosswise intersection.

Deferens. L., *de,* away from, + *ferens,* carrying.

Dendr-. G., *dendros,* tree.

Dens (gen. **dentis**). L., tooth.

Dermal. G., *derma,* skin

-deum. G., *odaios,* on the way.

Di-. G., twice; hence, twofold or double.

Di-, dia-. G., through, between, apart, across.

Diaphragm. G., *dia,* between, + *phragma,* partition.

Diarthrosis. G., *dia,* through, + *arthron,* joint.

Diastema (pl. **-ata**). G., *dia,* apart, + *histemi,* stand.

-didymis. G., *didymos,* twin.

Digit. L., *digitus,* finger.

Diphycercal. G., *diphyes,* twofold, + *kerkos,* tail.

Diplospondylous. G., *diploos,* double, + *spondyle,* vertebra.

Dipnoi. G., *di,* double, + *pneuma,* breath.

Distal. L., *distare,* to stand apart.

Dorsal. L., *dorsum,* back.

Duct. L., *ductus,* leading.

Duodenum. L., *duodecem,* twelve.

Dura. L., hard.

E-, ex-. L., out, out of, from.

Ectepicondyle. G., *ektos,* outside, + *epi,* upon, + *kondylos,* knuckle.

Ectoderm. G., *ektos,* outside, + *derma,* skin.

Effector. L., *efficere,* to bring to pass.

Efferent. L., *ex,* out, + *ferre,* to bear.

Ejaculatory. L., *e,* out, + *jaculor,* thrower.

Ek-, ekto-. G., out of, from, outside.

Elasmo-. G., *elasma,* a flat plate or strap.

Embolomerous. G., *en,* in, + *bolos,* lump, + *meros,* portion or part.

En-, endo-, ento-. G., *entos,* in, within.

Entepicondyle. G., *entos,* in, + *epi,* upon, + *kondylos,* knuckle.

Enteron. G., gut.

Enzyme. G., *en,* in, + *zyme,* leaven.

Epaxial. G., *epi,* upon, + L., *axis,* center line.

Edendyma. G., (*epi,* upon, + *en,* on, + *dyo,* put on), upper garment.

Epi-. G., upon, above.

Epibole. G., (*epi,* upon, + *bolos,* lump), layer.

Epiploic. G., *epiploon,* caul, omentum.

Epithelium. G., *epi,* upon, + *thele,* nipple.

Erythrocyte. G., *erythros,* red, + *kytos,* cell.

Esophagus. G., *oisein,* to carry, + *phagein,* to eat.

Ethmoid. G., *ethmos,* sieve, + *eidos,* form.

Eu-. G., adv., well; the better of alternates.

Eury-. G., broad.

Excretion. L., *ex,* out, + *cernere,* (p.p. *excretus*), to sift.

Exocrine. G., *ex,* out, + *krinein,* to separate.

Extensor. L., *ex,* out, + *tendere,* to stretch.

Extrinsic. L., *extrinsecus,* on the outside.

Facialis. L., *facies,* face.

Falciform. L., *falx* (gen. *falcis*), sickle, + *forma,* shape.

Fascia (pl. **-iae**). L., band.

Fauces. L., throat.

Fenestra. L., window.

Fer. L., *fero,* carry.

Fiber. L., *fibra,* string, thread.

Fibula. L., brooch, pin.

Filoplume. L., *filum,* thread, + *pluma,* soft feather, down.

Fimbria. L., thread, fringe.

Firmisternal. L., *firmus,* steadfast, strong, + G., *sternon,* chest.

Fissure. L., *findere* (p.p. *fissurus, -a, -um*), a cleave.

Flagellum. L., *flagrum* (dim. *flagellum*) a whip.

Flexor. L., *flexus,* bent.

Flocculus. NL., *floccus* (dim. *flocculus*), a tuft of wool.

Folium. L., leaf.

Follicle. L., *follis* (dim. *folliculus*), bag.

Fontanelle. F., a small fountain.

Foramen. L., hole.

Fornix. L., arch or vault.

Fossa. L., ditch, channel, something dug.

Fovea. L., small pit.

Frenulum. L., *frenum* (dim. *frenulum*), bridle.

Frontal. L., *frons* (gen. *frontis*), forehead, brow.

Fundus. L., bottom.

Funiculus. L., *funis* (dim. *funiculus*), a rope.

Furcula. L., *furca* (dim. *furcula*), fork.

Gametes. G., husband.

Gan-. G., *ganos*, brightness, sheen.

Ganglion. G., swelling under the skin.

Gastralium. G., *gaster*, belly.

Gastrula. NL. dim. (from G., *gaster*, belly).

Gem-. L., *geminus*, twin-born.

Gen-. G., *genos*, race; L., *genus*, race; L., *gena*, cheek; G., *genys*, cheek; L., *genu*, knee.

Genesis. G., birth, creation.

Geniculate. L., *geniculatus*, with bent knee.

Genital. L., *gignere* (p.p. *genitalis*), to beget.

Geo-. G., *ge*, the earth.

Germinal. L., *germen* (gen. *germinis*), bud germ.

Gladius. L., sword, blade.

Glans. L., acorn.

Glenoid. G., *glene*, socket, + *eidos*, form.

Gli-, **glia**. G., *gloia*, glue.

Glomus (pl. **glomera**). L., ball.

Gloss-, **glott-**. G., *glottos*, tongue.

Gluteus. G., *gloutos*, rump.

Gnathos. G., jaw.

Gonad. G., *gone*, seed.

Gracilis. L., slender.

Grad. L., *gradus*, step.

Granulosus. L., full of grains.

Granulum. L., small grain.

Guanin(e). Sp., *guano*, dung of sea fowl.

Guanophore. Sp. from Peruvian, *huanu*, dung, + *phoros*, bearing.

Gubernaculum. L., helm, rudder.

Gular. L., *gula*, throat.

Gymno-. G., naked.

Gyrus. G., *gyros*, a turn.

Habenula. L., *habena* (dim. *habenula*), strap.

Haemal, **hemal**. G., *haima*, blood.

Hamate. L., *hamatus*, hook shaped.

Helico-. G., *helix* (gen. *helikos*), twisted.

Hemi-. G., signifying half.

Hemopoietic. G., *haima*, blood, + *poietikos*, creative.

Hepatic. G., *hepar* (gen. *hepatos*), liver.

Hetero-. G. combining form signifying other, different.

Heterotopic. G., *heteros*, other, + *topos*, place.

Hilus. L., *hilum*, trifle, eye of seed.

Hippocampus. G., *hippos*, horse, + *kampos*, sea monster.

Histology. G., *histo*, web, + *logos*, discourse, ratio.

Holo-. G., *holos*, whole.

Homo-. G., *homos*, equal, same.

Homology. G., *homos*, same, + *logos*, discourse, ratio.

Hormone. G., *hormaein*, to excite.

Humor. L., moisture, fluid.

Hyaline. G., *hyalos*, glass.

Hyo-. G., *hyoeides* (*hy*, epsilon, + *eidos*, form), Y shaped.

Hypaxial. G., *hypo*, under, + L., *axis*, center line.

Hyper-. G., above, over.

Hypo-. G., signifying under, below.

Hypsodont. G., *hypso*, height, + *odous* (gen. *odontos*), tooth.

Hypural. G., *hypo*, under, + *oura*, tail.

Ichthy-. G., *ichthys*, fish.

Ileum. G., *eilein*, to wind or turn *or* L., *ileum*, groin.

Ilium. L., flank.

In-. L., not.

In-. L., in, into, within, toward, on.

Incisor. L., *incisus*, cut.

Incus. L., anvil.

Inductor. L., *inducere*, to lead on, excite.

Infra-. L., signifying below, lower than.

Infundibulum. L., a funnel.

Inguinal. L., *inguina,* groin.

Integument. L., *in,* over, + *tegumen,* a cover.

Inter-. L., signifying between, among.

Interstitial. L., *inter,* between, + *sistere,* to set.

Intestine. L., *intestinus,* internal or *intestinum,* intestine.

Intra-. L., inside.

Intrinsic. L., *intrinsecus,* inward.

Iris. G., (gen. *iridos*) rainbow.

Ischium. G., *ischion,* hip.

Iso-. G., *isos,* equal.

Isomer. G., *isos,* equal, + *meros,* part.

Iter. L., way, passage.

Jejunum. L., *jejunus,* empty.

Jugal. L., *jugum,* yoke.

Jugular. L., *jugulum,* pertaining to the neck.

Keratin. G., *keras* (adj. *keratinos*), horn.

Kinetic. G., *kinesis* (adj. *kinetikos*), movement.

Labial. L., *labia,* a lip.

Lacerate. L., *lacerare,* to tear.

Lacrimal. L., *lacrima,* tear.

Lacuna. L., hole, cavity.

Lagena. L., flask.

Lamina. L., thin plate.

Larva. L., ghost, mask.

Larynx. G., upper part of windpipe.

Lateral. L., *latus* (adj. *lateralis*), side.

-lecithal. G., *lecithos,* yolk.

Lemma. G., skin of a fruit.

Lepidotrichia. G., *lepis* (gen. *lepidos*), scale, + *thrix* (gen. *trichos*), hair.

Leuco-. G., *leukos,* white.

Levator. L., *levator,* lifter.

Lieno-. L., *lien,* spleen.

Ligamentum. L., a bandage.

Ling-. L., *lingua,* tongue.

Lipo-. G., *lipos,* fat.

Liss-. G., *lissos,* smooth.

Lith-. G., *lithos,* stone.

Lobus. G., *lobos,* lobe.

Lophodont. G., *lophos,* ridge, + *odous* (gen. *odontos*), tooth.

Lumbar. L., *lumbus* (n., *lumbaris*), loin.

Lumen. L., light, opening.

Luteus, -a, -um. L., yellow.

Lymph. L., *lympha,* water.

Macrophage. G., *makros,* large, + *phagein,* to eat.

Macula. L., spot, stain.

Magnus, -a, -um. L., large.

Malar. L., *mala,* upper jaw.

Malleus. L., hammer.

Mammillary. L., *mamma* (dim. *mammilla*), breast.

Man-. L., *manus,* hand.

Marsupium. L., pouch.

Mastoid. G., *mastos,* breast, + *eidos,* form.

Matrix. L., womb, groundwork, or mold.

Maxilla. L., bone.

Meatus. L., passage.

Medial. L., *medius* (adj. *medialis*), middle.

Mediastinum. L., *mediastinus,* servant, drudge.

Medulla. L., marrow, pith.

Mega-. G., *megas,* great.

Melanin. G., *melas,* black.

Meninx (pl. **meninges**). G., membrane.

Meniscus. G., *menos* (dim. *meniskos*), little moon.

Mental. L., *mentum,* chin.

-mere. G., *meros,* part.

Mes-, meso-. G., *mesos,* middle.

Met-, meta-. G., *meta,* next, after; denoting change of time or situation.

Metabolic. G., *meta,* change, + *bolos,* lump.

-metrium. G., *metra,* womb.

Micro-. G., *mikros,* small.

Mirabile. L., *mirus* (adj. *mirabilis*), wonderful.

Mito-. G., *mitos,* thread.

Molar. L., *mola* (gen. *molaris*), millstone.

-morph. G., *morpha,* shape.

Mucus. L., snivel, slippery secretion.

Muscle. L., *mus* (dim. *musculus*), mouse.

Myel-. G., *myelos,* marrow, spinal cord.

Myo-, mys-. G., *myos,* muscle.

Myodome. G., *myos,* muscle, + L., *domus,* house.

Naris. L., nostril.

Nas-. L., *nasus,* nose.

Neos. G., youthful, new.

Neoteny. G., *neos,* youthful, + *tenein,* to stretch.

Nephro-. G., *nephros,* kidney.

Neur-. G., *neuron,* sinew, tendon, nerve.

Neuromast. G., *neuron,* nerve, + *mastos,* round hill.

Neuropil. G., *neuron,* nerve, + *pilos,* felt.

Nictitating. L., *nictitatio,* to wink.

Nidamental. L., *nidamentum,* materials for a nest.

Nidus. L., nest.

Nodo-. L., *nodus,* knot.

Nuchal. L., *nucha,* nape of the neck.

Obliquus. L., slanting.

Obturator. L., *obturare,* to close.

Occipital. L., *occiput,* back of head.

Ocul-. L., *oculus,* eye.

Odon-. G., *odous* (gen. *odontos*), tooth.

-oid. G., *eidos,* form.

Olecranon. G., *olene,* lower arm, + *kranion,* head.

Olfactory. L., *olor,* a smell, + *facere,* to make.

Oligo-. G., *oligos,* few.

Omasum. L., paunch.

Omentum. L., fatty membrane, bowels.

Omni-. L., *omnis,* all.

Omo-. G., *omos,* shoulder.

Omphalo-. G., *omphalos,* the navel.

Ontogeny. G., *onta,* existing things, + *gennan,* to beget.

Oo-. G., *oon,* egg.

Operculum. L., lid.

Ophthalmic. G., *ophthalmos,* eye.

Opistho-. G., *opisthe,* backward, behind.

Optic. G., *opsis* (adj. *optikos*), sight.

Or-. L., *os* (gen. *oris*), mouth.

Ortho-. G., *orthos,* straight, direct.

Os, oss-. L., *os* (gen. *ossis*), bone.

Ossicle. L., *os* (dim. *ossiculum*), bone.

Ostium. L., door.

Ost-. G., *osteon,* bone.

Ostraco-. G., *ostracon,* pot, shell.

Otic. G., *otikos,* belonging to the ear.

Otolith. G., *ous* (gen. *otos*), ear, + *lithos,* stone.

Ovum. L., egg.

Oxyphil. G., *oxys,* acid, + *philos,* friend.

Paedo-. G., *pais* (gen. *paidos*), child.

Palatum. L., roof of the mouth.

Paleontology. G., *palaios,* ancient, + *onta,* existing things, + *logos,* discourse, ratio.

Pallium. L., cloak.

Palma. L., the open hand.

Palpebra. L., eyelid.

Pancreas. G., *pan,* all, + *kreas,* flesh.

Panniculus. L., *pannus* (dim. *panniculus*), cloth.

Papilla. L., pimple.

Para-. G., beside, near.

Parietal. L., *paries* (gen. *parietalis*), wall.

Parotid. G., *para,* near, + *ous* (gen. *otos*), ear.

Patella. L., *patera* or *patina* (dim. *patella*), dish or plate.

Pecten. L., comb.

Pectoral. L., *pectus* (gen. *pectoralis*), breast.

Pedunculus. L., *pes* (gen. *pedis;* dim. *pediculus*), foot.

Pelvis. L., basin.

Peri-. G., around.

Perineum. G., *perinaion,* part between anus and scrotum.

Peristalsis. G., *peristaltikos,* clasping and compressing.

Peritoneum. G., *peri,* around, + *tenein,* stretch.

Peroneal. G., *perone,* pin, fibula.

Petr-. G., *petros,* rock.

Phalanx (pl. **phalanges**). G., line of soldiers.

Pharynx. G., (gen. *pharyngos*), throat.

-philic. G., *philos,* love.

-phor, -phore. G., *phoros,* bearing.

Photophore. G., *phos* (gen. *photos*), light, + *phoros,* bearing.

Phragm. G., fence, partition.

Phrenic. G., *phren,* diaphragm.

Phyllo-. G., *phyllon,* leaf.

Phylogeny. G., *phylon,* race, + *gennan,* to beget.

-physis. G., *phyein,* to generate.

Pineal. L., *pinea,* pine cone.

Pinna. L., *penna, pinna,* feather; hence, wing.

Pir-. L., *pirum,* pear.

Pisiform. L., *pisum,* pea, + *forma,* shape.

Pituitary. L., *pituita,* slime, phlegm.

Pius, -a, -um. L., soft.

Placenta. L., a flat cake.

Placode. G., *plax*, plate, + *eidos*, form.

Planta. L., sole of the foot.

Plasma. G., form.

Plastron. F., breastplate.

Platy-. G., *platys*, flat.

Plectrum. G., *plektron*, hammer.

Pleuro-. G., *pleuros*, the side.

Plexus (pl. **plexus**). L., plaiting, braid.

Pneumatic. G., *pneumatikos*, pertaining to breath.

Pod. G., *pous* (gen. *podos*), foot.

Poikilothermous. G., *poikilos*, changeful, + *thermos*, heat.

Pons. L., (gen. *pontis*) bridge.

Popliteal. L., *poples* (gen. *poplitis*), ham.

Portal. L., *porta*, gate.

Porus. G., *poros*, passage.

Post-. L., behind, after.

Prae-, **pre-**. L., before, in front.

Prim-. L., *primus*, first.

Pro-. G., or L., before, in front of, or prior.

Proctodeum. G., *proktos*, anus, + *odaios*, on the way.

Profundus, **-a**, **-um**. L., deep.

Pronator. L., *pronare*, to bend forward.

Proprioceptor. L., *proprius*, special, + *capere*, to take.

Prostate. L., *pro*, in front, + *stare*, to stand.

Prot-. G., *protos*, first.

Proximal. L., *proximus*, next.

Psalterium. L. (from G., *psallein*, pluck the harp), stringed instrument, book of psalms.

Pseudo-. G., *pseudes*, false.

Pterygoid. G., *pteryx* (gen. *pterygos*), wing, + *eidos*, form.

Pteryla. G., *pteron*, feather, + *hyle*, a wood.

Pubis. L., mature.

Pudendum. L., *pudendus*, *-a*, *-um*, shameful.

Pulmonary. L., *pulmo*, lung.

Pygal. G., *pyge*, rump.

Pylorus. G., *pylouros* (*pyle*, gate, + *oresthai*, keep watch), gate-keeper.

Pyriform. L., *pirum*, pear, + *forma*, shape.

Quadriceps. L., *quattuor*, four, + *caput*, head.

Quadrigeminus. L., fourfold, four.

Quadro-. L., *quattuor*, four.

Radial. L., *radius*, rod, spoke.

Ramus. L., branch.

Re-. L., again, backward.

Receptor. L., *recipere*, to take back, receive.

Rectus, **-a**, **-um**. L., straight.

Remix (pl. **remiges**). L., rower.

Renal. L., *ren*, kidney.

Rest-. L., *restis*, rope.

Rete. L., network.

Reticulum. L., *rete* (dim. *reticulum*), a net.

Retina. L., *rete*, a net.

Retrices. L., *retro*, back, + *cedere*, to go.

Rhabdo-. G., *rhabdos*, rod.

Rhachitomous. G., *rhachis*, spine, + *tenmein*, to cut.

Rhinal. G., *rhis* (gen. *rhinos*), nose.

Rhombencephalon. G., *rhombos*, kind of parallelogram, + *enkephalos* (*en*, in, + *kephale*, head).

Rhyncho-. G., *rhynchos*, snout, beak.

Rodentia. L., *rodens* (gen. *rodentis*), gnawing.

Rostrum. L., beak.

Ruga. L., fold.

Ruminate. L., *ruminare*, to chew the cud.

Sacculus. L., *saccus* (dim. *sacculus*), a bag.

Sacrum. L., *sacer*, sacred.

Sagittal. L., *sagitta*, arrow.

Salpinx. G., trumpet.

Sarcolemma. G., *sarx* (gen. *sarkos*), flesh, + *lemma*, husk, skin.

Sartorius. L., *sartor*, tailor.

-saur. G., *sauros*, lizard; hence reptile.

Scala. L., staircase.

Scale. G., *skalenos*, a triangle with three unequal sides.

Schisme. G., cleft.

Sclera. G., *skleros*, hard.

Scrotum. L., pouch.

Sebum. L., tallow, grease.

Selenodont. G., *selene*, moon, + *odous* (gen. *odontos*), tooth.

Sella. L., chair, seat, saddle.

Seminiferous. L., *semen*, seed, + *ferre*, to bear.

Septum. L., *saeptum*, fence.

Ser-. L., *serum*, whey, watery liquid.

Serra. L., saw.

Sesamoid. G., *sesa,* sesame-seed, + *eidos,* form.

Sinus (pl. **sinus**). L., curve, cavity, bosom.

Soma (pl. **somata**). G., body.

Somite. G., *soma,* body, + *-ite,* indicating origin.

Sole. L., solea, sandal.

Spermatozoon (pl. **spermatozoa**). G., *sperma* (gen. *spermatos*), seed, + *zoon,* animal.

Sphenoid. G., *sphen,* wedge, + *eidos,* form.

Sphincter. G., *sphinkter,* that which binds tight (from *sphingein,* to bind tightly).

Spina. L., thorn.

Spiracle. L., *spiraculum,* air hole.

Splanchnic. G., *splanchna,* viscera.

Splenial. L., *splenium,* patch.

Squamo-. L., *squama,* scale.

Stapes. L., stirrup.

Stato-. G., *status,* standing.

Stego-. G., *stege,* roof.

Stereospondylous. G., *stereos,* solid, + *spondylos,* vertebra.

Sternum. G., *sternon,* chest.

Stomodeum. G., *stoma,* mouth, + *odaios,* on the way.

Stratum. L., layer.

Striatus, -a, -um. L., grooved, streaked.

Stylos. G., pillar.

Sub-. L., signifying under, beneath, near.

Sulcus. L., furrow.

Supinator. L., *supinare,* to bend forward.

Supra-. L., above.

Sym-, syn-. G., with or together.

Synapse. G., *syn,* together, + *haptein,* to fasten.

Syrinx. G., pipe.

Tabular. L., *tabula,* board, table.

Talonid. L., *talus,* heel, + G., *eidos,* form.

Tapetum. L., *tapete,* carpet.

Tarsus. G., *tarsos,* sole of the foot.

Tectum. L., roof.

Tegmentum. L., *tegumentum,* a covering.

Tela. L., web.

Tel-. G., *tele,* far off, distant.

Temporal. L., *temporalis,* belonging to time.

Temporal. L., *tempus* (gen. *temporis*), the temples.

Tendon. L., *tendere,* to stretch.

Tentorium. L., tent.

Teres. L., round, smooth.

Terminalis. L., *terminare,* to limit.

Testis. L., witness, hence (?) testicle.

Thalamus. G., *thalamos,* chamber or couch.

Theco-. G., *theke,* case.

-thelium. G., *thele,* nipple.

Ther-, -there. G., *ther,* beast, animal.

Thorax (pl. **thoraces**). G., breastplate, breast.

Thrombocytes. G., *thrombos,* clot, + *kytos,* cell.

Thymus. G., *thymos,* sweetbread.

Thyroid. G., *thyreos,* shield, + *eidos,* form.

-tome. G., *tomos,* to cut.

Trabecula. L., a little beam.

Trachea. G., *tracheia,* windpipe.

Trans-. L., across.

Trema (pl. **-ata**). G., hole, cavity.

Triceps. L., *tres,* three, + *caput,* head.

Tricho-, -trichia. G., *thrix* (gen. *trichos*), hair.

Trigeminus. L., born three together.

Triplo-. G., *triploos,* triple.

Triquetrum. L., *triquetrus,* three cornered, triangular.

Trochanter. G., *trochos,* wheel, pulley.

Trochlea. L. (from G., *trochilia,* pulley), pulley.

Trophoblast. G., *trophe,* nourishment, + *blastos,* shoot, germ.

Tropibasic. G., *trope,* a turning, + L., *basis,* base.

Tuberculum. L., a small hump.

Tunica. L., undergarment.

Turbinal. L., *turbo* (gen. *turbinis*), a top, anything that spins or shows turning.

Tympanum. L., drum.

-ula, -ulus. L., diminutive.

Ultimo-. L., *ultimus,* furthest, last.

Umbilical. L., *umbilicus,* navel.

Umbo-. L., shield.

Unciform. L., *uncus,* hook, + *forma,* shape.

Uncinate. L., *uncinatus,* furnished with a hook.

Unguli-. L., *unguis,* fingernail, toenail.

Urea. G., *ouron,* urine.

Uro-. G., *oura,* tail.

Urodele. G., *oura,* tail, + *delos,* evident.

Urodeum. G., *ouron,* urine, + *odaios,* on the way.

Urogenital. G., *ouron,* urine, + L., *genitalis,* genital.

Uropygial. G., *oura,* tail, + *pyge,* rump.

Urostyle. G., *oura,* tail, + *stylos,* pillar.

Uterus. L., *uterus,* womb.

Utriculus. L., small skin or leather bottle.

Vagina. L., sheath.

Vagus. L., wandering.

Vallate. L., *vallare,* to surround with a rampart.

Valvula. L., a little fold or valve.

Vas (pl. **vasa**). L., vessel or bowl.

Vascular. L., *vasculum,* small vessel.

Velum. L., veil.

Ventral. L., *venter* (gen. *ventris*), belly.

Ventricle. L., *ventriculus,* little cavity, loculus.

Vermiform. L., *vermis,* worm, + *forma,* shape.

Vertebra. L., joint.

Vesicle. L., *vesica* (dim. *vescicula*), bladder.

Vestibulum. L., entrance court.

Vibrissa. L., hair in the nostril.

Villus. L., shaggy hair.

Visceral. L., *viscus* (pl. *viscera*), entrails, bowels.

Vitelline. L., *vitellus,* yolk of egg.

Vitreus, -a, -um. L., glass.

Viviparous. L., *vivus,* living, + *parere,* to beget.

Vomer. L., ploughshare.

Vore, -vorous. L., *vorare,* to eat.

Xiphiplastron. G., *xiphos,* sword, + F., *plastron,* breastplate.

Zygapophysis. G., *zygon,* yoke, + *apophysis* (*apo,* away, + *physis,* growth), process of a bone.

Zygomatic. G., *zygon,* yoke.

Latin Word Endings

Although scientific terms are often used in English form, some knowledge of the use of these words in Latin form is desirable. Latin is a highly inflected language, with a variable series of terminations for nouns and adjectives expressing not only singular and plural numbers but also genders (of an artificial nature) and a variety of cases; still further, there are several different systems of forming such terminations ("declensions"). Fortunately, however, almost all use of scientific terms involves only two cases: nominative and genitive. Fewer than a score of endings affixed to the root word will cover most instances.

Adjectives (which must agree in gender, number, and case with their nouns) are "declined" according to one of the two following schemes, for each of which a common adjective is used as an example (the ending, attached to the root, is in boldface).

Most nouns follow one of these same schemes. Thus, *fibula* is a feminine noun of the first declension and is declined *fibula, fibulae, fibulae, fibularum; humerus* is a masculine noun of the second declension, declined *humerus, humeri, humeri, humerorum; sternum, sterna, sterni, sternorum,* a neuter noun of the second declension; *cutis, cutes, cutis, cutium* (skin), a feminine noun of the third declension.

There are, however, two complications: (1) in the third declension, most nouns have a short form for the nominative singular, a longer root for the other case endings. Thus *femur* (third declension neuter) becomes *femora,* and so on, in other cases; other typical examples are *meninx, meninges; foramen, foramina; caput, capita.* (2) A few nouns used anatomically belong to a further declension—a fourth declension. Of masculine words of this gender—*plexus* and *meatus* are examples— the plural spelling is the same as the singular; hence the English form is preferable for common use. A common neuter noun of this declension is *cornu* (horn), declined *cornu, cornua, cornus, cornuum.*

FIRST AND SECOND DECLENSION (Combined)

	Masculine	Neuter	Feminine
Nominative singular	magnus	magnum	magna
Nominative plural	magni	magna	magnae
Genitive singular	magni		magnae
Genitive plural	magnorum		magnarum

THIRD DECLENSION

	Masculine and Feminine	Neuter
Nominative singular	grandis	grande
Nominative plural	grandes	grandia
Genitive singular	grandis	
Genitive plural	grandium	

Appendix 3

References

A few of the more useful general works, or works on special topics or animal types, review articles, and a limited number of original research papers and monographs are listed here. To look further into the literature of any special topic, these two publications are most useful:

Zoological Record, 1864–date. London. Each annual volume lists all papers published during the year concerning each class of vertebrates and follows this with classified lists of those papers that deal with various topics in anatomy, embryology, and so forth.

Biological Abstracts, 1926–date. Philadelphia. A voluminous journal that attempts to abstract and index all papers published in any field of biology.

General

Alexander, R. McN. 1975. The Chordates. London, Cambridge University Press.

Bolk, L., et al., eds. 1931–1939. Handbuch der vergleichenden Anatomie der Wirbeltiere. 6 vols. Berlin and Vienna, Urban und Schwartzenberg. The classic reference for comparative anatomy, with summaries of all that was known to its date.

Bronn, H. G., et al. 1874–date. Klassen und Ordnungen des Thier-Reichs. Leipzig and Heidelberg, Winter. A voluminous work by various authors, published in parts, some old, some new, some as yet incomplete, which gives great attention to the anatomy of the various vertebrate groups as well as to classification and distribution.

Cuvier, G. 1805. Leçons d'Anatomie Comparée. 5 vols. Paris, Baudouin. The first great comparative anatomy.

Gaunt, A. S., et al. 1980. Morphology and analysis of adaptation. Am. Zool. 20:213–314. A recent symposium dealing with a variety of topics.

Giersberg, H., and Rietschel, P. 1967–1968. Vergleichende Anatomie der Wirbeltiere. 2 vols. Jena, Gustav Fischer. The first two volumes of a proposed three; an excellent and comprehensive work.

Goodrich, E. S. 1930. Studies on the Structure and Development of Vertebrates. London, The Macmillan Company. A stimulating discussion of many anatomic problems by a first rate authority. Reprinted by Dover Publications, New York, 1958.

Grassé, P.-P., ed. 1948–date. Traité de Zoologie, Anatomie, Systématique, Biologie. Paris, Masson et Cie. Not yet complete. Vol. XI treats of lower chordates; vols. XII-XVII of vertebrates.

Gregory, W. K. 1951. Evolution Emerging. 2 vols. New York, The Macmillan Company. Extremely valuable illustrations, although now somewhat out of date. Reprinted by Arno Press, New York, 1974.

Hildebrand, M. 1968. Anatomical Preparations. Berkeley and Los Angeles, University of California Press. An excellent book on how to make all sorts of preparations.

Kükenthal, W., and Krumbach, T., eds. 1923–date. Handbuch der Zoologie. Berlin and Zeipzig, W. de Gruyter and Company. A huge compendium like the Bronn, also incomplete.

Marinelli, W., and Strenger, A. 1954–1973.

Vergleichende Anatomie und Morphologie der Wirbeltiere. Wien, Franz Deuticke. Parts so far issued treat of cyclostomes, *Squalus,* and *Acipenser.*

Orr, R. T. 1982. Vertebrate Biology. 5th ed. Philadelphia, Saunders College Publishing. More natural history than anatomy.

Owen, R. 1866–1868. On the Anatomy of Vertebrates. 3 vols. London, Longmans, Green. Reprinted by AMS Press, New York, 1982. A classic, full of original observations.

Young, J. Z. 1981. The Life of Vertebrates. 3rd ed. Oxford, Clarendon Press. An excellent, group-by-group account, not only of structure but also of life habits and functions of the vertebrates.

Young, J. Z., and Hobbs, M. J. 1975. The Life of Mammals: Their Anatomy and Physiology. Oxford, Oxford University Press. A companion to the preceding, stressing anatomy and physiology, and including material on lower vertebrates.

Willson, M. E. 1984. Vertebrate Natural History. Philadelphia, Saunders College Publishing.

Functional Morphology

Alexander, R. McN. 1982. Locomotion of Animals. London, Chapman and Hall.

Alexander, R. McN., and Goldspink, G., eds. 1977. Mechanics and Energetics of Animal Locomotion. New York, Halsted Press.

Day, M. H., ed. 1981. Vertebrate Locomotion. Symp. Zool. Soc. Lond., No. 48. London and New York, Academic Press.

Denison, R. H., et al. 1961. Evolution and dynamics of vertebrate feeding mechanisms. Am. Zool. *1*:177–234.

Gans, C. 1974. Biomechanics. An Approach to Vertebrate Biology. Philadelphia, J. B. Lippincott Company. Reprinted by University of Michigan Press, Ann Arbor, 1980. A brief description of well-selected examples by a leader in the field.

Gans, C., and Bock, W. J. 1965. The functional significance of muscle architecture—a theoretical analysis. Ergeb. Anat. Entwickl. *38*:116–142.

Gray, J. 1968. Animal Locomotion. London, Weidenfeld and Nicolson. An extensive and detailed summary of the work done in this area by the most active group.

Hildebrand, M. 1981. Analysis of Vertebrate Structure. 2nd ed. New York, John Wiley and Sons. A basic comparative anatomy text with extensive discussions of functional morphology. especially of locomotor mechanisms.

Howell, A. B. 1944. Speed in Animals. Their Specializations for Running and Leaping. Chicago, University of Chicago Press. Reprinted by Hafner, New York, 1965.

Liem, K. F. 1970. Comparative functional anatomy of the Nandidae (Pisces: Teleostei). Fieldiana: Zool. *56*: 1–166.

Nursall, J. R., et al. 1962. Vertebrate locomotion. Am. Zool. *2*:127–208.

Smith, J. M., and Savage, R. J. C. 1956. Some locomotory adaptations in mammals. J. Linnean Soc. (London) *42*:603–622. An older paper, but a classic in the field.

Thompson, D'A. W. 1942. On Growth and Form. 2nd ed. Cambridge, England, Cambridge University Press. (An abbreviated edition was published in 1961.) A classic by a master mathematician, zoologist, and Greek scholar.

Tricker, R. A. R., and Tricker, B. J. K. 1967. The Science of Movement. New York, American Elsevier.

Physiology

Brobeck, J. K., ed. 1979. Best and Taylor's Physiological Basis of Medical Practice. 10th ed. Baltimore, Williams & Wilkins.

Buddenbrock, W. von. 1950–1961. Vergleichende Physiologie. 5 vols. Basel, Birkhäusen.

Field, J., et al., eds. 1959–date. Handbook of Physiology. Washington, American Physiological Society. A huge, still incomplete survey in many volumes; largely but far from entirely human.

Florey, E. 1977. General and Comparative Animal Physiology. 2nd ed. Philadelphia, W. B. Saunders Company.

Gordon, M. S., et al. 1982. Animal Physiology: Principles and Adaptations. 4th ed. New York, Macmillan, Inc.

Guyton, A. C. 1981. Textbook of Medical Physiology. 6th ed. Philadelphia, W. B. Saunders Company.

Hoar, W. S. 1975. General and Comparative Physiology. 2nd ed. Englewood Cliffs, N.J., Prentice-Hall, Inc. This and the following include much data on vertebrate physiology and anatomy.

Prosser, C. L. 1973. Comparative Animal Physiology. 3rd ed. Philadelphia, W. B. Saunders Company.

Ruch, T. C., and Patton, H. D., eds. 1973–1979. Medical Physiology and Biophysics. (20th edition of Howell's Textbook of Physiology.) Philadelphia, W. B. Saunders Company. Essentially human physiology alone.

Vander, A. J., et al. 1980. Human Physiology. The Mechanisms of Body Function. 3rd ed. New York, McGraw-Hill.

Evolutionary and Taxonomic Theory

The references here are only a few of the major sources in this field; there are many other textbooks, including several in paperback editions.

Cracraft, J., and Eldredge, N., eds. 1979. Phylogenetic Analysis and Paleontology. New York, Columbia University Press.

Darwin, C. 1859. On the Origin of Species. London, John Murray. (Reprinted in many versions by many companies.) The original classic, and still well worth reading.

Dobzhansky, T. 1970. Genetics of the Evolutionary Process. New York, Columbia University Press.

Dobzhansky, T., et al. 1977. Evolution. San Francisco, W. H. Freeman.

Futuyma, D. J. 1979. Evolutionary Biology. Sunderland, Mass., Sinauer Associates.

Mayr, E. 1963. Animal Species and Evolution. Cambridge, The Belknap Press of Harvard University Press.

Minkoff, E. C. 1983. Evolutionary Biology. Reading, Mass., Addison-Wesley Publishing Company.

Rensch, B. 1960. Evolution above the Species Level. New York, Columbia University Press.

Simpson, G. G. 1953. The Major Features of Evolution. New York. Columbia University Press.

Simpson, G. G. 1961. Principles of Animal Taxonomy. New York, Columbia University Press.

Sneath, P. H. A., and Sokal, R. R. 1973. Numerical Taxonomy, the Principles and Practice of Numerical Classification. San Francisco, W.H. Freeman.

Paleontology

Colbert, E. H. 1980. Evolution of the Vertebrates. 3rd ed. New York, John Wiley and Sons.

Piveteau, J. 1952–1968. Traité de Paléontologie. 7 vols. Paris, Masson et Cie. A comprehensive work, four volumes of which treat of vertebrates.

Romer, A. S. 1966. Vertebrate Paleontology. 3rd ed. Chicago, University of Chicago Press. The standard text on this topic.

Romer, A. S. 1971. The Vertebrate Story. (Rev. ed.) Chicago, University of Chicago Press. An elementary account of vertebrate evolution.

Stahl, B. J. 1974. Vertebrate History: Problems in Evolution. New York, McGraw-Hill. An interesting book that stresses disagreements and unsettled questions rather than simply presenting the "party line."

Lower Chordates

Barrington, E. J. W. 1965. The Biology of the Hemichordata and Protochordata. Edinburgh, Oliver and Boyd; San Francisco, W. H. Freeman and Company.

Barrington, E. J. W., and Jefferies, R. P. S., eds. 1975. Protochordates. Symposia of the Zoological Society of London, No. 36.

Berrill, N. J. 1950. The Tunicata. London, The Ray Society.

Berrill, N. J. 1955. The Origin of the Vertebrates. London, Oxford University Press.

Garstang, W. 1928. The morphology of Tunicata. Quart. J. Microsc. Sci. 72:51–187.

Grassé, P. -P., ed. 1948. Traité de Zoologie, Tome XI. Échinodermes-Stomochordés-Prochordés. Paris, Masson et Cie. Contains a comprehensive account of lower chordates by Dawydoff, Brien, Drach, and others.

Fishes

Alexander, R. McN. 1974. Functional Design in Fishes. 3rd ed. London, Hutchinson and Company.

Allis, E. P., Jr. 1897. The cranial muscles and cranial and first spinal nerves in *Amia calva.* J. Morphol. *12*:487–808. This and further works by Allis cited below are well-illustrated accounts of cranial anatomy.

Allis, E. P., Jr. 1903. The skull and cranial and first spinal muscles and nerves in *Scomber scomber.* J. Morphol. *18*:45–328.

Allis, E. P., Jr. 1909. The cranial anatomy of the mail-cheeked fishes. Zoologica (Stuttgart) *22*:1–219.

Allis, E. P., Jr. 1922. Cranial anatomy of *Polypterus.* J. Anat. *56*:189–294.

Allis, E. P., Jr. 1923. The cranial anatomy of *Chlamydoselachus anguineus.* Acta Zoologica *4*:123–221.

Berg, L. S. 1947. Classification of Fishes, Both Recent and Fossil. Ann Arbor, Edwards Brothers. A translation of a Russian original. A new edition in Russian published in 1949, translated into German in 1958.

Bond, C. 1979. The Biology of Fishes. Philadelphia, Saunders College Publishing.

Brodal, A., and Fänge, R., eds. 1963. The Biology of *Myxine.* Oslo, Universitetsforlaget.

Brown, M. E., ed. 1957. The Physiology of Fishes. 2 vols. New York, Academic Press. Despite the limitation of the title, it gives in the main a comprehensive account of fish biology and anatomy, although now largely replaced by Hoar and Randall (see below).

Budker, P. 1971. The Life of Sharks. London, Weidenfeld and Nicolson.

Daniel, J. F. 1934. The Elasmobranch Fishes. 3rd ed. Berkeley, University of California Press. Shark anatomy.

Dean, B. 1895. Fishes, Living and Fossil. New York, Macmillan. Old, but still valuable for Recent forms.

Dean, B. 1906. Chimaeroid fishes. Carnegie Institution of Washington, Publication 32.

Dean, B. 1916–1923. A Bibliography of Fishes. 3 vols. New York, American Museum of Natural History. Reprinted by Lubrecht and Cramer, Monticello, New York, 1973.

Gans, C., and Parsons, T. S. 1964. A photographic Atlas of Shark Anatomy. New York. Academic Press. Reprinted by University of Chicago Press, Chicago, 1981.

Gilbert, P. W., Mathewson, R. F., and Rall, D. P. 1967. Sharks, Skates, and Rays. Baltimore, Johns Hopkins Press.

Goodrich, E. S. 1909. A Treatise on Zoology, edited by E. Ray Lankester. Part IX. Vertebrata Craniata, Fascicule I. "Cyclostomes and Fishes." London, The Macmillan Company. A mine of data on fish anatomy; badly indexed, however.

Greenwood, P. H., and Norman, J. R. 1976. A History of Fishes. 3rd ed. New York, Halsted Press. Life history, habits, as well as structure.

Greenwood, P. H., et al., eds. 1973. Interrelationships of Fishes. (Supplement No. 1 to the Zoological Journal of the Linnean Society, Vol. 53, 1973.) London, Academic Press. A fairly recent survey with papers by many leading workers.

Harder, W. 1975. Anatomy of Fishes. Parts I and II. Stuttgart, E. Schweizerbart.

Hardisty, M. W., and Potter, I. C., eds. 1971–date. The Biology of Lampreys. New York and London, Academic Press.

Hoar, W. S., and Randall, D. J., eds. 1969–1979. Fish Physiology. 8 vols. New York, Academic Press. Mainly physiologic, but a mine of information on many aspects of the biology of fishes.

Jarvik, E. 1980. Basic Structure and Evolution of Vertebrates. 2 vols. New York and London, Academic Press. Primarily a detailed study of certain extinct or primitive fishes.

Lagler, K. F., et al. 1977. Ichthyology. 2nd ed. New York, John Wiley & Sons, Inc.

Liem, K. F., and Lauder, G. V., eds. 1982. Evolutionary morphology of the actinopterygian fishes. Am. Zool. *22*:237–345.

Marshall, N. B. 1965. The Life of Fishes. London, Weidenfeld and Nicolson.

Millot, J. 1954. Le troisième coelacanthe. Le Naturaliste Malagache. 1er Supplement. Superficial structures of *Latimeria.*

Millot, J. 1955. The coelacanth. Sci. Am. *193*(6):34–39.

Millot, J., and Anthony, J. 1958–date. Anatomie de *Latimeria chalumnae.* Paris, Centre National de la Recherche Scientifique.

Moy-Thomas, J. A., and Miles, R. S. 1971. Palaeozoic Fishes. 2nd ed. Philadelphia, W. B. Saunders Company. An excellent and up-to-date summary.

Northcutt, R. G., ed. 1977. Recent advances in the biology of sharks. Am. Zool. *27*:287–515.

Romer, A. S. 1946. The early evolution of fishes. Quart. Rev. Biol. *21*:33–69.

Thomson, K. S. 1969. The biology of lobe-finned fishes. Biol. Rev. *44*:91–154.

Amphibia

Ecker, A., Wiedersheim, R., and Gaupp, E. 1888–1904. Anatomie des Frosches. 3 vols., 2nd ed. Braunschweig, Friedrich Viewig und Sohn. A thorough account of frog anatomy, which has passed through the hands of three successive authors.

Francis, E. T. B. 1934. The Anatomy of the Salamander. London and New York, Oxford University Press.

Goin, C. J., and Goin, O. B. 1978. Introduction to Herpetology. 3rd ed. San Francisco, W. H. Freeman and Company.

Holmes, S. J. 1927. The Biology of the Frog. 4th ed. New York. Macmillan Company.

Lofts, B., ed. 1974–1976. Physiology of the Amphibia. 3 vols. New York and London, Academic Press.

Moore, J. A., ed. 1964. Physiology of the Amphibia. New York, Academic Press.

Noble, G. K. 1931. The Biology of the Amphibia. New York, McGraw-Hill. Reprinted by Dover Publications, New York, 1954. Still probably the best general reference on the modern amphibians.

Panchen, A. L., ed. 1981. The Terrestrial Environment and the Origin of Land Vertebrates. London and New York, Academic Press. A recent collection of papers on many aspects of this problem.

Parsons, T. S., and Williams, E. E. 1963. The relationships of the modern Amphibia: a re-examination. Quart. Rev. Biol. *38*:26–53.

Porter, K. R. 1972. Herpetology. Philadelphia, W. B. Saunders Company.

Romer, A. S. 1947. Review of the Labyrinthodontia. Bulletin, Museum of Comparative Zoology, Harvard, *99*:1–368.

Schmalhausen, I. I. 1968. The Origin of Terrestrial Vertebrates. New York, Academic Press. Translated from the Russian by L. Kelso; edited by K. S. Thomson.

Špinar, Z. V. 1972. Tertiary Frogs from Central Europe. The Hague, Dr. W. Junk N. V. Some of these excellent fossils have preserved soft parts.

Vial, J. L., ed. 1973. Evolutionary Biology of the Anurans. Columbia, University of Missouri Press. An excellent summary of many aspects.

Wiedersheim, R. 1879. Die Anatomie der Gymnophionen. Jena, Gustav Fischer.

Reptiles

See also the references by Goin and Goin and by Porter, listed above under Amphibia.

Bellairs, A. d'A. 1969. The Life of Reptiles. 2 vols. London, Weidenfeld and Nicolson.

Bellairs, A. d'A., and Underwood, G. 1951. The origin of snakes. Biol. Rev. *26*:193–237.

Bojanus, L. H. 1819–1821. Anatome Testudinis Europaeae. Vilnae, Josephi Zawadzki. Reprinted by the Society for the Study of Amphibians and Reptiles. Essentially a series of plates, probably the best yet made of any submammalian form.

Colbert, E. H. 1961. Dinosaurs, Their Discovery and Their World. New York, E. P. Dutton and Company.

Gans, C., et al., eds. 1969–date. Biology of the Reptilia. London, Academic Press. A large series, still far from complete, covering almost all aspects of reptilian biology.

Kemp, T. S. 1982. Mammal-like Reptiles and the Origin of Mammals. London and New York, Academic Press.

McGowan, C. 1983. The Successful Dragons. Toronto and Sarasota, Samuel Stevens. An interesting account of biomechanical and other functional aspects of the structure of extinct reptiles.

Romer, A. S. 1956. Osteology of the Reptiles. Chicago, University of Chicago Press. The basic reference for reptilian classification, as well as osteology.

Thomas, R. D. K., and Olson, E. C., eds. 1980. A Cold Look at the Warm-Blooded Dinosaurs. Boulder, Co., Westview Press. A collection of papers discussing many aspects and all sides of this interesting debate.

Underwood, G. 1967. A Contribution to the Classification of Snakes. London, Trustees of the British Museum (Natural History).

Williston, S. W. 1914. Water Reptiles of the

Past and Present. Chicago, University of Chicago Press.

Birds

Baumel, J. J. et al., eds. 1979. Nomina Anatomica Avium. London, Academic Press. Includes some discussion as well as a list of names.

Bowman, R. I. 1961. Morphological differentiation and adaptation in the Galápagos finches. University of California Publications in Zoölogy, Vol. 58.

Bradley, O. C. 1960. The Structure of the Fowl. 4th ed. Edinburgh and London, Oliver and Boyd, Ltd.

Chamberlain, I. W. 1943. Atlas of Avian Anatomy. East Lansing, Michigan State College, Agricultural Experiment Station.

De Beer, G. 1954. *Archaeopteryx lithographica.* London, British Museum (Natural History).

Dorst, J. 1974. The Life of Birds. 2 vols. New York, Columbia University Press. Mainly natural history.

Farner, D. S., et al., eds. 1971–1982. Avian Biology. 6 vols. New York, Academic Press. The major modern summary of anatomy and physiology.

Feduccia, A. 1980. The Age of Birds. Cambridge, Mass., Harvard University Press. A recent and readable account of avian origins, paleontology, and evolution.

Fürbringer, M. 1888. Untersuchungen zur Morphologie und Systematik der Vögel. Zugleich ein Beitrag zur Anatomie des Stütz- und Bewegungsorgane. 2 vols. Amsterdam and Jena, Gustav Fischer. Old but still basic work, including excellent comparative anatomic data.

Heilmann, G. 1926. The Origin of Birds. New York, D. Appleton-Century Company. Reprinted by Dover Publications.

King, A. S., and McLelland, J., eds. 1979–date. Form and Function in Birds. London and New York, Academic Press.

Koch, T. 1973. Anatomy of the Chicken and Domestic Birds. Ames, Iowa State University Press.

Lucas, A. M., and Stettenheim, P. R. 1972. Avian Anatomy. Integument. Washington, D.C., U.S. Dept. of Agriculture Handbook, 362.

Marshall, A. J., ed. 1960–1961. Biology and Comparative Physiology of Birds. 2 vols. New York, Academic Press. Still very useful, though partly replaced by Farner, et al. (see above).

Newton, A., and Gadow, H. 1893–1896. A Dictionary of Birds. London, Adam and Charles Black. Old but still useful.

Nickel, R., et al. 1977. Anatomy of the Domestic Birds. Berlin, Springer-Verlag.

Pycraft, W. P. 1910. A History of Birds. London, Methuen and Company. Includes anatomy.

Strong, R. M. 1939–1959. A bibliography of birds. Publication Field Museum of Natural History, Zoology, 25.

Sturkie, P. D. 1975. Avian Physiology, 3rd, ed. Ithaca, N.Y., Comstock Publishing Associates.

Thomson, A. L., ed. 1964. A New Dictionary of Birds. London, Thomas Nelson & Sons.

Webb, M. 1957. The ontogeny of the cranial bones, cranial peripheral, and cranial parasympathetic nerves, together with a study of the visceral muscles *of Struthio.* Acta Zoologica *38*:81–203.

Welty, J. C. 1982. The Life of Birds. 3rd ed. Philadelphia, Saunders College Publishing. Probably the best single volume on all aspects of birds.

Mammals

Anderson, S., and Jones, J. K., Jr., eds. 1967. Recent Mammals of the World—A Synopsis of Families. New York, The Ronald Press.

Baum, H., and Zietzschmann, O. 1936. Handbuch der Anatomie des Hundes. Berlin, P. Parey.

Bensley, B. A., and Craigie, E. H. 1938. Practical Anatomy of the Rabbit. 8th ed. Philadelphia, Blakiston Company.

Bourlière, F. 1954. The Natural History of Mammals. New York, Alfred A. Knopf.

Bradley, O. C., and Grahame, T. 1943. Topographical Anatomy of the Dog. 5th ed. New York, The Macmillan Company.

Crouch, J. E. 1969. Text-Atlas of Cat Anatomy. Philadelphia, Lea & Febiger.

Davis, D. D. 1964. The giant panda. A study

of evolutionary mechanisms. Fieldiana, Zoology, Mem. 3.

Davison, A., and Stromsten, F. A. 1937. Mammalian Anatomy, with Special Reference to the Cat. 7th ed. Philadelphia, Blakiston Company.

Ewer, R. F. 1973. The Carnivores. Ithaca, N.Y., Cornell University Press.

Field, H. E., and Taylor, M. E. 1969. An Atlas of Cat Anatomy. (Rev. ed.) Chicago, University of Chicago Press.

Flower, W. H., and Lydekker, R. 1891. An Introduction to the Study of Mammals, Living and Extinct. London, Adam and Charles Black. Reprinted by Arno, New York, 1978. Old, but still useful.

Gerhardt, U. 1909. Das Kaninchen. Leipzig, H. E. Ziegler, and R. Woltereck.

Getty, R. 1975. Sisson & Grossman's Anatomy of the Domestic Animal. 5th ed., 2 vols. Philadelphia, W. B. Saunders Company. Comprehensive accounts of horse, ox, sheep, goat, dog, cat, and domestic birds.

Goffart, M. 1971. Function and Form in the Sloth. Oxford, Pergamon Press.

Goss, C. M., ed. 1973. Gray's Anatomy. 29th ed. Philadelphia, Lea & Febiger. One of several standard human anatomies; others are those of Morris and Cunningham.

Greene, E. G. 1935. Anatomy of the rat. Tr. Am. Philosophical Soc. (n.s.) 27:1–370. Reprint by Hafner, New York, 1971.

Griffiths, M. 1968. Echidnas. Oxford, Pergamon Press.

Griffiths, M. 1978. The Biology of the Monotremes. New York and London, Academic Press.

Gunderson, H. L. 1976. Mammalogy. New York, McGraw-Hill.

Harrison, R. J., ed. 1972–date. Functional Anatomy of Marine Mammals. London and New York, Academic Press.

Hartman, C. G., and Straus, W. L., Jr., eds. 1933. The Anatomy of the Rhesus Monkey. Baltimore. Williams & Wilkins.

Hebel, R., and Stromberg, M. W. 1976. Anatomy of the Laboratory Rat. Baltimore, Williams & Wilkins.

Hill, W. C. O. 1953–1974. Primates: Comparative Anatomy and Taxonomy. 8 vols. Edinburg, University Press.

Hofer, H., et al., eds. 1956–date. Primatologia. Handbook of Primatologia. New York, S. Karger.

Howell, A. B. 1930. Aquatic Mammals. Springfield, Charles C. Thomas. Reprinted by Dover, 1970.

LeGros Clark, W. E. 1926. On the anatomy of the pen-tailed tree-shrew (Ptilocercus lowii). Proc. Zool. Soc. London 1926:1179–1309.

LeGros Clark, W. E. 1934. Early Forerunners of Man. Baltimore, Williams & Wilkins. A discussion of the anatomy of lower primates.

LeGros Clark, W. E. 1958. History of the Primates. 6th ed. London, British Museum (Natural History).

LeGros Clark, W. E. 1971. The Antecedents of Man. Rev. ed. New York, Time Books.

Lillegraven, J. A., et al., eds. 1979. Mesozoic Mammals: The First Two Thirds of Mammalian History. Berkeley and Los Angeles, University of California Press.

Matthews, L. H. 1969. The Life of Mammals. 2 vols. London, Weidenfeld and Nicolson.

Miller, M. E., et al. 1964. Anatomy of the Dog. Philadelphia, W. B. Saunders Company. One of the most detailed descriptions of a mammal other than man.

Nickel, R., Schummer, A., and Seiferle, E. 1960, 1961. Lehrbuch der Anatomie der Haustiere. Bd. 1; Bewegungsapparat. Bd. 2: Eingeweide. 2nd ed. Berlin and Hamburg, Parey.

Norris, K. 1966. Whales, Dolphins, and Porpoises. Berkeley, University of California Press.

Reighard, J. E., and Jennings, H. S. 1935. Anatomy of the Cat. 3rd ed. New York, Henry Holt and Company, Inc.

Schultz, A. H. 1969. The Life of Primates. London, Weidenfeld and Nicolson.

Sharman, G. B. 1970. Reproductive physiology of marsupials. Science 167:1221–1228.

Simpson, G. G. 1945. The principles of classification and a classification of mammals. Bulletin, American Museum of Natural History 88:1–350. The standard reference, currently being revised.

Slijper, E. J. 1962. Whales. London, Hutchinson and Company.

Thenius, E. 1972. Grundzüge der Verbreitungsgeschichte der Säugetiere. Jena, Gustav Fischer.

Vaughan, T. A. 1978. Mammalogy. 2nd ed. Philadelphia, W. B. Saunders Company.

Weber, M., Burlet, H. M. de, and Abel, O. 1927–1928. Die Säugetiere. 2 vols., 2nd ed. Jena, Gustav Fischer. A standard work on mammalian anatomy and classification.

Wimsatt, W. A., ed. 1970. Biology of Bats. 2 vols. New York, Academic Press.

Woollard, H. H. 1936. The anatomy of *Tarsius spectrum*. Proc. Zool. Soc. London *70*:1071–1184.

Zietzschmann, H. C. O., et al., eds. 1943. Ellenberger-Baum: Handbuch der vergleichenden Anatomie der Haustiere. 18th ed. Berlin, Springer-Verlag. The standard anatomy of domestic animals on which almost all others are based.

Cells and Tissues

Andrew, W., and Hickman, C. P. 1974. Histology of the Vertebrates. A Comparative Text. St. Louis, C. V. Mosby Company.

Bloom, W., and Fawcett, D. 1975. Textbook of Histology. 10th ed. Philadelphia, W. B. Saunders Company. One of the standard medical texts.

Brachet, J., and Mirsky, A. E., eds. 1964. The Cell: Biochemistry, Physiology, Morphology. 6 vols. New York, Academic Press.

Greep, R. O., and Weiss, L. 1973. Histology. 3rd ed. New York, McGraw-Hill.

Ham, A. W. and Cormack, D. H. 1979. Histology. 8th ed. Philadelphia, J. B. Lippincott Co.

Hammerson, F. 1980. Sobotta/Hammerson, Histology. 2nd ed. Baltimore, Urban and Schwarzenberg.

Kurtz, S. M., ed. 1964. Electron Microscopic Anatomy. New York, Academic Press.

Leeson, C. R., and Leeson, T. S. 1976. Histology. 3rd ed. Philadelphia, W. B. Saunders Company.

LeGros Clark, W. E. 1971. The Tissues of the Body. 6th ed. London, Oxford University Press. A text that stresses aspects of interest to the gross anatomist.

Patt, D. I., and Patt, G. R. 1969. Comparative Vertebrate Histology. New York, Harper & Row.

Embryology

Arey, L. B. 1974. Developmental Anatomy. 7th ed.(rev.). Philadelphia, W. B. Saunders Company. Entirely human.

Balinsky, B. I. 1981. An Introduction to Embryology. 5th ed. Philadelphia, Saunders College Publishing. Probably the best standard text.

Bellairs, R. 1971. Developmental Processes in Higher Vertebrates. London, Logo Press Ltd.

Bodemer, C. W. 1968. Modern Embryology. New York, Holt, Rinehart and Winston.

Conklin, E. G. 1932. The embryology of amphioxus. J. Morphol. *54*:69–151.

Damas, H. 1944. Recherches sur le developpement de *Lampetra fluviatilis* L. Contribution à l'etude de la céphalogenèse des vertébrés. Arch. Biol. *55*:1–284.

DeBeer, G. R. 1958. Embryos and Ancestors. 3rd ed. London and New York, Oxford University Press. A masterful and classic statement of the relationships among anatomy, embryology, and phylogeny.

De Haan, R. L., and Ursprung, H., eds. 1965. Organogenesis. New York, Holt, Rinehart and Winston.

Gould, S. J. 1977. Ontogeny and Phylogeny. Cambridge, Harvard University Press.

Hamilton, W. J., Boyd, J. D., and Mossman, H. W. 1972. Human Embryology. 4th ed. Baltimore, Williams & Wilkins.

Hopper, A. F., and Hart, N. H. 1980. Foundations of Animal Development. New York and Oxford, Oxford University Press.

Hörstadius, S. 1950. The Neural Crest. London, Oxford University Press.

Lillie, F. R. 1952. Development of the Chick. Revised and edited by H. L. Hamilton. New York, Holt.

Nelsen, O. 1953. Comparative Embryology of Vertebrates. New York, Blakiston Company.

Northcutt, R. G., and Gans, C. 1983. The genesis of neural crest and epidermal placodes: A reinterpretation of vertebrate origins. Quart. Rev. Biol. *54*:1–28.

Patten, B. M. 1948. Early Embryology of the Pig. 3rd ed. New York, McGraw-Hill.

Patten, B. M. 1971. Early Embryology of the Chick. 5th ed. New York, McGraw-Hill.

Patten, B. M., and Carlson, B. M. 1974. Foundations of Embryology. 3rd ed. New York, McGraw-Hill.

Romanoff, A. L. 1960. The Avian Embryo. New York, The Macmillan Company.

Saunders, J. W. 1982. Developmental Biology: Patterns, Problems and Principles. New York, Macmillan.

Starck, D. 1955. Embryologie. Stuttgart, Georg Thieme Verlag. Some emphasis on man but includes vertebrates generally.

Torrey, T. W., and Feduccia, A. 1979. Morphogenesis of the Vertebrates. 4th ed. New York, John Wiley & Sons.

Waddington, C. H. 1952. The Epigenetics of Birds. London, Cambridge University Press.

Willier, B. H., Weiss, P. A., and Hamburger, V., eds. 1955. Analysis of Development. Philadelphia, W. B. Saunders Company. Reprinted by Hafner, New York, 1971.

Skin

Bagnara, J. T., and Hadley, M. E. 1973. Chromatophores and Color Change. Englewood Cliffs, N.J., Prentice-Hall, Inc.

Elias, H., and Bortner, S. 1957. On the phylogeny of hair. Am. Mus. Novit. No. 1820:1–15.

Fingerman, M. 1965. Chromatophores. Physiol. Rev. 45:296–339.

Harvey, E. N. 1952. Bioluminescence. New York, Academic Press.

Kon, S. K., and Cowie, A. T. 1961. Milk: The Mammary Gland and its Secretion. New York, Academic Press. 2 vols.

Lillie, F. R. 1942. On the development of feathers. Biol. Rev. 17:247–266.

Maderson, P. F. A., ed. 1972. The vertebrate integument: Symposium. Am. Zool. 12:12–171.

Montagna, W., and Parakkal, P. F. 1974. The Structure and Function of Skin. 3rd ed. New York, Academic Press. (See also Sci. Am. 212[2]:56–69, 1965.)

Paris, P. 1914. Recherches sur la gland uro-pygienne des oiseaux. Arch. Zool. Exp. Gén. 53:132–276.

Rook, A. J., and Walton, G. S., eds. 1965. Comparative Physiology and Pathology of the Skin. Philadelphia, F. A. Davis Company.

Schaffer, J. 1940. Die Hautdrüsenorgane der Säugetiere. Berlin and Wien, Urban und Schwarzenberg.

Sokolov, V. E. 1982. Mammal Skin. Berkeley and Los Angeles, University of California Press.

Skeleton

Brien, P. 1962. Etude de la formation, de la structure des écailles des Dipneustes actuels et de leur comparaison avec les autres types d'écailles des poissons. Annalen Koninklijk Mus. Midden-Afrika, Ser. 8°, Zool. Wetensch., No. 108, pp. 53–125.

Denison, R. H. 1951. The exoskeleton of early Osteostraci. Fieldiana, Geol. 11:197–218.

Fisher, H. I. 1946. Adaptations and comparative anatomy of the locomotor apparatus of New World vultures. Am. Midland Nat. 35:545–727.

Flower, W. H. 1885. An Introduction to the Osteology of the Mammalia. 3rd ed. London, The Macmillan Company. An old but useful little book. Reissued by Dover Press, New York, 1962.

Goodrich, E. S. 1904. On the dermal fin-rays of fishes—living and extinct. Quart. J. Microsc. Sci. 47:465–522.

Goodrich, E. S. 1908. On the scales of fish, living and extinct, and their importance in classification. Proc. Zool. Soc. London, 1908:751–774.

Gregory, W. K., Miner, R. W., and Noble, G. K. 1923. The carpus of Eryops and the primitive cheiropterygium. Bull. Am. Mus. Nat. Hist. 48:279–288.

Gregory, W. K., and Raven, H. C. 1944. Studies on the origin and early evolution of paired fins and limbs. Ann. N. Y. Acad. Sci. 42:273–360.

Haines, R. W. 1942. The evolution of epiphyses and of endochondral bone. Biol. Rev. 17:267–292.

Jayne, H. 1898. Mammalian Anatomy. Part I. The Skeleton of the Cat. Philadelphia, J. B. Lippincott Co.

Kerr, T. 1952. The scales of primitive living actinopterygians. Proc. Zool. Soc. London *122*:55–78.

Kummer, B. 1959. Bauprinzipien des Säugerskeletes. Stuttgart, Georg Thieme Verlag.

McLean, F. C., and Urist, M. R. 1968. Bone. 3rd ed. Chicago, University of Chicago Press.

Moss, M. L. 1968. Comparative anatomy of vertebrate dermal bone and teeth. Acta Anat. (Basel) *71*:178–208.

Murray, P. D. F. 1936. Bones. A Study of the Development and Structure of the Vertebrate Skeleton. London, Cambridge University Press.

Ørvig, T. 1951. Histological studies of placoderms and fossil elasmobranchs. I. The endoskeleton, with remarks on the hard tissues of lower vertebrates in general. Arkiv f. Zool. *2(2)*:321–454.

Ørvig, T. 1977. A survey of odontodes ('dermal teeth') from developmental, structural, functional, and phyletic points of view. *In* Problems in Vertebrate Evolution, Linn. Soc. Symp. Ser., No. 4. London and New York, Academic Press.

Panchen, A. L. 1977. The Origin and Early Evolution of Tetrapod Vertebrae. *In* Problems in Vertebrate Evolution, Linn. Soc. Symp. Ser., No. 4. London and New York, Academic Press.

Parker, W. K. 1868. A Monograph on the Structure and Development of the Shoulder Girdle and Sternum. London, Ray Society.

Patterson, C. 1977. Cartilage bones, dermal bones, and membrane bones, or the exoskeleton versus the endoskeleton. *In* Problems in Vertebrate Evolution, Linn. Soc. Symp. Ser., No. 4. London and New York, Academic Press.

Reynolds, S. H. 1913. The Vertebrate Skeleton. 2nd ed. Cambridge, England, Cambridge University Press.

Romer, A. S. 1942. Cartilage: An embryonic adaptation. Am. Nat. *76*:394–404.

Schaeffer, B. 1941. The morphological and functional evolution of the tarsus in am-phibians and reptiles. Bull. Am. Mus. Nat. Hist. *78*:395–472.

Schaeffer, B. 1977. The dermal skeleton in fishes. *In* Problems in Vertebrate Evolution, Linn. Soc. Symp. Ser., No. 4. London and New York, Academic Press.

Schmalhausen, J. J. 1912–1913. Zur Morphologie der unpaaren Flossen. Ztschr. Wissensch. Zool. *100*:509–587; *104*:1–80.

Shufeldt, R. W. 1909. Osteology of birds. Bull. New York State Mus. *130*:5–381.

Wake, D. B. 1970. Aspects of vertebral evolution in the modern Amphibia. Forma et Functio *3*:33–60.

Wake, D. B., and Lawson, K. 1973. Developmental and adult morphology of the vertebral column in the plethodontid salamander *Eurycea bislineata*, with comments on vertebral evolution in the amphibia. J. Morph. *139*:251–300.

Watson, D. M. S. 1917. The evolution of the tetrapod shoulder girdle and fore-limb. J. Anat. *52*:1–63.

Westoll, T. S. 1958. The lateral fin-fold theory and the pectoral fins of ostracoderms and early fishes. *In* Westoll, T. S., ed.: Studies on Fossil Vertebrates. London, University of London, pp. 180–211.

Wilder, H. H. 1903. The skeletal system of *Necturus maculatus* Rafinesque. Mem. Boston Soc. Nat. Hist. *5*:357–439.

Williams, E. E. 1959. Gadow's arcualia and the development of tetrapod vertebrae. Quart. Rev. Bio. *34*:1–32.

Skull

Bellairs, A. d'A. 1949. The anterior braincase and interorbital septum of Sauropsida with a consideration of the origin of snakes. J. Linnean Soc. London, Zool. *41*:482–512.

Brock, G. T. 1929. On the development of the skull of *Leptodeira hotamboia*. Quart. J. Microsc. Sci. *73*:289–334.

Crompton, A. W. 1953. The development of the chondrocranium of *Spheniscus demersus* with special reference to the columella auris of birds. Acta Zoologica *34*:71–146.

Crompton, A. W. 1963. The evolution of the mammalian jaw. Evolution *17*:431–439.

DeBeer, G. R. 1937. The Development of the

Vertebrate Skull. London and New York, Oxford University Press. Publication preceded by a series of detailed papers on various forms by DeBeer and colleagues. Good bibliography.

Frazzetta, T. H. 1962. A functional consideration of cranial kinesis in lizards. J. Morphol. *111*:287–319.

Frazzetta, T. H. 1968. Adaptive problems and possibilities in the temporal fenestration of tetrapod skulls. J. Morphol. *125*:145–158.

Gaupp, E. 1900. Das Chondrocranium von *Lacerta agilis.* Anat. Hefte (Arb.) *15*:433–595.

Gregory, W. K. 1933. Fish skulls: A study of the evolution of natural mechanisms. Tr. Am. Philosophical Soc. *23*:75–481.

Hofer, H. 1955. Neuere Untersuchungen zur Kopfmorphologie der Vögel. Basel, Acta 11th Congrès International d'Ornithologie 104–137.

Jollie, M. T. 1957. The head skeleton of the chicken and remarks on the anatomy of this region in other birds. J. Morphol. *100*:389–436.

Jollie, M. T. 1960. The head skeleton of the lizard. Acta Zoologica *41*:1–64.

Kampen, P. N. van. 1905. Die Tympanalgegend des Säugetierschädels. Morphol. Jahrbuch *34*:321–722.

Lakjer, T. 1927. Studien über die Gaumenregion bei Sauriern im Vergleich mit Anamniern und primitiven Sauropsiden. Zool. Jahrbücher (Anat.) *49*:57–356.

Lang, C. 1956. Das Cranium der Ratiten mit besonderer Berücksichtigung von *Struthio camelus.* Ztschr. Wissensch. Zool. *159*:165–224.

Moore, W. J. 1981. The Mammalian Skull. Cambridge, Cambridge University Press.

Parker, W. K.: Structure and development of the skull. A long series of papers on the following forms: Ostrich, Fowl, *Rana,* Batrachia, Salmon, Pig, Urodela, *Tropidonotus,* Lacertilia, *Acipenser, Lepidosteus,* Edentata and Insectivora, Birds, Sharks and Skates, Crocodilia, *Opisthocomus,* in the following journals: Philos. Trans. Roy. Soc. London *(B) 156, 159, 161, 163, 164, 167, 169, 170, 173, 176,* 1866–1885; Tr. Zool. Soc. London *9, 10, 11, 13,* 1875–1891; Tr.

Linnean Soc. London, Zool. *1, 2,* 1875–1888. Old but well illustrated and valuable.

Peyer, B. 1912. Die Entwicklung des Schädelskelettes von *Vipera aspis.* Morphol. Jahrbuch *44*:563–621.

Starck, D. 1941. Zur Morphologie des Primordialcraniums von *Manis javanica* Desm. Morphol. Jahrbuch *86*:1–122. One of a series of mammal skull studies by Starck and his students.

Versluys, J. 1912. Das Streptostylie-Problem und die Bewegungen im Schädel bei Sauropsiden. Zool. Jahrbücher Suppl. *15*(2):545–714.

Muscles

Bourne, G. H., ed 1972–date. The Structure and Function of Muscle. 4 vols. New York, Academic Press.

Braus, H. 1901. Die Muskeln und Nerven der Ceratodusflosse. Semon's Zoologische Forschungsreisen in Australien *1*:137–300.

Cheng, C. C. 1955. The development of the shoulder region of the opossum, *Didelphys virginiana,* with special reference to the musculature. J. Morphol. *97*:415–471.

Dunlap, D. G. 1960. The comparative myology of the pelvic appendage in the Salientia. J. Morphol. *106*:1–76.

Edgeworth, F. H. 1935. The Cranial Muscles of Vertebrates. London, The Macmillan Company.

Elliott, D. H. 1965. Structure and function of mammalian tendon. Biol. Rev. *40*:392–421.

Ellsworth, A. H. F. 1974. Reassessment of muscle homologies and nomenclature. Huntington, New York, Krieger.

Fürbringer, M. Zur vergleichenden Anatomie des Brustschulterapparates und der Schultermuskeln. Jena. Ztschr. Naturwiss. 7:237–320, 1873; *8*:175–280, 1874; *34*:215–718, 1900; *36*:289–736, 1902; Morphol. Jahrbuch *1*:636–816, 1876.

George, J. C., and Berger, A. J. 1966. Avian Myology. New York, Academic Press.

Gilbert, P. W. 1957. The origin and development of the human extrinsic ocular muscles. Carnegie Inst. Washington, Contrib. Embryol. *36*:59–78.

Grundfest. H. 1960. Electric fishes. Sci. Am. *203*:(4):115–124.

Haas, G. 1931. Die Kiefermuskulatur und die Schädelmechanik der Schlangen in vergleichender Darstellung. Zool. Jahrbücher (Anat.) *53*:127–198. See also *ibid.* *52*:1–218, 1930.

Hofer, H. 1950 Zur Morphologie der Kiefermuskulatur der Vögel. Zool. Jahrbücher (Anat.) *70*:427–556.

Howell, A. B. 1937. Morphogenesis of the shoulder architecture: Aves. Auk *54*:363–375.

Huber, E. 1931. Evolution of Facial Musculature and Facial Expression. Baltimore, Johns Hopkins University Press.

Hudson, G. E. 1937. Studies on the muscles of the pelvic appendage in birds. Am. Midland Nat. *18*:1–108.

Jones, C. L. 1979. The morphogenesis of the thigh of the mouse with special reference to tetrapod muscle homologies. J. Morph. *162*:275–310.

Konigsberg, I. R. 1964. The embryological origin of muscle. Sci. Am. *211*(2):61–66.

Lakjer, T. 1926. Studien über die Trigeminusversorgte Kaumuskulatur der Sauropsiden. Copenhagen, C. A. Reitzel.

Lissmann, H. W. 1958. On the function and evolution of electric organs in fish. J. Exper. Biol. *35*:151–191.

Maurer, F. 1898. Die Entwicklung der ventralen Rumpfmuskulatur bei Reptilien. Morphol. Jahrbuch *26*:1–60.

McGowan, C. 1979. The hind limb musculature of the brown kiwi, *Apteryx australis mantelli.* J. Morph. *160*:33–74.

Nursall, J. R. 1956. The lateral musculature and the swimming of fish. Proc. Zool. Soc. London *126*:127–143.

Romer, A. S. 1927. The development of the thigh musculature of the chick. J. Morphol. *43*:347–385.

Romer, A. S. 1944. The development of tetrapod limb musculature—the shoulder region of *Lacerta.* J. Morphol. *74*:1–41.

Schumacher, G. H. 1961. Funktionelle Morphologie der Kaumuskulatur. Jena, Gustav Fischer.

Sewertzoff, A. N. 1907. Studien über die Entwickelung der Muskeln, Nerven und des Skeletts der Extremitäten der niederen Tetrapoda. Bull. Soc. Impériale Naturalistes Moscou (n.s.) *21*:1–430.

Shufeldt, R. W. 1890. The Myology of the Raven. London, Macmillan and Company.

Starck, D., and Barnikol, A. 1954. Beiträge zur Morphologie der Trigeminusmuskulatur der Vögel (besonders der Accipitres, Cathartidae, Striges und Anseres). Morphol. Jahrbuch *94*:1–64.

Stein, B. R. 1981. Comparative limb myology of two opossums, *Didelphis* and *Chironectes.* J. Morph. *169*:113–140.

Straus, W. L., and Rawles, M. E. 1953. An experimental study of the origin of the trunk musculature and ribs in the chick. Am. J. Anat. *92*:471–510.

Sullivan, G. E. 1962. Anatomy and embryology of the wing musculature of the domestic fowl *(Gallus).* Aust. J. Zool. *10*:458–518.

Sy, M. 1936. Funktionell-anatomische Untersuchungen am Vogelflügel. J. Ornithologie *84*:199–296.

Wilder, H. H. 1912. The appendicular muscles of *Necturus maculosus.* Zool. Jahrbücher Suppl. *15*(2):383–424.

Coelom

Butler, G. W. 1889, 1892. On the subdivision of the body cavity in lizards, crocodiles, and birds. Proc. Zool. Soc. London *1889*:452–474; snakes, *1892*:477–498.

Keith, A. 1905. The nature of the mammalian diaphragm and pleural cavities. J. Anat. Physiol. *39*:243–284.

Mall, F. P. 1897. Development of the human coelom. J. Morphol. *12*:395–453.

Wells, L. J. 1954. Development of the human diaphragm and pleural sacs. Carnegie Inst. Washington, Contrib. Embryol. *35*:107–134.

Mouth, Pharynx, Lungs

Adams, W. E. 1939. The cervical region of the Lacertilia. J. Anat. *74*:57–71.

Ames, P. L. 1971. The morphology of the syrinx in passerine birds. Yale Peabody Mus. Bull., No. 37.

Ballantyne, F. M. 1927. Air bladder and lungs; a contribution to the morphology of the air bladder of fish. Tr. Roy. Soc. Edinburgh 55:371–394.

Bijtel, J. H. 1949. The structure and the mechanism of movement of the gill filaments in Teleostei. Arch. Néerl. Zool. 8:267–288.

Brackenbury, J. H. 1980. Respiration and production of sounds by birds. Biol. Rev. 55:363–378.

Butler, P. M., and Joysey, K. A., eds. 1978. Development, Function, and Evolution of Teeth. New York and London, Academic Press.

Comroe, J. H., Jr. 1966. The lung. Sci. Am. 214(2):57–68.

Dahlberg, A. A., ed. 1971. Dental Morphology and Evolution. Chicago, University of Chicago Press.

Edmund, A. G. 1960. Tooth Replacement Phenomena in the Lower Vertebrates. Contribution 52, Life Sciences Division, Royal Ontario Museum, Toronto.

Gans, C. 1970. Strategy and sequence in the evolution of the external gas exchangers of ectothermal vertebrates. Forma et Functio 3:61–104.

Gaunt, A. S., ed. 1973. Vertebrate sound production. Am. Zool. 13:1139–1255.

Gibbs, S. P. 1956. The anatomy and development of the buccal glands of the lake lamprey (Petromyzon marinus L.) and the histochemistry of their secretion. J. Morphol. 98:429–470.

Gregory, W. K. 1934. A half century of trituberculy. The Cope-Osborn theory of dental evolution, with a revised summary of molar evolution from fish to man. Proc. Am. Philosophical Soc. 73:169–317.

Hughes, G. M. 1963. Comparative Physiology of Vertebrate Respiration. Cambridge, Mass., Harvard University Press.

Hughes, G. M., ed. 1976. Respiration of Amphibious Vertebrates. New York and London, Academic Press.

Jones, F. R. H., and Marshall, N. B. 1953. The structure and functions of the teleostean swimbladder. Biol. Rev. 28:16–83.

King, A. S. 1966. Avian lungs and air sacs. Int. Rev. Gen. Exp. Zool. 2:171–267.

Klapper, C. E. 1946. The development of the pharynx of the guinea pig with special emphasis on the fate of the ultimobranchial body. Am. J. Anat. 79:361–397.

Locy, W. A., and Larsell, O. 1916. The embryology of the birds' lung. Am. J. Anat. 19:447–501.

Marshall, N. B. 1960. Swimbladder structure of deep-sea fishes in relation to their systematics and biology. Discovery Reports 31:1–122.

Miles, A. E. W., ed. 1967. Structural and Chemical Organization of Teeth. 2 vols. New York, Academic Press.

Miller, W. S. 1947. The Lung. 2nd ed. Springfield, Ill., Charles C Thomas.

Moss, M. 1970. Enamel and bone in shark teeth; with a note on fibrous enamel in fishes. Acta Anat. 77:161–187.

Müller, B. 1908. The air sacs of the pigeon. Smithsonian Miscellaneous Collections 1:365–414.

Owen, R. 1840–1845. Odontography—A Treatise on the Comparative Anatomy of the Teeth. London, Hippolyte Bailliere. Despite its antiquity, a valuable comprehensive account; cf. the next.

Peyer, B. 1968. Comparative Odontology. Chicago, University of Chicago Press.

Poll, M. 1962. Étude sur la structure adulte et la formation des sacs pulmonaires des protoptères. Annalen Koninklijk Mus. Midden-Afrika, Ser. 8°, Zool. Wetensch., No. 108, pp. 131–171.

Schmidt-Nielsen, K. 1971. How birds breathe. Sci. Am. 225(6):72–79.

Scott, J. H., and Symons, N. B. 1977. Introduction to Dental Anatomy. 8th ed. Edinburgh and London, Churchill Livingstone.

Sonntag, C. F. 1920–1924. The comparative anatomy of the tongues of the Mammalia. Proc. Zool. Soc. London 1920:115–129; 1921:1–29, 277–322, 497–521, 741–767; 1922:639–657; 1923:129–153, 515–529; 1924:725–755.

Woodland, W. N. F. 1911. On the structure and function of the gas glands and retia mirabilia associated with the gas bladder of some teleostean fishes. Proc. Zool. Soc. London 1911:183–248.

Woskoboinikoff, M. 1932. Der Apparat der Kiemenatmung bei den Fischen. Zool. Jahrbücher (Anat.) 55:315–488.

Digestive System

Barrington, E. J. W. 1945. The supposed pancreatic organs of *Petromyzon fluviatilis* and *Myxine glutinosa*. Quart. J. Microsc. Sci. *85*:391–417.

Blake, I. H. 1930, 1936. Studies on the comparative histology of the digestive tube of certain teleost fishes. J. Morphol. *50*:39–70; *60*:77–102.

Burger, J. W. 1962. Further studies on the function of the rectal gland in the spiny dogfish., Physiol. Zool. *35*:205–217.

Calhoun, M. L. 1954. Microscopic Anatomy of the Digestive System of the Chicken. Ames, Iowa State College Press.

Cornselius, C. 1925. Morphologie, Histologie und Embryologie des Muskelmagens der Vögel. Morphol. Jahrbuch *54*:507–559.

Elias, H. 1955. Liver morphology. Biol. Rev. *30*:263–310.

Elias, H., and Sherrick, J. C. 1969. Morphology of the Liver. New York, Academic Press.

Gorham, F. W., and Ivy, A. C. 1938. General function of the gall bladder from the evolutionary standpoint. Field Museum of Natural History, Zoology Series, *22*:159–213.

Greene, C. W. 1912. Anatomy and histology of the alimentary tract of the king salmon. Bull. U.S. Bureau Fisheries, *32*:73–100.

Hill, W. C. O. 1926. A comparative study of the pancreas. Proc. Zool. Soc. London *1926*:581–631.

Hirsch, G. C. 1950. Magenlose Fische. Zool. Anzeiger, Ergänz. *145*:302–326.

Hopkins, G. S. 1895. On the enteron of American ganoids. J. Morphol. *11*:411–442.

Jacobshagen, E. 1913. Untersuchungen über das Darmsystem der Fische und Dipnoer. II. Jena. Ztschr. Naturw. *49*:373–810.

Kaden, L. 1936. Über Epithel und Drüsen des Vogelschlunds. Zool. Jahrbücher (Anat.) *61*:421–466.

Mitchell, P. C. 1901. On the intestinal tract of birds; with remarks on the valuation and nomenclature of zoological characters. Tr. Linnean Soc. London, Zool. *8*:173–275.

Mitchell, P. C. 1906. On the intestinal tract of mammals. Tr. Zool. Soc. London *17*:437–536.

Neumayer, L. 1930. Die Entwicklung des Darms von *Acipenser*. Acta Zoologica *11*:39–150.

Oguri, M. 1964. Rectal glands of marine and fresh water sharks, comparative histology. Science *144*:1151–1152.

Pernkopf, E. 1930. Beiträge zur vergleichende Anatomie des vertebraten Magens. Ztschr. Anat. *91*:329–390.

Peterson, H. 1908. Beiträge zur Kenntniss des Baues und der Entwickelung des Selachierdarmes. Jena. Ztschr. Naturw. *43*:619–652; *44*:123–148.

Preuss, F., and Fricke, W. 1979. Comprehensive schemata on the histology of the liver with consequences in terminology. J. Morph. *162*:211–220.

Rogers, T. A. 1958. The metabolism of ruminants. Sci. Am. *198*(2):34–38.

Rouiller, C., ed. 1963–1964. The Liver. 2 vols. New York, Academic Press.

Slijper, E. J. 1946. Die physiologische Anatomie der Verdauungsorgane bei den Vertebraten. Tabulae Biologicae *21*:1–81.

Excretory and Reproductive Systems

Bentley, P. J., and Follett, B. K. 1963. Kidney function in a primitive vertebrate, the cyclostome *Lampetra fluviatilis*. J. Physiol. *169*:902–918.

Borcea, J. 1906. Recherches sur le système urogenital des elasmobranchs. Arch. Zoologie exp. et gén. *4*(4):199–484.

Boyden, E. A. 1922. The development of the cloaca in birds. Am. J. Anat. *30*:163–201.

Buchanan, G., and Fraser, E. A. 1918. The development of the urinogenital system in the Marsupialia with special reference to *Trichosurus vulpecula*. Part I. J. Anat. *53*:35–95.

Chase, S. W. 1923. The mesonephros and urogenital ducts of *Necturus maculosus* Rafinesque. J. Morphol. *37*:457–532.

Conel, J. L. 1917. The urogenital system of myxinoids. J. Morphol. *29*:75–164.

Edwards, J. G. 1928, 1929. Studies on aglomerular and glomerular kidneys. Am. J. Anat. *42*:75–108; Anat. Rec. *44*:15–28.

Everett, H. B. 1945. The present status of the germ-cell problem in vertebrates. Biol. Rev. *20*:45–55.

Fox, H. 1963. The amphibian pronephros. Quart. Rev. Biol. *38*:1–25.

Fraser, E. A. 1950. The development of the vertebrate excretory system. Biol. Rev. *25*:159–187.

Grady, H. G., and Smith, D. E., eds. 1963. The Ovary. Baltimore, Williams & Wilkins.

Gray, P. 1930–1936. The development of the amphibian kidney. Quart. J. Microsc. Sci. *73*:507–546; *75*:425–466; *78*:445–473.

Holmgren, N. 1950. On the pronephros and the blood in *Myxine glutinosa.* Acta Zool. *31*:233–348.

Huber, G. C. 1917. On the morphology of the renal tubules of vertebrates. Anat. Rec. *13*:305–339.

Kempton, R. T. 1943, 1953. Studies on the elasmobranch kidney. J. Morphol. *73*:247–263; Biol. Bull. *104*:45–56.

Kindahl, M. 1938. Zur Entwicklung der Exkretionsorgane von Dipnoërn und Amphibien. Acta Zool. *19*:1–190.

Leigh-Sharpe, W. H. 1920–1926. The comparative morphology of the secondary sexual characters of elasmobranch fishes. J. Morphol. *34*:245–265; *35*:359–380; *36*:221–243; *42*:307–308.

Maschkowzeff, A. 1934–1935. Zur Phylogenie der Geschlechtsdrüsen und der Geschlechtsausfuhrgänge bei den Vertebrata auf Grund von Forschungen betreffend die Entwicklung des Mesonephros und der Geschlechtsorgane bei den Acipenseriden, Salmoniden und Amphibien. Zool. Jahrbücher (Anat.) *59*:1–68, 201–276.

Meyer, D. B. 1964. The migration of primordial germ cells in the chick embryo. Develop. Biol. *10*:154–190.

Moore, C. R. 1926. The biology of the mammalian testis and scrotum. Quart. Rev. Biol. *1*:4–50.

Schmidt-Nielsen, K. 1959. Salt glands. Sci. Am. *200*(1):101–116.

Semon, R. 1892. Studien über den Bauplan des Urogenitalsystems der Wirbeltiere. Dargelegt an der Entwickelung dieses Organsystems bei *Ichthyophis glutinosus.* Jena. Ztschr. Naturw. *26*:89–203.

Smith, C. L., et al. 1975. *Latimeria,* the living coelacanth, is ovoviviparous. Science. *190*:1105–1106.

Smith, H. W. 1932. Water regulation and its evolution in fishes. Quart. Rev. Biol. 7:1–26.

Smith, H. W. 1951. The Kidney. London and New York, Oxford University Press.

Smith, H. W. 1953. From Fish to Philosopher. Boston, Little, Brown and Company. Vertebrate evolution with kidney evolution as the leitmotif.

Witschi, E. 1948. Migration of the germ cells of human embryos from the yolk sac to the primitive gonadal folds. Carnegie Inst. Washington, Contrib. Embryol. *32*:67–80.

Young, W. C., ed. 1961. Sex and Internal Secretion. Baltimore, Williams & Wilkins.

Circulatory System

Abramson, D. I., ed. 1962. Blood Vessels and Lymphatics. New York, Academic Press.

Adolph, E. F. 1967. The heart's pacemaker. Sci. Am. *216*(3):32–37.

Barclay, A. K., Franklin, K. J., and Pritchard, M. M. L. 1945. The Foetal Circulation. Springfield, Illinois, Charles C Thomas.

Barnett, C. H., Harrison, R. J., and Tomlinson, J. D. W. 1958. Variations in the venous systems of mammals. Biol. Rev. *33*:442–487.

Benninghoff, A. 1921. Beiträge zur vergleichenden Anatomie und Entwicklungsgeschichte des Amphibienherzens. Morphol. Jahrbuch *51*:354–412.

Bugge, J. 1961. The heart of the African lungfish, *Protopterus.* Vidensk. Medd. fra Dansk. Naturhist. Foren. *123*:193–210.

Burnet, M. 1962. The thymus gland. Sci. Am. *207*(5):50–57.

Butler, E. G. 1927. The relative role played by the embryonic veins in the development of the mammalian vena cava posterior. Am. J. Anat. *39*:267–353.

Chardon, M. 1962. Contribution à l'étude du système circulatoire lié à la respiration des Protopteridae. Annalen Koninklijk Mus. Midden-Afrika, Ser. 8°, Zool. Wetensch., No. 108, pp. 53–99.

Chèvremont, M. 1948. Le système histiocytaire ou réticulo-endothélial. Biol. Rev. *23*:267–295.

Chiodi, V., and Bortolami, R. 1966. The Conducting System of the Vertebrate Heart. Bologna, Edizioni Calderini.

Congdon, E. D. 1922. Transformation of the

aortic arch during the development of the human embryo. Carnegie Inst. Washington, Contrib. Embryol. *14*:47–110.

Cooper, E. L., ed. 1975. Developmental immunology. Am. Zool. *15*:1–213.

Davis, D. D., and Storey, H. E. 1943. The carotid cirulation in the domestic cat. Publ. Field Museum of Natural History, Zool. *28*:5–47.

DeLong, K. T. 1962. Quantitative analysis of blood circulation through the frog heart. Science *138*:693–694.

Foxon, G. E. H. 1955. Problems of the double circulation in vertebrates. Biol. Rev. *30*: 196–228.

Greil, A. 1903. Beiträge zur vergleichenden Anatomie und Entwicklungsgeschichte des Herzens und des Truncus arteriosus der Wirbeltiere. Morphol. Jahrbuch *31*:123–310.

Greil, A. 1908–1913. Entwickelungsgeschichte des Kopfes und des Blutgefässsystems von *Ceratodus forsteri*. Semon's Zoologische Forschungsreise in Australien *1*:661–1492.

Heuser, C. H. 1923. The branchial vessels and their derivatives in the pig. Carnegie Inst. Washington, Contrib. Embryol. *15*: 121–139.

Hill, W. C. O. 1953. The blood-vascular system of *Tarsius*. Proc. Zool. Soc. London, *123*:655–692.

Hochstetter, T. 1906. Beiträge zur Anatomie und Entwickelungsgeschichte des Blutgefässsystems der Krokodile. Voeltzkow, A., Reise in Ostafrika, *4*:1–139.

Holmes, E. B. 1975. A reconsideration of the phylogeny of the tetrapod heart. J. Morph. *147*:209–228.

Huntington, G. S., and McClure, C. F. W. 1920. The development of the veins in the domestic cat. Anat. Rec. *20*:1–31.

Kampmeier, O. F. 1969. Evolution and Comparative Morphology of the Lymphatic System. Springfield, Illinois, Charles C Thomas.

Kern, A. 1926. Das Vogelherz. Morphol. Jahrbuch *56*:264–315.

Krogh, A. 1929. The Anatomy and Physiology of Capillaries. New Haven, Yale University Press.

Mathur, P. N. 1944. The anatomy of the reptilian heart. Part I. *Varanus monitor* (Linné). Proc. Ind. Acad. Sci., B, *20*:1–29.

Mayerson, H. S. 1963. The lymphatic system. Sci. Am. *208*(6):80–90.

Miller, J. F. A. P. 1964. The thymus and the development of immunological responsiveness. Science *144*:1544–1551.

Mossman, H. W. 1948. Circulatory cycles in the vertebrates. Biol. Rev. *23*:237–255.

O'Donoghue, C. H. 1912. The circulatory system of the common grass snake *(Tropidonotus natrix)*. Proc. Zool. Soc. London *1912*:612–647.

O'Donoghue, C. H. 1920. The blood-vascular system of the tuatara, *Sphenodon punctatus*. Philos. Tr. Roy. Soc. London *(B)210*: 175–252.

O'Donoghue, C. H., and Abbott, E. 1928. The blood-vascular system of the spiny dogfish, *Squalus acanthias* Linné, and *Squalus sucklii* Gill. Tr. Roy. Soc. Edinburgh *55*:823–890.

Padget, D. H. 1957. The development of the cranial venous system in man, from the viewpoint of comparative anatomy. Carnegie Inst. Washington, Contrib. Embryol. *36*:79–140.

Parsons, T. S., organizer. 1968. Functional morphology of the heart of vertebrates. Am. Zool. *8*:177–229.

Quiring, D. P. 1949. Collateral Circulation. Philadelphia, Lea and Febiger.

Regan, F. P. 1929. A century of study upon the development of the eutherian vena cava inferior. Quart. Rev. Biol. *4*:179–212.

Robertson, J. I. 1913. The development of the heart and vascular system of *Lepidosiren paradoxa*. Quart. J. Microsc. Sci. *59*:53–132.

Rusznyak, I., et al. 1967. Lymphatics and Lymph Circulation: Physiology and Pathology. 2nd ed. Oxford, Pergamon Press Ltd.

Satchell, G. H. 1971. Circulation in Fishes. Cambridge, England, Cambridge University Press.

Shearer, E. M. 1930. Studies on the embryology of circulation in fishes. Am. J. Anat. *46*:393–459.

Shindo, T. 1914. Zur vergleichenden Anatomie der arteriellen Kopfgefässe der Reptilien. Anat. Hefte *51*:267–356.

Webb, G. J. W. 1979. Comparative cardiac

anatomy of the reptilia. III. The heart of crocodilians and an hypothesis on the completion of the interventricular septum of crocodilians and birds. J. Morph. *161*:221–240.

Wiggers, C. J. 1957. The heart. Sci. Am. *196*(5):75–87.

Wood, J. E. 1968. The venous system. Sci. Am. *218*(1):86–96.

Yoffey, J. M., and Coortice, F. C. 1971 Lymphatics, Lymph and the Lymphomyeloid Complex. London and New York, Academic Press.

Sense Organs

Allin, E. F. 1975. Evolution of the Mammalian Middle Ear. J. Morph. *147*:403–438.

Allis, E. P., Jr. 1934. Concerning the course of the laterosensory canals in recent fishes, pre-fishes, and *Necturus*. J. Anat. *68*:361–415.

Allison, A. C. 1953. The morphology of the olfactory system in vertebrates. Biol. Rev. *28*:195–244.

Atz, J. W. 1952. Narial breathing in fishes and the evolution of internal nares. Quart. Rev. Biol. *27*:366–377.

Autrum, H. 1971–date. Handbook of Sensory Physiology. Berlin, Springer-Verlag. Numerous volumes with various authors.

Baldauf, R. J., and Heimer, L., eds. 1967. Vertebrate olfaction. Am. Zool. 7:385–432.

Baradi, A. F., and Bourne, G. H. 1953. Gustatory and olfactory epithelia. Internat. Rev. Cytol. *2*:289–330.

Bellairs, A. d'A., and Boyd, J. D. 1947, 1950. The lachrymal apparatus in lizards and snakes. Proc. Zool. Soc. London *117*:81–101; *120*:269–309.

Burne, R. H. 1909. The anatomy of the olfactory organ of teleostean fishes. Proc. Zool. Soc. London *1909*:610–662.

Cahn, P. H. 1967. Lateral Line Detectors. Bloomington, Ind., Indiana University Press.

Chranilov, N. S. 1927, 1929. Beiträge zur Kenntniss der Weber'schen Apparates der Ostariophysi. Zool. Jahrbücher (Anat.) *49*:501–597; *51*:323–462.

Dempster, W. T. 1930. The morphology of the amphibian endolymphatic organ. J. Morphol. *50*:71–120.

Dijkgraaf, S. 1952. Bau und Funktionen der Seitenorgane und des Ohrlabyrinths der Fische. Experientia *8*:205–216.

Disler, N. N. 1960. Lateral Line Sense Organs and Their Importance in Fish Behaviour. Moskva, Izdatelstvo Akademii Nauk SSSR.

Eakin, R. M. 1973. The Third Eye. Berkeley, University of California Press.

Fänge, R., Schmidt-Nielsen, K., and Osaki, H. 1958. The salt gland of the herring gull. Biol. Bull. *115*:162–171.

Frisch, K. von. 1936. Über den Gehörs in der Fische. Biol. Rev. *11*:210–243.

Gaupp, E. 1913. Die Reichertsche Theorie (Hammer- Amboss- und Kieferfrage). Arch. Anat. Physiol. Supplement *V*:1–417. On evolution of middle ear ossicles.

Guggenheim, L. 1948. Phylogenesis of the Ear. Culver City, Calif., Murray and Gee.

Holmgren, N. 1942. General morphology of the lateral sensory line system of the head in fishes. Kungl. Svenska Vetenskapsakad. Handl. (3)*20*(1):1–46.

Kleerekoper, H. 1969. Olfaction in Fishes. Bloomington, Ind., Indiana University Press.

Lowenstein, O. 1936. The equilibrium function of the vertebrate labyrinth. Biol. Rev. *11*:113–145.

Noble, G. K., and Schmidt, A. 1937. The structure and function of the facial and labial pits of snakes. Proc. Am. Philosophical Soc. 77:263–288.

Parrington, F. R. 1979. The evolution of the mammalian middle and outer ears: A personal review. Biol. Rev. *54*:369–387.

Parsons, T. S. 1959. Studies on the comparative embryology of the reptilian nose. Bull. Mus. Comp. Zool., Harvard, *120*:104–277.

Parsons, T. S., ed. 1966. The vertebrate ear. Am. Zool. *6*:368–466.

Polyak, S. 1957. The Vertebrate Visual System. Chicago, University of Chicago Press.

Pumphrey, R. J. 1948. The sense organs of birds. Ibis, *90*:171–199; Annual Report of the Smithsonian Institution 305–330.

Ralph, C. L., et al. 1979. The pineal complex and thermoregulation. Biol. Rev. *54*:41–72.

Retzius, G. 1881–1884. Das Gehörorgan der Wirbelthiere. Morphologischhistologische Studien. 2 vols. Stockholm, Samson and Wallin.

Rochon-Duvigneaud, A. 1943. Les Yeux et la Vision des Vertébrés. Paris, Masson et Cie.

Stensiö, E. A. 1947. The sensory lines and dermal bones of the cheek in fishes and amphibians. Kungl. Svenska Vetenskapsakad. Handl. (3)*24*(3):1–195.

Thomson, K. S. 1977. On the individual history of Cosmine and a possible electroreceptive function of the pore-canal system in fossil fishes. *In* Problems in Vertebrate Evolution, Linn. Soc. Symp. Ser., No. 4. London and New York, Academic Press.

Versluys, J. 1898. Die mittlere und äussere Ohrsphäre der Lacertilia und Rhynchocephalia. Zool. Jahrbücher (Anat.) *12*:161–406. See also *ibid. 18*:107–188, 1902.

Walls, G. L. 1942. The vertebrate eye and its adaptive radiation. Cranbrook Institute of Science, Bulletin No. 19.

Werner, S. C. 1960. Das Gehörorgan der Wirbeltiere und des Menschen. Leipzig, George Thieme.

Wever, E. G. 1978. The Reptile Ear. Its Structure and Function. Princeton, New Jersey, Princeton University Press.

Nervous System

Addens, J. L. 1933. The motor nuclei and roots of the cranial and first spinal nerves of vertebrates. Part I. Introduction, Cyclostomes. Zeitschr. Ges. Anat. Abt. I, *101*: 307–410.

Aronson, L. R. 1963. The central nervous system of sharks and bony fishes, with special reference to sensory and integrative mechanisms. *In* Sharks and Survival, P. W. Gilbert, ed., 165–241. Boston, D. C. Heath.

Auen, E.L., and Langebartel, D.A. 1977. The cranial nerves of the colubrid snakes *Elaphe* and *Thamnophis*. J. Morph. *154*: 205–222.

Bartelmez, G. W. 1915. Mauthner's cell and the nucleus motorius tegmenti. J. Comp. Neurol. *25*:87–128.

Bass, A. D. 1959. Evolution of nervous control from primitive organisms to man. Am. Assoc. Adv. Sci. Publ. 52.

Boeke, J. 1935. The autonomic (enteric) nervous system of *Amphioxus lanceolatus*. Quart. J. Microsc. Sci. 77:623–658.

Bullock, T. H. 1945. The anatomical organization of the nervous system of Enteropneusta. Quart. J. Microsc. Sci. *86*:55–111.

Campenhout, E., van. 1930. Historical survey of the development of the sympathetic nervous system. Quart. Rev. Biol. 5:23–50, 217–234.

Causey, G. 1960. The Cell of Schwann. Edinburg and London, E. and S. Livingstone Ltd.

Cobb, S. 1963. Notes on the avian optic lobe. Brain *86*:363–372.

Cobb, S. 1966. The brain of the emu, *Dromaeus novaehollandiae*. II. Anatomy of the principal nerve cell ganglia and tracts. Breviora. Mus. Comp. Zool., No. *250*:1–27.

Cole, F. J. 1897. On the cranial nerves of *Chimaera monstrosa*. Tr. Roy. Soc. Edinburgh *38*:631–680.

Conel, J. L. 1929, 1931. The development of the brain of *Bdellostoma stouti*. I. External growth changes. II. Internal growth changes. J. Comp. Neurol. *47*:343–403; *52*:365–499.

Crosby, E. C., and Schnitzlein, H. N. 1982. Comparative, Correlative Neuroanatomy of the Vertebrate Telencephalon. New York, Macmillan.

Franz, V. 1923. Nervensystem der Akranier. Jena. Ztschr. Naturw. *59*:401–526.

Gans, C., and Crosby, E. C., eds. 1964. Recent advances in neuroanatomy. Am. Zool. *4*: 4–96.

Goodrich, E. S. 1937. On the spinal nerves of the Myxinoidea. Quart. J. Microsc. Sci. *80*:153–158.

Hassler, F., and Stephan, H., eds. 1967. Evolution of the Forebrain. New York, Plenum Publishers.

Herrick, C. J. 1899. The cranial nerves of the bony fishes. J. Comp. Neurol. 9:153–455. Cf. also *10*:265–322, 1900; *11*:177–249, 1901.

Herrick, C. J. 1926. Brains of Rats and Men. Chicago, University of Chicago Press.

Herrick, C. J. 1931. An Introduction to Neurology. 5th ed. Philadelphia, W. B. Saunders Company.

Herrick, C. J. 1948. The Brain of the Tiger Salamander. Chicago, University of Chicago Press.

Hughes, A. 1960. The development of the peripheral nerve fiber. Biol. Rev. *35*:283–323.

Igarashi, S. and Kamiya, T. 1972. Atlas of the Vertebrate Brain. Tokyo, University of Tokyo Press.

Jerison, H. J. 1973. Evolution of the Brain and Intelligence. London, Academic Press.

Johnels, A. G. 1956. On the peripheral autonomic system of the trunk region of *Lampetra planeri.* Acta Zool. *37*:251–285.

Johnston, J. B. 1901. The brain of *Acipenser;* a contribution to the morphology of the vertebrate brain. Zool. Jahrbücher (Anat.) *15*:59–263.

Johnston, J. B. 1906. The Nervous System of Vertebrates. Philadelphia, Blakiston Company.

Kappers, C. U. A., Huber, G. C., and Crosby, E. C. 1936. The Comparative Anatomy of the Nervous System of Vertebrates, Including Man. 2 vols. New York, The Macmillan Company. A mine of information on comparative neurology, but difficult to work with for one not a neurologist.

Krieg, W. J. S. 1942. Functional Neuroanatomy. Philadelphia, Blakiston Company.

Kuhlenbeck, H. 1967–1978. The Central Nervous System of Vertebrates. 5 vols. New York, Academic Press. A rather idiosyncratic series with much interpretation.

Kuntz, A. 1953. The Autonomic Nervous System. 4th ed. Philadelphia, Lea & Febiger.

Larsell, O. 1967–1972. The Comparative Anatomy and Histology of the Cerebellum. 3 vols. Minneapolis, University of Minnesota Press.

Lindström, T. 1949. On the cranial nerves of the cyclostomes with special reference to the N. trigeminus. Acta Zool. *30*:315–458.

Llinas, R., ed. 1969. Neurobiology of Cerebellar Evolution and Development: Proceedings. Chicago, American Medical Association.

Mitchell, G. A. G. 1953. Anatomy of the Autonomic Nervous System. Edinburgh and London, E., and S. Livingstone, Ltd.

Nicol, J. A. C. 1952. Autonomic nervous systems in lower chordates. Biol. Rev. *27*: 1–49.

Nieuwenhuys, R. 1962. Trends in the evolution of the actinopterygian forebrain. J. Morphol. *111*:69–88.

Norris, H. W. 1913. Cranial nerves of *Siren lacertina.* J. Morphol. *24*:245–338.

Norris, H. W., and Hughes, S. P. 1920. The cranial, occipital and anterior spinal nerves of the dogfish. J. Comp. Neurol. *31*: 293–395.

Papez, J. W. 1929. Comparative Neurology. New York, Thomas Y. Crowell Company. Reprinted in 1961. Out of date, especially on fiber connections, but still useful for comparative approach.

Pearson, R., and Pearson, L. 1976. The Vertebrate Brain. London, Academic Press.

Petras, J. M., and Noback, C. R., eds. 1969. Comparative and evolutionary aspects of the vertebrate central nervous system. Ann. N.Y. Acad. Sci. *167*(1):1–513. Papers by many of the leading workers in one of the few truly comparative surveys.

Pick, J. 1970. The Autonomic Nervous System. Morphological, Comparative, Clinical and Surgical Aspects. Philadelphia, J. B. Lippincott Company.

Portmann, A. 1946, 1947. Etudes sur la cérebralisation chez les oiseaux. Alauda *14*:2–20; *15*:1–15.

Ransom, S. W., and Clark, S. L. 1959. The Anatomy of the Nervous System. 10th ed. Philadelphia, W. B. Saunders Company.

Retzlaff, E. 1957. A mechanism for excitation and inhibition of the Mauthner's cells in teleosts. J. Comp. Neurol. *107*:209–225.

Silén, L. 1950. On the nervous system of *Glossobalanus marginatus* Meek. Acta Zool. *31*:149–175.

Stefanelli, A. 1951. The mauthnerian apparatus in the Ichthyopsida. Quart. Rev. Biol. *21*:17–34.

Stensiö, E. A. 1963. The brain and the cranial nerves in fossil, lower craniate vertebrates. Skrifter Norske Videnskaps-Akademi 1 Olso I. Mat.-Naturv. Klasse. Ny Serie. No. *13*:1–120.

Stettner, L. J., and Matyniak, K. A. 1968. The brain of birds. Sci. Am. *218*(6):64–77.

Strong, O. S. 1895. The cranial nerves of Amphibia. J. Morphol. *10*:101–230.

Tretjakoff, D. 1927. Das periphere Nervensystem des Flussneunauges. Ztschr. Wiss. Zool. *129*:359–952.

Watkinson, G. B. 1906. The cranial nerves of *Varanus bivittatus.* Morphol. Jahrbuch *35*:450–472.

Weed, L. W. 1917. The development of the

cerebrospinal spaces in pig and in man. Carnegie Inst. Washington, Contrib. Embryol. *5*:3–116.

Weiss, P. A. 1934. In vitro experiments on the factors determining the course of the outgrowing nerve fiber. J. Exp. Zool. *68*:393–448.

Willard, W. A. 1915. The cranial nerves of *Anolis carolinensis*. Bull. Mus. Comp. Zool., Harvard *59*:17–116.

Worthington, J. 1906. The descriptive anatomy of the brain and cranial nerves of *Bdellostoma dombeyi*. Quart. J. Microsc. Sci. *57*:137–181.

Yntema, C. L., and Hammond, W. S. 1947. The development of the autonomic nervous system. Biol. Rev. *22*:344–359.

Young, J. Z. 1931. On the autonomic nervous system of the teleostean fish, *Uranoscopus scaber*. Quart. J. Microsc. Sci. *74*:492–525.

Young, J. Z. 1933. The autonomic system of selachians. Quart. J. Microsc. Sci. *75*:571–624.

Zeman, W., and Innes, J. R. M. 1963. Craigie's Neuroanatomy of the Rat. New York, Academic Press.

Endocrine Organs

Bargmann, W. 1960. The neruosecretory system of the diencephalon. Endeavour *19*:125–133.

Barrington, E. J. W. 1975. An Introduction to General and Comparative Endocrinology. 2nd ed. Oxford, Oxford University Press.

Barrington, E. J., ed. 1979–date. Hormones and Evolution. London and New York, Academic Press.

Bern, H. A. 1967. Hormones and endocrine glands of fishes. Science *158*:455–462.

Boyd, J. D. 1950. The development of the thyroid and parathyroid glands and the thymus. Ann. R. Coll. Surg. Engl. 7:455–471.

Chester Jones, I., and Henderson, I. W., eds. 1976–1978. General, Comparative and Clinical Endocrinology of the Adrenal Cortex. 3 vols. New York and London, Academic Press.

Dodd, J. M., and Kerr, T. 1963. Comparative morphology and histology of the hypoth-

alamo-neurohypophysial system. Symposium 9, Zool. Soc. London, 9–27.

Fields, W. S., Guillemin, R., and Carton, C. A., eds. 1956. Hypothalamic-hypophysial Interrelationships. A Symposium. Houston, Baylor University College of Medicine.

Goldsmith, E. D. 1949. Phylogeny of the thyroid: descriptive and experimental. Ann. N. Y. Acad. Sci. *50*:282–316.

Gorbman, A., ed. 1967. Recent developments in endocrinology. Am. Zool. 7:81–169.

Gorbman, A., et al. 1983. Comparative Endocrinology. New York, John Wiley & Sons.

Green, J. D. 1951. The comparative anatomy of the hypophysis, with special reference to its blood supply and innervation. Am. J. Anat. *88*:225–311.

Grossman, M. I. 1950. Gastrointestinal hormones. Physiol. Rev. *30*:33–90.

Holmes, R. L., and Ball, J. N. 1974. The Pituitary Gland: A Comparative Account. Cambridge, England, Cambridge University Press.

Lynn, G. W., and Wachowski, H. E. 1951. The thyroid gland and its functions in cold-blooded vertebrates. Quart. Rev. Biol. *26*:123–168.

Macchi, I. A., and Gapp, D. A., eds. 1973. Comparative aspects of the endocrine pancreas. Am. Zool. *13*:565–709.

Marshall, F. H. A. 1960. The Physiology of Reproduction. London, Longmans, Green.

Matty, A. J. 1966. Endocrine glands in lower vertebrates. Int. Rev. Gen. Exp. Zool. *2*:44–138.

Norris, H. W. 1941. The Plagiostome Hypophysis, General Morphology and Types of Structure. Lancaster, Science Press.

Pickford, G. E., et al. 1973. The current status of fish endocrine systems. Am. Zool. *13*:710–936.

Scharrer, E., and Scharrer, B. 1963. Neuroendocrinology. New York, Columbia University Press.

Turner, C. D., and Bagnara, J. T. 1976. General Endocrinology. 6th ed. Philadelphia, W. B. Saunders Company.

Villee, D. B. 1975. Human Endocrinology: A Developmental Approach. Philadelphia, W. B. Saunders Company.

Von Euler, U. S., and Heller, H. S., eds. 1963. Comparative Endocrinology. New York, Academic Press.

Watzka, M. 1933. Vergleichende Untersuchungen über den ultimobranchialen Kör-per. Z. Mikrosk. Anat. Forsch. *34*:485–533.

Wilkins, L. 1960. The thyroid gland. Sci. Am. *202*(3):119–129.

Wurtman, R. J., and Axelrod, J. 1965. The pineal gland. Sci. Am. *213*(1):50–60.

Index

Page numbers in *italics* indicate illustrations; those in roman indicate text or both text and illustrations.